Lecture Notes in Artificial Intelligence 11103

Subseries of Lecture Notes in Computer Science

More information about this series at http://www.springer.com/series/1244

Hung Son Nguyen · Quang-Thuy Ha
Tianrui Li · Małgorzata Przybyła-Kasperek (Eds.)

Rough Sets

International Joint Conference, IJCRS 2018
Quy Nhon, Vietnam, August 20–24, 2018
Proceedings

 Springer

Editors
Hung Son Nguyen
University of Warsaw
Warsaw
Poland

Quang-Thuy Ha (iD)
Faculty of Information Technology
Vietnam National University
Hanoi
Vietnam

Tianrui Li (iD)
School of Information Science
Southwest Jiaotong University
Chengdu
China

Małgorzata Przybyła-Kasperek (iD)
Institute of Computer Science
University of Silesia
Sosnowiec
Poland

ISSN 0302-9743   ISSN 1611-3349   (electronic)
Lecture Notes in Artificial Intelligence
ISBN 978-3-319-99367-6   ISBN 978-3-319-99368-3   (eBook)
https://doi.org/10.1007/978-3-319-99368-3

Library of Congress Control Number: 2018951242

LNCS Sublibrary: SL7 – Artificial Intelligence

This Springer imprint is published by the registered company Springer Nature Switzerland AG
The registered company address is: Gewerbestrasse 11, 6330 Cham, Switzerland

Preface

The proceedings of the 2018 International Joint Conference on Rough Sets (IJCRS 2018) contain the results of the meeting of the International Rough Set Society held at the International Centre for Interdisciplinary Science and Education (ICISE) and the University of Quy Nhon in Quy Nhon, Vietnam, during August 2018.

Conferences in the IJCRS series are held annually and comprise four main tracks relating the topic rough sets to other topical paradigms: rough sets and data analysis covered by the RSCTC conference series from 1998, rough sets and granular computing covered by the RSFDGrC conference series since 1999, rough sets and knowledge technology covered by the RSKT conference series since 2006, and rough sets and intelligent systems covered by the RSEISP conference series since 2007. Owing to the gradual emergence of hybrid paradigms involving rough sets, it was deemed necessary to organize Joint Rough Set Symposiums, first in Toronto, Canada, in 2007, followed by symposiums in Chengdu, China in 2012, Halifax, Canada, 2013, Granada and Madrid, Spain, 2014, Tianjin, China, 2015, where the acronym IJCRS was proposed, continuing with the IJCRS 2016 conference in Santiago de Chile and IJCRS 2017 in Olsztyn, Poland.

The IJCRS conferences aim at bringing together experts from universities and research centers as well as from industry representing fields of research in which theoretical and applicational aspects of rough set theory already find or may potentially find usage. They also become a place for researchers who want to present their ideas to the rough set community, or for those who would like to learn about rough sets and find out if they can be useful for their problems.

This year's conference, IJCRS 2018, celebrated the 20th anniversary of the first international conference on rough sets called RSCTC, which was organized by Lech Polkowski and Andrzej Skowron during June 22–26, 1998, in Warsaw, Poland. On this occasion, we listened to a retrospective talk delivered by Andrzej Skowron, who summarized the successes of this field and showed directions for further research and development.

IJCRS 2018 attracted 61 submissions (not including invited contributions), which underwent a rigorous reviewing process. Each accepted full-length paper was evaluated by three to five experts on average. The present volume contains 45 full-length regular and workshop submissions, which were accepted by the Program Committee, as well as six invited articles.

The conference program included five keynotes and plenary talks, a fellow talk, eight parallel sessions, a tutorial, the 6th International Workshop on Three-way Decisions, Uncertainty, and Granular Computing, and a panel discussion on rough sets and data science.

The chairs of the Organizing Committee also prepared the best paper award and the best student paper award. From all research papers submitted, the Program Committee nominated five papers as finalists for the award and, based on the final presentations during the conference, selected the winners.

We would like to express our gratitude to all the authors for submitting papers to IJCRS 2018, as well as to the members of the Program Committee for organizing this year's attractive program.

We also gratefully thank our sponsors: Vietnam National University in Ho Chi Minh City, for providing the technical support and human resources for the conference; the University of Quy Nhon, for sponsoring the reception and the conference facilities during the first day and the last day; Ton Duc Thang University, for sponsoring the pre-conference workshops on rough sets and data mining.

The conference would not have been successful without support received from distinguished individuals and organizations. We express our gratitude to the IJCRS 2018 honorary chairs, Andrzej Skowron, Huynh Thanh Dat, and Do Ngoc My, for their great leadership. We appreciate the help of Dinh Thuc Nguyen, Nguyen Tien Trung, Quang Vinh Lam, Quang Thai Thuan, Thanh Tran Thien, Luong Thi Hong Cam, Giang Thuy Minh, Phung Thai Thien Trang, Dao Thi Hong Le, Hung Nguyen-Manh, and all other representatives of Vietnam National University in Ho Chi Minh City and Quy Nhon University, who were involved in the conference organization. We would also like to thank Marcin Szeląg, Sinh Hoa Nguyen, and Dang Phuoc Huy, who supported the conference as tutorial, workshop, and special session chairs. We acknowledge the significant help from Khuong Nguyen-An, Tran Thanh Hai, Ly Tran Thai Hoc, and Marcin Szczuka provided at various stages of the conference publicity, website, and material preparation.

We are grateful to Tu Bao Ho, Hamido Fujita, Hong Yu, Andrzej Skowron, Piero Pagliani, and Mohua Banerjee for delivering excellent keynote and plenary talks and fellow talks. We thank Dominik Ślęzak and Arkadiusz Wojna for the tutorial. We are thankful to Hong Ye, Mohua Banerjee, Mihir Chakraborty, Bay Vo, and Le Thi Thuy Loan for the organization of workshops and special sessions.

Special thanks go to Alfred Hofmann of Springer, for accepting to publish the proceedings of IJCRS 2018 in the LNCS/LNAI series, and to Anna Kramer for her help with the proceedings. We are grateful to Springer for the grant of 1,000 Euro for the best paper award winners. We would also like to acknowledge the use of EasyChair, a great conference management system.

We hope that the reader will find all the papers in the proceedings interesting and stimulating.

August 2018

Hung Son Nguyen
Quang-Thuy Ha
Tianrui Li
Małgorzata Przybyła-Kasperek

Organization

Honorary Chairs

Andrzej Skowron   University of Warsaw, Poland
Thanh Dat Huynh   VNU-HCMC, Vietnam
Ngoc My Do   Quy Nhon University, Vietnam

General Chairs

Davide Ciucci   University of Milano-Bicocca, Italy
Dan Thu Tran   VNU-HCMC, Vietnam

Organizing Committee Chairs

Dinh Thuc Nguyen   VNU-HCMC, Vietnam
Tien Trung Nguyen   Quy Nhon University, Vietnam
Quang Vinh Lam   VNU-HCMC, Vietnam
Quang Thai Thuan   Quy Nhon University, Vietnam
Thanh Thien Tran   Quy Nhon University, Vietnam

Program Committee

Program Committee Chairs

Hung Son Nguyen   University of Warsaw, Poland
Quang-Thuy Ha   College of Technology, VNU-Hanoi, Vietnam
Tianrui Li   Southwest Jiaotong University, Chengdu, China
Małgorzata   University of Silesia, Poland
    Przybyła-Kasperek

Workshop, Special Sessions, and Tutorial Chairs

Marcin Szeląg   Poznań University of Technology, Poland
Sinh Hoa Nguyen   Polish-Japanese Academy of IT, Poland
Phuoc Huy Dang   Dalat University, Vietnam

Program Committee

Mani A.   Calcutta University, India
Piotr Artiemjew   University of Warmia and Mazury, Poland
Jaume Baixeries   Universitat Politecnica de Catalunya, Spain

Mohua Banerjee	Indian Institute of Technology Kanpur, India
Jan Bazan	University of Rzeszów, Poland
Rafael Bello	Universidad Central de Las Villas, Cuba
Nizar Bouguila	Concordia University, Canada
Jerzy Baszczyski	Poznań University of Technology, Poland
Mihir Chakraborty	Jadavpur University, India
Shampa Chakraverty	Netaji Subhas Institute of Technology, India
Chien-Chung Chan	University of Akron, USA
Mu-Chen Chen	National Chiao Tung University, Taiwan
Costin-Gabriel Chiru	Technical University of Bucharest, Romania
Victor Codocedo	INSA Lyon, France
Chris Cornelis	University of Granada, Spain
Zoltan Erno Csajbok	University of Debrecen, Hungary
Jianhua Dai	Hunan Normal University, China
Rafal Deja	WSB, Poland
Dayong Deng	Zhejiang Normal University, China
Thierry Denoeux	Université de Technologie de Compiegne, France
Fernando Diaz	University of Valladolid, Spain
Pawel Drozda	University of Warmia and Mazury, Poland
Didier Dubois	IRIT/RPDMP, France
Ivo Dntsch	Brock University, Canada
Zied Elouedi	Institut Superieur de Gestion de Tunis, Tunisia
Rafael Falcon	Larus Technologies Corporation, Canada
Victor Flores	Universidad Catolica del Norte, Chile
Wojciech Froelich	University of Silesia, Poland
Brunella Gerla	University of Insubria, Italy
Piotr Gny	Polish-Japanese Academy of IT, Poland
Anna Gomolinska	University of Białystok, Poland
Salvatore Greco	University of Catania, Italy
Rafal Gruszczynski	Nicolaus Copernicus University in Toruń, Poland
Jerzy Grzymala-Busse	University of Kansas, USA
Bineet Gupta	Shri RamSwaroop Memorial University, India
Christopher Henry	University of Winnipeg, Canada
Christopher Hinde	Loughborough University, UK
Qinghua Hu	Tianjin University, China
Van Nam Huynh	JAIST, Japan
Dmitry Ignatov	National Research University HSE, Russia
Masahiro Inuiguchi	Osaka University, Japan
Ryszard Janicki	McMaster University, Canada
Richard Jensen	Aberystwyth University, UK
Xiuyi Jia	Nanjing University of Science and Technology, China
Michal Kepski	University of Rzeszów, Poland
Md. Aquil Khan	Indian Institute of Technology Indore, India
Yoo-Sung Kim	Inha University, South Korea
Marzena Kryszkiewicz	Warsaw University of Technology, Poland
Yasuo Kudo	Muroran Institute of Technology, Japan

Yoshifumi Kusunoki	Osaka University, Japan
Sergei O. Kuznetsov	National Research University HSE, Russia
Xuan Viet Le	Quy Nhon University, Vietnam
Huaxiong Li	Nanjing University, China
Jiye Liang	Shanxi University, China
Churn-Jung Liau	Academia Sinica, Taipei, Taiwan
Tsau Young Lin	San Jose State University, USA
Pawan Lingras	Saint Mary's University, Canada
Caihui Liu	Gannan Normal University, China
Guilong Liu	Beijing Language and Culture University, China
Pradipta Maji	Indian Statistical Institute, India
Benedetto Matarazzo	University of Catania, Italy
Jess Medina	University of Cadiz, Spain
Ernestina Menasalvas	Universidad Politecnica de Madrid, Spain
Claudio Meneses	Universidad Catolica del Norte, Chile
Marcin Michalak	Silesian University of Technology, Poland
Tams Mihlydek	University of Debrecen, Hungary
Fan Min	Southwest Petroleum University, China
Pabitra Mitra	Indian Institute of Technology Kharagpur, India
Sadaaki Miyamoto	University of Tsukuba, Japan
Mikhail Moshkov	KAUST, Saudi Arabia
Michinori Nakata	Josai International University, Japan
Amedeo Napoli	Inria, France
Hoang Son Nguyen	Hue University, Vietnam
Loan T. T. Nguyen	TDTU, Vietnam
Long Giang Nguyen	Institute of Information Technology, VAST, Vietnam
M. C. Nicoletti	FACCAMP and UFSCar, Brazil
Vilem Novak	University of Ostrava, Czech Republic
Agnieszka Nowak-Brzezińska	University of Silesia, Poland
Piero Pagliani	Research Group on Knowledge and Information, Italy
Sankar Pal	Indian Statistical Institute, India
Krzysztof Pancerz	University of Rzeszów, Poland
Vladimir Parkhomenko	SPbPU, Russia
Andrei Paun	University of Bucharest, Romania
Witold Pedrycz	University of Alberta, Canada
Tatiana Penkova	Institute of Computational Modelling SB RAS, Russia
Georg Peters	Munich University of Applied Sciences and Australian Catholic University, Germany
Alberto Pettorossi	Università di Roma Tor Vergata, Italy
Jonas Poelmans	Clarida Technologies, UK
Lech Polkowski	Polish-Japanese Academy of IT, Poland
Henri Prade	IRIT - CNRS, France
Mohamed Quafafou	Aix-Marseille University, France
Elisabeth Rakus-Andersson	Blekinge Institute of Technology, Sweden
Sheela Ramanna	University of Winnipeg, Canada

Zbigniew Ras	University of North Carolina at Charlotte, USA
Grzegorz Rozenberg	Leiden University, The Netherlands
Henryk Rybiski	Warsaw University of Technology, Poland
Wojciech Rzasa	Rzeszów University, Poland
Hiroshi Sakai	Kyushu Institute of Technology, Japan
Guido Santos	Universittsklinikum Erlangen, Germany
Gerald Schaefer	Loughborough University, UK
Zhongzhi Shi	Institute of Computing Technology Chinese Academy of Sciences, China
Marek Sikora	Silesian University of Technology, Poland
Bruno Simões	Vicomtech-IK4, Spain
Roman Słowinski	Poznan University of Technology, Poland
John Stell	University of Leeds, UK
Jaroslaw Stepaniuk	Bialystok University of Technology, Poland
Zbigniew Suraj	University of Rzeszów, Poland
Paul Sushmita	Indian Institute of Technology Jodhpur, India
Piotr Synak	Security On Demand, Poland
Andrzej Szałas	University of Warsaw, Poland
Marcin Szczuka	University of Warsaw, Poland
Ryszard Tadeusiewicz	AGH University of Science and Technology in Kraków, Poland
Bala Krushna Tripathy	VIT University, India
Li-Shiang Tsay	North Carolina A&T State University, USA
Dmitry Vinogradov	Federal Research Center for Computer Science and Control, RAS, Russia
Bay Vo	Ho Chi Minh City University of Technology, Vietnam
Alicja Wakulicz-Deja	University of Silesia, Poland
Guoyin Wang	Chongqing University of Posts and Telecommunications, China
Szymon Wilk	Poznan University of Technology, Poland
Arkadiusz Wojna	Security On Demand, Poland
Marcin Wolski	Maria Curie-Skłodowska University, Poland
Wei-Zhi Wu	Zhejiang Ocean University, China
Yan Yang	Southwest Jiaotong University, China
Jingtao Yao	University of Regina, Canada
Yiyu Yao	University of Regina, Canada
Dongyi Ye	Fuzhou University, China
Hong Yu	Chongqing University of Posts and Telecommunications, China
Mahdi Zargayouna	Université Paris Est, France
Nan Zhang	Yantai University, China
Qinghua Zhang	Chongqing University of Posts and Telecommunications, China
Yan Zhang	University of Regina, Canada
Bing Zhou	Sam Houston State University, USA
Wojciech Ziarko	University of Regina, Canada
Beata Zielosko	University of Silesia, Poland

Additional Reviewers

Azam, Nouman
Benítez Caballero, María José
Bui, Huong
Chen, Chun-Hao
Czołombitko, Michał
Jankowski, Dariusz
Le, Tuong
Li, Jinhai
Mai, Son
Nguyen, Dan

Nguyen, Duy Ham
Nguyen, Hoang Son
Nguyen, Van Du
Nguyen, Viet Hung
Pham, Thi-Ngan
Ramírez Poussa, Eloisa
Shah, Ekta
Son, Le Hoang
Su, Ja-Hwung
Vluymans, Sarah

Introducing Histogram Functions into a Granular Approximate Database Engine (Industry Talk)

Dominik Ślęzak[1] and Arkadiusz Wojna[2]

[1] Institute of Informatics, University of Warsaw, Poland
[2] Security On-Demand, USA/Poland

Abstract. We discuss an approximate database engine that we started designing at Infobright, and now we continue its development for Security On-Demand (SOD). At SOD, it is used in everyday data analytics, allowing for fast approximate execution of ad-hoc queries over tens of billions of data rows [1]. In our engine, queries are run against collections of histograms that represent domains of single columns over groupings of consecutively loaded data rows (so-called packrows). Query execution process corresponds to transformation of such granulated summaries of the input data into summaries reflecting query results [2].

We compare our algorithms that generate histogram descriptions of the original data with data quantization methods that are widely used in data mining. We also introduce a new idea of extending SQL with function *hist(a)* that produces quantized representation of column *a* by means of merging *a*'s histograms corresponding to particular packrows into a unified *a*'s histogram over the whole data. We refer to our recent works on summary-based data visualization [3] and machine learning [4] in order to illustrate several scenarios of utilizing *hist* in practice.

Keywords: Big data analytics · Data granulation · Data quantization

References

1. Ślęzak, D., Chądzyńska-Krasowska, A., Holland, J., Synak, P., Glick, R., Perkowski, M.: Scalable cyber-security analytics with a new summary-based approximate query engine. In: Proceedings of BigData, pp. 1840–1849 (2017)
2. Ślęzak, D., Glick, R., Betliński, P., Synak, P.: A new approximate query engine based on intelligent capture and fast transformations of granulated data summaries. J. Intell. Inf. Syst. **50**(2), 385–414 (2018)
3. Chądzyńska-Krasowska, A., Stawicki, S., Ślęzak, D.: A metadata diagnostic framework for a new approximate query engine working with granulated data summaries. In: Polkowski, L., et al. (eds.) IJCRS 2017. LNCS, vol. 10313, pp. 623–643. Springer, Cham
4. Ślęzak, D., Borkowski, J., Chądzyńska-Krasowska, A.: Ranking mutual information dependencies in a summary-based approximate analytics framework. In: Proceedings of HPCS (2018)

Contents

Subjective Analysis of Price Herd Using Dominance Rough Set Induction: Case Study of Solar Companies

Hamido Fujita[1($\boxtimes$)] and Yu-Chien Ko[2]

[1] Faculty of Software and Information Science, Iwate Prefectural University,
Takizawa 020-0693, Japan
HFujita-799@acm.org
[2] Department of Information Management, Chung Hua University,
Hsinchu 30012, Taiwan
eugene@chu.edu.tw

Abstract. Herd behavior depends on subjectivity and objectivity combination. Usually the former over controls the latter and makes a special distinction from others. Especially, herd could regard itself as objective thus sacrificing all differences. Getting insight of the subjectivity appears more and more important in economics. However, the combination of subjectivity and objectivity varies with time evolution. To illustrate subjective analysis, we propose an inferential model to distinguish special enterprises from price herds. It assumes public finance as intrinsic self of subjectivity and the herding behavior as objective expectation of majority then identifies subjective actions.

Keywords: Subjectivity · Price herd · Decision making
Dominance-based rough set · Induction

1 Introduction

In the stock market, the majority's behavior represents the expectation of most investors. Deviation from the majority often rises from subjective decision like Fig. 1 where ph is a price herd which has two sets. One requires its elements to move higher prices when the majority decline; the other behaves in the opposite way when the majority increase their prices. Figure 1 presents subjective k or k' holds the pressure from majority and assumes a risk against majority's wisdom. For the judgment of rationality, subjectivity is usually assigned to non rational. In this research, the subjectivity behaves against the majority, not mattering about rationality. Contrarily, subjectivity can be objective if most expectations are not rational.

Theoretically, ph can be expressed with the characteristics of financial information. Its behavior is coded with expectation and hesitance cascading of most investors. We are motivated to identify the subjective enterprises by taking

© Springer Nature Switzerland AG 2018
H. S. Nguyen et al. (Eds.): IJCRS 2018, LNAI 11103, pp. 1–12, 2018.
https://doi.org/10.1007/978-3-319-99368-3_1

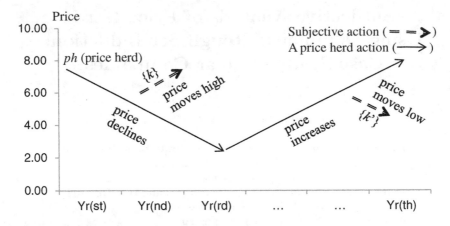

Fig. 1. The subjective actions vs a price herd's behavior

advantage of price herd model (*ph*) [14,20]. It will identify the behavior of a price herd through variables of Altman Z-Score. However, the timing points of identifying subjective enterprises in Fig. 1 have a problem, i.e. too many choices. To solve this, we propose a subjective clustering (*SC*) to distinguish special enterprises from the price herd. To illustrate its operation, we will apply *SC* on a solar industry through Taiwan Economic Journal (TEJ) database which provides public information of financial market.

The context of this article includes the innovative notions of *SC*, the model of *SC*, application of *SC* on a solar industry, discussion, and concluding remarks.

2   The Innovative Notions and Literatures of *SC*

In this research, *SC* is designed to identify the subjective enterprises distinguished from price herds. The induction of *SC* is updated from PH model [14] which is extended from dominance-based rough set approach (DRSA), rough set theory (RST). Its innovative notions and literatures are described below.

2.1   Subjectivity

Since "I think, therefore I am" [7,8] was proposed, subjectivity can be explored with inference. Recently, it is divided into two categories [16]. One adopts framework composed of conceptual consciousness to illustrate subjective behavior. The other adopts self-organizing power to rethink about ethics. In the information field, Bayesian probability is used in quantitative measures to construct a subjectivity framework [17]. This also builds a conceptual model for inference. Its subject concept is defined to comprise true distribution, probability space, hypotheses, observations, actions, and causal intervention [17]. The human behavior is regarded as correspondence of subjective consciousness. The corresponding inference about subjectivity thus can be expressed with scientific languages. The followings are its technical components.

2.2   Variables of Altman Z-Score

The financial information is an intrinsic part of companies. Therefore, management or decision underpinned by finance is a common sense. Altman's Z-Score has been playing a headship of discriminating survivals and failures, achieving up thirteen thousand citations in Google survey on 29 April 2017 and 75%–90% reliability [1,2]. The relevance between Altman variables and financial health has a highly positive correlation [15]. One of Altman variable, market values of equity (V), provides another expression of price. It is used as the price of PH and the price variation is treated as a herd behavior. Enterprises taking the opposite direction from price herds is regarded as subjective in this paper.

2.3   Granule and Evidence of PH

The idea of PH originates from identifying objectivity composed of characteristic granules. This paper designs a granule with three types of information: the inferential relevance between prices and herds, the herd's characteristic composed of objects' properties, and the decision preference [5]. The inferential relevance was proposed by Keynes [13] who expresses a rational belief about the inferential relevance based on the probability-relation. The objects' properties proposed by RST are expressed by indiscernibility [18], similarity [29], preference [10], etc. These properties are further formulated by relations [19], approximations (observable or unobservable) [19], classes (dominating or dominated) [10], etc. The decision preference of stakeholders in this paper is expressed by classified prices. Combining these three types of information can make the granules operated in mathematical sets to express a herding characteristic of objectivity, i.e. the majority. A granule in approximations verified to have certain relevance is defined as evidence, symbolized as $e_{j,k}$ in Eq. (1).

$$e_{j,k} = 1 \text{ or } 0 \qquad (1)$$

where 1 means a certain evidence, 0 means not a certain evidence, j indexes a variable, and k indexes an object. $e_{j,k}$ will comprises the induced PH.

2.4   Approximations in RST

In general, the objects' properties based on attributes cannot clearly specify a vague set. Therefore, approximations are used to express and estimate the vagueness by RST. The approximations are a pair of sets, i.e. $\underline{P}(X)$ and $\overline{P}(X)$ [19]. In this paper, a vague set X is designed as a simple herd containing $\underline{P}(X)$ and belonging to $\overline{P}(X)$, expressed in Eq. (2).

$$\underline{P}(X) \subseteq X \subseteq \overline{P}(X) \quad \text{where} \quad \overline{P}(X) = \bigcup_{X \cap X_k \neq \emptyset} X_k \qquad (2)$$

where P represents an inference function about the approximations of X based on attributes, $\underline{P}(X)$ is named the lower approximation, and $\overline{P}(X)$ is named the upper approximation. According to RST, $\underline{P}(X)$ has the certain characteristic of X by requiring all its elements in X. $\overline{P}(X)$ has the relevant characteristic of X by requiring $X \cap X_k \neq \emptyset$ where X_k is an equivalence class of X.

The objective characteristics of PH is designed as a set based on $\underline{P}(X)$. There are two types of estimations, i.e. the priori (hypothetical) $\underline{S}(X)$ and the posteriori (resolved) $\underline{S}'(X)$. $\underline{S}(X)$ is same as $\underline{P}(X)$ before making an inference. $\underline{S}'(X)$ is an induced $\underline{S}(X)$ by classifying X with the decision preference which is the dominance set in DRSA. Technically, $\underline{S}'(X)$ satisfies Eq. (2) and dominance induction next.

2.5 Induction of DRSA

PH is updated from DRSA. The induction of DRSA is a backward inference to classify objects with the determined preference. DRSA can disclose the cause of dominance thus able to support multi–criteria decision making (MCDM) [3, 6, 10, 11, 21, 22]. Its binary induction, i.e. dominance or non-dominance, is presented in Eq. (3) which can resolve the certain objects in $\underline{P}(X)$ by the dominance set $Cl_t^{\geq}$.

$$\underline{P}(X) \mapsto Cl_t^{\geq} \quad \text{where} \quad \underline{P}(X) = \{x | x \in Cl_t^{\geq}, D_P^+(x) \subseteq Cl_t^{\geq}\} \tag{3}$$

where t represents the number of the objects in the dominance set, $\underline{P}(X)$ is a lower approximation of Eq. (2), P is an inferential function covering a set of attributes, and $D_P^+(x)$ is a set having elements whose preferences are at least as x. The induction of Eq. (3) can find out $\underline{P}(X)$ from $Cl_t^{\geq}$. The constrains of induction contain membership ($\frac{|\underline{P}(X) \cap [x]|}{|[x]|}$), coverage ($\frac{|\underline{P}(X) \cap Cl_t^{\geq}|}{|Cl_t^{\geq}|}$), and accuracy ($\frac{\underline{P}(X)}{\overline{P}(X)}$) where $| \cdot |$ is the cardinality of a set and $[x]$ is an equivalence class. Mathematically, the membership and coverage degrees are expressed by Bayesian's conditional probabilities, like information cascade described next.

2.6 The Expectation of PH

In the information cascading, each cascade is predicted from an objective estimation and an observable value; the preference decision appears like H or L expressing high or low information about gain or loss; the observable value for adoption or rejection appears like V_H or V_L. Its estimation probability takes an action on the value, presented as $e'_{j,t}$ for attribute j at time t. In another word, $e'_{j,t}$ is an indicator of ph. It has three expectant rates with H for adopting V_H formulated as, $E_{j,2i}^H$ (up cascade), $E_{j,2i}^{non}$ (no cascade), and $E_{j,2i}^L$ (down cascade) at the sequence positions $2i$ ($i = 1, 2, ...$). These expectancy formulas are presented as Eq. (4) where $e'_{j,t}$ is an objective probability at initiation t of a price herd. The reason lies in $e'_{j,t}$ expresses the expectation of majority.

$$\begin{cases} E_{j,2i}^{non} = \left(E_{j,2}^{non}\right)^i \\ E_{j,2i}^H = E_{j,2}^H \times \left(1 + E_{j,2}^{non} + \left(E_{j,2}^{non}\right)^2 + ... + \left(E_{j,2}^{non}\right)^{i-1}\right) \\ E_{j,2i}^L = E_{j,2}^L \times \left(1 + E_{j,2}^{non} + \left(E_{j,2}^{non}\right)^2 + ... + \left(E_{j,2}^{non}\right)^{i-1}\right) \end{cases} \tag{4}$$

where

$$
\begin{cases}
E_{j,2}^{non} = e'_{j,t}(1 - e'_{j,t}) \\
E_{j,2}^{H} = \dfrac{e'_{j,t}(1 + e'_{j,t})}{2} \\
E_{j,2}^{L} = \dfrac{(e'_{j,t} - 1)(e'_{j,t} - 2)}{2}
\end{cases}
$$

3 A Subjective Enterprises, $\{k\}$

The behavior of subjective herd is designed to satisfy Eq. (5) which signifies a price herd through the objective probability $(e'_{j,t})$ where $0.4 < e'_j < 0.5$, j indexes an Altman variable, and t means a timing tag on years. The subjectiveness can be distinguished due to its opposite direction from majority.

$$
\begin{cases}
\text{Price herd:} & \{x_{k,t'-t} \mid V_{k,t'} \approx V_{k,t} \times E_{j,2}^{H}\} \\
\text{Subjective } k: & x_{k,t'-t} \text{ where } V_{k,t'} - V_{k,t} > 0
\end{cases}
\quad \text{where } 0.4 < e'_{j,t} < 0.5 \quad (5)
$$

where V means companies' stock price, j means a herding attribute, k represents a company, and t is the initial timing tags, and t' indicates the equilibrium time of herding movement. All companies satisfying Eq. (5) have unique subjectivity. Usually they keep their way opposite to herding movement.

3.1 The Information Table of PH and SC

The information table of ESPH is a data set containing all companies of the solar energy industry. It is mathematically defined as $IS = \{X, Q, f, R, V_H\}$, where $X = \{y \mid y = 1, 2, ..., n\}$ is a set of companies supposed to have securities' interests same as investors, $Q = \{q_1, q_2, ..., q_m\}$ represents a set of variables (defined by Altman in Table 1), m is the number of variables, $f : X \times Q \to R$ is a function transforming a variable's value of some company into a rank within

Table 1. Altman variables

Variables	Formula
q_1	Working capital/Total asset
q_2	Retained Earnings/Total assets; $q_2 = q_{21} + q_{22} + q_{23}$ where q_{21} is undistributed surplus earnings, q_{22} is special reserve, q_{23} is legal reserve. These sub items are defined as income tax and owners' equity in Taiwan
q_3	Earnings before interest and taxes/Total assets
q_4	Market value of equity (V)/Book value of total debt;
q_5	Sales/Total assets
V	Market value of equity; $V = $ (unchanged) security price $\times$ outstanding shares

Table 2. Data set and indicators of the price herd in 2010

k	q_1	q_2	q_3	q_4	q_5	V	s_1'	$s_2'(\downarrow)$	$s_3'(\downarrow)$	s_5'
1	0.048	−0.032	−0.035	4.388	0.381	0.060	0	0	0	0
2	0.003	0.212	0.092	2.594	0.629	0.321	1	1	1	1
3	0.056	0.157	0.082	1.917	0.889	0.916	1	1	1	1
4	0.094	0.293	0.158	12.227	0.328	3.413	1	1	1	1
5	0.103	0.101	0.071	2.339	0.642	0.114	1	1	1	1
6	0.038	−0.154	−0.067	2.693	0.467	0.097	0	0	0	0
7	0.038	−0.030	−0.020	1.495	0.374	0.228	1	0	0	1
8	0.024	0.183	0.096	1.633	0.955	0.073	0	0	0	0
9	0.000	0.051	0.041	1.975	1.015	0.650	1	1	1	1
10	0.159	0.137	0.102	3.236	0.639	0.354	1	1	1	1
11	0.066	0.077	0.038	5.444	0.601	0.087	0	0	0	0
12	−0.019	−0.049	−0.018	3.721	0.454	0.022	0	0	0	0
13	−0.045	−0.439	0.001	0.571	0.097	0.005	0	0	0	0
14	0.053	0.155	0.099	4.376	0.414	0.074	0	0	0	0
15	−0.073	0.011	0.064	0.622	0.184	0.004	0	0	0	0
16	−0.055	0.213	0.088	2.655	0.521	0.040	0	0	0	0
17	0.006	0.153	0.075	3.043	0.956	0.118	1	1	1	1
18	0.122	0.230	0.123	4.373	0.791	0.166	1	1	1	1
19	0.134	0.095	0.097	7.042	0.396	0.180	1	1	1	1
20	0.036	−0.372	0.009	8.376	0.392	0.019	0	0	0	0
21	0.019	−0.025	0.015	1.185	1.710	0.009	0	0	0	0
22	0.243	0.052	0.053	5.922	0.350	0.100	1	1	1	1
23	0.011	−0.153	−0.141	1.136	0.989	0.119	1	0	0	1
24	0.234	0.226	0.190	3.843	1.151	0.269	1	1	1	1
25	0.059	0.300	0.133	5.759	1.154	0.126	1	1	1	1
26	0.088	0.108	0.081	1.809	0.746	0.188	1	1	1	1
27	−0.115	−0.655	−0.467	3.003	0.455	0.013	0	0	0	0
28	0.083	0.119	0.106	2.570	0.769	0.144	1	1	1	1
29	0.108	−0.119	−0.111	2.916	1.525	0.016	0	0	0	0
30	0.052	0.231	0.135	11.294	0.679	0.132	1	1	1	1
31	0.141	0.088	0.103	3.420	0.659	0.080	0	0	0	0
32	0.202	0.319	0.340	83.134	0.771	0.251	1	1	1	1
33	0.071	0.176	0.106	1.855	1.616	0.058	0	0	0	0
34	0.018	0.114	0.074	3.843	0.632	0.025	0	0	0	0
35	0.049	0.255	0.178	4.112	0.581	0.127	1	1	1	1
36	0.163	0.260	0.221	14.430	0.682	0.047	0	0	0	0
37	0.234	0.310	0.168	5.320	0.501	0.083	0	0	0	0
38	0.196	0.185	0.131	11.196	0.852	0.182	1	1	1	1
39	0.033	0.105	0.077	2.042	1.460	0.128	1	1	1	1
40	0.128	−0.673	−0.024	3.516	0.339	0.030	0	0	0	0
41	−0.113	−0.134	−0.132	6.100	0.418	0.005	0	0	0	0
42	0.103	0.216	0.133	4.462	0.704	0.120	1	1	1	1
43	0.047	0.007	0.006	0.784	0.609	0.016	0	0	0	0
44	−0.006	0.002	0.020	3.057	0.360	0.031	0	0	0	0
45	−0.365	0.114	0.108	12.823	2.102	0.127	0	1	1	1

Note: k means index of companies.

the variable, $R = \{1st, 2nd, ..., nth\}$ is a ranking set, the ranking orders follow $1st \succeq 2nd \succeq ... \succeq nth$, and V_H represents the higher prices at the upper half securities such that $|V_H| \approx 0.5 \times |X|$. Our design takes half-half classification to estimate herding characteristics like pessimism, equilibrium, and optimism. IS adopts all companies instead of individuals to reduce the interference from the stock speculation of few companies.

4 Application of *SC*

The solar energy industry in Taiwan has been facing challenges like the global crisis in 2008 [26], oversupply in 2010 [4, 25], anti-dumping 2011–2016 [9], etc. On the time line, the stock price involved a turnaround, downward before upward during 2010–2014. Our case study empirically applies *SC* on TEJ to solve the price herd *ph* and subjective enterprises.

4.1 The Evidential Evidence of *ph*

The left part of Table 2 contains the dataset from TEJ. It is used to check the herding evidence composed of 1 or 0 in the right part. The first column represents the id (k) of companies. These evidence theoretically gives a quantitative measure about herding. Its embedded knowledge is illustrated in the followings.

4.2 The Behavior of *ph*

Figure 2 displays *ph* with the macro behavior of a solar industry in Taiwan during 2010–2011. Figure 3 is the behavior of *ph* within 2010–2011. As seen, the real prices were same as the expectation marked with circles, o.

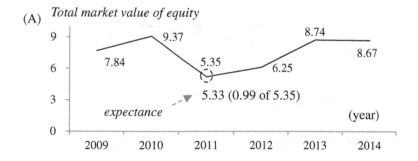

Fig. 2. The macro behavior of *ph* and its expectancy in 2011

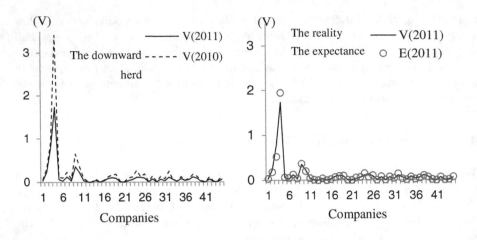

Fig. 3. The micro behavior of ph and their expectancy in 2011

4.3   The Behavior of a Subjective Enterprise, S

Figure 4 shows an subject deviated from ph in price variation. Because there is only one subjective enterprise, we use S to represent it. S assumed risk pressure and financial losses at the same time. With this result, no right or mistake is available for herding. The revealed knowledge is that about half companies had stock prices higher than their financial underpinning. Investors had no confidence in the stock price and most of them were apt to sell stocks at lower price.

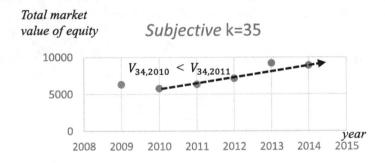

Fig. 4. The action of subjective enterprise during 2010–2011

5   Discussion

The subjective enterprises are very few in the analysis result. Their behavior appears not only deviated from the majority but unique. Followings are discussions.

5.1   Subjective Arguments

The behavioral correspondence from individual nature is objectively rational thus expressible by scientific languages. However, the behavior of individual consciousness containing emotion, hesitance, etc. leans to expectation instead of rationality. Therefore, Freud argues subjective psyche is composed of three parts, i.e. Id comprises the intrinsic nature, Ego involves consciousness of a subject who has expectancy, experience, emotion, information, etc., and Superego covers beyond Id and Ego [12, 27].

By the inferential framework of Fig. 5, we argue that Ego part is dynamic and variable with individual's consciousness combinations; Its variation might make Ego huge or nothing. For an individual company, evidential Ego is not suitably expressed by Bayesian theory because there might be not enough evidence to assure its subjectivity. Therefore, our research applies inference on Ego with all companies then distinguishes special ones from the majority. Price herd is used to disclose the subjective enterprises.

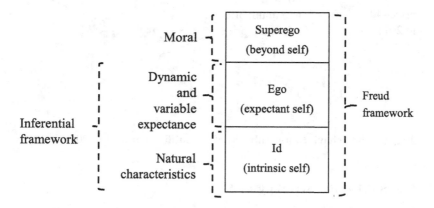

Fig. 5. A proposed inferential framework of subjectivity

5.2   Theoretical S in Herd Movement

We adopt financial information to comprise each companies as Freud's Id. The herd movement during 2010–2011 represents Ego behavior with expectant consciousness which is regarded as objective due to covering majority. In this period, only one company, i.e. S, in the solar industry is identified as unique and subjective. It is numbered as 35 in 45 companies. Its stock price from 2009 to 2014 is presented in Fig. 4. Based on its price behavior, S seems confident not impacted during herding movement.

5.3   Practical S in Herd Movement

S was one of twenty companies with the same financial health in 2010. Its business mission is set as the provider of total solution. It emphasis on future prediction and trust relationship [23]. In our observations, its actions during herding movement include reinvesting Topco Lancaster Investment in USA and DIO Energy GmbH in Germany [24]. At the worst situation in the financial market, it is still prominently subjective.

5.4 S Behavior After Herd Movement

The identified S is supposed to behave subjectively, i.e. not going down in price with others' expectation. In our trace on S during 2011 to 2018, it gradually climbs up in the financial market as Fig. 6, captured from Yahoo website [28]. Its price doubled in 6 years. Figure 6 has bar charts for each month with maximum and minimum prices.

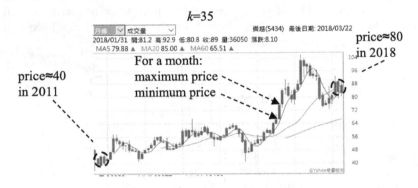

Fig. 6. The behavior after subjectivity identification during 2011–2018

5.5 The Subjective Strategy of S

S was established in Taiwan since 1990. From the very beginning, it continues reinvesting in related companies as many as twelve times [24]. In average, it has one reinvestment in two years, no matter economic environment bad or good. Currently, it is not only a manufacturer but an equipment supplier.

6 Concluding Remarks

Herd behavior is analyzed by extending from a price herd model to disclose subjective enterprises in this research [14]. Its result shows only one subjective enterprise deviated from the majority declining their prices. This subject is theoretically identified with the proposed subjective clustering, SC. By practical observations, it has a subjective characteristic, i.e. never stopping reinvestment even the stock market seemed to be crashing down.

In this research, SC solves the subjective enterprise by financial inference and discloses its characteristics with actions and observations. The former distinguishes the resulted enterprises with unique behavior. The latter gives supports about its subjectivity. Freud's framework is applied as the base of inference. With the herding expectations among companies, the subjectiveness is identified with clustering operations of Eq. (4). The dynamic and variable Ego thus can be resolved indirectly.

The subjective analysis is bigger than the general imagination. In the future work, it deserves more efforts and time to get deeper insight about subjective intelligence.

References

1. Altman, E.I., Iwanicz-Drozdowska, M., Laitinen, E.K., Suvas, A.: Financial distress prediction in an international context: a review and empirical analysis of Altman's Z-score model. J. Int. Financ. Manage. Acc. **28**, 131–171 (2016). https://doi.org/10.1111/jifm.12053
2. Altman, E.I.: Financial ratios, discriminant analysis and the prediction of corporate bankruptcy. J. Finance **23**(4), 589–609 (1968)
3. Augeri, M.G., Cozzo, P., Greco, S.: Dominance-based rough set approach: an application case study for setting speed limits for vehicles in speed controlled zones. Knowl. Based Syst. **89**, 288–300 (2015). https://doi.org/10.1016/j.knosys.2015.07.010
4. Cowen, T.: Why solar panel prices are falling (2011). http://marginalrevolution.com/marginalrevolution/2011/11/why-solar-panel-prices-are-falling.html
5. Denoeux, T.: Analysis of evidence-theoretic decision rules for pattern classification. Pattern Recognit. **30**(7), 1095–1107 (1997). https://doi.org/10.1016/S0031-3203(96)00137-9, http://www.sciencedirect.com/science/article/pii/S0031320396001379
6. Denoeux, T., Zouhal, L.M.: Handling possibilistic labels in pattern classification using evidential reasoning. Fuzzy Sets Syst. **122**(3), 409–424 (2001). https://doi.org/10.1016/S0165-0114(00)00086-5, http://www.sciencedirect.com/science/article/pii/S0165011400000865
7. Descartes, R.: A Discourse on the Method. Oxford University Press, Oxford (2006). http://www.rlwclarke.net/Theory/SourcesPrimary/DescartesDiscourseonMethod.pdf
8. Descartes, R.: Discourse on the method of rightly conducting ones reason and seeking truth in the sciences. In: Bennett, J (2017). http://www.earlymoderntexts.com/assets/pdfs/descartes1637.pdf
9. Trade of European Commission DG: The european union's measures against dumped and subsidised imports of solar panels from china (2016). http://trade.ec.europa.eu/doclib/docs/2015/july/tradoc_153587.pdf
10. Greco, S., Matarazzo, B., Slowinski, R.: Rough sets theory for multicriteria decision analysis. Eur. J. Oper. Res. **129**(1), 1–47 (2001)
11. Greco, S., Matarazzo, B., Slowinski, R.: Rough approximation by dominance relations. Int. J. Intell. Syst. **17**(2), 153–171 (2002)
12. Jacoby, J.: Is it rational to assume consumer rationality? some consumer psychological perspectives on rational choice theory **6**, 81 (2013)
13. Keynes, J.M.: A Treatise on Probability. Macmillan, London (1921)
14. Ko, Y.C., Fujita, H.: Evidential probability of signals on a price herd predictions: case study on solar energy companies. Int. J. Approximate Reasoning **92**, 255–269 (2018). https://doi.org/10.1016/j.ijar.2017.10.015, http://www.sciencedirect.com/science/article/pii/S0888613X1730213X

15. Ko, Y.C., Fujita, H., Li, T.: An evidential analysis of altman z-score for financial predictions: case study on solar energy companies. Appl. Soft Comput. **52**, 748–759 (2017). https://doi.org/10.1016/j.asoc.2016.09.050, http://www.sciencedirect.com/science/article/pii/S1568494616305099
16. Mansfield, N.: Subjectivity: Theories of the Self from Freud to Haraway. NYU Press, New York (2000). https://books.google.com.tw/books?id=qBVh5gVTlC4C
17. Ortega, P.A.: Subjectivity, bayesianism, and causality. Pattern Recognit. Lett. **64**, 63–70 (2015). https://doi.org/10.1016/j.patrec.2015.04.018, http://www.sciencedirect.com/science/article/pii/S016786551500135X. philosophical Aspects of Pattern Recognition
18. Pawlak, Z.: Granularity of knowledge, indiscernibility and rough sets. In: The 1998 IEEE International Conference on Fuzzy Systems Proceedings - IEEE World Congress on Computational Intelligence, vol. 1, pp. 106–110 (1998)
19. Pawlak, Z.: Rough probability. Bull. Polish Acad. Sci. Math **32**(9–10), 607–612 (1984)
20. Shiller, R.J.: Chapter 20 human behavior and the efficiency of the financial system. In: Handbook of Macroeconomics, vol. 1, pp. 1305–1340. Elsevier, New York (1999). https://doi.org/10.1016/S1574-0048(99)10033-8, http://www.sciencedirect.com/science/article/pii/S1574004899100338
21. Slowinski, R., Greco, S., Matarazzo, B.: Rough sets in decision making. In: Meyers, A.R. (ed.) Encyclopedia of Complexity and Systems Science, pp. 7753–7787. Springer, New York (2009). https://doi.org/10.1007/978-1-4614-1800-9
22. Slowinski, R., Stefanowski, J.: Rough classification in incomplete information systems. Math. Comput. Modell. **12**(10–11), 1347–1357 (1989)
23. TOPCO: About (2018). www.topco-global.com/webfront/pages/About.aspx
24. TOPCO: Company milestones (2018). www.topco-global.com/webfront/pages/About.aspx
25. Wang, U.: Report: solar panel supply will far exceed demand beyond 2012 (2012). http://www.forbes.com/sites/uciliawang/2012/06/27/report-solar-panel-production-will-far-exceed-demand-beyond-2012/#22ff115c6a19
26. Wiki: Financial crisis of 2007-2008. https://en.wikipedia.org/wiki/Financial_crisis_of_2007%E2%80%9308
27. Wikipedia: Freud's psychoanalytic theories (2018). https://en.wikipedia.org/wiki/Freud%27s_psychoanalytic_theories
28. yahoo: Stock (2018). https://tw.stock.yahoo.com/q/ta?s=5434
29. Yao, Y.: Rough sets, neighborhood systems and granular computing. In: 1999 IEEE Canadian Conference on Electrical and Computer Engineering, vol. 3, pp. 1553–1558 (1999)

Three-Way Decisions and Three-Way Clustering

Hong Yu$^{(\boxtimes)}$

Chongqing Key Laboratory of Computational Intelligence,
Chongqing University of Posts and Telecommunications,
Chongqing 400065, People's Republic of China
yuhong@cqupt.edu.cn

Abstract. A theory of three-way decisions is formulated based on the notions of three regions and associated actions for processing the three regions. Inspired by the theory of three-way decisions, some researchers have further investigated the theory of three-way decisions and applied it in different domains. After reviewing the recent studies on three-way decisions, this paper introduces the three-way cluster analysis. In order to address the problem of the uncertain relationship between an object and a cluster, a three-way clustering representation is proposed to reflect the three types of relationships between an object and a cluster, namely, belong-to definitely, uncertain and not belong-to definitely. Furthermore, this paper reviews some three-way clustering approaches and discusses some future perspectives and potential research topics based on the three-way cluster analysis.

Keywords: Three-way decisions · Three-way clustering · Uncertain
Soft clustering

1 Introduction

To model a particular class of human ways of problem solving and information processing, Professor Yao [55] proposed a theory of three-way decisions. The basic ideas of three-way decisions are to divide a universal set into three pairwise disjoint regions, or more generally a whole into three distinctive parts, and to act upon each region or part by developing an appropriate strategy [57].

The essential ideas of three-way decisions are commonly used in everyday life and widely applied in many fields and disciplines including medical decision-making, social judgement theory, hypothesis testing in statistics, management sciences and peer review process. In the last few years, we have witnessed a fast growing development and applications of three-way approaches in areas of decision making, email spam filtering, clustering analysis and so on [10,24,28,64].

The term "three-way decisions" embraces all aspects of a decision-making process, including tasks such as data and evidence collection and analysis for supporting decision making, reasoning, computing in order to arrive at a particular decision, justification and explanation of a decision. The unique feature

© Springer Nature Switzerland AG 2018
H. S. Nguyen et al. (Eds.): IJCRS 2018, LNAI 11103, pp. 13–28, 2018.
https://doi.org/10.1007/978-3-319-99368-3_2

of three-way decisions is a type of three-way approaches (i.e., the division of a whole into three parts) to problem solving and information processing. We may replace "decisions" in "three-way decisions" by other words to have specific interpretations such as three-way computing, three-way processing, three-way classification, three-way analysis, three-way clustering, three-way recommendation, and many others [59].

2   Reviews on Three-Way Decisions

The main idea of three-way decisions is to divide a universe into three disjoint regions and to process the different regions by using different strategies. By using notations and terminologies of rough set theory [38,39,58], we give a brief description of three-way decisions as follows [64].

Suppose U is a finite nonempty set of objects or decision alternatives and D is a finite set of conditions. Each condition in D may be a criterion, an objective, or a constraint. The problem of three-way decisions is to divide, based on the set of conditions in D, U into three pair-wise disjoint regions by a mapping f:

$$f : U \longrightarrow \{RI, RII, RIII\}. \tag{1}$$

The three regions are called Region I, Region II, and Region III, respectively.

Depending on the construction and interpretation of the mapping f, there are qualitative three-way decisions and quantitative three-way decisions. In qualitative three-way decision models, the universe is divided into three regions based on a function f that is of a qualitative nature. Quantitative three-way decision models are induced by that is of a quantitative nature. An evaluation-based three-way decision model uses an evaluation function that measures the desirability of objects with reference to the set of criteria.

It should be pointed out that we can have a more general description of three-way decisions by using more generic labels and names. For example, in an evaluation-based model of three-way decisions [55], we can use a pair of thresholds to divide a universe into three regions. If we arrange objects in an increasing order with lower values at left, then we can conveniently label the three regions as the left, middle, or right regions, respectively, or simply L, M, and R regions [64]. In a similar way, strategies for processing three regions can be described in more generic terms [5,6,57].

Originally, the concept of three-way decisions was proposed and used to interpret probabilistic rough set three regions. Further studies show that a theory of three-way decisions can be developed by moving beyond rough set theory. In fact, many recent studies go far beyond rough sets. In order to go further insights into three-way decisions and promote further research, this paper gives a brief review on the studies of three-way decisions from the following respects.

- Cost-sensitive sequential three-way decisions. Three-way decisions originate from the studies on the decision-theoretic rough set (DTRS) model. The DTRS presents a semantics explanation on how to decide a concept into positive, negative and boundary regions based on the minimization of the decision cost, rather than decision error. Li et al. [11] incorporated the three-way decisions into cost-sensitive learning and proposed a three-region cost-sensitive classification. It is evident that the boundary decision may achieve lower cost/risk than positive and negative decisions do, if available information for immediate decision is insufficient, which is consistent with human decision process [9,11]. Based on the DTRS, Ju et al. [26] constructed a generalized framework of cost-sensitive rough set with test cost and decision cost simultaneously, and further introduced multi-granulation DTRS into this field and proposed the cost-sensitive multi-granulation rough set model by considering two different costs [27], which enriches the semantics interpretation of cost-sensitive models based on the DTRS.

In real-world applications, the available information is always insufficient, or it may associate with extra costs to get available information, which leads to frequent boundary decision. However, if the available information continuously increases, the previous boundary decisions may be converted to positive or negative decisions, which forms a sequential decision process [54,56]. Li et al. proposed a cost-sensitive sequential three-way decision strategy [12,15], and introduced the method to handle the imbalance of misclassification cost and the insufficient of image information [13], and further investigated deep neural networks based on sequential granular feature extraction [14]. Considering the multilevel granular structure of real-world problems, Yang et al. proposed a unified model of sequential three-way decisions and multilevel incremental processing for complex problem solving [74].

- Determining the thresholds. Compared to two-way decisions approaches, three-way decisions approaches introduce deferment decision through a pair of thresholds (α, β). Therefore, for the three-way decision models, a great challenge is acquirement of a set of pairs of thresholds (α, β). Thus, Shang and Jia [23,25] studied this problem from an optimization viewpoint, in which the thresholds and corresponding cost functions for making three-way decisions can be learned from given data without any preliminary knowledge [17]. Zhang and Zou et al. [82] proposed a cost-sensitive three-way decisions model based on constructive covering algorithm (CCA); Zhang and Xing et al. [83] introduced CCA to the three-way decisions procedure and proposed a new three-way decisions model based on CCA to obtain POS, NEG and BND automatically.

Yao and his group explored the use of game-theoretic rough set (GTRS) model to handle thresholds determination issue. Afridi et al. [1] constructed a three-way clustering approach for handling missing data by introducing a method of thresholds determination based on a tradeoff game between the properties of accuracy and generality of clusters. Besides, Zhang and Yao applied GTRS in multi-criteria based three-way classification problem [76]. By considering probabilistic rough sets based models of game-theoretic rough sets for inducing three-way decisions, Rehman et al. [44] proposed an architecture of protein functions classification with probabilistic rough sets based three-way decisions.

- Three-way decisions with DTRS. Considering that incomplete data with missing values are very common in many data-intensive applications. Luo et al. [32] proposed an incremental approach for updating probabilistic rough approximations, with the variation of objects in an incomplete information system. Yang et al. [71] proposed the notions of weighted mean multi-granulation decision-theoretic rough set, optimistic multi-granulation decision-theoretic rough set, and pessimistic multi-granulation decision-theoretic rough set in an incomplete information system. Based on the DTRS, Liu et al. [30] proposed a novel three-way decision model by defining a new relation to describe the similarity degree of incomplete information.

Recently, Yao [60] have extended the theory of three-way decisions to the framework of interval sets and the corresponding three-way concept analysis in incomplete contexts. Li et al. [16] studied three-way cognitive concept learning via multi-granularity, and designed a three-way cognitive computing system which is in fact a dynamic process to update three-way granular concepts. Li et al. [13] simulated the human decision-making process, and proposed a dynamic sequential three-way decision method for cost-sensitive face recognition, by considering available information increases continuously. To deal with the problem of incremental overlapping clustering, Yu et al. [65] designed a dynamic three-way decision strategy to update the clustering when the data increase. Liu et al. [29] considered the dynamic change of loss functions in the DTRS with the time, and further proposed the dynamic three-way decision model. Zhang et al. [80] introduced a new three-way decision model based on dynamic decision making with the updating of attribute values.

- Three-way attribute reduction. The combination of three-way decisions and attribute reducts has theoretical significance and applicable prospects. In this regard, Chen et al. [3] discussed reduction issue based on three-way decisions in neighborhood rough sets. By utilizing double-quantitative measure, Zhang et al. [81] established a hierarchical reduct system, including qualitative/quantitative reducts, tolerant/approximate reducts. Furthermore, Zhang et al. [79] introduced three-way decisions into attribute reducts, and constructed a novel framework of three-way attribute reducts, aiming to directly quantify the final reduction action. Ren and Wei [45] studied three-way concept analysis, and proposed an approach for attribute reductions of three-way concept lattices. Ma and Yao [36] gave a general definition of class-specific attribute reducts, and thus, introduced the class-specific attribute reducts framework on the perspective of three-way decision.
- Three-way decisions and other theories. There are lots of excellent results on the combination of three-way decisions and other theories such as Dempster-Shafer theory, fuzzy sets, formal concept analysis and so on.

Wang et al. [51] proposed a Dempster-Shafer theory based intelligent three-way group sorting method. Zhao and Hu [75]investigated fuzzy and interval-valued fuzzy probabilistic rough sets and proposed their corresponding three-way decisions models, which are appropriate for fuzzy events. Hu [21] established the framework of three-way decisions spaces based on partially order sets and studied three-way decisions based on hesitant fuzzy sets [21,22]. In order to generate decision rules in incomplete information systems, Yang and Tan [72] constructed the evaluation function by combining the intuitionistic fuzzy set and the three-way decisions. To overcome the limitation of the existing three-way decisions models in uncertainty environment, Zhai et al. [78] extended the rough fuzzy set to tolerance rough set, thus, proposed the three-way decisions model based on tolerance rough fuzzy sets. Based on linguistic information-based decision-theoretic rough fuzzy sets, Sun et al. [49] established the corresponding three-way decisions approach to solve multiple attribute group decision problem. To mine three-way concepts to support three-way decisions in formal context, Li et al. [16] studied three-way cognitive concept learning via multi-granularity. Qi et al. [42] proposed the three-way concept analysis based on combining three-way decisions [55] and formal concept analysis [7]. Besides, Ren and Wei investigated the attribute reductions method over three-way concept lattices [45]. Aimed at analyzing the uncertainty and incompleteness in single-valued neutrosophic set, Singh [47] proposed three-way formal fuzzy concept lattice representation. With the issue of three-way concept lattices construction, Qian et al. [43] proposed approaches to create the three-way concept lattices based on the concept lattices of Type I-combinatorial context and Type II-combinatorial context. Yu et al. [73] made efforts on characterizing three-way concept lattices and three-way rough concept lattices, which enriched the theory of three-way concept lattices.

- Applications on three-way decisions. Since the theory of three-way decisions has been proposed, scholars have applied the idea to different applications. Yu and her group studied overlapping clustering [61], determining the number of clusters [62], incremental clustering [65] and so on, based on the three-way decision theory. They also applied the idea to refine and detect social community [66]. Min and his group applied three-way decisions to the incremental mining of frequent itemsets [18,37]. Shang and Jia combined the three-way decisions solution with text sentiment analysis to improve the performance of sentiment classification [85]. Miao and his group applied three-way decision into Chinese emotion recognition [50], and achieved an excellent result. Zhang and Wang studied the issue of sentiment uncertainty analysis, and applied three-way decisions to sentiment classification with sentiment uncertainty [77], with considering the scenarios of context dependent sentiment classification and topic-dependent sentiment classification. In order to solve multi-label sentiment classification, Ren and Wang [46] proposed the method of three-way decisions to recognize the multi-label sentiment orientation of Chinese text. Li and his group utilized cost-sensitive sequential three-way decision to face recognition [13]. Miao and his group proposed a novel algorithm for image segmentation with noise in the framework of decision-theoretic rough set model [8]. Three-way decisions also have been adapted to solve group decision making problem, by combining with theories of decision-theoretic rough sets [35], two universes fuzzy decision-theoretic rough set [48], cloud model [20] and prospect theory [31]. Moreover, the theory of three-way decisions has also been used in other fields such as email spam filtering [84] and recommender system [19].

3 Clustering Approaches for Uncertain Relationships Between Objects and Clusters

The task of cluster analysis or clustering is to group similar objects into the same cluster and dissimilar objects into different clusters. Obviously, there are three relationships between an object and a cluster: (1) the object certainly belongs to the cluster, (2) the object certainly does not belong to the cluster, and (3) the object might or might not belong to the cluster. It is a typical three-way decision processing to decide the relationship between an object and a cluster. Such relationships will inspire us to introduce the three-way decisions into the cluster analysis problem.

In the existing clustering approaches, some approaches such as fuzzy clustering, rough clustering and interval clustering, have been proposed to deal with this kind of uncertain relationship between objects and clusters. Sometimes, we also say that these approaches are soft clustering or overlapping clustering based on the meaning that an object can belong to more than one cluster. In other words, soft clustering technologies aim to relax the hard boundary of clusters by soft constraints, so that it can deal with problems such as overlapping clusters, outliers and uncertain objects [41].

Fuzzy c-means (FCM) is a method of clustering which allows an object to belong to more than one cluster. In the FCM, similarities between objects and each cluster are described by membership degrees based on the fuzzy sets theory, and all objects are assigned to k fuzzy clusters. However, it cannot get an exact representation of clusters by fuzzy sets. To solve this issue, Lingras and Peters [34] applied the rough sets theory to clustering, they presented a new cluster representation that an object can belong to multiple clusters with the concepts of lower and upper approximations. In rough clustering, every cluster might have the fringe region (boundary region) to decrease cluster errors. Objects in fringe regions need more information so that they can be assigned to certain clusters eventually. Next, they combined rough sets to k-means and proposed the rough k-means clustering which each cluster is described by a lower and upper approximation. Since changes in general lead to uncertainty, the appropriate methods for uncertainty modeling are needed in order to capture, model, and predict the respective phenomena considered in dynamic environments, Peters et al. [40] proposed the dynamic rough clustering to detect changing data structures. In addition, Lingras and Yan [33] developed fuzzy clustering by combining rough clustering, in which a cluster is represented by a lower and upper approximation and two thresholds α and β are used to divide the two approximations.

Considering clusters presented as interval sets with lower and upper approximations in rough k-means clustering are not adequate to describe clusters, Chen
and Miao [2] proposed an interval set clustering based on decision theory.

The rough sets theory has played an important role in dealing with
uncertainty. Yao introduced the Bayes risk decision-making into rough sets and
proposed the decision-theoretic rough set model, then proposed the concept
of three-way decisions [53]. The theory of three-way decisions extends binary-
decisions in order to overcome some drawbacks of binary-decisions. Inspired by
the three-way decisions, Yu [68] proposed a framework of three-way cluster analysis. The three-way clustering redefines the clustering representation and has been
applied to dealing with some problems such as overlapping incremental clustering [65], community detection [66] and high-dimensional data clustering [67].
Similar to rough clustering using a pair of lower and upper approximations to
represent a cluster, three-way clustering describes a cluster by a pair of sets.
Generally speaking, rough clustering usually restricts to the rough k-means and
its extension algorithms. The intersections between any two core regions do not
have to be empty in the three-way clustering, it is different to that the intersection between any two lower approximations is empty in rough clustering. For
example, we have shown some real-world cases in the reference [66], in which
some objects are core elements of two communities. Usually, uncertain objects
in fringe regions need further treatment in three-way clustering when further
information can be obtained.

In the above, we have discussed the existing approaches for dealing with
uncertain relationships. Rough clustering and interval clustering can also be
regarded as the approaches of three-way decisions in some sense, in which the
fringe objects are described well.

4   Three-Way Cluster Analysis

In cluster analysis, we need to solve two essential problems. One is how to represent a cluster. Another one is how to obtain the clusters, namely, how to
develop clustering algorithms. In this section, this paper will introduce a novel
framework of three-way cluster analysis. The basic idea of three-way clustering concludes two aspects: (1) the result of clustering is three-way, and (2) the
three-way decision strategy is used during the process of clustering.

4.1   Representation of Three-Way Clustering

Let $U = \{\mathbf{x}_1, \cdots, \mathbf{x}_n, \cdots, \mathbf{x}_N\}$ be a finite set, called the universe or the reference
set. $\mathbf{x}_n$ is an object which has D attributes, namely, $\mathbf{x}_n = (x_n^1, \cdots, x_n^d, \cdots, x_n^D)$.
x_n^d denotes the value of the d-th attribute of the object $\mathbf{x}_n$, where $n \in \{1, \cdots, N\}$,
and $d \in \{1, \cdots, D\}$.

The result of clustering scheme $\mathbf{C} = \{C^1, \cdots, C^k, \cdots, C^K\}$ is a family of
clusters of the universe, in which K means this universe is composed of K clusters. According to Vladimir Estivill-Castro, the notion of a "cluster" cannot be

precisely defined, which is one of the reasons why there are so many clustering algorithms [4]. There is a common denominator: a group of data objects. In the existing works, a cluster is usually represented by a single set, namely, $C^k = \{\mathbf{x}_1^k, \cdots, \mathbf{x}_i^k, \cdots, \mathbf{x}_{|C^k|}^k\}$, abbreviated as C without ambiguous.

From the view of making decisions, the representation of a single set means, that the objects in the set belong to this cluster definitely and the objects not in the set do not belong to this cluster definitely. This is a typical result of two-way decisions. For hard clustering, one object just belongs to one cluster; for soft clustering, one object might belong to more than one cluster. However, this representation cannot show which objects might belong to this cluster, and it cannot intuitively show the influence degree of the object during the processing of forming the cluster. Obviously, the use of three regions to represent a cluster is more appropriate than the use of a crisp set, which also directly leads to three-way decisions based interpretation of clustering.

In contrast to the general crisp representation of a cluster, we represent a three-way cluster C as a pair of sets:

$$C = (Co(C), Fr(C)). \tag{2}$$

Here, $Co(C) \subseteq U$ and $Fr(C) \subseteq U$. Let $Tr(C) = U - Co(C) - Fr(C)$. Then, $Co(C)$, $Fr(C)$ and $Tr(C)$ naturally form the three regions of a cluster as Core Region, Fringe Region and Trivial Region respectively. If $\mathbf{x} \in Co(C)$, the object $\mathbf{x}$ belongs to the cluster C definitely; if $\mathbf{x} \in Fr(C)$, the object $\mathbf{x}$ might belong to C; if $\mathbf{x} \in Tr(C)$, the object $\mathbf{x}$ does not belong to C definitely. These subsets have the following properties.

$$\begin{aligned}
U &= Co(C) \cup Fr(C) \cup Tr(C), \\
Co(C) \cap Fr(C) &= \emptyset, \\
Fr(C) \cap Tr(C) &= \emptyset, \\
Tr(C) \cap Co(C) &= \emptyset.
\end{aligned} \tag{3}$$

If $Fr(C) = \emptyset$, the representation of C in Eq. (2) turns into $C = Co(C)$; it is a single set and $Tr(C) = U - Co(C)$. This is a representation of two-way decisions. In other words, the representation of a single set is a special case of the representation of three-way cluster.

Furthermore, according to Formula (3), we know that it is enough to represent expediently a cluster by the core region and the fringe region.

In another way, for $1 \leq k \leq K$, we can define a cluster scheme by the following properties:

$$\begin{aligned}
&(i) \text{ for } \forall k, \ Co(C^k) \neq \emptyset; \\
&(ii) \ \bigcup_{k=1}^{K} (Co(C^k) \cup Fr(C^k)) = U.
\end{aligned} \tag{4}$$

Property (i) implies that a cluster cannot be empty. This makes sure that a cluster is physically meaningful. Property (ii) states that any object of U must definitely belong to or might belong to a cluster, which ensures that every object is properly clustered.

With respect to the family of clusters, $\mathbf{C}$, we have the following family of clusters formulated by three-way representation as:

$$\mathbf{C} = (\{Co(C^1), Fr(C^1)), \cdots, (Co(C^k), Fr(C^k)), \cdots, (Co(C^K), Fr(C^K))\}. \tag{5}$$

Obviously, we have the following family of clusters formulated by two-way decisions as:

$$\mathbf{C} = \{Co(C^1), \cdots, Co(C^k), \cdots, Co(C^K)\}. \tag{6}$$

Under the representation, we can formulate the soft clustering and hard clustering as follows. For a clustering, if there exists $k \neq t$, such that

$$\begin{aligned}
&(1) \ Co(C^k) \cap Co(C^t) \neq \emptyset, \ or \\
&(2) \ Fr(C^k) \cap Fr(C^t) \neq \emptyset, \ or \\
&(3) \ Co(C^k) \cap Fr(C^t) \neq \emptyset, \ or \\
&(4) \ Fr(C^k) \cap Co(C^t) \neq \emptyset,
\end{aligned} \tag{7}$$

we call it is a soft clustering; otherwise, it is a hard clustering.

As long as one condition of Eq. (7) is satisfied, there must exist at least one object belonging to more than one cluster.

Obviously, the representation of three-way brings the following advantages: the representation of a single set is a special case of the representation of three-way cluster; it intuitively shows that which objects are core of the cluster, and which ones are fringe of the cluster; it diversifies the type of overlapping; and it reduces the searching space when focusing on the overlapping/fringe objects.

4.2   An Evaluation-Based Three-Way Cluster Model

In this subsection, we will introduce an evaluation-based three-way cluster model, which produces three regions by using an evaluation function and a pair of thresholds on the values of the evaluation function. The model partially addresses the issue of trisecting a universal set into three regions.

Suppose there are a pair of thresholds (α, β) and $\alpha \geq \beta$. Although evaluations based on a total order are restrictive, they have a computational advantage. One can obtain the three regions by simply comparing the evaluation value with a pair of thresholds. Based on the evaluation function $v(\mathbf{x})$, we get the following three-way decision rules:

$$\begin{aligned}
Co(C^k) &= \{x \in U | v(\mathbf{x}) > \alpha\}, \\
Fr(C^k) &= \{x \in U | \beta \leq v(\mathbf{x}) \leq \alpha\}, \\
Tr(C^k) &= \{x \in U | v(\mathbf{x}) < \beta\}.
\end{aligned} \tag{8}$$

Yao proposed an evaluation-based three-way decisions model in the reference [57]. Naturally, an similar evaluation-based three-way cluster model is depicted in Fig. 1. We can divide the universe U according to Eq. 8 and design different strategies to process the three regions.

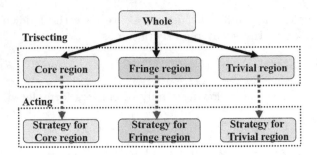

Fig. 1. An Evaluation-based Three-way Cluster Model

Based on the model, we have to pay attention to the following three points.

– About the evaluation function $v(\mathbf{x})$. It will be specified accordingly when an algorithm is devised. In fact, in order to devise the evaluation function, we can refer to the similarity measures or distance measures, probability, possibility functions, fuzzy membership functions, Bayesian confirmation measures, subsethood measures and so on.

– About the three-way thresholds α and β. For an evaluation-based model, we need to investigate ways to compute and to interpret a pair of thresholds. An optimization framework can be designed to achieve such a goal. That is, a pair of thresholds should induce a trisection that optimizes a given objective function. By designing different objective functions for different applications, we gain a great flexibility.

– The three-way decision strategy used during the process of clustering. Shortly, it concludes two aspects such as how to get the three-regions of a cluster and how to act on the three regions.

Of course, the previous two items serve to the third item. In other words, the basic research issues of three-way clustering are about how to obtain the three regions and how to act on the three regions, which is similar to the researches on three-way decisions.

4.3   Some Researches on Three-Way Clustering

In this subsection, I will summarize and discuss some issues and research points about the three-way clustering.

• Representation of three-way clustering. As discussed in Sect. 4.1, we can use a pair of sets to represent a cluster in three-way representation. Some works have been proposed in view of rough sets [34], interval sets [2], decision-theoretic rough sets [62] and mathematical morphology [52]. We can also represent the model of three-way clustering by using fuzzy set, shadow sets and other models. Different interpretations of three-way clustering could give different solutions to different kinds of clustering problems.

- How to get the three-way clustering. It is a good way to extend from the classical two-way decisions clustering approaches. The following properties are important to the efficiency and effectiveness of a novel algorithm: how to decide the thresholds, how to know the truth number of clusters. Yu et al. [69] proposed a method to determine the thresholds automatically based on gravitational search during the processing of clustering.
- Developing new clustering approaches for more uncertainty situations such as dynamic, incomplete data or multi-source data. For example, we had proposed a tree-based three-way clustering method for incremental overlapping clustering [65], a three-way decisions clustering algorithm for incomplete data based on attribute significance and miss rate [63], a semi-supervised three-way clustering framework for multi-view data [70], a three-way decision clustering approach for high dimensional data [67], and so on [68].
- Application of three regions. We can put forward the three-way clustering strategy to the application fields such as social network services, cyber marketing, E-commerce, recommendation service and other fields. Through the further work on the fringe region, we can know the influence degree of the object during the processing of forming the cluster, which is very helpful in some practical applications. For example, Yu et al. [66] have presented a method to detect and refine overlapping regions in complex networks by three-way clustering.

5   Conclusions

The notion of three-way decisions was introduced for meeting the needs to properly explain three regions of probabilistic rough sets. The theory of three-way decisions moves far beyond this original goal. We have seen a more general theory that embraces ideas from many fields and disciplines. This paper introduces most of recent studies on three-way decisions, in order to demonstrate the value and power as well as the great potentials of three-way decisions. For purpose of giving an example of researches related to three-way decisions, a three-way cluster analysis approach is introduced in this paper, which mainly addresses the problem that the uncertain relationship between an object and a cluster.

Acknowledgements. I am grateful to Professor Yiyu Yao for the discussions. In addition, this work was supported in part by the National Natural Science Foundation of China under grant No. 61533020, 61672120 and 61379114.

References

1. Afridi, M.K., Azam, N., Yao, J.T., Alanazi, E.: A three-way clustering approach for handling missing data using GTRS. Int. J. Approximate Reasoning **98**, 11–24 (2018)
2. Chen, M., Miao, D.Q.: Interval set clustering. Expert Syst. Appl. **38**(4), 2923–2932 (2011)

3. Chen, Y.M., Zeng, Z.Q., Zhu, Q.X., Tang, C.H.: Three-way decision reduction in neighborhood systems. Appl. Soft Comput. **38**, 942–954 (2016)
4. Estivill-Castro, V.: Why so many clustering algorithms: a position paper. ACM SIGKDD Explor. Newslett. **4**(1), 65–75 (2002)
5. Gao, C., Yao, Y.Y.: Actionable strategies in three-way decisions. Knowl.-Based Syst. **133**, 141–155 (2017)
6. Gao, C., Yao, Y.: Actionable strategies in three-way decisions with rough sets. In: Polkowski, L., et al. (eds.) IJCRS 2017. LNCS (LNAI), vol. 10314, pp. 183–199. Springer, Cham (2017). https://doi.org/10.1007/978-3-319-60840-2_13
7. Ganter, B., Wille, R.: Formal Concept Analysis: Mathematical Foundations. Springer, Heidelberg (1999). https://doi.org/10.1007/978-3-642-59830-2
8. Li, F., Miao, D.Q., Liu, C.H., Yang, W.: Image segmentation algorithm based on the decision-theoretic rough set model. CAAI Trans. Intell. Syst. **9**(2), 143–147 (2014)
9. Li, H.X., Zhou, X.Z.: Risk decision making based on decision-theoretic rough set: a three-way view decision model. Int. J. Comput. Intell. Syst. **4**(1), 1–11 (2011)
10. Li, H.X., Zhou, X.Z., Li, T.R., Wang, G.Y., Miao, D.Q., Yao, Y.Y.: Decision-Theoretic Rough Set Theory and Recent Progress. Science Press, Beijing (2011). (In Chinese)
11. Li, H., Zhou, X., Zhao, J., Huang, B.: Cost-sensitive classification based on decision-theoretic rough set model. In: Li, T., et al. (eds.) RSKT 2012. LNCS (LNAI), vol. 7414, pp. 379–388. Springer, Heidelberg (2012). https://doi.org/10.1007/978-3-642-31900-6_47
12. Li, H., Zhou, X., Huang, B., Liu, D.: Cost-sensitive three-way decision: a sequential strategy. In: Lingras, P., Wolski, M., Cornelis, C., Mitra, S., Wasilewski, P. (eds.) RSKT 2013. LNCS (LNAI), vol. 8171, pp. 325–337. Springer, Heidelberg (2013). https://doi.org/10.1007/978-3-642-41299-8_31
13. Li, H.X., Zhang, L.B., Huang, B., Zhou, X.Z.: Sequential three-way decision and granulation for cost-sensitive face recognition. Knowl.-Based Syst. **91**, 241–251 (2016)
14. Li, H.X., Zhang, L.B., Zhou, X.Z., Huang, B.: Cost-sensitive sequential three-way decision modeling using a deep neural network. Int. J. Approximate Reasoning **85**, 68–78 (2017)
15. Li, H.X., Zhou, X.Z., Huang, B.: Cost-sensitive sequential three-way decisions. In: Liu, D., Li, T.R., Miao, D.Q., Wang, G.Y., Liang, J.Y. (eds.): Three-Way Decisions and Granular Computing, pp. 42–59. Science Press, Beijing (2013). (In Chinese)
16. Li, J.H., Huang, C.C., Qi, J.J., Qian, Y.H., Liu, W.Q.: Three-way cognitive concept learning via multi-granularity. Inf. Sci. **378**, 244–263 (2016)
17. Li, W., Huang, Z., Jia, X.: Two-phase classification based on three-way decisions. In: Lingras, P., Wolski, M., Cornelis, C., Mitra, S., Wasilewski, P. (eds.) RSKT 2013. LNCS (LNAI), vol. 8171, pp. 338–345. Springer, Heidelberg (2013). https://doi.org/10.1007/978-3-642-41299-8_32
18. Li, Y., Zhang, Z.H., Chen, W.B., Min, F.: TDUP: an approach to incremental mining of frequent itemsets with three-way-decision pattern updating. Int. J. Mach. Learn. Cybernet. **8**(2), 441–453 (2017)
19. Huang, J.J., Wang, J., Yao. Y.Y., Zhong, N.: Cost-sensitive three-way recommendations by learning pair-wise preferences. Int. J. Approximate Reasoning **86**, 28–40 (2017)
20. Hu, J.H., Yang, Y., Chen, X.H.: Three-way linguistic group decisions model based on cloud for medical care product investment. J. Intell. Fuzzy Syst. **33**(6), 3405–3417 (2017)

21. Hu, B.Q.: Three-way decision spaces based on partially ordered sets and three-way decisions based on hesitant fuzzy sets. Knowl.-Based Syst. **91**, 16–31 (2016)
22. Hu, B.Q.: Three-way decisions based on semi-three-way decision spaces. Inf. Sci. **382**, 415–440 (2017)
23. Jia, X., Li, W., Shang, L., Chen, J.: An optimization viewpoint of decision-theoretic rough set model. In: Yao, J.T., Ramanna, S., Wang, G., Suraj, Z. (eds.) RSKT 2011. LNCS (LNAI), vol. 6954, pp. 457–465. Springer, Heidelberg (2011). https://doi.org/10.1007/978-3-642-24425-4_60
24. Jia, X.Y., et al.: Theory of Three-Way Decisions and Application. Nanjing University Press, Nanjing (2012). (In Chinese)
25. Jia, X.Y., Tang, Z.M., Liao, W.H., Shang, L.: On an optimization representation of decision-theoretic rough set model. Int. J. Approximate Reasoning **55**, 156–166 (2014)
26. Ju, H.R., Yang, X.B., Yu, H.L., Li, T.J., Yu, D.J., Yang, J.Y.: Cost-sensitive rough set approach. Inf. Sci. **355**, 282–298 (2016)
27. Ju, H.R., Li, H.X., Yang, X.B., Zhou, X.Z., Huang, B.: Cost-sensitive rough set: a multi-granulation approach. Knowl.-Based Syst. **123**, 137–153 (2017)
28. Liu, D., Li, T.R., Miao, D.Q., Wang, G.Y., Liang, J.Y.: Three-Way Decisions and Granular Computing. Science Press, Beijing (2013). (In Chinese)
29. Liu, D., Li, T., Liang, D.: Three-way decisions in dynamic decision-theoretic rough sets. In: Lingras, P., Wolski, M., Cornelis, C., Mitra, S., Wasilewski, P. (eds.) RSKT 2013. LNCS (LNAI), vol. 8171, pp. 288–299. Springer, Heidelberg (2013). https://doi.org/10.1007/978-3-642-41299-8_28
30. Liu, D., Liang, D.C., Wang, C.C.: A novel three-way decision model based on incomplete information system. Knowl.-Based Syst. **91**, 32–45 (2016)
31. Liu, S.L., Liu, X.W., Qin, J.D.: Three-way group decisions based on prospect theory. J. Oper. Res. Soc. **69**(1), 25–35 (2018)
32. Luo, C., Li, T., Chen, H.: Dynamic maintenance of three-way decision rules. In: Miao, D., Pedrycz, W., Ślęzak, D., Peters, G., Hu, Q., Wang, R. (eds.) RSKT 2014. LNCS (LNAI), vol. 8818, pp. 801–811. Springer, Cham (2014). https://doi.org/10.1007/978-3-319-11740-9_73
33. Lingras, P., Yan, R.: Interval clustering using fuzzy and rough set theory. In: Annual Meeting of the North American Fuzzy Information Processing Society, vol. 2, pp. 780–784. IEEE (2004)
34. Lingras, P., Peters, G.: Applying rough set concepts to clustering. In: International Conference on Rough Sets, Fuzzy Sets, Data Mining and Granular Computing, pp. 23–37. Springer, London (2012). https://doi.org/10.1007/978-1-4471-2760-4_2
35. Liang, D.C., Liu, D., Kobina, A.: Three-way group decisions with decision-theoretic rough sets. Inf. Sci. **345**, 46–64 (2016)
36. Ma, X.A., Yao, Y.Y.: Three-way decision perspectives on class-specific attribute reducts. Inf. Sci. **450**, 227–245 (2018)
37. Min, F., Zhang, Z.H., Zhai, W.J., Shen, R.P.: Frequent pattern discovery with tri-partition alphabets. Inf. Sci. **000**, 1–18 (2018)
38. Pawlak, Z.: Rough sets. Int. J. Comput. Inform. Sci. **11**, 341–356 (1982)
39. Pawlak, Z.: Rough Sets: Theoretical Aspects of Reasoning About Data. Kluwer Academic Publishers, Dordrecht (1991)
40. Peters, G., Weber, R., Nowatzke, R.: Dynamic rough clustering and its applications. Appl. Soft Comput. **12**(10), 3193–3207 (2012)
41. Peters, G., Crespo, F., Lingras, P., Weber, R.: Soft clustering-fuzzy and rough approaches and their extensions and derivatives. Int. J. Approximate Reasoning **54**(2), 307–322 (2013)

42. Qi, J., Wei, L., Yao, Y.: Three-way formal concept analysis. In: Miao, D., Pedrycz, W., Ślęzak, D., Peters, G., Hu, Q., Wang, R. (eds.) RSKT 2014. LNCS (LNAI), vol. 8818, pp. 732–741. Springer, Cham (2014). https://doi.org/10.1007/978-3-319-11740-9_67

43. Qian, T., Wei, L., Qi, J.J.: Constructing three-way concept lattices based on apposition and subposition of formal contexts. Knowl.-Based Syst. **116**, 39–48 (2017)

44. Rehman, H.U., Azam, N., Yao, J.T., Benso, A.: A three-way approach for protein function classification. PLoS ONE **12**(2), e0171702 (2017)

45. Ren, R.S., Wei, L.: The attribute reductions of three-way concept lattices. Knowl.-Based Syst. **99**, 92–102 (2016)

46. Ren, F.J., Wang, L.: Sentiment analysis of text based on three-way decisions. J. Intell. Fuzzy Syst. **33**(1), 245–254 (2017)

47. Singh, P.K.: Three-way fuzzy concept lattice representation using neutrosophic set. Int. J. Mach. Learn. Cybernet. **8**(1), 69–79 (2017)

48. Sun, B.Z., Ma, W.M., Xiao, X.: Three-way group decision making based on multi-granulation fuzzy decision-theoretic rough set over two universes. Int. J. Approximate Reasoning **81**, 87–102 (2017)

49. Sun, B.Z., Ma, W.M., Li, B.J., Li, X.N.: Three-way decisions approach to multiple attribute group decision making with linguistic information-based decision-theoretic rough fuzzy set. Int. J. Approximate Reasoning **93**, 424–442 (2018)

50. Wang, L., Miao, D., Zhao, C.: Chinese emotion recognition based on three-way decisions. In: Ciucci, D., Wang, G., Mitra, S., Wu, W.-Z. (eds.) RSKT 2015. LNCS (LNAI), vol. 9436, pp. 299–308. Springer, Cham (2015). https://doi.org/10.1007/978-3-319-25754-9_27

51. Wang, B., Liang, J.: A novel intelligent multi-attribute three-way group sorting method based on dempster-shafer theory. In: Miao, D., Pedrycz, W., Ślęzak, D., Peters, G., Hu, Q., Wang, R. (eds.) RSKT 2014. LNCS (LNAI), vol. 8818, pp. 789–800. Springer, Cham (2014). https://doi.org/10.1007/978-3-319-11740-9_72

52. Wang, P.X., Yao, Y.Y.: CE3: a three-way clustering method based on mathematical morphology. Knowl.-Based Syst. **155**, 54–65 (2018). https://doi.org/10.1016/j.knosys.2018.04.029

53. Yao, Y.: Three-way decision: an interpretation of rules in rough set theory. In: Wen, P., Li, Y., Polkowski, L., Yao, Y., Tsumoto, S., Wang, G. (eds.) RSKT 2009. LNCS (LNAI), vol. 5589, pp. 642–649. Springer, Heidelberg (2009). https://doi.org/10.1007/978-3-642-02962-2_81

54. Yao, Y.Y., Deng, X.F.: Sequential three-way decisions with probabilistic rough sets. In: Proceedings of 10th IEEE International Conference on Cognitive Informatics & Cognitive Computing, pp. 120–125. IEEE (2011)

55. Yao, Y.: An outline of a theory of three-way decisions. In: Yao, Y.T., et al. (eds.) RSCTC 2012. LNCS (LNAI), vol. 7413, pp. 1–17. Springer, Heidelberg (2012). https://doi.org/10.1007/978-3-642-32115-3_1

56. Yao, Y.: Granular computing and sequential three-way decisions. In: Lingras, P., Wolski, M., Cornelis, C., Mitra, S., Wasilewski, P. (eds.) RSKT 2013. LNCS (LNAI), vol. 8171, pp. 16–27. Springer, Heidelberg (2013). https://doi.org/10.1007/978-3-642-41299-8_3

57. Yao, Y.: Rough sets and three-way decisions. In: Ciucci, D., Wang, G., Mitra, S., Wu, W.-Z. (eds.) RSKT 2015. LNCS (LNAI), vol. 9436, pp. 62–73. Springer, Cham (2015). https://doi.org/10.1007/978-3-319-25754-9_6

58. Yao, Y.Y.: The two sides of the theory of rough sets. Knowl.-Based Syst. **80**, 67–77 (2015)

59. Yao, Y.Y.: Three-way decisions and cognitive computing. Cogn. Comput. **8**(4), 543–554 (2016)
60. Yao, Y.Y.: Interval sets and three-way concept analysis in incomplete contexts. Int. J. Mach. Learn. Cybernet. **8**(1), 3–20 (2017)
61. Yu, H., Wang, Y.: Three-way decisions method for overlapping clustering. In: Yao, J.T., et al. (eds.) RSCTC 2012. LNCS (LNAI), vol. 7413, pp. 277–286. Springer, Heidelberg (2012). https://doi.org/10.1007/978-3-642-32115-3_33
62. Yu, H., Liu, Z.G., Wang, G.Y.: An automatic method to determine the number of clusters using decision-theoretic rough set. Int. J. Approximate Reasoning **55**(1), 101–115 (2014)
63. Yu, H., Su, T., Zeng, X.: A three-way decisions clustering algorithm for incomplete data. In: Miao, D., Pedrycz, W., Ślęzak, D., Peters, G., Hu, Q., Wang, R. (eds.) RSKT 2014. LNCS (LNAI), vol. 8818, pp. 765–776. Springer, Cham (2014). https://doi.org/10.1007/978-3-319-11740-9_70
64. Yu, H., Wang, G.Y., Li, T.R., Liang, J.Y., Miao, D.Q., Yao, Y.Y.: Three-Way Decisions: Methods and Practices for Complex Problem Solving. Science Press, Beijing (2015). (In Chinese)
65. Yu, H., Zhang, C., Wang, G.Y.: A tree-based incremental overlapping clustering method using the three-way decision theory. Knowl.-Based Syst. **91**, 189–203 (2016)
66. Yu, H., Jiao, P., Yao, Y.Y., Wang, G.Y.: Detecting and refining overlapping regions in complex networks with three-way decisions. Inf. Sci. **373**, 21–41 (2016)
67. Yu, H., Zhang, H.: A three-way decision clustering approach for high dimensional data. In: Flores, V., et al. (eds.) IJCRS 2016. LNCS (LNAI), vol. 9920, pp. 229–239. Springer, Cham (2016). https://doi.org/10.1007/978-3-319-47160-0_21
68. Yu, H.: A framework of three-way cluster analysis. In: Polkowski, L., et al. (eds.) IJCRS 2017. LNCS (LNAI), vol. 10314, pp. 300–312. Springer, Cham (2017). https://doi.org/10.1007/978-3-319-60840-2_22
69. Yu, H., Chang, Z.H., Li, Z.X., Wang, G.Y.: An efficient three-way clustering algorithm based on gravitational search. In: ACM SIGKDD International Conference on Knowledge Discovery and Data Mining. ACM (2018)
70. Yu, H., Wang, X.C., Wang, G.Y., Zeng, X.H.: An active three-way clustering method via low-rank matrices for multi-view data. Inf. Sci. (2018). https://doi.org/10.1016/j.ins.2018.03.009
71. Yang, H.L., Guo, Z.L.: Multigranulation decision-theoretic rough sets in incomplete information systems. Int. J. Mach. Learn. Cybernet. **6**(6), 1005–1018 (2015)
72. Yang, X., Tan, A.: Three-way decisions based on intuitionistic fuzzy sets. In: Polkowski, L., et al. (eds.) IJCRS 2017. LNCS (LNAI), vol. 10314, pp. 290–299. Springer, Cham (2017). https://doi.org/10.1007/978-3-319-60840-2_21
73. Yu, H.Y., Li, Q.G., Cai, M.J.: Characteristics of three-way concept lattices and three-way rough concept lattices. Knowl.-Based Syst. **146**, 181–189 (2018)
74. Yang, X., Li, T.R., Fujita, H., Liu, D., Yao, Y.Y.: A unified model of sequential three-way decisions and multilevel incremental processing. Knowl.-Based Syst. **134**, 172–188 (2017)
75. Zhao, X.R., Hu, B.Q.: Fuzzy probabilistic rough sets and their corresponding three-way decisions. Knowl.-Based Syst. **91**, 126–142 (2016)
76. Zhang, Y., Yao, J.T.: Multi-criteria based three-way classifications with game-theoretic rough sets. In: Kryszkiewicz, M., Appice, A., Rybinski, H., Skowron, A., Ślęzak, D., Raś, Z.W. (eds.) ISMIS 2017. LNCS (LNAI), vol. 10352, pp. 550–559. Springer, Cham (2017). https://doi.org/10.1007/978-3-319-60438-1_54

77. Zhang, Z., Wang, R.: Applying three-way decisions to sentiment classification with sentiment uncertainty. In: Miao, D., Pedrycz, W., Ślęzak, D., Peters, G., Hu, Q., Wang, R. (eds.) RSKT 2014. LNCS (LNAI), vol. 8818, pp. 720–731. Springer, Cham (2014). https://doi.org/10.1007/978-3-319-11740-9_66

78. Zhai, J.H., Zhang, Y., Zhu, H.Y.: Three-way decisions model based on tolerance rough fuzzy set. Int. J. Mach. Learn. Cybernet. **8**(1), 35–43 (2017)

79. Zhang, X.Y., Miao, D.Q.: Three-way attribute reducts. Int. J. Approximate Reasoning **88**, 401–434 (2017)

80. Zhang, Q.H., Lv, G.X., Chen, Y.H., Wang, G.Y.: A dynamic three-way decision model based on the updating of attribute values. Knowl.-Based Syst. **142**, 71–84 (2018)

81. Zhang, X.Y., Miao, D.Q.: Double-quantitative fusion of accuracy and importance: systematic measure mining, benign integration construction, hierarchical attribute reduction. Knowl.-Based Syst. **91**, 219–240 (2016)

82. Zhang, Y., Zou, H., Chen, X., Wang, X., Tang, X., Zhao, S.: Cost-sensitive three-way decisions model based on CCA. In: Cornelis, C., Kryszkiewicz, M., Ruiz, E.M., Bello, R., Ślęzak, D., Shang, L. (eds.) RSCTC 2014. LNCS (LNAI), vol. 8536, pp. 172–180. Springer, Cham (2014). https://doi.org/10.1007/978-3-319-08644-6_18

83. Zhang, Y., Xing, H., Zou, H., Zhao, S., Wang, X.: A three-way decisions model based on constructive covering algorithm. In: Lingras, P., Wolski, M., Cornelis, C., Mitra, S., Wasilewski, P. (eds.) RSKT 2013. LNCS (LNAI), vol. 8171, pp. 346–353. Springer, Heidelberg (2013). https://doi.org/10.1007/978-3-642-41299-8_33

84. Zhou, B., Yao, Y.Y., Luo, J.G.: Cost-sensitive three-way email spam filtering. J. Intell. Inf. Syst. **42**, 19–45 (2014)

85. Zhou, Z., Zhao, W., Shang, L.: Sentiment analysis with automatically constructed lexicon and three-way decision. In: Miao, D., Pedrycz, W., Ślęzak, D., Peters, G., Hu, Q., Wang, R. (eds.) RSKT 2014. LNCS (LNAI), vol. 8818, pp. 777–788. Springer, Cham (2014). https://doi.org/10.1007/978-3-319-11740-9_71

Some Foundational Aspects of Rough Sets Rendering Its Wide Applicability

Andrzej Skowron[1,2(✉)] and Soma Dutta[3,4]

[1] Faculty of Mathematics, Informatics and Mechanics,
University of Warsaw, Banacha 2, 02-097 Warsaw, Poland
skowron@mimuw.edu.pl
[2] Systems Research Institute, Polish Academy of Sciences,
Newelska 6, 01-447 Warsaw, Poland
[3] Vistula University, Stokłosy 3, 02-787 Warsaw, Poland
somadutta9@gmail.com
[4] Department of Mathematics and Computer Science, University of Warmia and
Mazury, Sloneczna str. 54, 10-710 Olsztyn, Poland

Abstract. This paper aims to discuss about the reasons behind the wide applicability of the rough set approach in real-life projects. The rough set-based approximations of (vague) concepts is one among the most central notions, available in the literature, for dealing with imperfect data and/or information. Moreover, as the approach based on rough sets is directly driven from data it turns out to be advantageous for real life projects where data plays a crucial role. Besides, using rough set approach one can deal efficiently with algorithmic issues, especially in the context of searching for relevant computational building blocks (granules) for approximation of complex vague concepts. In this paper, we would focus on these few aspects of rough sets, in order to explain its wide applicability in real-life projects.

1 Introduction

The rough set (RS) approach was proposed by Professor Zdzisław Pawlak in 1982 [51,53][1] as a tool for dealing with imperfect knowledge and/or vague concepts. Many applications and methods based on rough set theory, alone or in combination with other approaches, have been developed.

The philosophy of rough set is grounded on the assumption that every object of a universe of discourse is associated with some information (data, knowledge). Objects characterized by the same information are indiscernible (similar) with respect to the available data. The *indiscernibility relation* generated in this way is the mathematical basis of rough set theory. A set of all indiscernible (similar) objects is called an elementary set, and this forms a *basic information granule (atom)* of knowledge about the universe. An arbitrary union of some elementary

---

[1] For more information readers are referred to some survey papers [55–57,67], books, *e.g.*, [19,57,71] and to the rough set database rsds.univ.rzeszow.pl.

© Springer Nature Switzerland AG 2018
H. S. Nguyen et al. (Eds.): IJCRS 2018, LNAI 11103, pp. 29–45, 2018.
https://doi.org/10.1007/978-3-319-99368-3_3

sets, called definable set, is referred to as *crisp* (precise) set. If a set is not crisp then it is called *rough* (imprecise, vague). A definable set is considered to be an *information granule*.

Thus, each rough set has *borderline cases* (*boundary–line*), *i.e.*, objects which cannot be classified with certainty as members of either the set or its complement. This means that borderline cases are those which cannot be properly classified by employing available information. Rough set theory can be viewed as a specific implementation of Frege's idea of vagueness, *i.e.*, imprecision in this approach is expressed by a boundary region of a set.

So, the assumption that objects can be "seen" only through the information available about them leads to the view that knowledge has granular structure. Due to the granularity of knowledge some objects of interest cannot be discerned, and thus they appear as the same (or similar). As a consequence, vague concepts, in contrast to precise concepts, cannot be characterized in terms of (information about) their elements. Therefore, in the proposed approach, it is assumed that any vague concept is replaced by a pair of precise concepts – called the *lower and the upper approximation* of the vague concept. The lower approximation consists of all objects which definitely belong to the concept and the upper approximation contains all objects which possibly belong to the concept. The difference between the upper and the lower approximation constitutes the *boundary region* of the vague concept. These approximation operations are the basic operations in rough set theory. Hence, rough set theory addresses vagueness not by means of membership to a set/concept, but by employing a boundary region to a set/concept. If the boundary region of a set is empty it means that the set is *crisp*, otherwise the set is *rough* (inexact). A nonempty boundary region of a set indicates the possibility that our knowledge about the set is not sufficient to define the set precisely.

In the development of rough set theory and its applications, one can distinguish three main stages. (i) During the first stage, the focus was based on the assumption that objects are perceived by means of partial information represented by attributes. (ii) In the second stage[2], the focus changed to looking at the strategies through which the concepts, given only on samples of objects, are approximated; as the strategies are different, finding relevant attributes as well as methods of selecting those attributes become the central notions of rough set literature. During this stage, approximation spaces and searching strategies for relevant approximation spaces have been considered to be the central point of interest in the study of rough sets. Many important achievements both in the theory and the applications were obtained. (iii) Nowadays, a new stage for rough sets has emerged based on the notion of interactive granular computations, in which how a relevant strategy for constructing an approximation space can be learned through interactions is also emphasized. As an example, one can consider perception based computing.

---

[2] This stage started a few years after the first paper by Pawlak on rough sets was published.

The rough set approach seems to be of fundamental importance in artificial intelligence and cognitive sciences. Relationship of rough sets with many other approaches such as fuzzy set theory, granular computing, evidence theory, formal concept analysis, (approximate) Boolean reasoning, multicriteria decision analysis, statistical methods, decision theory, matroids have been clarified by different researchers. There are reports on many hybrid methods obtained by combining rough sets with other approaches such as soft computing, statistical methods, natural computing, mereology, principal component analysis, singular value decomposition and support vector machines.

The main advantage of rough set theory in data analysis is that it does not necessarily need any additional information about data, other than some properties of objects. Whereas one needs additionally probability distribution function in statistics, basic probability assignments in evidence theory, a grade of membership or the value of possibility in fuzzy set theory, which are basically estimated from data. One can observe that the following application oriented aspects have emerged as a natural outcome of the fact that the theory of rough sets is grounded in data. Among many such a few are (i) introduction of efficient algorithms for finding hidden patterns in data, (ii) determination of optimal sets of data (data reduction) and evaluation of the significance of data, (iii) generation of sets of decision rules from data, (iv) easy-to-understand formulation of decision rules, (v) straightforward interpretation of obtained results, and (vi) suitability of many of its algorithms for parallel processing.

This paper aims to explain why the rough set approach leads to so many real-life applications. In this regard, we select the aspect related to 'close association' of the approach with data, and the basic notions of the approach for approximating concepts, as important reasons behind its wide applicability. We have already mentioned the importance of finding relevant searching strategies for the process of constructing approximation space, in application. In this regard, in Sect. 2, we would outline the rough set approach to searching for computational building blocks for cognition (*e.g.*, for approximation of vague concepts) based on parametrized approximation spaces. Here, we would try to touch the issues of the second and third stage of the development in the study of rough sets. An illustrative example related to discovery of relationships of rough sets with other approaches for dealing with uncertainty is presented in Sect. 3; the example, in particular, concerns to Dempster-Shafer theory. In Sect. 4, some comments on combination of rough sets with other soft computing approaches, such as fuzzy sets or neural networks, leading to improving the quality of constructed computational building blocks, are presented. Lastly, there is a concluding section listing some further possibilities to be explored.

2 Parametrized Approximation Spaces

In this section, we would concentrate on the two other stages of development in the rough set study mentioned in the introduction. One of them is the emergence of parametrized approximation spaces, and the other is introduction of interactions within a family of approximation spaces, parametrized by different purposes, contexts, or constraints.

In this regard, we put forward a discussion about importance of the rough set approach in searching for *computational building blocks for cognition*, as considered by Leslie Valiant as a fundamental question for Artificial Intelligence[3]. We emphasize on the necessity for a constructive search of the relevant components of approximation spaces from a given family of approximation spaces. It should be noted that these components need to be constructed from the available data.

The original approach by Pawlak was based on the notion of indiscernibility. Any such indiscernibility relation, generated from an equivalence relation, defines a partition of the universe of objects. Over the years, many generalizations of this approach are introduced; some of them are based on coverings rather than partitions (see, *e.g.*, [67]).

One should note that for dealing with covering based rough set approach, it first requires solving several new algorithmic problems, such as selection of family of definable sets and/or selection of relevant definition of approximation of sets among many possible ones. In the context of application, finding the relevant definition/strategy for approximation space is important as it is not given a priori, rather should be learned from data.

Let us first list down some of the foundational aspects for building the theory based on rough sets that need to be focused on in the context of applications (i) One of the key problems is that for a given problem (*e.g.*, classification problem) one needs to first discover the relevant covering for the target classification task. In the literature, there are numerous papers dedicated to theoretical aspects of the covering based rough set approach. However, still much more work should be done on, rather hard, algorithmic issues for discovering the relevant covering for a particular data. (ii) Another issue to be emphasized is related to *inclusion measures*. Parameters of such measures, for the purpose of application, sometimes need to be tuned so that they can induce high quality approximations. Usually, this is realized using the *minimum description length principle* (MDL) [63] for the constraints of the measures. In particular, approximation spaces with rough inclusion measures have been investigated. This approach was further extended to rough mereological approach. More general cases of approximation spaces with rough inclusion were also discussed in the literature including approximation spaces in Granular Computing (GrC). Finally, the approach for ontology approximation, used in hierarchical learning of complex vague concepts [71], is also worth to be mentioned here.

In the section below, we would show how different components of a generalized approximation space can be constructed from the perspective of application.

2.1 Some Examples for Generalized Approximation Space Parametrized by Different Constraints

Several generalizations of the classical rough set approach based on approximation spaces defined as pairs of the form (U, R), with an equivalence relation R,

---

[3] Leslie Valiant: https://people.seas.harvard.edu/~valiant/researchinterests.htm.

have been reported in the literature. These generalizations have emerged focusing on different application oriented views regarding the basic concepts used in the definition of rough sets. Searching strategies for relevant approximation spaces are crucial for real-life applications. They include discovery of uncertainty functions, inclusion measures as well as selection of methods for approximations of decision classes, and strategies for inductive extension of approximations from samples to relatively larger sets of objects.

Let us consider some examples of generalizations of the notions such as indiscernibility relation, inclusion relation and approximation space, following the requirements from the perspective of applications listed above.

A *generalized approximation space* [73] can be defined by a tuple $\mathcal{AS} = (U, I, \nu)$ where I is the *uncertainty function* defined on U with values in the power set $\mathcal{P}(U)$ of U. $I(x)$ is considered to be a *neighborhood* of x, and ν, the *inclusion function*, is defined on the Cartesian product $\mathcal{P}(U) \times \mathcal{P}(U)$ taking values in the interval $[0, 1]$; $\nu(X, Y)$ represents the degree of inclusion of the set X to the set Y. Then the lower and upper approximation operations are defined in $\mathcal{AS}$ in the following way.

$$LOW(\mathcal{AS}, X) = \{x \in U : \nu(I(x), X) = 1\} \text{ and } UPP(\mathcal{AS}, X) = \{x \in U : \nu((I(x), X) > 0\}.$$

In Pawlak's original definition [51], for a given information system (U, A)[4], $I(x)$ is equal to the equivalence class $A(x)$ generated from the indiscernibility relation $IND(A) = \{(x, y) \in U \times U : a(x) = a(y) \text{ for all } a \in A\}$, where $A(x) = \{y \in U : xIND(A)y\}$. In case of tolerance (or similarity) relation $T \subseteq U \times U$ one can consider $I(x) = \{y \in U : x \ T \ y\}$. That is, here $I(x)$ is equal to the tolerance class of x defined with respect to the relation T. For $X, Y \subseteq U$, the standard rough inclusion relation ν_{SRI}, available in the literature, is defined as follows[5].

$$\nu_{SRI}(X, Y) = \begin{cases} \dfrac{|X \cap Y|}{|X|}, & \text{if } X \neq \emptyset, \\ 1, & \text{otherwise.} \end{cases}$$

For the purpose of applications it is important to have some constructive definitions of I and ν.

One can consider another way to define $I(x)$. Usually together with $\mathcal{AS}$ we can associate a set $\mathcal{F}$ of formulae describing sets of objects of the universe U of $\mathcal{AS}$; $\mathcal{AS}$ basically gives the semantics ($\| \cdot \|_{\mathcal{AS}}$) such that for any formula α, $\|\alpha\|_{\mathcal{AS}} \subseteq U$[6]. Now, one can consider the following set $N_{\mathcal{F}}(x) = \{\alpha \in \mathcal{F} : x \in \|\alpha\|_{\mathcal{AS}}\}$, and construct $I(x) = \{\|\alpha\|_{\mathcal{AS}} : \alpha \in N_{\mathcal{F}}(x)\}$. Hence, more general uncertainty functions having values in $\mathcal{P}(\mathcal{P}(U))$ can be defined, and as a consequence different definitions of approximations can come up. For example,

---

[4] where U is a finite set and A is a set of attributes (*i.e.*, for any $a \in A$, $a : U \longrightarrow V_a$, where V_a is the set of values of a).

[5] $|X|$ denotes the cardinality of the set X.

[6] If $\mathcal{AS} = (U, A)$ then we will also write $\|\alpha\|_U$ instead of $\|\alpha\|_{\mathcal{AS}}$.

one can consider the following definitions of approximation operations over this approximation space $\mathcal{AS}$:

$$LOW(\mathcal{AS}, X) = \{x \in U : \nu(Y, X) = 1 \text{ for some } Y \in I(x)\} \text{ and}$$
$$UPP(\mathcal{AS}, X) = \{x \in U : \nu(Y, X) > 0 \text{ for any } Y \in I(x)\}.$$

An illustrative example of a set of formulas $\mathcal{F}$ can be based on a tolerance relation τ over U. Formulas from $\mathcal{F}$ defined over vectors of attribute values (or signatures of objects [40]) are used for defining tolerance classes. Then $I(x)$ consists of all tolerance classes of τ including the objects x. The family $\{\tau(x) : x \in U\}$ is a covering of U. Another example of covering of U can be obtained if, for a given tolerance relation τ over U, we take a family $\mathcal{C}(\tau)$ of all maximal (with respect to set theoretical inclusion) sets $Y \subseteq U$ satisfying the following condition: $\forall x, y \in Y$ $(y \in \tau(x))$. Then one can assign to $x \in U$ a family $\{Y \in \mathcal{C}(\tau) : x \in Y\}$. Certainly, $I(x)$ can be tuned by selecting relevant attributes taken for the definition of tolerance relation and/or parameters used to specify closeness of values and value vectors of attributes. It should be noted that the above presented scheme of approximation is not unique. In particular, the relationships (e.g., degrees of inclusion) of neighborhoods from $I(x)$ with the concept X and its complement may lead to other forms of approximation. Let us consider an illustrative example related to inducing classifiers. We assume that (U, A) is an information system and $X \subseteq U$ is a concept over U. However, we have only a partial information about this concept, i.e., we have a training set in the form of a decision system[7] (U_{tr}, A_{tr}, d), where $U_{tr} \subseteq U$, $A_{tr} = \{\bar{a} : a \in A\}$, $\bar{a}(x) = a(x)$ for $x \in U_{tr}$, and $d(x) = 1$ if $x \in X_{tr} = X \cap U_{tr}$, and $d(x) = 0$ if $x \in U_{tr} \setminus X_{tr}$. On the basis of this partial information an approximation of X over U should be induced. One of the approaches can be based on decision rules generated from (U_{tr}, A_{tr}, d). Let us assume that such a set $Rule$, of decision rules (for the decision 1 and 0, in our example), is obtained in the form of so called minimal decision rules [39, 55][8]. Now, for an arbitrary object x from U one can define $I(x)$ as a family of subsets of U_{tr} from (U_{tr}, A_{tr}, d) defined by the left hand sides of some rules belonging to the set $Rule$. For each such rule the object x should match the left hand side of the rule. In this way we obtain a subset of rules from $Rule$. The sets of objects from U_{tr} which satisfy the left hand sides of the selected rules create $I(x)$. We calculate the degrees of inclusion of sets from this family into X and its complement. The obtained degrees are used as arguments 'for' and 'against' membership of $x \in U$ to X. At this point, generally, a voting strategy is selected for resolving conflicts between these arguments to assign the tested object to the lower approximation of X or to the lower approximation

---

[7] Let us recall that a decision system is a triplet (U, A, d), where (U, A) is an information system and $d : U \longrightarrow V_d$ is the decision attribute with the set of values V_d such that $d \notin A$ [51].

[8] A rule of the form $lh(r) \longrightarrow d = i$, where $lh(r)$ is a conjunction of descriptors of the form $a = v$ for some $a \in A_{tr}$ and $i \in \{0, 1\}$ is minimal if this rule is true in U_{tr} but if we drop an arbitrary descriptor from $lh(r)$ the obtained rule will be no longer true in U_{tr} [39, 55].

of its complement. In the case, when the 'difference' between votes 'for' and 'against' is very 'small' the object is assigned to the boundary region.

The neighborhoods are defined relative to a given set of attributes (features) which can be tuned in the process of searching for more relevant features for the classification (e.g., using different reducts (see, e.g., [9,21,22]). For more complex vague concepts, this can be realized by hierarchical learning used for the ontology approximation discussed shortly below.

There are also different forms of rough inclusion functions. Let us consider two such examples. In the first example of a rough inclusion function, a threshold $t \in (0, 0.5)$ is used to relax the degree of inclusion of sets. The rough inclusion function ν_t is then defined by

$$
\nu_t (X,Y) = \begin{cases} 1 & \text{if } \nu_{SRI} (X,Y) \geq 1 - t, \\ \frac{\nu_{SRI}(X,Y)-t}{1-2t} & \text{if } t \leq \nu_{SRI} (X,Y) < 1 - t, \\ 0 & \text{if } \nu_{SRI} (X,Y) < t. \end{cases}
$$

Now, considering ν_t in place of ν in the above definitions of lower and upper approximations, one can obtain the approximations considered in the variable precision rough set approach (VPRSM) where Y is assumed to be a decision class and $I(x) = B(x)$ for any object x and a given set of attributes B. Another example of application of the standard inclusion was developed by using probabilistic decision functions. The rough inclusion relation can be also used for approximation of functions and relations [73].

Based on inclusion functions the *rough mereological approach* has also been generalized [61]. The inclusion relation $x\mu_r y$ with the intended meaning that x *is a part of y to a degree at least r*, has been taken as the basic notion of the rough mereology, a generalization of Leśniewski's notion of mereology [26].

As we already know, there can be families of approximation spaces for a particular purpose. We can think of that these families of approximation spaces are labeled by some parameters. Examples of a few such parameters are conditional attributes or formulas over these attributes, parametrized similarity relations used for description of neighborhoods, as well as different thresholds used to specify inclusion degrees of neighborhoods among different approximated concepts etc. By tuning such parameters, according to the chosen criteria (e.g., MDL principle), one can search for the optimal approximation space for describing/approximating concepts.

Thus, our knowledge about the approximated concepts is constrained by different parameters, and hence it is often partial and uncertain. So, it is reasonable to consider approximation of a concept based on both examples and counterexamples for the concepts [17] from the universe of objects. Hence, concept approximations constructed from a given sample of objects are extended, using inductive reasoning, on objects which are not yet observed. The rough set approach for dealing with concept approximation under such partial knowledge is now well developed.

2.2   Parametrized Approximation Space in a Complex Environment of Interacting Agents

Approximations of concepts should also take care of the constraints pertaining to dynamically changing environments. This leads to a more complex situation where the boundary regions are not crisp sets. This is also consistent to the postulate of the higher order vagueness considered by the philosophers (see, *e.g.*, [25]).

It is worthwhile to mention that a rough set approach for approximation of compound vague concepts has also been developed. For such concepts, it is hardly possible to expect that they can be approximated with high quality using the traditional methods [7,75]. In this context one first needs to consider the approximation of the domain-ontology of the concepts based on hierarchical learning. In several papers, the problem of ontology approximation (see, *e.g.*, [5]) has been discussed together with the possible applications in approximation of compound concepts or in knowledge transfer. In this case, a hierarchy of approximation spaces may need to be discovered for approximation of different concepts from the domain ontology. It is to be noted that in this approach different kinds of computational building blocks, called information granules, [70] work together, in parallel or in association. This involves interactions among different parts of the complex network of information granules. In any ontology [72], (vague) concepts and local dependencies between them are specified. Global dependencies can be derived from local dependencies. Such derivations can be used as hints in searching for relevant compound patterns (information granules) in approximation of more compound concepts from the ontology. The ontology approximation problem is one of the fundamental problems related to approximate reasoning. One should construct (in a given language that is different from the language in which the ontology is specified) not only approximations of concepts from ontology but also vague dependencies specified in the ontology. It is worthwhile to mention that an ontology approximation should be induced on the basis of incomplete information about concepts and dependencies specified in the ontology. Any method of approximation of vague dependency between two concepts X and Y should allow us to induce the arguments "for" and "against" that an object belongs to the concept Y on the basis of the arguments "for" and "against" that the object belongs to the concept X. Information granule calculi based on rough sets are capable to solve such problems. The approach towards approximation of a vague dependency between two concepts X and Y, based on only degrees of closeness (estimated from samples of objects) of X with Y and their extensions with respect to the approximation, is not satisfactory for approximate reasoning. Hence, more advanced approach should be developed. For complex vague dependencies, this can be performed in hierarchical way rather than in one step. Any argument can be thought of as a compound information granule (compound pattern). Arguments are fused by local schemes (production rules) discovered from data. Further fusions are possible through composition of local schemes, called approximate reasoning schemes (AR schemes) [49]. To estimate the degree to which (at least) an object belongs to a concept from ontology, the arguments "for" and "against" the membership to that concept are collected. Then a conflict resolution strategy is applied to aggregate the "for" and "against" degrees.

There are some other well established or emerging domains, not covered in this paper, where some generalizations of rough sets are proposed as the basic tools. These are often used in combination with other existing approaches. Among them rough sets based on (see references in [71]): (i) incomplete information and/or decision systems, (ii) non-deterministic information and/or decision systems, (iii) rough set model on two universes, (iv) dynamic information and/or decision systems, (v) dynamic networks of information and/or decision systems, are a few to name.

We know that rough sets play a crucial role in the development of *granular computing* (GrC) [58]. As in a complex network of information granules interactions play a natural role, *Interactive Granular Computing* (IGrC) comes in. So, in the study of parametrized approximation space one more dimension is added. The extension to IGrC [20] requires generalization of the basic concepts such as information and decision systems as well as methods for inducing hierarchical structures of information and decision systems interacting among themselves as well as with the environment. In the existing rough set approach, we assume that the results of computations of attribute values are given and are represented in data tables. In IGrC it is important also to resolve problems related to the process of perceiving values of attributes, *e.g.*, how these values of attributes are acquired through interaction with the environment and how to control this process to obtain data relevant for the target goals. Understanding interactive computations is one of the key problems for developing high quality intelligent systems working in complex environments [16]. In IGrC, computations are based on interactions of complex granules (c-granules, for short). Any c-granule consists of a physical part and a mental part linked in a special way [20]. IGrC is treated as the basis for (see, *e.g.*, [71] and references in this book): (i) Wistech Technology, in particular for approximate reasoning, called adaptive judgment, about properties of interactive computations, (ii) context inducing, (iii) reasoning about changes, (iv) process mining (this research was inspired by [54]), (v) perception based computing (PBC), (vi) risk management in computational systems [20] etc.

3   Rough Sets and Dempster-Shafer Theory

We know that the Dempster-Shafer theory [64,77] (see also http://www.science direct.com/journal/international-journal-of-approximate-reasoning/special-issue/10BG01ZSM7P) is widely used in decision support. In this section, our aim is only to give an illustrative example showing how the basic component of the Dempster-Shefer theory can be designed using the rough set notions [64–66][9].

---

[9] The readers are referred to the literature for other relationships of rough sets and Dempster-Shafer theory (see, *e.g.*, [11,12,76,79], [10]). For example, new methods of inducing rules were developed for searching rules with the large support for unions of few decision classes and eliminating many other decision classes (see, e.g., [33]).

In order to do that, first from the available data in the form of decision (information) systems the basic concepts of rough set theory such as generalized decision, lower approximation, upper approximation, and boundary region as well as aggregation of decision systems are defined. Next, on the basis of that one can define in a very simple way the basic concepts of the Dempster-Shafer theory.

Let us first recall the basic functions used in Dempster-Shafer theory [64].

By Θ we denote a nonempty finite set called the *frame of discernment*.

A function $m : \mathcal{P}(\Theta) \longrightarrow [0,1]$, where $\mathcal{P}(\Theta)$ is the powerset of Θ, is called the *mass function* if $m(\emptyset) = 0$ and $\sum_{\Delta \subseteq \Theta} m(\Delta) = 1$.

There are two more functions important in this theory. These are the *belief function* $Bel : \mathcal{P}(\Theta) \longrightarrow [0,1]$ and the *plausibility function* $Pl : \mathcal{P}(\Theta) \longrightarrow [0,1]$. They are defined as follows.

$$Bel(\Delta) = \sum_{\Gamma \subseteq \Delta} m(\Gamma) \quad \text{and} \quad Pl(\Delta) = \sum_{\Gamma \cap \Delta \neq \emptyset} m(\Gamma), \quad \text{where } \Delta \subseteq \Theta.$$

These functions have a simple intuitive interpretation in the rough set framework over decision systems [66].

Let $\mathbb{A} = (U, C, d)$ be a decision system [51,53,57]. We associate with the decision system $\mathbb{A}$ an approximation space $\mathcal{AS} = (U, I, \nu_{SRI})$, where $I(x) = C(x)$ for $x \in U$. We identify the set of decisions V_d with the frame of discernment Θ. By ∂_A we denote the generalized decision of $\mathbb{A}$, i.e., $\partial_A(x) = d(C(x)) = \{v \in V_d : \exists_{y \in C(x)} d(y) = v\}$. Now we can define the mass function $m_{\mathbb{A}}$ of the decision system $\mathbb{A}$ by

$$m_{\mathbb{A}}(\Delta) = \frac{|\{x \in U : \partial_C(x) = \Delta\}|}{|U|},$$

where $\Delta \subseteq V_d$. In fact, one can easily check that the function $m_{\mathbb{A}}$ satisfies the requirements for the mass function.

Now, one can obtain the following two facts for the belief function $Bel_{\mathbb{A}}$ and the plausibility function $Pl_{\mathbb{A}}$ defined on the basis of the mass function $m_{\mathbb{A}}$ [66]:

$$Bel_{\mathbb{A}}(\Delta) = \frac{|LOW(\mathcal{AS}, \bigcup_{i \in \Delta} : X_i)|}{|U|} \quad \text{and} \quad PL_{\mathbb{A}}(\Delta) = \frac{|UPP(\mathcal{AS}, \bigcup_{i \in \Delta} : X_i)|}{|U|},$$

where $X_i = \{x \in U : d(x) = i\}$ is the decision class related to the decision i, and $\Delta \subseteq V_d$.

In this way we obtain a very intuitive interpretation of the functions $Bel_{\mathbb{A}}$ and $Pl_{\mathbb{A}}$ in terms of the lower approximation and the upper approximation of unions of (relevant for Δ) decision classes.

Moreover, one can also obtain an interpretation of the so called Dempster-Shafer rule of combination using a relevant operation on decision tables. The Dempster-Shafer rule of combination aggregates two mass functions m_1 and m_2 to a new mass function $m_1 \otimes m_2$ defined by

$$m_1 \otimes m_2(\emptyset) = 0 \quad \text{and} \quad m_1 \otimes m_2(\Delta) = \frac{\sum_{A \cap B = \Delta} m_1(A) m_2(B)}{1 - \sum_{A \cap B = \emptyset} m_1(A) m_2(B)}, \quad \text{where } \emptyset \neq \Delta \subseteq V_d.$$

In the case when the mass functions m_1 and m_2 are defined by the decision systems $\mathbb{A}_1$ and $\mathbb{A}_2$, respectively, one can define a natural operation $\odot$ on these decision systems such that [66] $m_{\mathbb{A}_1} \otimes m_{\mathbb{A}_2} = m_{\mathbb{A}_1 \odot \mathbb{A}_2}$.

The presentation of the above basic notions of Dempster-Shafer theory appears natural using the basic concepts of rough sets. The presented approach allows us to design these definitions on the basis of the available data and to ground the basic concepts of Dempster-Shafer theory on data.

4   Combination of Rough Sets with Soft Computing Approaches Improving the Quality of the Constructed Granules

The main reason behind the success in developing methods with a high quality of approximating concepts is that these are based on combination of the rough set approach with other approaches. Relevant combinations of different languages for dealing with borderline cases, which these methods are using, lead to the improvement of their performance in searching for relevant granules as computational building blocks for approximation of complex vague concepts, especially the boundary regions.

Both fuzzy and rough set theory represent two different approaches to vagueness. Fuzzy set theory addresses *gradualness* of knowledge, expressed by the fuzzy membership, whereas rough set theory addresses *granularity* of knowledge, expressed by the indiscernibility relation. Both the theories are not competing but are rather complementary. In particular, the rough set approach provides tools for approximate construction of fuzzy membership functions.

Let us mention briefly two simple cases illustrating possible combination of methods based on rough sets and fuzzy sets.

In the first example, one can consider the rough-set methods for generation of decision rules for preliminary recognition of some regions (corresponding to some decisions). The left hand sides of obtained decision rules define crisp sets of objects. To resolve membership conflicts for objects close to boundaries of these sets one can use more 'elastic' approach based on fuzzy sets. In this elastic approach, fuzzy sets are spread over these crisp sets defined using the rough set approach.

In the second example, let us consider a situation when a fuzzy membership function μ_X for a concept X is given and we would like to modify this function to cover the fact that objects are perceived using attributes from a set C. This leads to considering for each indiscernibility class $C(x)$, its image obtained by μ_X, i.e., the set $\mu_X(C(x)) = \{\mu_X(y) : y \in C(x)\}$, instead of the particular value $\mu_X(x)$. So, $\mu_X(C(x))$ represents the possible set of values, that x and all elements similar to it with respect to the set of attributes C, can assume under μ_X. Then one can consider a pair of two fuzzy sets obtained by combination of the rough set and fuzzy set approaches. The combination is based on a rough-fuzzy model including μ_X and the approximation space $AS = (U, IND(C)$. Now, the pair, called the rough-fuzzy set, can be defined consisting of the lower approximation defined by the following fuzzy membership function:

$$LOW(\mathcal{AS}, \mu_X)(x) = inf\ \mu_X(C(x)),$$

and the upper approximation defined by the following fuzzy membership function

$$UPP(\mathcal{AS}, \mu_X)(x) = sup\ \mu_X(C(x)).$$

These two simple strategies of combination of rough sets and fuzzy sets are only illustrations of numerous other successful strategies in applications. More detailed discussion on relationships of rough sets and fuzzy sets, the reader can find, e.g., in [8,13,52,60,62,78]. Rough sets and fuzzy sets can work synergistically, often with other soft computing approaches. The developed systems exploit the tolerance for imprecision, uncertainty, approximate reasoning and partial truth under soft computing framework, and is capable of achieving tractability, robustness, and close resemblance with human like (natural) decision making for pattern recognition in ambiguous situations [69,80]. The developed methods have found applications in different domains such as bioinformatics and medical image processing. The objective of the rough-fuzzy integration is to provide a stronger paradigm of uncertainty handling in decision-making. Over the years, many methods and applications, in particular in pattern recognition, were developed on the basis of rough sets, fuzzy sets, and on their combination. The methods based on combination of the approaches exploit different abilities of the mixed languages used for generation as well as expressing patterns. This makes it possible to discover patterns of the higher quality, and have better approximation of the boundary region of a vague concept, in comparison to the situations when they are used in isolation. One should note that in this case the searching space for relevant patterns becomes larger in comparison to the cases when single approach is used. Similarly, developing efficient heuristics for searching relevant patterns is more challenging. These methods concern discovery of patterns such as decision rules, clusters and processes of feature selection. Readers can find more details in the literature (see, e.g., [29,38,68]) for the rough set based methods and [6,14,15,22,24,30–32,41,44,48,50,59,69]) for the methods based on combination of rough sets and fuzzy sets.

The characteristics of rough-fuzzy granulation have been further exploited in designing various neural network models for their efficient and speedy learning, and enhanced performance (see, e.g., [3,4,15,28,35,36,42,46,47,49]). This seems to be strongly promising to big data analysis. There are hybrid methods combining rough sets with methods using others statistical tools, e.g., kernel functions, case-based reasoning, wavelets, EM method, independent component analysis, principal component analysis etc. (see, e.g., [1,2,18,27,34,37,43,45,74]). We end this section, emphasizing the opinion, envisaged by other researchers [23] too, that the theoretical foundations of soft computing should be based on combination of rough sets, fuzzy sets genetic algorithm, and neural networks.

5   Conclusions

In this paper, we have discussed some aspects of the rough set approach which lead to its wide applicability in real-life projects. There are some other issues to be discussed such as the relevance of the rough set approach to the development of the foundations of different areas including machine learning, data mining, and data science. In particular, the role of the rough set approach in further development of IGrC as the basis for perception based computing, seems to be promising. Moreover, more work on extending the existing tools of mathematical logic should be done towards satisfying the requirement of '*a reconciliation between two contradictory characteristics–the apparent logical nature of reasoning and the statistical nature of learning*' as formulated by Leslie Valiant[10].

Acknowledgments. The authors would like to thank Professor Mihir Chakraborty for suggesting the problem considered in this paper.

References

1. Albanese, A., Sankar, F., Pal, K., Petrosino, A.: Rough sets, kernel set, and spatiotemporal outlier detection. IEEE Trans. Knowl. Data Eng. **26**, 194–207 (2014)
2. An, S., Shi, H., Hu, Q., Li, X., Dang, J.: Fuzzy rough regression with application to wind speed prediction. Inf. Sci. **282**, 388–400 (2014)
3. Banerjee, M., Mitra, S., Pal, S.K.: Rough-fuzzy MLP. IEEE Trans. Neural Nets **9**, 1203–1216 (1998)
4. Banerjee, M., Pal, S.K.: Roughness of a fuzzy set. Inf. Sci. **93**(3–4), 235–246 (1996)
5. Bazan, J.G.: Hierarchical classifiers for complex spatio-temporal concepts. In: Peters, J.F., Skowron, A., Rybiński, H. (eds.) Transactions on Rough Sets IX. LNCS, vol. 5390, pp. 474–750. Springer, Heidelberg (2008). https://doi.org/10.1007/978-3-540-89876-4_26
6. Bello, R., Falcón, R., Pedrycz, W.: Granular Computing: At the Junction of Rough Sets and Fuzzy Sets, Studies in Fuzziness and Soft Computing, vol. 234. Springer, Heidelberg (2010). https://doi.org/10.1007/978-3-540-76973-6
7. Breiman, L.: Statistical modeling: the two cultures. Statis. Sci. **16**(3), 199–231 (2001)
8. Chakraborty, M.K.: Membership function based rough set. Int. J. Approximate Reasoning **55**(1), 402–411 (2014)
9. Cornelis, C., Jensen, R., Martín, G.H., Ślęzak, D.: Attribute selection with fuzzy decision reducts. Inf. Sci. **180**(2), 209–224 (2010)
10. Denoeux, T.: Dempster-Shafer theory. Introduction, connections with rough sets and application to clustering. slides from letures at RSKT 2014, Shanghai, China, 25 October 2014. https://www.hds.utc.fr/~tdenoeux/dokuwiki/_media/en/rskt2014.pdf
11. Denoeux, T., Li, S., Sriboonchitta, S.: Evaluating and comparing soft partitions: an approach based on Dempster-Shafer theory. IEEE Trans. Fuzzy Syst. **26**(3), 1231–1244 (2017)
12. Dubois, D., Prade, H.: Rough fuzzy sets and fuzzy rough sets. Int. J. Gen. Syst. **17**, 191–208 (1990)

---

[10] https://people.seas.harvard.edu/~valiant/researchinterests.htm.

13. Dubois, D., Prade, H.: Rough fuzzy sets and fuzzy rough sets. Int. J. Gen. Syst. **17**(2–3), 191–209 (1990)
14. Ganivada, A., Ray, S., Pal, S.: Fuzzy rough sets, and a granular neural network for unsupervised feature selection. Neural Netw. **48**, 91–108 (2013)
15. Ganivada, A., Ray, S.S., Pal, S.: Fuzzy rough granular self-organizing map and fuzzy rough entropy. Theoret. Comput. Sci. **466**, 37–63 (2012)
16. Goldin, D., Smolka, S., Wegner, P. (Eds.): Interactive Computation: The New Paradigm. Springer, Heidelberg (2006). https://doi.org/10.1007/3-540-34874-3
17. Hastie, T., Tibshirani, R., Friedman, J.H.: The Elements of Statistical Learning: Data Mining, Inference, and Prediction. Springer, Heidelberg (2001). https://doi.org/10.1007/978-0-387-84858-7
18. Hu, Q., Yu, D., Pedrycz, W., Chen, D.: Kernelized fuzzy rough sets and their applications. IEEE Trans. Knowl. Data Eng. **23**, 1649–1667 (2011)
19. Chikalov, I., et al.: Three Approaches to Data Analysis. Test Theory, Rough Sets and Logical Analysis of Data, Series Intelligent Systems Reference Library, vol. 41. Springer, Heidelberg (2012). https://doi.org/10.1007/978-3-642-28667-4
20. Jankowski, A.: Interactive Granular Computations in Networks and Systems Engineering: A Practical Perspective. Springer, Heidelberg (2017). https://doi.org/10.1007/978-3-319-57627-5
21. Janusz, A., Ślęzak, D.: Rough set methods for attribute clustering and selection. Appl. Artif. Intell. **28**(3), 220–242 (2014)
22. Jensen, R., Shen, Q.: Computational Intelligence and Feature Selection: Rough and Fuzzy Approaches. IEEE Press Series on Cmputationa Intelligence. IEEE Press and Wiley, Hoboken (2008)
23. Joshi, M., Bhaumik, R.N., Lingras, P., Patil, N., Salgaonkar, A., Slezak, D.: Rough set year in India 2009. In: Sakai, H., Chakraborty, M.K., Hassanien, A.E., Slezak, D., Zhu, W. (eds.) RSFDGrC 2009. LNCS (LNAI), vol. 5908, pp. 67–68. Springer, Heidelberg (2009). https://doi.org/10.1007/978-3-642-10646-0_7
24. Joshi, M., Lingras, P., Rao, C.R.: Correlating fuzzy and rough clustering. Fundamenta Informaticae **115**(2–3), 233–246 (2012)
25. Keefe, R.: Theories of Vagueness. Cambridge Studies in Philosophy. Cambridge University Press, Cambridge (2000). Kindly check the edit made in Ref. [25]
26. Leśniewski, S.: Grungzüge eines neuen Systems der Grundlagen der Mathematik. Fundamenta Mathematicae **14**, 1–81 (1929)
27. Li, Y., Shiu, S.C.-K., Pal, S.K., Liu, J.N.-K.: A rough set-based case-based reasoner for text categorization. Int. J. Approximate Reasoning **41**(2), 229–255 (2006)
28. Lingras, P.: Fuzzy - rough and rough - fuzzy serial combinations in neurocomputing. Neurocomputing **36**(1–4), 29–44 (2001)
29. Lingras, P., Peters, G.: Rough clustering. Wiley Interdisc. Rev.: Data Min. Knowl. Disc. **1**(1), 64–72 (2011)
30. Maji, P., Pal, S.: RFCM: a hybrid clustering algorithm using rough and fuzzy sets. Fundamenta Informaticae **80**(4), 477–498 (2007)
31. Maji, P., Pal, S.: Rough set based generalized fuzzy c-means algorithm and quantitative indices. IEEE Trans. Syst. Man Cybern. Part B Cybern. **37**(6), 1529–1540 (2007)
32. Maji, P., Pal, S.K.: Rough-Fuzzy Pattern Recognition: Application in Bioinformatics and Medical Imaging. Wiley Series in Bioinformatics. Wiley, Hoboken (2012)
33. Marszal-Paszek, B., Paszek, P.: Classifiers based on nondeterministic decision rules. Rough Sets Intell. Syst. **2**, 445–454 (2013)
34. Mehera, S.K., Pal, S.K.: Rough-wavelet granular space and classification of multispectral remote sensing image. Applied Soft Comput. **11**, 5662–5673 (2011)

35. Mitra, P., Mitra, S., Pal, S.K.: Modular rough fuzzy MLP: evolutionary design. In: Zhong, N., Skowron, A., Ohsuga, S. (eds.) RSFDGrC 1999. LNCS (LNAI), vol. 1711, pp. 128–136. Springer, Heidelberg (1999). https://doi.org/10.1007/978-3-540-48061-7_17

36. Mitra, P., Mitra, S., Pal, S.K.: Evolutionary modular design of rough knowledge-based network using fuzzy attributes. Neurocomputing 36, 45–66 (2001)

37. Mitra, P., Pal, S., Siddiqi, M.A.: Nonconvex clustering using expectation maximization algorithm with rough set initialization. Pattern Recogn. Lett. 24, 863–873 (2003)

38. Nguyen, H.S.: Approximate boolean reasoning: foundations and applications in data mining. In: Peters, J.F., Skowron, A. (eds.) Transactions on Rough Sets V. LNCS, vol. 4100, pp. 334–506. Springer, Heidelberg (2006). https://doi.org/10.1007/11847465_16

39. Nguyen, H.S., Skowron, A.: Rough sets: from rudiments to challenges. In: Skowron, A., Suraj, Z. (eds.), vol. 71, pp. 75–173 (2013). https://doi.org/10.1007/978-3-642-30344-9_3

40. Nguyen, S., Skowron, A., Synak, P.: Discovery of data patterns with applications to decomposition and classification problems. In: Rough Sets in Knowledge Discovery 2: Applications, Case Studies and Software Systems, pp. 55–97 (1998). https://doi.org/10.1007/978-3-7908-1883-3_4

41. Pal, S., Meher, S., Dutta, S.: Class-dependent rough-fuzzy granular space, dispersion index and classification. Pattern Recogn. 45, 2690–2707 (2012)

42. Pal, S., Ray, S.S., Ganivada, A.: Granular Neural Networks, Pattern Recognition and Bioinformatics. Studies in Computational Intelligence, vol. 712. Springer, Heidelberg (2017). https://doi.org/10.1007/978-3-319-57115-7

43. Pal, S., Shiu, S.: Foundations of Soft Case-Based Reasoning. Wiley, Hoboken (2004)

44. Pal, S.K.: Soft data mining, computational theory of perceptions, and rough-fuzzy approach. Inf. Sci. 163(1–3), 5–12 (2004)

45. Pal, S.K., Mitra, P.: Multispectral image segmentation using the rough-set-initialized EM algorithm. IEEE Trans. Geosci. Remote Sens. 40, 2495–2501 (2002)

46. Pal, S.K., Mitra, P.: Case generation using rough sets with fuzzy representation. IEEE Trans. Knowl. Data Eng. 16(3), 292–300 (2004)

47. Pal, S.K., Pedrycz, W., Skowron, A., Swiniarski, R. (Eds.): Special volume: Rough-neuro computing. Neurocomputing 36(1–4), 1–262 (2001)

48. Pal, S.K., Peters, J.F. (eds.): Rough Fuzzy Image Analysis Foundations and Methodologies. Chapman & Hall/CRC, Boca Raton (2010)

49. Pal, S.K., Polkowski, L., Skowron, A. (eds.): Rough-Neural Computing: Techniques for Computing with Words. Cognitive Technologies. Springer, Heidelberg (2004). https://doi.org/10.1007/978-3-642-18859-6

50. Pal, S.K., Skowron, A. (eds.): Rough Fuzzy Hybridization: A New Trend in Decision-Making. Springer, Singapore (1999)

51. Pawlak, Z.: Rough sets. Int. J. Comput. Inf. Sci. 11, 341–356 (1982)

52. Pawlak, Z.: Rough sets and fuzzy sets. Fuzzy Sets Syst. 17, 99–102 (1985)

53. Pawlak, Z.: Rough Sets: Theoretical Aspects of Reasoning about Data, System Theory, Knowledge Engineering and Problem Solving, vol. 9. Kluwer Academic Publishers, Dordrecht (1991)

54. Pawlak, Z.: Concurrent versus sequential - the rough sets perspective. Bull. EATCS 48, 178–190 (1992)

55. Pawlak, Z., Skowron, A.: Rough sets and boolean reasoning. Inf. Sci. 177(1), 41–73 (2007)

56. Pawlak, Z., Skowron, A.: Rough sets: some extensions. Inf. Sci. **177**(1), 28–40 (2007)
57. Pawlak, Z., Skowron, A.: Rudiments of rough sets. Inf. Sci. **177**(1), 3–27 (2007)
58. Pedrycz, W., Skowron, S., Kreinovich, V. (eds.): Handbook of Granular Computing. Wiley, Hoboken (2008)
59. Peters, G., Crespo, F., Lingras, P., Weber, R.: Soft clustering - fuzzy and rough approaches and their extensions and derivatives. Int. J. Approximate Reasoning **54**(2), 307–322 (2013)
60. Polkowski, L.: Rough mereology as a link between rough and fuzzy set theories. a survey. In: Peters, J.F., Skowron, A., Dubois, D., Grzymała-Busse, J.W., Inuiguchi, M., Polkowski, L. (eds.) Transactions on Rough Sets II. LNCS, vol. 3135, pp. 253–277. Springer, Heidelberg (2004). https://doi.org/10.1007/978-3-540-27778-1_13
61. Polkowski, L. (Ed.): Approximate Reasoning by Parts. An Introduction to Rough Mereology, Intelligent Systems Reference Library, vol. 20. Springer, Heidelberg (2011). https://doi.org/10.1007/978-3-642-22279-5
62. Polkowski, L., Skowron, A.: Rough mereology: a new paradigm for approximate reasoning. Int. J. Approximate Reasoning **15**(4), 333–365 (1996)
63. Rissanen, J.: Minimum-description-length principle. In: Kotz, S., Johnson, N. (eds.) Encyclopedia of Statistical Sciences, pp. 523–527. Wiley, New York (1985)
64. Shafer, G.: Mathematical Theory of Evidence. Princeton University Press, Princeton (1976)
65. Skowron, A.: Boolean reasoning for decision rules generation. In: Komorowski, J., Raś, Z.W. (eds.) ISMIS 1993. LNCS, vol. 689, pp. 295–305. Springer, Heidelberg (1993). https://doi.org/10.1007/3-540-56804-2_28
66. Skowron, A., Grzymała-Busse, J.W.: From rough set theory to evidence theory. In: Yager, R., Fedrizzi, M., Kacprzyk, J. (eds.) Advances in the Dempster-Shafer Theory of Evidence, pp. 193–236. Wiley, New York (1994)
67. Skowron, A., Jankowski, A., Swiniarski, R.W.: Foundations of rough sets. In: Kacprzyk, J., Pedrycz, W. (eds.) Springer Handbook of Computational Intelligence, pp. 331–348. Springer, Heidelberg (2015). https://doi.org/10.1007/978-3-662-43505-2_21
68. Skowron, A., Pal, S.K. (Eds.): Special volume: rough sets, pattern recognition and data mining. Pattern Recogn. Lett. **24**(6), 829–831 (2003)
69. Skowron, A., Pal, S.K., Nguyen, H.S. (Eds.): Special issue on rough sets and fuzzy sets in natural computing. Theoret. Comput. Sci. **412**(42), 5816–5819 (2011)
70. Skowron, A., Stepaniuk, J.: Rough sets and granular computing: toward rough-ranular computing. In: Pedrycz, et al., vol. 58, pp. 425–448
71. Skowron, A., Suraj, Z. (eds.): Rough Sets and Intelligent Systems, Professor Zdzislaw Pawlak in Memoriam. Series Intelligent Systems Reference Library. Springer, Heidelberg (2013). https://doi.org/10.1007/978-3-642-30344-9
72. Staab, S., Studer, R. (eds.): Handbook on Ontologies. International Handbooks on Information Systems. Springer, Heidelberg (2004). https://doi.org/10.1007/978-3-540-92673-3
73. Stepaniuk, J. (ed.): Rough-Granular Computing in Knowledge Discovery and Data Mining. Springer, Heidelberg (2008). https://doi.org/10.1007/978-3-540-70801-8
74. Świniarski, R.W., Skowron, A.: Independent component analysis, principal component analysis and rough sets in face recognition. In: Peters, J.F., Skowron, A., Grzymała-Busse, J.W., Kostek, B., Świniarski, R.W., Szczuka, M.S. (eds.) Transactions on Rough Sets I. LNCS, vol. 3100, pp. 392–404. Springer, Heidelberg (2004). https://doi.org/10.1007/978-3-540-27794-1_19

75. Vapnik, V.: Statistical Learning Theory. Wiley, New York (1998)
76. Wu, W.-Z., Leung, Y., Zhang, W.-X.: Connections between rough set theory and Dempster-Shafer theory of evidence. Int. J. Gen. Syst. **31**(4), 405–430 (2002)
77. Yager, R., Liu, L. (eds.): Classic Works of the Dempster-Shafer Theory of Belief Functions, vol. 219. Springer, Heidelberg (2008). https://doi.org/10.1007/978-3-540-44792-4
78. Yao, Y.Y.: A comparative study of fuzzy sets and rough sets. Inf. Sci. **109**(1–4), 227–242 (1998)
79. Yao, Y.Y., Lingras, P.J.: Interpretations of belief functions in the theory of rough sets. Inf. Sci. **104**, 81–106 (1998)
80. Zadeh, L.A.: Fuzzy logic, neural networks, and soft computing. Commun. ACM **37**, 77–84 (1994)

What's in a Relation? Logical Structures of Modes of Granulation

Piero Pagliani[✉]

Research Group on Knowledge and Communication Models, Rome, Italy
pier.pagliani@gmail.com

1 Towards a Position Paper

1.1 What's a Granulation and What's an Approximation?

Granulation can be though of as a conceptual grid based on given knowledge, while approximation is the process of forming new knowledge through an available conceptual grid. In a wider sense, approximating is an operation required when a "scale" is used to determine something which does not fit exactly with the "precision" enabled by that scale. One can find instances of this dialectic between granulation and approximation in different fields spanning from data mining to story understanding, from pattern recognition to machine learning. We use the term "scale" in a general sense. Granulation is a sort of "conceptual scale". Granules are groups of items (or points) of a given universe of discourse formed by means of knowledge which has been acquired or hypothesized and stored, that is, an established knowledge. From now on, we use the terms "granule" and "neighbourhood", as well as "granulation" and "neighbourhood system", interchangeably.

Typically, items are grouped together if their share *to some extent* some well-established properties. But they could be grouped together also as a result of empirical evidences with little reference to any (at least apparent) rule. Therefore, the way in which granules are formed spans from the application of well-defined relations, up to "anarchical" grouping. To put it in another way, on the one extreme we deal with well-defined granules in which the logical structure is recognizable (for instance equivalence or order relations), while on the other extreme one deals with the breakup of the universe in parts which cannot be interpreted as neighbourhoods induced by any kind of relation, that is, non-structured granules in which it might be even difficult to understand why items are linked together.

However, pointless topology and the logical structure underlying its basic concepts, enable us to zoom-in and zoom-out different modes of granulation and understand their logical and geometrical properties even in some apparently unstructured cases.

1.2 What's in a Relation?

In the original formulation of Rough Set Theory, granules are formed by means of *equivalence relations*, that are very structured relations: reflexive, transitive

© Springer Nature Switzerland AG 2018
H. S. Nguyen et al. (Eds.): IJCRS 2018, LNAI 11103, pp. 46–60, 2018.
https://doi.org/10.1007/978-3-319-99368-3_4

and symmetric. Immediately from inception, other kinds of binary relations have been used, such as preorders (reflexive and transitive), partial orders (antisymmetric preorders) and tolerance relations (reflexive and symmetric). Also arbitrary binary relations have been taken into account.

Arbitrary granulations resulting in coverings of the universe of discourse, were pioneered by Zakowski and Pomikała. But since the beginning of the XXI century researches on covering-based rough sets started growing rapidly (we suggest to search on the Internet for an appropriate bibliography; a non-exhaustive list of works can be found in the References of [10]).

We call a granulation *pre-topological* respectively *topological*, if it induces approximation operators with properties close, respectively equal, to those of topological *interior* and *closure* operators. Since the adverb "close" means many a thing, we shall deal, actually, with different notions of "pre-topological" operators. The starting point is any relation $R \subseteq U \times U'$, where the elements of the sets U and U' receive a variety of interpretations. If $U = U'$ then R simply connects items on the basis of some criteria which is not embedded in the triple $\langle U, U, R \rangle$ itself, or are recoverable from it just formally, but not semantically. We denote this structure by $\langle U, R \rangle$ and call it a *square relational system*, SRS.

In this case and in other cases in which a binary relation is acting to form granules, a number of results are provided for free by topology and/or Modal Logic, because approximation operators are modal operators.

In a sense, this is the classical approach in Rough Set Theory and we shall see that it is a special case of more general approaches. Consider a relational system $\langle U, U', R \rangle$[1].

- **Property system interpretation**: U is a set of items and U' a set of properties, so that the relational structure is called a *property system*. This is a classical interpretation. If a property system is given, the elements of U can be grouped on the basis of the properties they fulfil, in order to form granules of knowledge. The geometry of the set of granules will depend on R. In other terms, R will induce one or more relations R^* on U with particular properties.
- **Pointless (or formal) topology interpretation**: U is a set of points and U' a set of formal (or abstract) neighbourhoods. Otherwise stated, U' is a set of abstract granules. In this case one point of interest are the relations R^* which are induced by R between abstract neighbourhoods.
- **Concrete neighbourhood interpretation**: an intermediate case is given when $U' = \wp(U)$, so that U' is a set of "real" granules of elements of U, that is, U' is a set of subsets of U. Modal Logic semantic based on neighbourhoods systems deals with similar relational structures. Moreover, covering-based approximations come from this situation.

Therefore, given a relational space $\langle U, U', R \rangle$, granules are formed in different ways. Basically, there are an *indirect* way and two *direct* ways.

---

[1] In formal topology, it is called a *basic pair* or a *Chu space*.

- **Indirect way**: $\langle a, b \rangle \in R^*$ because both $\langle a, u' \rangle$ and $\langle b, u' \rangle$ are in R, for *some* $u' \in U'$, where the meaning of "some" must be specified further. This means, for instance, that a and b are in the same granule because they share some property.
- **Direct way 1**: when $A \subseteq U$ and $\langle a, A \rangle \in R$ then A is a granule associated to a. Anyway, notice that A can be considered as (the extension of) a property, so that also in this case one can form granules in the indirect way. Actually, it is a two-face case.
- **Direct way 2**: when $U' = U$, so that the granule associated to a is the set of all the $b \in U$ such that $\langle a, b \rangle \in R$. We denote it by $R(a)$ and call it the $R - neighbourhood$ of a. The indirect way leads to this direct way by means of an induced relation R^*.

One main point of interest is to study the relationships between the properties of R and those of the induced relations R^* between items or between granules. Classical and generalised approximation operators from SRSs are within this case. Another point to be investigated concerns the relations between the operators definable within the concrete neighbourhood interpretation and those definable within the formal approach provided by pointless topology. We shall see that in some particular cases the three approaches give exactly the same result. That is, although one can think to deal with different situations, actually the inner logic is the same. However, the formal and the concrete approaches do not correspond exactly. Their ability to describe the properties of granulations are different and sometimes the language of one approach does not have any equivalent in the other language. The main aim of this survey is introducing a logical and mathematical tool-kit to be used in the researches about approximations and Rough Set Theory at large. Therefore, we will mention just a few new results (namely those in Sect. 3) but discuss a set of open problems.

2   Galois Adjunctions and Galois Connections

We shall study all the above cases starting with a small set of operators provided by pointless topology. These operators are defined by means of combinations of logical operators. Their inner logical structure make them into Galois adjunctions. From that, a number of result are easily deduced for free[2]. Assume $\mathbf{A} = \langle A, \leq_A \rangle$ and $\mathbf{B} = \langle B, \leq_B \rangle$ are partially ordered sets and let $\iota : \mathbf{A} \longmapsto \mathbf{B}$ and $\sigma : \mathbf{B} \longmapsto \mathbf{A}$ be two monotonic functions such that the following holds:

$$\iota(x) \leq_B y \quad \textit{if and only if} \quad x \leq_A \sigma(y) \tag{1}$$

---

[2] In Rough Set Theory the following operartors have been introduced by [4] and independently in [11].

We say that $\langle \iota, \sigma \rangle$ is a *Galois adjunction* between $\mathbf{A}$ and $\mathbf{B}$, where ι is the *lower adjoint* of σ and σ is the *upper adjoint* of ι. The pair $\langle \iota, \sigma \rangle$ is also called an *axiality*. The following facts are well-known:

Proposition 1. *If $\langle \iota, \sigma \rangle$ is a Galois adjunction between two partially ordered sets $\mathbf{A}$ and $\mathbf{B}$, then: (a) $\sigma\iota(x) \geq_A x$, any $x \in A$; (b) $\iota\sigma(y) \leq_B y$, any $y \in B$; (c) $\iota\sigma$ and $\sigma\iota$ are monotonic; (d) $\iota\sigma\iota\sigma = \iota\sigma$ and $\sigma\iota\sigma\iota = \sigma\iota$. Therefore: (e) $\iota\sigma$ is an interior operator; (f) $\sigma\iota$ is a closure operator.*

However, they both fail to be topological, because $\iota\sigma$ fails to be multiplicative and co-normal and $\sigma\iota$ fails to be additive and normal (where co-normal means $\iota\sigma(\top) = \top$ and normal $\sigma\iota(\bot) = \bot$, for $\top$ the maximal element of $\mathbf{B}$ and $\bot$ the minimal element of $\mathbf{A}$). In other terms, they are pre-topological operators.

A *Galois connection* is the antitone version of a Galois adjunction:

$$\iota(x) \leq_B y \quad \textit{if and only if} \quad x \geq_A \sigma(y) \tag{2}$$

In this case $\langle \iota, \sigma \rangle$ is called a *polarity*[3].

In a Galois connection, both $\iota\sigma$ and $\sigma\iota$ are pre-topological closure operators. Given a relational system, *constructors* which form Galois adjunctions and Galois connections can be defined by means of straightforward logical definitions:

Definition 1. *Let $\langle U, U', R \rangle$ be a relational system. Then:*

- $\langle e \rangle : \wp(U') \longmapsto \wp(U); \langle e \rangle(Y) = \{a \in U : \exists b(b \in Y \wedge a \in R^{\smile}(b))\};$
- $[e] : \wp(U') \longmapsto \wp(U); [e](Y) = \{a \in U : \forall b(a \in R^{\smile}(b) \Longrightarrow b \in Y)\};$
- $\langle i \rangle : \wp(U) \longmapsto \wp(U'); \langle i \rangle(X) = \{b \in U' : \exists a(a \in X \wedge b \in R(a))\};$
- $[i] : \wp(U) \longmapsto \wp(U'); [i](X) = \{b \in U' : \forall a(b \in R(a) \Longrightarrow a \in X)\};$
- $[[e]] : \wp(U') \longmapsto \wp(U); [[e]](Y) = \{a \in U : \forall b(b \in Y \Longrightarrow a \in R^{\smile}(b))\};$
- $[[i]] : \wp(U) \longmapsto \wp(U'); [[i]](X) = \{b \in U' : \forall a(a \in X \Longrightarrow b \in R(a))\}.$

$R^{\smile}$ is the inverse of R. Therefore $b \in R(a)$ if and only if $a \in R^{\smile}(b)$, so that the reader may interpret the above definitions according to her/his own intuition. The decorations "e" and "i" means *extensional* and, respectively, *intensional*. Formally, they just remember the direction, R or $R^{\smile}$, of the relation, but in many applications U' is a set of properties which may be fulfilled by the elements of a set U of objects. For this reason we keep the above decorations.

Further, the symbols $\langle e \rangle$ and $\langle i \rangle$, or collectively $\langle \cdot \rangle$, remind us that these are *possibility* operators. For instance, if $X = \langle e \rangle(Y)$ and $b \in Y$ then it is possible that b is in relation R (more precisely $R^{\smile}$) with the elements of X because there is *at least one* $b' \in Y$ such that aRb' for some $a \in X$. In turn, the symbol $[e]$ and $[i]$, or collectively $[\cdot]$, denote *necessity*: for instance, if $X = [e](Y)$ then in order to be in relation R with an element $a \in X$, it is necessary to be in Y because *at most all* the elements of Y are in relation with the elements of X. Notice, incidentally, that this is the "correct" relational reading of the clauses for possibility and necessity in Kripke models for modal logic, while the usual

---

[3] Sometimes this term denotes what we call a relational system.

reading runs in the opposite direction. For instance, Y is necessary in a if all the elements R-related to a belongs to Y.

Finally, $[[i]]$ and $[[e]]$ means *sufficiency*: for instance, if $X = [[e]](Y)$ and $a \in X$, then it is sufficient to belong to Y to be in relation $R^{\smile}$ with a, because *at least all* the members of Y are in relation with all the elements of X.

Moreover, it is worth noticing that the *logical core* of the constructors $\Diamond$-shaped is the pair $\langle \exists, \wedge \rangle$ (set-theoretically: they are built by means of non empty intersection), while that of $\Box$-shaped constructors is $\langle \forall \implies \rangle$ (set-theoretically they are built by means of the inclusion). Finally, because of their very logical core, these constructors fulfil the strategic properties we are looking for. In fact, $\langle \langle i \rangle, [e] \rangle$ and $\langle \langle e \rangle, [i] \rangle$ form Galois adjunctions, while $\langle [[e]], [[i]] \rangle$ forms a Galois connection between $\langle \wp(U), \subseteq \rangle$ and $\langle \wp(U'), \subseteq \rangle$. Hence, from Proposition 1 one obtains that $\langle i \rangle [e]$ and $\langle e \rangle [i]$ are pre-topological interior operators, while $[i] \langle e \rangle$, $[e] \langle i \rangle$, $[[i]][[e]]$ and $[[e]][[i]]$ are pre-topological closure operators, on $\wp(U)$ and $\wp(U')$, respectively. Thus we set:

Definition 2. *Let $\langle U, U', R \rangle$ be a relational system. Then:*

- $int : \wp(U) \longmapsto \wp(U); int(X) = \langle e \rangle([i](X))$ *(logical structure: $\exists \forall$).*
- $cl : \wp(U) \longmapsto \wp(U); cl(X) = [e](\langle i \rangle(X))$ *(logical structure: $\forall \exists$).*
- $\mathcal{A} : \wp(U') \longmapsto \wp(M); \mathcal{A}(Y) = [i](\langle e \rangle(Y))$ *(logical structure: $\forall \exists$).*
- $\mathcal{C} : \wp(U') \longmapsto \wp(M); \mathcal{C}(Y) = \langle i \rangle([e](Y))$ *(logical structure: $\exists \forall$).*
- $\mathcal{ITS} : \wp(U') \longmapsto \wp(U'); \mathcal{ITS}(Y) = [[i]][[e]](Y)$ *(logical structure: $\forall \forall$).*
- $est : \wp(U) \longmapsto \wp(U); est(X) = [[e]][[i]](X)$ *(logical structure: $\forall \forall$).*

Obviously, the symbols *int* and *cl* mean "interior" and "closure", respectively ($\mathcal{A}$ and $\mathcal{C}$ are their counterparts on the "formal" - that is, pointless - side)[4]. We have seen that this use is justified by the theory of adjointness relations. $\mathcal{ITS}$ and *est* give the intensional and, respectively, extensional sides of formal concepts in Formal Concept Analysis (see [17]). Moreover, *int* and *cl* fit the usual topological definitions. In fact, we know that for any subset X of U, a point a belongs to the interior of X if and only if there is a neighbourhood of a included in X. If the members of U' are interpreted as formal neighbourhoods (pointless neighbourhoods) we cannot verify directly if a neighbourhood b of a is included in a set of points X. However, we can check: first whether b is a formal neighbourhood of a, that is, whether $b \in R(a)$ or, equivalently, $a \in \langle e \rangle(b)$; second, whether the extension of b, that is, $R^{\smile}(b)$ or, equivalently, $\langle e \rangle(b)$, is included in X. From the adjunction property (1), $\langle e \rangle(b) \subseteq X$ if and only if $\{b\} \subseteq [i](X)$. The conclusion is that a belongs to the formal interior of X if and only if $a \in \langle e \rangle(b)$, for b belonging to $[i](X)$. To sum up, the interior of X is given by:

$$\{a : \exists b (a \in \langle e \rangle(b) \ \& \ b \in [i](X))\} = \langle e \rangle([i](X)) = int(X) \tag{3}$$

Similarly for closure. In fact for any subset X of U, a belongs to the closure of X if and only if the extension of any neighborhood of a has non empty

---

[4] The combination of quantifiers suggests an investigation of the relationships between the formal properties of the above operators and those in the hexagon of opposition which are obtained by similar combinations (see [2]).

intersection with X. Thus a belongs to the closure of X if and only if for all $b \in \langle i \rangle(a)$, $\langle e \rangle(b) \cap X \neq \emptyset$. That happens if $\langle i \rangle(a)$ is included in $\langle i \rangle(X)$, but from the adjunction property (1), $\langle i \rangle(a) \subseteq \langle i \rangle(X)$ if and only if $\{a\} \subseteq [e]\langle i \rangle(X)$ if and only if $a \in [e]\langle i \rangle(X)$. Finally, one can easily observe that

$$int(X) \subseteq X \subseteq cl(X), \quad any \ X \subseteq U. \tag{4}$$

Therefore, Galois adjunctions make it possible to define a pair of approximation operators which are mathematically sound and with a fair intuitive meaning. Symmetrically for $\mathcal{A}$ and $\mathcal{C}$. Anyhow, we again underline that int and cl are not topological, in general. If $R(U) = U'$ and $R^\smile(U') = U$, then int is co-normal and cl is normal (in this case we shall say that the property system is $normal$[5]). But generally int fails to be multiplicative because $\langle e \rangle$ is just granted to be additive, and cl fails to be additive because $[e]$ is just multiplicative.

In the next section we shall analyse the properties of SRSs, property systems and neighbourhood systems (both formal and concrete) which progressively make a plain set into a topological space, in order to identify their connections.

3   Granulation and Approximations

Given a relational system $\langle U, U'R \rangle$, $X \subseteq U$ and $Y \subseteq U'$, the relational definition of the constructors and operators are:

$$\langle e \rangle(Y) = \{u : u \in R^\smile(Y)\}; \quad \langle i \rangle(X) = \{u' : u' \in R(X)\} \tag{5}$$

$$[e](Y) = \{u : R(u) \subseteq Y\}; \quad [i](X) = \{u' : R^\smile(u') \subseteq X\} \tag{6}$$

$$\mathcal{A}(Y) = \{u' : R^\smile(u') \subseteq R^\smile(Y)\}; \quad cl(X) = \{u : R(u) \subseteq R(X)\} \tag{7}$$

$$\mathcal{C}(Y) = \bigcup\{R(u) : R(u) \subseteq Y\}; \quad int(X) = \bigcup\{R^\smile(u') : R^\smile(u') \subseteq X\} \tag{8}$$

From now on we usually will deal with $[\cdot]$, int and $\mathcal{C}$. The results for the other constructors and operators come by duality.

So, let us consider the classical definition of upper and lower approximation. Given an indiscernibility space $\langle U, E \rangle$ with E an equivalence relation[6]:

$$(lE)(X) = \bigcup\{E(Z) : E(Z) \subseteq X\}, \quad (uE)(X) = \bigcup\{E(Z) : E(Z) \cap X \neq \emptyset\} \tag{9}$$

When we come to arbitrary binary relations R, the definitions turn into:

$$(lR)(X) = \bigcup\{Z : R(Z) \subseteq X\}, \quad (uR)(X) = \bigcup\{Z : R(Z) \cap X \neq \emptyset\} \tag{10}$$

which are formally different from the literal translation of 9:

$$(lR)(X) = \bigcup\{R(Z) : R(Z) \subseteq X\}, \quad (uR)(X) = \bigcup\{R(Z) : R(Z) \cap X \neq \emptyset\} \tag{11}$$

---

[5] If $R(U) = U'$ then R is said to be $right$-$total$, or $surjective$, or that $R^\smile$ is $serial$. $R^\smile(U') = U$ means that R is $left$-$total$ or $serial$.

[6] From now on the interested reader is addressed to [12] and its bibliography.

Indeed, (10) coincides with (11) only under certain conditions. Since $u \in R^{\smile}(Y)$ if and only if $R(u) \cap Y \neq \emptyset$ and R-neighbouring is additive, one immediately notices that in a SRS the constructor $[e]$ coincides with the operator (lR) of (10), while the operator $\mathcal{C}$ coincides with the operator (lR) of (11).

Therefore we come to a couple of questions with the same answer: (i) when the definitions (10) and (11) coincide? (ii) when $\mathcal{C}$ and $[e]$ coincide? The latter question amounts also to the following row of questions: when $\mathcal{A}$ and $\langle e \rangle$, int and $[i]$, cl and $\langle i \rangle$ coincide, respectively? The answer is: when R is a preorder.

But when the above coincidences occur, a particular property emerges. Indeed, we know that $\mathcal{C}$ is an interior operator, which, however, is not multiplicative. But $[e]$ do is multiplicative (it is an upper adjoint). Similarly, the additivity of the lower adjoint $\langle e \rangle$ meets the closure properties of $\mathcal{A}$. The overall result must be split in two parts. In what follows given two operators $op1$ and $op2$ we set $op1 = op2$ if and only if for any argument x, $op1(x) = op2(x)$. If the operator op is defined by means of a relation R, we eventually write op_R, if needed.

Proposition 2. *Let $\langle U, R \rangle$ be a SRS. The following are equivalent: (i) R is a preorder, (ii) $[e]_R = \mathcal{C}_R$, $[i]_R = int_R$, $\langle e \rangle_R = \mathcal{A}_R$, $cl_R = \langle i \rangle_R$.*

Proposition 3. *Let $\langle U, R \rangle$ be a SRS. If R is a preorder, then $int_R, [i]_R, \mathcal{C}_R$ and $[e]_R$ are topological interior operators; cl_R, $\langle i \rangle_R$, $\mathcal{A}_R$ and $\langle e \rangle_R$ are topological closure operators.*

The converse of Proposition 3 holds just partially:

Corollary 1. *Let $\langle U, R \rangle$ be a SRS. If $[\cdot]_R$ and $\langle \cdot \rangle_R$ are topological interior, respectively closure, operators, then R is a preorder.*

The proof follows from Proposition 2. However, the converse of Corollary 1 does not hold for int, cl, $\mathcal{A}$ and $\mathcal{C}$. This is an important point which means that, for instance, there are relations R which are not preorders but such that $\mathcal{C}_R$ is a topological interior operator, nevertheless. Similarly for the topological properties of the other operators[7].

This means that not only $\mathbf{L}_{int_R}(U) = \{int_R(X) : X \subseteq U\}$ and $\mathbf{L}_{\mathcal{A}_R}(U') = \{\mathcal{A}_R(Y) : Y \subseteq U'\}$, $\mathbf{L}_{cl_R}(U) = \{cl_R(X) : X \subseteq U\}$ and $\mathbf{L}_{\mathcal{C}_R}(U') = \{\mathcal{C}_R(Y) : Y \subseteq U'\}$ but also $\mathbf{L}_{\langle e \rangle_R}(U) = \{\langle e \rangle_R(X) : X \subseteq U\}$ and $\mathbf{L}_{[i]_R}(U') = \{[i]_R(Y) : Y \subseteq U'\}$, $\mathbf{L}_{[e]_R}(U) = \{[e]_R(X) : X \subseteq U\}$ and $\mathbf{L}_{\langle i \rangle_R}(U') = \{\langle i \rangle_R(Y) : Y \subseteq U'\}$ are distributive lattices of sets.

About this fact, we have proved that if $\langle U, R \rangle$ is such that, say, $\mathcal{C}_R$ is a topological interior operator, then there is a transformation of $\langle U, R \rangle$ into a preorder $\langle U, R^* \rangle$ which amounts to a permutation of the rows $R(x)$, for some $x \in U$. Moreover, this transformation can be described by means of the operation of residuation between binary relations. An open issue is determining R^* by means of Galois connections and unities (see [1]). Some hints come from the fact that for $u \in U$, $x \in est(u)$ if and only if $u \in cl(x)$.

---

[7] Proposition 3 amends point (iv) of Corollary 1 of [9] and point (ii) of Facts 3 of [10], which state also the converse implication, erroneously.

Another open issue is determining the properties of int_R (in particular its pre-topological or topological ones) from the features of a property system $\langle U, U', R\rangle$. About this issue, we know that if $\langle U, U', R\rangle$ is a dichotomic system then cl_R induces an equivalence relation, hence a topological interior operator of a 0-dimensional topological space, that is, a space in which the elements are both closed and open (or *clopen*) - see [12]. However, this is just a first step and a more comprehensive understanding of the topic is required.

4   Concrete and Formal Neighbourhood Systems

Under the formal topology interpretation, in a relational system $\langle U, U', R\rangle$ the members of U' are formal neighbourhoods. Thus, a first move towards a concrete neighbourhood interpretation is replacing any formal neighbour u' with the set $R^\smile(u')$ of points associated with it. One obtains what follows:

Proposition 4. *Let $\langle U, U', R\rangle$ be a relational system, $\mathcal{Z} = \{R^\smile(u') : u' \in U'\}$. Then, for all $A \subseteq U, B \subseteq U'$:*
(1) $int(A) = \bigcup\{X \in \mathcal{Z} : X \subseteq A\}$, (2) $\mathcal{C}(B) = \bigcup\{Y \in \mathcal{W} : Y \subseteq B\}$,
(3) $cl(A) = \bigcap\{-X \in \mathcal{Z} : X \cap A = \emptyset\}$, (4) $\mathcal{A}(B) = \bigcap\{-Y \in \mathcal{W} : Y \cap B = \emptyset\}$.

More in general, let us now consider an association between points from a set U and subsets from $\wp(U)$ (possibly the elements of $\mathcal{Z}$)[8]. Thus, we work with the following ingredients:

Definition 3. *Let* $\mathbf{N} = \langle U, \wp(U), R\rangle$ *be a relational system, $X \subseteq U$ and $u \in U$. Let $x \in N \in R(u)$. Then $R(u)$ is called a* concrete neighbourhood family *of u; N is called a* concrete neighbourhood *of u; x is called a* concrete neighbour *of u; $\mathcal{N}(U) = \{R(u) : u \in U\}$ is called a* concrete neighbourhood system; *the pair $\langle U, \mathcal{N}(U)\rangle$ is called a* concrete neighbourhood space.

Given a concrete neighbourhood space, the following operators are definable:

$$G(X) = \{u : X \in R(u)\}, \quad F(X) = -G(-X) = \{u : -X \notin R(u)\}. \tag{12}$$

G is called a *core map* and F a *vicinity map* (induced by $\mathcal{N}(U)$).

The properties of these operators depends on the properties satisfied by the neighbourhood system. Indeed, consider the following conditions on $\mathcal{N}(U)$, for any $x \in U, A, N, N' \subseteq U$:
1. $U \in R(x)$; **0.** $\emptyset \notin R(x)$; **Id.** if $x \in G(A)$ then $G(A) \in R(x)$;
N1. $x \in N$, for all $N \in R(x)$;
N2. if $N \in R(x)$ and $N \subseteq N'$, then $N' \in R(x)$;
N3. if $N, N' \in R(x)$, then $N \cap N' \in R(x)$;
N4. there is an $N \neq \emptyset$ such that $R(x) =\uparrow N$ (the $\subseteq$ order filter of N).

---

[8] Notice that if $U' \neq U$ and one substitutes $\wp(U')$ for $\wp(U)$ then a more general picture is obtained. However, the result of the more specific case can be translated into the more general case by means of a map from U to U'.

Lemma 1. *For any $X, Y \subseteq U$, $x \in U$ the following correspondences hold:*

Condition	Equivalent properties of G	Equivalent properties of F
1	$G(U) = U$	$F(\emptyset) = \emptyset$
0	$G(\emptyset) = \emptyset$	$F(U) = U$
Id	$G(X) \subseteq G(G(X))$	$F(F(X)) \subseteq F(X)$
N1	$G(X) \subseteq X$	$X \subseteq F(X)$
N2	$X \subseteq Y \Rightarrow G(X) \subseteq G(Y)$	$X \subseteq Y \Rightarrow F(X) \subseteq F(Y)$
	$G(X \cap Y) \subseteq G(X) \cap G(Y)$	$F(X \cup Y) \supseteq F(X) \cup F(Y)$
N3	$G(X \cap Y) \supseteq G(X) \cap G(Y)$	$F(X \cup Y) \subseteq F(X) \cup F(Y)$

But $\langle U, \wp(U), R \rangle$ is a relational system, too so that we can define also the (abstract) operators int and cl alongside the (concrete) operators G and F. Therefore, a first question arises as to the conditions which make int and G (cl and F) coincide. We can immediately notice that if one substitutes int for G and cl for F, it is possible to verify that in any property system int and cl satisfy the equivalent properties of conditions **Id**, **N1** and **N2**. Moreover, int satisfies the equivalent property of **0** and cl the equivalent property of **1**. If the property system is normal, then int satisfies the equivalent property of **1** and cl satisfies that of **0**. Systems satisfying these conditions will be classified as $\mathcal{N}_{2Id}$ neighbourhood systems. Indeed, we have a precise result consistent with this scrutiny[9]:

Proposition 5. *Let $\langle U, \wp(U), R \rangle$ be a relational system. Then, for all $X \in \wp(U)$, $G(X) = int(X)$ if and only if $\mathcal{N}(U)$ is of type $\mathcal{N}_{2Id}$.*

Notice that according to (8), $int(X) = \{a \in U : \exists X'(X' \in R(a) \land R^{\smile}(\{X'\}) \subseteq X)\}$. Since $R^{\smile}(\{X'\}) = G(X)$, Proposition 5 shows when the recursive equation

$$G(X) = \{a : \exists X'(X' \in R(a) \land G(X') \subseteq X)\}$$

has a solution.

From the table above, we see that topological spaces are $\mathcal{N}_{2Id}$ spaces which fulfil **N3** in addition.

If an operator op depends on a system $\mathbf{S}$, we shall eventually write $op^{\mathbf{S}}$.

Let now $\mathbf{P} = \langle U, U', R \rangle$ be a formal neighbourhood system and $\mathbf{P}^{\mathbf{R}^{\smile}} = \langle U, \mathcal{Z}, \in \rangle$, where $\mathcal{Z}$ is the concrete counterpart of U' as defined in Proposition 4. Since $\langle u, u' \rangle \in R$ if and only if $u \in R^{\smile}(u')$ the relation R coincides with $\in$ if we replace u' with $R^{\smile}(u')$. It follows that for any $X \subseteq U$, $int^{\mathbf{P}}(X) = int^{\mathbf{P}^{\mathbf{R}^{\smile}}}(X)$.

---

[9] Details may be found in [12]. Pay attention that in that book $R(x)$ is denoted as $\mathcal{N}_x$ and property systems are called *"basic neighbourhood pairs"*, in the context of pre-topological formal spaces. A simplified proof can be found in [9].

Moreover, $\mathcal{Z}$ and $\in$ induce a concrete neighbourhood system by putting for any $u \in U$, $\mathcal{N}_u^{R^\smile} = \{X \in \mathcal{Z} : u \in X\} = \{R^\smile(u') : u' \in R(u)\}$. The family $\mathcal{N}_{R^\smile}(\mathbf{P}) = \{\mathcal{N}_u^{R^\smile} : u \in U\}$ will be called the *normal neighbourhood system,* NNS, induced by $\mathbf{P}$. Clearly $\mathcal{Z} = \bigcup(\mathcal{N}_{R^\smile}(\mathbf{P}))$.

Since $\langle U, \mathcal{N}(U) \rangle$ where $\mathcal{N}(U) = \{\in (u) : u \in U\}$ is a concrete neighbourhood system, an obvious question arises as to the connection between $int^{\mathbf{P}}$ (i.e. $int^{\mathbf{P}^{R^\smile}}$) and the operator $G^{\mathcal{N}_{R^\smile}(\mathbf{P})}$. The answer is: "no connections", because in a NNS only the properties $\mathbf{0}$, $\mathbf{N1}$ and the following weaker form of $\mathbf{Id}$:

$$if\ N \in \mathcal{N}_x,\ then\ \exists N' \in \mathcal{N}_x\ such\,that\ for\,any\ y \in N', N \in \mathcal{N}_y \qquad (\tau)$$

are granted. Thus, NNSs are poorly structured and one has:

$$G^{\mathcal{N}_{R^\smile}(\mathbf{P})}(X) = \begin{cases} \emptyset & if\ \neg\exists u\ s.\ t.\ X \in \mathcal{N}_u^{R^\smile} \\ X & otherwise \end{cases}$$

To obtain more structure, another class of concrete neighbourhood systems has to be defined out of $\mathbf{P}$:

Definition 4. *Let* $\mathcal{N}_g^{\uparrow R^\smile} = \bigcup\{\uparrow R^\smile(m) : m \in R(g)\}$. *The family* $\mathcal{N}_{\uparrow R^\smile}(\mathbf{P}) = \{\mathcal{N}_g^{\uparrow R^\smile} : g \in U\}$ *will be called* principal neighbourhood system, PNS, *induced by* $\mathbf{P}$. *Let us set* $\mathbf{P}^{\uparrow \mathbf{R}^\smile} = \langle U, \bigcup(\mathcal{N}_{\uparrow R^\smile}(\mathbf{P})), \in \rangle$.

PNSs enjoy more properties: $\mathbf{0}$, $\mathbf{N1}$, $\mathbf{N2}$ and $\mathbf{Id}$ (indeed, $\mathbf{N2}$ plus τ give $\mathbf{Id}$). As a consequence of this fact and Proposition 5, one obtains that for any $X \subseteq U$, $int^{\mathbf{P}}(X) = int^{\mathbf{P}^{R^\smile}}(X) = G^{\mathcal{N}_{\uparrow R^\smile}(\mathbf{P})}(X)$.

On the contrary, $int^{\mathbf{P}^{\uparrow \mathbf{R}^\smile}}$ has a poor behaviour:

$$int^{\mathbf{P}^{\uparrow \mathbf{R}^\smile}}(X) = \begin{cases} X & if\ \exists u'\ s.\ t.\ X \supseteq R^\smile(u') \\ \emptyset & otherwise \end{cases}$$

Other concrete neighbourhood systems defined on the basis of a relational system can be found in [10]. Notice that associating an item with more then one neighbourhood, is a way for describing the points of view of different knowledge subjects or different points of view of the same knowledge subject[10].

5 An Application: Covering-Based Rough Sets

The set $\mathcal{Z}$ of subsets of U defined in Proposition 4 is a covering of U, provided $R^\smile$ is serial. If $\mathcal{Z}$ is induced by a relational system $\mathbf{P}$, we denote it by $\mathbf{C}(\mathbf{P})$. Conversely, one can transform a covering of a set U into a relational system:

---

[10] This approach was pioneered in [5]. In [8] neighbourhood systems result from families of relational systems and two approximation operators "according to n relations" were introduced. Neighbourhood systems not fulfilling $\mathbf{N1}$ were investigated in [6].

Definition 5. *Let* $\mathbf{C} = \{K_i\}_{i \in I}$ *be a covering of a set* U, *with both* U *and* $\mathbf{C}$ *at most countable. Let us set for all* $x \in U, \langle x, K_i \rangle \in R$ *iff* $x \in K_i$. *The resulting relational system* $\mathbf{P}(\mathbf{C}) = \langle U, \mathbf{C}, R \rangle$ *will be called the* covering relational system, *CRS, induced by* $\mathbf{C}$.

Clearly, $\mathbf{P}(\mathbf{C})$ is a concrete neighbourhood system. Moreover, $\mathbf{C} = \bigcup(\mathcal{N}_{R^\smile}(\mathbf{P}(\mathbf{C})))$ so that $\mathbf{C} = \mathbf{C}(\mathbf{P}(\mathbf{C}))$. In [3,7,15] the following lower approximation operator is defined from a covering $\mathbf{C}$ of a set U: $(lC)_1(X) = \bigcup\{K_i : K_i \subseteq X\}$ (in [3] it is denoted by CL, in [7] by L_5 and in [15] by $\underline{C_1}$.

It is a natural way to define a lower approximation. What are its properties? If we work on the covering relational system $\mathbf{P}(\mathbf{C})$ we easily obtain: for any $X \subseteq U$, $(lC)_1(X) = int(X)$. From this equation one immediately has that $(lC)_1$ is decreasing, monotone and idempotent. It is neither multiplicative, nor additive. Therefore, it cannot have either a lower or an upper adjoint. However, it has a dual upper approximation operator, which is $(uC)_1(X) = \bigcap\{-K_i : K_i \cap X = \emptyset\}$. To my knowledge, this operator has not been taken into account in the literature on rough sets, but in [10].

A different approximation operator is the following: $(lC)_0(X) = \bigcup\{n(x) : n(x) \subseteq X\}$, where $n(x) = \bigcap\{K_i : x \in K_i\}$. It has been introduced in [7] with the symbol L. In order to understand its properties it must be noticed that for any $x \in U$, $n(x) = R_{\mathbf{C}}(x)$, where $R_{\mathbf{C}}$ is a preorder defined as: $R_{\mathbf{C}}(x) = \{\langle x, y \rangle : \forall K_i \in \mathbf{C}(x \in K_i \Longrightarrow y \in K_i)\}$. Therefore, now we have to work on the relational system $\mathbf{P}(R_{\mathbf{C}}) = \langle U, U, R_{\mathbf{C}} \rangle$. Using the above machinery, it is not difficult to show that $(lC)_0(X) = [e](X) = \mathcal{C}(X)$. Therefore, $(lC)_0$ is a topological interior operator. Moreover it coincides with the operators CL of [3], L_1 of [7] and $\underline{C_1}$ of [15]. Further, the dual operator of (lC_0) is $(uC)_2(X) = \{x : n(x) \cap X \neq \emptyset\}$, simply because $(uC)_2(X) = \langle e \rangle(X) = \mathcal{A}(X)$. The operator $(uC)_2$ has been introduced as XH in [3], U_1 in [7] and $\overline{C_2}$ in [15].

Since $(lC)_0$ is multiplicative, it has a lower adjoint, which is $(uC)_0(X) = \bigcup\{n(x) : x \in X\}$. It has been introduced as U in [7]. So we see that U is not the dual of L (which is U_1), but its lower adjoint. It is U_4 in [7] and IH in [3].

6   Granulation, Relations and Intuitionistic Formal Spaces

The concepts of granulation and approximation are strictly connected to the topological notion of an *adherence* and a *closure*. Therefore, let X be e set and K any increasing, monotone and idempotent (i.e. closure) operator on $\wp(X)$. Then we can introduce a sort of covering relation $\blacktriangleleft$ between points x and subsets A, by setting $x \blacktriangleleft A$ if and only if $x \in K(A)$, and extend this definition to subsets of X: $A \blacktriangleleft B$ if and only if all $x \in A$ are such that $x \blacktriangleleft B$. We, therefore, arrive at the following definition:

Definition 6. *Let* $\langle U, U', R \rangle$ *be a relational system. Then for any* $b \in U'$ *and* $Y, Y' \subseteq U'$, *the following relation is called a* formal semi-cover *or, shortly, a* semi-cover:

$$(basis)\ b \blacktriangleleft Y\ iff\ b \in \mathcal{A}(Y)\ \blacktriangleleft, \qquad (step)\ Y \blacktriangleleft Y'\ iff\ \forall y \in Y, y \blacktriangleleft Y'.$$

The relation ◄ is called "formal semi-covering" because, in view of the symmetry of $\mathcal{A}$ and cl, ◄ is the formal counterpart of the "concrete" concept of "adherence". Remember that the elements of U' now are to be thought of as formal neighbourhoods. Since concrete neighbourhoods can be combined by means of the set-theoretical intersection, we need a formal counterpart of this operation. So, let us assume that U' is equipped with a binary operation "·" which is associative, commutative and with a unity 1. Otherwise stated, $\langle U', \cdot, 1 \rangle$ is a commutative monoid. Now we lift the operation · from U' to $\wp(U')$ in the following way: $X \cdot Y = \{x \cdot y : x \in X \wedge y \in Y\}$, for $X, Y \subseteq U'$.

Since $\mathcal{A}$ is pre-topological, so is ◄. Let $\perp$ be any subset of U'. Then we call $\langle U', \cdot, 1, ◄, \perp \rangle$ a *pre-topological formal system*.

The difference between pre-topological formal systems and topological formal systems may be described as follows. Let us put $\Omega_{\mathcal{A}}(U') = \{X \subseteq U' : \mathcal{A}(X) = X\}$. We call the members of $\Omega_{\mathcal{A}}(U')$, $\mathcal{A}$-*saturated sets*. Since the operation · does not preserve saturation, let us set $X \bullet Y = \mathcal{A}(X \cdot Y)$. A pre-topological formal system is topological if $\langle \Omega_{\mathcal{A}}(U'), \bullet, \vee, U', \mathcal{A}(\perp) \rangle$, is a complete lattice with complete distributivity and ordering $\subseteq$. Since $\vee$ is $\cup$, $\bullet$ coincides with $\cap$, thus.

In terms of the covering relation ◄ the following properties are fundamental to obtain topological formal systems:

$$(left) \quad \frac{b ◄ Y}{b \cdot b' ◄ Y}; \quad (right) \quad \frac{b ◄ Y \quad b ◄ Y'}{b ◄ Y \cdot Y'}.$$

In general, both principles fail to hold even if · is idempotent. The same happens for the following important property:

$$(stability) \quad \frac{b ◄ Y \quad b ◄ Y'}{b \cdot b' ◄ Y \cdot Y'}$$

Proposition 6. *A pre-topological formal system is topological if (left) and (right) hold.*

Definition 7. *A pre-topological formal system in which the operation · is idempotent is called a* quasi-topological formal system[11].

Quasi-topological formal systems are abstraction of concrete neighbourhood systems. Indeed, we can build quasi-topological formal systems in the following way:

Definition 8. *Let* $\langle U, \wp(U), R \rangle$ *be a relational system. Set* $\perp = \{X \in \wp(U) : \langle e \rangle(X) = \emptyset\}$. *Then* $\langle \wp(U), \cap, U, ◄, \perp \rangle$ *is called a* formal neighbourhood system.

Proposition 7. *Any formal neighbourhood system is a quasi-topological formal system.*

---

[11] Cf. [12], where a more complete notion of a pre-topological formal system is defined, together with a classification of such systems.

But from $\langle U, \wp(U), R \rangle$ one obtains $\mathcal{N}(U) = \{R(x)\}_{x \in U}$ which is a concrete neighbourhood system. Indeed, it is another double-face system. Therefore, we say that then above formal neighbourhood system and this concrete neighbourhood system are *homogeneous*. Thus, the final question is obvious: is there any connection between the properties **1**, **0**, **Id**, **N1**, **N2**, **N3** and **N4**, which are definable on a concrete neighbourhood systems and the properties (*left*) and (*right*) definable on its homogeneous formal neighbourhood system? The answer is just partial:

Proposition 8. *Let* $\langle \wp(U), \cap, U, \blacktriangleleft, \perp \rangle$ *be a formal neighbourhood system induced by a relational system* $\langle U, \wp(U), R \rangle$. *If* $\{R(x)\}_{x \in U}$ *fulfils* **N3**, *then* (*right*) *holds.*

But the converse does not hold. On the contrary **N2** and (*left*) are equivalent:

Proposition 9. *Let* $\langle \wp(U), \cap, U, \blacktriangleleft, \perp \rangle$ *be a formal neighbourhood system induced by a relational system* $\langle U, \wp(U), R \rangle$. *Then,* $\{R(x)\}_{x \in U}$ *is a neighbourhood system fulfilling* **N2** *if and only if* (*left*) *holds.*

The above results is what has been established in [12]. Further achievements are not known to the author. In particular it is likely that there are no formal properties representing **N1** and **Id**. In a sense, it is hard to find the formal counterpart of these two conditions because they are defined by means of the membership relation between elements of U and subsets of U, which does not have any role in the formal, that is, pointless, framework. On the contrary, **N2**, **N3** and **N4** are defined by means of relations between subsets of U.

However, in a sense **Id** and **N1** are embedded in $\blacktriangleleft$, *via* the closure properties of $\mathcal{A}$. In particular they lead to the following properties of $\blacktriangleleft$:

$$\frac{b \in Y}{b \blacktriangleleft Y} \text{ (reflex)}; \ (i) \ \frac{b \blacktriangleleft Y \quad Y \blacktriangleleft Y'}{b \blacktriangleleft Y'}, \ (ii) \ \frac{b \blacktriangleleft b' \quad b' \blacktriangleleft Y}{b \blacktriangleleft Y} \text{ (trans)} \qquad (13)$$

We know that we can substitute any closure operator K on $\wp(U')$ for $\mathcal{A}$ in Definition 6. Conversely, if $\blacktriangleleft$ is a relation between a set X and its powerset $\wp(X)$ such that transitivity and reflexivity hold, than the operator K defined by $K(Y) = \{x : x \blacktriangleleft Y\}$ is a closure operator on $\wp(X)$ (see [16]).

About quasi-topological formal systems (hence formal neighbourhood systems) we know what follows: (i) (*stability*) gives (*right*), hence (ii) (*stability*) plus (*left*) implies that the system is topological.

Further investigations are required in order to better understand the connections between the formal and the concrete frameworks and the precision and accuracy of the descriptions enabled by the two approaches.

A case in point is the "dissonance" between topological formal systems and topological concrete spaces. For instance, the following cases are notable: (i) Topological formal neighbourhood systems in which **N3** does not hold. Actually, it is curious but not really a surprise because **N3** implies (*right*) but the opposite does not hold. (ii) Topological formal neighbourhood systems in which **N1** does not hold. Also this case is not really a surprise, in view of the above discussion. (iii) Formal neighbourhood systems such that $\mathbf{L}_{int}(U)$ and $\mathbf{L}_{\mathcal{A}}(\wp(U))$

are distributive lattices of sets, hence topological spaces, but in which (*left*) fails, so that they are not topological formal systems. This case is really tricky, actually, also because (*left*) and **N2** are equivalent. Thus, are there connections between the relations R such that R are not preorders but $\mathbf{L}_{int_R}(U)$ are distributive lattices (see Sect. 3) and the properties of the formal neighbourhood systems induced by the concrete neighbourhood systems $\mathcal{N}_{R^\smile}(\mathbf{P})$ or $\mathcal{N}_{\uparrow R^\smile}(\mathbf{P})$? Or are the formal and concrete approaches non commensurable, in a sense?

References

1. Crapo, H.: Unities and negation: on the representation of finite lattices. J. Pure Appl. Algebra **23**, 109–135 (1982)
2. Dubois, D., Prade, H.: From blanché's hexagonal organization of concepts to formal concept analysis and possibility theory. Log. Univers. **6**(1–2), 149–169 (2012)
3. Huang, A., Zhu, W.: Topological characterizations for three covering approximation operators. In: Ciucci, D., Inuiguchi, M., Yao, Y., Slezak, D., Wang, G. (eds.) RSFDGrC 2013. LNCS (LNAI), vol. 8170, pp. 277–284. Springer, Heidelberg (2013). https://doi.org/10.1007/978-3-642-41218-9_30
4. Järvinen, J.: Knowledge Representation and Rough Sets. Thesis. University of Turku (1999)
5. Lin, T.Y.: Granular computing on binary relations i: data mining and neighborhood systems, II: rough set representations and belief functions. In: Skowron, A., Polkowski, L. (eds.) Rough Sets in Knowledge Discovery Physica, pp. 107–140 (1998)
6. Lin, T.Y., Liu, G., Chakraborty, M.K., Slezak, D.: From topology to anti-reflexive topology. In: Proceedings of 2013 IEEE International Conference on Fuzzy Systems (FUZZ-IEEE), pp. 1–7 (2013)
7. Kumar, A., Banerjee, M.: Definable and rough sets in covering-based approximation spaces. In: Li, T., Nguyen, H.S., Wang, G., Grzymala-Busse, J., Janicki, R., Hassanien, A.E., Yu, H. (eds.) RSKT 2012. LNCS (LNAI), vol. 7414, pp. 488–495. Springer, Heidelberg (2012). https://doi.org/10.1007/978-3-642-31900-6_60
8. Pagliani, P.: Pretopologies and dynamic spaces. Fundamenta Informaticae **59**(2), 221–239 (2004)
9. Pagliani, P.: The relational construction of conceptual patterns - tools, implementation and theory. In: Kryszkiewicz, M., Cornelis, C., Ciucci, D., Medina-Moreno, J., Motoda, H., Raś, Z.W. (eds.) RSEISP 2014. LNCS (LNAI), vol. 8537, pp. 14–27. Springer, Cham (2014). https://doi.org/10.1007/978-3-319-08729-0_2
10. Pagliani, P.: Covering rough sets and formal topology. a uniform approach through intensional and extensional constructors. Trans. Rough Sets **XX**, 109–145 (2017)
11. Pagliani, P., Chakraborty, M.K.: Information quanta and approximation spaces (I and II). In: Hu, X., Liu, Q., Skowron, A., Lin, T.S., Yager, R.R., Zhang, B. (eds.) Proceedings of the IEEE International Conference on Granular Computing, Beijing, China, vol. 2, pp. 605–610 (2005). 611–616
12. Pagliani, P., Chakraborty, M.K.: A Geometry of Approximation. Trends in Logic, 27th edn. Springer, Netherlands (2008). https://doi.org/10.1007/978-1-4020-8622-9
13. Pawlak, Z.: Rough Sets: A Theoretical Approach to Reasoning About Data. Kluwer, Dordrecht (1991)

14. Pomykała, J.A.: Approximation operations in approximation space. Bull. Pol. Acad. Sci. Math. **35**, 653–662 (1987)
15. Qin, K., Gao, Y., Pei, Z.: On covering rough sets. In: Yao, J.T., Lingras, P., Wu, W.-Z., Szczuka, M., Cercone, N.J., Slezak, D. (eds.) RSKT 2007. LNCS (LNAI), vol. 4481, pp. 34–41. Springer, Heidelberg (2007). https://doi.org/10.1007/978-3-540-72458-2_4
16. Sambin, G.: Intuitionistic formal spaces and their neighbourhood. In: Ferro, R., Bonotto, C., Valentini, S., Zanardo, A. (eds.) Logic Colloquium 1988, pp. 261–285. Elsevier, North-Holland (1989)
17. Wille R.: Restructuring lattice theory: an approach based on hierarchies of concepts. In: Rival, I. (eds.) Ordered Sets. NATO Advanced Study Institutes Series, vol. 83, pp. 445–470. Springer, Dordrecht (1982). https://doi.org/10.1007/978-94-009-7798-3_15
18. Zakowski, W.: Approximations in the space (U, Π). Demonstratio Mathematica **XVI**, 761–769 (1983)

Multi-granularity Attribute Reduction

Shaochen Liang[1,2], Keyu Liu[1(✉)], Xiangjian Chen[1], Pingxin Wang[3],
and Xibei Yang[1,2]

[1] School of Computer, Jiangsu University of Science and Technology,
Zhenjiang 212003, Jiangsu, People's Republic of China
lshc940302@163.com, just_liukeyu@163.com
[2] Intelligent Information Processing Key Laboratory of Shanxi Province,
Shanxi University, Taiyuan 030006, Shanxi, People's Republic of China
[3] School of Science, Jiangsu University of Science and Technology, Zhenjiang 212003,
Jiangsu, People's Republic of China

Abstract. It is known that different parameters used in Gaussian kernel will provide us different granularities of information granulations. Therefore, kernel based fuzzy rough set has the characteristic of multi-granularity. From this point of view, a multi-granularity attribute reduction strategy is developed in this paper. Different from traditional reduction process that produces reduct by a fixed granularity, our strategy aims to derive reduct which is suitable for fuzzy rough approximations in terms of multi-granularity. To reduce the time consumption in reduction process and to avoid the consideration of all granularities may lead to the difficulty in eliminating attributes, the fuzzy rough approximations derived from the coarsest and the finest granularities are used to design constraint in multi-granularity attribute reduction. The experimental results show that compared with the traditional approach, not only the multi-granularity reduct may bring us almost the same performances for characterizing uncertainties, but also the multi-granularity reduction process is faster since only one reduct is required to be obtained for a set of the fuzzy rough approximations.

Keywords: Approximation quality · Attribute reduction
Conditional entropy · Fuzzy rough set · Multi-granularity

1 Introduction

Attribute reduction [10] plays a crucial role in the development of the rough set theory [1,4,17]. Different from the feature selections, most of the attribute reductions have clear semantic explanations with respect to different requirements. Presently, to derive reducts from data, exhaustive searching [19,20] and heuristic searching have been widely explored. Nevertheless, note that though the exhaustive searching can find all reducts in a given data, it is time-consuming and then the heuristic searching captures our attention. In the following, some state of the art results will be addressed, which aim to further speed up the reduction process in heuristic searching.

© Springer Nature Switzerland AG 2018
H. S. Nguyen et al. (Eds.): IJCRS 2018, LNAI 11103, pp. 61–72, 2018.
https://doi.org/10.1007/978-3-319-99368-3_5

1. Chen et al. [2] proposed an algorithm for computing reducts through a parallel way. Such approach can be interpreted as a "divide-and-conquer" strategy, it follows that less time is required for deriving a reduct.
2. Qian et al. [14–16] proposed an accelerator in the iteration of the heuristic searching. Such approach is based on the theoretical result: the rank of attributes will be preserved if some samples have been eliminated [16] through using the accelerator. Therefore, not only the process of computing reduct is accelerated, but also less memory is required.
3. Xu et al. [18] developed a heuristic algorithm based on sample selection technique. Different from the traditional heuristic searching on the whole data, the data is compressed by sample selection and then the efficiency of searching can be significantly improved.

Though the above methods have been demonstrated to be useful in speeding up the reduction process, they are only suitable for the definitions of attribute reductions constructed by one and only one granularity, i.e., those attribute reductions are defined by the rough sets based on one fixed information granulation. For example, neighborhood rough set attribute reduction is frequently designed with a given radius, such radius only provides a fixed result of neighborhood system or the so-called information granulation in Granular Computing.

Nevertheless, compared with single granularity, multi-granularity is more worthy to be addressed in many real-world applications. For instance, to evaluate the data distribution in a spatio-temporal space, Ji et al. [9] proposed a hierarchical entropy which is derived by different spatio-temporal granularities; the pseudo amino acid composition and the position-specific scoring matrix [21] provide us two different views of granularities for evaluating the performances of predictions; multi-granulation rough sets have been widely studied in References [3,8,11–13,22], the relationship among these different sizes of information granulations can be reflected by a multi-granularity technique. All of these results tell us that multi-granularity is commonly seen and then the re-consideration of the attribute reduction from multi-granularity is possible to provide us a new direction.

Take the model of fuzzy rough set as an example, the multi-granularity can be naturally formed if a set of the Gaussian kernel parameters is used [5,7]. A lesser parameter will generate a finer information granulation while a greater parameter may derive a coarser information granulation. The sizes of these different information granulations offer us the multi-granularity based results of fuzzy rough approximations. Therefore, to explore the multi-granularity attribute reduction, the constraint should be re-designed by using a set of the Gaussian kernel parameters instead of only one single parameter.

A simple way to design a multi-granularity attribute reduction is to fuse all the constraints in terms of all the considered parameters. However, it will bring us two challenges: 1. the complexity of the fused constraint will lower the speed of reduction process; 2. too many constraints will result in the difficulty of eliminating attributes. Therefore, we will develop a quick reduction process which is based on the computation of the coarsest and the finest granularities.

2   Preliminary Knowledge

2.1   Fuzzy Rough Set

Without loss of generality, a decision system is represented as $DS = <U, A, d>$, in which U is the set of samples, A is the set of condition attributes, and d is a decision attribute. $\forall x \in U$, $a(x)$ denotes the value of x over condition attribute $a \in A$, and $d(x)$ shows the label of x.

Given a decision system, an equivalence relation over d can be defined as $\mathrm{IND}(d) = \{(x, y) \in U \times U : d(x) = d(y)\}$. Immediately, a partition is obtained such that $U/\mathrm{IND}(d) = \{X_1, X_2, \cdots, X_q\}$, $X_k \in U/\mathrm{IND}(d)$ can be called the k-th decision class. Specially, the decision class which contains sample x is denoted by $[x]_{\mathrm{IND}(d)}$.

Moreover, $\forall B \subseteq A$, a Gaussian kernel based fuzzy relation [7] is denoted by R_B^σ, where σ is the Gaussian kernel parameter. By R_B^σ, $\forall x, y \in U$, the similarity between x and y is characterized as $\mathrm{R}_B^\sigma(x, y) = \exp\left(-\frac{\|x-y\|_B^2}{2\sigma^2}\right)$, in which $\|x - y\|_B$ is the Euclidean distance between x and y, i.e., $\|x - y\|_B = \sqrt{\sum_{a \in B} (a(x) - a(y))^2}$. Consequently, the fuzzy rough set of $X_k \in U/\mathrm{IND}(d)$ is defined as follows.

Definition 1. *Given a decision system $DS = <U, A, d>$, $\forall B \subseteq A$, the fuzzy rough lower and upper approximations of X_k are denoted by $\underline{\mathrm{R}_B^\sigma}(X_k)$ and $\overline{\mathrm{R}_B^\sigma}(X_k)$, respectively. $\forall x \in U$, the memberships that x belongs to them are*

$$\underline{\mathrm{R}_B^\sigma}(X_k)(x) = \min\{1 - \mathrm{R}_B^\sigma(x, y) : \forall y \notin X_k\}, \tag{1}$$

$$\overline{\mathrm{R}_B^\sigma}(X_k)(x) = \max\{\mathrm{R}_B^\sigma(x, y) : \forall y \in X_k\}. \tag{2}$$

2.2   Some Measurements

Approximation quality is a measurement in rough set theory, which reflects the percentage of the samples that belong to one of the decision classes determinately. The corresponding definition in fuzzy rough set [5] is presented as follows.

Definition 2. *Given a decision system $DS = <U, A, d>$, $\forall B \subseteq A$, the approximation quality with respect to B is defined as*

$$\gamma_B^\sigma(d) = \frac{|\bigcup_{k=1}^q \underline{\mathrm{R}_B^\sigma}(X_k)|}{|U|} = \frac{\sum_{x \in U} \max\{\underline{\mathrm{R}_B^\sigma}(X_k)(x) : \forall X_k \in U/\mathrm{IND}(d)\}}{|U|}, \tag{3}$$

where $|X|$ denotes the cardinality of the set X.

Conditional entropy is another measurement which characterizes the discriminating ability of $B \subseteq A$ relative to d. Presently, many definitions of conditional entropies have been proposed in terms of different requirements [6,7]. A typical representation of conditional entropy is shown in Definition 3 [23].

Definition 3. *Given a decision system* $DS = <U, A, d>$, $\forall B \subseteq A$, *the conditional entropy with respect to* B *is defined as*

$$\text{ENT}_B^\sigma(d) = -\frac{1}{|U|} \sum_{x \in U} |[x]_{\text{R}_B^\sigma} \cap [x]_{\text{IND}(d)}| \log \frac{|[x]_{\text{R}_B^\sigma} \cap [x]_{\text{IND}(d)}|}{|[x]_{\text{R}_B^\sigma}|}, \quad (4)$$

in which $[x]_{\text{R}_B^\sigma} = \sum_{y \in U} \text{R}_B^\sigma(x, y)/y$ *is the fuzzy information granule of* x.

3   Attribute Reduction

3.1   Heuristic Algorithm

By the above measurements, we can present the corresponding definitions of attribute reductions.

Definition 4. *Given a decision system* $DS = <U, A, d>$, $\forall B \subseteq A$,
   (1) B *is an approximation quality reduct* (γ-reduct) *if and only if* $\gamma_B^\sigma(d) = \gamma_A^\sigma(d)$ *and* $\forall C \subset B$, $\gamma_C^\sigma(d) \neq \gamma_B^\sigma(d)$;
   (2) B *is a conditional entropy reduct* (CE-reduct) *if and only if* $\text{ENT}_B^\sigma(d) = \text{ENT}_A^\sigma(d)$ *and* $\forall C \subset B$, $\text{ENT}_C^\sigma(d) \neq \text{ENT}_B^\sigma(d)$.

Approximation quality reduct and conditional entropy reduct are minimal subsets of A, which preserve the approximation quality and conditional entropy, respectively. These semantic explanations show the constraints of attribute reductions.
   To derive reducts by heuristic algorithm [10], different significance functions are required for different measurements.

Definition 5. *Given a decision system* $DS = <U, A, d>$, *if* $B \subset A$, *then* $\forall a \in A \backslash B$, *its significances with respect to different measurements are*

$$\text{Sig}_\gamma^\sigma(a, B, d) = \gamma_{B \cup \{a\}}^\sigma(d) - \gamma_B^\sigma(d); \quad (5)$$

$$\text{Sig}_{\text{ENT}}^\sigma(a, B, d) = \text{ENT}_B^\sigma(d) - \text{ENT}_{B \cup \{a\}}^\sigma(d). \quad (6)$$

$\text{Sig}_\gamma^\sigma(a, B, d)$ and $\text{Sig}_{\text{ENT}}^\sigma(a, B, d)$ reflect the variation of approximation quality and the variation of conditional entropy when attribute a is added into set B, respectively. Therefore, the higher the value of the significance function is, the more significant the condition attribute a will be in terms of the corresponding measurement.

Take approximation quality reduct as an example, it can be generated by Algorithm 1.

Algorithm 1. Heuristic Algorithm to Compute γ-reduct

Inputs: $DS =< U, A, d >$, Gaussian kernel parameter σ, threshold $\epsilon \in [0, 1)$;
Outputs: An approximation quality reduct B.
1. $B \leftarrow \emptyset$;
2. Compute $\gamma_A^\sigma(d)$;
3. **Do**
    1) $\forall a \in A \backslash B$, compute $\mathrm{Sig}_\gamma^\sigma(a, B, d)$; // $\gamma_\emptyset^\sigma(d) = 0$
    2) Select b such that $\mathrm{Sig}_\gamma^\sigma(b, B, d) = \max \{\mathrm{Sig}_\gamma^\sigma(a, B, d) : \forall a \in A \backslash B\}$;
    3) $B \leftarrow B \cup \{b\}$;
    4) Compute $\gamma_B^\sigma(d)$;
    Until $\gamma_A^\sigma(d) - \gamma_B^\sigma(d) \leq \epsilon \cdot \gamma_A^\sigma(d)$;
4. **Return** B.

In Algorithm 1, the most significant attribute b is selected and added into set B in each iteration until the constraint is satisfied. To avoid that the strict constraint may cause that no attribute can be eliminated, the threshold ϵ is employed.

In addition, in order to obtain the similarities between each two samples, we need to previously compute the Euclidean distance between each two samples that cost $O(|AT| \times |U|^2)$. And the overall time complexity of Algorithm 1 is at most $O(|AT|^2 \times |U|^2)$.

3.2 Multi-granularity Heuristic Algorithm

In real-world applications, it is not rare that several Gaussian kernel parameters should be considered instead of only one [8]. Without loss of generality, $T = \{\sigma_1, \sigma_2, \cdots, \sigma_m\}$ contains all the considered Gaussian kernel parameters, and they have been sorted in ascending order. In this case, Algorithm 1 will be executed m times to generate all the reducts.

It is time-consuming to generate m reducts. To solve the problem, a solution is to reduce the times of computing reducts. Therefore, the following definition of multi-granularity attribute reductions will be proposed.

Definition 6. *Given a decision system $DS = <U, A, d>$, $\forall B \subseteq A$,*
    (1) *B is a multi-granularity approximation quality reduct (MG-γ-reduct) if and only if $\forall \sigma \in T$, $\gamma_B^\sigma(d) = \gamma_A^\sigma(d)$ and $\forall C \subset B$, $\gamma_C^\sigma(d) \neq \gamma_B^\sigma(d)$;*
    (2) *B is a multi-granularity conditional entropy reduct (MG-CE-reduct) if and only if $\forall \sigma \in T$, $\mathrm{ENT}_B^\sigma(d) = \mathrm{ENT}_A^\sigma(d)$ and $\forall C \subset B$, $\mathrm{ENT}_C^\sigma(d) \neq \mathrm{ENT}_B^\sigma(d)$.*

Take the multi-granularity approximation quality reduct as an example, it can be generated by the multi-granularity heuristic algorithm presented below.

Algorithm 2. Heuristic Algorithm to Compute MG-γ-reduct

Inputs: $DS =< U, A, d >$, $T = \{\sigma_1, \sigma_2, \cdots, \sigma_m\}$, threshold $\epsilon \in [0, 1)$;
Outputs: A multi-granularity approximation quality reduct B.
1. $B \leftarrow \emptyset$;
2. Compute $\gamma_A^{\sigma_1}(d)$ and $\gamma_A^{\sigma_m}(d)$;
3. **Do**
    1) $\forall a \in A\backslash B$, $\mathrm{Sig}_\gamma^\sigma(a, B, d) = \frac{1}{2} \cdot \left(\mathrm{Sig}_\gamma^{\sigma_1}(a, B, d) + \mathrm{Sig}_\gamma^{\sigma_m}(a, B, d)\right)$;
    // $\gamma_\emptyset^{\sigma_1}(d) = 0$, $\gamma_\emptyset^{\sigma_m}(d) = 0$
    2) Select b such that $\mathrm{Sig}_\gamma^\sigma(b, B, d) = \max\{\mathrm{Sig}_\gamma^\sigma(a, B, d) : \forall a \in A\backslash B\}$;
    3) $B \leftarrow B \cup \{b\}$;
    4) Compute $\gamma_B^{\sigma_1}(d)$ and $\gamma_B^{\sigma_m}(d)$;
    Until $\gamma_A^{\sigma_1}(d) - \gamma_B^{\sigma_1}(d) \leq \epsilon \cdot \gamma_A^{\sigma_1}(d)$ and $\gamma_A^{\sigma_m}(d) - \gamma_B^{\sigma_m}(d) \leq \epsilon \cdot \gamma_A^{\sigma_m}(d)$;
4. Return B.

In Algorithm 2, the following two cases should be carefully noticed.

1. If all of the constraints in Definition 6 are considered, e.g., all of the approximation qualities in terms of all parameters should be preserved, then it will take too much time to generate the reduct. And the time complexity of such strategy is $O(m \times |AT|^2 \times |U|^2)$. This is consistent with our intuition: more constraints indicate more attributes are required, which will increase the iteration times. From this point of view, only two constraints are used in Algorithm 2, they are derived by the fuzzy rough approximations coming from the coarsest and the finest granularities, i.e., the approximation qualities derived by the maximal and the minimal parameters should be preserved. Therefore, the time complexity of Algorithm 2 is at most $O(|AT|^2 \times |U|^2)$.
2. In Algorithm 2, the most significant attribute b is determined by the mean value of the significances derived from σ_1 and σ_m. Then, b is added into set B in each iteration. Finally, when B satisfies $\gamma_A^{\sigma_1}(d) - \gamma_B^{\sigma_1}(d) \leq \epsilon \cdot \gamma_A^{\sigma_1}(d)$ and $\gamma_A^{\sigma_m}(d) - \gamma_B^{\sigma_m}(d) \leq \epsilon \cdot \gamma_A^{\sigma_m}(d)$, B is considered as the multi-granularity approximation quality reduct.

4 Experiments

To verify the effectiveness of our proposed algorithm, 12 data sets from UCI machine learning repository have been employed. Table 1 illustrates the details of them.

Table 1. Data set description

ID	Data sets	Samples	Attributes	Classes
1	Breast Cancer Wisconsin (Diagnostic)	569	30	2
2	Contraceptive Method	1473	9	3
3	Dermatology	366	34	6
4	Forest Type Mapping	523	27	4
5	Glass Identification	214	9	6
6	Libras Movement	360	90	15
7	Parkinsons	195	23	7
8	Pima Indians Diabetes	768	8	2
9	QSAR Biodegradation	1055	41	2
10	SPECTF Heart	267	44	2
11	Statlog (German Credit Data)	1000	24	5
12	Wine	178	13	3

The experiments are conducted on a personal computer with Intel i7-6700HP CPU (2.60 GHz) and 8 GB memory. In addition, the adopted software is Matlab R2014b.

4.1 Experimental Results and Discussions

Two groups of experiments have been designed. In these experiments, the threshold ϵ is set by $\epsilon = 0.05$, and Gaussian kernel parameters σ are set by $\sigma = 0.60, 0.65, 0.70, 0.75, 0.80$. Then Algorithm 1 and Algorithm 2 can be executed for generating reducts, respectively. Though Algorithms 1 and 2 have been presented for computing approximation quality reducts in this paper, they can also be used to compute conditional entropy reduct if the significance function is changed.

4.1.1 Time Consumptions of Reducts

In this experiment, the four reducts (see Definitions 4 and 6) are all calculated based on the whole samples. The time consumptions of these reducts are compared. The experimental results are shown in Table 2, in which better performances are highlighted in italic.

In Table 2, γ-reduct and CE-reduct are derived from Algorithm 1; MG-γ-reduct and MG-CE-reduct are derived from Algorithm 2. Note that 5 kernel parameters have been considered in this experiment, the time consumptions of γ-reduct/CE-reduct refer to the sum of time consumptions of computing 5 different reducts.

Table 2. Comparisons of time consumptions of reducts (seconds)

ID	γ-reduct	MG-γ-reduct	CE-reduct	MG-CE-reduct
1	39.2302	16.4294	58.5751	14.7688
2	29.2140	11.2767	23.3273	8.1779
3	16.5307	13.9467	16.1756	7.0508
4	28.5197	13.1976	25.2555	8.1192
5	0.4278	0.1716	0.3744	0.1118
6	99.5611	92.8676	137.4301	49.3479
7	2.0364	1.0152	1.9538	0.6087
8	5.2655	2.0594	5.3891	1.7293
9	239.2054	89.9737	294.0404	71.2781
10	5.3244	3.3556	11.3215	3.7685
11	50.7762	28.8793	71.2355	25.1949
12	0.3801	0.1941	0.5713	0.1868

With a careful observation of Table 2, it is not difficult to observe the following.

1. The computation of MG-γ-reduct requires less time than that of γ-reduct. For example, for the 9-th data set, 239.2054 s is required to generate γ-reduct while only 89.9737 s is needed to generate MG-γ-reduct.
2. The computation of MG-CE-reduct requires less time than that of CE-reduct. For example, for the 11-th data set, it takes 71.2355 s to generate CE-reduct, while it takes only 25.1949 s to generate MG-CE-reduct.

4.1.2   Performances of Reducts

Although the reducts can be generated faster by Algorithm 2, the performances of the obtained reducts are more important and should be deeply compared.

In this experiment, 10–fold cross validation is employed, which means the following progress is repeated 10 times: 90% of the samples in data are considered as the training samples for computing reducts, and the rest of the 10% samples are regarded as the test samples for evaluations, i.e., use reducts to compute approximation quality or conditional entropy over test samples, respectively.

Finally, the mean values of approximation quality and conditional entropy are recorded, which are displayed in Tables 3 and 4, respectively.

In Table 3, with a careful observation, we can detect the following. Compared with the approximation quality reduct, the multi-granularity approximation quality reduct may bring us similar values of approximation qualities. Take the 6-th data set as an example, if $\sigma = 0.70$, then by Algorithm 1, the approximation quality is 0.8139 over test data, whereas the approximation quality is 0.8181 for test data when Algorithm 2 is executed.

Table 3. Comparisons of approximation qualities

ID	Algorithms	$\sigma = 0.60$	$\sigma = 0.65$	$\sigma = 0.70$	$\sigma = 0.75$	$\sigma = 0.80$
1	Algorithm 1	0.4770	0.4330	0.3977	0.3622	0.3304
	Algorithm 2	0.4829	0.4376	0.3973	0.3617	0.3301
2	Algorithm 1	0.2007	0.1774	0.1576	0.1408	0.1264
	Algorithm 2	0.2007	0.1774	0.1576	0.1408	0.1264
3	Algorithm 1	0.9100	0.8892	0.8672	0.8469	0.8233
	Algorithm 2	0.9295	0.9063	0.8807	0.8531	0.8240
4	Algorithm 1	0.2897	0.2581	0.2306	0.2071	0.1869
	Algorithm 2	0.2903	0.2581	0.2306	0.2072	0.1870
5	Algorithm 1	0.1517	0.1374	0.1225	0.1097	0.0987
	Algorithm 2	0.1550	0.1374	0.1225	0.1097	0.0987
6	Algorithm 1	0.8734	0.8448	0.8139	0.7817	0.7489
	Algorithm 2	0.8821	0.8510	0.8181	0.7842	0.7499
7	Algorithm 1	0.3187	0.2839	0.2554	0.2297	0.2077
	Algorithm 2	0.3205	0.2856	0.2557	0.2299	0.2077
8	Algorithm 1	0.1388	0.1203	0.1052	0.0927	0.0822
	Algorithm 2	0.1388	0.1203	0.1052	0.0927	0.0822
9	Algorithm 1	0.3510	0.3183	0.2910	0.2657	0.2432
	Algorithm 2	0.3530	0.3202	0.2913	0.2658	0.2432
10	Algorithm 1	0.5893	0.5465	0.5079	0.4717	0.4393
	Algorithm 2	0.5913	0.5475	0.5079	0.4722	0.4399
11	Algorithm 1	0.8663	0.8371	0.8021	0.7645	0.7286
	Algorithm 2	0.8745	0.8402	0.8038	0.7663	0.7286
12	Algorithm 1	0.5903	0.5358	0.4856	0.4460	0.4100
	Algorithm 2	0.5972	0.5425	0.4930	0.4486	0.4090

With a careful observation of Table 4, we can detect the following.

Compared with the conditional entropy reduct, the multi-granularity conditional entropy reduct may bring us similar values of conditional entropy. Take the 11-th data set as an example, if $\sigma = 0.70$, then by Algorithm 1, the conditional entropy is 0.6986 over test data, whereas the conditional entropy is 0.6973 for test data when Algorithm 2 is executed.

Table 4. Comparisons of conditional entropies

ID	Algorithms	$\sigma = 0.60$	$\sigma = 0.65$	$\sigma = 0.70$	$\sigma = 0.75$	$\sigma = 0.80$
1	Algorithm 1	3.9835	4.7708	5.5784	6.3019	7.0091
	Algorithm 2	3.9835	4.7386	5.4837	6.2068	6.8996
2	Algorithm 1	10.9398	13.1812	15.1799	17.1674	19.1134
	Algorithm 2	10.9398	12.9029	14.8963	16.8817	18.8287
3	Algorithm 1	0.1795	0.2684	0.3805	0.5176	0.6852
	Algorithm 2	0.1762	0.2642	0.3774	0.5176	0.6852
4	Algorithm 1	7.7454	8.6768	9.4487	10.1994	10.8798
	Algorithm 2	7.7687	8.6103	9.3760	10.0695	10.6958
5	Algorithm 1	4.4620	4.7047	4.9680	5.1669	5.3244
	Algorithm 2	4.4643	4.7017	4.9106	5.0944	5.2561
6	Algorithm 1	0.1672	0.2358	0.3227	0.4223	0.5358
	Algorithm 2	0.1672	0.2349	0.3177	0.4155	0.5278
7	Algorithm 1	1.8201	2.0586	2.2946	2.5323	2.7413
	Algorithm 2	1.8201	2.0586	2.2886	2.5084	2.7168
8	Algorithm 1	14.0660	15.0780	15.9568	16.7202	17.5897
	Algorithm 2	14.0660	15.0780	15.9568	16.7202	17.3845
9	Algorithm 1	10.5192	11.9243	13.3234	14.6432	15.8110
	Algorithm 2	10.5296	11.9327	13.2680	14.5264	15.7034
10	Algorithm 1	1.5783	1.8406	2.0706	2.3034	2.5020
	Algorithm 2	1.5787	1.8220	2.0516	2.2663	2.4661
11	Algorithm 1	0.3171	0.4841	0.6986	0.9742	1.3033
	Algorithm 2	0.3203	0.4831	0.6973	0.9676	1.2962
12	Algorithm 1	1.6547	2.0331	2.4072	2.7064	2.9819
	Algorithm 2	1.6547	1.9916	2.3131	2.6143	2.8927

5   Conclusions

In this paper, we proposed the concept of multi-granularity attribute reduction in terms of fuzzy rough set. Such multi-granularity is realized by considering the multi-granulation generated by a set of Gaussian kernel parameters instead of only one parameter. Furthermore, to compute the multi-granularity reduct, traditional heuristic algorithm is modified by using the information provided by the coarsest and the finest granularities. Compared with traditional approach, the revised algorithm to compute multi-granularity reduct can significantly reduce the time consumptions, while the performance of characterizing uncertainties is preserved.

The following topics deserve our further investigations.

1. It may not be optimal to only use the coarsest and the finest granularities to determine the most significant attribute. Whether the multi-granularity heuristic algorithm can be further improved will be carefully analyzed.
2. Multi-granularity reducts will be employed in the classification learning task, and then the classification performance will be explored.
3. The quick reduct processes shown in Sect. 1 can also be introduced into the computations of multi-granularity reducts.

Acknowledgments. This work is supported by the Natural Science Foundation of China (No. 61572242, 61502211, 61503160), Open Project Foundation of Intelligent Information Processing Key Laboratory of Shanxi Province (No. 2014002), the Postgraduate Research & Practice Innovation Program of Jiangsu Province (KYCX18_2333).

References

1. An, S., Shi, H., Hu, Q.H., Li, X.Q., Dang, J.W.: Fuzzy rough regression with application to wind speed prediction. Inf. Sci. **282**, 388–400 (2014)
2. Chen, H.M., Li, T.R., Cai, Y., Luo, C., Fujita, H.: Parallel attribute reduction in dominance-based neighborhood rough set. Inf. Sci. **373**, 351–368 (2016)
3. Dai, J.H., Gao, S.C., Zheng, G.J.: Generalized rough set models determined by multiple neighborhoods generated from a similarity relation. Soft Comput. (2017). https://doi.org/10.1007/s00500-017-2672-x
4. Dai, J.H., Xu, Q.: Attribute selection based on information gain ratio in fuzzy rough set theory with application to tumor classification. Appl. Soft Comput. **13**, 211–221 (2013)
5. Dubois, D., Prade, H.: Rough fuzzy sets and fuzzy rough sets. Int. J. Gen. Syst. **17**, 191–209 (1990)
6. Hu, Q.H., Yu, D.R., Xie, Z.X., Liu, J.F.: Fuzzy probabilistic approximation spaces and their information measures. IEEE Trans. Fuzzy Syst. **16**, 549–551 (2006)
7. Hu, Q.H., Zhang, L., Chen, D.G., Pedrycz, W., Yu, D.R.: Gaussian kernel based fuzzy rough sets: model, uncertainty measures and applications. Int. J. Approx. Reasoning **51**, 453–471 (2010)
8. Hu, Q.H., Zhang, L.J., Zhou, Y.C., Pedrycz, W.: Large-scale multi-modality attribute reduction with multi-kernel fuzzy rough sets. IEEE Trans. Fuzzy Syst. (2017). https://doi.org/10.1109/TFUZZ.2017.2647966
9. Ji, S.G., Zheng, Y., Li, T.R.: Urban sensing based on human mobility. In: Proceedings of the 2016 ACM International Joint Conference on Pervasive and Ubiquitous Computing, pp. 1040–1051. ACM, New York (2016)
10. Jia, X.Y., Shang, L., Zhou, B., Yao, Y.Y.: Generalized attribute reduct in rough set theory. Knowl. Based Syst. **91**, 204–218 (2016)
11. Jing, Y.G., Li, T.R., Fujita, H., Yu, Z., Wang, B.: An incremental attribute reduction approach based on knowledge granularity with a multi-granulation view. Inf. Sci. **411**, 23–38 (2017)
12. Ju, H.R., Li, H.X., Yang, X.B., Zhou, X.Z., Huang, B.: Cost-sensitive rough set: a multi-granulation approach. Knowl. Based Syst. **123**, 137–153 (2017)

13. Liang, J.Y., Wang, F., Dang, C.Y., Qian, Y.H.: An efficient rough feature selection algorithm with a multi-granulation view. Int. J. Approx. Reasoning **53**, 912–926 (2012)
14. Qian, Y.H., Liang, J.Y., Pedrycz, W., Dang, C.Y.: An efficient accelerator for attribute reduction from incomplete data in rough set framework. Pattern Recognit. **44**, 1658–1670 (2011)
15. Qian, Y.H., Liang, J.Y., Pedrycz, W., Dang, C.Y.: Positive approximation: an accelerator for attribute reduction in rough set theory. Artif. Intell. **174**, 597–618 (2010)
16. Qian, Y.H., Wang, Q., Cheng, H.H., Liang, J.Y., Dang, C.Y.: Fuzzy-rough feature selection accelerator. Fuzzy Sets Syst. **258**, 61–78 (2014)
17. Vluymans, S., D'eer, L., Saeys, Y., Cornelis, C.: Applications of fuzzy rough set theory in machine learning: a survey. Fundamenta Informaticae **142**, 53–86 (2015)
18. Xu, S.P., Yang, X.B., Yu, H.L., Yu, D.J., Yang, J.Y., Tsang, E.C.C.: Multi-label learning with label-specific feature reduction. Knowl. Based Syst. **104**, 52–61 (2016)
19. Yao, Y.Y., Zhao, Y.: Discernibility matrix simplification for constructing attribute reducts. Inf. Sci. **179**, 867–882 (2009)
20. Yang, X.B., Qi, Y.S., Song, X.N., Yang, J.Y.: Test cost sensitive multigranulation rough set: model and minimal cost selection. Inf. Sci. **250**, 184–199 (2013)
21. Yu, D.J., Hu, J., Wu, X.W., Shen, H.B., Chen, J., Tang, Z.M., Yang, J., Yang, J.Y.: Learning protein multi-view features in complex space. Amino Acids **44**, 1365–1379 (2013)
22. Yue, X.D., Cao, L.B., Miao, D.Q., Chen, Y.F., Xu, B.: Multi-view attribute reduction model for traffic bottleneck analysis. Knowl. Based Syst. **86**, 1–10 (2015)
23. Zhang, X., Mei, C.L., Chen, D.G., Li, J.H.: Feature selection in mixed data: a method using a novel fuzzy rough set-based information entropy. Pattern Recognit. **56**, 1–15 (2016)

Tolerance Methods in Graph Clustering: Application to Community Detection in Social Networks

Vahid Kardan and Sheela Ramanna[⊠]

Department of Applied Computer Science, University of Winnipeg, Winnipeg,
Manitoba R3B 2E9, Canada
kardan-v@webmail.uwinnipeg.ca, s.ramanna@uwinnipeg.ca

Abstract. This article introduces a novel approach to graph clustering based on tolerance spaces. From a graph theory perspective, a community is considered as a group or cluster of nodes with interconnections between them. The proposed approach to community detection uses a tolerance relation which provides a mechanism for clustering objects (nodes or vertices of a graph) into groups termed as tolerance classes inspired by near set theory. The proposed tolerance-based community detection (TCD) algorithm uses the shortest path as the distance function for creating tolerance classes, where a tolerance class represents members of the same community. For parameter selection, an objective function based on two well-known quality functions, modularity and coverage, is used. To demonstrate the robustness of the proposed method, sensitivity analysis of the parameters is given. The effectiveness of the TCD algorithm has been demonstrated by testing it on four real-world data sets. Experimental results include the comparison of the TCD algorithm with four other methods. TCD was able to achieve the best results with two data sets. The contribution of this work is a new tolerance-based method for community detection in social networks.

Keywords: Community detection · Graph clustering
Near set theory · Tolerance spaces

1 Introduction

Research in discovering community structures has a deep and rich history and is of tremendous importance in sociology, biology and computer science disciplines where systems are often represented as graphs [6]. In most studies, a community is found by analyzing connections (edges) of the network, but other studies also include node attributes [17]. From a graph theory perspective, a community is considered as a group or cluster of nodes with interconnections between

This research has been supported by NSERC Discovery grant 194376 and Univ. of Winnipeg Major Research Grant.

© Springer Nature Switzerland AG 2018
H. S. Nguyen et al. (Eds.): IJCRS 2018, LNAI 11103, pp. 73–87, 2018.
https://doi.org/10.1007/978-3-319-99368-3_6

them. A popular real-world application of community detection can be found in social networks, where networked communities are fundamental structures for understanding social behavior [20].

There is intense interest in community detection algorithms based on network structures both in overlapping and non-overlapping communities, as evidenced by the most recent work found in [8]. In [25], overlapping communities are detected by a local-expansion-based method using rough set theory. Fuzzy granular theory was used to represent a social network where a vertex (node) can be part of several communities with different memberships of their association with each community [10].

In this paper, the focus is on discovering *non-overlapping* community structures in graphs with a novel approach based on tolerance spaces [24]. Also, only the edges of the network are considered for detecting communities. The proposed tolerance-based community detection (TCD) algorithm was inspired by near set theory [16]. The tolerance relation [21] provides us with a mechanism for clustering objects (nodes or vertices of a graph) into groups referred to as *tolerance classes*. The motivation for using tolerance classes is that the tolerance relation defines *similarity* rather than equivalence where nodes of the same community are highly similar, while nodes between communities have lower similarity. In case of graphs, community detection is considered as identifying subgraphs of a graph which are more densely connected within the subgraph than to the rest of the graph [13]. The TCD Algorithm uses the shortest path as the distance function for forming the tolerance classes in an *undirected graph*.

The effectiveness of our method has been demonstrated by testing it on four real-world data sets and benchmarked with four well-known algorithms. TCD was able to achieve the best results on two data sets. Also, sensitivity analysis has been performed to demonstrate the robustness of the proposed method. The contribution of this work is a new tolerance-based method for community detection in social networks.

This paper is organized as follows: We present research related to *non-overlapping* community methods in Sect. 2 due to space limitations. The theoretical framework for this research is given in Sect. 3. In Sect. 4 we present our method for combining graph theory and tolerance classes as well as defining the objective function to assess the quality of clustering. Tolerance-based Community Detection (TCD) algorithm is described in Sect. 5. In Sect. 6, the description of the four data sets, sensitivity analysis and the results is presented. Finally, in Sect. 7 suggestions for future work are given.

2   Related Works

In [7], the property of community structures was first explored in the context of social and biological networks. A divisive algorithm that uses *edge betweenness* as a metric to identify the boundaries of communities rather than on the cores was proposed. Subsequently, the authors proposed a new set of algorithms and an objective measure to choose the number of communities to partition the network in [13]. The *Louvain Method* is a heuristic method that is based on modularity optimization [1]. In 2014, a generalized version of this method was introduced that utilizes other quality functions for optimization instead of the original modularity function [2]. The *Infomap Method* is based on an information theoretic approach that reveals community structures in weighted and directed networks [19]. The *Label Propagation method (LPA)* detects a community of a node based on the labels of its neighbors. The algorithm first assigns a unique label to each node. Subsequently, these labels are propagated based on the majority labels of its neighbors. The latest version of the LPA algorithm is the Semi Synchronous Constrained Label Propagation Algorithm (SSCLPA) introduced by Chin and Ratnavelu in [3]. The *Fluid Communities method* uses the idea of expansion and contraction of fluids interacting in an environment [14].

3   Preliminaries: Tolerance Classes and Graphs

The algorithms presented in this paper are based on the concepts of neighborhoods and tolerance classes. Here, we recall their definitions.

Definition 1 Tolerance Relation [15,22,24]. Let O be a set of sample objects, and let τ be a binary relation (called a tolerance relation) on O ($\tau \subseteq O \times O$) that is reflexive (for all $x \in O$, $x\tau x$) and symmetric (for all $x, y \in O$, if $x\tau y$, then $y\tau x$) but transitivity of τ is not required.

Definition 2 Tolerance Space [15,22,24]. Then a tolerance space is defined as $\langle O, \tau \rangle$.

In this work, the sample space O is comprised of nodes and edges of the graph. Based on Zeeman [24], every pseudometric space determines tolerance relations with respect to some positive real threshold ε.

Definition 3 Neighbourhood. A neighborhood is defined as:

$$N(x) = \{y \in O : p(x, y) < \varepsilon\}.$$

In other words, all objects satisfy the tolerance relation with a single object in a neighborhood. Note that we do not use the $\tau_{p,\epsilon}$ neighborhood of x which is just an open ball in the pseudometric space $\langle O, p \rangle$ with the center x and radius ε [5].

Definition 4 Pre-class. A set $A \subseteq O$ is a τ-*preclass* (or briefly *preclass* when τ is understood) if and only if for any $x, y \in A$, $(x, y) \in \tau$.

Definition 5 Tolerance Class. The family of all preclasses of a tolerance space is naturally ordered by set inclusion and preclasses that are maximal with respect to a set inclusion are called τ-*classes* or just *classes*, when τ is understood. A maximal pre-class with respect to inclusion is called a tolerance class.

In other words, tolerance class is a pre-class where no additional element can be added to the pre-class.

Definition 6 Undirected Graph. A graph G is defined as a pair of (V, E), in which V is a set of vertexes, and $E \subseteq V \times V$ is a set of edges and in case of undirected graphs this pair is unordered or if edge $(u, v) \in E$ then $(v, u) \in E$.

The *degree* of a vertex v is defined as the number of edges containing v. Two vertexes are adjacent when they are both in a common edge.

Definition 7 Path. A path is a sequence of vertexes $P = (v_1, v_2, ..., v_n) \in V^n$ where $\forall i, 1 < i < n \ v_i$ is adjacent to v_{i+1}.

The length of the path P is defined as the number of vertexes in the sequence minus one, $n - 1$. The shortest path P between vertex s and z is the path with minimum length which $v_1 = s$ and $v_n = z$. This concept is utilized for defining a distance function for finding tolerance classes.

4   Combining Graph Theory and Tolerance Classes

In this research, tolerance classes are derived from graph components such as vertexes (nodes) or edges. Using Definition 5 for tolerance classes, we partition the graph to the final clusters. For applying the concept of tolerance classes, we will define a metric space in graphs as follows:

Consider a graph $G(V, E)$, where V is the set of vertexes, E is the set of edges and $d : V^2 \longrightarrow \mathbb{R}$ is defined as the number of edges in the shortest path (SP) between two vertexes $v, u \in V$. The metric space (V, d) in graph G is defined as:

$$d(v, u) = \begin{cases} \infty & \text{if no SP exists} \\ |SP| & \text{else} \end{cases} \tag{1}$$

where $|SP|$ denotes the number of edges in the shortest path between vertex v and u.

Next consider A and B as two non empty subset of V, and $\wp(V)$ as the power set of V, the closeness measure $c : (\wp(V))^2 \longrightarrow [0, 1]$ between A and B is defined as:

$$c(A, B) = \frac{|A \cap B|}{min(|A|, |B|)} \tag{2}$$

This parameter represents the percentage of members that the smaller set shares with the larger set.

Parameter β used in merging tolerance classes and clusters is defined as:

Definition 8 Merge Minimum Closeness Parameter (β). The minimum value of closeness measure (c) between a tolerance class (T) and a cluster (C) so that they can be merged together or:

$$c(T, C) > \beta \Rightarrow C = T \cup C$$

Another parameter used for controlling the sizes of the clusters is α which is defined as:

Definition 9 Minimum Cluster Size Parameter (α). This parameter is defined as the minimum size of each cluster (C) in the set of all clusters (L), to ensure that cluster size will not go below α or:

$$\forall C \in L \Rightarrow |C| > \alpha$$

The intuition behind this parameter is that if this condition is not met by a cluster member, then this member will join a different community based on majority voting of the cluster members. We now define the *objective function* for assessing the quality of clusters. This new objective function (O) introduced in Eq. 5 is a combination of a modularity function (Q) (given in Eq. 3) and a regularization parameter (S) (given in Eq. 4).

4.1 Objective Function

One of the challenges of TCD algorithm is to find a proper way for parameter selection. This problem is addressed by using an objective function based on two widely used quality functions introduced here. The first one (Q) is based on the well known modularity function introduced in [13]. Let L, w and v represent the set of clusters, number of edges inside the ith cluster, and total amount of edges where at least one end of the edge is inside the ith cluster respectively. Then:

$$Q(L) = \frac{1}{2|E|} \sum_{i=1}^{|L|} (w_i - \frac{v_i^2}{2.|E|}), \qquad Q : \wp(V) \longrightarrow [\frac{-1}{2}, 1) \qquad (3)$$

The second quality function (C) is based on the notion of coverage which is defined as the ratio of edges within clusters and the total number of edges or:

$$C(L) = \frac{\sum_{i=1}^{|L|} w_i}{|E|}, \qquad Q : \wp(V) \longrightarrow [0, 1] \qquad (4)$$

Finally, the objective function (O) is defined as follows:

$$O(L) = \eta_0.Q + (1 - \eta_0).C, \qquad O : \wp(V) \longrightarrow [\frac{-\eta_0}{2}, 1] \qquad (5)$$

with constant η_0 which is used to weight Q and C. In our case we set $\eta_0 = 0.5$ for balancing the weights. This objective function is used later in Sect. 5.6 for parameter selection.

5 Tolerance-Based Community Detection (TCD) Algorithm

In this section, we describe the proposed novel method for community detection based on tolerance classes given in Algorithm 4 which has three main functions. We begin the presentation by giving a detailed walk-through of each of these functions which form the basis of the TCD algorithm.

5.1   Get Tolerance Class Function

The pseudo code for the first function is given in Algorithm 1. This function has three inputs: G the graph, v the seed vertex for forming the tolerance class, and ϵ the maximum distance parameter (or distance threshold), and three variables: *root* representing the vertex which the Breadth-First Search (BFS) will be applied to, T the tolerance class, and set N containing the vertexes reached during the BFS.

In lines 2 and 3, *root* and T variables are initialized. Starting in line 4, the neighborhood (see Definition 3) of the *root* is found by doing a BFS on the *root*, and returning the set of vertexes in the range of distance threshold ϵ. It is important to note that the depth of BFS performed is not higher than ϵ. Therefore, all the vertexes are *not visited* during the BFS. Then the *root* node will be marked so that it will not get selected later as the root for subsequent Breadth-First Searches.

In lines 7 to 10, in the first iteration, T will be set to N since the variable T is empty. However, in subsequent iterations, T will be set to the intersection of T and N. In other words, after the first iteration, the vertexes which are not in the intersection will be removed from the set T. When all the remaining vertexes in T are selected as root, the loop will terminate resulting in the formation of a tolerance class.

Algorithm 1. Get Tolerance Class Function

1: **procedure** GETTOLERANCECLASS(G, v, ϵ)
2:     $root \leftarrow v$
3:     $T \leftarrow \emptyset$
4:     **while** $root \neq NULL$ **do**
5:         $N \leftarrow BFS(G, root, \epsilon)$
6:         $root.selected \leftarrow true$
7:         **if** $T \neq \emptyset$ **then**
8:             $T \leftarrow T \cap N$
9:         **else**
10:             $T \leftarrow N$
11:         $root \leftarrow NULL$
12:         **for each node** u **in** T **do**
13:             **if** $u.selected = false$ **then**
14:                 $root \leftarrow u$
15:                 break
         return T

5.2   Get Close Clusters Function

The pseudo code for the second function is given in Algorithm 2. This function has three inputs: L the set of the current clusters, T the tolerance class for which we are seeking to find the close clusters, and β the merge minimum closeness parameter. The output H represents the set of *close* clusters. In the *for* loop, the closeness measure of all current existing clusters in L with the tolerance class T will be calculated based on the Eq. 2. If the closeness measure satisfies the minimum threshold β, then the cluster will be added to the output H.

Algorithm 2. Get Close Clusters Function

1: **procedure** GETCLOSECLUSTERS(L, T, β)
2:    $H \leftarrow \emptyset$
3:    **for each cluster** $C \in L$ **do**
4:        $m \leftarrow calcCloseness(T, C)$
5:        **if** $m > \beta$ **then**
6:            $H.add(C)$
7:        **return** H

5.3 Find Nearest Cluster Function

The pseudo code for the third function is given in Algorithm 3. This function has four inputs: G the graph, L the set of detected clusters, C the intended cluster where the number of members in the cluster is less than parameter α, and ϵ the maximum distance parameter. The goal of this function is to find the nearest cluster to C as a candidate for merging in later steps. Furthermore, it is based on majority voting of all vertexes with respect to their neighborhoods. The variable *label* is the label of the nearest cluster. Also, array *counter* keeps track of the frequency of members of each cluster in the neighborhood of C.

Algorithm 3. Find Nearest Cluster Function

1: **procedure** FINDNEARESTCLUSTER(G, L, C, ϵ)
2:    $label \leftarrow NULL$
3:    **array** $counter[L.size()]$
4:    **for each** a **in** $counter$ **do**
5:        $a \leftarrow 0$
6:    **for each node** v **in** C **do**
7:        $N \leftarrow BFS(G, v, \epsilon)$
8:        **for each node** u **in** N **do**
9:            $lb \leftarrow u.clusterLabel$
10:           **if** $lb \neq C.label$ **then**
11:              $counter[lb] \leftarrow counter[lb] + 1$
12:              **if** $counter[lb] > counter[label]$ **then**
13:                $label \leftarrow lb$
14:    **return** $L.getClusterByLabel(label)$

In lines 4 and 5 all the frequencies are set to zero. Next, in lines 6 to 13, for every member of cluster C, first we find the vertexes in the neighborhood of that member by calling BFS function with the maximum depth of ϵ. Then for every vertex in the neighborhood N which is not a member of cluster C, based on the cluster membership label, the corresponding frequency in *counter* will be incremented. Next, if the new frequency is larger than the current cluster label with the largest frequency, the variable *label* will be updated to the new cluster's label. Finally, the corresponding cluster of *label* will be returned by this function.

5.4   TCD Function

We now discuss the main function TCD presented in Algorithm 4. This function has 4 inputs: G the graph, ϵ the maximum distance parameter, β the merge minimum closeness parameter and α the minimum cluster size. Also, variable L will contain the set of all the clusters when the algorithm is finished.

In line 2, we first *sort* the vertexes of the graph based on the degree of a vertex. By this step, we ensure that vertexes with a lower degree are accessed first to form the *tolerance classes*. The intuition here is that these vertexes have a lower chance of connecting with two different clusters. In other words, these vertexes are more likely to be inside a cluster rather than on the border.

Starting in line 4, for each vertex v in G, which is not yet clustered, we first find the corresponding tolerance class by calling the *getToleranceClass* function discussed in Sect. 5.1. Therefore, in line 5, variable T will contain the tolerance class of v.

In line 6, we have to find the set of clusters that can be merged with T. This is done by calling *getCloseClusters* function presented previously in Sect. 5.2. Then in lines 7 to 9, all of these clusters and T will be merged together to form a new cluster. Finally, in line 10, the new cluster will be added to the set L.

In the final stage of this algorithm starting from line 11, clusters with size less than α will be merged into the nearest cluster, which is found by calling *findNearestCluster* function discussed in Sect. 5.3.

5.5   Time Complexity

For the graph G(V, E) and Algorithm 4, the sorting will take $O(|V|.log(|V|))$ where $|V|$ represents the number of vertexes. In case of the *getToleranceClass* function given in Algorithm 1, the number of iterations in the *while* loop will not exceed the number of vertexes in the output of first Breadth-First Search (BFS). This number is not higher than b^ϵ where b represents the branching factor of the graph and ϵ is the maximum depth of BFS. Also, time complexity of a BFS with limited depth ϵ, is $O(b^\epsilon)$. Therefore, the overall time complexity for this function will be $O(b^{2\epsilon})$. For the *getCloseClusters* function given in Algorithm 2, since the number of clusters can not go beyond the number of vertexes the time complexity will be $O(|V|)$. Finally, for the *findNearestCluster* function given in Algorithm 3, since the intended cluster size is less than parameter α, the time complexity will be $O(\alpha.b^\epsilon)$. But considering that α is a small number (in this paper it is less than 14), time complexity can be assumed as $O(b^\epsilon)$. The overall

---

Algorithm 4. Tolerance Community Detection

```
 1: procedure TCD(G, ε, β, α )
 2:     sort(G)
 3:     L ← ∅
 4:     for  each node v ∈ V do
 5:         T ← getToleranceClass(G, v, ε)
 6:         H ← getCloseClusters(L, T, β)
 7:         for  each cluster C ∈ H do
 8:             T ← T ∪ C
 9:             L.remove(C)
10:         L.add(T)
11:     for  each cluster C ∈ L do
12:         if C.size() < α then
13:             K ← findNearestCluster(G, L, C, ε)
14:             K ← K ∪ C
15:             L.remove(C)
16:     return L
```

---

time complexity of the function given in Algorithm 4 is given in Eq. 6 where $|C|$ represents the number identified clusters:

$$O(|V|.log(|V|)) + |V|.(O(b^{2\epsilon}) + O(|V|)) + |C|.(O(b^{\epsilon})) = O(|V|^2 + |V|.b^{2\epsilon}) \quad (6)$$

5.6 Parameter Selection

As mentioned in the earlier sections, for parameter selection a new objective function was introduced in Eq. 5. The value of this function will be calculated for different instances of each parameter. In other words, all possible combinations will be examined. In our experiments, the values of maximum distance ϵ vary between 2 to 5. The range of merge minimum closeness parameter β is $[0.25, 0.95]$ and we increment this parameter by 0.05 in each iteration. In case of minimum cluster size α, the range is from 3 to 14. All possible combinations of these parameters are used. Then, the set of parameters with the highest value of the objective function is selected. In Sect. 6.1 the method's output sensitivity for parameters ϵ, α and β are discussed.

6 Results and Analysis

In this paper, we have compared the quality of our method with the latest version of Louvain [2], Infomap [19], Asynchronous Fluid Communities (AFC) [14] and Semi-Synchronous Label Propagation Algorithm (SSLPA) [4]. For comparison, four real data sets are used. The descriptions of these data sets are presented in Table 1.

For comparing the results obtained by different algorithms, two entropy-based measures are used; Normalized Mutual Information (NMI) which is a well known

Table 1. Summary of the real-world networks considered in this study.

Networks	Nodes	Edges	Clusters
Zachary [23]	34	78	2
Dolphins [12]	62	159	2
Pol-books [9]	105	441	3
Football [7]	115	613	12

measure and V-measure which is based on two concepts: completeness and homogeneity. Readers can refer to [18] for more information on these measures.

6.1   Parameter Sensitivity Analysis

For showing the robustness of our method, a generated graph with 2000 vertexes, 15045 edges and 99 clusters is used. This graph is generated by the benchmark generator described in [11]. The input parameters used for this generator were: $N = 2000$, $k = 15$, $k_{max} = 50$, $\gamma = -2$, $\beta = -1$, $s_{min} = 5$, $s_{max} = 50$, and $\mu = 0.1$. We have used the parameters with the *best* objective function score as the reference. These parameters are $\epsilon = 2$, $\beta = 0.9$, and $\alpha = 7$. In other words, for each parameter, all the other parameters will be fixed with the values selected based on the objective function.

In Fig. 1 from left to right, the scores of NMI measure for different values of minimum cluster size (α), merge minimum closeness (β), and maximum distance parameter (ϵ) are given. It is observed that in all cases the fluctuation of NMI measure is insignificant and it is around 0.9.

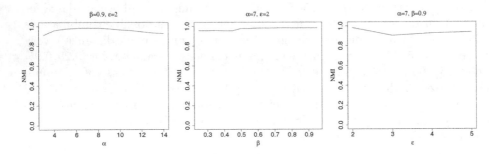

Fig. 1. NMI scores for different values of minimum cluster size (α), merge minimum closeness (β) and maximum distance (ϵ) parameters, based on the clusters obtained from the generated graph.

6.2   Complete Results

We now present the complete set of results for all the four data sets starting from Tables 2, 3, 4 and 5. In all of the experiments, parameters ϵ, β and α were selected by utilizing the parameter selection method discussed in Sect. 5.6.

Since AFC and Louvain methods generate different results for each run, we have used average and standard deviation over 100 runs to show the results. Also, for the AFC method, the number of clusters should be set as the input parameter. In our experiments, this parameter is set to the actual number of clusters for each data set.

In Table 2, the values for different quality measures with Zachary data set are presented. In Fig. 2, a visual comparison of the two main measures of NMI and V-Measure is shown. Also, Fig. 3 shows the clusters obtained by our method. This experiment shows that TCD obtains the second best result after AFC. It is worth noting that the AFC method will generate different results on different runs and number of clusters has to be set as the input parameter in advance. The parameters used in TCD are: $\epsilon = 2$, $\beta = 0.5$ and $\alpha = 3$.

Table 2. Completeness (c), Homogeneity (h), V-measure, and NMI scores for different algorithms based on the clusters obtained on **Zachary** network.

Method	Measure								No. Clusters	
	c		h		V		NMI			
	Avg.	SD	Avg.	SD	Avg.	SD	Avg.	SD	Avg.	SD
TCD	0.58	0.00	0.58	0.00	0.58	0.00	0.58	0.00	2.00	0.00
AFC	**0.69**	0.21	**0.69**	0.22	**0.69**	0.22	**0.69**	0.22	2.00	0.00
SSLPA	0.22	0.00	0.17	0.00	0.19	0.00	0.19	0.00	3.00	0.00
Louvain	0.41	0.04	0.77	0.07	0.54	0.05	0.56	0.05	4.00	0.00
Infomap	0.48	0.00	0.69	0.00	0.57	0.00	0.58	0.00	3.00	0.00

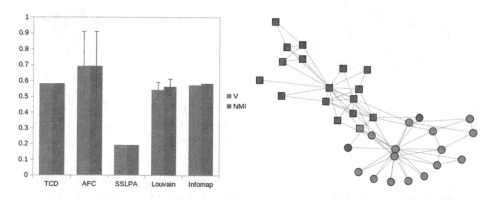

Fig. 2. Comparing V-measure, and NMI scores for **Zachary** network.

Fig. 3. TCD on **Zachary** network, ground-truth clusters are shown by different shapes

In Table 3, the values for different quality measures with Dolphins data set are presented. Also, in Fig. 4, NMI and V-Measure is illustrated visually, while

Fig. 5 shows the clusters obtained by our method. Here, our method got the best results by a large margin in comparison to the 4 other methods. It is interesting to note that the proposed TCD method gets the best scores in three out of four measures. Also, the homogeneity measure values with TCD is close to the best score acquired by Louvain method. The input parameters values for TCD are: $\epsilon = 5$, $\beta = 0.25$ and $\alpha = 3$.

Table 3. Completeness (c), Homogeneity (h), V-measure, and NMI scores for different algorithms based on the clusters obtained in **Dolphins** network.

Method	Measure										No. Clusters	
	c		h		V		NMI					
	Avg.	SD	Avg.	SD	Avg.	SD	Avg.	SD	Avg.	SD		
TCD	**0.80**	0.00	0.83	0.00	**0.81**	0.00	**0.81**	0.00	2.00	0.00		
AFC	0.61	0.21	0.64	0.21	0.62	0.21	0.62	0.21	2.00	0.00		
SSLPA	0.32	0.00	**0.91**	0.00	0.48	0.00	0.54	0.00	7.00	0.00		
Louvain	0.37	0.03	0.89	0.05	0.52	0.03	0.57	0.03	5.03	0.26		
Infomap	0.48	0.00	0.69	0.00	0.57	0.00	0.58	0.00	6.00	0.00		

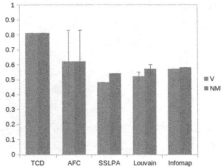

Fig. 4. Comparing V-measure, and NMI scores for **Dolphins** network.

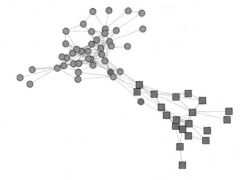

Fig. 5. TCD on **Dolphins** network, ground-truth clusters are shown by different shapes

In Table 4, the values for different quality measures with College Football data set are presented. The input parameters values for TCD are: $\epsilon = 2$, $\beta = 0.7$ and $\alpha = 7$.

In Table 5, the values for different quality measures with Pol-books data set are presented. For this data set the proposed TCD method shows the best result. Also we have to note that the proposed TCD method gets the best scores in three out of four measures. The input parameters values for TCD are: $\epsilon = 4$, $\beta = 0.45$ and $\alpha = 3$.

Table 4. Completeness (c), Homogeneity (h), V-measure, and NMI scores for different algorithms based on the clusters obtained in **Football** network.

Method	Measure										No. Clusters	
	c		h		V		NMI					
	Avg.	SD	Avg.	SD	Avg.	SD	Avg.	SD	Avg.	SD		
TCD	0.85	0.00	0.69	0.00	0.76	0.00	0.77	0.00	8.00	0.00		
AFC	0.89	0.02	0.88	0.03	0.89	0.03	0.89	0.03	12.00	0.00		
SSLPA	0.94	0.00	0.79	0.00	0.86	0.00	0.86	0.00	9.00	0.00		
Louvain	0.92	0.00	0.84	0.03	0.88	0.02	0.88	0.02	9.69	0.48		
Infomap	**0.93**	0.00	**0.92**	0.00	**0.92**	0.00	**0.92**	0.00	12.00	0.00		

Table 5. Completeness (c), homogeneity (h), V-measure, and NMI scores for different algorithms based on the clusters obtained in **Pol-books** network.

Method	Measure										No. Clusters	
	c		h		V		NMI					
	Avg.	SD	Avg.	SD	Avg.	SD	Avg.	SD	Avg.	SD		
TCD	**0.73**	0.00	0.51	0.00	**0.60**	0.00	**0.61**	0.00	2.00	0.00		
AFC	0.47	0.07	0.50	0.07	0.48	0.07	0.48	0.07	3.00	0.00		
SSLPA	0.34	0.00	**0.62**	0.00	0.44	0.00	0.46	0.00	8.00	0.00		
Louvain	0.48	0.03	0.63	0.02	0.54	0.02	0.55	0.02	4.71	0.45		
Infomap	0.48	0.00	0.62	0.00	0.54	0.00	0.54	0.00	5.00	0.00		

7   Conclusion

In this paper, we have presented a novel approach for community detection in social networks based on the concept of tolerance classes adapted from tolerances spaces and near set theory. We have proposed a tolerance-based community detection (TCD) algorithm that was tested against four well-known methods. For parameter selection, an objective function based on the popular modularity and coverage function has been used. The effectiveness of our method was tested on four real-world data sets. A detailed analysis of experiments and results is given using the standard measures of completeness, homogeneity, V-measure, and NMI. In addition, to demonstrate the robustness of the proposed method, sensitivity analysis of its parameters is presented. For the future work, we propose to extend the TCD algorithm for detecting overlapping communities and experimenting with large networks. Another potential extension is for directed graphs. Also we can explore other distance functions for forming the tolerance classes.

References

1. Blondel, V.D., Guillaume, J.L., Lambiotte, R., Lefebvre, E.: Fast unfolding of communities in large networks. J. Stat. Mech. Theory Exp. **2008**(10), P10008 (2008)
2. Campigotto, R., Céspedes, P.C., Guillaume, J.L.: A generalized and adaptive method for community detection. ArXiv preprint arXiv:1406.2518 (2014)
3. Chin, J.H., Ratnavelu, K.: A semi-synchronous label propagation algorithm with constraints for community detection in complex networks. Sci. Rep. **7**, 45836 (2017)
4. Cordasco, G., Gargano, L.: Community detection via semi-synchronous label propagation algorithms. ArXiv e-prints, March 2011
5. Engelking, R.: General Topology. Revised & Completed Edition. Heldermann Verlag, Berlin (1989)
6. Fortunato, S.: Community detection in graphs. Phys. Rep. **486**(3), 75–174 (2010)
7. Girvan, M., Newman, M.E.: Community structure in social and biological networks. Proc. Natl. Acad. Sci. **99**(12), 7821–7826 (2002)
8. Hajiabadi, M., Zare, H., Bobarshad, H.: IEDC: an integrated approach for overlapping and non-overlapping community detection. Knowl. Based Syst. **123**, 188–199 (2017)
9. Krebs, V.: Books about us politics. http://networkdata.ics.uci.edu/data.php?d=polbooks
10. Kundu, S., Pal, S.K.: Fuzzy-rough community in social networks. Pattern Recognit. Lett. **67**, 145–152 (2015)
11. Lancichinetti, A., Fortunato, S.: Benchmarks for testing community detection algorithms on directed and weighted graphs with overlapping communities. Phys. Rev. E **80**(1), 016118 (2009)
12. Lusseau, D., Newman, M.E.: Identifying the role that animals play in their social networks. Proc. R. Soc. London B Biol. Sci. **271**(Suppl 6), S477–S481 (2004)
13. Newman, M.E.J., Girvan, M.: Finding and evaluating community structure in networks. Phys. Rev. E **69**, 026113, February 2004. https://link.aps.org/doi/10.1103/PhysRevE.69.026113
14. Parés, F., et al.: Fluid communities: a competitive, scalable and diverse community detection algorithm. In: Cherifi, C., Cherifi, H., Karsai, M., Musolesi, M. (eds.) Complex Networks & Their Applications VI, pp. 229–240. Springer, Cham (2018). https://doi.org/10.1007/978-3-319-72150-7_19
15. Peters, J.F., Wasilewski, P.: Tolerance spaces: origins, theoretical aspects and applications. Inf. Sci. **195**, 211–225 (2012)
16. Peters, J.: Near sets. Special theory about nearness of objects. Fundamenta Informaticae **75**(1–4), 407–433 (2007)
17. Reihanian, A., Feizi-Derakhshi, M.R., Aghdasi, H.S.: Community detection in social networks with node attributes based on multi-objective biogeography based optimization. Eng. Appl. Artif. Intell. **62**, 51–67 (2017)
18. Rosenberg, A., Hirschberg, J.: V-measure: a conditional entropy-based external cluster evaluation measure. In: EMNLP-CoNLL, vol. 7, pp. 410–420 (2007)
19. Rosvall, M., Bergstrom, C.T.: Maps of random walks on complex networks reveal community structure. Proc. Natl. Acad. Sci. **105**(4), 1118–1123 (2008)
20. Wasserman, S., Faust, K.: Social Network Analysis. Cambridge University Press, Cambridge (1994)
21. Schroeder, M., Wright, M.: Tolerance and weak tolerance relations. J. Comb. Math. Comb. Comput. **11**, 123–160 (1992)

22. Wasilewski, P., Peters, J.F., Ramanna, S.: Perceptual tolerance intersection. In: Peters, J.F., Skowron, A., Chan, C.-C., Grzymala-Busse, J.W., Ziarko, W.P. (eds.) Transactions on Rough Sets XIII. LNCS, vol. 6499, pp. 159–174. Springer, Heidelberg (2011). https://doi.org/10.1007/978-3-642-18302-7_10
23. Zachary, W.W.: An information flow model for conflict and fission in small groups. J. Anthropol. Res. **33**(4), 452–473 (1977)
24. Zeeman, E.: The topology of the brain and visual perception. In: Fort, Jr., M.K. (ed.) Topology of 3-Manifolds and Related Topics, Conference Proceedings, pp. 240–256. University of Georgia Institute, Prentice-Hall Inc. (1962)
25. Zhang, Z., Zhang, N., Zhong, C., Duan, L.: Detecting overlapping communities with triangle-based rough local expansion method. In: Ciucci, D., Wang, G., Mitra, S., Wu, W.-Z. (eds.) RSKT 2015. LNCS (LNAI), vol. 9436, pp. 446–456. Springer, Cham (2015). https://doi.org/10.1007/978-3-319-25754-9_39

Similarity Based Rough Sets
with Annotation

Dávid Nagy$^{(\boxtimes)}$, Tamás Mihálydeák, and László Aszalós

Department of Computer Science, Faculty of Informatics, University of Debrecen,
Egyetem tér 1, Debrecen 4010, Hungary
{nagy.david,mihalydeak.tamas,aszalos.laszlo}@inf.unideb.hu

Abstract. In the authors' previous research the possible usage of the correlation clustering in rough set theory was investigated. Correlation clustering relies on a tolerance relation. Its result is a partition. From the similarity point of view singleton clusters have no information. A system of base sets can be generated from the partition, and if the singleton clusters are left out, then it is a partial approximation space. This way the approximation space focuses on the similarity (the tolerance relation) itself and it is different from the covering type approximation space relying on the tolerance relation. In this paper the authors examine how the partiality can be decreased by inserting the members of some singletons into an arbitrary base set and how this annotation affects the approximations. The authors provide software that can execute this process and also helps to select the destination base set and it can also handle missing data with the help of the annotation.

Keywords: Rough set theory · Correlation clustering
Set approximation

1 Introduction

In our previous study we examined whether the clusters, generated by correlation clustering, can be understood as a system of base sets. Correlation clustering is a clustering method in data mining which creates a partition. The groups, defined by this partition, contain the similar objects. In our previous paper (presented at IJCRS 2017) we showed that it is worth to generate the system of base sets from the partition. This way the base sets contain objects that are typically similar to each other and they are pairwise disjoint. There can be some clusters which have only one member. These singletons represent very little information regarding the similarity. This is they reason why they are not considered as base sets. This way we gained a partial approximation space. In practice there is always an expert who uses the systems. This user may have a background knowledge. We would like to offer a possibility to the user to implement this knowledge into the system by inserting a member of a singleton into a base set. We would like to show some situation where this annotation could be useful.

H. S. Nguyen et al. (Eds.): IJCRS 2018, LNAI 11103, pp. 88–100, 2018.
https://doi.org/10.1007/978-3-319-99368-3_7

The structure of the paper is the following: A theoretical background about the classical rough set theory comes first. In Sect. 3 we present our previous work. In Sect. 4 we define correlation clustering mathematically, shortly present the contraction method which finds a quasi-optimal partition, and how the representative member of a cluster can be chosen. In Sect. 5 the annotation process is described. In Sect. 6 our software is shown with a possible output. Finally we conclude the results.

2   Theoretical Background

From the theoretical point of view a Pawlakian approximation space (see in [10–12]) can be characterized by an ordered pair $\langle U, \mathcal{R} \rangle$ where U is a nonempty set of objects and $\mathcal{R}$ is an equivalence relation on U. In order to approximate an arbitrary subset S of U the followings have to be introduced:

- *the set of base sets*: $\mathfrak{B} = \{B \mid B \subseteq U,$ and $x, y \in B$ if $x\mathcal{R}y\}$, the partition of U generated by the equivalence relation $\mathcal{R}$;
- *the set of definable sets*: $\mathfrak{D}_{\mathfrak{B}}$ is an extension of $\mathfrak{B}$, and it is given by the following inductive definition:
  1. $\mathfrak{B} \subseteq \mathfrak{D}_{\mathfrak{B}}$;
  2. $\emptyset \in \mathfrak{D}_{\mathfrak{B}}$;
  3. if $D_1, D_2 \in \mathfrak{D}_{\mathfrak{B}}$, then $D_1 \cup D_2 \in \mathfrak{D}_{\mathfrak{B}}$.
- *the functions* l, u form a Pawlakian approximation pair $\langle \mathsf{l}, \mathsf{u} \rangle$, i.e.
  1. $Dom(\mathsf{l}) = Dom(\mathsf{u}) = 2^U$
  2. $\mathsf{l}(S) = \bigcup\{B \mid B \in \mathfrak{B}$ and $B \subseteq S\}$;
  3. $\mathsf{u}(S) = \bigcup\{B \mid B \in \mathfrak{B}$ and $B \cap S \neq \emptyset\}$.

3   Similarity Based Rough Sets

When we would like to define the base sets, we use the background knowledge embedded in an information system. The base sets represent background knowledge (or its limit). In a Pawlakian system we can say that two objects are indiscernible if all of their known attribute values are identical. The indiscernibility relation defines an equivalence relation. In some cases we have only a similarity (tolerance) relation. If we change the negativity of indiscernible relations to positivity of similarity (based on background knowledge), then we may rely on a tolerance relation. Some covering systems are based on a tolerance relation. It emphasizes the similarity to a given object and not the similarity of objects 'in general'. Using correlation clustering, we obtain a (quasi optimal) partition of the universe (see in [2–4]). The clusters contain such elements which are typically similar to each other and not just to a distinguished member. In our previous research we investigated whether the partition can be understood as a system of base sets (see in [9]). By our experiments, it is worth to generate a partition with correlation clustering. The base sets, generated from the partition, have several good properties:

- the similarity of objects relying on their properties (and not the similarity to a distinguished object) plays a crucial role in the definition of base sets;
- the system of base sets consists of disjoint sets, so the lower and upper approximation are closed in the following sense: Let S be a set and $x \in U$. If $x \in \mathsf{l}(S)$, then we can say, that every $y \in U$ object which are in the same cluster as x is in $l(S)$. If $x \in \mathsf{u}(S)$, then we can say, that every $y \in U$ object which are in the same cluster as x is in $u(S)$.
- only the necessary number of base sets appears (in applications we have to use an acceptable number of base sets);
- the size of base sets is not too small, or too big.

4   Correlation Clustering

Cluster analysis is a well-known method in data mining. The goal is to group the objects so that the objects in the same group are more similar to each other than to those which are in other groups. In many cases the similarity is based on the attribute values of the objects. Although, there are some cases when these values are not numbers, but we can still say something about their similarity or dissimilarity. Let's take the humans for example. We cannot describe someone's looks by a number, but we still make statements whether two persons are similar or dissimilar. These opinions are dependent on the person who makes the statements. Someone can say that two random persons are similar while others treat them as dissimilar. If we want to formulate the similarity and dissimilarity by using mathematics, we need a tolerance relation (i.e. a reflexive and symmetric relation). If this relation holds for two objects, we can say that they are similar. If this relation does not hold, then they are dissimilar. This relation is reflexive because every object is similar to itself. It is also symmetric because if some object is similar to another one, then the second object is also similar to the first object. However, the transitivity does not hold necessarily. If we take a human and a mouse, then due to their inner structure they are similar. This is the reason why mice are used in drug experiments. A human and a Paris doll are also similar due to their shape. This is why these dolls are used in show-windows. Although a mouse and a doll are dissimilar (except that both are similar to the same object). Correlation clustering is a clustering technique based on a tolerance relation (see in [6,7,14]).

The task is to find an $R \subseteq V \times V$ equivalence relation which is *closest* to the tolerance relation. A (partial) tolerance relation $\mathcal{R}$ (see in [8,13]) can be represented by a matrix M. Let matrix $M = (m_{ij})$ be the matrix of the partial relation $\mathcal{R}$ of similarity: $m_{ij} = 1$ if objects i and j are similar, $m_{ij} = -1$ if objects i and j are dissimilar, and $m_{ij} = 0$ otherwise.

A relation is called partial if there exist two elements (i,j) such that $m_{ij} = 0$. It means that if we have an arbitrary relation $R \subseteq V \times V$ we have two sets of pairs.

Let R_{true} be the set of those pairs of elements for which the R holds and R_{false} be the one for which R does not hold. If R is partial, then $R_{true} \cup R_{false} \subseteq V \times V$. If R is total, then $R_{true} \cup R_{false} = V \times V$.

A partition of a set S is a function $p : S \to \mathbb{N}$. The object classes, defined by the partition, are called clusters. Objects $x, y \in S$ are in the same cluster at partitioning p, if $p(x) = p(y)$. We call the following two cases conflicts:

– Two dissimilar objects end up in the same cluster
– Two similar objects end up in different clusters

The cost function is the number of these conflicts. The formal definition can be seen in [9]. For a relation the partition with the minimal cost function value is called *optimal*. Solving a correlation clustering problem is equivalent to minimizing its cost function, for the fixed relation. If the cost function value is 0, the partition is called *perfect*. Given the $\mathcal{R}$ and R we call the value f the distance of the two relations. The partition given this way, generates an equivalence relation. This relation can be considered as the closest to the tolerance relation.

It is easy to check that we cannot necessarily find a perfect partition for an arbitrary similarity relation. In Fig. 1 we can see a very simple example for the problem. Take the relation on the left. The dashed line denotes dissimilarity and the normal line similarity. On the right, Fig. 1 shows all the possible partition of these objects, where rectangles indicate the clusters. The thick lines denote the pairs which are counted in the cost function. In the upper row the value of the cost function is 1 (in each case), while in the two other cases it is 2 and 3, respectively.

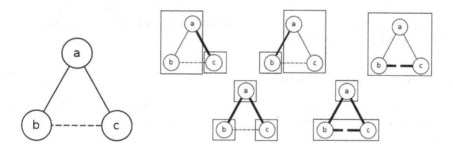

Fig. 1. Minimal frustrated similarity graph and its partitions

The number of partitions can be given by the Bell number (see in [1]), which grows exponentially. So the optimal partition cannot be determined in reasonable time. In a practical case a quasi optimal partition can be sufficient so a search algorithm can be used. We used an algorithm described in the next subsection.

4.1   Correlation Clustering by Contraction

We can define a *force* between objects based on a tolerance relation $\mathcal{R}$ as follows:

$$f_\mathcal{R}(i, S) = \sum_{j \in S} m_{ij}, \quad f_\mathcal{R}(R, S) = \sum_{i \in R} \sum_{j \in S} m_{ij}. \tag{1}$$

Based on the force $f_\mathcal{R}$ we can define two transformations of a partition:

- if $f_\mathcal{R}(R, S) > 0$, we can replace clusters R and S with cluster $R \cup S$ by *contracting them* into one cluster,
- if $f_\mathcal{R}(i, R) = \max_S f_\mathcal{R}(i, S)$ and $i \notin R$, then *move* object i from its cluster into cluster R.

We leave it to the reader to check that these two steps decrease the number of conflicts, so with them we can construct a greedy algorithm. This algorithm stops when we cannot apply either step to get to a better state.

The contraction method is just repeating these steps in the *right* order. We conducted many experiments to find the right order: The movement step alone is almost enough to generate a good partition. It groups the objects into several clusters but unfortunately this step is not able to join these clusters. If we have thousands of objects, then determining their most attractive cluster is a long task, although the process can be parallelized. In some rare cases, if we execute these movement steps in parallel, we could get into an infinite loop because some objects move back and forth between two clusters. If we only enable independent (i.e. no common cluster) movement steps, this problem disappears.

The contraction step is a big change, and—based on our experiments—it is not worth repeating, but worth following up with a movement step to liberate the objects which got into a worse relation with the contraction.

Different kinds of tolerance relations demand different variants of contraction methods (see in [5]).

4.2   Representative Member

We call a member representative if it is similar to most of the members and different from the least of the members in the same group. For any member m two values have been stored:

- α - the number of elements that are similar to m and are in the same group.
- β - the number of elements that are different from m and are in the same group.

Figure 2 shows a very simple example to the method. For the member A the two values are:

- $\alpha = 2$. Because there are two members (B and C) that are similar to A and are in the same group.
- $\beta = 2$. Because there are two members (F and E) that are different from A and belong in the same group.

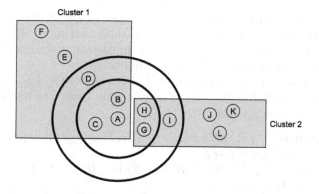

Fig. 2. α and β values for member A

In this example the similarity relation is based on the Euclidean distance of the objects. The smaller circle denotes the similarity threshold and the greater one denotes the difference threshold.

A member can be considered a possible representative if the following fraction is maximal:

$$r = \frac{\alpha^w - \beta^v}{\alpha + \beta + 1} \; v, w \in \mathbb{R}, v, w > 1, w > v \qquad (2)$$

The v and w are some weights. In our research we used 2 as both of their values.

For any group there can be more than one possible representative members. Although, only one member is chosen to be the actual representative. In this paper it is chosen randomly from the set of possible representative members.

5   Similarity Based Rough Sets with Annotation

Singleton clusters represent very little information because the system could not consider its member similar to any other objects without increasing the value of the cost function (see in Sect. 4). As they mean little information, we can leave them out. If we do not consider the singleton clusters, then we can generate partial system of base sets from the partition. Sometimes it can happen that an object does not belong to a cluster because the system could not consider it similar to any other objects based on the background information. This does not mean that this object is only similar to itself, but without proper information the system could not insert it into any cluster in order to decrease the number of conflicts. In medical applications it can occur that a patient has a similar disease as some other patients but has different data in the information system. In this case the search algorithm would consider this patient different from the others and so the patient does not belong to any non-singleton cluster. Although, a doctor or an expert could recognize that the patient could belong to a non-singleton cluster. The original partial system was defined by the correlation clustering. However, the user has some background knowledge. They can use this knowledge to help the system by inserting the members of some singletons into base sets (non-singleton clusters). With the help of the annotation process the user can put their own knowledge into the system. It also decreases the partiality by decreasing the number of singletons. After the annotation a new approximation space appears.

Let S be the set to be approximated, $\{x\}$ a singleton gained from the correlation clustering and B a base set. The following cases can happen with the base set B after the annotation if $B \subseteq \mathsf{l}(S)$:

- If $x \in S$, then $B' = \{x\} \cup B$ and $B' \subseteq \mathsf{l}(S)$ This way the approximation of the set S becomes more precise.
- If $x \notin S$, then $B' = \{x\} \cup B$ and $B' \subseteq \mathsf{u}(S)$ but $B' \not\subseteq \mathsf{l}(S)$ This increases the uncertainty relative to the set S.

The following cases can happen with the base set B after the annotation if $B \subseteq \mathsf{u}(S)$:

- If $x \in S$, then $B' = \{x\} \cup B$ and $B' \subseteq \mathsf{u}(S)$
- If $x \notin S$, then $B' = \{x\} \cup B$ and $B' \subseteq \mathsf{u}(S)$

The following cases can happen with the base set B after the annotation if $B \subseteq \mathsf{u}(S) \setminus \mathsf{l}(S)$:

- If $x \in S$, then $B' = \{x\} \cup B$ and $B' \subseteq \mathsf{u}(S) \setminus \mathsf{l}(S)$
- If $x \notin S$, then $B' = \{x\} \cup B$ and $B' \subseteq \mathsf{u}(S) \setminus \mathsf{l}(S)$

In both cases the upper approximation and the boundary region becomes larger. We can say that the annotation depends on the set to be approximated. It could be useful if:

- $x \in S$, then the user could only choose from those B base sets which are in $\mathsf{l}(S)$.
- $x \notin S$, then the user could only choose from those B base sets which are in $\mathsf{l}(\mathsf{u}(S)^c)$, where $\mathsf{u}(S)^c$ denotes the complement of the upper approximation.

This relative annotation looks very promising.

If there are more than one suitable base sets, then it can be useful if the user has some help to decide in which base set they should choose to put the member of a singleton into. The recommended base set is the one whose representative member is the most similar to the member of the given singleton. In this way, there is no need to compare it to each member of each base set.

The annotation process can be qualified as relevant or irrelevant regarding how it changes the representatives.

1. Relevant: After inserting a member of a singleton into a base set B, the representative member of the new base set B' is changed. In this case some real information is implemented into the system. Let us assume that the objects are members of political parties and the representative members are the leaders of these parties. The annotation process is when a new member is elected to a party. If the annotation is relevant, then it means that the balance of the party is changed, and a new leader is risen.

2. Irrelevant: After inserting a member of a singleton into a base set B, the representative member of the new base set B' is unchanged. In this case the implemented information is not relevant because it does not alter the base sets gained from the correlation clustering.

In either case the annotation can modify the set of possible representatives. As a conclusion we can say that, if after the annotation something was changed, then the user had some useful information which was not embedded in the similarity relation.

The order of the annotation is also worth to be checked. If we are to insert the members (O_1, O_2) of 2 different singletons into the same base set B, then the following question is needed to be answered. Is it still relevant to insert O_2 into B after putting O_1 into B?

– If the answer is yes, then the two members are interchangeable. This means that O_1, O_2 has some sort of similarity that was hidden in the similarity relation.
– If the answer is yes, then the two members are not interchangeable. This means that annotating O_1 makes it irrelevant to insert O_1 into B.

5.1   Dealing with Missing Data

In a real world application it can happen that an attribute value of an object is missing. This means that it can be unknown, unassigned or inapplicable (i.e. maiden name of a male). Coping with these data is usually a hard task. In many cases these values are often substituted. It is common to replace a missing value with the mean or the most frequent value. Typically this gives a rather good result in many situations. In early stage diabetes, it is not unusual that only the blood sugar level is higher than the normal level. If this value is missing for a patient, then it should not be replaced by the mean because the mean can be the normal blood sugar level. After the substitution this patient can be treated as a healthy one. This type of substitution does not consider the information of an object itself but the information of a collection of objects, therefore it can lead to a false conclusion. In this paper we propose another method to handle missing data. If an object has a missing attribute value, then it cannot be treated as similar to any other objects, so this entity forms a cluster alone. As mentioned earlier, these clusters cannot be treated as base sets. However, with the annotation the user has the possibility to decide whether an object with missing data is similar to other objects or not. The user has some background knowledge that can be used this way to cope with the missing values. In this case the information of an object itself is considered.

6 Program

The authors of this article wrote a program which helps us with the approximation and the annotation process. The software can be downloaded from: https://
github.com/lordimp88/NagyDavid. For giving the input datasets the user has
two options:

1. Generating random coordinate points
2. Reading continuous data from a file

1. *Random Points*
   The user gives the number of points, and then the points are generated in
   a 2 dimensional interval which is also given by the user In this option the
   base of the tolerance relation is the Euclidean distance of the objects (d).
   We defined a similarity (S) and a dissimilarity threshold (D). The tolerance
   relation $\mathcal{R}$ can be given this way for any objects O_1, O_2:

$$O_1 \mathcal{R} O_2 = \begin{cases} +1 & d(O_1, O_2) \leq S \\ -1 & d(O_1, O_2) > D \\ 0 & \text{otherwise} \end{cases} \tag{3}$$

2. *Continuous Data*
   Each row represents a single entity. In the software there is an option to
   normalize the data in the way described below. Let A be an attribute and v
   the value to be normalized. After the normalization:

$$v = \frac{v - min(A)}{max(A) - min(A)} \tag{4}$$

   The similarity is defined in two steps.
   (a) step: Let $A_1, A_2 \ldots A_n$ be the attributes, $t_1, t_2 \ldots t_n$ threshold values,
       O_1, O_2 two objects. Let $O_j(A_i)$ denote the attribute value of A_i for object
       O_j $(i = 1 \ldots n, j = 1, 2)$. If $\exists i \in \{1 \ldots n\} : |O_1(A_i) - O_2(A_i)| \geq t_i$, then
       the objects O_1 and O_2 are treated as different.
   (b) step: If the condition in the first step does not hold, then the tolerance
       relation $\mathcal{R}$ can be defined in the following way for any objects O_1, O_2
       using a similarity threshold S and a dissimilarity threshold D:

$$O_1 \mathcal{R} O_2 = \begin{cases} +1 & d(O_1, O_2) \leq S \\ -1 & d(O_1, O_2) > D \\ 0 & \text{otherwise} \end{cases} \tag{5}$$

   The d "distance" value is calculated for any objects O_1, O_2 by the following method:

$$d(O_1, O_2) = \sqrt{\sum_{i=1}^{n} (O_1(A_i) - O_2(A_i))^2} \tag{6}$$

The necessity of the first step can be explained by the following simple example. Let us assume that the objects are patients. It can happen that two patients are differ only in the blood pressure level and the other attribute values are relatively close to one another. So the distance between these two entities can be a small value. However, the patients cannot be treated as similar, because a high blood pressure level can indicate an illness. This fact remains hidden without the first step, because the similarity value can be small for the two patients. The same holds for normalized data.

After getting the input points the software runs a search algorithm which finds a quasi optimal partition. This algorithm is described in Subsect. 4.1. As mentioned earlier, the singleton clusters mean little information, so the software leaves them out and creates the system of base sets. After defining the base sets, the user can select a set of points for approximation.

6.1 Annotation

In the software the user has the option to insert the members of the left-out singleton clusters to any base set. Two singleton clusters cannot be merged together due to the similarity relation (their members are different). We mentioned earlier that there are two types of singletons:

- Its member is different from most of the objects so it forms a cluster alone.
- Due to the background knowledge the system decided that this object cannot be a member of any other group.

The software does not examine for a singleton which type it belongs, so there is no mandatory annotation for a singleton. It is up to the user to decide.

6.2 The Output of the Software

In this subsection we show a possible output generated by the software. In the following figures 20 points can be seen. The similarity relation is based on the

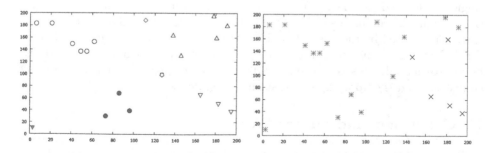

Fig. 3. Clusters (left) and the set to be approximated (right)

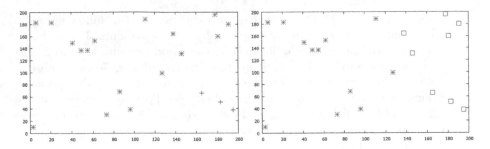

Fig. 4. The lower (left) and upper (right) approximation by clustering

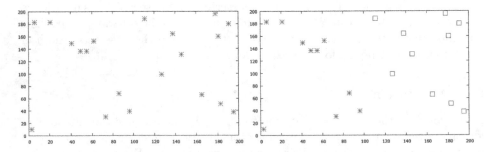

Fig. 5. The lower (left) and upper (right) approximation by clustering with annotation

Euclidean distance of the objects. Of course the software is capable of handling more points, but for better visibility only 20 points were used. The similarity threshold S was set to 50, and D was set to 90. In the left side of Fig. 3 the clusters generated by the correlation clustering can be seen. The singleton clusters contain the objects denoted by: the ◇ symbol, the △ symbol and the ▼ symbol. Some points were selected for approximation. The members of this set are denoted by the × symbols, and the other members are denoted by the star symbol. The members were chosen randomly. This set can be seen in the right side of Fig. 3. In Fig. 4 the reader can see the lower and upper approximation defined by the base sets gained from clustering after leaving the singletons out. The members of two singletons were inserted into two different base sets. The singleton denoted by the ◇ symbol was merged with the base set denoted by the △ symbol. The base set denoted by the ▽ symbol was extended with the singleton denoted by the △ symbol. The result of the annotation can be seen in Fig. 5. None of the members of the chosen singletons were members of the set to be approximated. This is the reason why the lower approximation became the empty set, and the upper approximation had more members.

7   Conclusion and Future Work

In [9] the authors introduced a partial approximation space relying on a similarity relation (a tolerance relation technically). The genuine novelty of approximation spaces is the systems of base sets: it is the result of correlation clustering, and so similarity is taken into consideration generally. Singleton clusters have no real information in approximation process, these clusters cannot be taken as base sets, therefore the approximation spaces are partial in general cases (the unions of base sets are proper subsets of universes.) In the present paper a new possibility appears in order to embed some information into the approximation spaces: a user may decide the status of a member of a singleton cluster: it can be put into a base set, and the approximation of a set changes according to the new system of base sets. This possibility is crucial in practical applications. The next step is to give up the pairwise disjoint property of base sets in the annotation process. This possibility helps a user a lot to make a decision about a member of singleton cluster: it may belong to more than one base sets, and so the user's decision is not so sharp. Another step to make in the near future is the investigation of influences of a similarity relation on valid logical consequences in a logical system relying on similarity based rough sets with or without annotation.

Acknowledgement. This work was supported by the construction EFOP-3.6.3-VEKOP-16-2017-00002. The project was co-financed by the Hungarian Government and the European Social Fund.

References

1. Aigner, M.: Enumeration via ballot numbers. Discrete Math. **308**(12), 2544–2563 (2008). http://www.sciencedirect.com/science/article/pii/S0012365X07004542
2. Aszalós, L., Mihálydeák, T.: Rough clustering generated by correlation clustering. In: Ciucci, D., Inuiguchi, M., Yao, Y., Ślęzak, D., Wang, G. (eds.) Rough Sets, Fuzzy Sets, Data Mining, and Granular Computing. RSFDGrC 2013. LNCS, vol. 8170, pp. 315–324. Springer, Heidelberg (2013). https://doi.org/10.1007/978-3-642-41218-9_34
3. Aszalós, L., Mihálydeák, T.: Rough classification based on correlation clustering. In: Miao D., Pedrycz W., Ślęzak, D., Peters, G., Hu, Q., Wang, R. (eds.) Rough Sets and Knowledge Technology. RSKT 2014. LNCS, vol. 8818, pp. 399–410. Springer, Cham (2014). https://doi.org/10.1007/978-3-319-11740-9_37
4. Aszalós, L., Mihálydeák, T.: Correlation clustering by contraction. In: 2015 Federated Conference on Computer Science and Information Systems (FedCSIS), pp. 425–434. IEEE (2015)
5. Aszalós, L., Mihálydeák, T.: Correlation clustering by contraction, a more effective method. In: Fidanova, S. (ed.) Recent Advances in Computational Optimization. SCI, vol. 655, pp. 81–95. Springer, Cham (2016). https://doi.org/10.1007/978-3-319-40132-4_6
6. Bansal, N., Blum, A., Chawla, S.: Correlation clustering. Mach. Learn. **56**(1–3), 89–113 (2004)
7. Becker, H.: A survey of correlation clustering. In: Advanced Topics in Computational Learning Theory, pp. 1–10 (2005)
8. Mani, A.: Choice inclusive general rough semantics. Inf. Sci. **181**(6), 1097–1115 (2011)
9. Nagy, D., Mihálydeák, T., Aszalós, L.: Similarity based rough sets. In: Polkowski, L. (ed.) Rough Sets. LNCS, vol. 10314, pp. 94–107. Springer, Cham (2017). https://doi.org/10.1007/978-3-319-60840-2_7

10. Pawlak, Z.: Rough sets. Int. J. Parallel Program. **11**(5), 341–356 (1982)
11. Pawlak, Z., Skowron, A.: Rudiments of rough sets. Inf. Sci. **177**(1), 3–27 (2007)
12. Pawlak, Z., et al.: Rough sets: theoretical aspects of reasoning about data. In: System Theory, Knowledge Engineering and Problem Solving, vol. 9. Kluwer Academic Publishers, Dordrecht (1991)
13. Skowron, A., Stepaniuk, J.: Tolerance approximation spaces. Fundamenta Informaticae **27**(2), 245–253 (1996)
14. Zimek, A.: Correlation clustering. ACM SIGKDD Explor. Newslett. **11**(1), 53–54 (2009)

Multidimensional Data Analysis for Evaluating the Natural and Anthropogenic Safety (in the Case of Krasnoyarsk Territory)

Tatiana Penkova[✉]

Institute of Computational Modelling of the Siberian Branch of the Russian
Academy of Sciences, Siberian Federal University, Krasnoyarsk, Russia
penkova_t@icm.krasn.ru

Abstract. This paper presents an approach to evaluating the natural and technogenic safety of the one of the largest regions in Siberia through the comprehensive analysis of territorial indicators. In order to explore geographical variations and patterns in occurrence of emergencies the multidimensional data analysis technique is applied to data of the Territory Safety Passports. For data modeling, principal components are selected and interpreted taking account of the contribution of the data attributes to the principal components. Data distribution on the principal components is analyzed at different levels of the territory detail: municipal areas and settlements. The results of this analysis have allowed to identify the high-risk areas and rank the territories according to danger degree of occurrence of the natural and technogenic emergencies. It gives the basis for decision making and makes it possible for authorities to allocate the forces and means for territory protection more efficiently and develop a system of measures to prevent and mitigate the consequences of emergencies in the large region.

Keywords: Multidimensional data analysis · Principal component analysis
Evaluating the natural and anthropogenic safety · Prevention of emergencies
Territorial management

1 Introduction

Prevention of natural and technogenic emergencies is a one of the major tasks of the territory management. Analytical support of decision-making processes based on modern technologies and efficient methods of data analysis is a necessary condition for improving the territorial safety system and management quality.

The Krasnoyarsk territory is the second largest federal subject of Russia and the third largest subnational governing body by area in the world. The Krasnoyarsk region lies in the middle of Siberia and occupies an area of 2,339,700 km^2, which is 13% of the country's total territory. This territory is characterised by heightened level of natural and technogenic emergencies which is determined by social-economic aspects, large resource potential, geographical location and climatic conditions. In the territory there are many accident prone technosphere objects including radiation-related objects, chemically-dangerous objects, fire-hazardous and dangerously explosive objects; hydraulic facilities; critically important objects; a lot of survival objects including

H. S. Nguyen et al. (Eds.): IJCRS 2018, LNAI 11103, pp. 101–109, 2018.
https://doi.org/10.1007/978-3-319-99368-3_8

boiler plants, power plants, pipelines and networks. Moreover, the territory is located in seven climatic zones. A number of large-scale natural emergencies, such as flood, forest fire, gale-strength wind and anomalously low temperature are recorded each year [1]. In order to improve the population and territory safety, a lot of monitoring systems and control tools for on-line observation are being actively introduced within the region [2–4]. The Ministry of Emergency has enacted the structure and order of conducting the Territory Safety Passport, which defines a system of indicators to assess the state of territory safety, the risk of emergencies and possible damages to create efficient prevention and mitigation actions [5]. At present, there are massive data collections about the state of controlled objects, occurred events and sources of emergencies. However, we have to admit that the processing stored data, aimed at obtaining the new and useful knowledge, is insufficient. The local databases remain unused, while the reasonable decisions, comprehensive analysis and emergencies prediction are sorely needed. Thus, identification of risk factors of emergencies based on monitoring data and investigation of their impact on key indicators of human safety are topical and important tasks in territorial management.

Data mining techniques provide the effective tool for discovering previously unknown, nontrivial, practically useful and interpreted knowledge needed to make decisions [6]. This paper presents the results of comprehensive multidimensional analysis of natural and technogenic safety indicators of the Krasnoyarsk territory in order to explore geographical variations and patterns in occurrence of emergencies by applying the data mining technique – principal component analysis – to data of the Territory Safety Passports.

The outline of this paper is as follows: Sect. 1 contains introduction. Section 2 describes the initial data. Section 3 presents results of principal component analysis: identification and interpretation of principal components; analysis of data distribution on the principal components at different levels of the territory detail. Section 4 draws the conclusion.

2   Data Description

Evaluating the natural and technogenic safety indicators is based on data of the Territory Safety Passports of the Krasnoyarsk territory collected in Center of Emergency Monitoring and Prediction (CEMP). Original dataset contains 1,690 objects, essentially discrete settlements-level geographical entities of the Krasnoyarsk territory, each with 12 measured attributes. Data attributes are listed in Table 1. One part of attributes characterizes the sensitivity of the territory to the risk factors effects (e.g. population density, the presence of industrial and engineering facilities) that is determined by the number of objects located on the territory (i.e. number of potential sources of emergencies), it is so-called "object attributes". The other part of attributes characterizes the presence of potential factor that can damage the health of people, can cause irreversible damage to the environment that is determined by the statistic of events occurred in the territory (i.e. number of emergencies), it is so-called "event attributes". In addition, some reference characteristics are used for data interpretation and map visualization. The preliminary correlation analysis of original data has shown a fairly strong relationship between "object" and "event" attributes, therefore for further analysis we will consider the attributes that characterize population and events. The correlation coefficients are presented in Table 2.

Table 1. List of the data attributes of Territory Safety Passports

No	Attributes	Description
1	Pop	Population
2	Soc_object	Number of important social facilities (e.g. educational, health, social, cultural and sports facilities)
3	Water_object	Number of dangerous water bodies
4	Indust_object	Number of potentially dangerous industrial objects (e.g. plants, factories, mines)
5	Oil_line	Number of pipeline sectors in 5 km radius from borders of settlement
6	Munic_object	Number of municipal facilities (e.g. power supply, water supply and heating facilities)
7	Flood_event	Number of floods
8	NFire_event	Number of natural fires
9	TFire_event	Number of technogenic fires
10	Munic_event	Number of accidents at municipal facilities
11	Nat_event	Number of natural events (excluding natural fires and floods)
12	Tech_event	Number of technogenic events (excluding technogenic fires and accidents at municipal facilities)

Table 2. Correlation coefficients between data attributes

No	2	3	4	5	6	7	8	9	10	11	12
1	**0.97**	0.39	**0.96**	0.04	0.28	0.29	0.08	**0.96**	**0.95**	0.08	**0.60**
2		0.36	**0.96**	0.01	0.25	0.25	0.05	**0.91**	**0.94**	0.06	**0.59**
3			0.39	−0.01	0.32	**0.60**	0.12	0.39	0.36	0.17	0.30
4				0.01	0.24	0.29	0.05	**0.91**	**0.91**	0.07	**0.56**
5					0.08	−0.02	0.06	0.07	0.02	0.05	0.14
6						0.29	0.08	0.31	0.43	0.13	0.48
7							0.06	0.33	0.30	0.13	0.28
8								0.10	0.06	−0.02	0.05
9									**0.93**	0.11	**0.63**
10										0.08	**0.58**
11											0.13

Within this research, the analysis and visualisation of multidimensional data are conducted using the ViDaExpert [7]. Data visualization on geographical maps is performed by applying the mapping tools «ArcGIS» [8].

3 Principal Component Analysis

Principal Component Analysis (PCA) is one of the most common techniques used to describe patterns of variation within a multi-dimensional dataset, and is one of the simplest and robust ways of doing dimensionality reduction. PCA is a mathematical procedure that uses an orthogonal transformation to convert a set of observations of possibly correlated variables into a set of values of linearly uncorrelated variables called principal components [9]. The number of principal components is always less than or equal to the number of original variables. This transformation is defined in such a way that the first principal component has the largest possible variance and each subsequent component, respectively, has the highest variance possible under the constraint that it is orthogonal to the preceding components.

3.1 Contribution of the Data Attributes to the Principal Components

One of the greatest challenges in providing a meaningful interpretation of multi-dimensional data using PCA is determining the number of principal components. In general, the method allows to identify k components based on k initial attributes. Table 3 shows the results of calculating the eigenvectors of the covariance matrix arranged in order of descending eigenvalues.

Table 3. Results of principal components calculation

Components	1	2	3	4	5	6	7
Eigenvalues	0.404	0.249	0.141	0.116	0.075	0.010	0.005
Accumulated dispersion	**0.504**	**0.652**	0.793	0.909	0.985	0.995	1
Pop	**0.509**	0.109	0.111	0.113	0.227	0.182	0.787
TFire_event	**0.513**	0.083	0.061	0.088	0.171	0.616	−0.557
NFire_event	0.060	**0.439**	−0.876	0.186	−0.022	−0.033	0.012
Munic_event	**0.503**	0.096	0.120	0.084	0.251	−0.764	−0.263
Flood_event	0.235	−0.314	−0.325	−0.853	0.109	−0.004	0.029
Nat_event	0.086	**−0.822**	−0.311	0.458	0.103	−0.015	0.010
Tech_event	**0.397**	−0.072	0.019	0.013	−0.913	−0.051	0.024

Based on combination of Kaiser's rule and the Broken-stick model [10], two principal components for data attributes were identified (PC1 and PC2) with 65% accumulated dispersion. Figure 1(a) illustrates the eigenvalues of components. As can be seen from Fig. 1(a), Kaiser's rule determines two principal components – eigenvalues of first two components are significantly greater than the average value and the Broken-stick model gives also two principal components – the line of Broken-stick model also cuts the eigenvalues of first two components. The contribution of the data attributes to principal components is presented in Fig. 1(b).

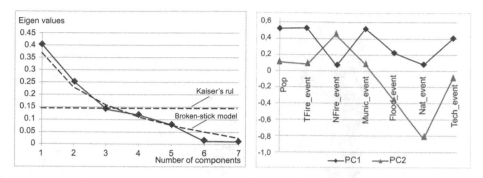

Fig. 1. (a) Eigenvalues of components. (b) Contribution of the data attributes to the first (PC1) and second (PC2) principal components

From Fig. 1(b) we can see that the first principal component (PC1) is characterised by the following attributes: a high level of population, high proportions of technogenic fires, accidents at municipal facilities and other technogenic events, a low percentage of natural events including natural fires and floods. In combination, these characteristics present the big settlements (e.g. cities) with high levels of technogenic hazards. The second principal component (PC2) is characterised by the following attributes: a low level of population, high proportion of natural fires, strong negative correlation with the percentage of natural events including floods and technogenic events including fires and accidents at municipal facilities. In combination, these characteristics present relatively small settlements (e.g. villages) with high levels of natural fires. This means that in comparison with other types of emergencies the technogenic and natural fires are the greatest threat for the Krasnoyarsk territory.

3.2   Data Distribution on the Principal Components

The data can be divided into groups according to where the settlements are located in terms of Territory Classifier. There are three levels of the territory detail: settlements, municipal areas and groups of municipal areas that give 1,690 objects, 65 objects and 8 objects respectively for the Krasnoyarsk territory. Figure 2 shows the visualisation of territorial groups (groups of municipal areas) on the geographic coordinates and the PCA plot, where: group 1 (green) – Angarsk Group; group 2 (rose) – Eastern Group; group 3 (purple) – Yeniseisk Group; group 4 (light blue) – Western Group; group 5 (yellow) – Central Group; group 6 (red) – Southern Group; group 7 (blue) – Taymyr Autonomous Okrug; group 8 (brown) – Evenk Autonomous Okrug. On a data map, the points in the form of triangles are settlements, and the color of these points corresponds to the color of the territorial group. Objects in the form of circles represent centroids of clusters of territorial groups.

As can be seen from Fig. 2, along the first principal component (PC1) the territorial groups are concentrated quite densely, it means that technogenic fires are general characteristic for all territorial groups of region, but along the second principal component (PC2) the territorial groups are distributed significantly and we can see that the natural fires are indicative of northern territorial groups.

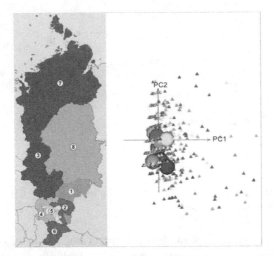

Fig. 2. Visualisation of territorial groups on the geographic map and the PCA plot (Color figure online)

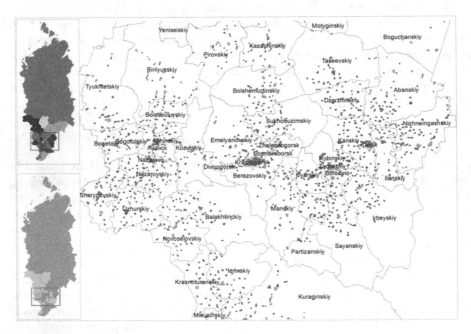

Fig. 3. Visualisation of the projections on the first principal component for municipal areas and settlements (Color figure online)

The visualisation of the projections on the first and second principal components on the geographic map is displayed in Figs. 3 and 4. On these figures, the negative values in range [−1, 0] correspond to Group 1 (blue), the positive values in range (0; 0.5] correspond to Group 2 (green) and the highest positive values in range (0.5; 1] correspond to Group 3 (red). The color intensity of municipal areas corresponds to the number of settlements in the group.

The lowest values of projections on the first principal component (Fig. 3, blue points) are observed for such settlements as: Ust-Kamo, Shigashet, Kasovo, Verhnekemskoe, Komorowskiy, Angutiha, Lebed. It can be explained by the fact that these settlements are very small villages and, at present, in these settlements there are no any socially significant objects and residents. The complete absence of the economic activity in these settlements leads to the lowest level (or absence) of technogenic fires. The highest values of the projections on the first principal component (Fig. 3, red points) are observed for such large settlements as Krasnoyarsk, Norilsk, Achinsk, Kansk, Minusinsk Lesosibirsk. These settlements present the big cities of the Krasnoyarsk territory where the population and number of socially significant and industrial facilities are above average level in region.

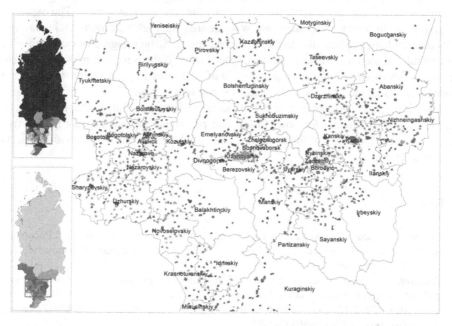

Fig. 4. Visualisation of the projections on the second principal component for municipal areas and settlements (Color figure online)

The lowest values of projections for the second principal component (Fig. 4, blue points) are observed for such settlements as: Turuhansk, Cheremshanka, Tanzybey, Emelyanovo, Ermakovskoe. Low levels of natural fires can be explained by the following facts: the absence of vegetation as a source of emergency in steppe areas (e.g. Western and Southern groups) and the absence of settlements in forest zone (e.g. Evenk Autonomous Okrug, Yeniseiysk and Turukhansky areas). The highest values of projections for the second principal component (Fig. 4, red points) are observed for such settlements as: Startsevo, Tilichet, Kuray, Baikal, Glinniy. The high risk of natural fires is observed in the large settlements that are located close to the forest zones. In addition, there is probability of natural fires in the big cities where the forests constitute the part of their territories.

4   Conclusion

In this paper the evaluating of natural and technogenic safety of the Krasnoyarsk territory in the context of settlements is carried out first time by applying the multidimensional data analysis technique – principal component analysis – to data of the Territory Safety Passports. The data analysis results show that the technogenic and natural fires are the greatest threat for territory of the Krasnoyarsk region. The explored geographical variations and patterns allow to identify the high-risk municipal areas and particular settlements, rank the territories according to danger degree of occurrence of the natural and technogenic emergencies. The results of this research make it possible for specialists of CEMP to develop a system of measures to prevent and mitigate the consequences of emergencies in the Krasnoyarsk territory.

The techniques and tools used in this paper make it easy to change the initial dataset (e.g. territories or threats) for other tasks. The presented approach to comprehensive multidimensional analysis of the territories can be adopted for different control objects in various areas.

References

1. The State of Natural and Anthropogenic Emergencies Protection of Territory and Population in the Krasnoyarsk Region: Annual Report of Ministry of Emergency, Krasnoyarsk, 230 p. (2016) (in Russian)
2. Penkova, T., Nicheporchuk, V., Metus, A.: Comprehensive operational control of the natural and anthropogenic territory safety based on analytical indicators. In: Polkowski, L., et al. (eds.) IJCRS 2017. LNCS (LNAI), vol. 10313, pp. 263–270. Springer, Cham (2017). https://doi.org/10.1007/978-3-319-60837-2_22
3. Shaparev, N.Y.: Environmental monitoring of the krasnoyarsk region in terms of sustainable environmental management. Inf. Anal. Bullet. (Sci. Tech. J.) **18**(12), 110–113 (2009) (in Russian)
4. Bryukhanova, E.A., Kobalinskiy, M.V., Shishatskiy, N.G., Sibgatulin, V.G.: Improvement of environmental monitoring information maintenance as an instrument for sustainable social and economic development (in the case of Krasnoyarsk Region). Inf. Commun. **1**, 43–47 (2014) (in Russian)

5. The Standard Territory Passport of Regions and Municipal Areas: The Regulation of Ministry of Emergency, No. 484, 25/10/2004 (in Russian)
6. Williams, G.J., Simoff, S.J. (eds.): Data Mining. LNCS (LNAI), vol. 3755, p. 329. Springer, Heidelberg (2006). https://doi.org/10.1007/11677437
7. Gorban, A., Pitenko, A., Zinovyev, A.: ViDaExpert: User-Friendly Tool for Nonlinear Visualization and Analysis of Multidimensional Vectorial Data. Cornell University Library. http://arxiv.org/abs/1406.5550
8. Using ArcViewGIS: The Geographic Information System of Everyone, 350 p. ESRI Press (1996)
9. Abdi, H., Williams, L.: Principal components analysis. Comput. Stat. **2**(4), 439–459 (2010)
10. Peres-Neto, P., Jackson, D., Somers, K.: How many principal components? Stopping rules for determining the number of non-trivial axes revisited. Comput. Stat. Data Anal. **49**(4), 974–997 (2005)

A Metaphor for Rough Set Theory: Modular Arithmetic

Marcin Wolski[1(✉)] and Anna Gomolińska[2]

[1] Department of Logic and Cognitive Science, Maria Curie-Skłodowska University,
Maria Curie-Skłodowska Sq. 4, 20-031 Lublin, Poland
`marcin.wolski@umcs.lublin.pl`
[2] Faculty of Mathematics and Informatics, University of Białystok,
Konstantego Ciołkowskiego 1M, 15-245 Białystok, Poland
`anna.gom@math.uwb.edu.pl`

Abstract. Technically put, a metaphor is a conceptual mapping between two domains, which allows one to better understand the target domain; as Lakoff and Núñes put it, the main function of a metaphor is *to allow us to reason about relatively abstract domains using the inferential structure of relatively concrete domains*. In the paper we would like to apply this idea of framing one domain through conceptual settings of another domain to rough set theory (RST). The main goal is to construe rough sets in terms of the following mathematical metaphor: *RST is a modular set-arithmetic*. That is, we would like to map/project modular arithmetic onto rough sets, and, as a consequence, to redefine the fundamental concepts/objects of RST. Specifically, we introduce new topological operators (which play a similar role as remainders in modular arithmetic), discuss their formal properties, and finally apply them to the problem of vagueness (which has been intertwined with RST since the 1980's).

Keywords: Rough set · Modular arithmetic · Remainder · Topolgy
Boundary · Vagueness

1 Introduction

Metaphors, as ambiguous as they are, have often provided us with deep insights into many fields of human activity; starting from very abstract theological problems of Trinity (e.g., the Tertullian's mataphor of *Sun*: Godfather is the star itself, Jesus is the light, and the Holy Spirit is the heat), to modern problems of cognitive science (e.g., the famous computer metaphor which has been dominating in the last 40 years in cognitive psychology). Technically speaking, a metaphor is a conceptual mapping between two domains, which allows one to better understand the target domain. Or, better still, as Lakoff and Núñes [6] put it: the main function of a metaphor is *to allow us to reason about relatively abstract domains using the inferential structure of relatively concrete domains.*

© Springer Nature Switzerland AG 2018
H. S. Nguyen et al. (Eds.): IJCRS 2018, LNAI 11103, pp. 110–122, 2018.
https://doi.org/10.1007/978-3-319-99368-3_9

E.g., the Sun is mapped on the Trinity, allowing one to concretely frame a seriously abstruse idea, or a computer is mapped on a human brain allowing one to frame how it functions.

A little bit more problematic is the role of metaphor in mathematics. The most vivid example seems to be the metaphor of *Divine Intellect*, which, although often very implicit, allowed most of mathematicians to (finally) accept the realm of infinite sets and non-constructive mathematics. But, as emphasised by the opponents, this framing is highly theological – e.g., the book *The Ghost in Turing's Machine. Taking God Out of Mathematics and Putting the Body Back In*, in which Rotman fights against the Platonism, as a way of framing mathematics.[1] In computer science the most well-known examples are given by *liquid*: e.g., the *flow* metaphor, which is the source of *information flow*, *memory leaks*, or the *law of conservation of memory*.

In the present paper we would like to apply the idea of a metaphor to rough set theory (RST). However, following suggestions by Lakoff and Núñes, we would like to do this in a *relatively concrete* way. As noted above, usually metaphors allow us merely to frame or conceptualise some very abstract ideas, bringing no concrete results. Yet, sometimes we are able to make one step further and materialise a given metaphor: e.g., the *liquid* metaphor has been *embodied* in computer science as a *liquid state machine* (LST). In the present paper we would like to follow this path, and apart from the conceptualisation/framing we would also like to deal with some *materialisation* of modular arithmetic within the conceptual body of RST. Under this view, we are interested in the set counterpart of remainders, which serve in modular arithmetic as the standard representatives of congruence classes. That is, we are going to enrich RST with new (topological) operators, the *remainder* **r** and *deficit* **d**, which assign to a given set some kind of remainders with respect to/modulo the underlying granularity of the universe.

Anyway, our hope is that the number metaphor will shed new light on the foundations of rough sets. More specifically, we shall address the problem of vague concepts, which have been intertwined with rough sets from the very beginning of this theory (the early 80s), mainly due to the existence of borderline cases. The second problem which we are going to address is the very nature/characteristic of a rough set itself.

Since the paper is the very first step in our project of redefining RST (as a kind of arithmetic), we cannot offer – apart from methodological considerations about vagueness – any discussion of (future) applications. Our next step is likely to focus upon the set remainder **r** and examine it against the background of another arithmetic system. We believe that this half of the boundary will finally lead us to some new results being both theoretically interesting and applicable.

2   Mathematical Preliminaries

In this section we shall recall basic definitions from rough set theory and modular arithmetic. We start with rough set theory, the motivations hidden under the

---

[1] A very interesting discussion of these problems may be found in Krajewski [5].

hood, and the methodological consequences of the original (Pawlak's) definitions. Then we shortly recall modular arithmetic. In the next section we shall use this arithmetic as a metaphor for rough sets.

2.1 Rough Set Theory

Let us start with the methodological assumptions staying behind rough set theory; as Pawlak observed [10]:

In the rough set approach vagueness is due to lack of information about some elements of the universe. If with some elements the same information is associated, in view of this information these elements are indiscernible. [...] It turns out that indiscernibility leads to the boundary-line cases, i.e., in view of the available information some elements cannot be classified to the concept or its complement and thus they form boundary-line cases.

The indiscernibility relation $E \subseteq U \times U$ between the elements of the universe U leads to the fundamental structures and operators of rough set theory. The detailed and extensive presentation of rough sets may be found in [9].

Definition 1 (Approximation Space). *A pair (U, E), where U is a nonempty set and E is an equivalence relation on U, is called an* approximation space. *A subset $X \subseteq U$ is called* definable *if $X = \bigcup \mathcal{Y}$ for some $\mathcal{Y} \subseteq U/E$, where U/E is the family of equivalence classes of E (the quotient set of E).*

As is well known, each equivalence relation E determines a partition U/E of the universe U, which is usually interpreted as a classification of objects (of course, each object x may be classified only to one equivalence class $[x]_E$). According to Z. Pawlak, *knowledge about a specific domain is construed as a classification of its elements* [10]. Thus, an approximation space expresses the information/knowledge encoded by the underlying information system. Any subset $X \subseteq U$ is called a *concept*, U/E is called a *knowledge basis*, and concepts build up from elements of the knowledge basis are called *definable concepts* or *exact concepts* (the set of all definable concept is denoted by $\mathcal{D}$). Since definable (exact) concept are supposed to form some algebraic structure (e.g., a topology or an algebra), usually the empty set $\emptyset$ is added to the knowledge basis. In the paper we always assume that $\emptyset \in \mathcal{D}$. An undefinable (not exact) concept is then approximated by a pair of exact concepts:

Definition 2 (Approximation Operators). *Let (U, E) be an approximation space. For every concept $X \subseteq U$, its E-lower and E-upper approximations are defined as follows, respectively:*

$$\underline{X} = \{a \in U : [a]_E \subseteq X\},$$

$$\overline{X} = \{a \in U : [a]_E \cap X \neq \emptyset\}.$$

By the usual abuse of language and notation, the operator $\underline{} : \mathcal{P}(U) \to \mathcal{P}(U)$ sending X to $\underline{X}$ will be called the *lower approximation operator*, whereas the operator $\overline{}: \mathcal{P}(U) \to \mathcal{P}(U)$ sending X to $\overline{X}$ will be called the *upper approximation operator*. Of course, U/E gives rise – as a base – to a topological space (U, τ_E), whose interior operator Int is $\underline{}$ and closure operator Cl is $\overline{}$. Therefore we obtain the standard Kuratowski axioms valid for rough approximations (we restrict our attention only to these axioms which are relevant to our study in the next section).

Proposition 1. *For every subset X of an approximation (U, E) space it holds:*

1. $\overline{\emptyset} = \emptyset$,
2. $\overline{X \cup Y} = \overline{X} \cup \overline{Y}$,
3. $\overline{\overline{X}} = \overline{X}$.

Proposition 2. *For every subset X of an approximation (U, E) space it holds:*

1. $\underline{\emptyset} = \emptyset$,
2. $\underline{X \cap Y} = \underline{X} \cap \underline{Y}$,
3. $\underline{\underline{X}} = \underline{X}$.

In this paper a *rough set* is defined as a pair $(\underline{X}, \overline{X})$, for some $X \subseteq U$; as a consequence a definable set is also a rough set. It may seem (philosophically) unintuitive, however it is necessary due to mathematical reasons – otherwise rough sets would not form any interesting structure. An alternative and equally popular approach is to define a rough set as an equivalence class of the rough equality relation $\equiv_E \subseteq \mathcal{P}(U) \times \mathcal{P}(U)$ defined by: $X \equiv_E Y$ iff $\underline{X} = \underline{Y}$ and $\overline{X} = \overline{Y}$. This definition is much more philosophically justified, but mathematically inconvenient.

Definition 3 (Representations of Rough Sets). *For an approximation space (U, E) and $X \subseteq U$, a pair $(\underline{X}, \overline{X})$ is called an* increasing representation *of X, whereas a pair $(\underline{X}, U \setminus \overline{X})$ is called a* disjoint representation *of X.*

The set $U \setminus \overline{X}$ is often called an *exterior of X* and denoted by $Ext(X)$, whereas $b(X) = \overline{X} \setminus \underline{X}$ is the boundary region of X. Of course, the choice of representation depends on a context of application. In the context of modal systems the increasing representation is more useful. On the other hand, in the context of abstract algebras the disjoint representation is more preferable.

However, there is a *snake in the garden*. As Marek and Truszczyński explains [7]:

The emphasis on the set X present in the original definition of rough sets is what we strive here to free ourselves from. After all, in most (if not all) applications set X we want to reason about is unknown or incompletely specified.

Or, better still, following Chakraborty [3], one may ask: *If X is already known why to approximate at all?*[2]

---

[2] Although Chakraborty's question makes perfect sense for abstract approximation spaces, the case of decision tables is a bit different: here the set X represents a decision attribute, which – although well known – still needs to be approximated by means of conditional attributes.

2.2   Modular Arithmetic

Although elementary modular arithmetic needs no introduction, we present here some basic information, at least in order to establish the notation. The detailed exposition of modular arithmetic may be found in [2]. Let us start with the most fundamental definition by Gauss, given in his *Disquisitiones Arithmeticae* (Arithmetic Investigations).

Definition 4 (Equivalence Modulo). *Let $\mathbb{Z}$ denote the set of integers and m be an integer. Then for a, b in $\mathbb{Z}$ we write*

$$a \equiv b \mod m$$

which reads "a is equivalent b modulo m", if $m|(a-b)$, where $|$ stands for the divisibility relation.

The parameter m is called *modulus*. Usually we employ the standard representation of integers modulo m defined in terms of remainders (since modular arithmetic is regarded in the paper merely as a metaphor, we are going to use the simplified version of this theorem).

Proposition 3. *Let $0 < m$ be a non-zero positive integer. Then for each $a \in \mathbb{Z}$ there exists a unique remainder r such that $a \equiv r \mod m$ and $r < m$.*

For this reason, we often use mod as an operator taking an arithmetic term t

$$t \bmod m,$$

and returning the corresponding reminder r; e.g.,

$$(5 + 2) \bmod 4 \text{ is } 3.$$

The remainders (i.e., $0, 1, 2, \ldots, m-1$), are called *standard representatives* for integers modulo m. Actually, each standard representative n stands for the equivalence class $\bar{n}$ (called *residue class*) of integers which are equivalent to $n \bmod m$; e.g. for $m = 4$, the representative 3 stands for the class $\bar{3} = \{\ldots, -5, -1, 3, 7, 11, \ldots\}$. The set of all congruence classes (or, alternatively, standard representatives) of the integers for a modulus m is usually called the *ring of integers modulo m*, denoted by $\mathbb{Z}/m$, which it actually forms when equipped with the following operations:

$$\bar{a} + \bar{b} = \overline{a+b},$$

$$\bar{a} - \bar{b} = \overline{a-b},$$

$$\bar{a} * \bar{b} = \overline{a*b},$$

where $\bar{a}$ stands for the residue class. Let us also recall that the ring forms an abelian group under addition $+$, and a monoid under multiplication $*$, where multiplication has to distribute over addition; i.e.,

$$a * (b + c) = (a * b) + (a * c).$$

The identity elements for $+$ and $*$ are denoted 0 and 1, respectively. If the multiplication is commutative, i.e.

$$a * b = b * a,$$

then the ring is called *commutative*.

3   The Metaphor of Modular Arithmetic

In this section we "project" modular arithmetic onto RST, that is, our aim is to formalise some ideas from this arithmetic within the RST frame. Since the full projection is not possible, modular arithmetic may be used here merely as a metaphor: e.g., *rough set theory is a modular set arithmetic*. That is, RST resembles modular arithmetic, and this similarity allows us to reinterpret and redefine some concepts and assumptions laying behind RST. However, we are not able to retrieve all concepts introduced in the previous section; specifically, we are not going to build a ring of residue classes (which is not compatible with RST), yet we use some Boolean ring machinery. The main emphasis in this section is put upon the standard representation of integers modulo m and it's RST counterpart.

As is well known, an approximation space (U, E) may be conceptualised also as a topological space (U, τ_E), whose closure operator Cl is the upper approximation, and interior operator Int is the lower approximation. All results presented in this section are valid for any topological space (after replacing $\overline{X}$ and $\underline{X}$ by $Cl(X)$ and $Int(X)$, respectively).

3.1   Modular Set Theory

Let us now come back to modular arithmetic. As already noted, we usually use remainders as the standard representatives. Thus, for given $a, m \in \mathbb{Z}$, the notation

$$a \bmod m$$

denotes/stands for a remainder $\mathbf{r}$ from Proposition 3. Of course, it means that

$$a = km + \mathbf{r}, \quad \text{where} \quad k \in \mathbb{Z}. \tag{1}$$

It suggests that the remainder may be construed as an *excess* or *nimiety* in size of a with respect to the *quantisation* of $\mathbb{Z}$ by means of m. In rough set theoretic terminology we could regard numbers of the form km as *definable*, and $\mathbf{r}$ as an excess which must be erased from a in order to obtain a definable number.

In RST the *quantisation* of U is given by the family of definable sets with respect to U/E (denoted by $\mathcal{D}$). As is well known it forms a Boolean algebra $(\mathcal{D}, \cap, \cup, ', \emptyset, U)$, where $'$ denotes the set complementation. As observed by Bernstein in 1924 [1], each Boolean algebra gives rise to a group; in particular $(\mathcal{D}, \uplus)$ and $(\mathcal{P}(U), \uplus)$, where $\uplus$ stands for the symmetric difference[3], are groups.

---

[3] $X \uplus Y = (X \setminus Y) \cup (Y \setminus X)$.

Each of them is actually an (additive) abelian group, in which every element is it's own inverse. Generally, such groups are called *Boolen groups*.

Definition 5 (Boolean Ring). *A ring* $R = (U, +, *, 0)$ *is Boolean if* $a^2 = a$ *for every* $a \in U$.

As always, each Boolean algebra induces also a Boolean ring. Thus we have:

Proposition 4. $(\mathcal{D}, \uplus, \cap, \emptyset)$ *and* $(\mathcal{P}(U), \uplus, \cap, \emptyset)$ *are Boolean rings.*

Let us now write a set-version of (1):

$$X = Y \uplus \mathbf{r} \quad \text{where} \quad Y \in \mathcal{D} \quad \text{and} \quad \mathbf{r} \in \mathcal{P}(U).$$

Since we want to take the maximal definable set $Y \subseteq X$, we have:

$$X = \underline{X} \uplus \mathbf{r}.$$

Therefore:

$$\mathbf{r} = X \setminus \underline{X}. \tag{2}$$

Let us compare it to the standard RST approach, which is based on the boundary region:

$$X \subseteq X \cup \mathbf{b}(X) = \overline{X} \in \mathcal{D} \quad \text{and} \quad \mathbf{b}(X) = \overline{X} \setminus \underline{X}. \tag{3}$$

Since $\mathbf{r}(X) \subseteq \mathbf{b}(X)$, we may say that within the modular arithmetic approach we are interested in the half of the boundary region. Interestingly if we replace $\cup$ by $\uplus$ in (3), then we define the second part $\mathbf{d}$ of the boundary, which may be interpreted as *deficit*.

$$X \subseteq X \uplus \mathbf{d} = \overline{X} \in \mathcal{D}. \tag{4}$$

In contrast to the previous scenario of remainder, where the set X has got too much elements, in the context of (4) the set X has got a deficit of points, and that is why X is not a definable set.

The natural next step in *materialisation* of the modular arithmetic metaphor in RST, is to convert (2) and (4) into definitions of new set operators:

$$\mathbf{r}(X) = X \setminus \underline{X}, \quad \text{for every} \quad X \subseteq U,$$

$$\mathbf{d}(X) = \overline{X} \setminus X, \quad \text{for every} \quad X \subseteq U.$$

Obviously, two halves become one:

Corollary 1. *For every subset X of an approximation (U, E) space it holds that*

$$\mathbf{b}(X) = \mathbf{r}(X) \uplus \mathbf{d}(X).$$

Before we examine formal properties of the remainder and deficit operators, it is worth to recall the formal characterisation of the boundary. Most importantly, **b** is not as well-behaved as either the lower approximation/interior (Proposition 2) or upper approximation/closure (Proposition 1) operator. In words of Willard [15]: *it is possible, but unrewarding, to characterize a topology completely by its frontier [i.e., boundary] operation.* For Clark [4] *to do so is not entirely clear.* However, Pervin [11] states the following axioms for the boundary:

Proposition 5. *For a topological space* (U, τ) *and its boundary* $\boldsymbol{b} : \mathcal{P}(U) \rightarrow \mathcal{P}(U)$, *which is defined by* $\boldsymbol{b}(X) = Cl(X) \setminus Int(X)$, *it always holds that:*

1. $\boldsymbol{b}(\emptyset) = \emptyset$,
2. $\boldsymbol{b}(X) = \boldsymbol{b}(X')$,
3. $\boldsymbol{b}(\boldsymbol{b}(X)) \subseteq \boldsymbol{b}(X)$,
4. $X \cap Y \cap \boldsymbol{b}(X \cap Y) = X \cap Y \cap (\boldsymbol{b}(X) \cup \boldsymbol{b}(Y))$.

for all $X, Y \subseteq U$.

3.2   Set Modular Remainder

Surprisingly, the remainder $\mathbf{r}(X)$ regarded as a set-operator is much better behaved than the boundary. However, before we discuss its behaviour, let us retrieve the original conceptualisation of RST.

Proposition 6. *For every subset* X *of an approximation* (U, E) *space it holds:*

1. $\underline{X} = X \setminus \boldsymbol{r}(X)$,
2. $\overline{X} = X \uplus \boldsymbol{r}(U \setminus X)$,
3. $\boldsymbol{b}(X) = \boldsymbol{r}(X) \cup \boldsymbol{r}(U \setminus X)$.

As a set operator the reminder behaves quite *smoothly*.

Proposition 7. *Let* (U, E) *be an approximation space and* $\boldsymbol{r}$ *be the induced reminder operator. Then the following conditions hold:*

1. $\boldsymbol{r}(\emptyset) = \emptyset$,
2. $\boldsymbol{r}(X \cap Y) = (\boldsymbol{r}(X) \cap Y) \cup (\boldsymbol{r}(Y) \cap X)$,
3. $\boldsymbol{r}(\boldsymbol{r}(X)) = \boldsymbol{r}(X)$.

In sheer contrast to the deficit operator (discussed in the next subsection), the reminder is idempotent.

Proof.

$$\mathbf{r}(\mathbf{r}(X)) = \mathbf{r}(X) \setminus \underline{\mathbf{r}(X)} = (X \setminus \underline{X}) \setminus \underline{(X \setminus \underline{X})} = (X \cap \underline{X}\,') \cap (X \cap \underline{X}\,')\,' =$$

$$(X \cap \underline{X}\,') \cap \overline{(X \cap \underline{X}\,')'} = (X \cap \underline{X}\,') \cap \overline{(X' \cup \underline{X})} = (X \cap \underline{X}\,') \cap (\overline{X'} \cup \overline{\underline{X}}) =$$

$$((X \cap \underline{X}\,') \cap \overline{X'}) \cup ((X \cap \underline{X}\,') \cap \overline{\underline{X}}) = ((X \cap \underline{X}\,') \cap \underline{X}\,') \cup ((X \cap \underline{X}\,') \cap \overline{\underline{X}}).$$

Thus we have $\mathbf{r}(\mathbf{r}(X)) = \mathbf{r}(X) \cup (\mathbf{r}(X) \cap \overline{\underline{X}})$ and for $(\mathbf{r}(X) \cap \overline{\underline{X}}) \subseteq \mathbf{r}(X)$, we obtain:

$$\mathbf{r}(\mathbf{r}(X)) = \mathbf{r}(X) \cup (\mathbf{r}(X) \cap \overline{\underline{X}}) = \mathbf{r}(X).$$

3.3   Set Modular Deficit

As in the previous subsection, before we discuss the formal behaviour of deficit operator $\mathbf{d}$, we shall define the RST conceptual body.

Proposition 8. *For every subset X of an approximation (U, E) space it holds:*

1. $\overline{X} = X \uplus \mathbf{d}(X)$,
2. $\underline{X} = X \setminus \mathbf{d}(U \setminus X)$,
3. $\mathbf{b}(X) = \mathbf{d}(X) \cup \mathbf{d}(U \setminus X)$.

The deficit operator is not as *smooth* as the remainder; most importantly, the deficit is not idempotent. Yet, it is still much better behaved than the boundary.

Proposition 9. *Let (U, E) be an approximation space and $\mathbf{d}$ be the induced deficit operator. Then the following conditions hold:*

1. $\mathbf{d}(\emptyset) = \emptyset$,
2. $\mathbf{d}(X \cup Y) = (\mathbf{d}(X) \setminus Y) \cup (\mathbf{d}(Y) \setminus X)$,
3. $\mathbf{d}(\mathbf{d}(X)) \neq \mathbf{d}(X)$,
4. $\mathbf{d}(\mathbf{d}(X)) \subseteq X$.

Interestingly, within this conceptualisation/metaphor, RST is not about approximations of undefinable (incompletely specified) sets; rather, RST – similarly like modular arithmetic – is primarily concerned with the remainder and deficit. Does it change much? Firstly, even if X is well specified (known), it still makes perfect sense to compute its value(s) modulo the underlying definable sets (quantisation). Secondly, we may introduce another representation of subsets of U – alternative to the increasing and disjoint representations introduced in Sect. 2.1.

Definition 6 (Modular Representation of Rough Sets). *For an approximation space (U, E) and $X \subseteq U$, a pair $(\mathbf{r}(X), \mathbf{d}(X))$ is called a* modular repre-*sentation of X.*

And thirdly, this new representation better shows the imperfectness of the set X. If we drop out X and put the specific values, e.g. $C, D \subseteq U$, such that $C \neq D$, then under the disjoint representation (C, D) the extent to which the underlying set (X) is unspecified or imperfect is hardly visible. In the increasing representation we may compute the boundary and have some rough knowledge about this problem. But under modular representation this issue is very clear: C is the set of elements of X which we have imperfect knowledge about, whereas D brings us elements outside X which, due to our imperfect knowledge, may be added to X[4].

---

[4] The modular representation is not – however – equivalent to a rough set, e.g., if $\underline{X} = \emptyset$, then $(\underline{X}, \overline{X})$ usually represents/approximates more than a single set. However, the modular representation is $(X, \mathbf{d}(X))$, which stands for X alone.

4   Vagueness: Set Modular Approach

In this section we discuss the set (modular) arithmetic against the background (of the problem) of vagueness (as discussed in philosophy and science). We also extend our conceptualisation on the case of topological spaces, which is more subtle and versatile.

Let us start with a small excerpt from the Stanford Encyclopedia of Philosophy:

Vagueness is standardly defined as the possession of borderline cases. [...] Borderline cases are inquiry resistant. Indeed, the inquiry resistance typically recurses. For in addition to the unclarity of the borderline case, there is normally unclarity as to where the unclarity begins. In other words 'borderline case' has borderline cases. This higher order vagueness shows that 'vague' is vague.

In other words, vagueness is defined as the possession of borderline cases which are inquiry resistant, in the sense that borderline cases have borderline cases (the so called *higher-order vagueness*).

As noted in the introductory section, in the (original) RST methodology, a set X, which is supposed to be approximated, is well-known or well-defined: in order to compute an approximation of X, for each object $x \in U$ we need to know how its equivalence class $[x]_E$ is related to X, e.g., if $[x]_E \subseteq X$ or $[x]_E \cap X \neq \emptyset$; thus, we must know all elements of X. That is why Chakraborty in [3] asks: *If X is already known why to approximate at all?* On the other hand, as observed by Pawlak [10], in RST *vagueness* occurs naturally as borderline cases, which result from the incompleteness of our knowledge; that is why X needs to be approximated.

Let us check the Encyclopedia of once again:

For instance, a boy may count as a borderline case of 'obese' because people cannot tell whether he is obese just by looking at him. A curious mother could try to settle the matter by calculating her boy's body mass index. The formula is to divide his weight (in kilograms) by the square of his height (in meters). If the value exceeds 30, this test counts him as obese. The calculation will itself leave some borderline cases. The mother could then use a weight-for-height chart. These charts are not entirely decisive because they do not reflect the ratio of fat to muscle, whether the child has large bones, and so on. The boy will only count as an absolute borderline case of 'obese' if no possible method of inquiry could settle whether he is obese. When we reach this stage, we start to suspect that our uncertainty is due to the concept of obesity rather than to our limited means of testing for obesity.

The main question here is whether our goal is to model or to deal with vagueness. On the one hand, the philosophical demands concerning vagueness are so high, that virtually any formal representation is prone to criticism. On the other hand, vague concepts are also used in hard sciences such as medicine. E.g., on the National Institute of Health Obesity Research web page once can find:

Obesity is a major contributor to serious health conditions in children and adults, including type 2 diabetes, cardiovascular disease, many forms of cancer, and numerous other diseases and conditions.

The solution here is to expel all borderline cases. As Weiner observes [14]:

As sometimes happens in such research, the decision is made to exclude borderline cases from the study. [...] For obvious reasons – the exclusion of borderline cases requires two sharp distinctions: a distinction between those who are obese and those who are borderline-obese and a distinction between those who are borderline-obese and those who are not obese.

Thus we have the two opposite approaches to borderline cases: philosophical (where these cases are *inquiry resistant*), and scientific (where these cases are well-defined and expelled). Interestingly, the metaphor of modular arithmetic allows us to *run with the hare and hunt with the hounds*.

Firstly, we would like to paraphrase the Wiener's distinctions as follows:

(I) a distinction between those who are obese and those who are borderline-obese;

(II) a distinction between those who are borderline-not-obese and those who are not obese.

If X is a set of obese people, then (I) may be modelled by $\mathbf{r}(X)$, and (II) may be represented by $\mathbf{d}(X)$. Now, we can generalise this approach and call $\mathbf{r}(X)$ a collection of borderline-members of X, whereas $\mathbf{d}(X)$ would be a set of borderline-non-members. Unfortunately, as long as we deal with approximation spaces, both (I) and (II) come in one *package*.

Corollary 2. *For every subset X of an approximation (U, E) space it holds that*

$$\mathbf{r}(X) = \emptyset \quad \textit{iff} \quad \mathbf{d}(X) = \emptyset.$$

As already discussed, any approximation space (U, E) might be viewed as a topological space (U, τ_E), whose base is given by U/E. Since any topology τ on a space U is uniquely determined by its closure operator or the collection $\mathcal{C}$ of all closed subset of U, we may assume that known (definable) sets of U, that is $\mathcal{D}$, is a sum: $\tau \cup \mathcal{C}$. For, as observed by Wiweger [16], in (U, τ_E) every open set is closed and every closed set is open, we have $\tau_E \cup \mathcal{C}_E = \tau_E = \mathcal{C}_E$. Hence, if X has a non-empty boundary, it is neither closed nor open, so both $\mathbf{r}(X)$ and $\mathbf{d}(X)$ are non-empty. If $\mathbf{d}(X)$ is empty, then X is closed, so it is also open, and $\mathbf{r}(X)$ must be empty. The case of $\mathbf{r}(X)$ is analogous.

Fortunately, the correspondence between binary relations and topologies on U can be generalised to the case of preorders R and Alexandrov topological spaces (U, τ_R). This time, as required, τ_R usually differs from $\mathcal{C}_R$, but our definitions of the remainder and deficit still make perfect mathematical sense – actually, all propositions from the previous section are valid for any topological space. However, our metaphor makes less (common)sense in this new settings. We may try correct it a bit by calling the members of τ_R *directly definable* and the elements of $\mathcal{C}_R$ *dually definable*. Then $\mathbf{r}$ could be related to directly definable sets, whereas $\mathbf{d}$ would relate to the dually definable ones.

Now let us come back to the comments given by (a) Marek and Truszczyński in [7] (X is unknown or unspecified) and (b) Chakraborty in [3] (X is well known by still needs approximations) – see the last paragraphs of Sect. 2.1. Concerning (a), if X is a plain subset of a topological space (U, τ_R), then we could call it *imperfect* if it includes borderline members: $\mathbf{r}(X) \neq \emptyset$. Concerning (b), if X is an open set, then it is well-defined, that is $\mathbf{r}(X) = \emptyset$, but we could call it *rugged* if there are borderline-non-members: $\mathbf{d}(X) \neq \emptyset$. Finally, a set X could be called *vague* if it is imperfect or rugged. As expected, under the modular representation it is directly visible if a set X is imperfect, rugged, or vague.

Let us go back to the fundamental question in the philosophy of vagueness, namely: *have borderline cases of X got borderline cases?* Because in our approach we distinguish borderline-members of X from borderline-non-members of X, we may only ask if the set of borderline-members (borderline-non-members) is imperfect, rugged, or vague. Let us consider, e.g., $\mathbf{d}(\mathbf{d}(X))$, which may be rugged (and thus vague). Therefore, we may model a phenomenon, which is similar to the second-order vagueness: *borderline-non-members may have borderline-non-members (vague may be vague)*. Interestingly, the non-empty set of borderline-members always stays vague, and hence it also stays inquiry resistant – as requested by the Stanford Encyclopedia of Philosophy; unfortunately, in a rather trivial way. Another solution to maintain the higher order vagueness was offered by Skowron [12,13], who discusses this problem within a dynamic settings, where the underlying set U or the knowledge/attributes are changing, which in turn makes the boundary to be in a state of flux. However, in such a case also crisp sets are unstable and may become vague. Needless to say, from purely philosophical point of view, both approaches are not (fully) adequate. On the bright side, our approach to the second order vagueness is consistent with the scientific methodology and practice [14].

5   Conclusions

In the paper we have discussed a metaphor within which rough set theory (RST) is regarded as a sort of modular set-arithmetic. To this end, we have mapped the conceptual domain of modular arithmetic (where a given number is assigned a remainder with respect to a given modulus) onto RST (where a set is given a remainder with respect to a given collection of definable sets). In result, we have introduced two new topological operators: the remainder $\mathbf{r}$ and the deficit $\mathbf{d}$, which may be roughly understood as halves of the boundary. We have presented their formal properties and discussed their application to the problem of vagueness. Interestingly, the idea of splitting the boundary in half allowed us to introduce a new representation for sets, which is philosophically more subtle. In particular, it has allowed us to address the methodological shortcomings of RST discussed by Marek and Truszczyński [7], and Chakraborty [3].

Acknowledgements. We are greatly indebted to anonymous referees for their valuable comments and corrections.

References

1. Bernstein, B.A.: Operations with respect to which the elements of a Boolean algebra form a group. Trans. Am. Math. Soc. **26**(2), 171–175 (1924)
2. Jones, W.B.: Modular Arithmetic. Blaisdell, New York (1964)
3. Chakraborty, M.: On some issues in the foundation of rough sets. Fundamenta Informaticae **148**(1–2), 123–132 (2016)
4. Clark, P.: Notes on general topology. https://pdfs.semanticscholar.org/fc7e/e8ebdfcec468f1317cf37673e2292e46ff6d.pdf
5. Krajewski, S.: Theological metaphors in mathematics. Stud. Log. Gramm. Rhetoric **44**(57), 13–30 (2016)
6. Lakoff, G., Núñez, R.E.: Where Mathematics Comes From. Basic Books, New York (2000)
7. Marek, V.M., Truszczyński, M.: Contributions to the theory of rough sets. Fundamenta Informaticae **39**(4), 389–409 (1999)
8. Pawlak, Z.: Rough sets. Int. J. Comput. Inf. Sci. **11**, 341–356 (1982)
9. Pawlak, Z.: Rough Sets: Theoretical Aspects of Reasoning about Data. Kluwer Academic Publisher, Dordrecht (1991)
10. Pawlak, Z.: An inquiry into vagueness and uncertainty. Institute of Computer Science report 29/94. Warsaw University of Technology (1994)
11. Pervin, W.J.: Foundations of General Topology. Academic Press, New York (1964)
12. Skowron, A.: Rough sets and vague concepts. Fundamenta Informaticae **64**(1–4), 417–431 (2005)
13. Skowron, A., Swiniarski, R.: Rough sets and higher order vagueness. In: Ślęzak, D., Wang, G., Szczuka, M., Düntsch, I., Yao, Y. (eds.) RSFDGrC 2005. LNCS (LNAI), vol. 3641, pp. 33–42. Springer, Heidelberg (2005). https://doi.org/10.1007/11548669_4
14. Weiner, J.: Science and semantics: the case of vagueness and supervaluation. Pac. Philos. Q. **88**(3), 355–374 (2007)
15. Willard, S.: General Topology. Addison-Wesley Publishing Co., Reading (1970)
16. Wiweger, A.: On topological rough sets. Bull. Pol. Acad. Scie. Math. **37**, 89–93 (1989)

A Method for Boundary Processing in Three-Way Decisions Based on Hierarchical Feature Representation

Jie Chen[1], Yang Xu[1], Shu Zhao[1(✉)], Yuanting Yan[1], Yanping Zhang[1],
Weiwei Li[2], Qianqian Wang[1], and Xiangyang Wang[3]

[1] School of Computer Science and Technology, Anhui University,
Hefei 230601, Anhui, People's Republic of China
zhaoshuzs2002@hotmail.com
[2] College of Astronautics, Nanjing University of Aeronautics and Astronautics,
Nanjing 210016, China
[3] Anhui Electrical Engineering Professional Technique College,
Hefei 230051, Anhui, People's Republic of China

Abstract. For binary classification problem, all samples can be divided into three regions based on the three-way decision theory: positive regions, negative regions and boundary regions. These samples in boundary regions may be impossible to make a definite decision for lacking of detailed information. More information obtained from positive and negative regions is crucial for boundary processing. In the real word, people may identify positive regions based on one rule, and identify negative regions on another. The samples in boundary regions are also divided to positive or negative regions based on different rules. In this paper, we propose a method for processing boundary regions in three-way decisions based on hierarchical feature representation ($HFR-TWD$), which can obtain hierarchical feature representation of positive and negative regions. Firstly, all samples are divided into three regions by $MinCA$, which builds the most accurate covers for each class. Then samples in positive regions and negative regions respectively construct hierarchical feature representation. Thirdly, the best feature representation of each class is selected by using boundary region validating. Finally, boundary samples in test set are divided according to best feature representation of each class. Experiments show that the proposed method $HFR - TWD$ improves classification accuracy.

Keywords: Boundary regions · Hierarchical Feature Representation
Three-way decision theory · MinCA

1 Introduction

In conventional two-way decision model, there are only two optional choices for a decision: positive decision or negative decision regardless of lacking of information or not. Thus, it may result in wrong decisions when the information is

© Springer Nature Switzerland AG 2018
H. S. Nguyen et al. (Eds.): IJCRS 2018, LNAI 11103, pp. 123–136, 2018.
https://doi.org/10.1007/978-3-319-99368-3_10

not enough. To address this issue, Yao proposed Three-way decision model [1–4], which extends two-way decision theory by incorporating an additional choice: boundary decision. Three-way decision theory presents the universe as positive, negative and boundary regions. Many researchers have done further research on it.

Yao et al. researched on the three-way decision semantically in $DTRS$ and proposed the three-way decision rough set model [2]. Liu et al. established a novel three-way decision model based on an incomplete system [5]. Yao proposed sequential three-way decisions to make a definite decision of acceptance or rejection for some uncertain samples [6]. Xu et al. proposed the single-object stream-computing based three-way decisions algorithm (SS3WD), it aims at solving challenges results from simultaneous addition and deletion of objects [7]. Gao and Yao introduced four actionable strategies to the trisecting-and-acting three-way decision model according to action benefit and action cost [8]. Qian and Dang solved the attribute reduction problem for sequential three-way decisions under dynamic granulation [9]. Cabitza et al. proposed two methods aiming at collective knowledge extraction from questionnaires with ordinal scales and dichotomous questions based on a three-way decision procedure and a statistical method [10].

In recent years, The three-way decision was widely used in the real life, such as spam filtering [11,12], text classification [13], rubust classification [14], medical decision-making [15], Parkinson's disease detection [16], management theory [17], risk preferences of decision-making [18], image data analysis [19,20], uncertainty management [21,22], oil exploration decision [23], sentiment analysis of text [24], cost-sensitive software defect prediction [25], cost-sensitive face recognition [26], conflict analysis [27], clustering analysis and covering reduction analysis [14,28], incomplete data analysis [5,29], malware analysis [30], social networks [31], recommendation systems [32] and etc.

The main superiority of three-way decision compared with two-way decision is the utility of the boundary decision. In three-way decision theory, both the positive and negative regions contain elements without uncertainty or fuzziness. The boundary decision is regarded as a feasible choice of decision when the available information for decision is too limited to make a proper decision. This is similar to the human decision strategy in the practical decision problems. In this case, how to reduce the boundary regions is a new problem [33].

Li et al. adopted the idea of tri-training algorithm [34] and put forward a tri-training algorithm based on three-way decisions to reduce the boundary regions [33]. We had proposed multi-view decision model based on constructive three-way decision theory, which mines the global information of all samples to classifying boundary samples [35]. Then, we had used three-way decision theory to multi-granular mining for boundary regions [36]. We also adopted a cost-sensitive method to deal with the boundary region [37].

Those researches mined new information to further investigate boundary regions. For a practical decision problem, we may find diverse characteristics between the types of decisions. People always take optimistic decision using these characteristics, while other decision may use different characteristics. Representative characters of a decision are unique. People will take different types of decision according to different representative characters. Namely, The feature representations of each decision are different and important. Because of deep architecture of human brain, the last few years have seen significant interest in "deep" learning algorithms that learn layered, hierarchical representations of high-dimensional data. The cognition process is hierarchical and abstract in layers. Making the right decision at the most optimal level is also a crucial issue.

In this paper, we propose a method to process boundary regions based on Hierarchical Feature Representation ($HFR - TWD$). We firstly divide all samples into three regions by $MinCA$. Then samples in positive regions and negative regions respectively construct hierarchical feature representation. The best feature representation of each class are selected using boundary region validating. Finally, boundary samples in test set are divided according to best feature representation of each class.

The paper is organized as follows: In Sect. 2, we introduce the related works. In Sect. 3, we introduce a method to process boundary regions based on Hierarchical Feature Representation ($HFR - TWD$) in detail. In Sect. 4, we analyze the experimental results. We draw our conclusion in Sect. 5.

2   Related Work

The three-way decisions model divides the universe into three regions according to two thresholds (α, β). One region represents the set of elements with membership grades are higher than α and these elements are accepted to be instances of the concept modeled by the fuzzy set. Another region represents the set of elements with membership grades are less than β and these elements are rejected. The third region represents the set of elements that are between α and β. These elements are neither accepted nor rejected to be instances of the concept modeled by the fuzzy set. Zhang and Xing [38] proposed three-way decisions model based on covering algorithm. Covering Algorithm (CA) is introduced to forming the covers, three regions are formed according to these covers and does not need any parameters. So, in this paper, we introduce $MinCA$ to process boundary regions. $MinCA$ builds the min covers, and we will get more accurate three regions according these min covers and does not need any parameters. The following describes the detail of CA and $MinCA$.

Covering algorithm (CA) is a constructively supervised learning algorithm that maps all samples in the data set to an n-dimensional sphere S^n. The sphere neighborhoods are utilized to divide the samples. The CA can construct to neural networks (NNs) based on the samples' own characteristics.

Definition 2.1. Cover Algorithm (CA): Given a training samples set $X = \{(x_1, l_1), (x_2, l_2), \cdots, (x_u, l_u)\}$, where l_i means Label(x_i) = l_i, which is the set in u-dimensional Euclidean space. $A_i = (A_i^1, A_i^2, \cdots, A_i^q)$ is q-dimensional characteristic attribute of the ith sample. We assume $C^j = \bigcup C_i^j$, $i \in [1, 2, \cdots]$. C^j represents all covers of the jth category samples. We can define the distance between sample i and the farthest similar point as $d_1(i)$ where the boundary does not have any dissimilar points, the distance between sample i and the nearest other as $d_2(i)$.

$$d_2(i) = \min\{d(x_i, x_k)\}, l_i \neq l_k, k \in [1, \cdots, u] \tag{1}$$
$$d_1(i) = \max\{d(x_i, x_k) | d(x_i, x_k) < d_2(i)\}, l_i = l_k, k \in [1, \cdots, u] \tag{2}$$
$$d(i) = (d_1(i) + d_2(i))/2 \tag{3}$$

Then, C_i^j is the ith cover of class j which is constructed by x^i and $d(i)$. The center of C_i^j is x^i, the radius is $d(i)$.

Definition 2.2. MinCA: We assume $C^j = \bigcup C_i^j$, $i \in [1, 2, \cdots]$. C^j represents all Min covers of the jth category samples, when

$$d(i) = d_1(i) \tag{4}$$

The center of Min cover C_i^j is x^i, the radius is $d(i)$. $C^1 = \bigcup C_i^1$ contains all covered samples of class 1 and $C^2 = \bigcup C_i^2$ contains all covered samples of class 2. So, C^1 is positive region (POS), C^2 is negative region (NEG). All uncovered samples are in boundary regions (BND).

In $MinCA$ model, it regards the max distance between the center and the similar points as the radius [38]. The covers of $MinCA$ model are smaller and more precise. The positive region POS and the negative region NEG accurately consist of those objects that we accept as satisfying the conditions and reject as satisfying the conditions. More uncertain samples are divided into BND regions for further precise decision. The difference between $MinCA$ and CA is shown in Fig. 1.

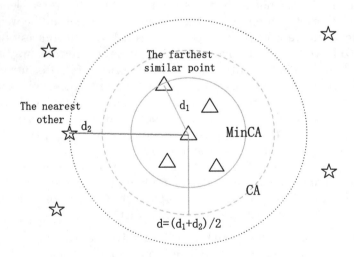

Fig. 1. The difference between $MinCA$ and CA

3   A Method to Process Boundary Regions Based on Hierarchical Feature Representation ($HFR-TWD$)

3.1   Hierarchical Feature Representation

Training samples are divided by $MinCA$ into three regions C^1, C^2 and uncovered samples, namely POS, NEG and BND. Then we will extract feature representation rules from POS and NEG regions. Mutual information is able to detect non-linear relationships among attributes. Therefore, we define mutual information relation metric to obtain feature representation.

Definition 3.1. Mutual Information Relation Metric: R_+/R_-
Given a training samples set $X = \{x_1, x_2, \cdots, x_u\}$, $A_i = (A^1, A^2, \cdots, A^q)$ is q-dimensional characteristic attribute of the sample. The information entropy of feature $A^s (s \in [1, \cdots, q])$ is defined as

$$H(A^s) = -\sum_{i=1}^{u} p(x_i) \log p(x_i) \tag{5}$$

The joint entropy of feature A^s and feature $A^t (t \in [1, \cdots, q])$ is defined as

$$H(A^s A^t) = -\sum_{i=1}^{u}\sum_{j=1}^{u} p(x_i x_j) \log p(x_i x_j) \tag{6}$$

The conditional entropy A^s to A^t is

$$H(A^s|A^t) = H(A^s A^t) - H(A^t) \tag{7}$$

The mutual information relationship between feature A^s and feature A^t are as follow:

$$I(A^s, A^t) = H(A^s) - H(A^s|A^t) = H(A^s) + H(A^t) - H(A^s A^t) \tag{8}$$

So, we can get mutual information relation metric R_+ using samples in POS region, and get mutual information relation metric R_- using samples in NEG region, where $r_{ij} = I(A_i, A_j)$ in R_+/R_- is the relationship between feature A_i and feature A_j. To eliminate self-influence, we set $r_{ii} = 1$.
R_+/R_- is a fuzzy equivalence relation on POS/NEG.

Definition 3.2. Quotient Space $A(\lambda)$: Define $d(\lambda)$ is a metric (or distance) function on R_+/R_-. Let

$$R_\lambda = \{R_+/R_- \geqslant \lambda\}, \lambda \geqslant 0 \tag{9}$$

R_λ is an equivalence relation on attribute A.
Let $A(\lambda)$ be a quotient space with respect to R_+/R_-.
Based on Quotient Space Theory, a family of quotient space $\{A(\lambda)|0 \leq \lambda \leq 1\}$ is an order-sequence under the inclusion relation of quotient sets. $A(\lambda)$ forms a

hierarchical structure with respect to attribute A. Thus, given fuzzy equivalence relations on attribute A, we have a corresponding hierarchical feature representation on attribute A.

Therefore, m levels feature representation of POS class and n levels feature representation of NEG class are obtained using mutual information relation metric R_+ and R_- based on $A(\lambda)$. Example 3.1 is the m levels feature representation of car dataset's [39] POS class.

Example 3.1. For car dataset, given attribute set $A = \{A^1, A^2, A^3, A^4, A^5, A^6\}$ and a fuzzy equivalence relation R_+ on POS. R_+ is represented by symmetric matrix as follows (Table 1):

Table 1. A symmetric matrix R_+ of A

$r_{ij}*100$	A^1	A^2	A^3	A^4	A^5	A^6
A^1	100	1.010	0.041	0.310	0.330	1.250
A^2	1.010	100	0.023	0.110	0.065	0.400
A^3	0.041	0.023	100	0.019	0.022	0.012
A^4	0.310	0.110	0.019	100	0.039	0.150
A^5	0.330	0.065	0.022	0.039	100	0.059
A^6	1.250	0.400	0.012	0.150	0.059	100

Let $r_{ij} = I(A_i, A_j)$. Based on the distance we construct the quotient space show below (Fig. 2).

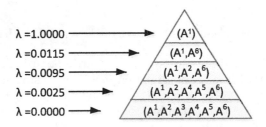

Fig. 2. Hierarchical Feature Representation of car dataset

3.2   The Selection of Best Representation

In this section, we obtain hierarchical feature representation based on mutual information relation metric of POS region and NEG region. Given m levels feature representation of POS class($sub + (i)$, $i \in [1, 2, \cdots, m]$) and n levels feature representation of NEG class($sub - (j)$, $j \in [1, 2, \cdots, n]$) consist of $m * n$ feature representation of all samples. Samples in boundary regions are validated based on each feature representation. The most accurate feature representation is best representation.

Define a function $L(x_i, C^1, C^2, sub)$: Computing the shortest distance between x_i and the center of cover in C^1 and C^2 using feature representation(sub):

$d^1 = (x_i, C^1, sub), d^2 = (x_i, C^2, sub)$

if $d^1 < d^2$, **return: 1**

else return: 2

The process of validation is presented as Algorithm 1.

Algorithm 1. Validation $(BND(X), m, n)$

Input: $BND(X) = \{(x_1, l_1), (x_2, l_2), \cdots, (x_s, l_s)\}$, where l_i means
    $Label(x_i) = l_i$,

Output: The best feature representation: $best+$ and $best-$

1 max=0;
2 **for** $i = 1; i \leq m; i++$ **do**
3   **for** $j = 1; j \leq n; j++$ **do**
4   int correct=0,error=0;
5   **for** $x_i, i = 1, \cdots, s$ **do**
6           $label1 = L(x_i, C^1, C^2, sub+(i));$
7           $label2 = L(x_i, C^1, C^2, sub-(j));$
8           $label3 = L(x_i, C^1, C^2, A_i);$
9   **if** $label1 == label2 \;||\; label1 == label3$ **then**
10             $label_i = label1;$
11   **else if** $label2 == label3$ **then**
12             $label_i = label3;$
13   **if** $Label(x_i) == label_i$ **then**
14   correct+1;
15   **else**
16   error+1;
17   **if** $max < correct$ **then**
18           $best+ = i, best- = j$, max= correct;
19 return $best+$ and $best-$;

3.3 Hierarchical Feature Representation Algorithm Based on Three Way Decision

In this section, we propose a Hierarchical Feature Representation algorithm based on Three Way Decision $(HFR - TWD)$.

We firstly divide train samples into three regions: POS, NEG and BND based on $MinCA$, which obtain mostly precise covers. Then we can get mutual information relation metric R_+ and R_- using samples in POS and NEG regions.

m levels feature representation of POS class and n levels feature representation of NEG class are obtained. The samples in BND will be validated using $m * n$ levels feature representation to select best representation $best+$ and $best-$. Finally, test samples are divided into three regions, and samples in boundary regions are examined by using $best+$ and $best-$.

The detail of $HFR - TWD$ is presented as Algorithm 2.

Algorithm 2. Hierarchical Feature Representation algorithm based on Three Way Decision($HFR - TWD$)

Input: train samples $X = \{(x_1, A_1), (x_2, A_2), \cdots, (x_u, A_u)\}$ and test samples
      $Y = \{y_1^1, y_2^1, \cdots, y_1^2, y_2^2, \cdots\}$, where y_i^j means $Lable(y_i) = j$

Output: POS samples($Lable(y_i) == 1$), NEG samples($Lable(y_i) == 2$)

1 //training:

2 train sample set X with attribute set A based on MinCA, generate Min cover set $C = \{c_1^1, c_2^1, c_3^1, \cdots, c_1^2, c_2^2, c_3^2, \cdots\}$;

3 delete covers where $coverednumber < Nmin$, then $POS = C^1 = \bigcup c^1$, $NEG = C^2 = \bigcup c^2$;

4 **for** $i=1$, A_i in POS regions, $i++$ **do**

5   **for** $j=1$, A_j in NEG regions, $j++$ **do**

6         $r_{ij} = I(A_i, A_j)$;

7   **if** $i==j$ **then**

8             $r_{ii} = 1$;

9 get m levels feature representation of POS class and n levels feature representation of NEG class based on **Definition 3.1**;

10 $\{best+, best-\} = Validation(BND(X), m, n)$;

11 //testing:

12 **for** *all test samples* $Y = \{y_i\}$ **do**

13   compute shortest distance d between y_i and cover center;

14   **if** $d <$ *radius of nearest cover* c^j **then**

15         $Lable(y_i) = j$;

16   **else**

17         $lable1 = L(y_i, C^1, C^2, best+)$;

18         $lable2 = L(y_i, C^1, C^2, best-)$;

19         $lable3 = L(y_i, C^1, C^2, A_i)$;

20   **if** $lable1 == lable2 \,||\, lable1 == lable3$ **then**

21             $Lable(y_i) = lable1$;

22   **else if** $lable2 == lable3$ **then**

23             $Lable(y_i) = lable3$;

---

4 Experiments

Our experiments are performed on two data sets from UCI Machine Learning Repository [39]. Table 2 shows the details of the data sets. All the samples used in experiment have complete attribute values. All comparative experiments results were published, we use these published results to evaluate our algorithm.

Table 2. Two data sets from UCI

Data	Number of data	Attributes	Classes
spambase	4601	58	2
chess	3196	36	2

4.1 Best Feature Representation Selection

We divide all samples into three regions through MinCA, Table 3 shows the number of three regions on *spambase* dataset and *chess* dataset. Figure 3 shows, we get 10 level feature representations of *POS* class and 10 level feature representations of *NEG* class, and the maximum correct number is 500 in *spambase* dataset, we conclude that the *best+* is sub+(9) and the *best−* is sub−(7). As for *chess* dataset, we get 5 level feature representations of *POS* class and 4 level feature representations of *NEG* class, the maximum correct number is 691 from Fig. 4, so the *best+* is sub+(4), the *best−* is sub−(2).

Table 3. The number of three regions

Data	Number of data	POS	NEG	BND
spambase	4601	2376	1432	793
chess	3196	1200	1087	909

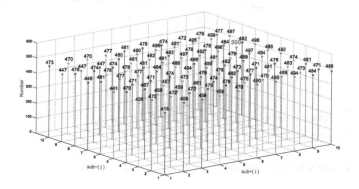

Fig. 3. The number of correct categories on *spambase* dataset

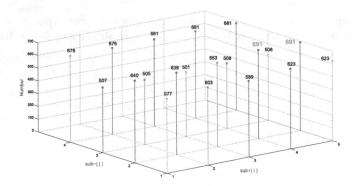

Fig. 4. The number of correct categories on *chess* dataset

4.2   Comparative Experiments

We firstly compare our algorithm with five algorithms on *spambase* dataset and *chess* dataset. Those comparative algorithms are three-way decisions model based on decision-theoretic rough set ($DTRS$) [1], Cost-sensitive three-way decisions model based on $CCA(CCTDM)$ [40], robustness three-way decisions model based on $CCA(R\text{-}TDM)$ [41] and Multi-granular three-way decision algorithm($MGTD$) [36]. All experiments are 10-fold cross-validation.

Comparative results are clearly shown in Fig. 5, Figure (a) indicts that the accuracy of our algorithm is up to 96.9% on average on *spambase* dataset, and it is superior to others. Figure (b) apparently indicts that the accuracy of our algorithm is better than those five algorithms on *chess* dataset.

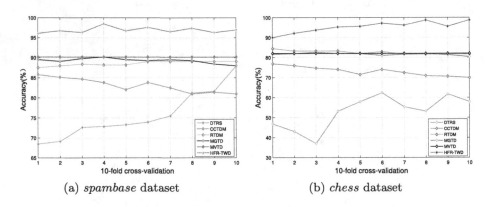

(a) *spambase* dataset   (b) *chess* dataset

Fig. 5. The whole classification accuracy of 6 three-way decision models

Then, we compare the performance of our algorithm with some latest algorithms on *spambase* dataset. Those comparative algorithms are integrated particle swarm optimization based J48 algorithm ($IPSO - J48$) [11], artificial bee-based decision tree $(ABBDT)$ [42], SVM and $SVM\&K-mean$ [43]. And we also compare the performance of our algorithm with other some latest algorithms on *chess* dataset. Those comparative algorithms are a weighted entropy frequent pattern mining $(WEFPM)$ [44], iterative sampling based frequent itemset mining $(ISbFIM)$ [45], an attributes similarity-based K-medoids clustering technique $(AS-KMC)$ [46], filter search strategy (relief-f) with an evolutionary search algorithm (differential evolution) $(RfDE)$ [47].

The comparative results are shown in Tables 4 and 5. From Table 4, we can see that the test accuracy of our algorithm is higher than others on *spambase* dataset. From Table 5, we can see that the best classification algorithm is $RfDE$ algorithm, but our algorithm is merely slightly lower than it.

Table 4. Classification accuracy on *spambase* dataset

Algorithm	Accuracy (%)
$ABBDT$ [42]	93.7
$IPSO-J48$ [11]	98.3
SVM [43]	96.1
$SVM\&K-mean$ [43]	98.0
$HFR-TWD$	98.9

Table 5. Classification accuracy on *chess* dataset

Algorithm	Accuracy (%)
$WEFPM$ [44]	93.2
$ISbFIM$ [45]	89.0
$AS-KMC$ [46]	94.8
$RfDE$ [47]	97.1
$HFR-TWD$	96.9

5   Conclusion

In this paper, we proposed a method $HFR-TWD$ to process boundary samples into a certain region. First of all, we utilize MinCA to divide all samples into three regions. Then, samples in POS and NEG respectively construct hierarchical feature representation, we use these hierarchical feature representations to handle BND region, and we will get the best feature representations. Finally, we use the best feature representations to handle boundary region in testing process. Compared with five three way decision models and other latest algorithms, the $HFR-TWD$ can find the best hierarchical feature representations to effectively handle samples from boundary region. So, we can conclude that the performance of $HFR-TWD$ algorithm is better.

Acknowledgments. This work was supported by National Natural Science Foundation of China (Grant Nos. 61602003, 61673020, and 61402006), National High Technology Research and Development Program (863 Plan)(Grant #2015A-A124102), Innovation Zone Project Program for Science and Technology of China's National Defense (Grant No. 2017-0001-863015-0009), the Provincial Natural Science Foundation of Anhui Province (Grant #1708085QF156), the Natural Science Foundation of Jiangsu Province (BK20170809), the China Postdoctoral Science Foundation (Grant No. 2018M632304).

References

1. Yao, Y.: The superiority of three-way decisions in probabilistic rough set models. Inf. Sci. **181**(6), 1080–1096 (2011)
2. Yao, Y.: Three-way decision: an interpretation of rules in rough set theory. In: Wen, P., Li, Y., Polkowski, L., Yao, Y., Tsumoto, S., Wang, G. (eds.) RSKT 2009. LNCS (LNAI), vol. 5589, pp. 642–649. Springer, Heidelberg (2009). https://doi.org/10.1007/978-3-642-02962-2_81
3. Yao, Y.: Three-way decisions with probabilistic rough sets. Inf. Sci. **180**(3), 341–353 (2010)
4. Yao, Y.: Two semantic issues in a probabilistic rough set model. Fundamenta Informaticae **108**(3), 249–265 (2011)
5. Liu, D., Liang, D., Wang, C.: A novel three-way decision model based on incomplete information system. Knowl. Based Syst. **91**(C), 32–45 (2016)
6. Yao, Y.: Granular computing and sequential three-way decisions. In: Lingras, P., Wolski, M., Cornelis, C., Mitra, S., Wasilewski, P. (eds.) RSKT 2013. LNCS (LNAI), vol. 8171, pp. 16–27. Springer, Heidelberg (2013). https://doi.org/10.1007/978-3-642-41299-8_3
7. Xu, J., Miao, D.Q.: A three-way decisions model with probablistic rough sets for stream computing. Int. J. Approx. Reason. **88**, 1–22 (2017)
8. Gao, C., Yao, Y.: Actionable strategies in three-way decisions. Knowl. Based Syst. **133**, 183–199 (2017)
9. Qian, J., Dang, C., Yue, X.: Attribute reduction for sequential three-way decisions under dynamic granulation. Int. J. Approx. Reason. **85**, 196–216 (2017)
10. Cabitza, F., Ciucci, D., Locora, A.: Exploiting collective knowledge with three-way decision theory: cases from the questionaire-based research. Int. J. Approx. Reason. **83**, 356–370 (2017)
11. Kaur, H., Sharma, A.: Novel email spam classification using integrated particle swarm optimization and J48. Int. J. Comput. Appl. **149**(7), 23–27 (2016)
12. Zhou, B., Yao, Y., Luo, J.: Cost-sensitive three-way email spam filtering. J. Intell. Inf. Syst. **42**(1), 19–45 (2014)
13. Li, Y.F., Zhang, L.B., Xu, Y., Yao, Y.Y.: Enhancing binary classification by modeling uncertain boundary in three-way decisions. IEEE Trans. Knowl. Data Eng. **29**(7), 1438–1451 (2017)
14. Yue, X., Chen, Y., Miao, D., Qian, J.: Tri-partition neighborhood covering reduction for robust classification. Int. J. Approx. Reason. **83**, 371–384 (2017)
15. Yao, J., Azam, N.: Web-based medical decision support systems for three-way medical decision making with game-theoretic rough sets. IEEE Trans. Fuzzy Syst. **23**(1), 3–15 (2015)

16. Liu, G., Zhang, Y., Hu, Z., et al.: Complexity analysis of electroencephalogram dynamics in patients with parkinson's disease. Parkinson's Dis. **2017**(6), Article no. 8701061 (2017)
17. Liu, D., Li, T., Liang, D.: Three-way decisions in dynamic decision-theoretic rough sets. In: Lingras, P., Wolski, M., Cornelis, C., Mitra, S., Wasilewski, P. (eds.) RSKT 2013. LNCS (LNAI), vol. 8171, pp. 291–301. Springer, Heidelberg (2013). https://doi.org/10.1007/978-3-642-41299-8_28
18. Li, H., Zhou, X.: Risk decision making based on decision-theoretic rough set: a three-way view decision model. Int. J. Comput. Intell. Syst. **4**(1), 1–11 (2011)
19. Li, H., Zhang, L.B.: Cost-sensitive sequential three-way decision modeling using a deep neural network. Int. J. Approx. Reason. **85**(C), 68–78 (2017)
20. Li, H., Zhang, L.: Sequential three-way decision and granulation for cost-sensitive face recognition. Knowl. Based Syst. **91**(C), 241–251 (2016)
21. Huang, B., Guo, C., Li, H.: Hierarchical structures and uncertainty measures for intuitionistic fuzzy approximation space. Inf. Sci. **336**(C), 92–114 (2015)
22. Ciucci, D., Dubois, D.: Three-valued logics, uncertainty management and rough sets. In: Peters, J.F., Skowron, A. (eds.) Transactions on Rough Sets XVII. LNCS, vol. 8375, pp. 1–32. Springer, Heidelberg (2014). https://doi.org/10.1007/978-3-642-54756-0_1
23. Yu, H., Chu, S., Yang, D.: Autonomous knowledge-oriented clustering using decision-theoretic rough set theory. Fundamenta Informaticae **115**(2–3), 141–156 (2012)
24. Ren, F., Wang, L.: Sentimental analysis of text based on three-way decisions. J. Intell. Fuzzy Syst. **33**(1), 245–254 (2017)
25. Li, W., Huang, Z., Li, Q.: Three-way decisions based software defect prediction. Knowl. Based Syst. **91**, 263–274 (2016)
26. Li, H., Zhang, L., Huang, B.: Sequential three-way decision and granulation for cost-sensitive face recognition. Knowl. Based Syst. **91**(C), 241–251 (2016)
27. Lang, G., Miao, D., Cai, M.: Three-way decisions approaches to conflict analysis using decision-theoretic rough set theory. Inf. Sci. **406**, 185–207 (2017)
28. Yu, H., Su, T., Zeng, X.: A three-way decisions clustering algorithm for incomplete data. In: Miao, D., Pedrycz, W., Ślęzak, D., Peters, G., Hu, Q., Wang, R. (eds.) RSKT 2014. LNCS (LNAI), vol. 8818, pp. 765–776. Springer, Cham (2014). https://doi.org/10.1007/978-3-319-11740-9_70
29. Wu, W.Z., Qian, Y., Li, T.J.: On rule acquisition in incomplete multi-scale decision tables. Inf. Sci. **378**(C), 282–302 (2017)
30. Nauman, M., Azam, N., Yao, J.T.: A three-way decision making approach to malware analysis using probabilistic rough sets. Inf. Sci. **374**, 193–209 (2016)
31. Peter, J.F., Ramanna, S.: Proximal three-way decisions: theory and applications in social networks. Knowl. Based Syst. **91**, 4–15 (2016)
32. Zhang, H., Min, F.: Three-way recommender systems based on random forests. Knowl. Based Syst. **91**(C), 275–286 (2016)
33. Li, P., Shang, L., Li, H.: A method to reduce boundary regions in three-way decision theory. In: Miao, D., Pedrycz, W., Ślęzak, D., Peters, G., Hu, Q., Wang, R. (eds.) RSKT 2014. LNCS (LNAI), vol. 8818, pp. 834–843. Springer, Cham (2014). https://doi.org/10.1007/978-3-319-11740-9_76
34. Zhou, Z.H., Li, M.: Tri-training: exploiting unlabeled data using three classifiers. IEEE Trans. Knowl. Data Eng. **17**(11), 1529–1541 (2005)

35. Chen, J., Zhao, S., Zhang, Y.: A multi-view decision model based on CCA. In: Ciucci, D., Wang, G., Mitra, S., Wu, W.-Z. (eds.) RSKT 2015. LNCS (LNAI), vol. 9436, pp. 266–274. Springer, Cham (2015). https://doi.org/10.1007/978-3-319-25754-9_24

36. Chen, J., Zhang, Y., Zhao, S.: Multi-granular mining for boundary regions in three-way decision theory. Knowl. Based Syst. **91**, 287–292 (2016)

37. Zhang, Y., et al.: Research on cost-sensitive method for boundary region in three-way decision model. In: Flores, V. (ed.) IJCRS 2016. LNCS (LNAI), vol. 9920, pp. 261–271. Springer, Cham (2016). https://doi.org/10.1007/978-3-319-47160-0_24

38. Zhang, Y., Xing, H., Zou, H., Zhao, S., Wang, X.: A three-way decisions model based on constructive covering algorithm. In: Lingras, P., Wolski, M., Cornelis, C., Mitra, S., Wasilewski, P. (eds.) RSKT 2013. LNCS (LNAI), vol. 8171, pp. 346–353. Springer, Heidelberg (2013). https://doi.org/10.1007/978-3-642-41299-8_33

39. UCI machine learning repository. http://archive.ics.uci.edu/ml/

40. Zhang, Y., Zou, H., Chen, X., et al.: Cost-sensitive three-way decisions model based on CCA. J. Nanjing Univ. (2015)

41. Zhang, Y., Zou, H., Chen, X., Wang, X., Tang, X., Zhao, S.: Cost-sensitive three-way decisions model based on CCA. In: Cornelis, C., Kryszkiewicz, M., Ślęzak, D., Ruiz, E.M., Bello, R., Shang, L. (eds.) RSCTC 2014. LNCS (LNAI), vol. 8536, pp. 172–180. Springer, Cham (2014). https://doi.org/10.1007/978-3-319-08644-6_18

42. Lee, Z.J., Lu, T.H., Huang, H.: A novel algorithm applied to filter spam e-mails for iPhone. Vietnam J. Comput. Sci. **2**(3), 143–148 (2015)

43. Elssied, N.O.F., Ibrahim, O., Osman, A.H.: Enhancement of spam detection mechanism based on hybrid k-mean clustering and support vector machine. Soft Comput. **19**, 3237–3248 (2015)

44. Devi, S.G., Sabrigiriraj, M.: Swarm intelligent based online feature selection (OFS) and weighted entropy frequent pattern mining (WEFPM) algorithm for big data analysis. Cluster Comput. **1**, 1–13 (2017)

45. Wu, X., Fan, W., Peng, J.: Iterative sampling based frequent itemset mining for big data. Int. J. Mach. Learn. Cybern. **1**(6), 1–8 (2015)

46. Narayana, G.S., Vasumathi, D.: An attributes similarity-based K-medoids clustering technique in data mining. Arab. J. Sci. Eng. **1**, 1–14 (2017)

47. Zainudin, M., Cheriet, M.: Feature selection optimization using hybrid relief-f with self-adaptive differential evolution. Int. J. Intell. Eng. Syst. **10**(2), 21–29 (2017)

Covering-Based Optimistic-Pessimistic Multigranulation Decision-Theoretic Rough Sets

Caihui Liu[1(✉)], Jin Qian[2], Nan Zhang[3], and Meizhi Wang[4]

[1] Department of Mathematics and Computer Science, Gannan Normal University,
Ganzhou 341000, Jiangxi, China
liu_caihui@163.com
[2] College of Computer Engineering, Jiangsu University of Technology,
Changzhou 213015, Jiangsu, China
qjqjlqyf@163.com
[3] School of Computer and Control Engineering, Yantai University,
Yantai 264005, Shandong, China
zhangnan0851@163.com
[4] Department of Physical Education, Gannan Normal University,
Ganzhou 341000, China
dei2002@163.com

Abstract. Multigranulation decision-theoretic rough sets (MDTRS) is a workable model for real-world decision making. The fruitful research achievements of the use of these models have been reported in different aspects. In most existing optimistic MDTRS models, the lower and upper approximations are defined based on the strategy *seeking commonality while preserving differences*, while pessimistic MDTRS models based on the strategy *Seeking commonality while eliminating differences* in the definitions of approximations. But in real life, one may need different strategies in defining lower approximation and upper approximation. This paper defines a new MDTRS approach in the frameworks of multi-covering approximation spaces by using different strategies in defining lower and upper approximation, namely, covering-based optimistic-pessimistic multigranulation decision-theoretic rough sets. We first explore a number of basic properties of the new model. Then, we elaborate on the relationship between the proposed models and the existing ones in literature and disclose the interrelationships of the new models.

Keywords: Covering · Multigranulation
Decision-theoretic rough sets · Optimistic · Pessimistic

1 Introduction

Since Yao and Wong [1] proposed the notion of decision-theoretic rough sets (DTRS), many researchers have been working on the theory. For example, Herbert and Yao [2] explored the game-theoretic rough set by combining game

© Springer Nature Switzerland AG 2018
H. S. Nguyen et al. (Eds.): IJCRS 2018, LNAI 11103, pp. 137–147, 2018.
https://doi.org/10.1007/978-3-319-99368-3_11

theory with DTRS. Liu et al. [3] discussed a multiple-category classification app-roach with decision-theoretic rough sets, which can effectively reduce misclassi-fication rate. Yu et al. [4] studied an automatic method of clustering analysis with the decision-theoretic rough set theory. Li et al. [5] studied an axiomatic characterization of decision-theoretic rough sets. Jia et al. [6] proposed an opti-mization representation of decision-theoretic rough set model and developed a heuristic approach and a particle swarm optimization approach for searching an attribute reduction with a minimum cost. Based on the DTRS, Yao [7,8] presented a new decision-making method known as three-way decisions, where a universe is divided into three pairwise disjoint regions, positive, negative and boundary regions by using an evaluation function and a pair of thresholds. Three-way decisions have been applied to many domains, such as email filtering [9], cost-sensitive face recognition [10], recommender system design [11], and so on.

The study on decision-theoretic rough set in a multigranulation environment is a new and interesting topic. Qian et al. [12] developed the multigranulation decision-theoretic rough set and proved that it is a general framework of many existing multigranulation rough set models. To tackle the problem of computa-tional cost in calculating the approximation of a target set with larger scale data, Qian et al. [13] proposed the combination of local rough sets with multigran-ulation decision-theoretic rough sets to obtain local multigranulation decision-theoretic rough sets (LMG-DTRSs) as a semi-unsupervised learning method. It is proved to be an excellent solution for dealing with data that have limited labels. However, those two models have their own limitations [14]: (1) All granu-lar structures in those models are based on equivalence relations, hence they are not suitable for coverings or neighborhoods based environments. (2) The models evaluate the multigranulation approximations in a quantitative way, so they are not suitable for the situations where general binary relations are considered. To tackle the above problems, Liu et al. [15] have proposed optimistic multigranu-lation decision-theoretic rough set model by employing the minimal descriptors of elements in a multi-covering space. The model may help to build a more reasonable and suitable decision environment for solving real world problems. Although, the successful fruits have been achieved on MDTRS, we found that in most existing optimistic MDTRS models [22–26], the lower and upper approxi-mations are defined based on the strategy *seeking commonality while preserving differences*, while pessimistic MDTRS models based on the strategy *Seeking com-monality while eliminating differences* in the definitions of approximations. But in real life, one may need different strategies in defining lower approximation and upper approximation [16]. In order to enlarge the usage scope of MDTRS, this paper proposed a new MDTRS model in the frameworks of multi-covering approximation spaces by using different strategy when defining lower and upper approximation, namely, covering-based optimistic-pessimistic multigranulation decision-theoretic rough sets (OP-CMDTRS). The motivation of this paper is outlined as follows.

– Two new fusion strategies are developed to deal with multi-source information systems.

– The models are constructed based on different strategies in defining lower approximations and upper approximations instead of using the same strategy adopted in the existing literatures.

The remainder of the paper is organized as follows. Section 2 reviews some basic notions and notations. Section 3 proposes the OP-CMDTRS model and discusses the interrelationships with the other generalized rough sets. Section 4 concludes the paper.

2   Preliminaries

In this section, some basic notions and notations will be reviewed.

2.1   Covering-Based Rough Sets

In this subsection, we will review some concepts related to the covering-based rough sets.

Definition 1 [17]. Let U be a universe of discourse and C a family of nonempty subsets of U. If $\cup C = U$, then C is called a covering of U. The ordered pair $\langle U, C \rangle$ is called a covering approximation space.

Definition 2 [19]. Let $\langle U, C \rangle$ be a covering approximation space, $x \in U$, then $md_C(x) = \{ K \in C_x | \forall S \in C_x (S \subseteq K \Rightarrow K = S) \}$ is called the minimal description of x, where $C_x = \{ K \in C | x \in K \}$.

2.2   Qian's MGRS

In this subsection, we will briefly outline the definition of optimistic multi-granulation rough sets.

Definition 3 [18]. Let $K = (U, \mathbf{R})$ be a knowledge base, where $\mathbf{R}$ is a family of equivalence relations on the universe U. Let $A_1, A_2, ..., A_m \in \mathbf{R}$, where m is a natural number. For any $X \subseteq U$, its optimistic lower and upper approximations with respect to $A_1, A_2..., A_m$ are defined as follows.

$$\sum_{i=1}^{m} A_i^O(X) = \{ x \in U | [x]_{A_1} \subseteq X \ or \ [x]_{A_2} \subseteq X \ or \ \cdots \ or \ [x]_{A_m} \subseteq X \}$$

$$\overline{\sum_{i=1}^{m} A_i^O}(X) = \neg \sum_{i=1}^{m} A_i(\neg X)$$

where $\neg X$ denotes the complement set of X. $(\sum_{i=1}^{m} A_i^O(X), \overline{\sum_{i=1}^{m} A_i^O}(X))$ is called the optimistic multi-granulation rough sets of X. Here, the word *optimistic* means that only a single granular structure is needed to satisfy the inclusion condition between an equivalence class and a target concept when multiple independent granular structures are available in the problem.

2.3 Decision-Theoretic Rough Sets

In [8], Yao proposed the theory of three-way decisions. Compared with two-way decisions, three-way decisions exhibit a third option, that is, non-commitment in addition to acceptance and rejection. The theory of three-way decisions can be described as follows.

Within the frame of three-way decisions, the set of states is given by $\Omega = \{X, \neg X\}$ (where $\neg X$ denotes the complement of X), the set of actions is given by $A = \{a_P, a_B, a_N\}$, where a_P, a_B and a_N represent the three actions in classifying an object x, namely, deciding $x \in POS(X)$, deciding x should be further investigated $x \in BND(X)$, and deciding $x \in NEG(X)$. $\lambda_{PP}, \lambda_{BP}$ and λ_{NP} denote the loss incurred for taking actions of a_P, a_B and a_N, respectively, when an object belongs to X. Similarly, $\lambda_{PN}, \lambda_{BN}$ and λ_{NN} denote the loss incurred for taking the correspondence actions when the object belongs to $\neg X$. By Bayesian decision procedure, for an object x, the expected loss $R(a_\bullet \mid [x])$ associated with taking the individual actions can be expressed as

$$R(a_P|[x]) = \lambda_{PP}P(X|[x]) + \lambda_{PN}P(\neg X|[x]),$$

$$R(a_N|[x]) = \lambda_{NP}P(X|[x]) + \lambda_{NN}P(\neg X|[x]),$$

$$R(a_B|[x]) = \lambda_{BP}P(X|[x]) + \lambda_{BN}P(\neg X|[x]).$$

Then the Bayesian decision procedure suggests the following three minimum-risk decision rules.

(P1) If $R(a_P|[x]) \leq R(a_B|[x])$ and $R(a_P|[x]) \leq R(a_N|[x])$, decide $x \in POS(X)$,
(N1) If $R(a_N|[x]) \leq R(a_P|[x])$ and $R(a_N|[x]) \leq R(a_B|[x])$, decide $x \in NEG(X)$,
(B1) If $R(a_B|[x]) \leq R(a_P|[x])$ and $R(a_B|[x]) \leq R(a_N|[x])$, decide $x \in BND(X)$.

By considering $0 \leq \lambda_{PP} \leq \lambda_{BP} < \lambda_{NP}$ and $0 \leq \lambda_{NN} \leq \lambda_{BN} < \lambda_{PN}$, (P1)–(B1) can be expressed concisely as:

(P2) If $P(X|[x]) \geq \alpha$ and $P(X|[x]) \geq \gamma$, decide $x \in POS(X)$,
(N2) If $P(X|[x]) \leq \gamma$ and $P(X|[x]) \leq \beta$, decide $x \in NEG(X)$,
(B2) If $P(X|[x]) \leq \alpha$ and $P(X|[x]) \geq \beta$, decide $x \in BND(X)$,

where:

$$\alpha = \frac{\lambda_{PN}-\lambda_{BN}}{(\lambda_{PN}-\lambda_{BN})+(\lambda_{BP}-\lambda_{PP})},$$

$$\beta = \frac{\lambda_{BN}-\lambda_{NN}}{(\lambda_{BN}-\lambda_{NN})+(\lambda_{NP}-\lambda_{BP})},$$

$$\gamma = \frac{\lambda_{PN}-\lambda_{NN}}{(\lambda_{PN}-\lambda_{NN})+(\lambda_{NP}-\lambda_{PP})}.$$

If $0 \leq \beta < \gamma < \alpha \leq 1$, (P2)–(B2) can be rewritten as follows:

(P3) If $P(X|[x]) \geq \alpha$, decide $x \in POS(X)$,
(N3) If $P(X|[x]) \leq \beta$, decide $x \in NEG(X)$,
(B3) If $\beta < P(X|[x]) < \alpha$, decide $x \in BND(X)$.

Based on the decision rules above, we obtain lower and upper approximations of the decision-theoretic rough sets as follows.

$\underline{PR}(X) = \{x \in U \mid P(X|[x]) \geq \alpha\}$ and $\overline{PR}(X) = \{x \in U \mid P(X|[x]) > \beta\}$.

3   Covering-Based Optimistic-Pessimistic Multigranulation Decision-Theoretic Rough Sets

In the MGRS theory, two kinds of strategies are used when approximating an observed concept. One is an optimistic strategy, i.e, *Seeking commonality while preserving difference* [18], and another one is pessimistic strategy, i.e., *Seeking commonality while eliminating differences* [18]. Here, we employ the optimistic strategy in the definition of lower approximation and pessimistic strategy in the definition of upper approximation of decision-theoretic rough sets in multi-covering approximation space $\langle U, \mathbf{C} \rangle$. We refer to this type of DTRS, covering-based optimistic-pessimistic multigranulation decision-theoretic rough sets (called OP-CMDTRS).

Definition 4.   Let $\langle U, \mathbf{C} \rangle$ be a multi-covering approximation space and $C_1, C_2, \cdots, C_n \in \mathbf{C}$, where n is a natural number. For any $X \subseteq U$, covering-based optimistic-pessimistic multigranulation decision-theoretic rough lower and upper approximations of X are defined as follows.

$$\underline{\sum\nolimits_{i=1}^{n} C_i^{OP,\alpha}}(X) = \{x \in U \mid \vee_{i=1}^{n}(P(X|\cap md_{C_i}(x)) \geq \alpha)\}$$

$$\overline{\sum\nolimits_{i=1}^{n} C_i^{OP,\beta}}(X) = U - \{x \in U \mid \vee_{i=1}^{n}(P(X|\cap md_{C_i}(x)) \leq \beta)\}$$
$$= \{x \in U \mid \wedge_{i=1}^{n}(P(X|\cap md_{C_i}(x)) > \beta)\}$$

The pair $(\underline{\sum_{i=1}^{n} C_i^{OP,\alpha}}(X), \overline{\sum_{i=1}^{n} C_i^{OP,\beta}}(X)$ is called a covering-based optimistic-pessimistic multigranulation decision-theoretic rough set.

Next, an example is given to explain the OP-CMDTRS models defined above.

Example 1.   Let $\langle U, \mathbf{C} \rangle$ be a multi-covering approximation space. $\mathbf{C}$ a family of coverings on U and $U = \{x_1, x_2, x_3, x_4\}$. $C_1, C_2 \in \mathbf{C}$ are two coverings on U such that $C_1 = \{\{x_1, x_2\}, \{x_2, x_3, x_4\}, \{x_3, x_4\}\}$, $C_2 = \{\{x_1, x_3\}, \{x_2, x_4\}, \{x_1, x_2, x_4\}\}$.

Suppose $X = \{x_1, x_4\}$. According to the above definitions, we have the following results.

First, we calculate the minimal descriptions for each element under granular structure C_1, C_2.

For C_1:

$$\cap md_{C_1}(x_1) = \{x_1, x_2\},$$
$$\cap md_{C_1}(x_2) = \{x_2\},$$
$$\cap md_{C_1}(x_3) = \cap md_{C_1}(x_4) = \{x_3, x_4\}.$$

For C_2:

$$\cap md_{C_2}(x_1) = \{x_1\},$$
$$\cap md_{C_2}(x_2) = \cap md_{C_2}(x_4) = \{x_2, x_4\},$$
$$\cap md_{C_2}(x_3) = \{x_1, x_3\}.$$

According to Definition 4, we obtain:

$$P(X \mid \cap md_{C_1}(x_1)) = \frac{P(X \cap (\cap md_{C_1}(x_1)))}{P(\cap md_{C_1}(x_1))} = \frac{1/4}{1/2} = \frac{1}{2} = 0.5$$

$$P(X \mid \cap md_{C_1}(x_2)) = 0,$$
$$P(X \mid \cap md_{C_1}(x_3)) = 0.5,$$
$$P(X \mid \cap md_{C_1}(x_4)) = 0.5,$$
$$P(X \mid \cap md_{C_2}(x_1)) = 1,$$
$$P(X \mid \cap md_{C_2}(x_2)) = 0.5,$$
$$P(X \mid \cap md_{C_2}(x_3)) = 0.5,$$
$$P(X \mid \cap md_{C_2}(x_4)) = 0.5.$$

If $\alpha = 0.6$ and $\beta = 0.3$, by Definition 4, the following result is formed.

$$\underline{\sum_{i=1}^{2} C_i^{OP, 0.6}}(X) = \{x_1\}, \quad \overline{\sum_{i=1}^{2} C_i^{OP, 0.3}}(X) = \{x_1, x_2, x_3, x_4\}.$$

Proposition 1. Let $\langle U, \mathbf{C} \rangle$ be a multi-covering approximation space and $C_1, C_2, \cdots, C_n \in \mathbf{C}$, where n is a natural number. Covering-based optimistic-pessimistic multigranulation decision-theoretic rough lower and upper approximations satisfy the following properties.

(1) $\underline{\sum_{i=1}^{n} C_i^{OP, \alpha}}(\emptyset) = \emptyset$, $\overline{\sum_{i=1}^{n} C_i^{OP, \beta}}(\emptyset) = \emptyset$;

(2) $\underline{\sum_{i=1}^{n} C_i^{OP, \alpha}}(U) = U$, $\overline{\sum_{i=1}^{n} C_i^{OP, \beta}}(U) = U$.

Remark 1. Let $\langle U, \mathbf{C} \rangle$ be a multi-covering approximation space and $C_1, C_2, \cdots, C_n \in \mathbf{C}$, for any $X \subseteq U$, the following properties may not hold.

(1) $\underline{\sum_{i=1}^{n} C_i^{OP, \alpha}}(X) \subseteq \overline{\sum_{i=1}^{n} C_i^{OP, \beta}}(X)$;

(2) $\underline{\sum_{i=1}^{n} C_i^{OP, \alpha}}(X) \subseteq X$;

(3) $X \subseteq \overline{\sum_{i=1}^{n} C_i^{OP, \beta}}(X)$.

Example 2 explains Remark 1.

Example 2 (Example 1 continued). If $\alpha = 0.6$ and $\beta = 0.51$, we have that

$$\overline{\sum_{i=1}^{2} C_i^{OP,0.6}}(X) = \{x_1\}, \quad \overline{\sum_{i=1}^{2} C_i^{OP,0.51}}(X) = \emptyset,$$

then

$$\overline{\sum_{i=1}^{n} C_i^{OP,\alpha}}(X) \subseteq \overline{\sum_{i=1}^{n} C_i^{OP,\beta}}(X) \text{ and } X \subseteq \overline{\sum_{i=1}^{n} C_i^{OP,\beta}}(X) \text{ are not}$$

hold.

If $\alpha = 0.5$, we have that $\overline{\sum_{i=1}^{2} C_i^{OP,0.5}}(X) = U$, then $\overline{\sum_{i=1}^{n} C_i^{OP,\alpha}}(X) \subseteq X$ is not satisfied.

Proposition 2. Let $\langle U, \mathbf{C} \rangle$ be a multi-covering approximation space and $C_1, C_2, \cdots, C_n \in \mathbf{C}$, where n is a natural number. For any $X \subseteq U$, covering-based optimistic-pessimistic multigranulation decision-theoretic rough lower and upper approximations satisfy the following two properties.

(1) $\overline{\sum_{i=1}^{n} C_i^{OP,\alpha}}(X) = \cup_{i=1}^{n} \underline{C_i^{\alpha}}(X)$;
(2) $\overline{\sum_{i=1}^{n} C_i^{OP,\beta}}(X) = \cap_{i=1}^{n} \overline{C_i^{\beta}}(X)$.

where $\underline{C_i^{\alpha}}(X) = \{x \in U \mid P(X \mid \cap md_{C_i}(x)) \geq \alpha\}$ and $\overline{C_i^{\beta}}(X) = \{x \in U \mid P(X \mid \cap md_{C_i}(x)) \geq \beta\}$ are defined in [20].

Proof: It is obvious according to Definition 4 and Definition 3.1 in [20].

4   Relationships of the Models

We discuss some interesting interrelationships between the proposed models and the existing ones.

Theorem 1. Let $\langle U, \mathbf{C} \rangle$ be a multi-covering approximation space and $C_1, C_2, \cdots, C_n \in \mathbf{C}$. For any $X \subseteq U$, we have that

(1) $\overline{\sum_{i=1}^{n} C_i^{OP,\alpha}}(X) = O_{\sum_{i=1}^{n} C_i}{}_{\alpha}(X)$;
(2) $\overline{\sum_{i=1}^{n} C_i^{OP,\beta}}(X) = \overline{P_{\sum_{i=1}^{n} C_i}{}^{\beta}}(X)$;

where $O_{\sum_{i=1}^{n} C_i}{}_{\alpha}(X) = \{x \in U \mid P(X \mid \cap md_{C_1}(x)) \geq \alpha \text{ or } \cdots \text{ or } P(X \mid \cap md_{C_n}(x)) \geq \alpha\}$

$\overline{P_{\sum_{i=1}^{n} C_i}{}^{\beta}}(X) = U - \{x \in U \mid P(X \mid \cap md_{C_1}(x)) \leq \beta \text{ or } \cdots \text{ or } P(X \mid \cap md_{C_n}(x)) \leq \beta\}$

Proof: It is straightforward.

Theorem 2. Let $\langle U, \mathbf{C} \rangle$ be a multi-covering approximation space and $C_1, C_2, \cdots, C_n \in \mathbf{C}$. For any $X \subseteq U$, we have

(1) If $\alpha = 1$, $\overline{\sum_{i=1}^{n} C_i^{OP,\alpha}}(X) = \underline{FR_{\sum_{i=1}^{n} C_i}}(X)$;
(2) If $\beta = 0$, $\overline{\sum_{i=1}^{n} C_i^{OP,\beta}}(X) = \overline{FR_{\sum_{i=1}^{n} C_i}}(X)$.

where $\underline{FR_{\sum_{i=1}^{n} C_i}}(X)$ and $\overline{FR_{\sum_{i=1}^{n} C_i}}(X)$ are defined in [21].

Proof: Here we only prove (1), the other parts can be proved in a similar way. According to Definition 4, we have

$$\sum_{i=1}^{n} \underline{C_i^{OP,1}}(X) = \{x \in U|\, \vee_{i=1}^{n}(P(X|\cap mdc_i(x)) \geq 1)\}$$
$$= \{x \in U|P(X|\cap mdc_1(x)) = 1 or \cdots or$$
$$P(X|\cap mdc_n(x)) = 1\}$$
$$= \{x \in U|\cap mdc_1(x) \subseteq X or \cdots or \cap mdc_n(x) \subseteq X\}$$
$$= \underline{FR_{\sum_{i=1}^{n} C_i}}(X).$$

This completes the proofs of Theorem 2.

Remark 2. Let $\langle U, \mathbf{C}\rangle$ be a multi-covering approximation space, $C_1, C_2, \cdots,$ $C_n \in \mathbf{C}$ and $C_1 = \{C_{11}, C_{12}, \cdots, C_{1p}\}$, $C_2 = \{C_{21}, C_{22}, \cdots, C_{2q}\}$, ..., $C_n = \{C_{n1}, C_{n2}, \cdots, C_{nl}\}$, where $p, q, \ldots, l$ are all natural numbers. For any $X \subseteq U$, the follows may not satisfied.

(1) $\sum_{i=1}^{n} \underline{C_i^{OP,\alpha}}(C_{ij}) = C_{ij}$,

(2) $\overline{\sum_{i=1}^{n} C_i^{OP,\beta}}(C_{ij}) = C_{ij}$.

Remark 2 shows that for any element C_{ij} in the coverings which construct the given DTRS model, the lower or upper approximation of C_{ij} in that model is may not itself anymore, which is true in classical multigranulation rough set model.

Example 3 is employed to explain Remark 2.

Example 3. Let $\langle U, \mathbf{C}\rangle$ be a multi-covering approximation space, where $U = \{1, 2, 3, 4\}$, $C_1, C_2 \in \mathbf{C}$, $C_1 = \{\{1, 2\}, \{2, 3, 4\}, \{3, 4\}\}$, $C_2 = \{\{1, 3\}, \{2, 4\}, \{1, 2, 4\}\}$. Let $X = C_{11} = \{1, 2\}$.

For C_1:
$$md_{C_1}(1) = \{\{1, 2\}\},$$
$$md_{C_1}(2) = \{\{1, 2\}, \{2, 3, 4\}\},$$
$$md_{C_1}(3) = md_{C_1}(4) = \{\{3, 4\}\}$$

For C_2:
$$md_{C_2}(1) = \{\{1, 3\}\{1, 2, 4\}\},$$
$$md_{C_2}(2) = md_{C_2}(4) = \{\{2, 4\}\},$$
$$md_{C_2}(3) = \{\{1, 3\}\}$$

Then
$$P(X|\cap md_{C_1}(1)) = \frac{P(X\cap(\cap md_{C_1}(1)))}{P(\cap md_{C_1}(1))} = \frac{1/4}{1/2} = \frac{1}{2} = 0.5,$$

$$P(X|\cap md_{C_1}(2)) = 0,$$
$$P(X|\cap md_{C_1}(3)) = 0.5,$$
$$P(X|\cap md_{C_1}(4)) = 0.5,$$
$$P(X|\cap md_{C_2}(1)) = 1,$$
$$P(X|\cap md_{C_2}(2)) = 0.5,$$
$$P(X|\cap md_{C_2}(3)) = 0.5,$$
$$P(X|\cap md_{C_2}(4)) = 0.5.$$

If $\alpha = 0.6$, $\beta = 0.3$, then

$$\underline{\sum_{i=1}^{2} C_i^{OP,0.6}}(C_{11}) = \{1,2\},$$

$$\overline{\sum_{i=1}^{2} C_i^{OP,0.3}}(C_{11}) = U$$

Obviously, we have

$$\overline{\sum_{i=1}^{2} C_i^{OP,0.3}}(C_{11}) = U \neq C_{11} = \{1,2\},$$

Theorem 3. Let $\langle U, \mathbf{C} \rangle$ be a multi-covering approximation space and $C_1, C_2, \cdots, C_n \in \mathbf{C}$. For any $X \subseteq U$ and $0 \leq \beta_2 \leq \beta_1 < \alpha_1 \leq \alpha_2 \leq 1$ we have

(1) $\underline{\sum_{i=1}^{n} C_i^{OP,\alpha_2}}(X) \subseteq \underline{\sum_{i=1}^{n} C_i^{OP,\alpha_1}}(X)$;

(2) $\overline{\sum_{i=1}^{n} C_i^{OP,\beta_1}}(X) \subseteq \overline{\sum_{i=1}^{n} C_i^{OP,\beta_2}}(X)$.

Proof: We only prove (1), the part (2) can be proved in a similar way.

According to Definition 5, we have

$$\underline{\sum_{i=1}^{n} C_i^{OP,\alpha_2}}(X) = \{x \in U \mid \vee_{i=1}^{n}(P(X \mid \cap md_{C_i}(x)) \geq \alpha_2)\}$$

$$\underline{\sum_{i=1}^{n} C_i^{OP,\alpha_1}}(X) = \{x \in U \mid \vee_{i=1}^{n}(P(X \mid \cap md_{C_i}(x)) \geq \alpha_1)\}$$

If $\alpha_1 \leq \alpha_2$, then for any $i \in \{1, 2, \ldots, n\}$, we have

$$P(X \mid \cap md_{C_i}(x)) \geq \alpha_2 \geq \alpha_1$$

Therefore,

$$\{x \in U \mid \vee_{i=1}^{n}(P(X \mid \cap md_{C_i}(x)) \geq \alpha_2)\} \subseteq$$
$$\{x \in U \mid \vee_{i=1}^{n}(P(X \mid \cap md_{C_i}(x)) \geq \alpha_1)\}$$

i.e. $\underline{\sum_{i=1}^{n} C_i^{OP,\alpha_2}}(X) \subseteq \underline{\sum_{i=1}^{n} C_i^{OP,\alpha_1}}(X)$.

Theorem 3 states that for the same concept with different values of α and β, the corresponding approximations are different, i.e., the higher the value of α, the lower the lower approximation, and the bigger the value of β, the bigger the upper approximation.

Example 4 (Example 3 continued). Suppose $\alpha = 0.7$ and $\beta = 0.2$, according to Definitions 4, we have

$$\underline{\sum_{i=1}^{2} C_i^{OP,0.7}}(C_{11}) = \{1\}, \overline{\sum_{i=1}^{2} C_i^{OP,0.2}}(C_{11}) = \{1,3,4\}.$$

Obviously, we have

$$\underline{\sum_{i=1}^{2} C_i^{OP,0.7}}(C_{11}) = \{1\} \subset \underline{\sum_{i=1}^{2} C_i^{OP,0.6}}(C_{11}) = \{1,2\};$$

$$\overline{\sum_{i=1}^{2} C_i^{OP,0.2}}(C_{11}) = \{1,3,4\} \subset \overline{\sum_{i=1}^{2} C_i^{OP,0.3}}(C_{11}) = U.$$

Theorem 4. Let $\langle U, \mathbf{C} \rangle$ be a multi-covering approximation space and $C_1, C_2, \cdots, C_n \in \mathbf{C}$. If $C_1, C_2, \cdots, C_n$ are all partitions, then for any $X \subseteq U$ and $0 \leq \beta \leq \alpha \leq 1$ we have

(1) $\overline{\sum_{i=1}^{n} C_{i}^{OP,\alpha}(X)} = \sum_{i=1}^{n} \overline{A_{i}^{OP,(\alpha,\beta)}}(X) = \sum_{i=1}^{n} \overline{A_{i}^{O,(\alpha,\beta)}}(X);$

(2) $\overline{\sum_{i=1}^{n} C_{i}^{OP,\beta}(X)} = \sum_{i=1}^{n} \overline{A_{i}^{OP,(\alpha,\beta)}}(X) = \sum_{i=1}^{n} \overline{A_{i}^{P,(\alpha,\beta)}}(X).$

where $\sum_{i=1}^{n} \underline{A_{i}^{OP,(\alpha,\beta)}}(X)$, $\sum_{i=1}^{n} \overline{A_{i}^{OP,(\alpha,\beta)}}(X)$ are defined in [16] and $\sum_{i=1}^{n} \overline{A_{i}^{O,(\alpha,\beta)}}(X)$,

$\sum_{i=1}^{n} \underline{A_{i}^{P,(\alpha,\beta)}}(X)$ are defined in [18].

5 Conclusion

In the present paper, we mainly discussed a kind of multigranulation decision-theoretic rough set model in the multi-covering space by employing the new strategy. We gave the properties of the proposed model. And we also found some interrelationships between the proposed model and other existing models.

Acknowledgements. This work was supported by the China National Natural Science Foundation of Science Foundation under Grant Nos.: 61663002, 61741309, 61403329, 61305052 and Jiangxi Province Natural Science Foundation of China under Grant No.: 20171BAB202034.

References

1. Yao, Y.Y., Wong, S.K.M.: A decision theoretic framework for approximating concepts. Int. J. Man Mach. Stud. **37**, 793–809 (1992)
2. Herbert, J.P., Yao, J.T.: Game-theoretic rough sets. Fundamenta Informaticae **108**(3–4), 267–286 (2011)
3. Liu, D., Li, T.R., Li, H.X.: A multiple-category classification approach with decision-theoretic rough sets. Fundamenta Informaticae **115**(2–3), 173–188 (2012)
4. Yu, H., Liu, Z.G., Wang, G.Y.: An automatic method to determine the number of clusters using decision-theoretic rough set. Int. J. Approx. Reason. **55**(1), 101–115 (2014)
5. Li, T.J., Yang, X.P.: An axiomatic characterization of probabilistic rough sets. Int. J. Approx. Reason. **55**(1), 130–141 (2014)
6. Jia, X.Y., Tang, Z.M., Liao, W.H., Shang, L.: On an optimization representation of decision-theoretic rough set model. Int. J. Approx. Reason. **55**(1), 156–166 (2014)
7. Yao, Y.Y.: Three-way decisions with probabilistic rough sets. Inf. Sci. **180**, 341–353 (2010)
8. Yao, Y.: An outline of a theory of three-way decisions. In: Yao, J.T., et al. (eds.) RSCTC 2012. LNCS (LNAI), vol. 7413, pp. 1–17. Springer, Heidelberg (2012). https://doi.org/10.1007/978-3-642-32115-3_1
9. Zhou, B., Yao, Y., Luo, J.: A three-way decision approach to email spam filtering. In: Farzindar, A., Kešelj, V. (eds.) AI 2010. LNCS (LNAI), vol. 6085, pp. 28–39. Springer, Heidelberg (2010). https://doi.org/10.1007/978-3-642-13059-5_6

10. Li, H.X., Zhang, L.B., Huang, B., Zhou, X.Z.: Sequential three-way decision and granulation for cost-sensitive face recognition. Knowl. Based Syst. **91**, 241–251 (2016)
11. Zhang, H.R., Min, F.: Three-way recommender systems based on random forests. Knowl. Based Syst. **91**, 275–286 (2016)
12. Qian, Y.H., Zhang, H., Sang, Y.L., Liang, J.L.: Multigranulation decision-theoretic rough sets. Int. J. Approx. Reason. **55**(1), 225–237 (2014)
13. Qian, Y.H., Liang, X.Y., Lin, G.P.: Local multigranulation decision-theoretic rough sets. Int. J. Approx. Reason. **82**, 119–137 (2017)
14. Liu, C., Wang, M., Zhang, N.: Covering-based optimistic multigranulation decision-theoretic rough sets based on maximal descriptors. In: Polkowski, L., et al. (eds.) IJCRS 2017. LNCS (LNAI), vol. 10314, pp. 238–248. Springer, Cham (2017). https://doi.org/10.1007/978-3-319-60840-2_17
15. Liu, C., Wang, M.: Optimistic decision-theoretic rough sets in multi-covering space. In: Flores, V., et al. (eds.) IJCRS 2016. LNCS (LNAI), vol. 9920, pp. 282–293. Springer, Cham (2016). https://doi.org/10.1007/978-3-319-47160-0_26
16. Qian, J.: Research on multigranulation decision-theoretic rough set models. J. Zhengzhou Univ. (Nat. Sci. Edn.). https://doi.org/10.13705/j.issn.1671-6841. 2017069. (in Chinese with English Abstract)
17. Zakowski, W.: Approximations in the space (U, Π). Demonstratio Mathematica **16**, 761–769 (1983)
18. Qian, Y.H., Liang, J.Y., Yao, Y.Y., Dang, C.Y.: MGRS: a multi-granulation rough set. Inf. Sci. **180**, 949–970 (2010)
19. Zhu, W., Wang, F.Y.: On three types of covering rough sets. IEEE Trans. Knowl. Data Eng. **19**, 1131–1144 (2007)
20. Gong, Z.T., Shi, Z.H.: On the covering probabilistic rough set models and its Bayes desicions. Fuzzy Syst. Math. **22**(4), 142–148 (2008). (Chinese with English abstract)
21. Liu, C.H., Miao, D.Q., Qian, J.: On multi-granulation covering rough sets. Int. J. Approx. Reason. **55**, 1404–1418 (2014)
22. Li, H.X., Zhang, L.B., Zhou, X.Z., Huang, B.: Cost-sensitive sequential three-way decision modeling using a deep neural network. Int. J. Approx. Reason. **85**, 68–78 (2017)
23. Feng, T., Fan, H.T., Mi, J.S.: Uncertainty and reduction of variable precision multi-granulation fuzzy rough sets based on three-way decisions. Int. J. Approx. Reason. **85**, 36–58 (2017)
24. Sun, B.Z., Ma, W.M., Xiao, X.: Three-way group decision making based on multi-granulation fuzzy decision-theoretic rough set over two universes. Int. J. Approx. Reason. **81**, 87–102 (2017)
25. Qian, Y.H., Liang, X.Y., Lin, G.P., Qian, G., Liang, J.Y.: Local multigranulation decision-theoretic rough sets. Int. J. Approx. Reason. **82**, 119–137 (2017)
26. Ju, H.R., Li, H.X., Yang, X.B., Zhou, X.Z., Huang, B.: Cost-sensitive rough set: a multi-granulation approach. Knowl. Based Syst. **123**, 137–153 (2017). https://doi.org/10.1016/j.knosys.2017.02.019

Studies on CART's Performance in Rule Induction and Comparisons by STRIM
In a Simulation Model for Data Generation and Verification of Induced Rules

Yuichi Kato[1(✉)], Shoya Kawaguchi[1], and Tetsuro Saeki[2]

[1] Shimane University, 1060 Nishikawatsu-cho, Matsue, Shimane 690-8504, Japan
ykato@cis.shimane-u.ac.jp
[2] Yamaguchi University, 2-16-1 Tokiwadai, Ube, Yamaguchi 755-8611, Japan
tsaeki@yamaguchi-u.ac.jp

Abstract. The tree based method is a conventional statistical method that involves constructing a tree structure for a classification model through recursively splitting a dataset by explanatory variables to minimize some impurity criteria for the response variable. This tree structure induces many if-then rules with product forms. In this paper, we study a basic tree based approach — the classification and regression trees (CART) method — based on a simulation model for data generation and verification for induced rules. We compare CART with the statistical test rule induction method (STRIM) to clarify its performance and problems. We also apply both methods to a real-world dataset and consider their performances based on the simulation results.

1 Introduction

Activities of modern society are based around various network systems, which produce massive datasets that are destroyed or stored with no use. Such datasets contain diverse patterns and features of human activities. Nowadays, as efficient and timely application of this information can inform business strategies, there has been rapid development and expansion of data mining research and technology, particularly in areas concerning e-business. This paper focuses on a tree based model often used among such data mining methods. The model is a statistical method and constructs a classification model through recursively splitting a dataset by explanatory variables to minimize some impurity criterion for the response variable. The aim of splitting the dataset is to visually arrange the dataset in a tree structure, which presents many if-then rules hidden in the dataset and can indicate business strategies or information.

We previously proposed an if-then rule induction method called the statistical test rule induction method (STRIM) [1–9] which statistically interprets the classical Rough Sets theory [10–13]. We studied the validity of STRIM based on a simulation model for data generation and verification of induced rules (SM for DG & VIR) and considered the differences and/or relationships between the rules

© Springer Nature Switzerland AG 2018
H. S. Nguyen et al. (Eds.): IJCRS 2018, LNAI 11103, pp. 148–161, 2018.
https://doi.org/10.1007/978-3-319-99368-3_12

induced by the classical method and STRIM [1,3,5,9]. Specifically, the simulation model was used to (1) generate the decision table using pre-specified rules in a rule box and hypotheses for deciding the decision attribute's value against the condition attributes' values generated by random numbers, (2) apply a chosen rule induction method to the generated decision table, and (3) confirm whether the applied method properly induced pre-specified if-then rules. That is, the simulation model can be used to examine the ability of any rule induction method and to study the features of the chosen method.

In this paper, we choose the classification and regression trees (CART) method [14] as the most basic tree-based approach. CART is usually used for classification problems after inducing the classification tree which consists of many if-then rules. We focus on the validity of its induced rules since their rules indicate diverse patterns and features of human activities hidden behind the analyzed dataset and their patterns and features are useful for gaining new business strategies. Specifically, we apply CART to the aforementioned simulation model and examine its performance in rule induction since its performance in a simulation has not yet been reported, except on a real-world dataset. The simulation results clarify the following:

(1) CART tends to induce only some of the pre-specified rules and their sub-rules with longer rule length due to the tree structure.

(2) CART cannot properly process the conflicting data and eliminate the indifferent data in the dataset of interest due to the bisection method for the dataset.

(3) The problems and features of (1) and (2) generate a large number of rules with longer rule length in some cases more than the size of the dataset (the decision table).

After the simulation experiment, CART and STRIM — which has been already validated in a simulation — are applied to a real-world dataset and the resulting rules are judged in consideration of the simulation results.

2   Simulation Model for Data Generation and Verification of Induced Rules

In statistics, a dataset $U = \{u(i)|i = 1, ..., N = |U|\}$ is collected from a population of interest to estimate and/or infer properties and features of the population. Here, $u(i)$ is an object with several attributes, whose properties and features contribute to the estimation and inference of the population. Let us denote an observation system by $S = (U, A, V)$. Here, A is the set of an attribute and V is the set of the attribute's values; that is, $V = \bigcup_{a \in A} V_a$ and V_a is the set of the value of attribute a. When randomly sampling $u(i)$ from the population, each attribute becomes a random variable with the respective attribute value as its outcome.

Here, there are two main types of dataset, with a division between the response and explanatory variables and those without it. In the former case, the set of attributes A is denoted $A = C \cup \{D\}$ to distinguish from the latter case. Here, D is a decision attribute and the response variable, and $C = \{C(j)|j = 1, ..., |C|\}$ is the set of condition attribute $C(j)$ and also $C(j)$ is an explanatory variable for the response variable. If D and $C(j)$ are qualitative variables, D denotes the random variable of the class of $u(i)$ and is affected by the set C of random variables $C(j)$. Note, however, that CART can deal with both qualitative and quantitative variables. This paper studies the CART's performance in rule induction dealing with qualitative variables and compares it to the results of STRIM based on the system $S = (U, A = C \cup D, V)$ called the decision table in Rough Sets theory.

In the classical Rough Sets theory, the decision table is denoted as $S = (U, A = C \cup \{D\}, V, \rho)$. Here, ρ: $U \times A \to V$ and ρ is called an information function. However, this paper does not need ρ since we recognize D and $C(j)$ as random variables and V as the set of their outcomes, that is, the sample space, as described above.

Figure 1 outlines a SM for DG & VIR. Randomly sampling $u(i)$ from the population, the outcome of $C = (C(1), ..., C(|C|))$; that is, $u^C(i) = (v_{C(1)}(i), ..., v_{C(|C|)}(i))$ is obtained and becomes the input into the rule box. The rule box transfers $u^C(i)$ to the output $u^D(i)$ using the rule box's pre-specified rules and hypotheses with regard to the output as shown in Table 1, which shows the following three cases: (1) the uniquely determined case, (2) the indifferent case (the rules are not specified at all the inputs), and (3) the conflicted case. Cases (2) and (3) often happen in the real-world. The observer in Fig. 1 records $u(i) = (u^C(i), u^D(i))$. NoiseC and NoiseD are introduced to adapt the model for the real-world dataset. NoiseC adjusts the value of $u^C(i) = (v_{C(1)}(i), ..., v_{C(|C|)}(i))$ or makes $v_{C(j)}(i)$ a missing value and NoiseD adjusts the value of $u^D(i)$.

Generating $u^C(i) = (v_{C(1)}(i), ..., v_{C(|C|)}(i))$ using random numbers and transforming it into $u^D(i)$ using the model shown in Fig. 1, $U = \{u(i) = (u^C(i), u^D(i)) |i = 1, ..., N = |U|\}$ can be obtained and applied to any rule induction method to investigate the extent to which the method induces the pre-specified rules. That is, the system can be used to investigate the performance for any rule induction method. To date, most conventional studies have applied rule induction methods to real-world datasets and judged the results only by the domain knowledge before studying the method's properties and features via the simulation with a white rule box like that shown in Fig. 1.

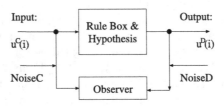

Fig. 1. A simulation model for data generation and verification of induced rules. The rule box contains if-then rules $R(d, k)$: if $CP(d, k)$ then $D = d$ ($d = 1, 2, ..., k = 1, 2, ...$).

Table 1. Hypotheses for the decision attribute value.

Hypothesis 1	$u^C(i)$ coincides with $R(d, k)$, and $u^D(i)$ is uniquely determined as $D = d$ (uniquely determined data)
Hypothesis 2	$u^C(i)$ does not coincide with any $R(d, k)$, and $u^D(i)$ can only be determined randomly (indifferent data)
Hypothesis 3	$u^C(i)$ coincides with several $R(d, k)$ $(d = d1, d2, ...)$, and their outputs of $u^C(i)$ conflict with each other. Accordingly, the output of $u^C(i)$ must be randomly determined from the conflicted outputs (conflicted data)

3 Examination of CART on SM for DG & VIR

CART is the most basic tree based model approach and can be adapted into other methods such as the multiple additive regression tree (MART) [15], bagging (bootstrap aggregating) [16], and random forests [17]. We now examine CART on SM for DG & VIR.

3.1 The CART Method

CART is implemented using a statistical software package and is used across various fields including medical science, environmental science and econometrics. We briefly describe the method with qualitative variables for use in the following section (see literature [14] for more detail).

CART recursively splits U in $S = (U, A = C \cup D, V)$ and constructs a binary tree T for classification as shown in Fig. 2, where T is a set of nodes t_l; that is, $T = \{t_l | l = 1, ..., l_e\}$. Here, t_l denotes a set of $u(i)$ labeled l. The root node $t = t_1$ includes the whole set of U. The split rule s_1 divides t_1 into two sections: the left node $t_L = t_2$ satisfying s_1 (y in Fig. 2) and the right node $t_R = t_3$ not satisfying s_1 (n in Fig. 2). Accordingly, $|t| = |t_L| + |t_R|$. The split rule s_t repeatedly divides t. A node t without s_t is called a terminal node and the set of terminal nodes is denoted as $\tilde{T}$. In Fig. 2, $l_e = 9$ and $\tilde{T} = \{t_4, t_5, t_7, t_8, t_9\}$.

The right size tree T is constructed using the following three procedures:

(1) A progression process to grow the tree:

A split rule s_t at t is selected based on an impurity criterion $r_c(t)$ such as a classification error rate, Gini coefficient or entropy measure. For example, an entropy measure is $r_c(t) = - \sum_{1 \le k \le |V_{a=D}|} p(k|t) \log p(k|t)$. Here, $p(k|t) = \sum_{u^D(i) \in t} 1(u^D(i) = k)/|t|$; $1(\bullet)$ is a function taking 1 if the given in parentheses is true, otherwise 0 and $p(k|t)$ is the probability of the event $u^D(i) = k$ in t. The impurity of t in T is $R(t) = p(t)r_c(t)$, where $p(t) = |t|/|U|$. Accordingly, the reduction amount of the impurity by s_t can be defined as $\Delta R(s_t, t) = R(t) - R(t_L) - R(t_R)$ and the following s_t should be selected to dominate the frequency of a specific $D = d$ in t: $s_t^* = \max \arg_{s_t \in S_t} \Delta R(s_t, t)$. This procedure is repeated until t satisfies the pre-specified stopping condition, such as $\Delta R(s_t^*, t) \leq R^*$. We denote the stopped tree $T_{\max}$.

(2) A receding process to prune the tree:

Let us define an adapting degree of T for the dataset as $R(T) = \sum_{t \in \tilde{T}} R(t)$. As T increases, $R(T)$ decreases. However, a too large T will produce overfitting and cause large classification errors for the future dataset. Thus, $R_\alpha(T)$ with a penalty of the complexity parameter α is defined as $R_\alpha(T) = R(T) + \alpha|\tilde{T}|$ and a subtree $T(\alpha)$ of $T_{\max}$ satisfying $T(\alpha) = \arg\min_{T \preceq T_{\max}} R_\alpha(T)$ can be found. Corresponding to the increasing α, $0 = \alpha_0 < \alpha_1 < \alpha_2 < ...$, the nesting sequence of subtree $T_{\max} = T_0 \succ T_1 \succ ... \succ T_J = \{t_1\}$ can be found.

(3) A process to select the best tree:

The cross-validation method can obtain the expectation of $R(T_j)$ ($j = 0, 1, ..., J$) $R^{CV}(T_j)$ and the minimum $R^{CV}(T_{j0})$: $R^{CV}(T_{j0}) = \min_j R^{CV}(T_j)$. However, the following T_{j1} satisfying the one standard error (1SE) rule is often used: $R^{CV}(T_{j1}) \leq R^{CV}(T_{j0}) + \hat{SE}(R^{CV}(T_{j0}))$ where $j1$ is the maximum tree number satisfying the inequality and $\hat{SE}(R^{CV}(T_{j0}))$ is the estimated standard deviation of $R^{CV}(T_{j0})$.

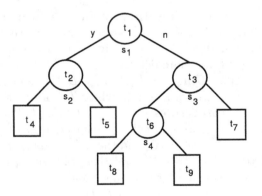

Fig. 2. An example of tree T.

3.2   Simulation Experiment with CART

We conducted a simulation experiment of CART on SM for DG & VIR in Fig. 1. Specifically, we specified the rules shown in Table 2 denoting, for example, $CP(1,1) = 110000$ with $CP(1,1) = (C(1) = 1) \wedge (C(2) = 1)$ as the condition part of the if-then rule, where $|C| = 6$, $V_a = \{1, 2, ..., 6\}$ ($a = C(j)$ ($j = 1, ..., |C|$), $a = D$). Then, we generated $v_{C(j)}(i)$ ($j = 1, ..., |C| = 6$) with a uniform distribution and formed $u^C(i) = (v_{C(1)}(i), ..., v_{C(6)}(i))(i = 1, ..., N = 10,000)$. Next, we transformed $u^C(i)$ into $u^D(i)$ using the pre-specified rules in Table 2 and the hypotheses in Table 1 without generating NoiseC and NoiseD for a plain

Table 2. An example of pre-specified rules in the rule box in Fig. 1.

$R(d,k)$	$CP(d,k)$	$D = d$
$R(1,1)$	110000	$D = 1$
$R(1,2)$	001100	$D = 1$
$R(2,1)$	220000	$D = 2$
$R(2,2)$	002200	$D = 2$
$R(3,1)$	330000	$D = 3$
$R(3,2)$	003300	$D = 3$
$R(4,1)$	440000	$D = 4$
$R(4,2)$	004400	$D = 4$
$R(5,1)$	550000	$D = 5$
$R(5,2)$	000500	$D = 5$
$R(6,1)$	660000	$D = 6$
$R(6,2)$	006600	$D = 6$

experiment. We randomly sampled $N_B = 5,000$ and formed a new dataset as the decision table. Finally, we applied the sampled dataset to CART, which was already implemented and freely presented as the function *rpart* in the R programming language [18].

Table 3 shows an example of the output by *rpart* in the list structure obtained through the procedures (1)–(3) mentioned in Sect. 3.1, although CART also outputs the tree structure. When the tree structure becomes too large and complicated, the list structure becomes easier to handle and understand the analyzed results. Table 3 shows the following:

(1) The node 1) at Line Number 1 (LN $= 1$) is the root t_1 and contains $5,000$ data points. If the node is represented by $D = 5$ which has the most frequent occurrence of $u^D(i)$ ($i = 1, ..., N_B$), the $4,139$ objects of $u(i)$ will be lost. The occurrence rates of $D = 1, ...6$ are (0.17 0.16 0.17 0.17 0.17 0.16) respectively.

(2) LN $= 2$ shows that node 2) which is obtained by splitting the parent node 1) with the condition $C(3) = 1 \lor 2 \lor 5 \lor 6$ ($= s_1$), holds $3,402$ objects of $u(i)$ satisfying the condition, and if the node is represented by the most frequent occurrence attribute value $D = 1$, then the node will lose $2,701$ objects of $u(i)$. The same applies hereafter.

(3) LN $= 5$ shows that node 16) is a terminal node obtained by splitting the parent node 8) with the condition $C(4) = 1$. It holds 143 objects of $u(i)$ satisfying the condition and can be represented by $D = 1$ permitting the loss of 10 objects of $u(i)$. By tracing the nodes 16) $\rightarrow$ 8) $\rightarrow$ 4) $\rightarrow$ 2) accumulating and arranging the split conditions, we obtain $(C(4) = 1) \land (C(3) = 1) \land (C(4) = 1 \lor 2 \lor 3 \lor 4 \lor 5) \land (C(3) = 1 \lor 2 \lor 5 \lor 6) = (C(3) = 1) \land (C(4) = 1)$. That is, the following product form of an if-then rule with rule length 2 ($RL = 2$) is obtained: if $(C(3) = 1) \land (C(4) = 1)$ then $D = 1$.

Table 3. An example of the output by *rpart*.

Line Number	Output Node Information (node), split, n, loss, yval, (yprob), * denotes terminal node
1	1) root 5000 4139 5 (0.17 0.16 0.17 0.17 0.17 0.16)
2	2) C3=1,2,5,6 3402 2760 1 (0.19 0.18 0.13 0.14 0.18 0.17)
3	4) C4=1,2,3,4,5 2827 2263 1 (0.2 0.2 0.14 0.14 0.19 0.13)
4	8) C3=1 676 443 1 (0.34 0.14 0.12 0.14 0.12 0.13)
5	16) C4=1 143 10 1 (0.93 0.014 0.014 0 0.021 0.021) *
6	17) C4=2,3,4,5 533 433 1 (0.19 0.18 0.15 0.18 0.14 0.16)
6	34) C1=2,3,5 276 209 2 (0.15 0.24 0.2 0.13 0.16 0.12)
7	68) C2=1,3,6 147 106 3 (0.14 0.2 0.28 0.15 0.088 0.14) *
8	69) C2=2,4,5 129 92 2 (0.16 0.29 0.1 0.12 0.25 0.093)
...	... ...
136	123) C2=6 7 2 6 (0 0.14 0 0.14 0 0.71) *
137	31) C3=4 130 11 4 (0.0077 0.015 0.023 0.92 0.015 0.023) *

Table 4. Arrangement of induced rules for each D by rule length.

$D = d$	Number of rules by rule length						
	1	2	3	4	5	6	Total
1	0	1	0	86	458	792	1,337
2	0	1	0	159	94	1,296	1,550
3	0	2	0	75	222	0	299
4	0	1	0	150	96	1,296	1,543
5	0	1	4	237	36	0	278
6	0	1	0	44	60	504	609
Total	0	7	4	751	966	3,888	5,616

(4) LN $= 7$ also shows a terminal node and derives the if-then rule: if $(C(1) = 2 \vee 3 \vee 5) \wedge (C(2) = 1 \vee 3 \vee 6) \wedge (C(3) = 1) \wedge (C(4) = 2 \vee 3 \vee 4 \vee 5)$ then $D = 3$ by tracing node 68) $\to$ 34) $\to$ 17) $\to$ 8) $\to$ 4) $\to$ 2), accumulating and arranging the split conditions. This rule contains 36 rules of the product form with $RL = 4$ such as if $(C(1) = 2) \wedge (C(2) = 1) \wedge (C(3) = 1) \wedge (C(4) = 2)$ then $D = 3$.

Arranging the if-then rules contained in Table 3 with the product form produces the amount of rules for each D by RL, as shown in Table 4, which is then compared with the specified rules in Table 2. Table 4 has the following implications:

(i) CART induced seven rules with $RL = 2$. Six of the seven coincided with the specified rules: $R(1,2)$, $R(2,2)$, $R(3,2)$, $R(4,2)$, $R(5,2)$, and $R(6,2)$ in Table 2. The other at $D = 3$ was the rule: if $(C(3) = 4) \wedge (C(4) = 3)$ then $D = 3$ and did not coincide with a pre-specified rule.

(ii) Excluding the six rules, CART induced unnecessary and/or partial rules with respect to the pre-specified rules, amounting to $5,610$ from the decision table of $|U| = 5,000$ each of which can be recognized as an if-then rule of $RL = 6$. That is, CART may create new unrelated rules from the decision table while arranging the decision table and inducing rules from it.

Implication (i) is inferred as follows. As mentioned in (3) of Table 3, CART split the root node and induced the rule $R(1,2)$: if $(C(3) = 1) \wedge (C(4) = 1)$ then $D = 1$. By contrast, the data not satisfying $(C(3) = 1)$, that were satisfying $C(3) = \bar{1} = 2 \vee 3... \vee 6$ were used for inducing the rule if $(C(3) = 2 \vee 3... \vee 6) \wedge ... \wedge (C(1) = 1) \wedge (C(2) = 1)$ then $D = 1$, which was included in the rule set of $D = 1$ of $RL = 3,...,6$ in Table 4. Thus, CART induces only the partial rule of $R(1,1)$. Figure 3 is a simplified illustration of this process. The same reasoning applies to the rules for $D = 2,...,6$.

Generally, tree based approaches, including CART carry the restriction that $U(R(j1,1)) \cap U(R(j2,2)) = \phi$, $(j1, j2 = 1,...,|V_{a=D}|, j1 \neq j2)$, where $U(R(j1,1))$ is the subset of U satisfying $R(j1,1)$. Accordingly, they cannot express the conflict rules. Real-world datasets include not only the conflicting data but also the indifferent data (see Table 1). In addition, this approach cannot eliminate the indifferent data.

From the above considerations of implication (i), the tree based approach will cause, for example, $C(1) = \bar{1} = 2 \vee 3... \vee 6, ..., C(6) = \bar{1} = 2 \vee 3... \vee 6$, which is why CART caused more rules than $|U| = 5,000$ (see implication (ii)).

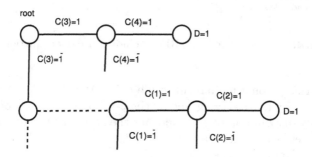

Fig. 3. Simplified Tree diagram derived from Table 3.

4   Experimental Studies of STRIM

We proposed STRIM [1–9] which statistically interprets the classical Rough Sets theory and we studied its validity based on the model shown in Fig. 1 before applying it to a real-world dataset to confirm its usefulness. The outline of the algorithm is shown in C-language style in Fig. 4 (details in [8,9]). At LN(Line Number) $= 8$–9, for each decision attribute value di, the statistically independent condition attributes against di are reduced. At LN $= 10$, the function rule_check() (the body is at LN $= 19$–33) systematically forms a trying rule by

Line   Algorithm to induce if-then rules by STRIM with a reduct function
Number

```
1     int main(void) {
2     int rdct_max[|CV|]={0,...,0}; //initialize maximum value of C(j)
3     int rdct[|CV|]={0,...,0}; //initialize reduct results by D=1
4     int rule[|C|]={0,...,0}; //initialize trying rules
5     int tail=-1; //initialize value set
6     input data; // set decision table
7     for (di=1; di<=|D|; di++) {// induce rule candidates every D=1
8     attribute_reduct(rdct_max)
9     set rdct[ck] ; // if (rdct_max[ck]==0) {rdct[ck]=0; }else {rdct[ck]=1; }
10    rule_check(rcdct, redct_max, tail, rule); // the first stage process
11    }// end di
12    arrange rule candidates // the second stage
13    }// end main
14    int attribute_reduct(int rdct_max[]) {
15    make contingency table for D=1 vs. C(j)
16    Test H0(j,l);
17    if H0(j,l) is rejected then set rdct_max[j,l]=jmax else rdct_max[j,l]=0; //
      jmax:the attribute value of the maximum frequency
18    }// end of attribute_reduct
19    int rule_check(int rdct[], int rdct_max[], int tail,int rule[]) {// the first stage
      process
20    for (ci=tail+1; cj<|C|; ci++) {
21    for (cj=1; cj<=rdct[ci]; cj++) {
22    rule[ci]=rdct_max[cj]; // a trying rule set for testing
23    count frequency of the trying rule; // count n1, n2, ...
24    if (frequency>=N0) {//sufficient frequency ?
25    if (|z|>3.0) {//sufficient evidence ?
26    add the trying rule as a rule candidate
27    }// end of if |z|
28    rule_check(ci,rule)
29    }// end if frequency
30    }// end cj
31    rule[ci]=0; // trying rules reset
32    }// end ci
33    }// end rule_check
```

Fig. 4. An algorithm for STRIM including a reduct function.

the dimension rule[] (condition part of a rule CP). At LN $= 25$, we examine the degree of the validity for the trying rule by the z-value which is the degree of bias in the frequency distribution of D supposing the standard normal distribution. This is used to select the rule as a candidate. The selected candidates are finally arranged into the induced rules at LN $= 12$.

Table 5 shows that STRIM with the dataset corresponding to Table 3 induced all the rules specified in Table 2. For example, $R(7)$ in Table 5 coincides with those at LN $= 5$ in Table 3. The frequency distribution of D $(n_1, ..., n_6) = (133, 2, 2, 0, 3, 3)$ is extremely biased at $D = 1$ representing the decision attribute although CART shows the rates in place of the frequency distribution. The other frequencies $(n_2, ..., n_6) = (2, 2, 0, 3, 3)$ may have been caused by the conflict. The accuracy, coverage and p-value corresponding to the z-value are also shown.

Let us consider why large differences exist between the rule induction results by CART and STRIM. As mentioned above, STRIM systematically explores the condition part of an if-then rule as CP, statistically tests whether the frequency distribution $(n_1, ..., n_6)$ of the $U(CP)$ has bias or not and induces the set of rules

Table 5. Induced rules by STRIM for the dataset corresponding to Table 3.

Induced rule $R(i)$	Condition part C	D	p-value(z)	$(n_1,..,n_6)$	Accuracy	Coverage
1	66000	6	1.38E−160(26.98)	(1, 2, 2, 3, 1, 149)	0.9430	0.1865
2	33000	3	5.45E−151(26.15)	(0, 0, 139, 1, 1, 1)	0.9789	0.167
3	00220	2	3.17E−145(25.63)	(2, 138, 1, 2, 1, 1)	0.952	0.1673
4	00660	6	1.27E−137(24.94)	(6, 1, 4, 2, 3, 137)	0.895	0.171
5	11000	1	8.81E−133(24.49)	(126, 1, 1, 0, 1, 0)	0.977	0.148
6	00550	5	2.16E−131(24.36)	(1, 1, 1, 2, 133, 3)	0.943	0.154
7	00110	1	1.24E−130(24.29)	(133, 2, 2, 0, 3, 3)	0.930	0.156
8	55000	5	1.35E−129(24.19)	(1, 0, 0, 2, 130, 4)	0.949	0.151
9	00330	3	1.59E−127(23.99)	(3, 3, 130, 2, 2, 3)	0.909	0.157
10	22000	2	6.93E−124(23.64)	(0, 116, 1, 1, 2, 1)	0.959	0.141
11	44000	4	4.35E−123(23.57)	(2, 2, 0, 123, 2, 4)	0.925	0.147
12	00440	4	1.74E−117(23.01)	(1, 2, 3, 119, 2, 3)	0.915	0.143

Table 6. Arrangement of induced rules by CART for the Rakuten travel dataset.

$D = d$	Number of rules by rule length						
	1	2	3	4	5	6	Total
1	1	0	2	20	22	0	45
2	0	0	20	95	112	306	533
3	0	0	3	43	24	324	394
4	0	0	5	43	75	48	171
5	0	1	0	17	24	0	42
Total	1	1	30	218	257	678	1185

with a high p-value. In this process, the conflicting data partially included in the $U(CP)$ hardly contributes to bias and the $U(CP)$ supported by the indifferent dataset can be easily removed as shown in Table 5, since such a dataset barely causes bias. Even in the case when NoiseC and NoiseD are contaminated in $u(i)$ with a high percentage, STRIM has a high rate of inducing the pre-specified rules. That is, in terms of rule induction, STRIM is robust against such noises [4].

On the other hand, tree based approaches, including CART divide the given dataset into many parts recursively thus splitting the parts based on the criterion of the impurity where each part of the dataset included in the terminal node induces rules. However, there is no way to properly remove the rules controlled by the indifferent data or to handle the conflicting data, which induce many meaningless rules as shown in Table 4 and Fig. 3. However, the criteria inducing rules by the bias of STRIM and the reduction of the impurity of CART use the same concept so CART partially induces the same rules as STRIM.

Table 7. Induced rules for the Rakuten dataset by STRIM.

Induced rule $R(i)$	Condition part C	D	p-value(z)	(n_1,..,n_6)	Accuracy	Coverage
1	005050	5	0.0(46.31)	(6, 4, 6, 74, 662)	0.88	0.65
2	055000	5	0.0(44.28)	(9, 9, 9, 46, 594)	0.89	0.59
3	005005	5	0.0(41.28)	(6, 3, 5, 27, 488)	0.92	0.48
4	000010	1	0.0(40.72)	(634, 175, 26, 4, 0)	0.76	0.64
5	040040	4	5.90E−183(28.82)	(7, 22, 39, 338, 60)	0.73	0.34
6	000044	4	3.92E−166(27.44)	(8, 35, 57, 380, 110)	0.64	0.39
7	030030	3	1.45E−164(27.31)	(43, 109, 407, 88, 2)	0.63	0.40
8	004004	4	6.28E−138(24.97)	(10, 19, 39, 296, 81)	0.67	0.30
9	020000	2	1.02E−84(19.47)	(146, 344, 168, 28, 1)	0.50	0.34
CART	050050	5	5.77E−281(35.80)	(1, 2, 24, 147, 511)	0.75	0.51

5   Studies with a Real-World Dataset

The Rakuten Institute of Technology provides an open dataset of Rakuten Travel [19]. This dataset contains about $6,200,000$ questionnaire survey ratings $A = \{C(1) = \text{Location}, C(2) = \text{Room}, C(3) = \text{Meal}, C(4) = \text{Bath (Hot Spring)}, C(5) = \text{Service}, C(6)=\text{Amenity}, D = \text{Overall} \}$ for about $130,000$ travel facilities using a set of categorical values $V_a = \{\text{Dissatisfied (1), Somewhat dissatisfied (2), Neither satisfied nor dissatisfied (3), Satisfied (4), Very Satisfied (5)}\}$, $\forall a \in A$, that is, $|V_{a=D}| = |V_{a=C(j)}| = 5$. We constructed a decision table of $N = 10,000$ surveys by randomly selecting $2,000$ samples, each with $D = m$ ($m = 1, ..., 5$), from about $400,000$ surveys of the 2013–2014 dataset because there were heavy biases with respect to the frequency of $D = m$. Finally we randomly sampled $N_B = 5,000$ from the $10,000$ surveys and re-constructed the decision table.

We applied CART to the decision table and arranged the results shown in Table 6 in the same way as Table 4. Table 6 shows the same tendency as Table 4. That is, CART induced many rules with long rule length although the specified rules of a real-world dataset were unknown.

We applied STRIM to the same dataset and obtained Table 7, which shows the following:

(1) The rule lengths of all induced rules were less or equal than two. $R(4)$ coincides with the rule at $D = 1$ of RL=1 in Table 6.
(2) The rule at $D = 5$ of $RL = 2$ in Table 6 corresponds to $R(1)$, $R(2)$ or $R(3)$ of $D = 5$ in Table 7 so that the rule is specifically written as CART in Table 7. Comparing with the z-value, accuracy and coverage of $R(1)$, $R(2)$ or $R(3)$, those of the CART rule in Table 7 was lower than the STRIM equivalents except the coverage of $R(3)$ whereas the other rules of $D = 5$ by CART were included in the rules with RL ≥ 4. Specifically, STRIM finds $R(3)$ independent of that of CART. Accordingly, the rules of $D = 5$ by STRIM seems more useful than those by CART.
(3) The rules induced by STRIM in Table 7 indicate that $C(1)$ (Location) and $C(4)$ (Bath (Hot Spring)) can be commonly reducted through $D = 1, ..., 5$ while CART commonly induced many rules with RL ≥ 5 due to the lack of the reduct function; all of these rules appear meaningless.

Combining these considerations with the simulation experiment results, we found that CART barely induces the proper or valid rules compared to those by STRIM; however, we note that the proper and/or valid rules are unknown when using a real-world dataset and can only be guessed using the domain knowledge.

6   Conclusion

Methods to extract and/or find knowledge and/or information from large datasets are actively researched. We focused on CART as a basic tree based approach with a response variable and explanatory variables. CART presents their relationships in a tree structure and is used for classification problems. Specifically, we investigated the validity and availability of CART for inducing a tree structure in a simulation model of data generation and verification of induced rules (SM for DG & VIR) under the condition that both variables took qualitative values (although CART can also handle quantitative variables). We then compared CART with STRIM expanding the classical Rough Sets theory. We examined CART's validity and performance in if-then rule induction in a simulation experiment since the tree structure is a kind of if-then rule structure and can be easily transformed into the form of the rules. The following list presents our key findings with respect to the SM for DG & VIR:

(1) CART is likely to induce a large number of rules with longer rule length than those of the pre-specified rules and most of these appear meaningless.
(2) CART cannot properly handle the conflicting data nor effectively remove the indifferent data.
(3) CART has no way of reducting the explanatory variables (the condition attributes) that have nothing to do with the response variable (the decision attribute).
(4) The cause of (1), (2) and (3) is that CART only recursively splits the subset of the given dataset U based on the criterion reducing the impurity until the subset meets the stop condition of the splitting. CART does not have a reproducing and/or reusing process of the split subset data, and thus the rules depend on the subset already split (see Fig. 3). By contrast, STRIM does not split U and uses the subset of U any number of times to test a trying rule with a clear criterion that the rule has bias.
(5) The criterion reducing the impurity by CART and the bias criterion by STRIM are based on a common concept for rule induction and thus both methods partially induce the same rules.

After considering the above findings, we applied CART to the real-world dataset of Rakuten Travel and confirmed the same tendency as in the simulation by comparing CART's induced rules with those of STRIM. Note that judging which rules induced by CART should be adopted would be very difficult without conducting the simulation.

Many tree based approaches derived from CART have been proposed such as MART [15], Bagging [16], and Random Forests [17] and so on [20–22]. Future work should study these methods in SM for DG & VIR to clarify their features before applying them to real-world datasets.

Acknowledgements. We truly thank Rakuten Inc. for presenting Rakuten Travel dataset [19].

References

1. Matsubayashi, T., Kato, Y., Saeki, T.: A new rule induction method from a decision table using a statistical test. In: Li, T., et al. (eds.) RSKT 2012. LNCS (LNAI), vol. 7414, pp. 81–90. Springer, Heidelberg (2012). https://doi.org/10.1007/978-3-642-31900-6_11
2. Kato, Y., Saeki, T., Mizuno, S.: Studies on the necessary data size for rule induction by STRIM. In: Lingras, P., Wolski, M., Cornelis, C., Mitra, S., Wasilewski, P. (eds.) RSKT 2013. LNCS (LNAI), vol. 8171, pp. 213–220. Springer, Heidelberg (2013). https://doi.org/10.1007/978-3-642-41299-8_20
3. Kato, Y., Saeki, T., Mizuno, S.: Considerations on rule induction procedures by STRIM and their relationship to VPRS. In: Kryszkiewicz, M., Cornelis, C., Ciucci, D., Medina-Moreno, J., Motoda, H., Raś, Z.W. (eds.) RSEISP 2014. LNCS (LNAI), vol. 8537, pp. 198–208. Springer, Cham (2014). https://doi.org/10.1007/978-3-319-08729-0_19
4. Saeki, T., Kato, Y., Mizuno, S.: Studies of rule induction by STRIM from the decision table with contaminated attribute values from missing data and noise – in the case of critical dataset size–. World Acad. Sci. Eng. Technol. Int. J. Comput. Electr. Autom. Control Inf. Eng. **19**(6), 1244–1249 (2015)
5. Kato, Y., Saeki, T., Mizuno, S.: Proposal of a statistical test rule induction method by use of the decision table. Appl. Soft Comput. **28**, 160–166 (2015)
6. Kato, Y., Saeki, T., Mizuno, S.: Proposal for a statistical reduct method for decision tables. In: Ciucci, D., Wang, G., Mitra, S., Wu, W.-Z. (eds.) RSKT 2015. LNCS (LNAI), vol. 9436, pp. 140–152. Springer, Cham (2015). https://doi.org/10.1007/978-3-319-25754-9_13
7. Kitazaki, Y., Saeki, T., Kato, Y.: Performance comparison to a classification problem by the second method of quantification and STRIM. In: Flores, V., et al. (eds.) IJCRS 2016. LNCS (LNAI), vol. 9920, pp. 406–415. Springer, Cham (2016). https://doi.org/10.1007/978-3-319-47160-0_37
8. Fei, J., Saeki, T., Kato, Y.: Proposal for a new reduct method for decision tables and an improved STRIM. In: Tan, Y., Takagi, H., Shi, Y. (eds.) DMBD 2017. LNCS, vol. 10387, pp. 366–378. Springer, Cham (2017). https://doi.org/10.1007/978-3-319-61845-6_37
9. Kato, Y., Itsuno, T., Saeki, T.: Proposal of dominance-based rough set approach by STRIM and its applied example. In: Polkowski, L., et al. (eds.) IJCRS 2017. LNCS (LNAI), vol. 10313, pp. 418–431. Springer, Cham (2017). https://doi.org/10.1007/978-3-319-60837-2_35
10. Pawlak, Z.: Rough sets. Int. J. Inf. Comput. Sci. **11**(5), 341–356 (1982)
11. Skowron, A., Rauser, C.M.: The discernibility matrix and functions in information systems. In: Słowiński, R. (ed.) Intelligent Decision Support. Handbook of Application and Advances of Rough Set Theory, vol. 11, pp. 331–362. Kluwer Academic Publishers, Dordrecht (1992). https://doi.org/10.1007/978-94-015-7975-9_21

12. Grzymala-Busse, J.W.: LERS – a system for learning from examples based on rough sets. In: Słowiński, R. (ed.) Intelligent Decision Support. Handbook of Applications and Advances of the Rough Sets Theory, vol. 11, pp. 3–18. Kluwer Academic Publishers, Dordrecht (1992). https://doi.org/10.1007/978-94-015-7975-9_1

13. Ziarko, W.: Variable precision rough set model. J. Comput. Syst. Sci. **46**, 39–59 (1993)

14. Brieman, L., Frieman, J.H., Olshen, R.A., Stone, C.J.: Classification and Regression Trees. Chapman & Hall, New York (1984)

15. Frieman, J.H.: Greedy function approximation: gradient boosting machine. Ann. Stat. **29**, 1189–1232 (2001)

16. Brieman, L.: Bagging predictions. Mach. Learn. **26**(2), 123–140 (1996)

17. Brieman, L.: Random forests. Mach. Learn. **45**(1), 5–23 (2001)

18. https://cran.r-project.org/web/packages/rpart/index.html

19. http://rit.rakuten.co.jp/opendataj.html

20. Zheng, Z., Wang, G., Wu, Y.: A rough set and rule tree based incremental knowledge acquisition algorithm. In: Wang, G., Liu, Q., Yao, Y., Skowron, A. (eds.) RSFDGrC 2003. LNCS (LNAI), vol. 2639, pp. 122–129. Springer, Heidelberg (2003). https://doi.org/10.1007/3-540-39205-X_16

21. Sikder, I.U., Munakata, T.: Application of rough set and decision tree for characterization of premonitory factors of low seismic activity. Expert Syst. Appl. **36**, 102–110 (2009)

22. Buregwa-Czuma, S., Bazan, J.G., Bazan-Socha, S., Rzasa, W., Dydo, L., Skowron, A.: Resolving the conflicts between cuts in a decision tree with verifying cuts. In: Polkowski, L., et al. (eds.) IJCRS 2017. LNCS (LNAI), vol. 10314, pp. 403–422. Springer, Cham (2017). https://doi.org/10.1007/978-3-319-60840-2_30

Rseslib 3: Open Source Library of Rough Set and Machine Learning Methods

Arkadiusz Wojna[1(✉)] and Rafał Latkowski[2]

[1] Security On-Demand, 12121 Scripps Summit Dr 320, San Diego, CA 92131, USA
[2] Loyalty Partner, Złota 59, 00-120 Warsaw, Poland
{wojna,rlatkows}@mimuw.edu.pl

Abstract. The paper presents a new generation of Rseslib library - a collection of rough set and machine learning algorithms and data structures in Java. It provides algorithms for discretization, discernibility matrix, reducts, decision rules and for other concepts of rough set theory and other data mining methods. The third version was implemented from scratch and in contrast to its predecessor it is available as a separate open-source library with API and with modular architecture aimed at high reusability and substitutability of its components. The new version can be used within Weka and with a dedicated graphical interface. Computations in Rseslib 3 can be also distributed over a network.

1 Introduction

Rough set theory [15] was introduced by Pawlak as a methodology for data analysis based on approximation of concepts in information systems. Discernibility is a key concept in this methodology, which is the ability to distinguish objects, based on their attribute values. Along with theoretical research rough sets were developed in practical directions as well. To facilitate applications software tools implementing rough set concepts and methods have been developed. This paper describes one of such tools.

Rseslib 3 is a library of rough set and machine learning algorithms and data structures implemented in Java. It is the successor of Rseslib 2 used in Rough Set Exploration System (RSES) [2]. The first version of the library started in 1993 and was implemented in C++. It was used as the core of Rosetta system [14]. Rseslib 2 was the first version of the library implemented in Java and it stands for the core of RSES. The third version of the library was entirely redesigned and all the methods available in this version were implemented from scratch. The following features are distinguishing the version 3 from its predecessor:

- available as a library with an API
- open source distributed under GNU GPL license
- modular component-based architecture
- easy-to-reuse data representations and methods
- easy-to-substitute components
- available in Weka.

© Springer Nature Switzerland AG 2018
H. S. Nguyen et al. (Eds.): IJCRS 2018, LNAI 11103, pp. 162–176, 2018.
https://doi.org/10.1007/978-3-319-99368-3_13

As open source library of rough set methods in Java Rseslib 3 fills in an uncovered gap in the spectrum of rough set software tools. The algorithms in Rseslib 3 can be used both by users who need to apply ready-to-use rough set methods in their data analysis tasks as well as by researchers interested in extension of the existing rough set methods who can use the source code of the library as the basis for their extended implementations. The library can be used also within the following external tools: Weka [1], the dedicated graphical interface Qmak and Simple Grid Manager distributing computations over a network of computers.

The library is not limited to rough sets, it contains and is open to concepts and algorithms from other areas of machine learning and data mining. That is related to another goal of the project which is to provide a universal library of highly reusable and substitutable components at a very elementary level unmet in open source data mining Java libraries available today.

Looking for analogous open source Java projects one can find Modlem[1] and Richard Jensen's programs[2]. Modlem is a Weka package providing a covering algorithm inducing decision rules. The algorithm contains some aspects of rough set theory. Richard Jensen developed a number of programs in Java providing various rough set methods, some of them are provided with their source code.

There are useful libraries of rough set methods developed in other programming languages: RoughSets [18] in R and NRough [23] in C#. RoughSets package was extended with RapidRoughSets [8] - an extension facilitating the use of the package in RapidMiner, a popular java platform for data mining, machine learning and predictive analytics. There are a number of tools providing rough set methods within graphical interface like RSES [2], Rosetta [14] or ROSE [17].

2 Data

The concept of the library is based on classical representation of data in machine learning. It is assumed that a finite set of objects U, a finite set of conditional attributes $A = \{a_1, \ldots, a_n\}$ and a decision attribute dec are given. Each object $x \in U$ is represented by a vector of values $(x_1, \ldots, x_n)$. The value x_i is the value of the attribute a_i on the object x belonging to the domain of values V_i corresponding to the attribute a_i: $x_i \in V_i$. The type of a conditional attribute a_i can be either numerical, if its values are comparable and can be represented by numbers $V_i \subseteq \mathbb{R}$ (e.g.: age, temperature, height), or nominal, if its values are incomparable, i.e., if there is no linear order on V_i (e.g.: color, sex, shape).

The library contains many algorithms implementing various methods of supervised learning. These methods assume that each object $x \in U$ is assigned with a value of the decision attribute $dec(x)$ called a decision class and they learn from the objects in U a function approximating the real function dec on all objects outside U. At present the algorithms in the library assume that the domain of values of the decision attribute dec is discrete and finite: $V_{dec} = \{d_1, \ldots, d_m\}$.

The library reads data from files in three formats: ARFF, CSV and RSES2.

---

[1] https://sourceforge.net/projects/modlem.
[2] http://users.aber.ac.uk/rkj/site/?page_id=79.

3   Discretizations

Some algorithms require data in form of nominal attributes, e.g. some rule based algorithms like the rough set based classifier. Discretization (known also as quantization or binning) is data transformation converting data from numeric attributes into nominal attributes. The library provides a number of discretization methods. Each method splits domain of a numerical attribute into a number of disjoint intervals. New nominal attribute is formed by encoding a numerical value into an identifier of an interval.

The following discretization methods are available in Rseslib:

- Equal width intervals
- Equal frequency intervals
- Holte's 1R algorithm [7]
- Entropy minimization (static and dynamic) [5]
- ChiMerge algorithm [10]
- Maximal discernibility (MD) heuristic (global and local) [13].

4   Discernibility Matrix

Computation of reducts is based on the concept of discernibility matrix [21]. The library provides 4 types of discernibility matrix. Each type is $|U| \times |U|$ matrix defined for all pairs of objects $x, y \in U$. The values of discernibility matrix $M(x, y)$ are defined as the subsets of the set of conditional attributes: $M(x, y) \subseteq A$. If a data set contains numerical attributes discernibility matrix can be computed using either the original or the discretized numerical attributes.

The first type of discernibility matrix M^{all} depends on the values of the conditional attributes only, it does not take the decision attribute into account:

$$M^{all}(x, y) = \{a_i \in A : x_i \neq y_i\}$$

In many applications, e.g. in object classification, we want to discern objects only if they have different decisions. The second type of discernibility matrix M^{dec} discerns objects from different decision classes:

$$M^{dec}(x, y) = \begin{cases} \{a_i \in A : x_i \neq y_i\} & \text{if } dec(x) \neq dec(y) \\ \emptyset & \text{if } dec(x) = dec(y) \end{cases}$$

If data are inconsistent, i.e. if there are one or more pairs of objects with different decisions and with equal values on all conditional attributes then $M^{dec}(x, y) = \emptyset$ like for pairs of objects with the same decision. To overcome this inconsistency the concept of generalized decision was introduced [16,20]:

$$\partial(x) = \{d \in V_{dec} : \exists y \in U : \forall a_i \in A : x_i = y_i \wedge dec(y) = d\}$$

If U contains inconsistent objects x, y they have the same generalized decision. The next type of discernibility matrix M^{gen} is based on generalized decision:

$$M^{gen}(x, y) = \begin{cases} \{a_i \in A : x_i \neq y_i\} & \text{if } \partial(x) \neq \partial(y) \\ \emptyset & \text{if } \partial(x) = \partial(y) \end{cases}$$

This type of discernibility matrix removes inconsistencies but discerns pairs of objects with the same original decision, e.g. an inconsistent object from a consistent object. The fourth type of discernibility matrix M^{both} discerns a pair of objects only if they have both the original and the generalized decision different:

$$M^{both}(x, y) = \begin{cases} \{a_i \in A : x_i \neq y_i\} & \text{if } \partial(x) \neq \partial(y) \wedge dec(x) \neq dec(y) \\ \emptyset & \text{if } \partial(x) = \partial(y) \vee dec(x) = dec(y) \end{cases}$$

Data can contain missing values. All types of discernibility matrix available in the library have 3 modes to handle missing values [11]:

- different value — an attribute a_i discerns x, y if the value of one of them on a_i is defined and the value of the second one is missing (missing value is treated as yet another value): $a_i \notin M(x, y) \Leftrightarrow x_i = y_i \vee (x_i = * \wedge y_i = *)$
- symmetric similarity — an attribute a_i does not discern x, y if the value of any of them on a_i is missing: $a_i \notin M(x, y) \Leftrightarrow x_i = y_i \vee x_i = * \vee y_i = *$
- nonsymmetric similarity — asymmetric discernibility relation between x and y: $a_i \notin M(x, y) \Leftrightarrow (x_i = y_i \wedge y_i \neq *) \vee x_i = *$.

The first mode treating missing value as yet another value keeps indiscernibility relation transitive but the next two modes make it intransitive. Such a relation is not an equivalence relation and does not define correctly indiscernibility classes in the set U. To eliminate that problem the library provides an option to transitively close an intransitive indiscernibility relation.

5   Reducts

Reduct [21] is a key concept in rough set theory. It can be used to remove some data without loss of information or to generate decision rules.

Definition 1. *The subset of attributes $R \subseteq A$ is a (global) reduct in relation to a discernibility matrix M if each pair of objects discernible by M is discerned by at least one attribute from R and no proper subset of R holds that property:*

$$\forall x, y \in U : M(x, y) \neq \emptyset \Rightarrow R \cap M(x, y) \neq \emptyset$$

$$\forall R' \subsetneq R \, \exists x, y \in U : M(x, y) \neq \emptyset \wedge R' \cap M(x, y) = \emptyset$$

If M is a decision-dependent discernibility matrix the reducts related to M are the reducts related to the decision attribute *dec*.

Reducts defined in Definition 1 called also global reducts are sometimes too large and generate too specific rules. To overcome this problem the notion of local reducts was introduced [26].

Definition 2. *The subset of attributes $R \subseteq A$ is a local reduct in relation to a discernibility matrix M and an object $x \in U$ if each object $y \in U$ discerned from x by M is discerned from x by at least one attribute from R and no proper subset of R holds that property:*

$$\forall y \in U : M(x,y) \neq \emptyset \Rightarrow R \cap M(x,y) \neq \emptyset$$

$$\forall R' \subsetneq R \exists y \in U : M(x,y) \neq \emptyset \wedge R' \cap M(x,y) = \emptyset$$

It may happen that local reducts are still too large. In the extreme situation there is only one global or local reduct equal to the whole set of attributes A. In such situations partial reducts [12] can be helpful.

Let P be the set of all pairs of objects $x, y \in U$ discerned by a discernibility matrix M: $P = \{\{x,y\} \subseteq U : M(x,y) \neq \emptyset\}$ and let $\alpha \in (0;1)$.

Definition 3. *The subset of attributes $R \subseteq A$ is a global α-reduct in relation to a discernibility matrix M if it discerns at least $(1 - \alpha)|P|$ pairs of objects discernible by M and no proper subset of R holds that property:*

$$|\{\{x,y\} \subseteq U : R \cap M(x,y) \neq \emptyset\}| \geq (1 - \alpha)|P|$$

$$\forall R' \subsetneq R : |\{\{x,y\} \subseteq U : R' \cap M(x,y) \neq \emptyset\}| < (1 - \alpha)|P|$$

Let $P(x)$ be the set of all objects $y \in U$ discerned from $x \in U$ by a discernibility matrix M: $P(x) = \{y \in U : M(x,y) \neq \emptyset\}$ and let $\alpha \in (0;1)$.

Definition 4. *The subset of attributes $R \subseteq A$ is a local α-reduct in relation to a discernibility matrix M and an object $x \in U$ if it discerns at least $(1 - \alpha)|P(x)|$ objects discernible from x by M and no proper subset of R holds that property:*

$$|\{y \in U : R \cap M(x,y) \neq \emptyset\}| \geq (1 - \alpha)|P(x)|$$

$$\forall R' \subsetneq R : |\{y \in U : R' \cap M(x,y) \neq \emptyset\}| < (1 - \alpha)|P(x)|$$

The following algorithms computing reducts are available in Rseslib:

– **All Global Reducts**
  The algorithm computes all global reducts from a data set. The algorithm is based on the fact that a set of attributes is a reduct if and only if it is a prime implicant of a boolean CNF formula generated from the discernibility matrix [19]. First the algorithm calculates the discernibility matrix and then it transforms the discernibility matrix into a boolean CNF formula. Finally it applies an efficient algorithm finding all prime implicants of the formula using well-known in the field of boolean reasoning advanced techniques accelerating computations [4]. All found prime implicants are global reducts.

– **All Local Reducts**
  The algorithm computes all local reducts for each object in a data set. Like the algorithm computing global reducts it uses boolean reasoning. The first step is the same as for global reducts: the discernibility matrix specified by parameters is calculated. Next for each object x in the data set the row of the discernibility matrix corresponding to the object x is transformed into a CNF formula and all local reducts for the object x are computed with the algorithm finding prime implicants.

- **One Johnson Reduct**

  The method computes one reduct with greedy Johnson algorithm [9]. The algorithm starts with the empty set of attributes called the candidate set and adds iteratively one attribute maximizing the number of discerned pairs of objects according to the semantics of a selected discernibility matrix. It stops when all objects are discerned and checks if any of the attributes in the candidate set can be removed. The final candidate set is a reduct.

- **All Johnson Reducts**

  A version of the greedy Johnson algorithm in which the algorithm branches and traverses all possibilities rather than selecting one of them arbitrarily when more than one attribute cover the maximal number of uncovered fields of the discernibility matrix. The result is the set of the reducts found in all branches of the algorithm.

- **Global Partial Reducts**

  The algorithm finding global α-reducts described in [12]. The value α is the parameter of the algorithm.

- **Local Partial Reducts**

  The algorithm finding local α-reducts described in [12]. The value α is the parameter of the algorithm.

The table below presents time (in seconds) of computing decision-related reducts by particular algorithms on some data sets. Numerical attributes were discretized with the local maximal discernibility method. The experiments were run on Intel Core i7-4790 3.60 GHz processor.

Dataset	Attributes	Objects	All global	All local	Global partial	Local partial
Segment	19	1540	0.6	0.9	0.2	0.2
Chess	36	2131	4.1	66.1	0.2	0.4
Mushroom	22	5416	2.9	4.9	0.8	1.5
Pendigits	16	7494	10.4	23.2	2.2	4.3
Nursery	8	8640	6.5	6.7	1.5	2.8
Letter	16	15000	44.6	179.7	9.7	20.5
Adult	13	30162	62.1	70.1	18.0	33.0
Shuttle	9	43500	91.8	92.5	22.7	48.4
Covtype	12	387342	8591.9	8859.0	903.7	7173.7

6　Rules Generated from Reducts

Reducts described in the previous section can be used in Rseslib to generate decision rules. As reducts can be generated from a discernibility matrix using generalized decision Rseslib uses generalized decision rules:

Definition 5. *A decision rule indicates the probabilities of the decision classes at given values of some conditional attributes:*

$$a_{i_1} = v_1 \wedge \ldots \wedge a_{i_p} = v_p \Rightarrow (p_1, \ldots, p_m)$$

where p_j is defined as $p_j = \dfrac{\left| \{ x \in U : x_{i_1} = v_1 \wedge \ldots x_{i_p} = v_p \wedge dec(x) = d_j \} \right|}{\left| \{ x \in U : x_{i_1} = v_1 \wedge \ldots x_{i_p} = v_p \} \right|}.$

A data object x is said to match a rule if the premise of the rule is satisfied by the attribute values of x: $x_{i_1} = v_1, \ldots, x_{i_p} = v_p$. Rseslib provides the option to allow the values v_k in the descriptors of a rule to be missing values: $a_{i_k} = *$. An object x satisfies a descriptor with missing value $a_{i_k} = *$ if the value of the attribute a_{i_k} on x is missing: $x_{i_k} = *$.

Each decision rule r: $a_{i_1} = v_1 \wedge \ldots \wedge a_{i_p} = v_p \Rightarrow (p_1, \ldots, p_m)$ in Rseslib is assigned with its support in the data set U used to generate rules:

$$support(r) = \left| \{ x \in U : x_{i_1} = v_1 \wedge \ldots x_{i_p} = v_p \} \right|$$

Rseslib provides two algorithms generating decision rules from reducts:

- **Rules from global reducts** (Johnson reducts are global reducts). Given a set of global reducts GR the algorithm finds all templates in the data set:

$$Templates(GR) = \left\{ \bigwedge_{a_i \in R} a_i = x_i : R \in GR, x \in U \right\}$$

For each template the algorithm generates one rule with the decision probabilities p_j as defined in Definition 5:

$$Rules(GR) = \{ t \Rightarrow (p_1, \ldots, p_m) : t \in Templates(GR) \}$$

- **Rules from local reducts.** For each object $x \in U$ the algorithm applies the selected algorithm $LR : U \mapsto \mathcal{P}(A)$ computing local reducts $LR(x)$ for x and generates the set of templates as the union of the sets of templates from all objects in U:

$$Templates(LR) = \left\{ \bigwedge_{a_i \in R} a_i = x_i : R \in LR(x), x \in U \right\}$$

The set of decision rules is obtained from the set of templates in the same way as in case of global reducts:

$$Rules(LR) = \{ t \Rightarrow (p_1, \ldots, p_m) : t \in Templates(LR) \}$$

7 Classification

7.1 Rough Set Classifier

Rough set classifier provided in Rseslib uses the Algorithms computing discernibility matrix, reducts and rules generated from reducts described in the previous sections. It enables to apply any of the discretization methods listed in Sect. 3 to transform numerical attributes into nominal attributes. A user of the classifier selects a discretization method, a type of discernibility matrix and an algorithm generating reducts. The classifier computes a set of decision rules and the support of each rule in the training set.

Let *Rules* denote the computed set of decision rules. The rules are used in classification to determine a decision value when provided with an object x to be classified. First, the classifier calculates the vote of each decision class $d_j \in V_{dec}$ for the object x:

$$vote_j(x) = \sum_{\{t \Rightarrow (p_1,\ldots,p_m) \in Rules:\, x\, \text{matches}\, t\}} p_j \cdot support(t \Rightarrow (p_1,\ldots,p_m))$$

Then the classifier assigns to x the decision with the greatest vote:

$$dec_{roughset}(x) = \max_{d_j \in V_{dec}} vote_j(x)$$

7.2 K Nearest Neighbors/RIONA

Rseslib provides an originally extended version of the k nearest neighbors (k-nn) classifier [24]. It can work with data containing both numerical and nominal attributes and implements fast neighbor search that make the classifier work in reasonable time for large data sets.

In the learning phase the algorithm induces a distance measure from a training set and constructs an indexing tree used for fast neighbor search. Optionally, the algorithm can learn the optimal number k of nearest neighbors from the training set. The distance measure is the weighted sum of distances between values of two objects on all conditional attributes. The classifier provides two metrics for nominal attributes: Hamming metric and Value Difference Metric (VDM), and three metrics for numerical attributes: the city-block Manhattan metric, Interpolated Value Difference Metric (IVDM) and Density-Based Value Difference Metric (DBVDM). IVDM and DBVDM metrics are adaptations of VDM metric to numerical attributes. For computation of the weights in the distance measure three methods are available: distance-based method, accuracy-based method and a method using perceptron.

While classifying an object the classifier finds k nearest neighbors in the training set according to the induced distance measure and it applies one of three methods of voting for the decision by the found neighbors: equally weighted, with inverse distance weights or with inverse square distance weights.

The algorithm has also the mode to work as RIONA algorithm [6]. This mode implements a classifier combining the k-nn method with rule induction where the nearest neighbors not validated by additional rules are excluded from voting.

7.3   K Nearest Neighbors with Local Metric Induction

K nearest neighbors with local metric induction is the k nearest neighbors method extended with an extra step - the classifier computes a local metric for each classified object [22]. While classifying an object, first the classifier finds a large set of the nearest neighbors (according to a global metric). Then it generates a new, local metric from this large set of neighbors. At last, the k nearest neighbors are selected from this larger set of neighbors according to the locally induced metric and used to vote for the decision.

In comparison to the standard k-nn algorithm this method improves classification accuracy particularly for the case of data with nominal attributes. It is reasonable to use this method rather for large data sets (2000 training objects or more).

7.4   Classical Classifiers

Rseslib delivers also implementations of classifiers well-known in the machine learning community (see [25] for more details):
C4.5 - decision tree developed by Quinlan
AQ15 - rule-based classifier with a covering algorithm
Neural network - classical backpropagation algorithm
Naive Bayes - simple Bayesian network
Support vector machine
PCA - classifier using principal component analysis
Local PCA - classifier using local principal component analysis
Bagging - metaclassifier combining a number of "weak" classifiers
AdaBoost - another popular metaclassifier.

8   Other Algorithms

Beside rough set and classification methods Rseslib provides many other machine learning and data mining algorithms. Each algorithm is available as separate class or method and easy to use as an independent component. That includes:
Data transformation: discretizations, missing value completion (non-invasive data imputation by Gediga and Duentsch), attribute selection, numerical attribute scaling, new attributes (radial, linear and arithmetic transformations)
Data filtering: missing values filter, Wilson's editing, Minimal Consistent Subset (MSC) by Dasarathy, universal boolean function based filter
Data sampling: with repetitions, without repetitions, with given class distribution
Data clustering: k approximate centers algorithm
Data sorting: attribute value related, distance related
Rule induction: from global reducts, from local reducts, AQ15 algorithm
Metric induction: Hamming and Value Difference Metric (VDM) for nominal attributes, city-block Manhattan, Interpolated Value Difference Metric (IVDM)

and Density-Based Value Difference Metric (DBVDM) for numerical attributes, attribute weighting (distance-based, accuracy-based, perceptron)

Principal Component Analysis (PCA): OjaRLS algorithm

Boolean reasoning: two different algorithms generating prime implicant from a CNF boolean formula

Genetic algorithm scheme: a user provides cross-over operation, mutation operation and fitness function only

Classifier evaluation: single train-and-classify test, cross-validation, multiple test with random train-and-classify split, multiple cross-validation (all types of tests can be executed on many classifiers).

9   Modular Component-Based Architecture

Providing a collection of rough set and machine learning algorithms is not the only goal of Rseslib. It is designed also to assure maximum reusability and substitutability of the existing components in new components of the library. Hence a strong emphasis is put on its modularity. The code is separated into loosely related elements as small as possible so that each element can be used independently of other elements. For each group of the elements of the same type a standardizing interface is defined so that each element used in an algorithm can be easily substituted by any other element of the same type. Code separation and standardization is applied both to the algorithms and to the objects.

The previous sections presented the range of algorithms available in Rseslib. Below there is a list of the objects in the library implementing various data-related mathematical concepts that can be used as isolated components:

Basic: attribute, data header, data object, boolean data object, numbered data object, data table, nominal attribute histogram, numeric attribute histogram, decision distribution

Boolean functions/operators: attribute value equality, numerical attribute interval, nominal attribute value subset, binary discrimination, metric cube, negation, conjunction, disjunction

Real functions/operators: scaling, perceptron, radius function, multiplication, addition

Integer functions: discrimination (discretization, 3-value cut)

Decision distribution functions: nominal value to decision distribution, numeric value to vicinity-based decision distribution, numeric value to interpolated decision distribution

Vector space: vector, linear subspace, principal components subspace, vector function

Linear order

Indiscernibility relations

Distance measures: Hamming, Value Difference Metric, city-block Manhattan, Interpolated Value Difference Metric, Density-Based Value Difference Metric, metric-based indexing tree

Rules: boolean function based, equality descriptors rule, partial matching rule

Probability: gaussian kernel function, hypercube kernel function, m-estimate.

The structure of rough set algorithms in Rseslib is one of the examples of the component-based architecture. Each of the six modules: *Discretization, Logic, Discernibility, Reducts, Rules* and *Rough Set Classifier* provides well-abstracted algorithms with clearly defined interfaces that allow algorithms from other modules to use them as their components. It is easy to extend each module with implementation of a new method and to add the new method as an alternative in all components using the module.

The component-based architecture of Rseslib makes it possible to implement unconventional combinations of data mining methods. For example, perceptron learning is used as one of the attribute weighting methods in the algorithm computing a distance measure between data objects. Estimation of value probability at given decision is another example of such combination: it uses k nearest neighbors voting as one of the methods defining conditional value probability.

10   Tools

10.1   Rseslib Classifiers in Weka

Weka [1] is a very popular machine learning and data mining software equipped with the system of packages updated independently of Weka core allowing people all over the world to contribute to Weka and maintain easily their extensions.

Rseslib is such an official Weka package available from Weka repository. Rseslib version 3.1.2 (the latest at the moment of preparing this paper) provides three Rseslib classifiers with full configuration in Weka: rough set classifier, k nearest neighbors/RIONA and k nearest neighbors with local metric induction. These three classifiers can be used, tested and compared with other classifiers within all Weka interfaces.

10.2   Graphical Interface Qmak

Qmak is a graphical user interface dedicated to Rseslib library. It is a tool for data analysis, data classification, classifier evaluation and interaction with classifiers. Qmak provides the following features:

- visualization of data, classifiers and single object classification
- interactive classifier modification by a user
- classification of test data with presentation of misclassified objects
- experiments on many classifiers: single train-and-classify test, cross-validation, multiple test with random train-and-classify split, multiple cross-validation.

Qmak 1.0.0 (the latest at the moment of preparing this paper) with Rseslib 3.1.2 provides visualization of 5 classifiers: rough set classifier, k nearest neighbors, C4.5 decision tree, neural network and principal component analysis classifier. Visualization of a rough set classifier presents the decision rules of the classifier (see Fig. 1). The rules can be filtered and sorted by attribute occurrence, attribute values, length, support and accuracy. Visualization of classification by

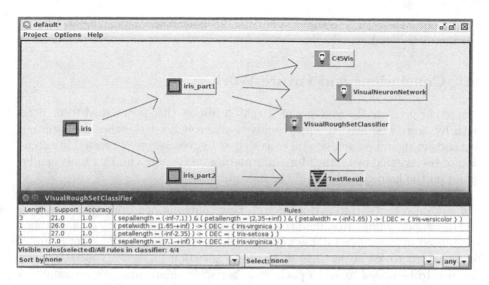

Fig. 1. Qmak project panel with instance of rough set classifier displayed

rough set classifier shows the decision rules matching a classified object enabling the same types of filtering and sorting criteria as visualization of the classifier.

Users can implement new classifiers and their visualization and add them easily to Qmak. It does not require any change in Qmak itself. A new classifier can be added using GUI or in the configuration file.

Qmak is available from Rseslib homepage. Help on Qmak can be found in the main menu of the application.

10.3   Computing in Cluster

Simple Grid Manager is a tool for running massive Rseslib-based experiments on all available computers. It is the successor of the previous version of software dedicated to Rseslib 2 [3]. Using SGM a user can create an ad-hoc cluster of computers by running server part on one machine and client part on all machines designated to run the experiments. The server reads experiment lists from script files, distributes tasks between all available client machines, collects results of executed tasks and stores them in a result file. The main features of the tool are:

- Executes train-and-test experiments with any set of classifiers from Rseslib library (or user written classifiers compatible with Rseslib standards)
- Allows ad-hoc cluster creation without any configuration and maintenance
- Automatically resumes failed jobs and skips completed jobs in case of restart
- Uses robust communication that allows creation of a cluster over non-reliable networks
- Enables utilizing multi-core architectures by executing many client instances on one machine.

Simple Grid Manager is available from Rseslib homepage. The guide on how to run the distributed experiments can be found in [25].

11   Conclusions and Future Work

The paper presents the contents of Rseslib 3 library that is designed to be used both by users who need to apply ready-to-use rough set or other data mining methods in their data analysis tasks as well as by researchers interested in extension of the existing methods. More information on Rseslib 3 and its tools can be found on the home page[3] and in the user guide [25].

The development of Rseslib 3 is continued. The repository of the library[4] is maintained by GitHub and is open to new contributions from all researchers and developers willing to extend the library. There is ongoing work on a classifier specialized in imbalanced data. The algorithms computing reducts are planned to be added to Weka package as attribute selection methods. Discretizations are also to be added to Weka package as separate algorithms. We are going to add Rseslib to Maven repository and to investigate the possibility of connecting Rseslib to RapidMiner.

Acknowledgment. We would like to thank Professor Andrzej Skowron for his mentorship over the project and for his advice on the development and Professor Dominik Ślęzak for his remarks to this paper. It must be emphasized that the library is the result of joint effort of many people and we express our gratitude to all the contributors: Jan Bazan, Rafał Falkowski, Grzegorz Góra, Wiktor Gromniak, Marcin Jałmużna, Łukasz Kosson, Łukasz Kowalski, Michał Kurzydłowski, Łukasz Ligowski, Michał Mikołajczyk, Krzysztof Niemkiewicz, Dariusz Ogórek, Marcin Piliszczuk, Maciej Próchniak, Jakub Sakowicz, Sebastian Stawicki, Cezary Tkaczyk, Witold Wojtyra, Damian Wójcik and Beata Zielosko.

References

1. Weka 3: Data Mining Software in Java. http://www.cs.waikato.ac.nz/ml/weka
2. Bazan, J.G., Szczuka, M.: The rough set exploration system. In: Peters, J.F., Skowron, A. (eds.) Transactions on Rough Sets III. LNCS, vol. 3400, pp. 37–56. Springer, Heidelberg (2005). https://doi.org/10.1007/11427834_2
3. Bazan, J.G., Latkowski, R., Szczuka, M.: DIXER – distributed executor for rough set exploration system. In: Ślęzak, D., Yao, J.T., Peters, J.F., Ziarko, W., Hu, X. (eds.) RSFDGrC 2005. LNCS, vol. 3642, pp. 39–47. Springer, Heidelberg (2005). https://doi.org/10.1007/11548706_5
4. Brown, F.M.: Boolean Reasoning: The Logic of Boolean Equations. Kluwer Academic Publishers, Dordrecht (1990)
5. Fayyad, U., Irani, K.: Multi-interval discretization of continuous-valued attributes for classification learning. In: Proceedings of the 13th International Joint Conference on Artificial Intelligence, pp. 1022–1027. Morgan Kaufmann (1993)

---

[3] http://rseslib.mimuw.edu.pl.
[4] https://github.com/awojna/Rseslib.

6. Góra, G., Wojna, A.: RIONA: a new classification system combining rule induction and instance-based learning. Fundam. Inform. **51**(4), 369–390 (2002)
7. Holte, R.C.: Very simple classification rules perform well on most commonly used datasets. Mach. Learn. **11**(1), 63–90 (1993)
8. Janusz, A., Stawicki, S., Szczuka, M., Ślęzak, D.: Rough set tools for practical data exploration. In: Ciucci, D., Wang, G., Mitra, S., Wu, W.-Z. (eds.) RSKT 2015. LNCS, vol. 9436, pp. 77–86. Springer, Cham (2015). https://doi.org/10.1007/978-3-319-25754-9_7
9. Johnson, D.S.: Approximation algorithms for combinatorial problems. J. Comput. Syst. Sci. **9**(3), 256–278 (1974)
10. Kerber, R.: Chimerge: discretization of numeric attributes. In: Proceedings of the 10th National Conference on Artificial Intelligence, pp. 123–128. AAAI Press (1992)
11. Latkowski, R.: Flexible indiscernibility relations for missing attribute values. Fundam. Inform. **67**(1–3), 131–147 (2005)
12. Moshkov, M., Piliszczuk, M., Zielosko, B.: Partial Covers, Reducts and Decision Rules in Rough Sets: Theory and applications. Studies in Computational Intelligence, vol. 145. Springer, Heidelberg (2008). https://doi.org/10.1007/978-3-540-69029-0
13. Nguyen, H.S.: Discretization of real value attributes: a Boolean reasoning approach. Ph.D. thesis, Warsaw University (1997)
14. Øhrn, A., Komorowski, J., Skowron, A., Synak, P.: The design and implementation of a knowledge discovery toolkit based on rough sets - the ROSETTA system. In: Polkowski, L., Skowron, A. (eds.) Rough Sets in Knowledge Discovery 2: Applications, Case Studies and Software Systems, pp. 376–399. Physica-Verlag (1998)
15. Pawlak, Z.: Rough Sets - Theoretical Aspects of Reasoning about Data. Kluwer Academic Publishers, Dordrecht (1991)
16. Pawlak, Z., Skowron, A.: Rudiments of rough sets. Inf. Sci. **177**(1), 3–27 (2007)
17. Prędki, B., Wilk, S.: Rough set based data exploration using ROSE system. In: Raś, Z.W., Skowron, A. (eds.) ISMIS 1999. LNCS, vol. 1609, pp. 172–180. Springer, Heidelberg (1999). https://doi.org/10.1007/BFb0095102
18. Riza, L.S., et al.: Implementing algorithms of rough set theory and fuzzy rough set theory in the R package "RoughSets". Inf. Sci. **287**, 68–89 (2014)
19. Skowron, A.: Boolean reasoning for decision rules generation. In: Komorowski, J., Raś, Z.W. (eds.) ISMIS 1993. LNCS, vol. 689, pp. 295–305. Springer, Heidelberg (1993). https://doi.org/10.1007/3-540-56804-2_28
20. Skowron, A., Grzymała-Busse, J.W.: From rough set theory to evidence theory. In: Yager, R.R., Kacprzyk, J., Fedrizzi, M. (eds.) Advances in the Dempster-Shafer Theory of Evidence, pp. 193–236. Wiley, New York (1994)
21. Skowron, A., Rauszer, C.: The discernibility matrices and functions in information systems. In: Slowinski, R. (ed.) Intelligent Decision Support, Handbook of Applications and Advances of the Rough Sets Theory, pp. 331–362. Kluwer Academic Publishers, Dordrecht (1992)
22. Skowron, A., Wojna, A.: K nearest neighbor classification with local induction of the simple value difference metric. In: Tsumoto, S., Słowiński, R., Komorowski, J., Grzymała-Busse, J.W. (eds.) RSCTC 2004. LNCS, vol. 3066, pp. 229–234. Springer, Heidelberg (2004). https://doi.org/10.1007/978-3-540-25929-9_27
23. Widz, S.: Introducing NRough framework. In: Polkowski, L., Yao, Y., Artiemjew, P., Ciucci, D., Liu, D., Ślęzak, D., Zielosko, B. (eds.) IJCRS 2017. LNCS, vol. 10313, pp. 669–689. Springer, Cham (2017). https://doi.org/10.1007/978-3-319-60837-2_53

24. Wojna, A.: Analogy-based reasoning in classifier construction. In: Peters, J.F., Skowron, A. (eds.) Transactions on Rough Sets IV. LNCS, vol. 3700, pp. 277–374. Springer, Heidelberg (2005). https://doi.org/10.1007/11574798_11
25. Wojna, A., Latkowski, R., Kowalski, Ł.: RSESLIB: User Guide. http://rseslib. mimuw.edu.pl/rseslib.pdf
26. Wróblewski, J.: Covering with reducts - a fast algorithm for rule generation. In: Polkowski, L., Skowron, A. (eds.) RSCTC 1998. LNCS, vol. 1424, pp. 402–407. Springer, Heidelberg (1998). https://doi.org/10.1007/3-540-69115-4_55

Composite Sequential Three-Way Decisions

Xin Yang[1,2], Ning Wang[3], Tianrui Li[1(✉)], Dun Liu[4], and Chuan Luo[5]

[1] School of Information Science and Technology, Southwest Jiaotong University,
Chengdu 611756, China
trli@swjtu.edu.cn
[2] School of Computer Science, Sichuan Technology and Business University,
Chengdu 611745, China
[3] Chongqing Normal University Foreign Trade and Business College,
Chongqing 401520, China
[4] School of Economics and Management, Southwest Jiaotong University,
Chengdu 610031, China
[5] College of Computer Science, Sichuan University, Chengdu 610065, China

Abstract. In this paper, we present a composite framework of sequential three-way decisions to deal with hybrid data based on the fusion of different granularities. According to the top-down manner, we construct a multilevel composite granular structure by the addition of a new attribute type, and define a general composite binary relation based on three kinds of fusion strategies. At each level, the particular regions including seven selections are considered to induce the acceptance, non-commitment, and rejection rules. Some uncertain objects may be further investigated by more types of attributes at the next level. In this way, such multilevel processing of hybrid data naturally leads to the composite sequential three-way decisions.

Keywords: Sequential three-way decisions · Hybrid data
Composite binary relation

1 Introduction

Compared to two-way decisions, three-way decisions provide three choices for decision-making, namely, the decisions of acceptance, non-commitment, and rejection [14]. As a useful tool to solve human problem and process information, the basic notion of three-way decisions can be interpreted as a two-step approach [16]. The first step with trisecting is to divide the objects into three pair-wise disjoint regions, denoted as Region I, Region II, and Region III, respectively. The second step with acting is to move objects among three regions by appropriate strategies. In the past few years, such framework has attracted a lot of researches associated with granular computing and rough sets [5,12].

More particularly, Yao and Deng [17] proposed sequential three-way decisions with probabilistic rough sets under a cost-accuracy trade off, and further Yao [15] introduced a sequential framework of three-way decisions with a high-level

© Springer Nature Switzerland AG 2018
H. S. Nguyen et al. (Eds.): IJCRS 2018, LNAI 11103, pp. 177–186, 2018.
https://doi.org/10.1007/978-3-319-99368-3_14

conceptual understanding of granular computing. Moreover, Yang et al. [11] presented a general sequential three-way decisions model with the incremental processing. Based on such multilevel granularity of framework, the objects are gradually assigned into different regions under the dynamic process of decision-making. For clear understanding, we illustrate the multilevel sequential processing in Pawlak rough sets [6]. As a special model of three-way decisions, rough sets divide the objects into three ordered regions, namely, positive, boundary, and negative regions. A multilevel granular structure can be constructed by a nested sequence of attributes. Subsequently, at a particular level, seven possible situations resulting from the combinations of regions may be adopted, e.g., the objects only in boundary region are further investigated at the next level of granularity due to the insufficiency of available information. In this way, we generate the acceptance and rejection rules and make the delayed decisions with the non-commitment rules at each level.

As an efficient and effective model, sequential three-way decisions have been researched in many real-world applications, e.g., face recognition [3], deep neural networks [4], attribute reduction [7], multi-class statistical recognition [9]. However, they seldom consider hybrid data, namely, so-called the composite decision table [18], which includes various of attribute types, e.g., categorical, numerical, interval-valued and set-valued, etc. With the advent of the era of Big Data, the information of objects may be collected by various types of attributes. Note that, the different types of data can provide us with a different specific descriptions on objects. Hence, it is desired to mine the valuable information from such hybrid data by the fusion strategy, e.g., the intersection composite relation [18] and the quantitative composite relation [10]. To tackle the complex problem-solving for hybrid data in granular computing, it is noteworthy that a multilevel granular structure may be constructed from hybrid data. Moreover, we need to pay more attention to the fusion of models associated with each type of attributes. In fact, the sequential strategy may be a suitable approach for the composite data. It leads to the multiple different models for the different hybrid data at the different levels. Therefore, the main motivation of this paper is to combine sequential three-way decision with hybrid data. We construct a multilevel composite granular structure by the addition of a new attribute type. Subsequently, we discuss the general composite binary relations by the optimistic, neutral, and pessimistic strategies. Finally, we propose a composite sequential three-way decisions model with DTRS for dynamic hybrid data decision-making.

The remainder of this paper is organized as follows. Section 2 introduces the basic notion of three-way decisions. In Sect. 3, a composite sequential three-way decisions model is proposed for dynamically addressing hybrid data. Finally, Sect. 4 concludes our proposal and points out the future work.

2   The Theory of Three-Way Decisions

In general, three-way decisions can be categorized into two classes, namely, the static and dynamic three-way decisions. In fact, as a dynamic multilevel framework associated with trisecting-and-acting, sequential three-way decisions are a natural extension of the former. In this section, we briefly introduce the notion of such framework [11,15,16].

2.1   The Static Three-Way Decisions

For the trisecting task in three-way decisions, we consider a decision table $DT = (U, AT)$. U is a finite nonempty set of objects; AT is a finite set of attributes. Suppose $v : U \longrightarrow L$ is an evaluation function, which estimates the decision states of objects in U. Let $(L, \succeq)$ denotes a totally ordered set, (α, β) denotes a pair of thresholds which satisfies $\alpha \succeq \beta$. Based on the evaluation function $v(x)$ in such a decision table DT, three pair-wise disjoint regions can be constructed as follows:

$$\text{Region I}(v) = \{x \in U \mid v(x) \succeq \alpha\},$$
$$\text{Region II}(v) = \{x \in U \mid \alpha \succ v(x) \succ \beta\}, \tag{1}$$
$$\text{Region III}(v) = \{x \in U \mid \beta \succeq v(x)\},$$

where Region I$(v) \cup$ Region II$(v) \cup$ Region III$(v) = U$. For simplicity, we denote this three regions as R_1, R_2, and R_3. Once we obtain a tri-partition based on the evaluation-based approach, three different strategies for the acting task are developed on R_1, R_2, and R_3, respectively [16].

Based on the trisecting-and-acting framework, many generalizations of sets can be explored with three-way decisions, such as rough sets [6]. As Yao stated in [16], there are three structures for three regions, namely, unordered three regions without any preferences, non-linearly ordered three regions, and linearly ordered three regions. It is reasonable to state that, corresponding to R_1, R_2, and R_3, respectively, the positive, boundary, and negative regions in rough sets belong to the third situation. By considering probabilistic rough sets, DTRS is an improved Bayesian approach to three-way decisions based on the overall minimum decision risk [13].

For categorical data, the equivalence relation R_E is utilized to divide the objects into the equivalence granules $[x]$. Given a concept $X \subseteq U$, we can regard the condition probability $\Pr(X|[x]) = \frac{|X \cap [x]|}{|[x]|}$ as the evaluation function $v(x)$ to measure the similarity between $[x]$ and X. Suppose a pair of thresholds (α, β) satisfied $0 \leqslant \beta < \alpha \leqslant 1$. We give the representation of three regions in DTRS model as follows:

$$\text{POS}(X) = \{x \in U \mid \Pr(X|[x]) \geqslant \alpha\},$$
$$\text{BND}(X) = \{x \in U \mid \beta < \Pr(X|[x]) < \alpha\}, \tag{2}$$
$$\text{NEG}(X) = \{x \in U \mid \Pr(X|[x]) \leqslant \beta\}.$$

In DTRS model, the positive, boundary, and negative regions generate three rules with a yes, delayed and no decision, respectively. Moreover, we can systematically and mathematically calculate two thresholds by the well-known Bayesian decision procedure. The details can be found in [13].

2.2 The Dynamic Three-Way Decisions

With a dynamic idea for practical decision-making, we provide a general framework of sequential three-way decisions by our previous work [11].

Suppose $GS = (GS_1, GS_2, \ldots, GS_{n-1}, GS_n)$ is the n levels granular structure, where $GS_i = (U_i, G_i, v_i(x), \alpha_i, \beta_i), i = 1, 2, \ldots, n$. At ith level, U_i denotes the processing objects, G_i denotes a particular granulation, $v_i(x)$ denotes an evaluation function and (α_i, β_i) denotes a pair of thresholds which satisfies $\alpha_i \succ \beta_i$ $(i.e., \alpha_i \succeq \beta_i \wedge \neg(\beta_i \succeq \alpha_i))$. From first level to nth level in sequential three-way decisions, the ith level of three regions can be expressed as:

$$\begin{aligned}
\mathrm{R}_1^i(v_i) &= \{x \in U_i \mid v_i(x) \succeq \alpha_i\}, \\
\mathrm{R}_2^i(v_i) &= \{x \in U_i \mid \alpha_i \succ v_i(x) \succ \beta_i\}, \\
\mathrm{R}_3^i(v_i) &= \{x \in U_i \mid \beta_i \succeq v_i(x)\},
\end{aligned} \tag{3}$$

where $\mathrm{R}_1^i(v_i) \cup \mathrm{R}_2^i(v_i) \cup \mathrm{R}_3^i(v_i) = U_i$, and U_i has seven selections resulting from the combinations of R_1, R_2, and R_3 depicted as follows:

(1) $U_i = \mathrm{R}_1^{i-1}(v_{i-1})$,

(2) $U_i = \mathrm{R}_2^{i-1}(v_{i-1})$,

(3) $U_i = \mathrm{R}_3^{i-1}(v_{i-1})$,

(4) $U_i = \mathrm{R}_1^{i-1}(v_{i-1}) \cup \mathrm{R}_2^{i-1}(v_{i-1})$,

(5) $U_i = \mathrm{R}_1^{i-1}(v_{i-1}) \cup \mathrm{R}_3^{i-1}(v_{i-1})$,

(6) $U_i = \mathrm{R}_2^{i-1}(v_{i-1}) \cup \mathrm{R}_3^{i-1}(v_{i-1})$,

(7) $U_i = U_{i-1} = \mathrm{R}_1^{i-1}(v_{i-1}) \cup \mathrm{R}_2^{i-1}(v_{i-1}) \cup \mathrm{R}_3^{i-1}(v_{i-1})$,

where $i \neq 1$, $U_1 = U$ is a set of original objects for processing. Seven situations may be selected by the objective. For example, in rough sets, for the most of binary classification problems, the boundary region attracts our more concern than other two regions. We need to investigate the objects of boundary region at the next level. In a similar case, for multi-class, we may pay attention to both boundary and negative regions. For above two different tasks, the second and sixth options can be adopted with our sequential three-way decisions, respectively. Moreover, the remaining five situations may also be carried out for complex problem solving.

To sum up, structures, models and granularity for each level is a key issue for the sequential hybrid data analysis. In next section, we will introduce a novel multilevel composite granular structure to make a sequence of three-way decisions by the fusion of granularities.

3 Composite Sequential Three-Way Decisions

In real-world decision-making, we may collect various types of data over time to enhance our evidence for making the definite decisions of acceptance or rejection. In other words, we may tackle these data at each level in the sequential decision procedure. The different types of attributes may naturally lead to a multilevel composite granular structure. We adopt the fusion strategy to obtain the granules of objects. In this way, the composite sequential three-way decisions are proposed for hybrid data sets.

3.1 The Multilevel Composite Granular Structure

Definition 1 *(Composite decision table). let $CDT = (U, AT = C \bigcup D, V, f)$ be a composite decision table, where U is a nonempty finite set of objects; AT is a nonempty finite set of attributes, C is a set of condition attributes, which consists of m different types of attributes, $C = \bigcup_{k=1}^{m} C^k$, where C^k is a subset of C with the same attribute type and m denotes the number of attribute types, D is a decision attribute set, $C \bigcap D = \varnothing$; V is a domain of the attributes, $V = \bigcup_{a \in AT} V_a$; $f = U \times AT \to V$ is an information function, $f(x_i, a_l)$ denotes the attribute value of object x_i under a_l, $i = 1, 2, \cdots, |U|$, $l = 1, 2, \cdots, |AT|$.*

Table 1. A composite decision table CDT

U	a_1	a_2	a_3	a_4	a_5	d
x_1	1	0.6	0.4	[0.4, 0.6]	{0}	0
x_2	2	0.3	0.5	[0.5, 0.7]	{0,1}	1
x_3	1	0.3	0.2	[0.6, 0.7]	{0,2}	1
x_4	2	0.5	0.3	[0.2, 0.4]	{1,3}	0
x_5	2	0.4	0.5	[0.2, 0.3]	{0,3}	1
x_6	2	0.3	0.4	[0.6, 0.7]	{1,2}	0

Example 1. Table 1 is a composite decision table, which includes four types of data, namely, categorical, numerical, interval-valued and set-valued data. Let $U = \{x_1, x_2, x_3, x_4, x_5, x_6\}$ be a set of objects, $C = \{a_1, a_2, a_3, a_4, a_5\}$ be a set of attributes on U. In Table 1, $C^1 = \{a_1\}$ is categorical data, $C^2 = \{a_2, a_3\}$ is numerical data, $C^3 = \{a_4\}$ is interval-valued data and $C^4 = \{a_5\}$ is set-valued data.

Based on a level-by-level addition strategy with the different attribute types, it is easy to construct a multilevel composite granular structure in a composite decision table.

Definition 2 *(Multilevel composite granular structure). Given a composite decision table $CDT = (U, AT = C \bigcup D, V, f)$, $C = \bigcup_{k=1}^{m} C^k$. Let $CDT_i = (U_i, AT_i = C_i \bigcup D, V_i, f_i)$ be the ith level of composite decision table CDT, where $C_i = \bigcup_{j=1}^{i} C^j \subseteq C$, $i = 1, 2, \ldots, m$. At the ith level, $CR_i^{C_i}$ is the composite binary relation and $[x]_{C_i}$ is the composite granules based on $CR_i^{C_i}$. The ith level of* *composite granular structure CGS_i and the multilevel composite granular structure CGS are denoted respectively as follows:*

$$CGS_i = (U_i, CR_i^{C_i}, C_i, [x]_{C_i}), \tag{4}$$
$$CGS = (CGS_1, \ldots, CGS_{m-1}, CGS_m). \tag{5}$$

Example 2. Given a composite decision table CDT shown in Table 1. According to Definition 2, we can construct a nested sequence of attributes and a multilevel composite granular structure as follows:

$$\{a_1\} \subset \{a_1, a_2, a_3\} \subset \{a_1, a_2, a_3, a_4\} \subset \{a_1, a_2, a_3, a_4, a_5\}$$
$$Level-1: CGS_1 = (U, CR_1^{C_1}, C_1 = \{a_1\}, [x]_{C_1}),$$
$$Level-2: CGS_2 = (U, CR_2^{C_2}, C_2 = \{a_1, a_2, a_3\}, [x]_{C_2}),$$
$$Level-3: CGS_3 = (U, CR_3^{C_3}, C_3 = \{a_1, a_2, a_3, a_4\}, [x]_{C_3}),$$
$$Level-4: CGS_4 = (U, CR_4^{C_4}, C_4 = \{a_1, a_2, a_3, a_4, a_5\}, [x]_{C_4}).$$

3.2 The Composite Binary Relation with the Fusion of Granularities

Under such multilevel composite granular structure CGS, the sequential three-way decisions can be used to make a faster decision by a less overall cost of decision process with the acceptance accuracy. Besides, the objects with various types of attributes can be granulated by the multiple binary relations separately. However, we should consider the solution of granulation for such hybrid data by a single composite relation. Consequently, the fusion strategy will be investigated with the composite binary relation $CR_i^{C_i}$ at each level.

Given a composite decision table $CDT = (U, AT = C \bigcup D, V, f)$, where C contains m types of attributes, $C = \bigcup_{k=1}^{m} C^k$. Suppose R_{C^k} is the binary relation for the kth type of attributes, and $[x]_{C^k}$ is the granules induced by R_{C^k}. In this case, we suggest three possible strategies by the different fusion approaches for the granules [10].

For instance, we have three types of attributes C^1, C^2, and C^3, and three binary relations R_{C^1}, R_{C^2}, and R_{C^3} at a particular level. For an object $x_1 \in U = \{x_1, x_2, x_3, x_4, x_5\}$. Suppose that we have three granules for x_1 as $[x_1]_{C^1} = \{x_1, x_2, x_4\}, [x_1]_{C^2} = \{x_1, x_2, x_3\}$, and $[x_1]_{C^3} = \{x_1, x_2\}$ by R_{C^1}, R_{C^2}, and R_{C^3}, respectively. By the union and intersection operations of sets, we have

$$[x]_{C^1 \cap C^2 \cap C^3} = [x_1]_{C^1} \cup [x_1]_{C^2} \cup [x_1]_{C^3}$$
$$= \{x_1, x_2, x_4\} \cup \{x_1, x_2, x_3\} \cup \{x_1, x_3\} = \{x_1, x_2, x_3, x_4\},$$
$$[x]_{C^1 \cup C^2 \cup C^3} = [x_1]_{C^1} \cap [x_1]_{C^2} \cap [x_1]_{C^3}$$
$$= \{x_1, x_2, x_4\} \cap \{x_1, x_2, x_3\} \cap \{x_1, x_3\} = \{x_1\}.$$

It is obvious to find that, the former is the optimistic strategy for the fusion of granules and the latter is pessimistic. This two ideas are similar with the optimistic and pessimistic multigranulation rough set proposed by Qian [8]. More particularly, they focused on the fusion of a series of lower and upper approximations associated with a family of indiscernibility relations. Furthermore, the optimistic and pessimistic approach may lead to two extreme directions. Specifically speaking, these two strategies may induce coarser or finer granules. Hence, to obtain a appropriate granule for object x_1, we may adopt a mixed operation as follows:

$$[x]_{C^1 \cup C^2 \cap C^3} = [x_1]_{C^1} \cup [x_1]_{C^2} \cap [x_1]_{C^3}$$
$$= \{x_1, x_2, x_4\} \cup \{x_1, x_2, x_3\} \cap \{x_1, x_3\} = \{x_1, x_3\},$$
$$[x]_{C^1 \cap C^2 \cup C^3} = [x_1]_{C^1} \cap [x_1]_{C^2} \cup [x_1]_{C^3}$$
$$= \{x_1, x_2, x_4\} \cap \{x_1, x_2, x_3\} \cup \{x_1, x_3\} = \{x_1, x_2, x_3\}.$$

We can observe that the neutral results with the granule $[x]$ is obtained as follows:

$$[x]_{C^1 \cap C^2 \cap C^3} \subseteq [x]_{C^1 \cup C^2 \cap C^3} \subseteq [x]_{C^1 \cap C^2 \cup C^3} \subseteq [x]_{C^1 \cup C^2 \cup C^3}.$$

Based on above analysis, three strategies, namely, union, mixed, and intersection operations, may be adopted to define the composite binary relation. To construct a reasonable granular structure by the top-down manner, a more compatible fusion method is used to define a general composite binary relation.

Definition 3 *(Composite binary relation). Let* $CDT_i = (U_i, AT_i = C_i \bigcup D, V_i, f_i)$ *be the ith level of composite decision table* $CDT = (U, AT = C \bigcup D, V, f)$, *where* $C = \bigcup_{k=1}^{m} C^k$, $C_i = \bigcup_{j=1}^{i} C^j \subseteq C, i = 1, 2, \ldots, m$. *Suppose* R_{C^j} *is the binary relation for the jth type of attributes. For* $x, y \in U$, *the composite binary relation* $CR_i^{C_i}$ *at the ith level is defined as:*

$$CR_i^{C_i} = \{(x, y) \in U \times U \mid |\{R_{C^j} : (x, y) \in R_{C^j}\}| \geq \lambda\}, \tag{6}$$

where $| * |$ *denotes the cardinality of a set,* $\lambda = 1, 2, \ldots, |C_i|$ *are the control parameters for the results of granulation, and* $|C_i|$ *denotes the number of attribute types at the ith level.*

In Definition 3, we can select different fusion strategies for the composite binary relation $CR_i^{C_i}$ by different λ. As we introduced before, the general fusion strategies are described as follows:

(1) If $\lambda = 1$, the optimistic (union) strategy is adopted.
(2) If $1 < \lambda < |C_i|$, the neutral (mixed) strategy is adopted.
(3) If $\lambda = |C_i|$, the pessimistic (intersection) strategy is adopted.

Example 3. To illustrate the fusion process of granulation by Definition 3, the equivalence relation R_E [6], the neighborhood relation R_N [2], the similarity relation R_I [19], and the tolerance relation R_S [1] are provided to deal with categorical, numerical, interval-valued, and set-valued attributes C_E, C_N, C_I, and C_S, respectively in Table 1. The results of granulation with respect to the composite decision table CDT are shown in Tables 2 and 3.

Table 2. The results of granulation by the single binary relation

U	$C_E = \{a_1\}$	$C_N = \{a_2, a_3\}$	$C_I = \{a_4\}$	$C_S = \{a_5\}$
x_1	$\{a_1, a_3\}$	$\{a_1, a_4, a_5\}$	$\{a_1, a_2\}$	$\{a_1, a_2, a_3, a_5\}$
x_2	$\{a_2, a_4, a_5, a_6\}$	$\{a_2, a_4, a_5, a_6\}$	$\{a_1, a_2, a_3, a_6\}$	$\{a_1, a_2, a_3, a_4, a_5, a_6\}$
x_3	$\{a_1, a_3\}$	$\{a_3, a_4, a_6\}$	$\{a_2, a_3, a_6\}$	$\{a_1, a_2, a_3, a_5, a_6\}$
x_4	$\{a_2, a_4, a_5, a_6\}$	$\{a_1, a_2, a_3, a_4, a_5, a_6\}$	$\{a_4, a_5\}$	$\{a_2, a_4, a_5, a_6\}$
x_5	$\{a_2, a_4, a_5, a_6\}$	$\{a_1, a_2, a_4, a_5, a_6\}$	$\{a_4, a_5\}$	$\{a_1, a_2, a_3, a_4, a_5\}$
x_6	$\{a_2, a_4, a_5, a_6\}$	$\{a_2, a_3, a_4, a_5, a_6\}$	$\{a_2, a_3, a_6\}$	$\{a_2, a_3, a_4, a_6\}$

Table 3. The results of granulation by the composite binary relation

U	$CR_4^{C_4}, \lambda = 1$	$CR_4^{C_4}, \lambda = 2$	$CR_4^{C_4}, \lambda = 3$	$CR_4^{C_4}, \lambda = 4$
x_1	$\{a_1, a_2, a_3, a_4, a_5\}$	$\{a_1, a_2, a_3, a_5\}$	$\{a_1\}$	$\{a_1\}$
x_2	$\{a_1, a_2, a_3, a_4, a_5, a_6\}$	$\{a_1, a_2, a_3, a_4, a_5, a_6\}$	$\{a_2, a_4, a_5, a_6\}$	$\{a_2, a_6\}$
x_3	$\{a_1, a_2, a_3, a_4, a_5, a_6\}$	$\{a_1, a_2, a_3, a_6\}$	$\{a_3, a_6\}$	$\{a_3\}$
x_4	$\{a_1, a_2, a_3, a_4, a_5, a_6\}$	$\{a_2, a_4, a_5, a_6\}$	$\{a_2, a_4, a_5, a_6\}$	$\{a_4, a_5\}$
x_5	$\{a_1, a_2, a_3, a_4, a_5, a_6\}$	$\{a_1, a_2, a_4, a_5, a_6\}$	$\{a_2, a_4, a_5\}$	$\{a_4, a_5\}$
x_6	$\{a_2, a_3, a_4, a_5, a_6\}$	$\{a_2, a_3, a_4, a_5, a_6\}$	$\{a_2, a_3, a_4, a_6\}$	$\{a_2, a_6\}$

In Table 2, it is easy to find that the different binary relations induce the different sizes of granules. To get the fusion granules in composite decision table CDT, four kinds of composite binary relations are implemented corresponding to the control parameters $\lambda = 1, 2, 3, 4$ in Table 3. Indeed, the impact of λ bring the different results by the control of granulation. Moreover, with the increase of λ, the sizes of granules with each object monotonicly decrease. In other words, we obtain the biggest size of granules when $\lambda = 1$ due to the optimistic fusion strategy. The pessimistic granulation with the smallest size of granules is adopted by setting λ equal to 4. $\lambda = 2$ and $\lambda = 3$ are our neutral strategy for the fusion of granulation.

3.3 Composite Sequential Three-Way Decisions with DTRS

With coarse-grained granules, one type of data is used to make an acceptance or rejection decision for some objects. However, more types of data may be considered due to the lack of information evidence. The non-commitment decisions is made for the rest of objects since we may have stronger support by other types of data. Based on this recognition, by the fusion of binary relations, the sequential approach to three-way decisions under the multilevel composite granular structure can be proposed to tackle hybrid data.

Suppose the second situation in Eq. (4) is adopted to construct composite sequential three-way decision with DTRS model. Given a composite decision table $CDT = (U, AT = C \bigcup D, V, f)$. Let $CGS_i = (U, CR_i^{C_i}, C_i, [x]_{C_i})$ be the *ith* level of composite granular structure, where $CR_i^{C_i}$ is the composite binary relation. At *ith* level, three-way regions in DTRS model are defined as follows:

$$\text{POS}_i \ (X_i) = \{x \in U_i \mid \Pr(X_i | [x]_{C_i}) \geqslant \alpha_i\},$$
$$\text{BND}_i \ (X_i) = \{x \in U_i \mid \beta_i < \Pr(X_i | [x]_{C_i}) < \alpha_i\}, \quad (7)$$
$$\text{NEG}_i \ (X_i) = \{x \in U_i \mid \Pr(X_i | [x]_{C_i}) \leqslant \beta_i\},$$

where X_i denotes the concept at the *ith* level, $X_i \subseteq U_i$, $U_i = \text{BND}_{i-1}(X_{i-1})$, $U_i \subseteq U(i \neq 1, U_1 = U)$, and α_i, β_i are two thresholds at the *ith* level, $0 \leqslant \beta_i < \alpha_i \leqslant 1$.

In such composite sequential three-way decisions model, our further investigation should consider three issues as follows:

(1) The selection of hybrid data at each level of granular structure. In sequential three-way decisions, a sequence of different types of attributes may be added into CDT successively according to their significance. To construct a reasonable and monotonic granular structure, we should determine which types of data are more important for decision-making at a particular level.
(2) The fusion of binary relations with various types of attributes. This paper presents a general composite binary relations by the control parameter λ. It is necessary to optimize λ by some objectives, such as the evaluation of granulation or the accuracy of decisions.
(3) The determination of thresholds for the different hybrid data. A sequence of thresholds should be calculated and interpreted by a reasonable way. In fact, each level contains the different hybrid data in our proposed composite granular structure. For instance, hybrid data at the *ith* level may consist of categorical and numerical data, or interval-valued and set-valued data. The former and the latter may need different thresholds to obtain a tri-partition in terms of three-way decisions.

4   Conclusions

To address the fusion of different attribute types in granular computing, this paper presented a composite framework of sequential three-way decisions. We proposed the multilevel composite granular structure, and investigated three fusion methods associated with different binary relations by the optimistic, neutral, and pessimistic strategies. Besides, we consider a general composite relation by the control parameter λ of granulation. To efficiently handle a huge hybrid data, incremental leaning and parallel computing may be introduced into composite sequential three-way decisions in our future work.

Acknowledgments. This work is supported by the National Science Foundation of China (Nos. 61573292, 61572406, 71571148).

References

1. Guan, Y.Y., Wang, H.K.: Set-valued information systems. Inf. Sci. **176**(17), 2507–2525 (2006)
2. Hu, Q.H., Yu, D.R., Xie, Z.X.: Neighborhood classifiers. Expert Syst. Appl. **34**(2), 866–876 (2008)
3. Li, H.X., Zhang, L.B., Huang, B., Zhou, X.Z.: Sequential three-way decision and granulation for cost-sensitive face recognition. Knowl. Based Syst. **9**, 1241–251 (2016)
4. Li, H.X., Zhang, L.B., Zhou, X.Z., Huang, B.: Cost-sensitive sequential three-way decision modeling using a deep neural network. Int. J. Approx. Reason. **85**, 68–78 (2017)
5. Li, J.H., Huang, C.C., Qi, J.J., Qian, Y.H., Liu, W.Q.: Three-way cognitive concept learning via multi-granularity. Inf. Sci. **378**, 244–263 (2017)
6. Pawlak, Z.: Rough sets. Int. J. Comput. Inf. Sci. **11**(5), 341–356 (1982)
7. Qian, J., Dang, C.Y., Yue, X.D., Zhang, N.: Attribute reduction for sequential three-way decisions under dynamic granulation. Int. J. Approx. Reason. **85**, 196–216 (2017)
8. Qian, Y.H., Liang, J.Y., Yao, Y.Y., Dang, C.Y.: MGRS: a multi-granulation rough set. Inf. Sci. **180**(6), 949–970 (2010)
9. Savchenko, A.V.: Fast multi-class recognition of piecewise regular objects based on sequential three-way decisions and granular computing. Knowl. Based Syst. **91**, 252–262 (2016)
10. Wang, L.N., Yang, X., Chen, Y., Liu, L., An, S.Y., Zhuo, P.: Dynamic composite decision-theoretic rough set under the change of attributes. Int. J. Comput. Intell. Syst. **11**, 355–370 (2018)
11. Yang, X., Li, T.R., Fujita, H., Liu, D., Yao, Y.Y.: A unified model of sequential three-way decisions and multilevel incremental processing. Knowl. Based Syst. **134**, 172–188 (2017)
12. Yang, X., Li, T.R., Liu, D., Chen, H.M., Luo, C.: A unified framework of dynamic three-way probabilistic rough sets. Inf. Sci. **420**, 126–147 (2017)
13. Yao, Y.Y.: Decision-Theoretic Rough Set Models. In: Yao, J.T., Lingras, P., Wu, W.-Z., Szczuka, M., Cercone, N.J., Ślęzak, D. (eds.) RSKT 2007. LNCS, vol. 4481, pp. 1–12. Springer, Heidelberg (2007). https://doi.org/10.1007/978-3-540-72458-2_1
14. Yao, Y.Y.: Three-way decisions with probabilistic rough sets. Inf. Sci. **180**(3), 341–353 (2010)
15. Yao, Y.: Granular computing and sequential three-way decisions. In: Lingras, P., Wolski, M., Cornelis, C., Mitra, S., Wasilewski, P. (eds.) RSKT 2013. LNCS (LNAI), vol. 8171, pp. 16–27. Springer, Heidelberg (2013). https://doi.org/10.1007/978-3-642-41299-8_3
16. Yao, Y.Y.: Three-way decisions and cognitive computing. Cogn. Comput. **8**(4), 543–554 (2016)
17. Yao, Y.Y., Deng, X.F.: Sequential three-way decisions with probabilistic rough sets. In: Proceedings of the 10th IEEE International Conference on Cognitive Informatics and Cognitive Computing, pp. 120–125 (2011)
18. Zhang, J.B., Li, T.R., Chen, H.M.: Composite rough sets for dynamic data mining. Inf. Sci. **257**, 81–100 (2014)
19. Zhang, Y.Y., Li, T.R., Luo, C., Zhang, J.B., Chen, H.M.: Incremental updating of rough approximations in interval-valued information systems under attribute generalization. Inf. Sci. **373**, 461–475 (2016)

NDER Attribute Reduction
via an Ensemble Approach

Huixiang Wen[1], Appiahmantey Eric[2], Xiangjian Chen[1],
Keyu Liu[1(✉)], and Pingxin Wang[3]

[1] School of Computer, Jiangsu University of Science and Technology,
Zhenjiang 212003, Jiangsu, People's Republic of China
JUST_liukeyu@163.com
[2] School of Computer Science and Communication Engineering, Jiangsu University,
Zhenjiang 212013, Jiangsu, People's Republic of China
[3] School of Science, Jiangsu University of Science and Technology,
Zhenjiang 212003, Jiangsu, People's Republic of China

Abstract. Traditional attribute reduction based on neighborhood decision error rate aims to reduce the decision errors through selecting valuable attributes. To further improve the performances of the selected
attributes in reducts, an ensemble selector is introduced into such framework. Different from the previous strategy, our approach is realized
through considering a set of the fitness functions instead of one and only
one fitness function, which makes the ensemble selecting of attribute is
possible. The experimental results on 10 UCI data sets and 2 KEEL
data sets demonstrate that our ensemble selector is effective in improving the stabilities of both reducts and classification results. In addition,
the classification accuracies can also be increased.

Keywords: Attribute reduction · Ensemble selector
Fitness function · Classification accuracy

1 Introduction

In the filed of rough set, to deal with data with continuous values or even mixed
values, Hu et al. [10] have proposed the concept of Neighborhood Rough Set
(NRS). Presently, NRS has been widely explored because the strong adaptability
to complex data.

Similar to other rough sets, attribute reduction [1,12,15,20] is also a key
topic in NRS. With respect to different requirements, many attribute reductions
have been studied in terms of NRS. A topic example is that Hu et al. [11] proposed the concept of Neighborhood Decision Error Rate (NDER) based attribute
reduction. Different from the previous measures such that approximation quality [6], conditional entropy [22] for defining attribute reductions, NDER provides
us a criterion from the perspective of the performance of classification learning.

© Springer Nature Switzerland AG 2018
H. S. Nguyen et al. (Eds.): IJCRS 2018, LNAI 11103, pp. 187–201, 2018.
https://doi.org/10.1007/978-3-319-99368-3_15

Therefore, attribute reduction is effective in reducing the incorrect neighborhood decisions, it follows that the classification performance [13, 16, 19, 21] of the neighborhood classifier [9] may be improved.

Given a definition of attribute reduction, the immediate problem is to find the reduct. Up to now, due to the lower time complexity, heuristic algorithm [7] is favored by the majority of researchers in rough set. For instance, Yao et al. [24] analyzed the structure of reduct constructions in heuristic searching. They pointed out that most of the searching strategies [26] posssess two similar structures: (1) "adding one attribute into the pool set step by step until the constraint is satisfied", it is referred to as the addition control strategy; (2) "deleting one attribute from the pool set for step by step until the constraint is satisfied", it is referred to as the deletion control strategy.

It should be noticed that no matter what kind of the searching strategy is selected, one and only one fitness function is used for evaluating the significance of attribute and then deciding which attribute should be added or deleted. However, one fitness function may be sensitive to the data perturbation. For example, if part of the samples have been changed, then the fitness value may be quite different from the reduct results that derived by raw data.

To overcome the limitations of one fitness function, we try to design a algorithm to compute NDER based reduct which aims to achieve higher stability. Since it has been reported that the ensemble strategy [17, 21, 23, 26] is an effective technique to improve the stability in the field of feature selection, our algorithm will then design an ensemble selector to evaluate the significance of attribute, i.e., a set of the fitness functions instead of only one fitness function is used.

The rest of the paper is organized as follows. In Sect. 2, we will review some basic concepts related to neighborhood relation. In Sect. 3, following the limitation of attribute reduction based on NDER, an ensemble approach is proposed to compute NDER reducts. Section 4 analyzes the effectiveness of our approach over 10 UCI data sets and 2 KEEL data sets. We then conclude with some remarks and perspectives for further work in Sect. 5.

2   Preliminary Knowledge

2.1   Neighborhood Relation

Without loss of generality, a decision system can be represented as $DS = <U, AT, d>$ in which U is the set of samples, AT is the set of condition attributes and d is a decision attribute. Furthermore, $\forall x \in U$, $d(x)$ expresses the label of sample x, and $a_i(x)$ denotes its value over condition attribute $a_i \in AT$.

Given a decision system, since the classification task is considered in this paper, an equivalence relation over d can be defined such that $\text{IND}_d = \{(x, y) \in U \times U : d(x) = d(y)\}$. By IND_d, a partition $U/\text{IND}_d = \{X_1, X_2, \ldots, X_q\}$ is induced, $X_k \in U/\text{IND}_d$ is referred to as the k-th decision class. Specially, the decision class which contains sample x is denoted by $[x]_d$.

Furthermore, a relation can also be defined in terms of condition attributes. For instance, $\forall A \subseteq AT$, Hu et al. [9] have defined a neighborhood relation such that $N_A = \{(x,y) \in U \times U : \triangle_A(x,y) \le \sigma\}$. In N_A, $\sigma \ge 0$, $\triangle_A(.,.)$ is the distance function [4] with respect to A.

In the context of this paper, Euclidean distance is employed, i.e., $\triangle_A(x,y) = \sqrt{\sum_{a_i \in A} (a_i(x) - a_i(y))^2}$. By N_A, the neighborhood of sample x is formed such that $N_A(x) = \{y \in U : (x,y) \in N_A\}$. To avoid that only the sample x belongs to the neighborhood of x, Hu et al. [9] modified σ for each $x \in U$ such that

$$\delta = \min_{y \in U \wedge y \ne x} \triangle_A(x,y) + \sigma \cdot \Big(\max_{y \in U \wedge y \ne x} \triangle_A(x,y) - \min_{y \in U \wedge y \ne x} \triangle_A(x,y) \Big). \quad (1)$$

Assuming that the neighborhood relation derived from δ with respect to A is denoted by δ_A, then the neighborhood of x is $\delta_A(x) = \{y \in U : \triangle_A(x,y) \le \delta\}$.

2.2 Neighborhood Rough Set and Classifier

Definition 1. *Given a decision system $DS =< U, AT, d >$, $\forall A \subseteq AT$, the neighborhood lower and upper approximations of d with respect to A are then defined as $\underline{\delta_A}(d) = \bigcup_{k=1}^{q} \underline{\delta_A}(X_k)$ and $\overline{\delta_A}(d) = \bigcup_{k=1}^{q} \overline{\delta_A}(X_k)$, where $\underline{\delta_A}(X_k) = \{x \in U : \delta_A(x) \subseteq X_k\}$ and $\overline{\delta_A}(X_k) = \{x \in U : \delta_A(x) \cap X_k \ne \emptyset\}$.*

Through further considering the partial inclusion between neighborhood and decision class, Hu et al. [9] proposed the following Neighborhood Classifier (NEC). Different from KNN [5,18] which specifies the number of neighbors, NEC uses σ to select neighbors.

Algorithm 1. Neighborhood Classifier (NEC)

Inputs: $DS =< U, AT, d >, A \subseteq AT$, test sample $y \notin U$, and parameter σ;
Outputs: Predicted decision label $\mathrm{Pre}_A(y)$.
1. $\forall x \in U$, compute $\triangle_A(y,x)$;
2. Compute δ, and obtain $\delta_A(y)$;
3. $\forall X_k \in U/\mathrm{IND}_d$, compute $\Pr(X_k|\delta_A(y)) = |\delta_A(y) \cap X_k|/|\delta_A(y)|$;
4. $X_j = \arg\max\{\Pr(X_k|\delta_A(y)) : \forall X_k \in U/\mathrm{IND}_d\}$;
5. Find the corresponding decision label $\mathrm{Pre}_A(y)$ in terms of X_j;
6. **Return** $\mathrm{Pre}_A(y)$.

By NEC, Neighborhood Decision Error Rate (NDER) is defined as follows.

Definition 2. *Given a decision system* $DS = < U, AT, d >$, $\forall A \subseteq AT$, *then NDER related to A is then defined as*

$$\text{NDER}_A(d) = \frac{|\{x \in U : \text{Pre}_A(x) \neq d(x)\}|}{|X|}. \tag{2}$$

in which $|X|$ *is the cardinal number of set* X.

In Definition (2), for each computation of $\text{Pre}_A(x)$, x is considered as a test sample. If the predicted label of x is obtained, then it can be compared with the true label of x. Obviously, NDER ia generated by a leave-one-out validation strategy.

It should be noticed that $\text{NDER}_A(d)$ is counted by predictions of all samples in a decision system, it does not highlight the decision errors occur in one of the specific decision classes. For such reason, a local strategy to compute neighborhood decision error rate can be obtained as Definition 3 shows.

Definition 3. *Given a decision system* $DS = < U, AT, d >$, $\forall A \subseteq AT$, $\forall X_k \in U/\text{IND}_d$, *then NDER of* X_k *with respect to A is defined as*

$$\text{NDER}_A^{X_k}(d) = \frac{|\{x \in X_k : \text{Pre}_A(x) \neq d(x)\}|}{|X_k|}. \tag{3}$$

3   Attribute Reduction

3.1   NDER Based Attribute Reduction

The definition of attribute reduction with the constraint of NDER is defined as follows.

Definition 4. *Given a decision system* $DS = < U, AT, d >$, $\forall A \subseteq AT$, *A is referred to as a Neighborhood Decision Error Rate Reduct (NDERR) if and only if*

1. $\text{NDER}_A(d) \leq \text{NDER}_{AT}(d)$;
2. $\forall B \subset A, \text{NDER}_B(d) > \text{NDER}_{AT}(d)$.

In the following, the addition strategy will be employed to compute NDERR. For each iteration in addition strategy, the most significant attribute can be determined by the following fitness function. $\forall A \subseteq AT$, then $\forall a_i \in AT - A$, its significance with respect to neighborhood decision error rate is:

$$\Phi(a_i) = \text{NDER}_A(d) - \text{NDER}_{A \cup \{a_i\}}(d). \tag{4}$$

The above fitness function indicates that if the value of $\Phi(a_i)$ is higher, then a_i will more important. This is mainly because higher value of $\Phi(a_i)$ implies that the lower NDER will be achieved if a_i is added into A.

Example 1. Suppose that the $\text{NDER}_A(d) = 0.8$. when a_1 is added into A, we obtained the $\text{NDER}_{A \cup \{a_1\}}(d) = 0.6$; Similarity, when a_2 is added into A, and then we compute the $\text{NDER}_{A \cup \{a_2\}}(d) = 0.5$; By the computation, we obtained that $\Phi(a_1) = 0.2$ and $\Phi(a_2) = 0.3$. So attribute a_2 is selected.

The following Algorithm 2 shows us the detailed process of computing NDERR by $\Phi(a_i)$.

Algorithm 2. Process to compute NDERR

Inputs: $DS = <U, AT, d>$, and parameter σ;
Outputs: One NDERR A.
1. $A \leftarrow \emptyset$, let $\text{NDER}_A(d) = 1$;
2.   Compute $\text{NDER}_{AT}(d)$;
3.   **Do**
       (1) $\forall a_i \in AT - A$, compute $\Phi(a_i)$;
       (2) Select $b \in AT - A$ such that $\Phi(b) = \max\{\Phi(a_i) : \forall a_i \in AT - A\}$;
       (3) $A \leftarrow A \cup \{b\}$;
       (4) Compute $\text{NDER}_A(d)$;
       Until $\text{NDER}_A(d) \leq \text{NDER}_{AT}(d)$
4.   **Return** A.

3.2   Ensemble Process

Algorithm 2 uses one and only one fitness function to determine the significance of the attribute. In this subsection, we will present an ensemble selector for determining the significance of the attribute through using a set of the fitness functions. Such set of fitness functions can be defined by the NDER of specific decision class, i.e., $\text{NDER}_A^{X_k}(d)$. $\forall A \subseteq AT$ and $\forall X_k \in U/\text{IND}_d$, then $\forall a_i \in AT - A$, the significance of a_i with respect to NDER of X_k is:

$$\Phi_{X_k}(a_i) = \text{NDER}_A^{X_k}(d) - \text{NDER}_{A \cup \{a_i\}}^{X_k}(d). \tag{5}$$

Since for supervised data, more than one decision classes can be obtained and then the set of the fitness functions is $\{\Phi_{X_1}, \ldots, \Phi_{X_q}\}$. Therefore, the following Algorithm is designed to compute NDERR.

Algorithm 3. Ensemble process to Compute NDERR

Inputs: $DS = <U, AT, d>$, parameter σ;
Outputs: One NDERR A.
1.　$A \leftarrow \emptyset$, let $\text{NDER}_A(d) = 1$;
2.　Compute $\text{NDER}_{AT}(d)$;
3.　**Do**
　　(1) Temporary pool $T \leftarrow \emptyset$;
　　(2) **For** $k = 1$ to q
　　　　(i) $\forall a_i \in AT - A$, compute $\Phi_{X_k}(a_i)$;
　　　　(ii) Select $b \in AT - A$ such that $\Phi_{X_k}(b) = \max\{\Phi_{X_k}(a_i) : \forall a_i \in AT - A\}$;
　　　　(iii) Add b into T;
　　End
　　(3) For each different attribute in T, compute the frequency
　　　　of occurrences;
　　(4) **If** Two or more attributes in T have the maximal frequency
　　　　of occurrences
　　Then
　　　　Select an attribute b which ranks high in the order of the raw
　　　　attributes;
　　Else
　　　　Select an attribute b in T with the maximal frequency of
　　　　occurrences;
　　　　`// Ensemble selector`
　　End
　　(5) $A \leftarrow A \cup \{b\}$;
　　(6) Compute $\text{NDER}_A(d)$;
　　Until $\text{NDER}_A(d) \leq \text{NDER}_{AT}(d)$
4.　**Return** A.

The step 3 is the main step in this attribute reduction process. For each iteration in step 3, the aim is to select a significant attribute and then add it into the pool set. The time complexity of this step is $O(n^2 \times m^2)$, where n is the numbers of attributes and m is the numbers of samples. The overall time complexity is $O(nr \times m^2)$ if there are n candidate attributes, and r attributes are selected. Similar to Algorithm 3, the Algorithm 2 also comes with a time complexity of $O(nr \times m^2)$.

Different from Algorithm 2, single fitness function Φ is replaced by a set of fitness functions $\{\Phi_{X_1}, \Phi_{X_2}, \ldots, \Phi_{X_q}\}$ in Algorithm 3.

The following Fig. 1 further shows us a detailed mechanism of ensemble strategy shown in Algorithm 3.

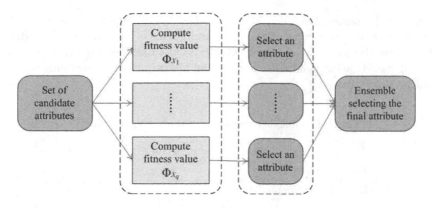

Fig. 1. Ensemble process.

Following Fig. 1, for each decision class X_k, we obtain the set of fitness values in terms of set of candidate attributes such that $\{\Phi_{X_k}(a_1), \ldots, \Phi_{X_k}(a_n)\}$ where $1 \leq k \leq q$, therefore, the attribute a_i with maximal fitness value $\Phi_{X_k}(a_i)$ is selected for decision class X_k. Similarity, different decision classes may generate a collection of the attributes and then the majority principle is regarded as the ensemble voting for deriving the final selected attribute, i.e., the attribute with maximal frequency of occurrence is selected. If two or more attributes have the maximal frequency of occurrence, then the attribute which ranks high in the order of the raw attributes is finally selected.

3.3   Measuring Stabilities

Following attribute reduction, a natural problem is to test the performances of reduct. In this paper, it is assumed that the stability indicates the degree of varying of reducts when sample variations happen. Therefore, the stability of reduct [20,25] can be defined as following.

Definition 5. *Given a decision system $DS =< U, AT, d >$, suppose that U is divided into t groups with the same size such that $U_1, U_2, \ldots, U_t$, then the stability of reduct is:*

$$\text{St}_{\text{reduct}} = \frac{2}{t \cdot (t-1)} \sum_{r=1}^{t-1} \sum_{r'=r+1}^{t} \frac{|A_r \cap A_{r'}|}{|A_r \cup A_{r'}|}, \tag{6}$$

in which A_r is the reduct obtained in $< U - U_r, AT, d >$.

The value of $\text{St}_{\text{reduct}}$ is used as an index to describe the stability of reduct. Obviously, $\text{St}_{\text{reduct}} \in [0, 1]$, if $\text{St}_{\text{reduct}} = 0$, it indicates that the same element does not exist between any two reducts, then the reduct obtained by the algorithm is completely unstable. If $\text{St}_{\text{reduct}} = 1$, it indicates that the results of any two reducts are the same, then the reduct obtained by the algorithm is completely stable. The greater the value of $\text{St}_{\text{reduct}}$, the higher the stability of the reduct.

Following the stability of reduct, we use NEC to further investigate the stabilities of classification results [8]. Firstly, the following joint distribution matrix should be used (Table 1).

Table 1. Joint distribution of classification results.

	$\text{NEC}_{A_r}(x) = d(x)$	$\text{NEC}_{A_r}(x) \neq d(x)$
$\text{NEC}_{A_{r'}}(x) = d(x)$	a	b
$\text{NEC}_{A_{r'}}(x) \neq d(x)$	c	d

$\text{NEC}_{A_r}(x)$ is the predicated label of sample x if classifier NEC is used over attribute sets A_r, a, b, c and d are numbers of samples which satisfy the corresponding conditions, respectively. Therefore, the agreement of classification results between reducts A_r and $A_{r'}$ is: $\text{Agg}(A_r, A_{r'}) = \frac{a+d}{a+b+c+d}$, it follows that the stability of classification result is:

$$\text{St}_{\text{classification}} = \frac{2}{t \cdot (t-1)} \sum_{r=1}^{t-1} \sum_{r'=r+1}^{t} \text{Agg}(A_r, A_{r'}). \tag{7}$$

4   Efficiency Analysis

To evaluate the performances of Ensemble process, 10 UCI data sets and 2 KEEL data sets have been selected, which are shown in Table 2. All the experiments have been carried out on a personal computer with Windows 10, Inter Core i5-6300HQ CPU (2.50 GHz) and 16.00 GB memory. The programming is Matlab R2016a. Moreover, for each data set, we have appointed 10 different parameters used in neighborhood relation such that $\sigma = \{0.05, 0.10, \ldots, 0.50\}$.

4.1   Comparison of Stabilities

In this subsection, we will compare the stabilities of two types of reducts which are obtained by Algorithms 2 and 3, respectively. Such stabilities are reflected by how data perturbation will influence the results of reducts.

From this point of view, 10-folder cross-validation has been adopted in this experiment. Therefore, the obtained stabilities of reducts are average values derived by cross-validation. The following Fig. 2 displays the detailed results of stabilities.

By Fig. 2, it is not difficult to observe the following.

1. In most cases, Algorithm 3 is superior to Algorithm 2 for improving the stabilities of reducts. From this point of view, the ensemble selector we proposed in Algorithm 3 does work.

2. In most cases, the stabilities of classification results based on reducts derived by Algorithm 3 are greater than those derived by Algorithm 2. Therefore, we know that the reduct with higher stability may help us to generate stable classification results.

Table 2. Data sets description. The full name of data set nr 10 is: Parkinson Multiple Sound Recording.

ID	Data sets	Samples	Attributes	Decision classes	Sources
1	Cardiotocography	2126	22	10	UCI
2	Contraceptive method	1473	10	3	UCI
3	Dermatology	366	35	6	UCI
4	Glass identification	214	10	6	UCI
5	Libras movements	360	90	15	UCI
6	Seeds	218	8	3	UCI
7	Statlog (Heart)	270	13	2	UCI
8	Steel plates faults	1941	34	2	UCI
9	Wine quality	6498	11	7	UCI
10	Parkinson	1208	26	2	UCI
11	Ringnorm	7400	21	2	KEEL
12	Twonorm	7400	21	2	KEEL

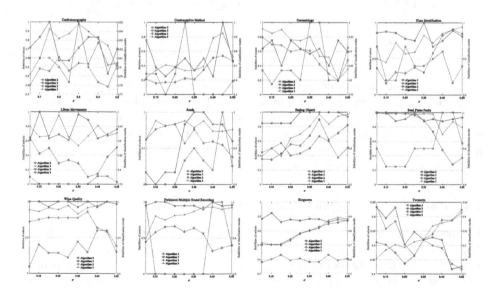

Fig. 2. Stabilities of reducts and classification results.

4.2   Statistical Comparisons of Reducts

In this section, we will make the statistical comparisons of algorithms considered in this paper. The Wilcoxon signed rank test is selected for comparing two algorithms. The purpose of this computation is trying to reject the null-hypothesis that the two algorithms perform equally well for computing reduct.

For each data set, we have appointed 10 different parameters used in neighborhood relation to obtain reducts, it follows that 10 stabilities will be derived with respect to each algorithm. Take the data "Cardiotocography" for instance, the 10 stabilities of reducts derived by Algorithm 2 are "0.6000, 0.6000, 0.6000, 0.5000, 0.3333, 1, 0.5000, 0.3333, 0.2857, 0.5252" while the 10 stabilities of reducts derived by Algorithm 3 are "0.7143, 0.8333, 1, 0.5714, 0.5714, 0.7143, 0.8333, 0.8333, 0.4286, 0.7071", the corresponding p-value (p-value is the probability of observing the given result, or one more extreme, by chance if the null hypothesis is true.) of Wilcoxon signed rank test is 0.0334. The detailed results of p-values are shown in Table 3.

Table 3. p-value of Wilcoxon signed rank test for comparing stabilities of reducts.

ID	Algorithm 2 and Algorithm 3	ID	Algorithm 2 and Algorithm 3
1	0.0334	7	0.0001
2	0.0094	8	0.0125
3	0.0363	9	0.0001
4	0.0002	10	0.0333
5	0.0001	11	0.0200
6	0.1567	12	0.0034

Suppose that the significance level is given by 0.05, that is, if p-value is less than 0.05, then we reject the null-hypothesis. Therefore, following the detailed p-value shown in Table 3, we can see that most of the p-values are less than 0.05, from which we can conclude that Algorithm 2 and 3 do not perform equally well from the viewpoint of the stability of the reduct. In other words, Algorithm 3 is so different from Algorithm 2 for computing reducts.

4.3   Comparisons of Classification Performances

To further test the classification performances of the reducts obtained by our Algorithm 3, classification accuracies are employed to evaluate classification performances. In this subsection, not only neighborhood classifier (NEC) has been employed, but also four types of fuzzy rough approaches [13,14] have been used, they are Fuzzy Rough Classifier (FRC) [2], three robust fuzzy rough classifiers include k-mean-FRC, k-median-FRC and k-trimmed-FRC [13,14,19]. We option to compare with the four types of fuzzy rough classifiers mainly because: (1) both Algorithm 2 and 3 are designed to derive based on neighborhood rough set theory; (2) the structure of fuzzy rough set is quite different from that of neighborhood rough set and then fuzzy rough classifier can also be regarded as the third-party classifier. Therefore, by using third-party classifier, the comparisons of the classification performances of the different reducts may be more objective.

In this experiment, we have selected three parameters such that $\sigma = \{0.1, 0.2, 0.3\}$. For each σ, we use 10-folder cross-validation to obtain 10 different reducts over training sets by both Algorithm 2 and 3. Immediately, we compute the classification accuracies of the five classifiers by using the reducts over testing sets. Similar to Ref. [14], the value of k in FRC is 3. The following Tables 4, 5, 6, 7 and 8 show us the average classification accuracies of each classifier.

With an investigation of above results, we can observe the following.

1. In most cases, Algorithm 3 provides us reducts which can generate higher classification accuracies in terms of five different classifiers. From this point of view, Algorithm 3 is superior Algorithm 2 since the induced reducts are more effective in classification learning.
2. Different from NEC, by considering four types of fuzzy rough classifiers, greater value of σ may help us to obtain reducts which are with higher classification accuracies. For example, in "Libras Movements" data set, if σ is set by 0.1, 0.2 and 0.3, then the classification accuracies of FRC based on the reducts generated by Algorithm 2 are 0.4722, 0.7611 and 0.7889, respectively; the classification accuracies of FRC based on the reducts generated by Algorithm 3 are 0.5611, 0.7922 and 0.8083, respectively.

Table 4. Mean values of classification accuracies (NEC).

ID	$\sigma = 0.1$		$\sigma = 0.2$		$\sigma = 0.3$	
	Algorithm 2	Algorithm 3	Algorithm 2	Algorithm 3	Algorithm 2	Algorithm 3
1	0.7888	0.7822	0.6844	0.7065	0.6204	0.6571
2	0.4807	0.4976	0.4745	0.4786	0.4440	0.4508
3	0.9290	0.9563	0.9290	0.9290	0.7706	0.7732
4	0.6125	0.5421	0.4255	0.4628	0.4069	0.4351
5	0.7417	0.7667	0.5028	0.4917	0.2372	0.2694
6	0.9190	0.9333	0.9048	0.9190	0.7667	0.7952
7	0.7593	0.7815	0.7481	0.7519	0.7000	0.7409
8	0.9987	0.9985	0.9794	0.9788	0.7372	0.7970
9	0.9440	0.9552	0.9492	0.9494	0.8937	0.9440
10	0.6631	0.6746	0.6506	0.6655	0.6258	0.6316
11	0.7285	0.7310	0.6472	0.6533	0.6283	0.6409
12	0.5580	0.5939	0.5022	0.5022	0.4649	0.5472
Average	0.7603	0.7677	0.6998	0.7074	0.6079	0.6402

Table 5. Mean values of classification accuracies (FRC).

ID	$\sigma = 0.1$		$\sigma = 0.2$		$\sigma = 0.3$	
	Algorithm 2	Algorithm 3	Algorithm 2	Algorithm 3	Algorithm 2	Algorithm 3
1	0.6035	0.6225	0.6130	0.6225	0.5844	0.6416
2	0.4311	0.4277	0.4334	0.4338	0.4612	0.4750
3	0.4892	0.4264	0.8661	0.8771	0.9154	0.9373
4	0.6206	0.6402	0.6777	0.6965	0.6917	0.6965
5	0.4722	0.5611	0.7611	0.7922	0.7889	0.8083
6	0.8905	0.9000	0.8905	0.9408	0.8905	0.9408
7	0.6333	0.7000	0.7000	0.7296	0.6963	0.7296
8	0.9541	0.9320	0.9897	0.9981	0.9985	0.9981
9	0.9211	0.8543	0.9210	0.9327	0.9210	0.9327
10	0.6259	0.6416	0.6349	0.6424	0.6349	0.6424
11	0.5812	0.5824	0.5487	0.5497	0.5487	0.5497
12	0.6035	0.6225	0.6130	0.6225	0.5844	0.6416
Average	0.6522	0.6676	0.7208	0.7365	0.7263	0.7495

Table 6. Mean values of classification accuracies (k-mean-FRC).

ID	$\sigma = 0.1$		$\sigma = 0.2$		$\sigma = 0.3$	
	Algorithm 2	Algorithm 3	Algorithm 2	Algorithm 3	Algorithm 2	Algorithm 3
1	0.5848	0.6126	0.5476	0.6216	0.5476	0.6403
2	0.4263	0.4284	0.4270	0.4291	0.4243	0.4318
3	0.4783	0.4234	0.8251	0.8775	0.8716	0.9181
4	0.6159	0.7007	0.6392	0.7146	0.6963	0.7146
5	0.4694	0.5417	0.7500	0.7557	0.7972	0.8194
6	0.9408	0.9190	0.9095	0.9286	0.9048	0.9381
7	0.6333	0.6481	0.6704	0.6889	0.7185	0.6444
8	0.7439	0.7733	0.9449	0.9402	0.9918	0.9995
9	0.8711	0.8763	0.9324	0.9370	0.9270	0.9157
10	0.6250	0.6267	0.6523	0.6399	0.6515	0.6747
11	0.5467	0.5467	0.5351	0.5355	0.5351	0.5355
12	0.5848	0.6126	0.5476	0.6216	0.5476	0.6403
Average	0.6267	0.6425	0.6984	0.7242	0.7177	0.7397

Table 7. Mean values of classification accuracies (k-median-FRC).

ID	$\sigma = 0.1$		$\sigma = 0.2$		$\sigma = 0.3$	
	Algorithm 2	Algorithm 3	Algorithm 2	Algorithm 3	Algorithm 2	Algorithm 3
1	0.5476	0.6238	0.5667	0.6429	0.5857	0.6524
2	0.4263	0.4325	0.4263	0.4325	0.4243	0.4325
3	0.4756	0.3196	0.7270	0.7706	0.9207	0.9399
4	0.6206	0.6311	0.6206	0.6404	0.6299	0.6869
5	0.1611	0.2889	0.6333	0.7028	0.7833	0.8083
6	0.9143	0.9286	0.9143	0.9143	0.9143	0.9190
7	0.6074	0.6111	0.6926	0.6993	0.6926	0.6630
8	0.6919	0.6816	0.8944	0.8983	0.9912	0.9892
9	0.8711	0.8203	0.9330	0.9551	0.9275	0.9348
10	0.6200	0.6333	0.6506	0.6738	0.6631	0.6647
11	0.5382	0.5367	0.5382	0.5497	0.5321	0.5497
12	0.5476	0.6238	0.5667	0.6429	0.5857	0.6524
Average	0.5852	0.5943	0.6803	0.7102	0.7209	0.7410

Table 8. Mean values of classification accuracies (k-trimmed-FRC).

ID	$\sigma = 0.1$		$\sigma = 0.2$		$\sigma = 0.3$	
	Algorithm 2	Algorithm 3	Algorithm 2	Algorithm 3	Algorithm 2	Algorithm 3
1	0.5203	0.5299	0.5944	0.5952	0.6325	0.6429
2	0.4318	0.4535	0.4318	0.4535	0.4750	0.4762
3	0.4729	0.3114	0.6883	0.7104	0.8962	0.9482
4	0.5508	0.6544	0.5787	0.6730	0.5508	0.6730
5	0.1917	0.1899	0.5361	0.5667	0.7611	0.7639
6	0.9095	0.9190	0.9190	0.9286	0.9190	0.9286
7	0.5741	0.6407	0.6926	0.7222	0.7222	0.7222
8	0.6770	0.6772	0.8413	0.8449	0.9758	0.9799
9	0.7306	0.7533	0.9217	0.9102	0.9217	0.9217
10	0.6168	0.6441	0.6656	0.6573	0.6656	0.6573
11	0.5353	0.5555	0.5232	0.5267	0.5232	0.5267
12	0.5203	0.5299	0.5944	0.5952	0.6325	0.6429
Average	0.5609	0.5716	0.6656	0.6820	0.7229	0.7403

5   Conclusion and Future Work

In this paper, an ensemble strategy has been introduced into the process of computing reduct. It uses a set of fitness functions instead of single one to determine which attribute should be selected in the process of computing reduct. The experiment results have demonstrated that the our approach cannot only improve the stabilities of both reducts and classification results, but also strength the classification performances. The future work will be focused on the following two aspects.

1. Only addition control strategy is employed in this paper. The deletion, addition-deletion control strategies will be further explored.
2. The weights of different fitness functions are also interesting issues to be addressed.
3. Such approach may also be considered in some other rough set models, such as decision-theoretic rough set [3], etc.

Acknowledgments. This work is supported by the Natural Science Foundation of China (Nos. 61572242, 61502211, 61503160).

References

1. Wang, C.Z., Shao, M.W., He, Q., Qian, Y.H., Qi, Y.L.: Feature subset selection based on fuzzy neighborhood rough sets. Knowl. Based Sys. **111**, 173–179 (2016)
2. Dubois, D., Prade, H.: Rough fuzzy sets and fuzzy rough sets. Int. J. Gener. Syst. **17**, 191–209 (1990)
3. Dou, H.L., Yang, X.B., Song, X.N., Yu, H.L., Wu, W.Z.: Decision-theoretic rough set: a multicost strategy. Knowl. Based Syst. **91**, 71–83 (2016)
4. Wilson, D.R., Martinez, T.R.: Improved heterogeneous distance functions. J. Artif. Intell. Res. **6**, 1–34 (1997)
5. Tsang, E.C.C., Hu, Q.H., Chen, D.G.: Feature and instance reduction for PNN classifiers based on fuzzy rough sets. Int. J. Mach. Learn. Cybern. **7**, 1–11 (2016)
6. Mi, J.S., Wu, W.Z., Zhang, W.X.: Approaches to knowledge reduction based on variable precision rough set model. Inf. Sci. **159**, 255–272 (2004)
7. Li, J.Y., Fong, S., Wong, R.K., Millham, R., Wong, K.K.L.: Elitist binary wolf search algorithm for heuristic feature selection in high-dimensional bioinformatics datasets. Sci. Rep. **254**, 19–28 (2017)
8. Kuncheva, L.I., Whitaker, C.J.: Measures of diversity in classifier ensembles and their relationship with the ensemble accuracy. Mach. Learn. **51**, 181–207 (2003)
9. Hu, Q.H., Yu, D.R., Xie, Z.X.: Neighborhood classifier. Expert Syst. Appl. **34**, 866–876 (2008)
10. Hu, Q.H., Liu, J.F., Wu, C.X.: Neighborhood rough set based heterogeneous feature subset selection. Inf. Sci. **18**, 3577–3594 (2008)
11. Hu, Q.H., Pedrycz, W., Yu, D.R., Liang, J.: Selecting discrete and continuous features based on neighborhood decision error minimization. IEEE Trans. Syst. Man Cybern. Part B **40**, 137–150 (2010)
12. Hu, Q.H., Yu, D.R., Xie, Z.X., Li, X.D., Ensemble Rough Subspaces: EROS. Pattern Recogn. **40**, 3728–3739 (2007)
13. Hu, Q.H., An, S., Yu, X., Yu, D.R.: Robust fuzzy rough classifiers. Fuzzy Sets Syst. **183**, 26–43 (2011)
14. Hu, Q.H., Zhang, L., An, S., Zhang, D., Yu, D.R.: On robust fuzzy rough set models. IEEE Trans. Fuzzy Syst. **20**, 636–651 (2012)

15. Xu, S.P., Yang, X.B., Yu, H.L., Tsang, E.C.C.: Multi-label learning with label-specific feature reduction. Knowl. Based Syst. **104**, 52–61 (2016)
16. Xu, S.P., Yang, X.B., Song, X.N., Yu, H.L.: Prediction of protein structural classes by decreasing nearest neighbor error rate. In: 2015 International Conference on Machine Learning and Cybernetics, Guangzhou, China, 12–15 July 2015, pp. 7–13 (2015)
17. Xu, S., Wang, P., Li, J., Yang, X., Chen, X.: Attribute reduction: an ensemble strategy. In: Polkowski, L. (ed.) IJCRS 2017. LNCS, vol. 10313, pp. 362–375. Springer, Cham (2017). https://doi.org/10.1007/978-3-319-60837-2_30
18. Li, S.Q., Harner, E.J., Adjeroh, D.A.: Random KNN feature selection-a fast and stable alternative to random forests. BMC Bioinform. **12**, 1–11 (2011)
19. Zhao, S.Y., Chen, H., Li, C.P., Du, X.Y., Sun, H.: A novel approach to building a robust fuzzy rough classifier. IEEE Trans. Fuzzy Syst. **23**, 769–786 (2015)
20. Yang, X.B., Qi, Y., Yu, H.L., Song, X.N., Yang, J.Y.: Updating multigranulation rough approximations with increasing of granular structures. Knowl. Based Syst. **64**, 59–69 (2014)
21. Yang, X.B., Xu, S.P., Dou, H.L., Song, X.N., Yu, H.L., Yang, J.Y.: Multigranulation rough set: a multiset based strategy. Int. J. Comput. Intell. Syst. **10**, 277–292 (2017)
22. Zhang, X., Mei, C.L., Chen, D.G., Li, J.H.: Feature selection in mixed data: a method using a novel fuzzy rough set-based information entropy. Pattern Recogn. **56**, 1–15 (2016)
23. Wang, X.Z., Xing, H.J., Li, Y., Hua, Q., Dong, C.R., Pedrycz, W.: A study on relationship between generalization abilities and fuzziness of base classifiers in ensemble learning. IEEE Trans. Fuzzy Syst. **23**(5), 1638–1654 (2014)
24. Yao, Y.Y., Zhao, Y., Wang, J.: On reduct construction algorithms. Trans. Comput. Sci. **2**, 100–117 (2008)
25. Qian, Y.H., Wang, Q., Cheng, H.H., Liang, J.Y., Dang, C.Y.: Fuzzy-rough feature selection accelerator. Fuzzy Sets Syst. **258**, 61–78 (2015)
26. Zhou, Z.H., Yu, Y.: Ensembling local learners through multimodal perturbation. IEEE Trans. Syst. Man Cybern. Part B **35**(4), 725–735 (2005)

Considerations on Rule Induction Methods by the Conventional Rough Set Theory from a View of STRIM

Tetsuro Saeki[1(✉)], Jiwei Fei[1], and Yuichi Kato[2]

[1] Yamaguchi University, 2-16-1 Tokiwadai, Ube, Yamaguchi 755-8611, Japan
tsaeki@yamaguchi-u.ac.jp
[2] Shimane University, 1060 Nishikawatsu-cho, Matsue, Shimane 690-8504, Japan
ykato@cis.shimane-u.ac.jp

Abstract. In this paper, the rule induction method STRIM, the classical Rough Sets (RS) theory and the notion of three-way decision rules are summarized and their performance is examined by applying them to a real-world dataset and a simulation dataset. From these experimental studies, the problems inherent in the rule induction method by the conventional RS theory based on the indiscernibility are pointed out and a comparison is made with STRIM. Specifically, the rule induction methods that are based on indiscernibility and do not consider the decision table which is only a sample of outcomes obtained by chance from a population of interest are highly dependent upon the samples in the decision table given. This paper states that such rule induction methods are thus problematic and need to be improved to create a more robust rule induction method.

1 Introduction

Extracting the properties and structures hidden in a large dataset is about discovering knowledge and/or information, and that is important for making good strategical decisions and acting consistently. For example, Rough Sets (RS) theory proposed by Pawlak [1] in 1982 is used for reducting a dataset, creating a decision table [2,3], and inducing if-then rules hidden in the decision table [4,5]. Here, the dataset is a set of objects each of which is featured by particular values: its condition attributes and its decision attribute. RS theory first focuses on an indiscernibility property of these objects and provides inclusion relationships of the target object set by defining lower and upper approximations. These approximate expressions provide two representative rules with necessity (accuracy $= 1.0$) and possibility (accuracy > 0.0) respectively. However, the necessity rule imposes a severe condition, i.e., accuracy $= 1.0$, on the rule induction. Therefore, Ziarko [6] proposed a variable precision rough set model (accuracy $= 1.0 - \varepsilon$) with an admissible error ($\varepsilon \in [0.0, 0.5)$).

Yao [7–9] divided the target set into positive, negative, and boundary regions using the lower and upper approximations and proposed three-way decision rules

H. S. Nguyen et al. (Eds.): IJCRS 2018, LNAI 11103, pp. 202–214, 2018.
https://doi.org/10.1007/978-3-319-99368-3_16

corresponding to those regions. Yao also suggested that the boundary parameters (α, β) of the three-way decision rules should be determined by considering accuracy as a type of conditional probability representation and introducing a cost function from a Bayesian decision perspective. This consideration extends Pawlak's and Ziarko's rule induction methods and corresponds to them in some special cases. However, Yao does not propose a new reduction method or a new rule induction method for the decision table and the new related algorithms.

As an alternative to RS theory, the statistical test rule induction method (STRIM) which considers the decision table as a sample dataset obtained from a population has been proposed [10–17]. STRIM uses a statistical reduct method on the decision table [14] and a statistical rule induction method from the reduced table [16]. Note that STRIM was studied independently of the conventional RS methods and was not based on the approximation concept. Specifically, STRIM recognizes the condition attributes and decision attributes of the decision table as random variables and the decision table as their outcomes. Moreover STRIM proposes a data generation model of the decision table by a system which generates input sets of condition attribute values and transforms them into the corresponding output of the decision attribute value through pre-specified if-then rules and hypotheses with regard to the decision attribute value based on causality. This system can also be used for confirming the validity of any rule induction method by applying the method to the dataset generated by the system and investigating whether the method can or cannot induce the pre-specified rules.

In this paper, we first summarize STRIM and give an example of testing its performance by applying it to a real-world dataset. We then state the basics of the if-then rule induction method by STRIM from the viewpoint of proof by contradiction in propositional logic. We then summarize the conventional RS theory based on indiscernibility, and point up the problem of its rule induction method based on indiscernibility in contrast to STRIM. We study this experimentally by applying the LEM2 algorithm, implementing the classical RS theory to the data generation model described above and comparing the results with those of the same experiment using STRIM. Lastly, the idea of three-way decision rules is summarized and we point out that the idea is fundamentally based on the concept of indiscernibility and will cause the same problems as does the classical RS theory. From three summarizations and studies of the conventional methods, this paper points out that the rule induction method based on the concept of indiscernibility of the given decision table needs to be improved as the decision table is merely a sample obtained from the population.

2   The Conventional STRIM

In RS theory, the decision table is expressed as: $S = (U, A = C \cup \{D\}, V, \rho)$. Here $U = \{u(i)|i = 1, ..., |U| = N\}$ is a sample set, A is an attribute set, $C = \{C(j)|j = 1, ..., |C|\}$ is a condition attribute set $C(j)$, a condition attribute, is a member of C, and D is a decision attribute. V is a set of attribute values denoted $V = \bigcup_{a \in A} V_a$ and characterized by the information function $\rho: U \times A \to V$.

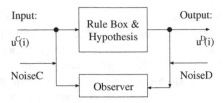

Fig. 1. Data generation model: The rule box contains if-then rules $R(d, k)$: if $CP(d, k)$ then $D = d$ ($d = 1, 2, ..., k = 1, 2, ...$).

Table 1. Hypotheses with regard to decision attribute value.

Hypothesis 1	$u^C(i)$ coincides with $R(k)$, and $u^D(i)$ is uniquely determined as $D = d(k)$ (uniquely determined data)
Hypothesis 2	$u^C(i)$ does not coincide with any $R(d)$, and $u^D(i)$ can only be determined randomly (indifferent data)
Hypothesis 3	$u^C(i)$ coincides with several $R(d)$ ($d = d1, d2, ...$), and their outputs of $u^C(i)$ conflict with each other. Accordingly, the output of $u^C(i)$ must be randomly determined from the conflicted outputs (conflicted data)

Generally, inducing if-then rules from a decision table implicitly assumes a causal relationship between the condition attributes and decision attributes. Therefore, in STRIM, we propose a model in which S is derived from the input/output relationships shown in Fig. 1. In other words, STRIM considers the decision table to be a sample dataset obtained from an input–output system that includes a rule box as shown in Fig. 1 and hypotheses regarding the decision attribute values, as shown in Table 1. A sample $u(i)$ consists of its condition attribute values $u^C(i)$ and decision attribute values $u^D(i)$. Here, $u^C(i)$ is an input to the rule box and is transformed to the output $u^D(i)$ using the rules (generally unknown) contained in the rule box and the hypotheses. The hypotheses consist of three cases corresponding to the nature of the input. The three cases are: uniquely determined, indifferent, and conflicted (see Table 1). In contrast, $u(i) = (u^C(i), u^D(i))$ is measured by an observer (Fig. 1). The existence of NoiseC and NoiseD causes missing values in $u^C(i)$ and changes $u^D(i)$ to create another $u^D(i)$ value. These noises bring the system closer to a real-world system. Differing from the conventional RS theory, STRIM includes the data generation model shown in Fig. 1. This data generation model suggests that the values $(u^C(i), u^D(i))$, i.e., a decision table is the outcome of the random variables $(C, D) = ((C(1), ..., C(|C|), D)$ observing the population. Therefore, in STRIM, $\rho(u(i), C(j))$ are the outcome of the random variables $C(j)$. Note that there is no concept of the information function in STRIM, i.e., $S = (U, A = C \cup \{D\}, V)$ is the decision table and V is the sample space in STRIM.

Table 2. STRIM rule induction results for Rakuten Travel dataset.

$CP(d,k)$	$C(1)C(2)$ $...C(6)$	D	p-value (z)	Accuracy	Coverage	$f = (n_1, n_2, n_3, n_4, n_5)$
(5,1)	005050	5	0.0 (64.08)	0.876	0.629	(11, 12, 9, 146, 1258)
(5,2)	005005	5	0.0 (58.31)	0.915	0.486	(17, 6, 5, 62, 972)
(1,1)	000010	1	0.0 (57.78)	0.766	0.639	(1277, 346, 40, 4, 1)
(4,1)	040040	4	0.0 (40.37)	0.719	0.348	(16, 37, 90, 695, 129)
(3,1)	030030	3	0.0 (38.12)	0.633	0.392	(73, 203, 784, 170, 9)
(2,1)	020000	2	3.0E−168 (27.62)	0.494	0.348	(303, 695, 351, 51, 6)

Given a dataset created by the data generation model in Fig. 1, five processes are carried out: (1) STRIM extracts significant pairs of condition attributes and their values, e.g., $C(j) = v_{j_k}$, for rules of $D = d$ using the local reduct [14,16,17]; (2) STRIM constructs a trying condition part of the rules, e.g., $CP(d,k) = \wedge_j(C(j_k) = v_j)$, using the reduct results; (3) STRIM investigates whether $U(CP(d,k))$ has caused a bias at n_d in the frequency distribution of the decision attribute values $f = (n_1, n_2, ..., n_{M_D})$. Here, $n_m = |U(CP(d,k)) \cap U(m)|$ $(m = 1, ..., |V_D| = M_D)$, $U(CP(d,k)) = \{u(i)|u^{C=CP(d,k)}(i)$, i.e., $u^C(i)$ sastifies $CP(d,k)\}$, and $U(m) = \{u(i)|u^{D=m}(i)\}$ since the $u^C(i)$ coinciding with $CP(d,k)$ in the rule box is transformed to $u^D(i)$ based on hypothesis 1 or 3 (Table 1). In other words, $CP(d,k)$ coinciding with one of the rules in the rule box creates bias in $f = (n_1, n_2, ..., n_{M_D})$. Specifically, STRIM uses a statistical test method for the investigation of the bias specifying a null hypothesis $H0$: f does not have any bias, i.e., $CP(d,k)$ is not a rule; the alternative hypothesis is $H1$: f has a bias, i.e., $CP(d,k)$ is a rule and has a proper significance level. Here, $H0$ is tested using the sample dataset, i.e., the decision table and the proper test statistics; for example,

$$z = \frac{(n_d + 0.5 - np_d)}{(np_d(1 - p_d))^{0.5}} \quad (d = 1, 2, ..., M_D), \tag{1}$$

where $p_d = P(D = d)$, $n = \sum_{j=1}^{5} n_j$, z obeys the standard normal distribution under a proper condition [18] and is considered an index of the bias of f; (4) If $H0$ is rejected, the assumed $CP(d,k)$ becomes a candidate for the rules in the rule box; (5) STRIM repeats processes (1–4) to obtain a set of rule candidates, then arranges the rule candidates and induces the final results [16,17].

Figure 2 shows a STRIM algorithm that includes a reduct function. Here, line nos. (LN) 8 and 9 are the reduct part of process (1), process (2) is executed at LN 10, where the dimension rule[] is used as the rule candidate, process (3) is executed at LN 25 in the rule_check() function, process (4) is executed at LN 26, and process (5) is executed from LN 7 to LN 11 and LN 12.

A rule induction example obtained by applying STRIM to the Rakuten Travel dataset, which is maintained by the Rakuten Institute of Technology follows

Line Algorithm to induce if-then rules by STRIM with a reduct function
No.

```
1    int main(void) {
2    int rdct_max[|CV|]={0,...,0}; //initialize maximum value of C(j)
3    int rdct[|CV|]={0,...,0}; //initialize reduct results by D=l
4    int rule[|C|]={0,...,0}; //initialize trying rules
5    int tail=-1; //initialize value set
6    input data; // set decision table
7    for (di=1; di<=|D|; di++) {// induce rule candidates every D=l
8    attribute_reduct(rdct_max)
9    set rdct[ck] ; // if (rdct_max[ck]==0) {rdct[ck]=0; }else {rdct[ck]=1; }
10   rule_check(rcdct, redct_max, tail, rule); // the first stage process
11   }// end di
12   arrange rule candidates // the second stage
13   }// end main
14   int attribute_reduct(int rdct_max[]) {
15   make contingency table for D=l vs. C(j)
16   Test H0(j,l);
17   if H0(j,l) is rejected then set rdct_max[j,l]=jmax else rdct_max[j,l]=0; //
     jmax:the attribute value of the maximum frequency
18   }// end of attribute_reduct
19   int rule_check(int rdct[], int rdct_max[], int tail,int rule[]) {// the first stage
     process
20   for (ci=tail+1; cj<|C|; ci++) {
21   for (cj=1; cj<=rdct[ci]; cj++) {
22   rule[ci]=rdct_max[cj]; // a trying rule set for test
23   count frequency of the trying rule; // count n1, n2, ...
24   if (frequency>=N0) {//sufficient frequency ?
25   if (|z|>3.0) {//sufficient evidence ?
26   add the trying rule as a rule candidate
27   }// end of if |z|
28   rule_check(ci,rule)
29   }// end if frequency
30   }// end cj
31   rule[ci]=0; // trying rules reset
32   }// end ci
33   }// end rule_check
```

Fig. 2. STRIM algorithm with reduct function.

[17] (for another example, see [16]). The dataset concerned contains approximately $6,200,000$ questionnaire surveys of ratings $A = \{ C(1) = $ "Location," $C(2) = $ "Room," $C(3) = $ "Meal," $C(4) = $ "Bath (Hot Spring)," $C(5) = $ "Service," $C(6) = $ "Amenity," and $D = $ "Overall" $\}$ of approximately $130,000$ travel facilities by using a set of categorical values $V_a = \{ $ "Dissatisfied $(DS(1))$," "Somewhat dissatisfied $(SD(2))$," "Neither satisfied nor dissatisfied $(NN(3))$," "Satisfied $(ST(4))$," and "Very Satisfied $(VS(5))$" $\}$, where $\forall a \in A$, i.e., $|V_{a=D}| = |M_D| = |V_{a=C(j)}| = M_{C(j)} = 5$. We constructed a decision table of $N = 10,000$ questionnaire surveys by randomly selecting $2,000$ samples, each of $D = m$ $(m = 1, ..., 5)$, from approximately $400,000$ surveys from the 2013–2014 dataset, choosing these surveys because they contained heavy biases with respect to the frequency of $D = m$. We applied STRIM to this decision table and obtained Table 2, which represents the following:

(1) $CP(d = 5, k = 1)$ represents a rule stating that if $(C(3) = VS(5)) \bigwedge (C(5) = VS(5))$ then $D = VS(5)$, and its accuracy and coverage are 0.876 and 0.639, respectively.

Table 3. Examples of rules induced by LEM2 for the first simulation dataset (Case1).

Rule no.	Rule	(accuracy, coverage)	$f = (n_1, n_2, n_3, n_4, n_5, n_6)$
1	(C1 = 1) & (C2 = 1) & (C4 = 6) = >(D = 1)	(1.0, 0.0296)	(15,0,0,0,0,0)
2	(C1 = 4) & (C3 = 1) & (C4 = 1) = >(D = 1)	(1.0, 0.0355)	(18,0,0,0,0,0)
3	(C1 = 1) & (C2 = 1) & (C3 = 1) = >(D = 1)	(1.0, 0.0197)	(10,0,0,0,0,0)
4	(C1 = 5) & (C2 = 6) & (C3 = 1) & (C4 = 1) = >(D = 1)	(1.0, 0.0138)	(7,0,0,0,0,0)
...	...	...	...
8	(C1 = 1) & (C2 = 1) & (C3 = 5) & (C5 = 6) = >(D = 1)	(1.0, 0.0099)	(5,0,0,0,0,0)
...	...	...	...
24	(C1 = 5) & (C2 = 6) & (C3 = 5) & (C4 = 4) & (C5 = 2) = >(D = 1)	(1.0, 0.002)	(1,0,0,0,0,0)
...	...	...	...
27	(C1 = 2) & (C3 = 2) & (C4 = 5) & (C5 = 6) & (C6 = 5) = >(D = 1)	(1.0, 0.002)	(1,0,0,0,0,0)
...	...	...	...

(2) This rule implies the frequency $f = (11, 12, 9, 146, 1258)$ of the decision attribute values, and the bias at $D = 5$ is $z = 64.08$ as calculated by Eq. (1) corresponding to the p-value= 0.0.

(3) STRIM suggests that $C(1) =$ "Location" and $C(4) =$ "Bath (Hot Spring)" can be reducted because no rules use those attributes.

3   Considerations on a Rule Induction Method by STRIM from the Viewpoint of Proof by Contradiction

In propositional logic, a logical expression Q is often derived from several logical expressions $P_1, P_2, ..., P_n$. It can be proved that Q is also true (T) from the interpretation that all P_j $(j = 1, ..., n)$ is T. Simultaneously, if $P_1 \wedge P_2 \wedge ... \wedge P_n = P$, $P \rightarrow Q$ is valid. Here, Q is referred to as a logical consequence from P. If $P \rightarrow Q$ is shown to be true, a reasoning result Q' for arbitrary P' can be obtained using reasoning rules by modus ponens. In propositional logic, to demonstrate that $P \rightarrow Q$ is true, the proof by contradiction is often used to indicate that $P \wedge \sim Q =$ false (F) because $P \rightarrow Q = \sim P \vee Q = \sim (P \wedge \sim Q) = $ T.

As described in Sect. 2, rules hidden in the decision table are derived by evaluating the condition part $CP(d, k) = \wedge_j (C(j_k) = v_j)$ of the if-then rule for $D = d$ by a hypothesis test. We propose an algorithm to estimate rule candidates by rejecting $H0$: f does not have any bias and $CP(d, k)$ is not a rule. Now, let $P_j = \text{T}$ when $C(j_k) = v_k$ and let $P_j = \text{F}$ when $C(j_k) \neq v_k$. In addition, let $Q = \text{T}$ when $D = d$ and $Q = \text{F}$ when $D \neq d$. For example, in $CP(d = 5, k = 1)$ in Table 2, the number of samples of U where $P = \text{T}$ is $11 + 12 + 9 + 146 + 1{,}258 = 1{,}436$, and among them the number of samples where $D \neq 5$ ($Q = \text{F}$, i.e., $\smallfrown Q = \text{T}$) is $11 + 12 + 9 + 146 = 178$. Therefore, under $H0$, the number of samples for $P \wedge \smallfrown Q = \text{T}$ is 178. Note that $(C, D) = ((C(1), ..., C(|C|)), D)$ are random variables. Under $P(D = 5) = 1/5$ and the judgment model in Table 1, the occurrence probability of such a distribution shows that the p-value is equal to or less than 0.0. Thus, $H0$ is rejected in this case, i.e., it is determined statistically that $P \wedge \smallfrown Q = \text{F}$. Therefore, it can be seen that $P \rightarrow Q = \text{T}$ is shown with critical p-value $= 0.0$. Here, since (C, D) are random variables it is necessary to consider the problem that the if-then rule induction method (Sect. 2) is rooted in the fact that the propositional logic $P \rightarrow Q$ is judged to be statistically true or false using proof by contradiction.

4   Considerations on Conventional RS Theory and Its Application to a Rule Induction Problem

Conventional RS theory focuses on the following equivalence relation and the equivalence set of indiscernibility within the decision table S of interest:

$$I_B = \{(u(i), u(j)) \in U^2 | \rho(u(i), a) = \rho(u(j), a), \forall a \in B \subseteq C\}.$$

Here, I_B is an equivalence relation in U and derives the quotient set, $U/I_B = \{[u_i]_B | i = 1, 2, ..., |U| = N\}$, and $[u_i]_B = \{u(i) \in U | (u(j), u_i) \in I_B, u_i \in U\}$. $[u_i]_B$ is an equivalence set with the representative element u_i. Let it be that $\forall X \subseteq U$, then X can be approximated as $B_*(X) \subseteq X \subseteq B^*(X)$ using the equivalence set:

$$B_*(X) = \{u_i \in U | [u_i]_B \subseteq X\}, \tag{2}$$

$$B^*(X) = \{u_i \in U | [u_i]_B \cap X \neq \phi\}. \tag{3}$$

$B_*(X)$ and $B^*(X)$ are the lower and upper approximations respectively of X by B. Note that the pair $(B_*(X), B^*(X))$ is typically referred to as a rough set of X by B.

Specifically, we let $X = \{u(i) | \rho(u(i), D) = d\} = U(d) = \{u(i) | u^{D=d}(i)\}$, and define a set of $u(i)$ as $U(CP) = \{u(i) | u^{C=CP}(i)\}$. If $U(CP) \subseteq U(d)$, then, with necessity, CP can be used as the condition part of the if-then rule of $D = d$. In other words, the following expression of if-then rules with necessity is obtained:

$$Rule(d, k) : \text{ if } CP = \wedge_j (C(j_k) = v_{j_k}) \text{ then } D = d. \tag{4}$$

Similarly, with possibility, $C^*(X)$ derives the condition part CP of the if-then rule of $D = d$. However, the approximations $B_*(X) \subseteq X \subseteq B^*(X)$ of $U(d)$ by lower/upper approximation are too severe or too loose, respectively, and, in many cases, it is impossible to induce effective rules due to the inclusion relationship. Ziarko then expanded the original RS by introducing an admissible error in two ways:

$$\underline{B}_\epsilon(U(d)) = \{u(i) | acc \geq 1 - \varepsilon\}, \tag{5}$$

Table 4. Examples of rules induced by STRIM for the first simulation dataset (Case1).

$CP(d,k)$	$C(1)$ ...$C(6)$	D	p-value(z)	Accuracy	Coverage	$f = (n_1, n_2, n_3, n_4, n_5, n_6)$
(6,1)	660000	6	5.91E−98(20.97)	0.938	0.1883	(1, 2, 1, 2, 0, 90)
(3,1)	330000	3	1.94E−97(20.92)	0.978	0.1778	(0, 0, 88, 1, 1, 0)
(2,1)	002200	2	2.70E−89(20.00)	0.942	0.1698	(,1 81, 1, 1, 1, 1)
(5,1)	550000	5	1.71E−81(19.08)	0.987	0.1477	(0, 0, 0, 0, 78, 1)
(6,2)	006600	6	2.99E−81(19.05)	0.889	0.1674	(6, 1, 1, 0, 2, 80)
(5,2)	005500	5	9.91E−81(18.99)	0.964	0.1515	(0, 1, 1, 1, 80, 0)
(1,1)	001100	1	2.42E−79(18.82)	0.920	0.1578	(80, 1, 2, 0, 3, 1)
(3,1)	003300	3	8.65E−77(18.50)	0.888	0.1596	(3, 2, 79, 2, 2, 1)
(4,1)	004400	4	1.50E−76(18.48)	0.949	0.1456	(1, 0, 1, 75, 1, 1)
(1,2)	110000	1	4.86E−74(18.17)	0.959	0.1381	(70, 1, 1, 0, 1, 0)
(2,2)	220000	2	9.07E−68(17.35)	0.938	0.1279	(0, 61, 1, 0, 2, 1)
(4,2)	440000	4	1.45E−65(17.06)	0.918	0.1301	(1, 1, 0, 67, 2, 2)
(6,3)	600600	6	6.82E−24(10.01)	0.532	0.1046	(8, 9, 11, 6, 10, 5)
(5,3)	500500	5	7.14E−08(7.08)	0.464	0.0739	(10, 10, 11, 5, 39, 9)
(3,3)	030300	3	2.33E−08(5.46)	0.390	0.0606	(11, 6, 30, 12, 10, 8)

Table 5. Comparison of the number of induced rules by rule length derived by using LEM2 and STRIM.

Case no.	Method	Number of rules by rule length						
		1	2	3	4	5	6	Total
Case1	LEM2	0	0	82	1073	623	0	1778
	STRIM	0	15	0	0	0		15
Case2	LEM2	0	0	72	1108	556	0	1736
	STRIM	0	14	0	0	0	0	14
Case3	LEM2	0	0	74	1106	616	0	1796
	STRIM	0	13	0	0	0	0	13

$$\overline{B}_\varepsilon(U(d)) = \{u(i)|acc > \varepsilon\}, \tag{6}$$

where $acc = |U(d) \cap U(CP(k))|/|U(CP(k))| = n_d/n$, $\varepsilon \in [0, 0.5)$. The pair $(\underline{B}_\varepsilon(U(d)), \overline{B}_\varepsilon(U(d)))$ is called an ε-lower and ε-upper approximation that satisfies the properties $B_*(U(d)) \subseteq \underline{B}_\varepsilon(U(d)) \subseteq \overline{B}_\varepsilon(U(d)) \subseteq B^*(U(d))$, $\underline{B}_{\varepsilon=0}(U(d)) = B_*(U(d))$, and $\overline{B}_{\varepsilon=0}(U(d)) = B^*(U(d))$. The ε-lower and/or ε-upper approximations induce if-then rules with admissible errors in the same manner as the lower and/or upper approximations.

As described above, in conventional RS theory, an equivalence relation I_B at a given U is first focused on. Then, based on this relation, an equivalence set at a given U is derived, and the target set is approximated by the equivalence set. Using these approximated sets, if-then rules are induced respectively, as described above. However, the outcome $\rho(u(i), C(k))$ of the random variable $C(k)$ is used for the equivalence relation $I_B = \{(u(i), u(j)) \in U^2 | \rho(u(i), a) = \rho(u(j), a), \forall a = \forall C(k) \in B \subseteq C\}$. Therefore, the equivalence event I_B is a probability event controlled by the conditional joint probability $P((C(k) = \rho(u(i), C(k)), C(k) = \rho(u(j), C(k))) | \rho(u(i), C(k)) = \rho(u(j), C(k)), \forall C(k) \in B \subseteq C)$.

Here, we confirm the rule induction performance using the conventional RS theory in a simulation experiment. First, we set the following rule in the Rule Box in Fig. 1:

$$R(d) : \text{ if } R_d \text{ then } D = d, \quad (d = 1, ..., M_D = 6) \tag{7}$$

$$R_d = (C(1) = d) \wedge (C(2) = d) \vee (C(3) = d) \wedge (C(4) = d).$$

Assume that random variables $C(j)$ $(j = 1, ..., |C| = 6)$ are distributed uniformly and generate inputs $u^C(i) = (v_{C(1)}(i), ..., v_{C(6)}(i))$ $(i = 1, ..., N = 10000)$. Then, using the pre-specified rule (7) and the hypothesis in Table 1, the output $u^D(i)$ is generated to create a decision table. We randomly selected samples by $N_B = 3,000$ from the decision table and formed a new decision table. Table 3 shows some of the 1,778 rules obtained by applying the LEM2 algorithm implementing the lower approximation in ROSE2 [18] to this decision table. In Table 3, by focusing on the rule for $D = 1$ as an example, two or three rules are shown for rule lengths 3 4, and 5. Table 4 shows the results of analyzing the same decision table by STRIM. This simulation experiment was repeated three times, and the numbers of rules induced by each method were arranged and compared according to the rule length in Table 5. We observe the following from these tables.

(1) LEM2 induced all rules for accuracy $= 1$. Some of the induced rules with rule length 3 or 4 shown in Table 3 are sub-rules of the pre-specified rules. If specifying admissible error ε for accuracy and estimating rules by use of VPRS, it is possible to induce the pre-specified rules shown in Table 4. However, in VPRS neither an induction algorithm nor a specifying method for ε has been proposed.

(2) As shown in Table 4, STRIM induced all 12 pre-specified rules and three extra rules. Statistical evidence (p-value or z-value) is shown in these rules. Although it seems that the pre-specified rules can be estimated using appropriate ε and VPRS, the main component of the induction in STRIM is the statistical test The induced rules are based on evidence, i.e., a sufficient number of data that can be used by the statistical test. On the other hand, the coverages of the rules induced in LEM2 are only small percentages, i.e., they include rules of length 5, and by any criterion that is not sufficiently restrictive to be accepted as a rule.

(3) The decision table can be considered a collection of many unarranged if-then rules. LEM2 and STRIM summarize those rules so that human beings can grasp and use the structure and/or features of the rules. From conducting the rule induction experiment three times by LEM2 and STRIM (Table 5), we see that LEM2 summarizes 3,000 rules in somewhat more than 1,700 rules; however, it is clear that LEM2 cannot adequately deal with the given decision table. On the other hand, STRIM induces all pre-specified rules (generally unknown). Note that STRIM induces several additional rules; however, the difference between STRIM and LEM2 can be clearly observed from the accuracy coverage and z-value (Table 4). The validity of the analyzed result by STRIM for the real-world dataset in Table 2 can be inferred to some extent from this simulation result. In any case, we can infer that the rule induction method by the conventional RS based on stochastically varying equivalence relations derives different rules for each decision table, and that the lower approximation rule based on such an equivalence relation cannot fully summarize the decision table.

5   Three-Way Decision Rules and Their Application to the Classification Problem

Yao proposed the concept of three-way decision rules as a new rule induction and decision-making method based on a new interpretation of the classical RS theory [7–9]. Specifically, using a classical RS, Yao proposed to divide U into three regions of X, i.e., the positive region $POS(X)$, the boundary region $BND(X)$, and the negative region $NEG(X)$:

$$POS(X) = B_*(X), \tag{8}$$

$$BND(X) = B^*(X) - B_*(X), \tag{9}$$

$$NEG(X) = U - POS(X) \cup BND(X) = U - B^*(X) = (B^*(X))^C. \tag{10}$$

Any element $x \in POS(X)$ certainly belongs to X, and any element $x \in NEG(X)$ does not belong to X. One cannot decide with certainty whether or not an element $x \in BND(X)$ belongs to X. Similar to the conventional RS theory, we let $X = U(d)$ and can obtain the following decision rules corresponding to (8), (9), and (10):

$$Des([x]) \rightarrow_P Des(U(d)), \text{ for } [x] \subseteq POS(U(d)), \tag{11}$$

$$Des([x]) \rightarrow_B Des(U(d)), \text{ for } [x] \subseteq BND(U(d)), \tag{12}$$

$$Des([x]) \rightarrow_N Des(U(d)), \text{ for } [x] \subseteq NEG(U(d)). \tag{13}$$

Here, $Des([x])$ denotes the logic formula defining the equivalence class $[x]$. For example, $[x]$ is defined by $\wedge_j (C(j_k) = v_{j_k})$.

Yao links (11), (12), and (13) to the rule accuracy (or confidence) based on the probability measure as follows:

$$acc(Des([x]) \rightarrow_\Lambda Des(U(d))) = Pr(U(d)|[x]) = \frac{|[x] \cap U(d)|}{|[x]|}. \qquad (14)$$

Here, $Pr(U(d)|[x])$ is the conditional probability of $U(d)$ given $[x]$. In other words, the probability that the element of $[x]$ exists in $U(d)$ is estimated by the cardinal number. According to accuracy, the positive, boundary, and negative rules are defined by the conditions: $acc = 1$, $0 < acc < 1$, and $acc = 0$, respectively. However, like the idea of VPRS, such approximation based on acc is impractical because the condition is too severe to handle real-world datasets. Therefore, Yao introduced tolerance, similar to VPRS, and proposed rules for the classification problem as follows:

(P1) If $Pr(U(d)|[x]) \geq \alpha$, decide $[x] \subseteq POS(U(d))$,
(B1) If $\beta < Pr(U(d)|[x]) < \alpha$, decide $[x] \subseteq BND(U(d))$,
(N1) If $Pr(U(d)|[x]) \leq \beta$, decide $[x] \subseteq NEG(U(d))$.

Here, $0 \leq \beta < \alpha \leq 1$. As described above, Yao associated the accuracy of the induced rule with the conditional probability. Furthermore, when applying this induced rule to the classification problem, Yao proposed determining boundary parameters (α, β) in accordance with a criterion that minimizes the costs and/or losses by errors based on Bayesian statistics [19]. A detailed discussion is given in the literature [8].

Ziarko did not report a method to specify a reasonable admissible error ε. Yao specified error ε based on Bayesian statistics and included previous studies as a special case. For example, Eqs. (5) and (6) correspond to $\alpha = 1 - \varepsilon$ and $\beta = \varepsilon$, respectively. However, Yao did not propose a specific rule induction method and/or algorithm, such as the decision matrix method [4] or LEM2 [5]. In addition, the three-way decision rules constructing three regions, i.e., the positive, boundary, and negative regions are based on the equivalence relation, which depends on the given decision table and will induce different rules for each sample dataset obtained from the same population similar to the results in classical RS theory.

6   Conclusion

This paper has summarized the concept and validity of a STRIM algorithm that induces rules without using RS theory but by using a statistical test. Furthermore, the rule induction performance of STRIM has been demonstrated through a real-world dataset analysis and a simulation experiment. STRIM has the following features.

(1) There is a data generation model in which the roles of input, output, input/output converting mechanism, observation, and noise generation are clear.

(2) The condition attributes (input) and the decision attribute (output) are considered random variables. Therefore, for example, $\rho(u(i), C(k))$ in the decision table are the outcomes of the random variables $C(k)$. In other words, the decision table is the set of outcomes randomly obtained from the population with condition attributes and decision attribute.

(3) The if-then rule is an input/output converting mechanism that causes bias in the output distribution under the decision attribute value hypothesis (Table 1).

(4) The judgment of bias in the output distribution is determined by a statistical test using a given decision table. Therefore, although STRIM uses a sample dataset, it has an objective criterion that satisfies the criteria for statistical testing with a significance level.

(5) The statistical test is rooted in the proof by contradiction, which is often used when demonstrating the logical consequences of propositional logic.

We have also summarized the conventional RS theory and the associated rule induction method, and pointed out problems there with shown by the results of the simulation experiment. Corresponding to points (1) to (4) above, the conventional RS theory and the rule inducing method are described as follows.

(i) There is no data generation model. Thus, there is no alternative to studying the given decision table at the starting point.

(ii) As there is no data generation model, such as the information function $\rho(u(i), C(k))$, $\rho(u(i), D)$ is needed for convenience. The information function is such that the function value is different for each sample for the same attribute $C(k)$.

(iii) The criterion for adopting a rule is accuracy, and the adoption criteria are not clear (coverage is very small e.g. only one sample satisfies the rule).

(iv) The induced rules are established using only the given decision table, and different rules are derived from different decision tables obtained from the same population because the equivalence class and lower and upper approximation sets differ for each decision table.

From the above, it is considered that the indiscernibility based on the equivalence class is not the essence of a good rule induction method and an improved rule induction method is needed.

References

1. Pawlak, Z.: Rough sets. Int. J. Inform. Comput. Sci. **11**(5), 341–356 (1982)
2. Skowron, A., Rauser, C.M.: The discernibility matrix and functions in information systems. In: Słowiński, R. (ed.) Intelegent Decision Support, Handbook of Application and Advances of Rough Set Theory, pp. 331–362. Kluwer Academic Publishers, Boston (1992)
3. Thangavel, K., Pethalakshmi, A.: Dimensional reduction based on rough set theory. Rev. Appl. Soft Comput. **9**, 1–2 (2009)
4. Shan, N., Ziarko, W.: Data-based acquisition and incremental modification of classification rules. Comput. Intell. **11**(2), 357–370 (1995)

5. Grzymala-Busse, J.W.: LERS – a system for learning from examples based on rough sets. In: Słowiński, R. (ed.) Intelligent Decision Support, Handbook of Applications and Advances of the Rough Sets Theory, pp. 3–18. Kluwer Academic Publishers, Boston (1992)
6. Ziarko, W.: Variable precision rough set model. J. Comput. Syst. Sci. **46**, 39–59 (1993)
7. Yao, Y.: Three-way decision: an interpretation of rules in rough set theory. In: Wen, P., Li, Y., Polkowski, L., Yao, Y., Tsumoto, S., Wang, G. (eds.) RSKT 2009. LNCS, vol. 5589, pp. 642–649. Springer, Heidelberg (2009). https://doi.org/10.1007/978-3-642-02962-2_81
8. Yao, Y.: Three-way decision with probabilistic rough sets. Inf. Sci. **180**, 341–353 (2010)
9. Yao, Y.: Rough sets and three-way decisions. In: Ciucci, D., Wang, G., Mitra, S., Wu, W.-Z. (eds.) RSKT 2015. LNCS (LNAI), vol. 9436, pp. 62–73. Springer, Cham (2015). https://doi.org/10.1007/978-3-319-25754-9_6
10. Matsubayashi, T., Kato, Y., Saeki, T.: A new rule induction method from a decision table using a statistical test. In: Li, T., et al. (eds.) RSKT 2012. LNCS (LNAI), vol. 7414, pp. 81–90. Springer, Heidelberg (2012). https://doi.org/10.1007/978-3-642-31900-6_11
11. Kato, Y., Saeki, T., Mizuno, S.: Studies on the necessary data size for rule induction by STRIM. In: Lingras, P., Wolski, M., Cornelis, C., Mitra, S., Wasilewski, P. (eds.) RSKT 2013. LNCS, vol. 8171, pp. 213–220. Springer, Heidelberg (2013). https://doi.org/10.1007/978-3-642-41299-8_20
12. Kato, Y., Saeki, T., Mizuno, S.: Considerations on rule induction procedures by STRIM and their relationship to VPRS. In: Kryszkiewicz, M., Cornelis, C., Ciucci, D., Medina-Moreno, J., Motoda, H., Raś, Z.W. (eds.) RSEISP 2014. LNCS (LNAI), vol. 8537, pp. 198–208. Springer, Cham (2014). https://doi.org/10.1007/978-3-319-08729-0_19
13. Kato, Y., Saeki, T., Mizuno, S.: Proposal of a statistical test rule induction method by use of the decision table. Appl. Soft Comput. **28**, 160–166 (2015)
14. Kato, Y., Saeki, T., Mizuno, S.: Proposal for a statistical reduct method for decision tables. In: Ciucci, D., Wang, G., Mitra, S., Wu, W.-Z. (eds.) RSKT 2015. LNCS (LNAI), vol. 9436, pp. 140–152. Springer, Cham (2015). https://doi.org/10.1007/978-3-319-25754-9_13
15. Kitazaki, Y., Saeki, T., Kato, Y.: Performance comparison to a classification problem by the second method of quantification and STRIM. In: Flores, V., et al. (eds.) IJCRS 2016. LNCS (LNAI), vol. 9920, pp. 406–415. Springer, Cham (2016). https://doi.org/10.1007/978-3-319-47160-0_37
16. Fei, J., Saeki, T., Kato, Y.: Proposal for a new reduct method for decision tables and an improved STRIM. In: Tan, Y., Takagi, H., Shi, Y. (eds.) DMBD 2017. LNCS, vol. 10387, pp. 366–378. Springer, Cham (2017). https://doi.org/10.1007/978-3-319-61845-6_37
17. Kato, Y., Itsuno, T., Saeki, T.: Proposal of dominance-based rough set approach by STRIM and its applied example. IJCRS 2017, Part I. LNCS (LNAI), vol. 10313, pp. 418–431. Springer, Cham (2017). https://doi.org/10.1007/978-3-319-60837-2_35
18. Walpole, R.E., Myers, R.H., Myers, S.L., Ye, K.: Probability and Statistics for Engineers and Scientists, 8th edn, pp. 187–191. Pearson Prentice Hall, Upper Saddle River (2007)
19. Dud, R., Hart, P.E.: Pattern Classification and Scene Analysis. Wiley, New York (1973)

Multi-label Online Streaming Feature Selection Based on Spectral Granulation and Mutual Information

Huaming Wang, Dongming Yu, Yuan Li, Zhixing Li$^{(\boxtimes)}$, and Guoyin Wang$^{(\boxtimes)}$

Chongqing Key Laboratory of Computational Intelligence,
Chongqing University of Posts and Telecommunications,
Chongqing 400065, People's Republic of China
{lizx,wanggy}@cqupt.edu.cn

Abstract. Instances in multi-label data sets are generally described as a high-dimensional feature vector, as brings the "curse of dimensionality" problem. To ease this problem, some multi-label feature selection algorithms have been proposed. However, they all handle feature selection problems with the assumption that all candidate features are available beforehand. While in some real applications, feature selection must be conducted in the online manner with dynamic features, for example, novel topics arise constantly with a set of features in social networks. Online streaming feature selection (OSFS), dealing with dynamic features, has attracted intensive interest in recent years. Some online feature selection methods are designed for single-label applications, They can not be directly applied in multi-label scenarios. In this paper, we propose a multi-label online streaming feature selection algorithm based on spectral granulation and mutual information (ML-OSMI), which takes high-order label correlations into consideration. Moreover, comprehensive experiments are conducted to verify the effectiveness of the proposed algorithm on twelve multi-label high-dimensional benchmark data sets.

Keywords: Multi-label feature selection · Streaming features
Mutual information · Granular computing

1   Introduction

Multi-label data emerge on various real-world domains, such as image processing, text classification, bioinformatics and information retrieval [1–5]. In these applications, each instance is associated with multiple labels simultaneously. For example, a document may belong to many topics and a gene could have several functions [5]. Moreover, multi-label data are generally represented by very high dimensional vectors, as brings a large number of features and most of them are irrelevant or redundant [6]. Unnecessary features may not only reduce the performance of classifiers but result in the increment of memory storage and computation time. To ease these problems, feature selection techniques have been wildly

© Springer Nature Switzerland AG 2018
H. S. Nguyen et al. (Eds.): IJCRS 2018, LNAI 11103, pp. 215–228, 2018.
https://doi.org/10.1007/978-3-319-99368-3_17

studied, which select a relative small subset of features from the original feature space to remove irrelevant and redundant features without losing discriminative information for later processing.

A number of feature selection methods dealing with multi-label data have been proposed [7–9]. However, they handle feature selection problems with the assumption that all candidate features are available before the learning starts and have to wait for the calculation of all the features, which is very deficient in practice. Online streaming feature selection [10], evaluating features dynamically with the arrival of new features, is a more time efficiency and intuitive way to solve such problems. Existing online feature selection methods [11–14]. But they are designed for single-label learning tasks and cannot be directly applied to multi-label tasks. One commonly encountered way is transforming the multi-label problems into single-label problems. Then single-label online feature selection methods can be adopted. Nevertheless, it ignores the correlation among labels which may carry useful information for learning task, or leads to extremely high and unbalanced label space [9,15].

In this paper, we analyze multi-label online streaming feature selection problem and design an online streaming feature selection algorithm based on spectral granulation and mutual information. The proposed algorithm first granulates labels using spectral clustering. Then it transforms label granules into new multi-class labels and performs feature selection on the new label space. The main contributions of this study are summarized as follows: (1) Although there are multi-label feature selection methods for constant features and single-label feature selection algorithms for dynamic features, we introduce dynamic feature selection into multi-label scenarios. (2) We designed a novel multi-label online streaming feature selection algorithm. (3) Comprehensive experiments are conducted to compare our proposed methods with traditional multi-label methods and single-label online streaming feature selection algorithms on various benchmark multi-label data sets.

2   Related Works

2.1   Multi-label Feature Selection

In multi-label learning tasks, each instance is associated with multiple labels and these labels are generally correlated, as makes multi-label feature selection tasks more complicated than single-label ones. Moreover, there are evidences showing that taking label correlations into consideration can benefit the learning model [7]. Hence, exploring label dependence is an important issue. Multi-label feature selection algorithms can be divided into three categorizes by the type of correlations they considered, first-order, second-order and high-order methods.

First-order ones, such as BR [15], consider each label independently and transform the multi-label feature selection task into several binary single-label sub-problems. LCFS [16] is a second-order algorithm. It builds new labels based on relations among the original labels to capture pair-wise label correlations and then conducts BR approach on the expanded label space to select a subset of informative features. First-order and second-order algorithms assume that labels are independent to each other or pair-wise correlated. However, correlations among labels in real applications are more complicated. LP transforms multi-label data set to a new single-label multi-class data set, then any single-label feature selection could be adopted [15]. However, when the number of labels is extreme large, LP based methods could suffer from terribly class-imbalance problems [6].

MDMR [17] defines mutual information based evaluations to guide feature selection procedure, considering multi-label feature selection problems in two aspects, namely feature dependency and feature redundancy. [18] implements a multi-label feature selection method similar to MDMR named MLMRMR based on the single-label feature selection algorithm mRMR [19]. [9] partitions labels into clusters according to their similarity using a balanced k-means methods and then undertakes feature selection based on mRMR viewing each cluster of labels as a new multi-label subtask. RFS [20] introduces $\ell_{2,1}$-norm on both loss function and regularization to eliminate unnecessary features. [21] solves multi-label feature selection with streaming labels by ranking features iteratively, where the labels arrive one at a time. [7] proposes a multi-label feature selection method called MIFS. The labels are first mapped to a low-dimensional space with less noisy. Then it conducts feature selection on the reduced label space.

2.2   Online Streaming Feature Selection

Online streaming feature selection focuses on the feature selection problems with dynamic features. Grafting [13], Alpha-investing [14], fast-OSFS [11] and SAOLA [22] are several state-of-the-art algorithms proposed to solve online streaming feature selection problems. Grafting treats the feature selection task as a stream-wise regularized risk minimization problem. New features are selected if the improvement of accuracy made by them is greater than a predefined threshold. However, it has no mechanism to remove redundant features selected previously, rendering it suffering from the nesting effect. Alpha-investing [14] uses a step-wise linear regression model and a p-value to determine new features which are selected or not. Furthermore, alpha-investing and Grafting used prior information about the structure of feature space, which is impossible to obtain on the original streaming tasks. Hence, they might not produce good performance in real applications. Wu [11] proposed the fast-OSFS algorithm, needing no prior knowledge about the feature space, which contains two major steps: online relevance analysis and online redundancy analysis. The first step discards irrelevant features and the second eliminates redundant features. SAOLA [22] is another online feature selection method dealing with dynamic features using mutual information based criterions to guide feature selection heuristically.

Though there are several online feature selection methods proposed, they are designed for single-label tasks and can not apply directly in multi-label scenarios. In this paper, we study the multi-label feature selection problems with dynamic (or streaming) features and propose a multi-label online streaming feature selection algorithm.

3   The Proposed Method

In this section, we first describe the multi-label online streaming feature selection problem. Then, we design a multi-label online streaming feature selection method. The proposed method applies spectral clustering which granulates labels into clusters and captures high-order label correlations. Moreover, the relevance and redundancy of features are redefined using mutual information to guide multi-label feature selection procedure.

3.1   Problem Statement

Definition 1 (Traditional Multi-label Feature Selection). Let X be the sample space and $x_i \in X$ is a feature vector. $Y = \{l_1, l_2, ..., l_m\}$ is a set of labels. Multi-label learning is objective to produce a function $H = \{X \rightarrow 2^L\}$ which assigns each instance with a set of relevant labels. Traditional multi-label feature selection holds the assumption that instances are represented with a fixed dimensional feature space $F = \{f_1, f_2, ..., f_d\}$. They aim to select an optimal subset of features $SF \subseteq F$ without harming the predictive performance.

Definition 2 (Streaming Features). Streaming features denote a feature space where features flow in one by one over time with fixed number of instances. With a dynamic feature space, the dimensionality may tend to very high or even infinite. Besides, each feature is required to be processed when its arrival. Hence, feature selection procedure should be conducted in the online manner.

Definition 3 (Multi-label Online Streaming Feature Selection). Multi-label online streaming feature selection copes with a streaming feature vector F_s^t, where $F_s^t = \{f_1, f_2, ..., f_t\}$ and f_t denotes the feature arrives at time t. As the features flow in continuously, multi-label streaming feature selection task is objective to remove irrelevant and redundant features from the available feature set F_s^t while holds discriminative information with more than one targets $Y = \{l_1, l_2, ..., l_m\}$.

There are three major challenges in the multi-label streaming feature selection scenario:

- **The dynamic and uncertain nature of the feature space.** The dimensionality of the feature space grows over time and may even tend to infinite.
- **The streaming nature of the feature space.** The subset of selected features should be updated timely with new features flow in one at a time.
- **The complex correlations among labels.** There are complex correlations among labels and evidences show that taking label correlations into consideration will benefit learning model.

3.2   The Framework of ML-OSMI

The framework of the proposed multi-label online streaming feature selection algorithm is shown in Fig. 1. To capture label correlations, the original label space is first transformed into a multi-class multi-target one with much lower dimensionality. Then, the new labels are used to select features. To conduct feature selection procedure with many labels and streaming features, we adopt relevance test and redundancy test to guide the online feature selection, motivated by single-label online streaming feature selection methods [11]. Section 3.3 gives the details of label space transformation and Sect. 3.4 redefines the relevance and redundancy of features.

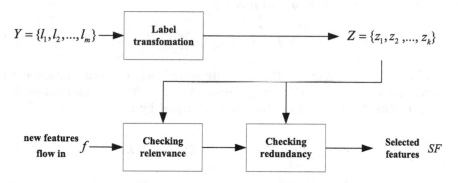

Fig. 1. Framework of the proposed algorithm

3.3   Capturing Label Correlations by Spectral Granulation

In multi-label data, a label is generally related to a small set of labels from the entire label space [9,23]. Hence, the label correlations can be explored as much as possible by dividing labels into partitions, where the labels in one partition are relevant to each other and the labels in different partitions are irrelevant. The partitions of labels are considered as granulas in this paper. The labels in the same granula are high correlated while the labels in different granula are mutually independent or weakly related. To generate the granulas, labels are clustered using spectral clustering with cosine similarity. Then, each label clusters is transformed into a multi-class label applying LP framework [6]. Finally, we get a new label space consists of multi-class labels with much lower dimensionality than the original label space. The new multi-class labels are used to steer feature selection processing taking label correlations into account.

3.4   Evaluations Based on Mutual Information

To perform multi-label feature selection, an algorithm must be able to measure the dependency between features and labels. Mutual information is often employed to characterize this dependency. Given two random variables x and y, their mutual information is defined in terms of probability density functions $p(x)$, $p(y)$ and $p(x, y)$:

$$mi(x, y) = \int \int p(x, y) log \frac{p(x, y)}{p(x)p(y)} dxdy . \qquad (1)$$

the normalized version of mutual information is:

$$nmi(x, y) = \frac{2 \times mi(x, y)}{h(x) + h(y)} . \qquad (2)$$

where $h(x) = \int p(x) \log p(x) dx$. Given conditional variable z, the conditional mutual information between x and y is

$$cmi(x, y|z) = \int \int p(x, y|z) log \frac{p(x, y)}{p(x)p(y)} dxdy . \qquad (3)$$

Given a finite set of features F and a finite set of labels L, mutual information based feature selection methods is objective to find the optimal subset of features $SF^* \subseteq F$ without reducing the information shared by features and labels, as can be written as:

$$SF^* = \arg \min_{SF \subseteq F} \{|SF| : mi(SF, L) = mi(F, L)\} . \qquad (4)$$

It can also be considered as removing every unnecessary feature from F. Using conditional mutual information, this formulation can be expressed as:

$$SF^* = \arg \min_{SF \subseteq F} \{|SF| : \forall f \epsilon F - SF, cmi(f, L|SF) = 0\} . \qquad (5)$$

The Eq. (5) indicates that an optimal reduction of the original feature set F should contain no irrelevant or redundant features. However, either Eqs. (4) or (5) is difficult to calculate. In the following, we redefine the relevance and redundancy of features based on mutual information to guide the feature selection procedure to achieve this target.

Definition 4 (Relevance) given a finite label set $L = \{l_1, l_2, ..., l_n\}$, the relevance of the feature f and the label set L is defined as:

$$rel(f, L) = \max\{nmi(f, l_i), l_i \epsilon L\} . \qquad (6)$$

The $rel(f, L)$ measures the relevance between feature f and the label set L. Moreover, it delivers in pairwise manner, as can be calculated with efficiency. Obviously, if $rel(f, L) = 0$, f shares little information with any label $l_i \in L$. In other words, f can be discarded without harming the predictive performance. However, 0 is a threshold which is too strict to use in real applications. A compromise choice is using a small positive relevance threshold α. If $rel(f, L) \leq \alpha$, f is considered to be an irrelevant feature.

Definition 5 (Redundancy). let F is a finite feature set, for any feature $g\epsilon F$, the significance of g on L given another $h\epsilon F$ is defined as

$$sig(g, L|h) = \max\{cmi(g, l_i|h), l_i\epsilon L\} . \tag{7}$$

which means that a feature g is redundant and can be removed from F if there exists a feature $h\epsilon F - g$ satisfying $sig(g, L|h) = 0$. This is a loosed and approximate version of the formulation $cmi(g; l_i|F - g) = 0$ described in Eq. 5. It only considers second-order conditional dependency but is much easier and efficient to calculate.

3.5   The Proposed Method

We propose a multi-label online feature selection algorithm named multi-label online streaming feature selection based on spectral granulation and mutual information (ML-OSMI) on the basis of Sects. 3.2, 3.3 and 3.4. The pseudo-code

---

Algorithm 1. ML-OSMI

Input: Feature stream F, label space L and the relevance threshold α
Output: selected features SF
1 granulating labels into $Z = \{z_1, z_2, ..., z_k\}$ using spectral clustering;
2 $SF = \emptyset$;
3 **repeat**
4 $\quad$ get f from the stream F;
5 $\quad$ /*checking relevance */
6 $\quad$ **if** $rel(f, L) \leq \alpha$ **then**
7 $\quad\quad$ | continue;
8 $\quad$ **end**
9 $\quad$ /*checking redundancy */
10 $\quad$ $added = 1$;
11 $\quad$ **for** a_j *in* SF **do**
12 $\quad\quad$ /*checking whether f is redundant*/
13 $\quad\quad$ **if** $sig(f, z_i, a_j) == 0$ **then**
14 $\quad\quad\quad$ | $added = 0$;
15 $\quad\quad\quad$ | break;
16 $\quad\quad$ **end**
17 $\quad\quad$ /*checking whether a_j is redundant*/
18 $\quad\quad$ **if** $sig(a_j, z_i, f) == 0$ **then**
19 $\quad\quad\quad$ | $SF = SF \backslash a_j$;
20 $\quad\quad$ **end**
21 $\quad$ **end**
22 $\quad$ **if** $added == 1$ **then**
23 $\quad\quad$ | $SF = SF \cup f$;
24 $\quad$ **end**
25 **until** *no new features or stopping criteria met*;
26 **return** SF

---

of ML-OSMI is shown in Algorithm 1. ML-OSMI delivers as follows. As a new feature f flows in, if $rel(f, L) \leq \alpha$ is satisfied, f is considered to be a irrelevant feature and discarded. The online feature selection waits for the next feature. If f passes the relevance checking at Step 6, the algorithm assesses two kinds of redundancy, the redundancy of f and the redundancy of selected features before time t. Suppose SF is the set of selected features before f arrives. Firstly, the algorithm checks the redundancy of f to determine whether there exists a feature $a_j \in SF$ making f conditionally independent to the label set. If there has no such a feature in SF, f is selected. Then, the algorithm removes all features made to be redundant by f from SF. If there has no new features, the algorithm terminates.

3.6 Analysis of Time Efficiency

The time complexity of the proposed algorithm consists of two parts: the complexity of conducting relevance analysis and the complexity of removing redundant features. In the analysis, the number of samples is omitted for simplicity. Let F_t be the features arrived before time t. F_t^r is a subset of F_t containing all features which are relevant to the label set. Suppose SF_t is the selected feature subset at time t and $r = |SF_t|$. Let $m = |F_t|$ be the number of features in F_t and $p = |F_t^r|$. When the number of feature is extremely high, it has $m \gg p \gg r$. Hence, the average time complexity of the proposed algorithm is $O(km+kpr)$, where k is the number of label granulas and $k \ll r$. If all features are discarded on the relevance test, the best time complexity is $O(km)$. While all features pass the independence test, the worst-case complexity is $O(kmr)$. Noticing that $k \ll n$ and $r \ll m$, where n is the cardinality of the original label set, one can concludes that $O(kmr) \ll O(nm^2)$.

4   Experiment Results

4.1 Experiment Settings

We use twelve multi-label high-dimensional benchmark data sets from various domains as our test beds. The details of data sets are shown in Table 1. The *scene* is from the image processing application. *emotions* and *CAL500* involve emotions classification of music. *genbase* and *yeast* are obtained in biology domain. The rest seven data sets are from text and natural language processing topics. All data sets are available at the MEKA website[1]. The experiments are conducted on a personal computer with Windows Server 2016, Inter(R) Core (TM) i7-6850K CPU and 64 GB memory employing MATLAB R2016a platform.

To illustrate the effectiveness of the proposed algorithm, we compare our algorithm with four state-of-the-art multi-label feature selection algorithms and two state-of-the-art single-label online feature selection algorithms. The comparisons contain the number of selected of features, running time and prediction

---

[1] http://meka.sourceforge.net/#datasets.

performances. The predictions are delivered by the multi-label k-nearest neighbors algorithm(ML-KNN) [24] trained with the selected features. ML-KNN is a well known multi-label classification method for its efficiency. In our experiments, the number of nearest neighbors is set to the recommended value 10 and the smoothing factor is 1. Five widely used evaluations are used to measure the predictive performances, namely Hamming Loss, Coverage, One Error, Ranking Loss and Average Precision [6]. The greater the value of Average Precision, the better the performance of the model. For the other four evaluations, the less their value are, the better the model is.

Table 1. Details of the benchmark data sets

Ind	Dataset	Instance	Feature	Label	Domain
1	emotions	593	72	6	music
2	bibtex	7395	1836	159	text
3	CAL500	502	68	174	music
4	delicious	16105	500	983	text
5	enron	1702	1001	53	text
6	genbase	662	1186	27	biology
7	languagelog	1460	1004	75	text
8	medical	978	1449	45	ext
9	scene	2407	294	6	images
10	tmc2007	28596	49060	22	text
11	20NG	19299	1006	20	text
12	yeast	2417	103	14	biology

4.2 Comparisons with Traditional Multi-label Feature Selection Methods

The comparative multi-label feature selection algorithms are F-Score [25], MLM-RMR [18,19], RFS [20] and MIFS [7]. Comparisons on running time and predictive performances are given. The implements of these algorithms can be found on Github[2] and the parameters such as the size of selected features are set as their default value. Moreover, the 5-fold validation mechanism is adopted on all data sets. Table 2 gives the running time and Fig. 2 shows the predictive performances of multi-label feature selection methods.

(1) ML-OSMI vs. F-Score. As is shown in Table 2, F-Score takes fewer time on 8 of 12 data sets except for *CAL500*, *enron*, *genbase* and *medical*. However, Fig. 2 shows that ML-OSMI achieves higher Average Precision on 11 of 12 except for the *bibtex*. There has no significant difference on Coverage among all feature selection methods. For Hamming Loss, ML-OSMI delivers better results on

---

[2] https://github.com/KKimura360/MLC_toolbox.

Table 2. Running time (Seconds)

Ind	F-Score	MIFS	RFS	MLMRMR	Proposed
1	**0.013**	0.117	0.795	0.026	0.102
2	**3.719**	29.751	603.866	30.022	18.655
3	0.141	0.897	0.529	**0.011**	0.041
4	**7.294**	424.569	3989.808	96.614	45.698
5	0.650	1.837	20.059	2.209	**0.388**
6	0.361	0.529	3.103	0.882	**0.025**
7	**1.074**	1.978	13.013	2.575	3.441
8	0.733	1.157	6.733	1.558	**0.532**
9	**0.029**	0.717	34.849	1.044	2.388
10	**0.499**	27.580	21669.986	20.382	1.142
11	**0.523**	19.950	6544.833	14.217	24.936
12	**0.027**	0.568	34.542	0.285	0.545

enron, genbase, languagelog, medical and *scene*. On other 7 data sets, ML-OSMI and F-Score perform equally well. Besides, ML-OSMI obtains better performance on 9 out of 12 data sets for One Error and 10 out of 12 data sets for Ranking Loss.

(2) ML-OSMI vs. MIFS. Table 2 says that ML-OSMI uses fewer time to select features on 9 out of 12 data sets than MIFS. Figure 2 shows that ML-OSMI performs better than MIFS on all data sets but the *scene* on Average Precision, Hamming Loss and Ranking Loss. For Coverage, neither of them shows superiority. Moreover, except for *scene* and *languagelog*, ML-OSMI gains better results of One Error than MIFS.

(3) ML-OSMI vs. RFS. The comparisons between ML-OSMI and RFS in Table 2 show that ML-OSMI achieves better time efficiency on all data sets. For the predictive performances, Fig. 2 indicates that ML-OSMI gets better results evaluated by Average Precision, One Error and Ranking Loss on all data sets except for the *emotions* and *languagelog*. Besides, ML-OSMI outperforms RFS on 8 out of 12 data sets on Hamming Loss and delivers the same results on 3 of the remaining 4 data sets. For Coverage, ML-OSMI and RFS perform almost equally well.

(4) ML-OSMI vs. MLMRMR. Table 2 shows that MLMRMR takes less time than ML-OSMI on *emotions, CAL500, languagelog, scene, 20NG* and *yeast*, while ML-OSMI takes less time than MLMRMR on the other 6 data sets. As Fig. 2 shows, ML-OSMI performs better than MLMRMR on *enron* and *scene* and MLMRMR performs better than ML-OSMI on *enron* and *bibtex*. On the remaining 9 data sets, ML-OSMI performs as good as MLMRMR.

4.3 Comparisons with OSFS Methods in Streaming Feature Scenario

We also compare ML-OSMI with two state-of-the-art OSFS algorithms, Alpha-investing [14] and SAOLA [22]. To evaluate the effectiveness of the proposed multi-label online streaming feature selection algorithm, we choose 8 data sets with extreme high dimensionality to simulate the streaming feature selection scenario. Average Precision and Hamming Loss are used as the criterions to demonstrate the performance of the algorithms. Figure 3 reports the performances of LP-SAOLA, LP-alpha-investing and ML-OSMI with the features flowing in continuously over time. Table 3 gives the running time.

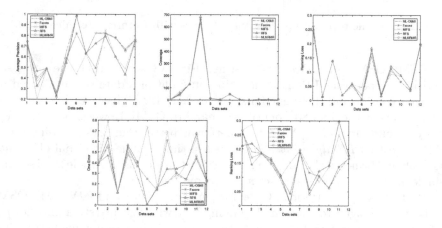

Fig. 2. Comparisons with multi-label feature selection methods

Table 3. Running time (Seconds)

Dataset	lp-alpha-investing	lp-saola	Proposed
emotions	**0.004**	0.154	0.102
bibtex	**15.331**	435.656	18.655
CAL500	**0.003**	0.193	0.041
delicious	**6.180**	43.281	45.698
enron	0.416	109.994	**0.388**
genbase	1.137	1.068	**0.025**
languagelog	**0.875**	106.720	3.441
medical	**0.481**	150.697	0.532
scene	**0.211**	1.153	2.388
tmc2007	18.408	52.927	**1.142**
20NG	41.580	157.684	**24.936**
yeast	**0.007**	0.027	0.545

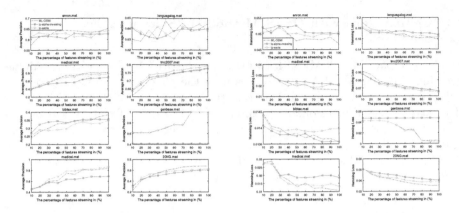

Fig. 3. The predictive performance changes with features streaming in

(1) ML-OSMI vs. LP-alpha-investing. Figure 3 shows that the proposed algorithm outperforms LP-alpha-investing on 6 out of 8 data sets evaluated by Average Precision and Hamming Loss. For *mc2007*, LP-alpha-investing generates better results on the prior 80% of features than ML-OSMI. However, with new features continuously flow in, ML-OSMI performs better than LP-alpha-investing. Table 3 says that *LP-alpha-investing* takes less time dealing with 8 out of 12 data sets. It should be noted that LP-alpha-investing transforms the whole label set into a single multi-class label, as makes it more time efficiency.

(2) ML-OSMI vs. LP-saola. On *enron*, *medical*, *bibtex* and *20NG*, ML-OSMI gets better Average Precision and Hamming Loss with features streaming flowing in. Besides, compared to LP-SAOLA, the proposed algorithm gains better time efficiency on 9 out of 12 data sets except for *delicious*, *scene* and *yeast*. Especially, on six relatively higher dimensional data sets with thousands of features, *bibtex*, *enron*, *genbase*, *medical*, *tmc2007* and *20NG*, the proposed algorithm shows better efficiency for taking relative less time.

5   Conclusion

In this paper, we propose a multi-label online streaming feature selection algorithm to address multi-label feature selection with dynamic features. The proposed method first granulates the labels. Labels in the same granula are high correlated and labels in different granula are mutually independent or weakly correlated. Then, transforming each granula of labels into a multi-class label, the original labels is converted into a new space with much lower dimensionality, taking high-order correlations into consideration. Moreover, the relevance and redundancy of features are redefine based on mutual information to guide feature selection procedure. Finally, the features are selected with the new label space in online manner. Comprehensive experiments are conducted to verify the effectiveness of the proposed method, comparing it with traditional multi-label feature selection methods and online streaming feature selection methods. Results have shown that the proposed multi-label online feature selection algorithm can effectively solve multi-label feature selection with dynamic features. In our future work, we will study how to deliver feature selection with features and labels flow in simultaneously.

Acknowledgements. This work was supported by the National Key Research and Development Program of China (Grant no. 2016YFB1000900), the National Natural Science Foundation of China (Grant nos. 61572091, 61772096), Chongqing Basic and Frontier Research Project (cstc2015jcyjA40018) and The Science and Technology Project Affiliated to the Education Department of Chongqing Municipality (KJ1500438).

References

1. Hua, X.S., Qi, G.J.: Online multi-label active annotation: towards large-scale content-based video search. In: International Conference on Multimedia 2008, Vancouver, British Columbia, Canada, pp. 141–150, October 2008
2. Lai, H., Yan, P., Shu, X., Wei, Y., Yan, S.: Instance-aware hashing for multi-label image retrieval. IEEE Trans. Image Process. **25**(6), 2469 (2016)
3. Trohidis, K., Tsoumakas, G., Kalliris, G., Vlahavas, I.P.: Multi-label classification of music into emotions. In: ISMIR 2008, 9th International Conference on Music Information Retrieval, Drexel University, Philadelphia, PA, USA, 14–18 September 2008, pp. 325–330 (2008)
4. Wu, B., Lyu, S., Hu, B.G., Ji, Q.: Multi-label learning with missing labels for image annotation and facial action unit recognition. Patt. Recogn. **48**(7), 2279–2289 (2015)
5. Zhang, M.L., Zhou, Z.H.: Multilabel neural networks with applications to functional genomics and text categorization. IEEE Trans. Knowl. Data Eng. **18**(10), 1338–1351 (2006)
6. Tsoumakas, G., Katakis, I., Vlahavas, I.P.: Mining multi-label data. In: Data Mining and Knowledge Discovery Handbook, 2nd edn., pp. 667–685 (2010)
7. Jian, L., Li, J., Shu, K., Liu, H.: Multi-label informed feature selection. In: Proceedings of the Twenty-Fifth International Joint Conference on Artificial Intelligence, IJCAI 2016, New York, NY, USA, 9–15 July 2016, pp. 1627–1633 (2016)
8. Lee, J., Kim, D.W.: Mutual information-based multi-label feature selection using interaction information. Expert Syst. Appl. **42**(4), 2013–2025 (2015)
9. Li, F., Miao, D., Pedrycz, W.: Granular multi-label feature selection based on mutual information. Patt. Recogn. **67**, 410–423 (2017)
10. Wu, X., Yu, K., Wang, H., Ding, W.: Online streaming feature selection. In: Proceedings of the 27th International Conference on Machine Learning (ICML 2010), 21–24 June 2010, Haifa, Israel, pp. 1159–1166 (2010)
11. Wu, X., Yu, K., Ding, W., Wang, H.: Online feature selection with streaming features. IEEE Trans. Patt. Anal. Mach. Intell. **35**(5), 1178 (2013)
12. Wang, J., et al.: Online feature selection with group structure analysis. IEEE Trans. Knowl. Data Eng. **27**(11), 3029–3041 (2016)
13. Perkins, S., Theiler, J.: Online feature selection using grafting. In: Machine Learning, Proceedings of the Twentieth International Conference (ICML 2003), 21–24 August 2003, Washington, DC, USA, pp. 592–599 (2003)

14. Zhou, J., Foster, D.P., Stine, R.A., Ungar, L.H.: Streaming feature selection using alpha-investing. In: Proceedings of the Eleventh ACM SIGKDD International Conference on Knowledge Discovery and Data Mining, Chicago, Illinois, USA, 21–24 August 2005, pp. 384–393 (2005)
15. Cherman, E.A., Monard, M.C., Lee, H.D.: A comparison of multi-label feature selection methods using the problem transformation approach. Electr. Notes Theor. Comput. Sci. **292**, 135–151 (2013)
16. Spolaôr, N., Monard, M.C., Lee, H.D.: Feature selection for multi-label learning. In: Proceedings of the 24th International Conference on Artificial Intelligence, Series, IJCAI 2015, pp. 4401–4402. AAAI Press (2015)
17. Lin, Y., Hu, Q., Liu, J., Duan, J.: Multi-label feature selection based on max-dependency and min-redundancy. Neurocomputing **168**, 92–103 (2015)
18. Kimura, K., Sun, L., Kudo, M.: MLC toolbox: A MATLAB/OCTAVE library for multi-label classification. CoRR, abs/1704.02592 (2017). http://arxiv.org/abs/1704.02592
19. Peng, H., Long, F., Ding, C.: Feature selection based on mutual information: criteria of max-dependency, max-relevance, and min-redundancy. IEEE Trans. Patt. Anal. Mach. Intell. **27**(8), 1226 (2005)
20. Nie, F., Huang, H., Cai, X., Ding, C.H.Q.: Efficient and robust feature selection via joint $l_{2,1}$-norms minimization. In: Advances in Neural Information Processing Systems 23: 24th Annual Conference on Neural Information Processing Systems 2010. Proceedings of a meeting held 6–9 December 2010, Vancouver, British Columbia, Canada, pp. 1813–1821 (2010)
21. Lin, Y., Hu, Q., Zhang, J., Wu, X.: Multi-label feature selection with streaming labels. Inf. Sci. **372**, 256–275 (2016)
22. Yu, K., Wu, X., Ding, W., Pei, J.: Towards scalable and accurate online feature selection for big data. In: 2014 IEEE International Conference on Data Mining, ICDM 2014, Shenzhen, China, 14–17 December 2014, pp. 660–669 (2014)
23. Sun, L., Kudo, M., Kimura, K.: Multi-label classification with meta-label-specific features. In: 23rd International Conference on Pattern Recognition, ICPR 2016, Cancún, Mexico, 4–8 December 2016, pp. 1612–1617 (2016)
24. Zhang, M.L., Zhou, Z.H.: ML-KNN: a lazy learning approach to multi-label learning. Patt. Recogn. **40**(7), 2038–2048 (2007)
25. Kong, D., Ding, C.H.Q., Huang, H., Zhao, H.: Multi-label reliefF and F-statistic feature selections for image annotation. In: 2012 IEEE Conference on Computer Vision and Pattern Recognition, Providence, RI, USA, 16–21 June 2012, pp. 2352–2359 (2012)

Bipolar Queries with Dialogue: Rough Set Semantics

Soma Dutta[1,2(✉)] and Andrzej Skowron[3,4]

[1] Vistula University, Stokłosy 3, 02-787 Warsaw, Poland
somadutta9@gmail.com
[2] Department of Mathematics and Computer Science,
University of Warmia and Mazury, Sloneczna str. 54, 10-710 Olsztyn, Poland
[3] Faculty of Mathematics, Informatics and Mechanics,
University of Warsaw, Banacha 2, 02-097 Warsaw, Poland
skowron@mimuw.edu.pl
[4] Systems Research Institute, Polish Academy of Sciences,
Newelska 6, 01-447 Warsaw, Poland

Abstract. This paper proposes an interpretation of characterizing required condition and desired condition of an user, that is a bipolar query, from the perspective of rough set semantics with an additional feature of learning the user's need through dialogue.

1 Introduction

Bipolar queries are meant to express human preferences and intentions by distinguishing the required and desired components. In the context of machine-driven search in response to an user's query, understanding this distinction between required and desired conditions, articulated to a machine through natural language, is a real challenge. In literature (see, *e.g.*, [5,8–10,12]), there are two ways of viewing this bipolar nature of a query given by a human user; one is bipolar univariate and other is unipolar bivariate. In the first case, one scale passing gradually from negative evaluation to positive evaluation via neutral cases is considered. In the latter, two more or less independent scales, which separately account for positive and negative evaluations for both required and desired conditions, is considered. In this paper, our approach will be inclined to the second way of viewing bipolar queries. The next important issue is to assess the query as a whole by aggregating its bipolar assessments. The methods for assessing each of the components of a bipolar query and aggregating them together are varying in the literature.

In [8] authors have presented a way to distinguish between an agent's requirement and desire in a formal set up so that an automated search engine can satisfy a particular objective of an user. They have posed the problem through an example that an user is looking for a house which is *cheap and possibly close to the public transport*. The task is to identify, among these two constraints, the required one and the desired one, and accordingly aggregate the preferences on houses

© Springer Nature Switzerland AG 2018
H. S. Nguyen et al. (Eds.): IJCRS 2018, LNAI 11103, pp. 229–242, 2018.
https://doi.org/10.1007/978-3-319-99368-3_18

giving more priority to the *required condition* than the *desired condition*. In [12], the issue is addressed by first selecting houses satisfying the attribute 'cheap', and then order them using the criterion 'close to public transport'. Kacprzyk and Zadrożny [8], on the other hand, have emphasized on an approach where the former condition has to be satisfied *necessarily*, and the latter only if *possible*. In this regard, they come up with an operator, defined by standard logical connectives, to capture the sense of *and possibly*, and the operator is named as *and possibly* operator. The semantics of this operator is then investigated by considering different many-valued logical connectives for 'conjunction' and 'implication'. Thus given a house *h*, *h is cheap and possibly close to the public transport* is translated to a value from a suitable value set. But the development of the theory ignores the following aspects.

- How to decide which operators would fit suitable to satisfy the user's choice?
- How the system is perceiving situations while searching for answers related to the user's queries?
- How the decision process reflects the natural aspect of learning from data?
- How the user's requirement and intention can be realized without a component of interaction or dialogue between the user and the system?

We focus on the key strategy taken in [8], which naturally leads towards two terms; the first condition has to be satisfied *necessarily* and then the second condition if *possible*. This perspective naturally brings in rough set theory [17] as a possible model for finding a suitable semantics for bipolar queries. Moreover, in [8] the notion of *cheap and possibly close to the public transport* is realized through the key rule that *if there are houses satisfying both then fine, otherwise, choose only the cases satisfying the first*, where the notions of *cheap* and *close to public transport* are represented by fuzzy sets. Thus, based on some price value and distance measure each house is identified with the degree to which it is cheap and the degree to which it is close to public transport. Then the aggregation of these two measures is nothing but a mere calculation of numbers, from where retrieving back the original semantics of cheap and close to public transport and refining the search by modifying the semantics a bit is impossible. In this context, the method proposed in this paper based on rough set would be advantageous as the rough set theoretic approximation of any vague concept remains grounded in the data.

Below we present a preliminary idea based on rough sets and information systems so that the process of obtaining a cluster of houses, suitable to user's choice, (i) be grounded in the available data, (ii) be flexible for refinement based on modification of data, and (iii) be sensitive to the user's feedback through initiation of dialogues.

In this regard, in Sect. 2, first we present the basic notions from the theory of rough sets. Section 3 discusses about our proposed method of addressing the notion of bipolar queries using rough set semantics, and a formal language corresponding to the proposed semantics. In Sect. 4, we present a proposal by introducing dialogue between the user and the system in order to understand the user's need better. Lastly, there is a concluding section listing some further possibilities to be explored.

2   Preliminary of Rough Sets and Decision Systems

The notion of rough sets was introduced by Pawlak [15,16] in order to address the concepts which have borderline cases apart from the cases which surely belong and surely do not belong to the concepts. The notion of rough set is defined based on a notion of information system which describes a set of objects of a universe with respect to certain attributes.

Definition 1. *An information system is a triple* $\mathbb{A} = (U, \mathcal{A}, V)$*, where* U *is a set of objects,* $\mathcal{A}$ *is a set of attributes, and* V *is a set of values such that for each* $a \in \mathcal{A}$*,* $a : U \mapsto V$*.*

Given an information system $\mathbb{A}$ for any $B \subseteq \mathcal{A}$ we can create an equivalence relation, known as indiscernibility relation, in the following way.

Definition 2. *Given an information system* $\mathbb{A} = (U, \mathcal{A}, V)$ *for* $B \subseteq \mathcal{A}$ *the indiscernibility relation with respect to* B*, denoted as* IND_B*, is defined as follows. For any* $x, y \in U$*,* $x \; IND_B \; y$ *iff* $a(x) = a(y)$ *for each* $a \in B$*.*

This relation IND_B partitions the whole universe into equivalence classes, and that generates an approximation space (U, IND_B).[1]

Definition 3. *Given an approximation space* (U, IND_B)*, for any set* $X \subseteq U$*, there are two approximations of* X *with respect to* IND_B*.*

- *The lower approximation of* X *with respect to the attributes of* B*, denoted as* $Low_B(X)$*, is given as* $Low_B(X) = \cup\{[u]_B : [u]_B \subseteq X\}$*.*
- *The upper approximation of* X *with respect to the set of attributes* B*, denoted as* $Upp_B(X)$*, is given as* $Upp_B(X) = \cup\{[u]_B : [u]_B \cap X \neq \phi\}$*.*

Definition 4. *Given an approximation space* (U, IND_B)*, any set* $X \subseteq U$ *is represented by a pair* $(Low_B(X), Upp_B(X))$*, called a rough set*[2]*.*

The lower approximation of a set X with respect to the set of attributes B represents those objects for which the whole equivalence class with respect to B is completely contained in X; that is, $Low_B(X)$ is the union of those equivalence

---

[1] Instead of calling (U, IND_B) as approximation space one may call it indiscernibility space. But as the notions of approximation, like the lower and upper approximation operators, are defined based on IND_B, we follow the prevalent practice of calling (U, IND_B) as approximation space, rather than calling (U, Low, Upp) as the approximation space generated from the indiscernibility space (U, IND_B).

[2] In literature (see, *e.g.*, [1,13–16,18,22–24]) there are different variant definitions of rough sets; interrelations among these different definitions and operations parallel to set theoretic union, intersection and complementation are also studied by different researchers (see, *e.g.*, [1–3,7]). Unlike ordinary sets, the set of all rough sets over a universe U with respect to intersection, union and complementation does not form a Boolean algebra, and so intersection, union and complementation operations of rough sets are a bit different than the usual ones. In this paper, instead of going into the detail we refer the readers to the cited above literature.

classes with respect to B for which it is sure that the concept X applies. The upper approximation of X with respect to B is the union of those equivalence classes which has non-empty intersection with X; that is, $Upp_B(X)$ contains those elements from the universe for which either they belong to X or they are equivalent to some elements belonging to X. So, $Upp_B(X)$ contains those elements which are possibly in X. So, if X represents a vague concept, then $Low_B(X)$ contains those elements which surely belong to X, $Upp_B(X)^c$ contains those elements which surely do not belong to X, and $Upp_B(X) \setminus Low_B(X)$ contains those elements which are the borderline instances of X.

One more important point is aggregation of information systems over the same universe and approximation of a set with respect to the individual information systems and their aggregated information system. Usually, if $X \subseteq U$ is included in the lower approximation with respect to one information system and $Y \subseteq U$ is included in the lower approximation of another, then $X \cap Y$ is also included in the lower approximation of the joint information system consisting of the union of both the sets of attributes of those two information systems. This is also an outcome of the nature of intersection operation of two rough sets. These properties of rough set theoretic operations and aggregation of rough sets would have an impact in Sect. 3 in the context of defining a semantics for a modal language.

In [18, 21], departing from the notion of equivalence class, a notion of generalized approximation space, based on a notion of neighbourhood of an element of the universe, is proposed. Then analogous to the notion that an equivalence class of an element u is contained in a set X or has non-empty intersection with X, *a neighbourhood of u is included in X to a degree* is introduced. Let us present a few basic definitions in this regard.

Definition 5. *A generalized approximation space is a tuple $(U, \mathcal{J}, v)$ where $\mathcal{J}$ is an uncertainty function given as $\mathcal{J} : U \mapsto P(P(U))$, and $v : P(U) \times P(U) \mapsto [0, 1]$.*

For any $x \in U$, $\mathcal{J}(x)$ can be considered to be a family of neighbourhoods of x, and v is a graded inclusion function determining how a subset of U is included in another subset. For X belonging to $\mathcal{J}(x)$, one can have different interpretations such as 'a neighbourhood of x', 'a cover of x', or even 'an equivalence class of x' in usual sense. When we are beyond the classical sense of partition, the clusters around x are not very crisp as that of an equivalence class as no condition of disjointness is imposed between two different clusters of two elements. So, $\mathcal{J}$ allows some uncertainty in the formation of a cluster around an element, and hence is called *uncertainty function*. In the context of ordinary approximation space each member of $\mathcal{J}(x)$ of the generalized approximation space represents an equivalence class of x based on IND_B with respect to some set B of attributes. There can be different definitions for a graded inclusion relation also; as an example the standard one is given as follows.

Example 1. For any $X, Y \subseteq U$, the standard rough set inclusion is given as

$$v_{SRI}(X,Y) = \begin{cases} \frac{|X \cap Y|}{|X|} & \text{if } X \neq \emptyset \\ 1 & \text{otherwise.} \end{cases}$$

Now based on the above notion of generalized approximation space, one can define the lower and upper approximation of any set $X \subseteq U$ in the following way.

Definition 6. *Given a generalized approximation space $(U, \mathcal{J}, v)$, for any set $X \subseteq U$,*

- $Low(X) = \{u \in U : v(Y, X) = 1 \text{ for some } Y \in \mathcal{J}(x)\}$
- $Upp(X) = \{u \in U : v(Y, X) > 0 \text{ for any } Y \in \mathcal{J}(x)\}.$

Instead of such crisp conditions for defining $Low(X)$ and $Upp(X)$ based on the neighbourhood function $\mathcal{J}$, one can also impose the conditions respectively $v(Y, X) \geq 1 - t$ and $0 < v(Y, X) < 1 - t$ (or $t \leq v(Y, X) < 1 - t$ [26]) for some very small positive number t, in the definitions of the lower and upper approximations of X.

Let us consider an information system $(U, \mathcal{A}, V)$. We can now create a generalized approximation space $(U, \mathcal{J}_{\mathcal{A}}, v_t)$ such that for each $x \in U$, $\mathcal{J}_{\mathcal{A}}(x) = \{[x]_B : B \subseteq A\}$, and $v_t : P(U) \times P(U) \mapsto [0,1]$ where $t \in [0, .5)$. Thus we have a family of generalized approximation spaces parametrized by the thresholds $t \in [0, .5)$ such that the lower approximation operator $Low_{B,t}$ and the upper approximation operator $Upp_{B,t}$ are defined in the following way.

Definition 7. *Given a generalized approximation space $(U, \mathcal{J}_{\mathcal{A}}, v_t)$, for any $X \subseteq U$ and $B \subseteq \mathcal{A}$,*

- $Low_{B,t}(X) = \cup\{[x]_B : v_t(X, [x]_B) \geq 1 - t\}$
- $Upp_{B,t}(X) = \cup\{[x]_B : 0 < v_t(X, [x]_B) < 1 - t\}.$

3   Information System Based Interpretation of Required and Desired Conditions

Let an information system have a database of houses characterized as cheap and expensive with respect to a set of amenities and price. The system also has a characterization of the same set of houses with respect to other parameters in terms of the decision values for closed to public transport. That is, in terms of rough set literature there are two decision tables [16] for a set of houses - one for the decision attribute 'cheap' (C) and the other for the decision attribute 'close to public transport' (P). Now following the basic notions of rough sets we can design the following simple method so that the system can select a set of houses as 'cheap and possibly close to the public transport'.

(i) Let C be the class of houses which belong to the decision class *cheap* based on a set of attributes $\mathcal{A}_1$, and P be the class of houses which belong to the decision class *close to public transport* based on a set of attributes $\mathcal{A}_2$. As the respective sets of attributes, viz., $\mathcal{A}_1$ and $\mathcal{A}_2$, for the lower and upper approximations for C and P are clear from the context, for simplicity of presentation let us use just $\underline{C}$, $\underline{P}$ and $\overline{P}^c$ in the present sequel.

(ii) As the first preference is to choose houses which are surely cheap, and then look for the houses which are possibly close to public transport too, we first focus on identifying $\underline{C}$. Now, if $\underline{C} \cap \overline{P} = \phi$ we would choose $\underline{C}$. If $\underline{C} \cap \overline{P} \neq \phi$, we can have the following possible relations of $\underline{P}$ (sure cases of P) and $\overline{P}^c$ (surely negative cases of P) with $\underline{C}$ (see Fig. 1).

- $\underline{C} \cap \underline{P} = \phi$, $\underline{C} \cap \overline{P}^c = \phi$.   • $\underline{C} \cap \underline{P} \neq \phi$, $\underline{C} \cap \overline{P}^c = \phi$.
- $\underline{C} \cap \underline{P} = \phi$, $\underline{C} \cap \overline{P}^c \neq \phi$.   • $\underline{C} \cap \underline{P} \neq \phi$, $\underline{C} \cap \overline{P}^c \neq \phi$.

(iii) Now in order to formalize C *and possibly* P we can simply consider the following interpretation: $\underline{C} \cap \underline{P}$ else $\underline{C} \cap \overline{P}$ else $\underline{C}$.

That is, the system would choose those houses which are both surely cheap and surely close to the public transport if such a non-empty set exists. If not, the system would prefer to select those houses which are surely cheap and still possible to be counted as close to public transport if such a non-empty set exists; otherwise it would choose only the set of surely cheap houses. One can notice, both from the Fig. 1 and from the cases listed in item (ii), that our target search criterion is such that the resultant cluster can never be $\underline{C} \cap \overline{P}^c$.

(iv) **Extension of above proposal for more than two constraints.** In the above case we have considered only two constraints C and P and a preference of the first over the second. Based on the same framework let us consider some possible ways of extending the idea for more than two constraints.

  (a) Let there be three constraints C_1, C_2, C_3 such that the user wishes to have $C_1 \succ C_2 \succ C_3$ where $C_i \succ C_j$ represents that C_i is preferred over C_j, and C_1, C_2, C_3 are perceived with respect to $\mathcal{A}_1, \mathcal{A}_2, \mathcal{A}_3$ respectively. So, in this case we can first look for the clusters obtained from the rule $\underline{C_1} \cap \underline{C_2}$ else $\underline{C_1} \cap \overline{C_2}$ else $\underline{C_1}$. Let H be the obtained cluster. Now we do not want to meet the constraint C_3 at the cost of deviating from H. So, our next step would be to look for the cluster following the rule $\underline{H} \cap \underline{C_3}$ else $\underline{H} \cap \overline{C_3}$ else $\underline{H}$.

  (b) Let $\{C_1, C_2, \ldots, C_n\}$ be a set of constraints which are equally required, and $\{P_1, P_2, \ldots, P_m\}$ be a set of constraints which are equally desired, and for each i, j, $C_i \succ P_j$. Then instead of each single table for the constraints, we can consider the joint table with decision attribute $C_1 \& \ldots \& C_n$ and the other decision table with decision attribute $P_1 \& \ldots \& P_m$, where $\&$ is interpreted as holding all the component decisions together. Then considering $C_1 \& \ldots \& C_n$ as C and $P_1 \& \ldots \& P_m$ as P we can proceed as above.

(v) **Refinement of search:** In the above proposal the main key behind the search is $\underline{C} \cap \underline{P}$ else $\underline{C} \cap \overline{P}$ else $\underline{C}$. It indicates that the best choice would be when $\underline{C} \cap \underline{P}$ is non-empty. In case if $\underline{C} \cap \underline{P}$ is empty the next choice would be $\underline{C} \cap \overline{P}$, and if that possibility does not work too, then the outcome would be $\underline{C}$. But if we keep practical aspects of a search in mind, then often we like to refine the search by making some adjustment in the set of parameters/attributes so that we can accommodate both the constraints surely C and possibly P. As the background database with description and decision about houses are available, designing the refinement of search following the above mentioned direction is not difficult. As we now will be directly dealing with adjusting the set of attributes in order to get a cluster of houses better fitting the user's intention, instead of using the notation $\underline{C}$ we would now switch to the notation $Low_{\mathcal{A}_1}(C)$.

- Suppose $Low_{\mathcal{A}_1}(C) \cap Low_{\mathcal{A}_2}(P) = \phi$.
- As the database for both the decision attributes are available, we can check whether by dropping a few attributes from $\mathcal{A}_2$ some houses from $Low_{\mathcal{A}_1}(C)$ get included in a refined lower approximation of P. So, we start with checking if for some $h \in Low_{\mathcal{A}_1}(C)$ there is some $\mathcal{A}'_2 \subseteq \mathcal{A}_2$ such that $v_t([h]_{\mathcal{A}'_2}, P) \geq 1-t$ for some $t \in [0, 0.5)$. In that case, $[h]_{\mathcal{A}'_2} \subseteq Low_{\mathcal{A}'_2, t}(P)$, and we thus have $Low_{\mathcal{A}_1}(C) \cap Low_{\mathcal{A}'_2, t}(P) \neq \phi$.
- As the next step it can be checked whether for $\mathcal{A}'_2 \subseteq \mathcal{A}_2$, one obtains a good overlap of the equivalence class of houses generated with respect to $\mathcal{A}'_2$ with $Low_{\mathcal{A}_1}(C)$. So, we can check if for some $h \in Low_{\mathcal{A}_2}(P)$, there is some $\mathcal{A}'_2 \subseteq \mathcal{A}_2$ such that $v([h]_{\mathcal{A}'_2}, Low_{\mathcal{A}_1}(C)) \geq 1 - t$ for some very small threshold $t \in [0, 0.5)$. For such case, $[h]_{\mathcal{A}'_2} \subseteq Low_{\mathcal{A}'_2, t}(Low_{\mathcal{A}_1}(C)) \cap Low_{\mathcal{A}_2}(P)$, and we obtain a modified cluster which may better satisfy the user.
- If the above options do not work, as a next possibility the system can drop some amenities from $\mathcal{A}_1$ and check if for $\mathcal{A}'_1 \subseteq \mathcal{A}_1$ and $[h]_{\mathcal{A}_1} \subseteq Low_{\mathcal{A}_1}(C)$ whether both $[h]_{\mathcal{A}'_1} \cap Low_{\mathcal{A}_2}(P) \neq \phi$ and $v([h]_{\mathcal{A}'_1}, Low_{\mathcal{A}_1}(C)) \geq 1-t$, for a negligibly small threshold $t > 0$, hold. In such case, $v([h]_{\mathcal{A}'_1}, C) \geq 1-t$ as $Low_{\mathcal{A}_1}(C) \subseteq C$. Thus we obtain a modified cluster $Low_{\mathcal{A}'_1, t} \cap Low_{\mathcal{A}_2}(P) \neq \phi$.

It is to be noted that for refining the search $Low_{\mathcal{A}_1}(C)$, the cluster corresponding to the required condition, always has been given a priority over $Low_{\mathcal{A}_2}(P)$, the cluster corresponding to the desired condition. In the first case, the system makes an attempt by checking if some houses from $Low_{\mathcal{A}_1}(C)$ can be considered as *surely close to public transport* to some degree if some attributes characterizing P are ignored. So, the search starts from some houses belonging to $Low_{\mathcal{A}_1}(C)$. In the second case, the search starts from the houses which are already considered as surly close to public transport. The target is to check if for some subset $\mathcal{A}'_2$ of the set of attributes characterizing P, $v([h]_{\mathcal{A}'_2}, Low_{\mathcal{A}_1}(C)) \geq 1 - t$ for some small positive quantity t. In case of positive result, the already obtained cluster $Low_{\mathcal{A}_1}(C)$ is tuned a bit by a modified cluster $Low_{\mathcal{A}'_2, t}(Low_{\mathcal{A}_1}(C))$. Thus, without affecting $Low_{\mathcal{A}_1}(C)$ much a set of houses can be obtained from

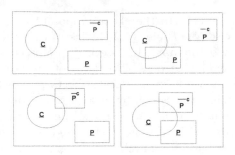

Fig. 1. Possible cases of $\underline{P}$ and $\overline{P}^c$ with $\underline{C}$.

$Low_{\mathcal{A}_2',t}(Low_{\mathcal{A}_1}(C)) \cap Low_{\mathcal{A}_2}(P)$. If dropping a few attributes of P does not work, as the third option the system goes for dropping a few attributes from $\mathcal{A}_1$, the set characterizing C. In this context, with respect to the smaller set of attributes $\mathcal{A}_1'$, for a class $[h]_{\mathcal{A}_1}$ which was already included in $Low_{\mathcal{A}_1}(C)$, it is checked if $[h]_{\mathcal{A}_1'} \cap Low_{\mathcal{A}_2}(P) \neq \phi$ and for a small positive number t, $v([h]_{\mathcal{A}_1'}, Low_{\mathcal{A}_1}(C)) \geq 1-t$ is still satisfied. In such case, $Low_{\mathcal{A}_1',t}(C) \cap Low_{\mathcal{A}_2}(P)$ is considered as the refined cluster as $v([h]_{\mathcal{A}_1'}, Low_{\mathcal{A}_1}(C)) \geq 1 - t$ implies $v([h]_{\mathcal{A}_1'}, C) \geq 1 - t$. As in $Low_{\mathcal{A}_1',t}(C)$ we consider a kind of neighbourhood of $Low_{\mathcal{A}_1}(C)$, we do not much move away from $Low_{\mathcal{A}_1}(C)$.

Thus, as a general scheme we may consider a collection of generalized approximation spaces $\mathbb{AS} = \{\mathbb{A}_{t,\mathcal{A}_i} : \mathcal{A}_i \subseteq \mathcal{A}\}_{t \in [0,0.5)}$ where $\mathbb{A}_{t,\mathcal{A}_i} = (U, \mathcal{J}_{\mathcal{A}_i}, v_t)$, and $\mathcal{J}_{\mathcal{A}_i}(x) = \{[x]_B : B \subseteq \mathcal{A}_i\}$. As dropping some attributes helps to generate a bigger equivalence class, the system can check whether dropping some attributes from $\mathcal{A}_2$ ($\subseteq \mathcal{A}$), the set of attributes characterizing P, and/or $\mathcal{A}_1$ ($\subseteq \mathcal{A}$), the set of attributes characterizing C, can include some common cases in the respective equivalence classes. As our target is not to deviate from the cluster $Low_{\mathcal{A}_1}(C)$, in each time we can check whether a newly obtained enlarged equivalence class, say $[h]_{\mathcal{A}_2'}$ or $[h]_{\mathcal{A}_1'}$ has a significantly good overlap with $Low_{\mathcal{A}_1}(C)$. Moreover, a tuning of the threshold t also can generate a bigger set of possibilities without deviating from the main target. For instance, let us choose $\mathcal{A}_2' \subseteq \mathcal{A}_2$. So, surely $[h]_{\mathcal{A}_2} \subseteq [h]_{\mathcal{A}_2'}$; but $[h]_{\mathcal{A}_2} \subseteq Low_{\mathcal{A}_2}(P)$ does not mean $[h]_{\mathcal{A}_2'} \subseteq Low_{\mathcal{A}_2}(P)$. Now let us consider that for some $t_1 \in [0, 0.5)$, for all $[h]_{\mathcal{A}_2} \subseteq Low_{\mathcal{A}_2}(P)$, $v_{t_1}([h]_{\mathcal{A}_2'}, Low_{\mathcal{A}_2'}(P)) \geq 1 - t_1$. So, $v_{t_1}([h]_{\mathcal{A}_2'}, P) \geq 1 - t_1$ and $[h]_{\mathcal{A}_2'} \subseteq Low_{\mathcal{A}_2',t_1}(P)$. Hence $Low_{\mathcal{A}_2}(P) \cap Low_{\mathcal{A}_2',t_1}(P) \neq \phi$. Tuning the threshold helps when for a prefixed threshold t_1, for some $[h]_{\mathcal{A}_2'} \subseteq Low_{\mathcal{A}_2}(P)$, $v_{t_1}([h]_{\mathcal{A}_2'}, Low_{\mathcal{A}_2}(P)) \geq 1 - t_1$ is not the case. Then we may slightly change the threshold, and consider a modified threshold $t_2 \in [0, 0.5)$ such that $t_1 \leq t_2$. With respect to this new threshold t_2 if $v_{t_2}([h]_{\mathcal{A}_2'}, Low_{\mathcal{A}_2}(P)) \geq 1 - t_2$, then as before we can claim $Low_{\mathcal{A}_2}(P) \cap Low_{\mathcal{A}_2',t_2}(P) \neq \phi$. So, without moving away from the initial cluster we can enlarge our possibilities by considering a modified cluster $Low_{\mathcal{A}_1} \cap Low_{\mathcal{A}_2',t_2}(P)$.

The above discussion on refining a search to serve an user better, reflects the need for introducing interactions/dialogues among the user and the system. We would attempt to throw light on this issue in Sect. 4.

3.1   A Modal Language Representing Above Semantics

Let us now present a syntax which can provide a language to express the basic ingredients and operational parts of the above semantics where we have two decision tables $(U, \mathcal{A}_1, C)$ and $(U, \mathcal{A}_2, P)$. More specifically, it is not needed to emphasize on this term 'two decision tables'. We can talk about a single decision table with an extended set $\mathcal{A}$ of finitely many conditional attributes including $\mathcal{A}_1 \cup \mathcal{A}_2$, and an extended set of finitely many decision attributes, combining all decision parameters that we would like to address. It is to be noted that the decision attributes are always of different status than that of the conditional attributes. Usually, a decision class, i.e. a particular value for a decision attribute, is described by different sets of possible values of the conditional attributes. Each possible combination of values for the conditional attributes can be represented by an equivalence class generated from the indiscernibility relation, obtained with respect to the set of all conditional attributes. But a single decision class may contain objects of different equivalence classes, and two different decision classes may contain objects from the same equivalence class. So, usually, decision classes are approximated with respect to the equivalence classes obtained from a set of conditional attributes.

The main aim of this section is to provide an outline of a formal language where we can express the proposed key rule of search as a well-formed formula. Having such a formal language would be advantageous as it may be used to express constraints for higher order aggregations of different information systems.

1. Atomic propositions: $a = v$ for $a \in \mathcal{A}_1 \cup \mathcal{A}_2 \cup \{P, C\}$ and v belonging to the set V_a of values of the attribute a
2. Logical Connectives: $\wedge, \vee \, \neg$
3. Modal operators: $\Box_\mathcal{B}, \Diamond_\mathcal{B}$ (finitely many modal operators suffixed by subsets of $\mathcal{A}$).
4. Formulas: Any atomic formula is a formula, and if α, β are formulas then formulas obtained from them by using logical connectives and modal operators are formulas too.

From the above alphabet we can have compound formula of the form $a \in V'$ for any $V' \subseteq V_a$, where $a \in V'$ represents the disjunction of the atomic formulas $a = v$ for all $v \in V'$.

Interpretation of the Above Language

Let us consider the decision system $(U, \mathcal{A} \cup \mathcal{D}, V \cup V_d)$, where U is a set of houses, $\mathcal{A}$ is a set of conditional attributes including $\mathcal{A}_1, \mathcal{A}_2$, and $\mathcal{D}$ is a set of decision attributes containing the decision attribute C (cheap) and P (close to the public transport). For the conditional attributes the value set $V = \{V_a : a \in \mathcal{A}\}$, and the same for the decision attributes is V_d which includes V_{d_1} and V_{d_2}, the set of values respectively for C and P. We now interpret the above language with respect to the given decision system.

1. $||a = v|| = \{x \in U : a(x) = v\}$.
2. $||\alpha \wedge \beta|| = ||\alpha|| \cap ||\beta||$, $||\alpha \vee \beta|| = ||\alpha|| \cup ||\beta||$, $||\neg\alpha|| = ||\alpha||^c$ with standard set theoretic intersection, union and complementation operations.
3. For any formula with the modal operator $\square$ in the front is interpreted as follows.
   $||\square_{\mathcal{A}_1}\alpha|| = Low_{\mathcal{A}_1}(||\alpha||)$.
4. For any formula with the modal operator $\diamond$ in the front is interpreted as follows.
   $||\diamond_{\mathcal{A}_2}\alpha|| = Upp_{\mathcal{A}_2}(||\alpha||)$

So, interpretation of a formula of the form $a \in V'$ would be $\cup_{v \in V'}\{x \in U : a(x) = v\}$.

Now let us concentrate on presenting the key rule C *and possibly* P so that it can capture the semantics proposed in Sect. 2. Let $\alpha = \square_{\mathcal{A}_1}(C \in D_1)$, $\beta = \square_{\mathcal{A}_2}(P \in D_2)$, and $\gamma = \diamond_{\mathcal{A}_2}(P \in D_2)$ where $D_1 \subseteq V_{d_1}$ and $D_2 \subseteq V_{d_2}$. Let α' represent the formula $\square_{\mathcal{A}_1 \cup \mathcal{A}_2}(C \in D_1 \wedge P \in D_2)$. Following the usual rough set semantics for intersection we know $Low_B(X) \cap Low_{B'}(Y) \subseteq Low_{B \cup B'}(X \cap Y)$ [1–3,7]. So, $||\alpha \wedge \beta|| \subseteq ||\alpha'||$. Now considering $\delta = (\alpha' \wedge (\alpha \wedge \beta)) \vee (\neg(\alpha \wedge \beta) \wedge (\alpha \wedge \gamma))$ we can notice that if there is a house belonging to $Low_{\mathcal{A}_1}(||C \in D_1||) \cap Low_{\mathcal{A}_2}(||P \in D_2||)$, then the result would be $Low_{\mathcal{A}_1}(||C \in D_1||) \cap Low_{\mathcal{A}_2}(||P \in D_2||)$; and if not, it would pick up the houses from the cluster $Low_{\mathcal{A}_1}(||C \in D_1||) \cap Upp_{\mathcal{A}_2}(||P \in D_2||)$. So, the formula $\neg\delta \vee \alpha$ has exactly the same semantics what the key rule $\underline{C} \cap \underline{P}$ else $\underline{C} \cap \overline{P}$ else $\underline{C}$ intends to have.

4   A Dialogue Based Approach to Bipolar Queries

In this section, we would make an attempt to introduce interactions or dialogues between the user and the system. This would help the system to better understand the user's need, and to initiate negotiations for providing alternative choices. In [6] we have presented a formal language for dialogues and that can be exploited for our present purpose. Let us now present a prototypical case of the user-system interactions to better understand the user's perspective of a specific query.

– First the dialogue is initiated when the user gives a description, say *houses that are cheap and possibly close to public transport*. The user's description is treated as a sequence of attributes $\langle C, P \rangle$ where the order of the appearance indicates the preference of the first attribute over the second.
– The system has a database of houses characterized with respect to amenities and prices. So, the system can forward a dialogue with a sequence of attributes representing amenities and a budget for price. The dialogue may be formalized as $\langle a_1, a_2, \ldots a_m, b \rangle$.

- In return the user's can change the ordering of the attributes representing her preference for particular amenities and put a value for b, the budget. Instead of changing order, the preference for the amenities can also be expressed in terms of values from a specific scale.
- In a similar fashion, the system would also enquire for parameters describing location, connectivity that specify the feature close to the public transport. In response the user returns a sequence of values and/or attributes describing her preference.
- With the given attributes for both *cheap* and *close to public transport* the system can compare with the available databases. Based on the constraint, given by the user, the system might need to drop some attributes and consider the decision classes namely, cheap and close to public transport, and their approximations based on the modified subsets of attributes. In that way, clusters for $Low_{\mathcal{A}_1}(C)$, $Upp_{\mathcal{A}_1}(C)$, $Low_{\mathcal{A}_2}(P)$, $Upp_{\mathcal{A}_2}(P)$ are generated, and the system looks for the cluster that satisfies the condition $Low_{\mathcal{A}_1}(C) \cap Low_{\mathcal{A}_2}(P)$ *else* $Low_{\mathcal{A}_1}(C) \cap Upp_{\mathcal{A}_2}(P)$ *else* $Low_{\mathcal{A}_1}(C)$.
- Now each house, from the obtained cluster of houses, can be individually identified with a sequence presenting their amenities and price budget, as well as descriptions pertaining to close to public transport. A typical such sequence may look like $\langle a_1, a_2, \ldots, v_b; b_1, b_2 \ldots v_p \rangle$ where a_i represents amenities, v_b represents budget price, the semicolon (;) represents the end of description for the required condition, b_i's represent the attributes corresponding to the desired condition, and v_p represents some values for the decision *close to public transport*. All such sequences can be forwarded to the user in the next round of the dialogue. So, the system as a dialogue would send a set of sequences to the user.
- If the user is satisfied with the result, she can send the acceptance feedback through a sequence $\langle a_1, a_2, \ldots, v_b; b_1, b_2 \ldots v_p; \boxtimes \rangle$; Otherwise, $\langle a_1, a_2, \ldots, v_b; b_1, b_2 \ldots v_p; \boxtimes \rangle$, representing her dissatisfaction, is forwarded.
- If the user is not satisfied, the system can explore different refinement strategies.

  • In this context, based on user's preference over the parameters for distance from public transport the system can search for a cluster $Low_{\mathcal{A}_2 \setminus \{b_j\}, t_1}(P)$ for $t_1 \in [0, 0.5)$ so that $Low_{\mathcal{A}_1}(C) \cap Low_{\mathcal{A}_2 \setminus \{b_j\}, t_1}(P)$ becomes non-empty (cf. Sect. 3). The dialogue can continue for finitely many rounds based on the system's output and the user's feedback. For instance, if $Low_{\mathcal{A}_2 \setminus \{b_j\}, t_1}(P) \cap Low_{\mathcal{A}_1}(C)$ does not satisfy the user, then the system can tune the threshold t_1 to t_2 such that $t_1 \leq t_2$, or drop $\{b_j, b_k\}$ from $\mathcal{A}_2$ based on the preference of parameters, described by the user at the beginning of the dialogue. Then a new search begins to check the possibility $Low_{\mathcal{A}_1}(C) \cap Low_{\mathcal{A}_2 \setminus \{b_j, b_k\}, t_2}(P)$.
  • The system also can drop some attributes from $\mathcal{A}_1$, and check if there is a non-empty cluster $Low_{\mathcal{A}_1 \setminus \{a_j\}, t}(C) \cap Low_{\mathcal{A}_2}(P)$ fitting to the user need. Feedback of the user collected at each round of answer may help to learn the system the more precise interval for the threshold t.

5 Conclusions

In this paper, we provide a semantics for bipolar queries, where intention of an user is understood in terms of the required condition and desired condition, from the context of rough sets. We tried to capture the priority of required condition over the desired condition following the proposal that *if both are satisfied then fine, otherwise the required condition has to be satisfied*. This was the approach taken in [8] too. But in our context, one more advantage is that this search for finding an outcome satisfying the user is based on the user's description of attributes defining the notion of 'cheap' and 'close to public transport'. Unlike fuzzy set theoretic approaches taken in [8], the notions of 'cheap' and 'close to public transport' are not given a priori by fuzzy membership functions; rather they are learnt by matching the available data with the user's descriptions. Moreover, we have also introduced interactions between the user and the system and the possibility of modifying the search based on the user's feedback. The proposal of dialogue between an user and a system can be extended among multiple sources of information systems. In the existing literature on multiple sources of information systems [11,19,20,25], usually the information collected from multiple sources are aggregated by some means; the incorporation of the user's feedback and tuning the search based on that have not been addressed.

This paper only addresses situations where different constraints of the user can be arranged in a linear order of preference. In practice, it can be a complex relation of preference among different constraints of the user. This needs further investigation. In this regard, a further point of reference could be the Belief-Desire-Intention (BDI) model of Casali et al. [4], where the desire of an user is described in terms of positive and negative preference relations, and intention of an user reflects practical necessities which cannot be violated. In the BDI model, the semantics for desire and intention are given by different models; they are kept connected by some bridging rules. This approach [4] along with the model of dialogue base [6], allowing interactions among different databases of a network of information systems, may help us to extend the research further.

References

1. Banerjee, M., Chakraborty, M.K.: A category for rough sets. Found. Comput. Decis. Sci. **18**(3–4), 167–180 (1993)
2. Banerjee, M., Chakraborty, M.K.: Rough algebra. Bull. Polish Acad. Sc. (Math.) **41**(4), 293–297 (1993)
3. Bonikowski, Z.: A certain conception of the calculus of rough sets. Notre Dame J. Formal Logic **33**, 412–421 (1992)
4. Casali, A., Godo, L., Sierra, C.: g-BDI: a graded intensional agent model for practical reasoning. In: Torra, V., Narukawa, Y., Inuiguchi, M. (eds.) MDAI 2009. LNCS (LNAI), vol. 5861, pp. 5–20. Springer, Heidelberg (2009). https://doi.org/10.1007/978-3-642-04820-3_2
5. Dubois, D., Prade, P.: An overview of the symmetric bipolar representation of positive and negative information in possibility theory. Fuzzy Sets Syst. **160**(10), 1355–1366 (2009)

6. Dutta, S., Wasilewski, P.: Dialogue in hierarchical learning of concepts using prototypes and counterexamples. Fundame. Informaticae **162**, 1–20 (2018)
7. Gehrke, M., Walker, E.: The structure of rough sets. Bull. Polish Acad. Sc. (Math.) **40**, 235–245 (1992)
8. Kacprzyk, J., Zadrożny, S.: Bipolar queries: some inspirations from intention and preference modeling. In: Trillas, E., Bonissone, P., Magdalena, L., Kacprzyk, J. (eds.) Combining Experimentation and Theory. Studies in Fuzziness and Soft Computing, vol. 271, pp. 191–208. Springer, Heidelberg (2012). https://doi.org/10.1007/978-3-642-24666-1_14
9. Kacprzyk, J., Zadrożny, S.: Compound bipolar queries: a step towards an enhanced human consistency and human friendliness. In: Matwin, S., Mielniczuk, J. (eds.) Challenges in Computational Statistics and Data Mining. SCI, vol. 605, pp. 93–111. Springer, Cham (2016). https://doi.org/10.1007/978-3-319-18781-5_6
10. Kacprzyk, J., Zadrożny, S.: Compound bipolar queries: the case of data with a variable quality. In: 2017 IEEE International Conference on Fuzzy Systems, FUZZ-IEEE 2017, Naples, Italy, 9–12 July 2017, pp. 1–6. IEEE (2017)
11. Khan, M.A., Banerjee, M.: A preference-based multiple-source rough set model. In: Szczuka, M.S., Kryszkiewicz, M., Ramanna, S., Jensen, R., Hu, Q. (eds.) RSCTC 2010. LNCS (LNAI), vol. 6086, pp. 247–256. Springer, Heidelberg (2010). https://doi.org/10.1007/978-3-642-13529-3_27
12. Lacroix, M., Lavency, P.: Preferences: putting more knowledge into queries. In: proceedings of the 13th International Conference on Very Large Databases, Brighton, UK, pp. 217–225 (1987)
13. Komorowski, J., Pawlak, Z., Polkowski, L., Skowron, A.: Rough sets: a tutorial. In: Pal, S.K., Skowron, A. (eds.) Rough Fuzzy Hybridization: A New Trend in Decision Making, p. 398. Springer, Singapore (1999)
14. Pagliani, P., Chakraborty, M.: A Geometry of Approximation: Rough Set Theory: Logic, Algebra and Topology of Conceptual Patterns. Trends in Logic, vol. 27. Springer, Heidelberg (2008)
15. Pawlak, Z.: Rough sets. Int. J. Comput. Inf. Sci. **11**, 341–356 (1982)
16. Pawlak, Z.: Rough Sets: Theoretical Aspects of Reasoning about Data, System Theory, Knowledge Engineering and Problem Solving, vol. 9. Kluwer Academic Publishers, Dordrecht (1991)
17. Pawlak, Z., Skowron, A.: Rudiments of rough sets. Inf. Sci. **177**(1), 3–27 (2007)
18. Pawlak, Z., Skowron, A.: Rough sets: some extensions. Inf. Sci. **177**(1), 28–40 (2007)
19. Rasiowa, H., Marek, W.: Mechanical proof systems for logic II, consensus programs and their processing. J. Intell. Inf. Syst. **2**(2), 149–164 (1993)
20. Rauszer, C.M.: Rough logic for multi-agent systems. In: Masuch, M., Pólos, L. (eds.) Logic at Work 1992. LNCS, vol. 808, pp. 161–181. Springer, Heidelberg (1994). https://doi.org/10.1007/3-540-58095-6_12
21. Skowron, A., Stepaniuk, J.: Tolerance approximation spaces. Fundamenta Informaticae **27**(2–3), 245–253 (1996)
22. Skowron, A., Jankowski, A., Swiniarski, R.W.: Foundations of rough sets. In: Kacprzyk, J., Pedrycz, W. (eds.) Springer Handbook of Computational Intelligence, pp. 331–348. Springer, Heidelberg (2015). https://doi.org/10.1007/978-3-662-43505-2_21
23. Nguyen, H.S., Skowron, A.: Rough sets: from rudiments to challenges. In: Skowron, A., Suraj, Z. (eds.) Rough Sets and Intelligent Systems - Professor Zdzisław Pawlak in Memoriam. Intelligent Systems Reference Library, vol. 42, pp. 75–173. Springer, Heidelberg (2013). https://doi.org/10.1007/978-3-642-30344-9_3

24. Yao, Y.: Two views of the theory of rough sets in finite universes. Int. J. Approx. Reasoning **15**, 291–317 (1996)
25. Qian, Y., Jiye, L.J., Yao, Y.Y., Dang, C.: MGRS: a multi-granulation rough set. Inf. Sci. **180**, 949–970 (2010)
26. Ziarko, W.: Variable precision rough set model. J. Comput. Syst. Sci. **46**, 39–59 (1993)

Approximation by Filter Functions

Ivo Düntsch[1,2(✉)], Günther Gediga[3], and Hui Wang[1,4]

[1] School of Mathematics and Informatics, Fujian Normal University,
Fuzhou, Fujian, China
[2] Brock University, St. Catharines, ON L2S 3A1, Canada
ivo@duentsch.net
[3] Institut für Evaluation und Marktanalysen, Brinkstr. 19, 49143 Jeggen, Germany
gediga@eval-institut.de
[4] School of Computing and Mathematics, Ulster University,
Newtownabbey, Northern Ireland
H.Wang@ulster.ac.uk

Abstract. In this exploratory article, we draw attention to the common formal ground among various estimators such as the belief functions of evidence theory and their relatives, approximation quality of rough set theory, and contextual probability. The unifying concept will be a general filter function composed of a basic probability and a weighting which varies according to the problem at hand. To compare the various filter functions we conclude with a simulation study with an example from the area of item response theory.

Keywords: Filter functions · Belief functions
Approximation quality · Contextual probability

1 Introduction

In order to classify a data point $x \in Q$ about which we have no precise knowledge, one may take into account information that is available in a neighbourhood of x and use this to classify x. Neighbourhoods can be defined in various ways; prominent examples are by distance functions in a numerical context or as equivalence or similarity classes with respect to a chosen relation in a nominal context [10].

The original rough set concept of neighbourhood of a point x is a class of an equivalence relation which contains x. This was generalized to consider the relationship of subsets of Q with $R(x)$, where R is a binary relation on Q and $R(x) = \{y \in Q : xRy\}$. From each of these neighbourhood concepts lower and upper approximations can be derived, and we invite the reader to consult [13] for an introduction to such generalization.

The ordering of authors is alphabetical and equal authorship is implied.

Ivo Düntsch gratefully acknowledges support by Fujiang Normal Univeristy, the Natural Sciences and Engineering Research Council of Canada Discovery Grant 250153, and by the Bulgarian National Fund of Science, contract DN02/15/19.12.2016.

H. S. Nguyen et al. (Eds.): IJCRS 2018, LNAI 11103, pp. 243–256, 2018.
https://doi.org/10.1007/978-3-319-99368-3_19

Even if we have decided in principle which type of neighbourhood of $E \subseteq Q$ should be considered, it is often still not clear which neighbourhood should be used. For example, one crucial issue in the k – nearest neighbour method is the choice of k. In other words, decisions have to be made which sets we allow to be neighbourhoods of a point or a set, and this is where filter functions come in useful.

The Oxford English Dictionary gives various definitions of *filter*, among others, [9]:

- A porous device for removing impurities or solid particles from a liquid or gas passed through it.
- A device for suppressing electrical or sound waves of frequencies not required.
- *Computing* A function used to alter the overall appearance of an image in a specific manner.
- *Computing* A piece of software that processes data before passing it to another application, for example to reformat characters or to remove unwanted types of material.

A filter function may be considered as a rule that tells us which sets are selected to serve as an approximation (or description) of a subset E of the universe Q, and how these "neighbourhoods" will be weighted.

Throughout, Q denotes a finite nonempty set with $|Q| = n$, and $\mathcal{N}$ is a family of subsets of Q.

At times, we will suppose that $\mathcal{N}$ is a – not necessarily proper – Boolean subalgebra of 2^Q with atom set $\mathrm{At}(\mathcal{N}) = \{A_1, \ldots, A_k\}$. In this case, if $Y \in \mathcal{N}$, we define $\mathrm{noa}(Y)$ as the number of atoms of $\mathcal{N}$ contained in Y.

A *probability measure* on a Boolean subalgebra $\mathcal{N}$ of 2^Q is an additive function p on $\mathcal{N}$, i.e. if $Q_1, \ldots, Q_k \in \mathcal{N}$, and the Q_i are pairwise disjoint, then $p(\bigcup\{Q_i : 1 \leq i \leq k\}) = \sum\{p(Q_i) : 1 \leq i \leq k\}$; we require furthermore that $p(Q) = 1$. This is the standard definition of measure theory.

The *sampling probability* on $\mathcal{N}$ is defined by

$$(1.1) \qquad p_{\mathcal{N}}(Y) := \begin{cases} \sum\{\frac{|A_i|}{n} : A_i \subseteq Y\}, & \text{if } Y \neq \emptyset, \\ \emptyset, & \text{otherwise.} \end{cases}$$

This assignment is based on the principle of indifference and assumes ignorance about the distribution within the atoms of $\mathcal{N}$.

A generalization of probability measures are *mass functions* or *basic probabilities* [11], or *basic belief functions* [16]: A *mass function on $\mathcal{N}$* is a function $m : \mathcal{N} \rightarrow [0, 1]$ such that $\sum\{m(Y) : Y \in \mathcal{N}\} = 1$. A *focal element* is a set $Y \in \mathcal{N}$ with $m(Y) \neq \emptyset$. Owing to the finiteness of Q, the restriction to the upper bound 1 for $m(Y)$ is one of convenience which may be obtained by appropriate weighting. Unlike the Dempster–Shafer model, we assume an open world situation, and do not require that $m(\emptyset) = 0$; here, we follow [14, Sect. 4.8].

If p is a probability measure on $\mathcal{N}$, then the function $m_p : \mathcal{N} \to [0,1]$ defined by

(1.2)
$$m_p(Y) := \begin{cases} p(Y), & \text{if } Y \in \text{At}(\mathcal{N}), \\ 0, & \text{otherwise.} \end{cases}$$

is a mass function. So, formally, probabilities are special mass functions (often called *Bayesian mass functions*).

2   Filter Functions

In general, a filter is a function which passes information that is pertinent to the application area, and reduces (or leaves out) information considered to be irrelevant. This concept of a filter originates with signal processing, but the same idea may be applied to elements of weighted structures. There is no relation to the filter concept in lattice theory.

We consider filter functions $F : 2^Q \to [0,1]$ of the general form

(2.1)
$$F(E) = \sum \{m(Y) \cdot w(E,Y), Y \in \mathcal{N}\}.$$

A filter consists of several parts:

- A set $\mathcal{N}$ of neighbourhoods which are often determined by an indicator function and, perhaps, other parameters. In such a way, the pool $\mathcal{N}$ of possible neighbourhoods is adjusted to the needs of the problem under consideration. How the initial $\mathcal{N}$ is chosen is a topic for further research.
- A weighting function $w : 2^Q \times \mathcal{N} \to [0,1]$ which re–scales the weights of the neighbourhoods in such a way that desired properties such as the value of an upper bound or the sum of the re–scaled values are guaranteed. In most cases, the values of w will be in $[0,1]$.

If $E \subseteq Q$ is an event (or a piece of evidence), and $Y \in \mathcal{N}$, it is reasonable to suppose that Y should not be considered a neighbourhood of E, if $E \cap Y = \emptyset$. On the other hand, any Y which contains E should be considered a neighbourhood of E; these are, in some sense, "boundary" situations.

In this spirit, we define our main indicator functions by

$\text{ind}^u(X,Y) = 1 \Longleftrightarrow$ if $X \cap Y \neq \emptyset$,	Upper indicator
$\text{ind}^l(X,Y) = 1 \Longleftrightarrow Y \subseteq X$,	Lower indicator.

Other indicators we use are

$\text{ind}^z(Y) = 1 \Longleftrightarrow \text{ind}^u(Y,Y) = 1 \Longleftrightarrow Y \neq \emptyset$,	
$\text{ind}^{sub}(X,Y) = 1 \Longleftrightarrow X \subseteq Y$,	Subset indicator,
$\text{ind}^{eq}(X,Y) = 1 \Longleftrightarrow \text{ind}^{sub}(X,Y) \cdot \text{ind}^{sub}(Y,X) = 1 \Longleftrightarrow X = Y$	Equality indicator.

We suppose, as is customary, that an indicator function takes values in $\{0,1\}$. Now we define the *upper* and the *lower filter*:

(2.2) $$F_m^u(E) := \sum\{m(Y) \cdot \text{ind}^u(E,Y) : Y \in \mathcal{N}\}, \qquad \text{Upper filter}$$

(2.3) $$F_m^l(E) := \sum\{m(Y) \cdot \text{ind}^l(E,Y) : Y \in \mathcal{N}\}, \qquad \text{Lower filter.}$$

Lower and upper filters as defined above are not the only one, which select a neighbourhood of some evidence E; they are, as we shall see, maximal filters of their type: For the upper filter and $E \neq \emptyset$, a set $Y \in \mathcal{N}$ is a neighbourhood of E, if they have at least one element in common. A simple way to sharpen this is the demand that they have at least $k \geq 1$ elements in common. If E has exactly one element, then the situation is unchanged, but if E consists of more than one element, the number of neighbourhood sets will be reduced. These considerations lead us to *upper* and *lower* $k - filters$ $(1 \leq k \leq |Q|)$ by first defining the indicators

(2.4) $$\text{ind}^{u,k}(X,Y) = 1 \Longleftrightarrow |X \cap Y| \geq k,$$

(2.5) $$\text{ind}^{l,k}(X,Y) = 1 \Longleftrightarrow Y \subseteq X \text{ and } |Y| \geq k.$$

A similar parametrization may be used to demand that a neighbourhood should cover more than $s\%$ of the event. So, we define the indicator functions

(2.6) $$\text{ind}^{u,s}(X,Y) = 1 \Longleftrightarrow X = Y \text{ or } |X \cap Y| \gtrsim s \cdot |X|,$$

(2.7) $$\text{ind}^{l,s}(X,Y) = 1 \Longleftrightarrow X = Y \text{ or } Y \subsetneq X \text{ and } |Y| \gtrsim s \cdot |X|.$$

The boundary values of the parameterized indicators are easily seen to be

$$\text{ind}^{u,k=1}(X,Y) = \text{ind}^{u,s=0}(X,Y) = \text{ind}^u(X,Y), \quad \text{ind}^{u,k=|Q|}(X,Y) = \text{ind}^{u,s=1}(X,Y) = \text{ind}^{sub}(X,Y)$$

$$\text{ind}^{l,k=1}(X,Y) = \text{ind}^{l,s=0}(X,Y) = \text{ind}^l(X,Y), \quad \text{ind}^{l,k=|Q|}(X,Y) = \text{ind}^{l,s=1}(X,Y) = \text{ind}^{eq}(X,Y).$$

The respectively weighted upper and lower filter are now defined by

(2.8) $$F_m^{u,s}(E) := \sum_{Y \in \mathcal{N}} m(Y) \cdot \text{ind}^{u,k}(E,Y),$$

(2.9) $$F_m^{l,s}(E) := \sum_{Y \in \mathcal{N}} m(Y) \cdot \text{ind}^{l,k}(E,Y),$$

(2.10) $$F_m^{u,s}(E) := \sum_{Y \in \mathcal{N}} m(Y) \cdot \text{ind}^{u,s}(E,Y),$$

(2.11) $$F_m^{l,s}(E) := \sum_{Y \in \mathcal{N}} m(Y) \cdot \text{ind}^{l,s}(E,Y).$$

The parameterized filters are antitone with respect to s:

Theorem 1. *Let $s, t \in [0, 1]$, and $s \leq t$. Then, $F_m^{l,t}(E) \leq F_m^{l,s}(E)$ and $F_m^{u,t}(E) \leq F_m^{u,s}(E)$.*

Proof. We show the claim only for the lower filter, as the remaining claim is proved similarly. First, consider

$$F_m^{l,t}(E) \leq F_m^{l,s}(E) \iff F_m^{l,s}(E) - F_m^{l,t}(E) \geq 0,$$
$$\iff \sum_{Y \in \mathcal{N}} m(Y) \cdot \mathrm{ind}^{l,t}(E, Y) - \sum_{Y \in \mathcal{N}} m(Y) \cdot \mathrm{ind}^{l,s}(E, Y) \geq 0,$$
$$\iff \sum_{Y \in \mathcal{N}} m(Y) \cdot (\mathrm{ind}^{l,t}(E, Y) - \mathrm{ind}^{l,s}(E, Y)) \geq 0.$$

Since $s \leq t$, we have $|Y| \geq t \cdot |X|$ implies $|Y| \geq s \cdot |X|$, and therefore, $\mathrm{ind}^{l,t}(E, Y) = 1$ implies $\mathrm{ind}^{l,s}(E, Y) = 1$. It follows that $\mathrm{ind}^{l,s}(E, Y) \geq \mathrm{ind}^{l,t}(E, Y)$, i.e. $\mathrm{ind}^{l,s}(E, Y) - \mathrm{ind}^{l,t}(E, Y) \geq 0$. Since $m(Y) \geq 0$, we conclude $F_m^{l,t}(E) \leq F_m^{l,s}(E)$.

The same proof shows that the parameterized filters are antitone as well.

3   Approximation and Estimation

In this section we show how commonly used belief and approximation measures fit into the scheme of filter functions as proposed in (2.1). For an overview of different interpretations of "belief" we refer the reader to [7].

3.1   Evidence Measures

Evidence theory has been widely studied as an alternative to classical probability theory, see the source book edited by Yager and Liu [21]. For a thoughtful discussion of belief and probability we invite the reader to consult [4,7], where, among others, it was shown that "a key part of the important Dempster-Shafer theory of evidence is firmly rooted in classical probability theory".

In evidence theory and related fields, two functions are obtained from a mass function $m : \mathcal{N} \to [0, 1]$:

$$(3.1) \qquad \mathrm{bel}_m(E) := \sum_{Y \in \mathcal{N}, Y \subseteq E} m(Y), \qquad \text{degree of belief,}$$

$$(3.2) \qquad \mathrm{pl}_m(E) := \sum_{Y \in \mathcal{N}, Y \cap E \neq \emptyset} m(Y), \qquad \text{degree of plausibility.}$$

These concepts were introduced by Dempster [1], who called them, respectively, *lower* and *upper probability*. A belief function assigns the total amount of belief supporting E without supporting $Q \setminus E$, and $\mathrm{pl}_m(E)$ quantifies the maximal amount of belief that might support E [15]. It is straightforward to show that $\text{pl}_m(E) = \text{bel}_m(Q) - \text{bel}_m(Q \setminus E)$.

Conversely, every mass function can be obtained from a function bel which satisfies certain conditions, see e.g. [11, Chap. 2].

Belief and plausibility are easily related to the upper and lower filter function as follows:

$$\text{bel}_m(E) = \sum \{m(Y) : Y \subseteq E, Y \in \mathcal{N}\} = \sum \{m(Y) \cdot \text{ind}^l(E, Y) : Y \in \mathcal{N}\} = F_m^l(E),$$

$$\text{pl}_m(E) = \sum \{m(Y) : E \cap Y \neq \emptyset, Y \in \mathcal{N}\} = \sum \{m(Y) \cdot \text{ind}^u(E, Y) : Y \in \mathcal{N}\} = F_m^u(E).$$

3.2   Rough Set Approximation Quality

Suppose that $X \subseteq Q$, and that $\mathcal{N}$ is a Boolean algebra with atoms $A_1, \ldots, A_k$. Then, $\text{At}(\mathcal{N})$ can be considered the partition of Q obtained from some equivalence relation θ on Q; in other words, we work with a rough set approximation space $\langle Q, \theta \rangle$. In rough set theory [10], the *upper approximation of X* is the set $\text{upp}(X) := \bigcup \{A_i : A_i \cap X \neq \emptyset\}$ and the *lower approximation of X* is the set $\text{low}(X) := \bigcup \{A_i : A_i \subseteq X\}$. These approximations lead to two statistics relative to $\mathcal{N}$:

$$(3.3) \qquad \qquad \mu^{\mathcal{N}*}(E) = \frac{|\text{upp}(E)|}{n},$$

$$(3.4) \qquad \qquad \mu_*^{\mathcal{N}}(E) = \frac{|\text{low}(E)|}{n}.$$

Inspection of the indices used in "classical rough set theory" such as α, γ, rough membership, other element counting etc. shows that these indices are valid only in case we assume the principle of indifference: Assuming no knowledge of the distribution within the equivalence classes, we let p be the sampling probability measure on $\mathcal{N}$ as defined in (1.1). There may be other assumptions within the frame of lower and upper set approximations, which consequently lead to other evaluation schemes. The principle of indifference is widely used in rough set theory – explicitly or implicitly. For example, the general rough membership function defined in [8, Definition 4.3.] is a special filter in our terminology for which the principle of indifference is a hidden assumption; otherwise the estimator of this index is biased and unsuitable for applications. In [8] only point estimators of indices or membership functions are addressed - but this is not the whole story: The reliability of the indices needs to be discussed as well. Assuming the principle of indifference, we are able to compute confidence intervals such as the reliability of the general rough membership function or other filters, as we demonstrate in the present work.

Using the mass function m determined by p as defined in (1.2) we can describe $\mu^{\mathcal{N}*}(E)$ and $\mu^{\mathcal{N}}_*(E)$ in terms of upper and lower filter:

$$\mu^{\mathcal{N}*}(E) = \sum\{\frac{|A_i|}{n} : E \cap A_i \neq \emptyset\},$$
$$= \sum\{m(Y) : E \cap Y \neq \emptyset, Y \in \mathcal{N}\},$$
$$= \sum\{m(Y) \cdot \text{ind}^u(E \cap Y), Y \in \mathcal{N}\},$$
$$= F_m^u(E),$$
$$\mu^{\mathcal{N}}_*(E) = \sum\{\frac{|A_i|}{n} : A_i \subseteq E)\},$$
$$= \sum\{m(Y) \cdot \text{ind}^l(E \cap Y), Y \in \mathcal{N}\},$$
$$= F_m^l(E).$$

This shows the close connection of rough set approximation to the estimators of evidence theory, observed first by Skowron [12].

The *approximation quality* is the function

$$(3.5) \qquad \gamma(E) := \frac{|\text{low}(E)|}{n} + \frac{|\text{low}(Q \setminus E)|}{n}..$$

$\gamma(E)$ is the relative frequency of all elements of Q which are correctly classified under the granulation of information by $\mathcal{N}$ with respect to being an element of E or not. In terms of filter functions, this becomes

$$(3.6) \qquad \gamma(E) = F_m^l(E) + F_m^l(Q \setminus E).$$

3.3   Pignistic Probability

According to Smets [15], decision making under uncertainty can (and should) be done in two steps. On a *credal level*, an assignment of beliefs is made to pieces of evidence. In order to be coherent on a *pignistic level* (decision level), the uncertainties quantified by the belief function must be turned into a probability measure. In such a way, the two levels of handling uncertainty and decision making are clearly separated unlike, as Smets claims, in Bayesian reasoning.

A *pignistic probability distribution* (with respect to the mass function m and the Boolean algebra $\mathcal{N}$) [16, Sect. 3] is a function $pp : \mathcal{N} \to [0,1]$ which is defined by

$$(3.7) \qquad pp_m(E) := \sum\{m(Y) \cdot \frac{|E \cap Y|}{\text{noa}(Y)} : Y \in \mathcal{N}^+\}$$

If E is an atom of $\mathcal{N}$, we obtain

$$(3.8) \qquad \mathrm{pp}_m(E) = \sum \{m(Y) \cdot \frac{|E|}{\mathrm{noa}(Y)} : E \subseteq Y \in \mathcal{N}\}.$$

Note that $E \subseteq Y$ implies that $Y \neq \emptyset$. It was shown in [15] that pp is indeed a probability measure, if $\mathcal{N} = 2^Q$. Setting

$$w(E,Y) := \begin{cases} \frac{|E \cap Y|}{\mathrm{noa}(Y)} & \text{if } Y \neq \emptyset, \\ 0, & \text{otherwise,} \end{cases}$$

we see that $\mathrm{pp}(E) = \sum \{m(Y) \cdot w(E,Y), Y \in \mathcal{N}\}$ as in (2.1).

3.4   Contextual Probability

Another two step procedure to reason under uncertainty, called *contextual probability* was first proposed in [17], and subsequently developed in [19]. It is a secondary probability, which is defined in terms of a basic (primary) function; it can be used to estimate the primary probability from a data sample through a process called *neighbourhood counting*; for details see [20].

Given a mass function m over 2^Q, we first define a weight function by

$$w(E,Y) := \begin{cases} \frac{|E \cap Y|}{|Y|} & \text{if } Y \neq \emptyset, \\ 0, & \text{otherwise.} \end{cases}$$

The *contextual probability* is the function $\mathrm{cp}^m : 2^Q \to [0,1]$ defined by

$$(3.9) \qquad \mathrm{cp}^m(E)) = \sum \{m(Y) \cdot w(E,Y) : Y \in \mathcal{N}\},$$

Wang [17] showed that cp^m is a probability distribution if $\mathcal{N} = 2^Q$.

This definition of contextual probability was found problematic when trying to find a simple relationship between the primary probability and the secondary probability, so the definition was refined in [18], and extended in [20]. The work on estimating contextual probability from data sample has spawned a series of papers exploring the various forms of neighbourhood counting for multivariate data, sequences, trees, and graphs. We give a somewhat simplified version of the revised definition, and also extend its range over 2^Q.

Suppose that p is a probability measure on $\mathcal{N}$, and let $K := \sum \{p(Y) \cdot |Y| : Y \in \mathcal{N}\}$ be a normalization factor. The *contextual probability with respect to p*, is defined by

$$(3.10) \qquad \mathrm{cp}^p(E) := \sum \{p(Y) \cdot \frac{|E \cap Y|}{K}, Y \in \mathcal{N}\}.$$

Setting $w(E,Y) := \frac{|E \cap Y|}{K}$ and using the mass function m_p of (1.2), we see that cp^p is an instance of a general filter function.

4   Probabilistic Knowledge Structures

In this section we apply some of the filter functions defined previously to a situation well known in the context of psychometric aspects of learning, in particular, knowledge structures [5,6]. Connections of knowledge structures to other concepts including rough sets were exhibited in [2].

Suppose that U is a set of students, Q is a set of problems, and $S \subseteq U \times Q$ is a binary relation between students and problems, called a *solving relation*; uSq means that student u solves problem q. For each $u \in U$, the set $S(u) := \{q \in Q : uSq\}$ is called the *empirical (observed) solving pattern* of u. The set $\{S(u) : u \in U\}$ is called an *empirical knowledge structure* (EKS) with respect to U and Q, denoted by $\widehat{\mathcal{K}}$. With each $X \subseteq Q$ we associate a number $\mathrm{obs}(X) = |\{u \in U : S(u) = X\}|$. Thus, $\mathrm{obs}(X)$ is the number of times that X was observed as a student's solving pattern.

A *probabilistic knowledge structure* (PKS) is a tuple $\langle \mathcal{N}, m \rangle$ where $\mathcal{N} \subseteq 2^Q$, and m is a mass function on $\mathcal{N}$. We interpret m as *item–pattern probability* in the sense that

(4.1) $m(X) = p(\text{each } x \in X \text{is solved, and no problem in } Q \setminus X \text{ is solved}).$

in other words $m(X)$ is the probability that X is an observed item pattern. $m(\emptyset)$ is the probability that no item in Q is solved, and $m(\{x\})$ is the probability that only x is solved.

Given a PKS, we estimate the probabilities by the relative frequencies of the observed item patterns by

(4.2)
$$\hat{m}(X) = \hat{p}(\text{each } x \in X \text{ is solved, and no problem in } Q \setminus X \text{is solved}) = \frac{\mathrm{obs}(X) \cdot |X|}{n}.$$

In this way we not only obtain insight into the probability nature of the mass function and its derivations, but we may use the empirical counterpart of relative frequencies as estimates and as a basis for statistical inference.

Using a PKS as a workhorse, we will explore which interpretation this context offers for different filter functions. First, consider F_m^l, which is just the belief function bel_m. Then, according to our interpretation,

$$\mathrm{bel}_m(E) = \sum \{m(Y) : Y \in \mathcal{N}, Y \subseteq E\},$$
$$= p_{\mathrm{bel}_m}((\text{some items in } E \text{ are solved or no item is solved})$$
$$\text{and no item outside } E \text{ is solved.})$$

Considering a solving path $\emptyset \subseteq \{x_1\} \subseteq \{x_1, x_2\} \subseteq \ldots \subseteq E$, we see that p_{bel_m} is a cumulative probability function with $p_{\mathrm{bel}_m}(Q) = 1$. A problem which may arise is that the condition "some item in E is solved or no item is solved" is not always acceptable. Thus, we may remove the latter condition – which corresponds to $m(\emptyset) \neq \emptyset$, and define

$$\mathrm{bel}_m^+(E) = \sum \{m(Y) : Y \in \mathcal{N}^+, Y \subseteq E\},$$
$$= p_{\mathrm{bel}_m^+} \text{ (some items in } E \text{ are solved and no item outside } E \text{ is solved.)}$$

bel_m^+ is also a cumulative function, but $\mathrm{bel}_m^+(Q) = 1 - m(\emptyset)$.

Turning to F_m^u, we recall that $F_m^U = \mathrm{pl}_m$. Then,

$$\mathrm{pl}_m(E) = \sum \{m(Y) : Y \cap E \neq \emptyset, Y \in \mathcal{N}\},$$
$$= p_{\mathrm{pl}} \text{(at least one problem in } E \text{ is solved)}.$$

If $E = \{x\}$, then $p_{\mathrm{pl}}(\{x\})$ is the *item solving probability* of x.

To estimate only the states in $\mathcal{N}$, we let $\mathrm{ind}_{\mathcal{N}}(E) := 1$ if and only if $E \in \mathcal{N}$, and define

(4.3)
$$\mathrm{bel}_m^{\min}(E) := \mathrm{ind}_{\mathcal{N}}(E) \cdot F_m^l(E) = \sum \{m(Y) \cdot \mathrm{ind}_{\mathcal{N}}(E) \cdot \mathrm{ind}^l(E, Y) : Y \in \mathcal{N}\}.$$

$F_m^{l,\min}$ may be regarded as some sort of minimal lower filter, as only elements of $\mathcal{N}$ are allowed to be approximated. Observe that the lower filter F_m^l coincides with $\mathrm{bel}_m^{\min}$ if and only if $\mathcal{N} = 2^Q$.

To parameterize the upper filter $F_m^u(E)$ to use only states in $\mathcal{N}$ that contain E we shall consider $\mathrm{pl}_m^{\min} := F_m^{u,1}$ as defined in (2.10) with $s = 1$.

Suppose we have a set of five questions $Q = \{1, 2, 3, 4, 5\}$ and $\mathcal{N}$ consisting of 12 item patterns, each supplied with a basic probability, as shown in Fig. 1.

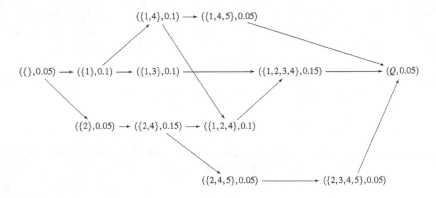

Fig. 1. A weighted knowledge structure

Given the PKS in Fig. 1, we have performed some empirical experiments to compute sampling distributions of the defined filter procedures. We use a multinomial sampling, and $N = 50, N = 100$, or $N = 1,000$ observations of item patterns. For 10,000 simulations of the sampling process, we computed the sampling distributions of the functions $\text{bel}_m, \text{pl}_m, \text{cp}^m, \text{bel}_m^{\min}, \text{pl}_m^{\min}$, and $pl[k = 2]$ for all subsets of $2^{\{1,2,3,4,5\}}$. We have computed the mean, bias, median, upper and lower quartile, and the 2.5%- and 97.5%-quantile of the sampling distributions of these functions for each subset Q.[1]

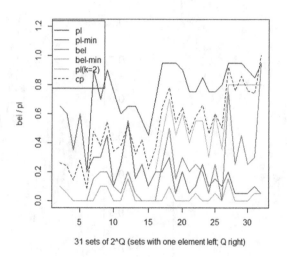

31 sets of 2^Q (sets with one element left; Q right)

Fig. 2. Simulation graph

Figure 2 shows the mean of the different filter functions on the nonempty subset of 2^Q. The left most is the value of $\{1\}$, followed by the values of the sets $\{2\}, \ldots, \{5\}$. The sets with two elements follow in lexicographical order, followed by the sets with 3, 4, and finally, 5 elements.

We observe that the values of the functions $\text{pl}_m, pl_m^{\min}$, and $\text{pl}_m^{k=2}$ are equal for sets with one element, and $\text{pl}_m^{\min}$ and $\text{pl}^{k=2}$ are identical for sets with two elements. The larger the number of elements, the larger the difference of pl_m and $\text{pl}_m^{\min}$. The same observations hold for bel_m and $bel_m^{\min}$. Furthermore, the graphs of $\text{pl}_m^{k=2}$ and cp^m are quite similar – up to events with 1 element.

By way of example, Fig. 3 shows the confidence intervals of cp^m for 50, respectively, 500 observations.

The organisation of the x-axis in Fig. 3 is the same as in Fig. 2. It can be see from Fig. 3 that – given a quite sparse PKS as our example of Fig. 1 – the 95% confidence bounds are quite narrow, even if we assume a small empirical basis of only 50 observations (left part of the figure). An empirical basis of 500 item

---

[1] The tables and the R-source of the simulation procedure are available for download at www.roughsets.net.

patterns allows us a precise estimate of the cp^m values. The same is true for the other measures; we omit the details for these which can be found in the archive.

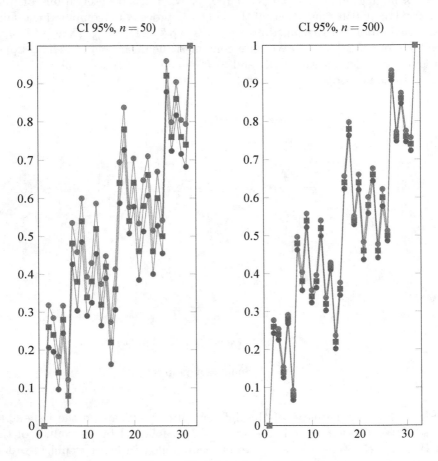

Fig. 3. CI and median of cp^m

5   Summary and Outlook

We have exhibited a common form of several estimators employed in reasoning under uncertainty. The novelty is not that connections exist among them – these have been known for some time –, but the interpretation as filter functions, a term we have borrowed from digital imaging. A filter, such as an edge detector, extracts salient features of a scene, or, as in our case, of a situation for further processing. A simulation study indicates how some filters behave in various situations.

In future work we shall explore whether and how the filter concept can be extended to other estimators, for example, to kernel functions such as k-nearest neighbour. We will also investigate a logical approach to filter functions applied in applications of theories of visual perception and digital imaging, following the path started in [3].

Acknowledgement. We are grateful to the referees for constructive comments.

References

1. Dempster, A.P.: Upper and lower probabilities induced by a multivalued mapping. Ann. Math. Stat. **38**(2), 325–339 (1967)
2. Düntsch, I., Gediga, G.: A note on the correspondences among entail relations, rough set dependencies, and logical consequence. J. Math. Psychol. **45**, 393–401 (2001). MR 1836895
3. Düntsch, I., Gediga, G.: On the gradual evolvement of things. In: Skowron, A., Suraj, Z. (eds.) Rough Sets and Intelligent Systems - Professor Zdzisław Pawlak in Memoriam, vol. 1, pp. 247–257. Springer, Heidelberg (2012). https://doi.org/10.1007/978-3-642-30344-9_8
4. Fagin, R., Halpern, J.: Uncertainty, belief, and probability. Comput. Intell. **7**(3), 160–173 (1991)
5. Falmagne, J.C., Doignon, J.P.: Learning Spaces. Springer, Heidelberg (2011). https://doi.org/10.1007/978-3-642-01039-2
6. Falmagne, J.C., Koppen, M., Villano, M., Doignon, J.P., Johannesen, J.: Introduction to knowledge spaces: how to build, test and search them. Psychol. Rev. **97**, 201 (1990)
7. Halpern, J.Y., Fagin, R.: Two views of belief: belief as generalized probability and belief as evidence. Artif. Intell. **54**, 275–317 (1992)
8. Mani, A.: Probabilities, dependence and rough membership functions. Int. J. Comput. Appl. **39**(1), 17–35 (2017)
9. Oxford English Dictionaries: Definition of "filter" (2018). https://en.oxforddictionaries.com/definition/filter. Accessed 20 Mar 2018
10. Pawlak, Z.: Rough Sets: Theoretical Aspects of Reasoning About Data, System Theory, Knowledge Engineering and Problem Solving, vol. 9. Kluwer, Dordrecht (1991)
11. Shafer, G.: A Mathematical Theory of Evidence. Princeton University Press, Princeton (1976)
12. Skowron, A.: The rough sets theory and evidence theory. Fundamenta Informaticae **13**, 245–262 (1990)
13. Słowiński, R., Vanderpooten, D.: Similarity relations as a basis for rough approximations. ICS Research Report 53, Polish Academy of Sciences (1995)
14. Smets, P.: Belief functions. In: Smets, P., Mandani, A., Dubois, D., Prade, H. (eds.) Non-standard Logics for Automated Reasoning. Academic Press, London (1988)
15. Smets, P.: Belief functions versus probability functions. In: Bouchon, B., Saitta, L., Yager, R.R. (eds.) IPMU 1988. LNCS, vol. 313, pp. 17–24. Springer, Heidelberg (1988). https://doi.org/10.1007/3-540-19402-9_51
16. Smets, P., Kennes, R.: The transferable belief model. Artif. Intell. **66**(2), 191–234 (1994)
17. Wang, H.: Contextual probability. J. Telecommun. Inf. Technol. **3**, 92–97 (2003)

18. Wang, H., Dubitzky, W.: A flexible and robust similarity measure based on contextual probability. In: Proceedings of the Nineteenth International Joint Conference on Artificial Intelligence (IJCAI), pp. 27–34 (2005)
19. Wang, H., Düntsch, I., Gediga, G., Guo, G.: Nearest Neighbours without k. In: Dunin-Keplicz, B., Jankowski, A., Skowron, A., Szczuka, M. (eds.) Monitoring, Security, and Rescue Techniques in Multiagent Systems. Advances in Soft Computing, pp. 179–189. Springer, Heidelberg (2006). https://doi.org/10.1007/3-540-32370-8_12
20. Wang, H., Murtagh, F.: A study of the neighborhood counting similarity. IEEE Trans. Knowl. Data Eng. **20**(4), 449–461 (2008)
21. Yager, R., Liu, L. (eds.): Classic Works of the Dempster-Shafer Theory of Belief Functions. Studies in Fuzziness and Soft Computing, vol. 219. Springer, Heidelberg (2008). https://doi.org/10.1007/978-3-540-44792-4

A Test Cost Sensitive Heuristic Attribute Reduction Algorithm for Partially Labeled Data

Shengdan Hu[1,2,3], Duoqian Miao[1,2]([✉]), Zhifei Zhang[1,4], Sheng Luo[1,2], Yuanjian Zhang[1,2], and Guirong Hu[2]

[1] Department of Computer Science and Technology, Tongji University, Shanghai 201804, China
dqmiao@tongji.edu.cn, hushengdan@163.com
[2] Key Laboratory of Embedded System and Service Computing, Ministry of Education, Tongji University, Shanghai 201804, China
[3] Department of Computer Science, Shanghai Normal University Tianhua College, Shanghai 201815, China
[4] State Key Laboratory for Novel Software Technology, Nanjing University, Nanjing 210023, China

Abstract. Attribute reduction is viewed as one of the most important topics in rough set theory and there have been many researches on this issue. In the real world, partially labeled data is universal and cost sensitivity should be taken into account under some circumstances. However, very few studies on attribute reduction for partially labeled data with test cost have been carried out. In this paper, based on mutual information, the significance of an attribute in partially labeled decision system with test cost is defined, and for labeled data, a heuristic attribute reduction algorithm TCSPR is proposed. Experimental results show the impact of test cost on reducts for partially labeled data and comparative experiments of classification accuracy indicate the effectiveness of the proposed method.

Keywords: Attribute reduction · Uncertainty · Rough set
Test cost sensitive · Partially labeled data

1 Introduction

Uncertainty is a common phenomenon in the world. Reasoning and knowledge acquisition with uncertain or incomplete information is always a core subproblem of artificial intelligence. There have been plenty of theories on the problem of uncertainty, for example, probability theory [2], possibility theory [4,5], fuzzy set [3], rough set [6,7], evidence theory [8,9], cloud model [1]. As an extension of set theory, rough set which was proposed by Polish computer scientist Zdzislaw Pawlak [6] in 1982, is a soft computing tool to model imperfect knowledge. In rough set, it is assumed that knowledge is based on the ability to classify

© Springer Nature Switzerland AG 2018
H. S. Nguyen et al. (Eds.): IJCRS 2018, LNAI 11103, pp. 257–269, 2018.
https://doi.org/10.1007/978-3-319-99368-3_20

objects and tabular representation of knowledge is often employed. Uncertainty in this theory is represented by the boundary region of a set, and the boundary region can be specified in terms of a pair of crisp sets which give the lower and the upper approximation of the original set.

Recently, a deluge of data from a variety of sources has reached an unprecedented volume. However, there may exist incomplete data due to various reasons. In decision systems of rough set, some values of the decision attributes may be missing and the systems are actually partially labeled data. This could often occur in reality. For example, for the information system of patients in a hospital, some diagnoses of diseases may be missing due to the patients stop doing further examinations. Knowledge acquisition from partially labeled data is akin to semi-supervised learning which attracts plenty of researchers. To deal with partially labeled data, some methods based on rough set have been proposed [10–13], and incremental methods in dynamic system are studied [14–17].

The existing rough set-based methods for partially labeled data mentioned above seek low classifying error rates or high accuracy and implicitly assume that all classes or features have the same cost, nevertheless, this assumption may not be suitable in real-world scenarios. For example, in a clinical diagnosis system, it may cause some damage to a patient who is misclassified as cancer class, but may result in serious damage if a patient who has cancer is misclassified as non-cancer class and could not get treatment timely. Also, a patient often needs to undertake a number of medical tests, in this case, money and/or time for these tests are regarded as test costs and the costs may be various according to different tests. From these two examples, we can infer that cost sensitivity should be considered in some problems. In the cost sensitive settings, it is aimed to minimize the total cost, rather than simply minimize the error rate. Turney [18] concluded nine types of costs in inductive concept learning and in decision systems, some researchers have done much research on decision cost [19–22] and test cost [23,24]. However, there have been few studies about cost sensitive in decision system with missing decision values. Motivated by these analysis, this paper focuses on tackling the problem of attributes reduction for partially labeled data with test cost sensitive. We first define the significance of an attribute in the partially labeled decision system with test cost based on mutual information. Next, for labeled data, a heuristic algorithm TCSPR for attribute reduction is proposed. Then some attribute reduction experiments are conducted on several data sets to find out the impact of test cost on reducts. In order to verify the effectiveness of the proposed method, the quality of reducts are compared.

The remainder of the paper is organized as follows. Some preliminary concepts and uncertainty measures based on information entropy in rough set are briefly reviewed in Sect. 2. In Sect. 3, the definition of partially labeled decision system with test cost is given and an attribute reduction algorithm TCSPR is proposed. Section 4 illustrates some experiments and results. Section 5 concludes the paper with some discussions.

2   Preliminary

In this section, we present a review of some basic rough set concepts related to this article. One can refer to references [6,7,25] for detail of the theory.

2.1   Rough Set

Definition 1. *An information system is a tuple $IS = (U, A, V, f)$, where $U = \{x_1, x_2, \ldots, x_n\}$ is a finite nonempty set of objects, and $A = \{a_1, a_2, \ldots, a_m\}$ is a finite nonempty set of attributes, V is a nonempty set of values of $a_i \in A(i = 1, 2, \ldots, m)$, $f : U \to V$ is a nonempty set of information functions each of which maps an object in U to a exact value in V.*

If $A = C \cup D$, where C is a set of condition attributes and D is a decision attributes set, the information system is called decision information system or decision table and denoted as $DS = (U, A = C \cup D, V, f)$.

Definition 2. *Let $DS = (U, A = C \cup D, V, f)$ be a decision information system, $B \subseteq A$ be an equivalence relation (also called B-indiscernibility relation), for an arbitrary set $X \subseteq U$, the lower approximation and upper approximation of X with respect to B respectively are defined as:*
$$\underline{B}(X) = \{x \in U | [x]_B \subseteq X\},$$
$$\overline{B}(X) = \{x \in U | [x]_B \cap X \neq \emptyset\},$$
where $[x]_B$ is the equivalence class including x with respect to B, and $[x]_B = \{y \in U | f(x, a) = f(y, a), \forall a \in B\}$. If $\underline{B}(X) = \overline{B}(X)$, X is B-definable, and if $\underline{B}(X) \neq \overline{B}(X)$, X is rough with respect to B.

Definition 3. *Let $DS = (U, A = C \cup D, V, f)$ be a decision information system, and the objects in U are partitioned into r disjoint crisp subsets by decision attributes set D, namely, $U/D = \{D_1, D_2, \ldots, D_r\}$, then C-positive region of D is defined as:*
$$POS_C(D) = \bigcup_{i=1}^{r} \underline{C}(D_i),$$
and the boundary region of D w.r.t. C is defined as:
$$BN_C(D) = \bigcup_{i=1}^{r} \overline{C}(D_i) - \bigcup_{i=1}^{r} \underline{C}(D_i).$$

For any $B \subseteq A$ and $X \subseteq U$, the positive region $POS_B(X)$ is the collection of the objects that can be certainly classified as members of X with respect to relation B. The boundary region $BN_B(X)$, in a sense, is the undecidable area of the universe and none of the objects in this region can be certainly classified into X or $\sim X$. In rough set theory, uncertainty can be represented by the boundary region of a set.

2.2   Uncertainty Measure Based on Entropy and Reduct

In rough set theory, there are some algebraic measurement methods to express the inexactness of object or set, such as accuracy, roughness, attribute dependency degree. Inspired by Shannon's information entropy, Miao gave the information representation of the concepts and operations about rough set theory, and proposed the heuristic reduction algorithm based on mutual information [26].

Definition 4. *Let $DS = (U, A = C \cup D, V, f)$ be a decision information system, $B \subseteq A$ and the objects in U are partitioned into m disjoint crisp subsets $\{B_1, B_2, \ldots, B_m\}$ by B, then rough entropy of B is defined as:*

$$H(B) = -\sum_{i=1}^{m} \frac{|B_i|}{|U|} log_2 \frac{|B_i|}{|U|},$$

where $|B|$ denotes the cardinality of B, and $\sum_{i=1}^{m} |B_i| = |U|$ holds.

Definition 5. *Let $DS = (U, A = C \cup D, V, f)$ be a decision information system, $U/C = \{X_1, X_2, \cdots, X_m\}$ and $U/D = \{Y_1, Y_2, \cdots, Y_n\}$, the entropy of D conditioned on C is defined as:*

$$H(D|C) = -\sum_{i=1}^{m} \frac{|X_i|}{|U|} \sum_{j=1}^{n} \frac{|X_i \cap Y_j|}{|X_i|} log_2 \frac{|X_i \cap Y_j|}{|X_i|}. \tag{1}$$

Let $I(x; y)$ be the mutual information of x and y, the increment of mutual information, which is defined as:

$$I(B \cup \{a\}; D) - I(B; D) = H(D|B) - H(D|B \cup \{a\}), \tag{2}$$

can be used to measure the attribute significance.

Definition 6. *Let $DS = (U, A = C \cup D, V, f)$ be a decision information system, $B \subseteq C$, for $\forall a \in C - B$, the significance measure of a on B can be defined by mutual information as:*
$SGF(a, B, D) = H(D|B) - H(D|B \cup \{a\}).$
If $B = \emptyset$, the significance measure of a is:
$SGF(a, D) = H(D) - H(D|\{a\}).$

$SGF(a, B, D)$ expresses the importance of attribute a to decision D conditioned on the given attributes B.

Reduct is a subset of attributes that maintains some particular properties as the original data. For a given decision table, there may be multiple reducts. Based on the definitions above, relative reduct can be defined as follows.

Definition 7. *Let $DS = (U, A = C \cup D, V, f)$ be a decision information system and $B \subseteq C$, B is a reduct of C relative to D iff:*
(1) $H(D|B) = H(D|C)$;
(2) $\forall a \in B, H(a|B - \{a\}) > 0$.

In a given decision table, the intersection of all attribute reducts is core, and each element of a core should be in every reduct. The core may be an empty set.

2.3 Test Cost Sensitive Rough Set

Definition 8 ([27]). *A test cost sensitive decision system is a tuple $TDS = (U, A = C \cup D, V, f, c)$, where U, A, C, D, V and f have the same meanings as in definition 1, $c : C \rightarrow R^+ \cup \{0\}$ is the test cost function and R^+ is the set of positive real numbers.*

Assuming that the test cost of every attribute is independent, test cost function can be represented by a vector $c = [c(a_1), c(a_2), \cdots, c(a_{|C|})]$, where $c(a_i)(i = 1, 2, \ldots, |C|)$ is the test cost of attribute a_i, and for $\forall B \subseteq C, c(B) = \sum_{a \in B} c(a)$.

3   Attribute Reduction for Partially Labeled Data

3.1   Partially Labeled Decision System

In a partially labeled decision system, some values of the decision attributes are missing. In the light of test cost, partially labeled decision system can be defined as:

Definition 9. *A partially labeled decision system with test cost is a tuple* $TPDS = (U = L \cup N, A = C \cup D, V, f, c)$, *where* U, A, C, D, V, f *and* c *have the same meanings as in definition 8. L denotes the set of labeled objects, and N denotes the set of unlabeled objects.*

Then, we can define the significance of attribute a on B in a partially labeled decision system with test cost as follows:

$$SGF(a, B, D, c(a), \lambda) = (H(D|B) - H(D|B \cup \{a\}))c(a)^\lambda, \qquad (3)$$

where $H(D|B)$ and $H(D|B \cup \{a\})$ are the entropy of D conditioned on B and $B \cup \{a\}$ respectively, and they can be calculated by equation (1) in which the number of objects $|U|$ should be replaced by the number of labeled objects $|L|$. $c(a)$ is the test cost of attribute a, and $c(a) \geq 0$. λ is a parameter that can adjust the weight of test cost and $\lambda \leq 0$. If $\lambda = 0$, the significance of attribute a on B is based on conditional entropy as shown in definition 6. $c(a_1), c(a_2), \cdots, c(a_{|C|})$ and λ can be specified in real application by domain experts.

3.2   Attribute Reduction Algorithm

It has been proved that finding a minimal reduct of a decision table with exhaustive algorithm is NP-hard in rough set [28], and correspondingly, computing the minimal test cost of a reduct will have the same complexity. Actually, some heuristic algorithms have been proposed, and most of them are greedy.

In this paper, a heuristic algorithm (TCSPR) for attribute reduction of partially labeled data based on test cost sensitive is as Algorithm 1. In the algorithm, based on the objects with labeled, we first find the core of the attributes set. Then in each iterative step of the while loop, after the computation of significance of every attribute in the unselected attributes subset, choose the attribute with highest significance and add it to the reduct set, until the end condition holds. The significance is computed based on Eq. (3).

Algorithm 1. A heuristic attribute reduction algorithm for partially labeled data with test cost sensitive, called TCSPR

Input: $TPDS = (U = L \cup N, A = C \cup D, V, f, c), \lambda$
Output: An attributes subset B as a relative reduct
$U \leftarrow L$;
$B \leftarrow \emptyset$;
for all $a \in C$ **do**
    if $POS_{C-\{a\}}(D)! = POS_C(D)$ **then**
        $B \leftarrow B \bigcup \{a\}$;
    end if
end for
$tempA \leftarrow C - B$;
while $H(D|B)! = H(D|C)$
    for all $a \in tempA$ **do**
        compute $SGF(a, B, D, c(a), \lambda)$;
    end for
    select a' with maximal $SGF(a', B, D, c(a'), \lambda)$;
    $B \leftarrow B \bigcup \{a'\}$;
    $tempA \leftarrow tempA - \{a'\}$;
end while
return B;

3.3   Complexity Analysis

If the core is a reduct of the attributes set, then it is the minimal reduct. Let m be the number of condition attributes, l be the number of labeled objects, namely $m = |C|$, $l = |L|$, the computational complexity of finding the core is $O(ml)$, and this is the best case of finding a reduct. In the worst case, the reduct is the whole condition attributes, correspondingly the computational complexity is $O(m^2 l^2)$.

4   Experiments

In this section, some experiments are conducted on several data sets from UCI repository [29] with the following purposes: (1) to find out the impact of parameter λ on the reducts of partially labeled data, (2) to find out the impact of test cost on the reducts of partially labeled data, (3) to compare the classification accuracy of classifiers trained from partially labeled data.

4.1   Data Sets and Experiment Environment

According to the experimental requirements of attribute reduction, we adopt 4 data sets with task of classification, as shown in Table 1. The datasets are preprocessed as follows: (1) we delete the eleventh attribute of dataset mushroom because of missing attribute values, (2) the continuous attributes in the dataset wine and ionosphere are discreted by Weka using 3 bins. All the attributes are

Table 1. Summary of datasets

Name	#attributes	#objects	#classes
wine	13	178	3
zoo	16	101	7
mushroom	21	8124	2
ionosphere	34	351	2

identified by natural numbers for convenience. All the experiments are implemented in MATLAB on a PC with CPU 2.60 GHz and 4 GB memory.

4.2 Impact of Parameter λ on Reduct

Because there are no existent test costs of datasets in Table 1, we first assume some values of them. The test costs can be produced by many methods, and here we adopt normal distribution with the mean is 0, and the variance is 1. Then we scale the costs between 1 and 100. Let the test costs of all the attributes be the numbers in Table 2.

Table 2. Test costs of datasets

dataset	#attributes	test cost
wine	13	42 49 54 59 64 71 1 61 100 46 93 79 71
zoo	16	61 100 67 41 1 52 52 28 70 55 60 35 31 36 21 34
mushroom	21	48 70 1 54 45 17 32 45 100 86 16 91 52 38 51 36 37 65 63 63 51
ionosphere	34	84 29 69 39 47 39 24 27 51 1 78 54 81 67 74 8 24 43 61 96 48 100 74 97 62 36 8 63 85 48 40 47 21 41

With the test costs in Table 2, we let labeled ratio be 0.2, 0.4, 0.6, 0.8, and 1.0 respectively, the reducts produced by Algorithm 1 with different λ ($\lambda = 0, -0.5, -1, -2, -4$) are shown in Table 3. When $\lambda = 0$, we do not consider the test costs of attributes and the significance of attribute is based on conditional entropy in reality. Here, we assume the objects of different class in the labeled data are of the same proportion as the objects of different class in the whole dataset. When the labeled ratio is 1.0, that is the dataset and there are no unlabeled data.

In the wine and zoo datasets, when the labeled ratio is up to 0.4, the changes of reducts based on different λ ($\lambda = -0.5, -1, -2, -4$) are small. In the mushroom dataset, the core of attributes is {5} when the labeled ratio is less than or equal to 0.8, however the core is {1, 3, 5, 9, 13, 14} when the labeled ratio is bigger than 0.85, owing to the huge difference between the core, the reducts are very different. In the ionosphere dataset, it seems that the difference between reducts are mainly caused by labeled ratio and λ has tiny impact on the reducts.

Table 3. Reducts on different λ with the same cost

dataset	ratio	core	reduct				
			λ = 0	λ = −0.5	λ = −1	λ = −2	λ = −4
wine	0.2	∅	{2,11,13}	{2,7,11,13}	{2,7,11,13}	{1,2,4,7,10}	{1,2,4,7,10}
	0.4	{13}	{2,5,10,11,12,13}	{1,2,7,9,11,13}	{1,2,4,7,9,10,13}	{1,2,3,4,7,8,10,13}	{1,2,3,4,7,8,10,13}
	0.6	{3,13}	{1,3,9,11,13}	{1,3,7,9,11,13}	{1,2,3,7,8,10,13}	{1,2,3,7,8,10,13}	{1,2,3,7,8,10,13}
	0.8	{1,3,13}	{1,3,4,9,11,12,13}	{1,3,7,8,10,11,13}	{1,2,3,4,7,8,10,13}	{1,2,3,4,7,8,10,13}	{1,2,3,4,7,8,10,13}
	1.0	{1,3,13}	{1,3,4,9,11,12,13}	{1,2,3,4,7,8,10,13}	{1,2,3,4,7,8,10,13}	{1,2,3,4,7,8,10,13}	{1,2,3,4,7,8,10,13}
zoo	0.2	∅	{1,6,13}	{4,5,6,13}	{4,5,6,13}	{4,5,6,13,14}	{4,5,6,13,14,15,16}
	0.4	{6}	{1,6,8,13}	{4,5,6,8,13}	{4,5,6,8,13}	{4,5,6,8,13}	{4,5,6,8,13,16}
	0.6	{6,13}	{3,4,6,11,13}	{4,5,6,8,12,13}	{4,5,6,8,12,13}	{4,5,6,8,12,13}	{4,5,6,8,12,13,16}
	0.8	{6,13}	{3,4,6,8,13}	{4,6,8,12,13}	{4,5,6,8,12,13}	{4,5,6,8,12,13}	{4,5,6,8,12,13,15,16}
	1.0	{6,13}	{3,4,6,8,13}	{4,6,8,12,13}	{4,5,6,8,12,13}	{4,5,6,8,12,13}	{4,5,6,8,12,13,16}
mushroom	0.2	{5}	{1,5}	{3,5}	{3,5}	{3,5}	{3,5}
	0.4	{5}	{5,19}	{3,5,11,21}	{3,5,7,8,11}	{3,5,7,8,11}	{3,5,7,8,11}
	0.6	{5}	{5,19}	{3,4,5,11}	{3,5,7,8,11}	{3,5,7,8,11}	{3,5,7,8,11}
	0.8	{5}	{3,5,19}	{1,3,4,5,11}	{3,5,7,11,12}	{1,3,5,7,8,11}	{1,3,5,7,8,11}
	1.0	{1,3,5,9,13,14}	{1,3,4,5,9,13,14,21}	{1,3,5,7,9,13,14,21}	{1,3,5,7,9,11,13,14,21}	{1,3,5,7,9,11,13,14,17}	{1,3,5,7,9,11,13,14,17}
ionosphere	0.2	∅	{4,15,20,22,34}	{4,7,8,10,16,27,33}	{4,8,10,16,17,27,33}	{4,7,8,10,16,27,33}	{4,7,8,10,16,27,33}
	0.4	{5}	{1,4,5,14,25,28,34}	{5,6,8,10,12,16,25,32,33}	{5,8,10,16,17,18,25,27,32,33}	{5,8,10,16,17,18,25,27,32,33}	{5,8,10,16,17,18,25,26,27,32,33,34}
	0.6	{4,5,6,18,23,26,34}	{1,4,5,6,8,9,14,18,23,25,26,29,34}	{4,5,6,10,12,16,18,23,25,26,27,32,33,34}	{4,5,6,10,12,16,17,18,23,25,26,27,32,33,34}	{4,5,6,10,12,16,17,18,23,25,26,27,32,33,34}	{4,5,6,7,10,12,16,17,18,23,25,26,27,32,33,34}
	0.8	{4,5,6,8,18,22,23,26,32,34}	{1,4,5,6,8,9,18,22,34}	{4,5,6,7,8,10,12,14,16,18,22,23,25,26,27,32,33,34}	{4,5,6,7,8,10,12,14,16,18,22,23,25,26,27,32,33,34}	{4,5,6,7,8,10,12,14,16,18,22,23,25,26,27,32,33,34}	{4,5,6,7,8,10,12,14,16,18,22,23,25,26,27,32,33,34}
	1.0	{4,5,6,8,18,22,23,26,32,34}	{3,4,5,6,8,9,10,11,18,22,23,24,26,27,29,31,32,34}	{4,5,6,7,8,10,12,14,16,18,22,23,25,26,27,32,33,34}	{4,5,6,7,8,10,12,14,16,18,22,23,25,26,27,32,33,34}	{4,5,6,7,8,10,12,14,16,18,22,23,25,26,27,32,33,34}	{4,5,6,7,8,10,12,14,16,18,22,23,25,26,27,32,33,34}

From Table 3, we can find that as the ratio raises, the core of a dataset expands, and this may result in the change of reduct. The changes are tiny in some datasets, such as wine and zoo, but huge in mushroom and ionosphere. The reducts are very different when $\lambda = 0$ compared to the reducts based on test costs ($\lambda = -0.5, -1, -2, -4$). Meanwhile, the attributes with the smallest test cost, namely the attribute 7 in wine dataset, attribute 5 in zoo dataset, attribute 3 in mushroom dataset and attribute 10 in ionosphere dataset respectively, are almost in all the reducts with different ratio and λ, but do not appear in the cores and reducts when $\lambda = 0$. The results in this table also indicates that the impact of λ on the reducts of partially labeled data may be limited, and the reducts almost the same when $\lambda = -1$ and $\lambda = -2$ in some datasets, however there may be some fluctuation when $\lambda = -0.5$ and $\lambda = -4$ compared to $\lambda = -1$.

4.3   Impact of Test Cost on Reduct

In the experiments above, we find the impact of λ on the reducts of partially labeled data based on the same test cost. Here, we let $\lambda = -1$, and conduct some experiments to show the impact of test costs on the reducts of partially labeled data. We let labeled ratio be 0.2, 0.4, 0.6, 0.8 and 1.0 respectively, test costs be produced randomly and satisfy normal distribution, the reducts produced by Algorithm 1 with different test costs are shown in Table 4.

In wine and ionosphere datasets, the reducts expand when labeled ratios raise in general. In zoo, the numbers of attributes in reduct change a little, mainly be 5 or 6. But the numbers change a lot in mushroom, for example, the reduct is {1,5} when the ratio is 0.2 and test cost is [1 40 51 76 1 78 17 63 85 73 100 97 38 70 27 14 94 61 59 94 82], and the reduct is {5,19} when the ratio is 0.6 and test cost is [31 53 57 77 39 1 26 70 22 62 45 71 4 39 19 100 59 70 22 34 38].

Table 4. Reducts on different test cost with the same λ

dataset	ratio	test cost	reduct
wine	0.2	48 70 1 54 45 17 32 45 100 86 16 91 52 41 68 36 39 95 92 93 67 1 68 100 60 79	{3,7,11,13} {3,4,8,9,12}
	0.4	1 100 58 73 14 38 45 60 14 61 24 52 65 69 100 97 29 66 17 1 94 55 53 93 80 70	{1,3,5,9,11,13} {4,5,6,7,9,11,13}
	0.6	85 97 64 63 100 1 52 28 56 89 94 36 53 70 57 75 77 37 60 1 100 48 29 58 22 65	{1,3,6,8,10,12,13} {2,3,5,7,10,11,12,13}
	0.8	42 100 78 1 63 24 48 57 70 48 87 44 61 68 52 74 59 29 1 100 33 53 65 53 25 37	{1,3,4,5,6,10,12,13} {1,3,5,6,9,11,12,13}
	1.0	46 100 21 77 61 42 15 41 47 1 57 23 47 51 23 37 52 1 68 100 66 36 54 20 87 68	{1,3,7,8,10,11,13} {1,2,3,5,9,10,11,12,13}
zoo	0.2	36 37 23 100 11 25 17 1 35 32 95 33 5 97 84 34 36 61 68 78 65 81 80 42 49 54 59 64 71 1 61 100	{3,7,8,13} {1,10,13,14}
	0.4	72 1 12 55 62 32 67 85 71 17 37 34 47 39 35 100 51 57 72 4 24 78 74 23 100 60 67 68 44 1 58 10	{2,3,6,10,14} {4,6,8,13,14}
	0.6	100 37 85 51 75 57 44 50 77 83 77 43 28 64 67 1 48 100 42 31 21 1 34 55 22 53 67 32 59 61 46 96	{4,6,8,12,13,16} {4,6,9,12,13}
	0.8	31 65 38 43 68 83 63 87 38 66 47 42 1 76 100 70 90 51 13 23 25 82 35 52 10 44 1 28 72 100 64 52	{1,3,6,9,12,13} {3,4,6,9,11,13}
	1.0	99 90 1 19 46 45 25 50 43 100 52 34 58 50 77 63 3 58 33 26 20 63 39 1 100 33 49 52 8 75 49 46	{3,4,6,9,13} {3,4,6,8,13}
mushroom	0.2	27 71 59 50 55 16 49 86 53 52 35 50 56 60 43 46 95 1 100 58 73 1 40 51 76 1 78 17 83 85 73 100 97 38 70 27 14 94 61 59 94 82	{1,5,18} {1,5}
	0.4	39 80 100 75 87 80 58 71 48 84 40 42 47 1 96 72 48 94 28 63 59 56 56 14 44 39 67 84 85 14 48 1 5 45 100 17 58 37 85 6 46 65	{5,14,19} {5,11,19}
	0.6	31 53 57 77 39 1 26 70 22 62 45 71 4 39 19 100 59 70 22 34 38 99 56 87 1 54 39 16 81 74 66 23 100 76 56 66 57 10 56 39 34 29	{5,19} {4,5,7,11,17}
	0.8	33 1 66 56 44 44 27 67 42 29 74 40 32 38 26 20 100 81 52 17 26 60 87 29 1 25 74 76 78 62 70 52 89 28 78 42 56 80 94 34 100 83	{2,5,10,20,21} {3,4,5,19}
	1.0	38 33 32 29 42 43 73 100 59 33 65 48 1 78 53 46 61 51 4 35 35 31 100 21 33 26 11 42 39 95 40 15 98 86 41 1 34 43 56 40 62 60	{1,3,5,9,13,14,19,21} {1,3,4,5,9,11,13,14}
ionosphere	0.2	76 100 92 73 49 50 98 24 66 15 94 90 67 65 1 82 9 5 69 42 78 85 90 49 79 70 89 81 91 63 63 94 11 54 27 46 68 71 32 44 57 48 60 82 34 50 9 78 41 28 49 24 53 42 100 78 1 63 24 48 57 70 48 87 44 61 68 56	{8,10,15,18,24,33} {8,13,16,18,20,23,34}
	0.4	13 28 54 17 75 27 100 14 52 1 25 29 58 74 55 35 70 70 50 63 58 17 69 27 28 51 37 42 64 77 40 56 39 53 55 34 51 58 48 80 34 32 12 64 63 45 63 50 52 48 55 48 100 23 10 24 25 38 43 42 57 54 61 80 70 21 1 64	{1,4,5,8,10,22,24,25,34} {5,8,9,12,20,21,24,25,32,33}
	0.6	22 66 44 53 67 46 57 61 75 55 82 42 40 90 26 41 68 45 82 71 49 19 57 1 48 64 47 66 100 68 49 60 53 3 45 87 54 70 30 57 71 21 64 51 31 55 7 1 25 30 25 11 63 15 58 76 55 58 46 18 54 80 100 44 14 60 65 50	{1,4,5,6,10,12,18,22,23,24, 25,26,34} {1,4,5,6,8,12,13,14,18,20,21, 23,25,26,31,34}
	0.8	26 90 1 39 22 36 55 52 35 65 62 62 40 70 62 100 71 73 50 23 41 34 38 60 38 31 41 46 57 30 55 51 76 45 41 11 5 54 38 60 19 100 49 75 39 40 65 61 13 1 14 4 47 33 35 6 68 58 43 74 20 57 55 55 24 54 30 19	{1,3,4,5,6,8,9,18,22,23,24, ,26,29,32,34} {3,4,5,6,8,11,12,16,18,20,22, 23,24,26,27,32,33,34}
	1.0	39 81 46 58 39 56 1 41 35 43 58 55 98 51 57 32 53 36 62 78 69 48 31 89 100 64 43 84 50 53 78 36 25 69 46 39 48 61 59 23 55 40 25 9 51 28 37 18 1 40 35 64 55 60 19 15 7 48 100 53 19 55 11 4 53 19 44 68	{1,4,5,6,7,8,9,10,11,14,18, 22,23,26,29,32,33,34} {4,5,6,8,9,10,11,14,15,18, 21,22,23,26,27,29,32,33,34}

From Table 4, we find that the attributes in reducts vary a lot according to different test costs except the attributes in core, which indicates that test cost has great impact of reduct. Furthermore, the attribute with low cost probably be a member of reduct and this is consistent with the conclusion in Sect. 4.2.

4.4 Quality of Reducts

From partially labeled data based on reduct, one can train classifier and use it to predict the classification of new objects. Here, some experiments are conducted to show the prediction performance of the classifiers. First, the numbers of attributes based on different reduction algorithms are shown in Fig. 1, where Pawlak stands for reduction based on attribute dependency degree, and Entropy stands for reduction based on entropy. Obviously, TCSPR gets more attributes than other algorithms in most cases.

Then, based on the reduced partially labeled data from three different methods (Pawlak, Entropy and TCSPR), we use decision tree model and CART algorithm to train classifier. To avoid randomness, 10-fold cross-validation is adopted

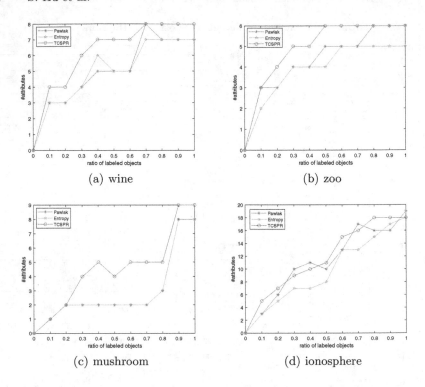

Fig. 1. Relationship between number of attributes and labeled ratio

and this is done 10 times. The relationships between classification accuracy of unlabeled data and ratio of labeled objects are shown in Fig. 2, where accuracy is the mean of the 100 experimental results, and "original" indicates the classification accuracy of the classifier trained by the whole dataset. Generally speaking, for all the three methods, there are some identical phenomenons in Fig. 2: in the wine and zoo datasets, the classification accuracy raises as the ratio of labeled data increases, and this accords with the common recognition; In the mushroom dataset, the classification accuracy is already near to 1 when the ratio is 0.1, and when the ratio increases, the accuracy decreases till the ratio is 0.8, then the accuracy goes up sharply when the ratio is 0.9; However, in the ionosphere dataset, the classification accuracy fluctuates according to the ratio. Figure 2 also shows that the accuracies of Pawlak and Entropy are closer, especially in the zoo and mushroom datasets, which indicates that the great impact of test cost on classification accuracy. In mushroom dataset, the classification accuracy of TCSPR is much superior than that of the other two methods, and in zoo dataset, the classification accuracy of TCSPR is higher than that of the other two methods when ratio is less than 0.6, however it approaches to other two in other settings. So, on the premise of not reducing the classification accuracy obviously, considering test cost for partially labeled data and finding the reduct with minimal test cost, which can be studied in future, are meaningful.

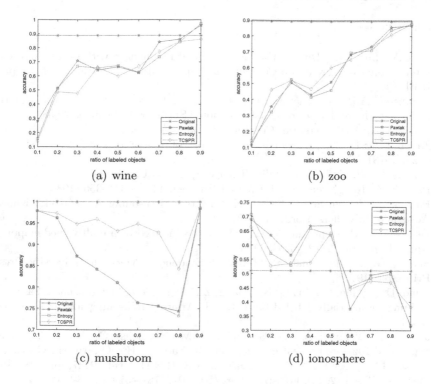

Fig. 2. Relationship between classification accuracy and labeled ratio

5   Conclusion

Based on rough set theory, this paper focuses on the attributes reduction of partially labeled data with test cost sensitive. Based on mutual information and the test cost of every condition attribute, we give the definition of attribute significance. Then a heuristic algorithm (TCSPR) for attribute reduction based on the significance is proposed. Experiments indicate the impact of labeled ratio and test cost on the reducts, and the effectiveness of our algorithm is verified too. In the future, more comparative experiments should be conducted to analyze the quality of the reducts, and further work can concentrate on incremental attribute reduction of partially labeled data with test cost sensitive.

Acknowledgements. The authors would like to thank the anonymous reviewers for their constructive comments that help improve the manuscript. This research was supported by the National Key R&D Program of China (213), National Natural Science Foundation of China (61673301), Major Project of Ministry of Public Security (20170004), and the Open Research Funds of State Key Laboratory for Novel Software Technology (KFKT2017B22).

References

1. Li, D.Y., Liu, C.Y., Du, Y., Han, X.: Artificial intelligence with uncertainty. Inn: International Conference on Computer and Information Technology, vol. 15, p. 2. IEEE (2008)
2. Dempster, A.P.: A Generalization of Bayesian Inference. Classic Works of the Dempster-Shafer Theory of Belief Functions. Springer, Heidelberg (2008)
3. Zadeh, L.A.: Fuzzy sets. Inf. Control **8**(3), 338–353 (1965)
4. Zadeh, L.A.: Fuzzy sets as a basis for a theory of possibility. Fuzzy Sets Syst. **1**(1), 3–28 (1978)
5. Dubois, D., Prade, H.: Possibility Theory. Springer, US (1988)
6. Pawlak, Z.: Rough sets. Int. J. Comput. Inf. Sci. **11**(5), 341–356 (1982)
7. Pawlak, Z.: Rough Sets: Theoretical Aspects of Reasoning About Data. Kluwer Academic Publishers, Dordrecht (1991)
8. Dempster, A.P.: Upper and lower probabilities induced by a multivalued mapping. Ann. Math. Stat. **38**(2), 325–339 (1967)
9. Shafer, G.: A Mathematical Theory of Evidence. Princeton University Press, Princeton (1976)
10. Miao, D.Q., Gao, C., Zhang, N., Zhang, Z.F.: Diverse reduct subspaces based co-training for partially labeled data. Int. J. Approx. Reasoning **52**(8), 1103–1117 (2011)
11. Jensen, R., Vluymans, S., Parthaláin, N.M., Cornelis, C., Saeys, Y.: Semi-supervised fuzzy-rough feature selection. In: Yao, Y., Hu, Q., Yu, H., Grzymala-Busse, J.W. (eds.) RSFDGrC 2015. LNCS (LNAI), vol. 9437, pp. 185–195. Springer, Cham (2015). https://doi.org/10.1007/978-3-319-25783-9_17
12. Zhang, W., Miao, D.Q., Gao, C., Li, F.: Rough set attribute reduction algorithm for partially labeled data. Comput. Sci. **44**(1), 25–31 (2017). (in Chinese)
13. Dai, J.H., Hu, Q.H., Zhang, J.H., Hu, H., Zheng, N.G.: Attribute selection for partially labeled categorical data by rough set approach. IEEE Trans. Cybern. **PP**(99), 1–12 (2017)
14. Ciucci, D.: Temporal dynamics in information tables. Fundamenta Informaticae **115**(1), 57–74 (2012)
15. Luo, C., Li, T.R., Chen, H.M., Fujita, H., Yi, Z.: Efficient updating of probabilistic approximations with incremental objects. Knowl.-Based Syst. **109**, 71–83 (2016)
16. Jing, Y.G., Li, T.R., Fujita, H., Yu, Z., Wang, B.: An incremental attribute reduction approach based on knowledge granularity with a multi-granulation view. Inf. Sci. **411**, 23–38 (2017)
17. Lang, G.M., Miao, D.Q., Yang, T., Cai, M.J.: Knowledge reduction of dynamic covering decision information systems when varying covering cardinalities. Inf. Sci. **346**(C), 236–260 (2016)
18. Turney, P.D.: Types of cost in inductive concept learning. In: 17th ICML Proceedings of the Cost-Sensitive Learning Workshop, California, pp. 1–7 (2000)
19. Yao, Y.Y., Wong, S.K.M.: A decision theoretic framework for approximating concepts. Int. J. Man-Mach. Stud. **37**, 793–809 (1992)
20. Yao, Y.Y., Zhao, Y.: Attribute reduction in decision-theoretic rough set models. Inf. Sci. **178**(17), 3356–3373 (2008)
21. Huang, J.J., Wang, J., Yao, Y.Y., Zhong, N.: Cost-sensitive three-way recommendations by learning pair-wise preferences. Int. J. Approx. Reasoning **86**(C), 28–40 (2017)

22. Li, H., Zhou, X., Zhao, J., Huang, B.: Cost-sensitive classification based on decision-theoretic rough set model. In: Li, T. (ed.) RSKT 2012. LNCS (LNAI), vol. 7414, pp. 379–388. Springer, Heidelberg (2012). https://doi.org/10.1007/978-3-642-31900-6_47

23. Yang, X.B., Qi, Y.S., Song, X.N., Yang, J.Y.: Test cost sensitive multigranulation rough set: model and minimal cost selection. Inf. Sci. **250**(11), 184–199 (2013)

24. Ju, H.J., Li, H.X., Yang, X.B., Zhou, X.Z., Hang, B.: Cost-sensitive rough set: a multi-granulation approach. Knowl.-Based Syst. **123**(1), 137–153 (2017)

25. Zhang, W.X., Wu, W.Z., Liang, J.Y.: Rough Sets Theory and Methods. Science Press, Beijing (2003). (in Chinese)

26. Miao, D.Q., Hu, G.R.: A heuristic algorithm for reduction of knowledge. J. Comput. Res. Dev. **36**(6), 681–684 (1999). (in Chinese)

27. Min, F., He, H.P., Qian, Y.H., Zhu, W.: Test-cost-sensitive attribute reduction. Inf. Sci. **181**(22), 4928–4942 (2011)

28. Wong, S.K.M., Ziarko, W.: On optimal decision rules in decision tables. Bull. Polish Acad. Sci. Math. **33**(11–12), 693–696 (1985)

29. http://archive.ics.uci.edu/ml/index.php

Logic on Similarity Based Rough Sets

Tamás Mihálydeák$^{(\boxtimes)}$

Department of Computer Science, Faculty of Informatics, University of Debrecen,
Egyetem tér 1, Debrecen 4010, Hungary
mihalydeak.tamas@inf.unideb.hu

Abstract. Pawlak's indiscernibility relation (which is an equivalence
relation) represents a limit of our knowledge embedded in an information system. Covering approximation spaces generated by tolerance relations treat objects which are similar to a given object in the same way.
Similarity based rough sets rely on the similarity of objects in general
and preserve the benefit of pairwise disjoint system of base sets. By
using correlation clustering not only a pairwise disjoint system of base
sets can be generated but representative members of base sets can be
defined. These representative members have an important logical usage.
The author shows that there is a logical system relying on similarity base
sets in which the truth values of first-order formulas can be counted in
an effective simple way.

Keywords: Rough set theory · Correlation clustering · Partial logic
Multivalued logic

1   Introduction

Pawlak's original theory of rough sets (see in e.g. [13,14,16]), covering systems
relying on tolerance relations [17], general covering systems [15,20], decision theoretic rough set theory [19], general partial approximation spaces [5] are different
systems of rough set theory. There is a very important common property: all
systems rely on given background knowledge and we cannot say more about an
arbitrary set (representing a 'new' property) or about its members then its lower
and upper approximations make possible. The base sets represent background
knowledge at least some regard:

- in Pawlak's system they represent the limit of background knowledge by indiscernibility relation;
- in covering systems relying on tolerance relation objects which are similar to
  a given one are treated in the same way;
- in general covering systems a base set corresponds to a property informally;
- general partial approximation spaces give up covering requirement in order
  to represent partiality appearing in information systems.

© Springer Nature Switzerland AG 2018
H. S. Nguyen et al. (Eds.): IJCRS 2018, LNAI 11103, pp. 270–283, 2018.
https://doi.org/10.1007/978-3-319-99368-3_21

The system of similarity based rough sets (see in [12]) focuses on similarity in general and shows a possibility to define partial and pairwise disjoint system of base sets. Similarity relations generate new systems of properties: those objects belong to the same base set which are similar to each other (not only to a given object). In the present paper a partial first–order logic is created in order to give a possibility to use logical tools therefore the consequences of background knowledge can be investigated.

After giving a general picture of approximation spaces the influences of background knowledge on membership relations are surveyed. Then the most important features of similarity based sets are given in order to show a possibility of creating base sets relying on similarity relations in general with preserving pairwise disjoint property of base sets. Finally a partial first-order logic relying on similarity base sets is presented.

2   Theoretical Background

The notion of general approximation spaces can represent the bases of the most important kinds of rough set theory:

Definition 1. *The ordered 5-tuple $\langle U, \mathfrak{B}, \mathfrak{D}_\mathfrak{B}, \mathsf{l}, \mathsf{u} \rangle$ is a general partial approximation space with a Pawlakian approximation pair if*

1. *U is a nonempty set;*
2. *$\mathfrak{B} \subseteq 2^U$, $\mathfrak{B} \neq \emptyset$ and if $B \in \mathfrak{B}$, then $B \neq \emptyset$;*
3. *$\mathfrak{D}_\mathfrak{B}$ is an extension of $\mathfrak{B}$, and it is given by the following inductive definition:*
    (a) $\mathfrak{B} \subseteq \mathfrak{D}_\mathfrak{B}$;
    (b) $\emptyset \in \mathfrak{D}_\mathfrak{B}$;
    (c) if $D_1, D_2 \in \mathfrak{D}_\mathfrak{B}$, then $D_1 \cup D_2 \in \mathfrak{D}_\mathfrak{B}$.
4. *the functions l, u form a Pawlakian approximation pair $\langle \mathsf{l}, \mathsf{u} \rangle$, i.e.*
    (a) $\mathsf{l}(S) = \bigcup \mathcal{C}^\mathsf{l}(S)$, where $\mathcal{C}^\mathsf{l}(S) = \{B \mid B \in \mathfrak{B} \text{ and } B \subseteq S\}$;
    (b) $\mathsf{u}(S) = \bigcup \mathcal{C}^\mathsf{u}(S)$, where $\mathcal{C}^\mathsf{u}(S) = \{B \mid B \in \mathfrak{B} \text{ and } B \cap S \neq \emptyset\}$.

Informally, the set U is the universe of approximation; $\mathfrak{B}$ is a nonempty set of base sets; $\mathfrak{D}_\mathfrak{B}$ (i.e. the set of definable sets) contains not only the base sets, but those which can be used to approximate any subset of U; the functions l, u (and b) determine the lower and upper approximation of any set.

The characteristic difference between the kinds of approximation spaces (with a Pawlakian approximation pair) appears in the base sets (members of $\mathfrak{B}$). Only four main kinds of approximation spaces are mentioned here: the original Pawlakian; covering generated by a tolerance relation; general covering; general (partial):

1. From the theoretical point of view an original Pawlakian approximation space (see in [13,16]) can be characterized by an ordered pair $\langle U, \mathcal{R} \rangle$ where U is a nonempty set of objects and $\mathcal{R}$ is an equivalence relation on U. $\mathcal{R}$ is called an indiscernibility relation and it determines a partition on U. The equivalence classes of generated partition are base sets and so they are the members of $\mathfrak{B}$.

2. Pawlakian approximation spaces (relying on an indiscernibility relation) have been generalized using tolerance relations (instead of equivalence ones), which are similarity relations and so they are symmetric and reflexive. Covering-based approximation spaces generated by tolerance relations (see e.g. in [17]) generalize Pawlakian approximation spaces in two points:

   (a) $\mathcal{R}$ is a tolerance relation;
   (b) if $[x] = \{y \mid y \in U, x\mathcal{R}y\}$, then $\mathfrak{B} = \{[x] \mid x \in U\}$.

3. General covering approximation spaces (see e.g. in [20]) do not rely on tolerance relations, any nonempty subset of U can be a base set. There is only one requirement: $\bigcup \mathfrak{B} = U$.

4. In the case of general (partial) approximation spaces (see e.g. in [5]) the last requirement is given up: any family $\mathfrak{B}$ of nonempty subsets of U can be a set of base sets.

3   Influences of Embedded Knowledge on Membership Relations

What is the importance of set of base sets from the theoretical point of view? It represents a sort of limit of our knowledge embedded in an information system. In some situation it makes our judgment of the membership relation uncertain – making the set vague – because a decision about a given object affects the decision about all other objects which are in a same base set.

The main source of uncertainty is in our background knowledge. Let S be a subset of U, and $x, y \in U$. What is the consequence of embedded and limited background knowledge? What can be said about y with respect to x?

1. In an original Pawlakian space relying on an equivalence relation $\mathcal{R}$:
   - if $x \in \mathsf{l}(S)$ (i.e. x is a member of S necessarily), then $y \in S$ for all $y, x\mathcal{R}y$;
   - if $x \in \mathsf{u}(S) \setminus \mathsf{l}(S)$ (i.e. x is a member of S possibly), then y may be a member of S for all $y, x\mathcal{R}y$ (it means that there are y_1, y_2 such that $x\mathcal{R}y_1, y_1 \in S$, and $x\mathcal{R}y_2, y_2 \notin S$);
   - if $x \in \mathsf{l}(\bar{S})(= U \setminus \mathsf{u}(S))$ (i.e. x is not a member of S necessarily), then $y \notin S$ for all $y, x\mathcal{R}y$.

2. In a covering space generated by a tolerance relation $\mathcal{R}$:
   - if $x \in \mathsf{l}(S)$ (i.e. x is a member of S necessarily), then $y \in S$ for all $y, y \in [x']$ where $x' \in [x]$ and $[x'] \in \mathcal{C}^{\mathsf{l}}(S)$;
   - if $x \in \bigcup(\mathcal{C}^{\mathsf{u}}(S) \setminus \mathcal{C}^{\mathsf{l}}(S))$ (i.e. x is a member of S possibly), then there is an x' and a base set $[x']$ such that $x \in [x']$, $[x'] \cap S \neq \emptyset$, $[x'] \nsubseteq S$ and y may be a member of S for all $y \in [x']$;
   - if $x \in \mathsf{l}(\bar{S})(= U \setminus \mathsf{u}(S))$ (i.e. x is not a member of S necessarily), then $y \notin S$ for all $y, x\mathcal{R}y$.

3. In a general covering space:
   - if $x \in \mathsf{l}(S)$ (i.e. x is a member of S necessarily), then there is a base set B, such that $x \in B$ and $B \in \mathcal{C}^{\mathsf{l}}(S)$) therefore $y \in S$ for all $y \in B$;

- if $x \in \bigcup(\mathcal{C}^u(S) \setminus \mathcal{C}^l(S))$ (i.e. x is a member of S possibly), then there is a base set B such that $x \in B$, $B \cap S \neq \emptyset$ and $B \not\subseteq S$ therefore y may be a member of the set S for all $y \in B$;
- if $x \in \mathsf{I}(\bar{S})(= U \setminus \mathsf{u}(S))$ (i.e. x is not a member of S necessarily), then there is a base set B such that $B \cap S = \emptyset$ therefore $y \notin S$ for all $y \in B$.

4. In a general partial space:
   - if $x \in \mathsf{I}(S)$ (i.e. x is a member of S necessarily), then there is a base set B, such that $x \in B$ and $B \in \mathcal{C}^l(S))$ therefore $y \in S$ for all $y \in B$;
   - if $x \in \bigcup(\mathcal{C}^u(S) \setminus \mathcal{C}^l(S))$ (i.e. x is a member of S possibly), then there is a base set B such that $x \in B$, $B \cap S \neq \emptyset$ and $B \not\subseteq S$ therefore y may be a member of the set S for all $y \in B$;
   - if $x \in \mathsf{I}(\bar{S})$ (i.e. x is not a member of S necessarily), then there is a base set B such that $B \cap S = \emptyset$ therefore $y \notin S$ for all $y \in B$;
   - otherwise we do not know anything about x (i.e. there is no any base set B such that $x \in B$), therefore we cannot say anything about y with respect to x.

Boundary regions play a crucial role in the representation of uncertainty coming from given background knowledge. In [4] the authors showed that theoretically different boundary regions can be introduced into a general partial approximation space $\langle U, \mathfrak{B}, \mathfrak{D}_{\mathfrak{B}}, \mathsf{I}, \mathsf{u} \rangle$:

1. $\mathsf{b}_1(S) = \mathsf{u}(S) \setminus \mathsf{I}(S)$;
2. $\mathsf{b}_2(S) = \bigcup(\mathcal{C}^u(S) \setminus \mathcal{C}^l(S))$;
3. $\mathsf{b}_3(S) = \bigcup \mathcal{C}^b(S)$, where $\mathcal{C}^b(S) = \{B \mid B \in \mathfrak{B}, \ B \cap S \neq \emptyset, \text{ and } B \not\subseteq S\}$.

In original Pawlakian spaces there is no difference between different types of boundary regions, i.e. if $\langle U, \mathfrak{B}, \mathfrak{D}_{\mathfrak{B}}, \mathsf{I}, \mathsf{u} \rangle$ is an original Pawlakian space characterized by an ordered pair $\langle U, \mathcal{R} \rangle$, then $\mathsf{b}_1(S) = \mathsf{b}_2(S) = \mathsf{b}_3(S)$ for all $S \subseteq U$. In general case the boundary regions defined according to the first point are not definable sets necessarily, therefore this definition cannot be used in general approximations spaces where we want to rely on only definable sets. If there are only finite number of base sets (i.e. $\mathfrak{B}$ is finite), then the sets $\mathsf{b}_2(S), \mathsf{b}_3(S)$ are definable for all $S \subseteq U$. Some important connections between different types of boundary regions were showed in [4,6]:

- $\mathsf{b}_1(S) \subseteq \mathsf{b}_2(S) \subseteq \mathsf{u}(S)$;
- $\mathsf{b}_1(S) = \mathsf{b}_2(S)$ if and only if $\mathsf{b}_2(S) \cap \mathsf{I}(S) = \emptyset$;
- if $\mathfrak{B}$ is one-layered (i.e. the base sets are pairwise disjoint), then there is no difference between different types of boundary regions, i.e.
  - $\mathsf{b}_1(S) = \mathsf{b}_2(S) = \mathsf{b}_3(S)$;
  - $\mathsf{b}_1(S)$ is definable;
  - $\mathsf{b}_i(S) \cap \mathsf{I}(S) = \emptyset$, where $i = 1, 2, 3$;
  - $\mathsf{u}(S) = \mathsf{I}(S) \cup \mathsf{b}_i(S)$, where $i = 1, 2, 3$.

Notice that only lower and upper approximations (and so only background and embedded knowledge represented by base sets) are used, and in a finite one-layered case there is no real difference between different types of boundary regions.

The next step is to make clear the 'nature', the usage and the influences of background (and embedded) knowledge.

1. In the original Pawlakian case the limit of our knowledge appears explicitly: base sets consist of indiscernible objects, there is no way to distinguish them from each other.
2. In covering structures generated by tolerance relations a base set contains objects which are similar to a given object, and therefore we treat them in the same way. Being similar to a given object is a property, but it is a very special (not a general) one, it is generated by the tolerance relation.
3. In general covering spaces base sets can be considered as the representations of real properties, and we suppose that all object have at least one (known, represented) property. Objects with the same property (members of a base set) are handled in the same way. (The system of base sets cannot be generated by tolerance relations in some cases.)
4. General partial spaces are similar to general covering ones, but it is not supposed that all objects have at least one property represented by a base set. In practical cases information systems are not total, there is no relevant information about an object: it may be in our database but some information is missing, and so it does not have any property represented by a base set.

Some problems appear in different cases. In practical applications indiscernibility relation (as an equivalence relation) may be too strong. In the case of huge number of objects if we have a reflexive and symmetric relation, then it may be difficult to decide whether it is transitive. Covering spaces generated by tolerance relations give possibilities to use only reflexive and symmetric relations, but too many base sets appear, (each object generate a base set). These base sets are not about similarity (in general), but only about similarity to given objects (to their generators). In general covering and partial spaces there is no room for similarity, these spaces rely on only common properties of objects. A pairwise disjoint system of base sets generated from a covering system (relying on a tolerance relation or a family of properties) or a general partial system is not a real solution: it is difficult to give any meaning represented by received base sets and too many small base sets appear, therefore the system may become very close to classical set theory.

The following question appears: is there any way to use similarity in general and to preserve the benefit of pairwise disjoint system? The system of similarity based rough sets gives a possible solution. The system was presented at IJCRS2017 [12].

4   Similarity Based Rough Sets

Suppose that there is a universe U, and a (not necessarily total) tolerance relation $\mathcal{R}$, which represents similarity among objects belonging to U. Of course the base of the similarity can be the properties of our object. If U is finite (as in practical cases) and we have an arbitrary fixed ordering of members of U, then a (partial) tolerance relation can be defined by a matrix M (see in [8,17]):

- $m_{ij} = 1$ whenever objects u_i and u_j are similar,
- $m_{ij} = -1$ whenever objects u_i and u_j are dissimilar,
- $m_{ij} = 0$ otherwise.

A relation is partial if there exist two elements (u_i, u_j) such that $m_{ij} = 0$. It means that if we have an arbitrary relation $R \subseteq U \times U$ we have two sets of pairs. Let R_{true} be the set of those pairs of elements for which the R holds, and R_{false} be the one for which R does not hold. If R is partial then $R_{true} \cup R_{false} \subseteq U \times U$. If R is total then $R_{true} \cup R_{false} = U \times U$.

The task given at the end of previous section is to find an $R \subseteq U \times U$ equivalence relation *closest* to the tolerance relation. Correlation clustering is a clustering technique based on a tolerance relation (see in [1–3]) and its result is a partition. A partition of a set U is a function $p : U \to \mathbb{N}$. Objects $u_i, u_j \in U$ are in the same cluster at partitioning p, if $p(u_i) = p(u_j)$.

The cost function counts the negative cases i.e. it gives the number of cases whenever two dissimilar objects are in the same cluster, or two similar objects are in different clusters. The cost function of a partition p and a relation $\mathcal{R}$ with matrix M is

$$f(p, M) = \frac{1}{2} \sum_{i<j} (m_{ij} + abs(m_{ij})) - \sum_{i<j} \delta_{p(u_i)p(u_j)} m_{ij},$$

where δ is the Kronecker delta symbol. For a fixed relation the partition with the minimal cost function value is called *optimal*. Solving a correlation clustering problem is equivalent to minimizing its cost function. The partition given this way, generates an equivalence relation. This relation can be considered as the closest to the tolerance relation. There are many different techniques for correlation clustering, here these methods are not analyzed because they depend on U.

There is a natural way to determine the representative members of a clusters: We call a member representative if it is similar to most of the members and different from the least of the members in its cluster.

From the approximation point of view the most important point is that the result of correlation clustering can give a system of base sets (with representative members): we use only non-singleton clusters as base sets. Singleton clusters are not able to represent any information connected with given similarity relation.

By applying a correlation clustering process on U connected with the similarity relation $\mathcal{R}$ the similarity based general approximation space $SBAP$ can be defined:

Definition 2. $SBAP = \langle U, \mathcal{R}, \mathfrak{B}, \mathsf{V} \rangle$ *is a similarity based general approximation space, where*

- U is a nonempty set;
- $\mathcal{R}$ is a tolerance relation on the set U;
- $\mathfrak{B}(= \{B_i \mid i = 1, \ldots, n\})$, where B_i is a non-singleton cluster received by the correlation clustering process on U, relying on the tolerance relation $\mathcal{R}$;
- $\mathsf{V} = \langle u_1, u_2, \ldots, u_n \rangle$ is the representative vector of $SBAP$, i.e. $u_i \in B_i$ is a representative object of B_i determined by the correlation clustering process.

5   Logic on Similarity Based Rough Sets (LSBRS)

Similarity and its logical properties are investigated in rough set theory extensively (see e.g. Ewa Orłowska's and Dimiter Vakerelov's papers [7,18]). In an information system different similarity relations can be defined and the mentioned papers introduce different (modal) logical systems and deal with different (logical) properties of relations.

LSRBS is *not* a logical system *of* similarity relations appearing in an information system, it is not about the logical connections of different relations. It is a logic *on* similarity based rough sets, i.e. in its semantics the system of similarity based rough sets (given by a universe, a tolerance relation and a process of correlation clustering) plays crucial role which is similar to the role of classical set theory in the semantics of classical first–order logic. LSBRS is a partial three–valued logic.[1]

5.1   Language of Logic on Similarity Based Rough Sets

The language of LSBRS is not independent from the given similarity based general approximation space which characterizes the 'word' relying on background knowledge.

Definition 3. Let $SBAP = \langle U, \mathcal{R}, \mathfrak{B}, \langle u_1, u_2, \ldots, u_n \rangle \rangle$ be a similarity based general approximation space. $L = \langle LC, Var, Con, Term, Rep, Form \rangle$ is a first order language relying on $SBAP$ with the set Rep of representatives , if

1. $LC = \{\neg, \wedge, \vee, \supset, \equiv, \forall, \exists, (,)\}$, LC is the set of logical constants.
2. $Var = \{x_i \mid i = 0, 1, 2, \ldots \}$, Var is the denumerable infinite set of individual variables.
3. $Con = \mathcal{N} \cup \bigcup_{n=1}^{\infty} \mathcal{P}(n))$, where $\mathcal{N}$ is a set of name parameters, and $\mathcal{P}(n)$ is the set of n–argument predicate parameters. Con is the denumerable set of non–logical constants.
4. The sets LC, Var, $\mathcal{N}$, $\mathcal{P}(n)$ $(n = 1, 2, \ldots)$ are pairwise disjoint.
5. $Term = Var \cup \mathcal{N}$, $Term$ is the set of terms.
6. $Rep = \{a_1, a_2, \ldots, a_n\} \subseteq \mathcal{N}$.
7. The set $Form$ (the set of formulas) is given by the following inductive definition:
   (a) If $P \in \mathcal{P}(n)(n = 1, 2, \ldots)$ and $t_1, t_2, \ldots, t_n \in Term$, then $P(t_1, t_2, \ldots, t_n) \in Form$;

---

[1] Different versions of partial first–order logic relying on rough sets are e.g. in [9–11].

(b) If $A, B \in Form$, then $\neg A, (A \circ B), \in Form$ where $\circ \in \{\wedge, \vee, \supset, \equiv\}$;
(c) If $A \in Form$, $x \in Var$, then $\forall x A, \exists x A \in Form$;

Remark 1. Later the members of the set *Rep* will be the names of representative objects.

5.2   Semantics of LSBRS

Definition 4. Let $SBAP = \langle U, \mathcal{R}, \mathfrak{B}, \langle u_1, \ldots, u_n \rangle \rangle$ be a similarity based general approximation space, and L be a language of first order logic relying on $SBAP$ with the set Rep of representatives. $\langle SBAP, \varrho \rangle$ is an interpretation relying on the similarity based general approximation space $SBAP$ if

– ϱ is an interpretation function such that
  1. $Dom(\varrho) = Con$;
  2. If $a_i \in Rep$, i.e. a_i is a representative, then $\varrho(a_i) = u_i$.
  3. If $b \in \mathcal{N} \setminus Rep$ (i.e. b is a non-representative name parameter), then $\varrho(b) \in U$.
  4. If $P \in \mathcal{P}(1)$ i.e. P is a one-argument predicate parameter, then
     $\varrho(P) = \langle \varrho(P)_1, \ldots, \varrho(P)_n \rangle$, where $\varrho(P)_1, \ldots, \varrho(P)_n \in \{-1, 0, 1\}$;
  5. If $P \in \mathcal{P}(m), (m > 1)$ i.e. P is an n-argument predicate parameter, then

$$\varrho(P) = \begin{pmatrix} \varrho(P)_{11}, & \cdots, & \varrho(P)_{1n} \\ \vdots & \ddots & \vdots \\ \varrho(P)_{m1}, & \cdots, & \varrho(P)_{mn} \end{pmatrix}$$

where $\varrho(P)_{ij}, \in \{-1, 0, 1\}$ $(1 \leq i \leq n, 1 \leq j \leq m)$.

The points 4, 5 show that the semantic value of a predicate parameter may and must be characterized only by lower, upper approximations and the boundary region. Lower approximation corresponds to positive region, upper approximation corresponds to the union of positive and boundary region.

Let $\langle SBAP, \varrho \rangle$ be an interpretation relying on the similarity based general approximation space $SBAP$. If P is a one argument predicate parameter, then

– the set $\varrho^+(P) = \cup\{B_i \mid \varrho(P)_i = 1\}$ is the positive region of P;
– the set $\varrho^\star(P) = \cup\{B_i \mid \varrho(P)_i = 0\}$ is the boundary region of P;
– the set $\varrho^-(P) = \cup\{B_i \mid \varrho(P)_i = -1\}$ is the negative region of P;
– there is no information about objects which do not belong to the set $\varrho^+(P) \cup \varrho^\star(P) \cup \varrho^-(P)$.

Similar positive, negative and boundary region can be constructed in the case of m-argument predicate parameter P, $(m > 1)$:

– $\varrho^+(P) = \cup\{B_i \mid \varrho(P)_{1i} = 1\} \times \cup\{B_i \mid \varrho(P)_{2i} = 1\} \times \cdots \times \cup\{B_i \mid \varrho(P)_{mi} = 1\}$
  is the positive region of P (where $1 \geq i \geq n$);
– $\varrho^\star(P) = \cup\{B_i \mid \varrho(P)_{1i} = 0\} \times \cup\{B_i \mid \varrho(P)_{2i} = 0\} \times \cdots \times \cup\{B_i \mid \varrho(P)_{mi} = 0\}$
  is the boundary region of P (where $1 \geq i \geq n$);

- $\varrho^-(P) = \cup\{B_i \mid \varrho(P)_{1i} = -1\} \times \cdots \times \cup\{B_i \mid \varrho(P)_{mi} = -1\}$ is the negative region of P (where $1 \geq i \geq n$);
- if an m-tuple $\langle u_1, \ldots, u_m \rangle$ does not belong to the set $\varrho^+(P) \cup \varrho^\star(P) \cup \varrho^-(P)$, then there is no information about at least one object of the m-tuple $\langle u_1, \ldots, u_m \rangle$.

Definition 5. Function v is an assignment relying on the interpretation $\langle SBAP, \varrho \rangle$ if $v : Var \to U$.

Definition 6. Let v be an assignment relying on the interpretation $\langle SBAP, \varrho \rangle$, $x \in Var$ and $u \in U$. $v[x : u]$ is a modified assignment of v, if $v[x : u]$ is an assignment, $v[x : u](y) = v(y)$ if $x \neq y$, and $v[x : u](x) = u$.

5.3   Semantic Rules of LSBRS

In the semantics of LSBRS the semantic value of an expression depends on a given interpretation $Ip = \langle SBAP, \varrho \rangle$, a given assignment v (relying on Ip). For the sake of simplicity in order to treat semantic paritiality (i.e. some formulas have no semantic value) a null entity is used. We use number 0 for falsity, number 1 for truth, number 1/2 for uncertainty and number 2 for null entity. The semantic value of an expression A with respect to $Ip = \langle SBAP, \varrho \rangle$, and the assignment v is denoted by $[\![A]\!]_v^{Ip}$ or $[\![A]\!]^{\langle SBAP, \varrho \rangle}_v$. For the sake of simplicity the superscripts are omitted.

Semantic rules are the followings:

1. If $x \in Var$, then $[\![x]\!]_v = v(x)$.
2. If $c \in \mathcal{N}$ i.e. c is a name parameter, then $[\![a]\!]_v = \varrho(a)$
3. If $P \in \mathcal{P}(1)$, i.e. P is a one-argument predicate parameter and $t \in Term$,
   then $[\![P(t)]\!]_v = \begin{cases} 1 & \text{if } [\![t]\!]_v \in \varrho^+(P) \\ 1/2 & \text{if } [\![t]\!]_v \in \varrho^\star(P) \\ 0 & \text{if } [\![t]\!]_v \in \varrho^-(P) \\ 2 & \text{otherwise} \end{cases}$
4. If $P \in \mathcal{P}(m)$, i.e. P is an m-argument predicate parameter and $t_1, t_2, \ldots, t_m \in Term$, then
   $$[\![P(t_1, \ldots, t_m)]\!]_v = \begin{cases} 1 & \text{if } \langle [\![t_1]\!]_v, \ldots, [\![t_m]\!]_v \rangle \in \varrho^+(P) \\ 1/2 & \text{if } \langle [\![t_1]\!]_v, \ldots, [\![t_m]\!]_v \rangle \in \varrho^\star(P) \\ 0 & \text{if } \langle [\![t_1]\!]_v, \ldots, [\![t_m]\!]_v \rangle \in \varrho^-(P) \\ 2 & \text{otherwise} \end{cases}$$
5. If $A \in Form$, then
   $[\![\neg A]\!]_v = \begin{cases} 2 & \text{if } [\![A]\!]_v = 2 \\ 1 - [\![A]\!]_v & \text{otherwise} \end{cases}$
6. If $A, B \in Form$, then
   $[\![(A \wedge B)]\!]_v = \begin{cases} 2 & \text{if } [\![A]\!]_v = 2, \text{ or } [\![B]\!]_v = 2; \\ \min\{[\![A]\!]_v, [\![B]\!]_v\} & \text{otherwise} \end{cases}$
   $[\![(A \vee B)]\!]_v = \begin{cases} 2 & \text{if } [\![A]\!]_v = 2, \text{ or } [\![B]\!]_v = 2; \\ \max\{[\![A]\!]_v, [\![B]\!]_v\} & \text{otherwise} \end{cases}$
   $[\![(A \supset B)]\!]_v = \begin{cases} 2 & \text{if } [\![A]\!]_v = 2, \text{ or } [\![B]\!]_v = 2; \\ \max\{[\![\neg A]\!]_v, [\![B]\!]_v\} & \text{otherwise} \end{cases}$

7. If $A \in Form, x \in Var$ and $\mathcal{V}(A) = \{u \mid u \in U$ such that $[\![A]\!]_{v[x:u]} \neq 2\}$, then

$$[\![\forall x A]\!]_v = \begin{cases} 2 & \text{if } \mathcal{V}(A) = \emptyset, \\ \min\{[\![A]\!]_{v[x:u]} \mid u \in \mathcal{V}(A)\} & \text{otherwise} \end{cases}$$

$$[\![\exists x A]\!]_v = \begin{cases} 2 & \text{if } \mathcal{V}(A) = \emptyset, \\ \max\{[\![A]\!]_{v[x:u]} \mid u \in \mathcal{V}(A)\} & \text{otherwise} \end{cases}$$

5.4 Central Logical Notions

Definition 7. Let $SBAP = \langle U, \mathcal{R}, \langle u_1, u_2, \ldots, u_n \rangle \rangle$ be a similarity based general approximation space, $L = \langle LC, Var, Con, Term, Rep, Form \rangle$ be a first order language relying on $SBAP$ with the set Rep of representatives and $\Gamma \subseteq Form, A, B \in Form$.

- The formula A is a strong consequence of the members of set Γ (in notation $\Gamma \vDash_s A$) over the similarity based general approximation space $SBAP$ if all members of Γ are true, then A is true with respect to all interpretations and assignments relying on $SBAP$.
- The formula A is a weak consequence of the members of set Γ (in notation $\Gamma \vDash_w A$) over the similarity based general approximation space $SBAP$ if all members of Γ are not false, then A is not false with respect to all interpretations and assignments relying on $SBAP$.
- The formula A is logically equivalent with the formula B (in notation $A \Leftrightarrow B$) over the similarity based general approximation space $SBAP$ if $[\![A]\!]_v = [\![B]\!]_v$ for all interpretations and assignments relying on $SBAP$.
- The formula A is degenerate with respect to an interpretations and assignment v relying on $SBAP$ if $[\![A]\!]_v = 2$

5.5 Theorems About LSBRS

Next three theorems show the sources of partiality. Their proofs are the trivial consequences of semantic rules.

Theorem 1. $[\![P(t_1, t_2, \ldots, t_n)]\!]_v = 2$ if and only if there is a t_i such that $[\![t_i]\!]_v \notin \bigcup \mathfrak{B}$.

Theorem 2. Let A be a formula, b be a non-representative name parameter and x be a variable.

- If b has an occurrence in A and $[\![b]\!]_v \notin \bigcup \mathfrak{B}$, then $[\![A]\!]_v = 2$.
- If x is a free variable of A and $[\![x]\!]_v \notin \bigcup \mathfrak{B}$, then $[\![A]\!]_v = 2$.

Theorem 3. If $[\![A]\!]_v = 2$, then there is a non-representative name parameter b in A such that $[\![b]\!]_v \notin \bigcup \mathfrak{B}$ or a free variable x of A such that $[\![x]\!]_v \notin \bigcup \mathfrak{B}$

Theorem 4. Let P be an n-argument predicate parameter, and $t_1, \ldots, t_n \in Term$. If $[\![t_i]\!]_v \in B_i, (i = 1, \ldots, n)$ (therefore $[\![t_i]\!]_v$ and u_i are in the same base set i.e. u_i is a representative object of $[\![t_i]\!]_v$), then $[\![P(t_1, \ldots, t_n)]\!]_v = [\![P(a_1, \ldots, a_n)]\!]_v$

Proof. It is a trivial consequence of interpretation of representatives.

Let $A \in Form$ be a formula, and $t_1, t_2 \in Term$ be terms. A new notation $[A]_{t_2}^{t_1}$ has to be introduced:

- Suppose that $t_2 \in Var$, and term t_1 is substitutable for variable t_2 in the formula A. Then the formula $[A]_{t_2}^{t_1}$ is the result of substitution of term t_1 for all free occurrences of variable t_2.
- Suppose that $t_1, t_2 \in \mathcal{N}$ (i.e. t_1, t_2 are name parameters). Then the formula $[A]_{t_2}^{t_1}$ is the result of substitution of name parameter t_1 for all occurrences of name parameter t_2.

Corollary 1. If u is a member of the base set represented by the object u_i (therefore $\varrho(a_i) = u_i = \llbracket a_i \rrbracket$), then $\llbracket A \rrbracket_{v[x:u]} = \llbracket [A]_x^{a_i} \rrbracket_v$

The next theorem is fundamental because it shows that in determining the truth value of a formula we have to take into consideration only the values of predicates on representatives. The proof is a direct consequence of Theorem 4 and Corollary 1.

Theorem 5. Let $A \in Form$, $x_1, \ldots, x_k \in Var$ such that there is at least one free occurrence of x_i in A $(i = 1, \ldots, k)$, and $b_1, \ldots, b_l \in \mathcal{N}$ such that there is at last one occurrence of b_j in A $(j = 1, \ldots, l)$.

- If there is an i or a j such that $\llbracket x_i \rrbracket_v \notin \bigcup \mathfrak{B}$ or $\llbracket b_j \rrbracket_v \notin \bigcup \mathfrak{B}$ then $\llbracket A \rrbracket_v = 2$.
- If $\llbracket x_i \rrbracket_v = \llbracket a_i^\star \rrbracket$ $(i = 1, \ldots, k)$, and $\llbracket b_j \rrbracket_v = \llbracket a_j^{\star\star} \rrbracket$ $(j = 1, \ldots, l)$, where $a_i^\star, a_j^{\star\star} \in Rep$ $(i = 1, \ldots, k)$, $(j = 1, \ldots, l)$, then

$$\llbracket A \rrbracket_v = \llbracket [A]_{x_1, \ldots, x_k, b1, \ldots, bl}^{a_1^\star, \ldots, a_k^\star, a_1^{\star\star}, \ldots, a_l^{\star\star}} \rrbracket$$

The next theorem shows that in quantified cases one has to take into consideration only the values of predicates on representatives.

Theorem 6. Let $SBAP$ be a similarity based general approximation space, L be a first-order language relying on $SBAP$, $A \in Form$, $x \in Var$, and $Rep = \{a_1, a_2, \ldots, a_n\}$. Then

$$\forall x A \Leftrightarrow [A]_x^{a_1} \wedge [A]_x^{a_2} \wedge \cdots \wedge [A]_x^{a_n}$$

$$\exists x A \Leftrightarrow [A]_x^{a_1} \vee [A]_x^{a_2} \vee \cdots \vee [A]_x^{a_n}$$

Proof. If $\llbracket \forall x A \rrbracket_v = 2$ or $\llbracket \exists x A \rrbracket_v = 2$, then $\mathcal{V}(A) = \emptyset$, i.e. $\llbracket A \rrbracket_{v[x:u]} = 2$ for all $u \in U$. Therefore according to Corollary 1 $\llbracket [A]_x^{a_i} \rrbracket_v = 2$ for all $i = 1, 2, \ldots, n$ and so $\llbracket [A]_x^{a_1} \wedge [A]_x^{a_2} \wedge \cdots \wedge [A]_x^{a_n} \rrbracket = 2$, and $\llbracket [A]_x^{a_1} \vee [A]_x^{a_2} \vee \cdots \vee [A]_x^{a_n} \rrbracket = 2$.

If $\llbracket [A]_x^{a_1} \wedge [A]_x^{a_2} \wedge \cdots \wedge [A]_x^{a_n} \rrbracket = 2$, then there is an i such that $\llbracket [A]_x^{a_i} \rrbracket = 2$. It means that there is at least one term t in A which is different from x and a_i and the source of semantic value gap, i.e. $\llbracket t \rrbracket_v \notin \bigcup \mathfrak{B}$. Therefore $\llbracket A \rrbracket_{v[x:u]} = 2$ for all $u \in U$, i.e. $\mathcal{V}(A) = \emptyset$, $\llbracket \forall x A \rrbracket_v = 2$ and $\llbracket \exists x A \rrbracket_v = 2$.

If $\llbracket \forall x A \rrbracket_v \neq 2$ or $\llbracket \exists x A \rrbracket_v \neq 2$, then

$$[\![\forall x A]\!]_v = \min\{[\![A]\!]_{v[x:u]} \mid u \in \mathcal{V}(A)\} = \min\{[\![A]_x^{a_1}]\!]_v, [\![A]_x^{a_2}]\!]_v, \ldots, [\![A]_x^{a_n}]\!]_v\} =$$
$$= [\![A]_x^{a_1} \wedge [A]_x^{a_2} \wedge \cdots \wedge [A]_x^{a_n}]\!]_v$$
$$[\![\exists x A]\!]_v = \max\{[\![A]\!]_{v[x:u]} \mid u \in \mathcal{V}(A)\} = \max\{[\![A]_x^{a_1}]\!]_v, [\![A]_x^{a_2}]\!]_v, \ldots, [\![A]_x^{a_n}]\!]_v\} =$$
$$= [\![A]_x^{a_1} \vee [A]_x^{a_2} \vee \cdots \vee [A]_x^{a_n}]\!]_v \qquad \square$$

The next theorem shows that in rough set theory we have to be careful when we use some generally accepted classical logical laws. For example the contraposition law of implication does not hold, the modus ponens holds but the modus tollens does not. It is enough to give the statements only for one-argument predicate parameters.

Theorem 7. Let $P, Q \in Con$ be two one-argument predicate parameters. Then

- $P(x) \supset Q(x) \not\equiv \neg Q(x) \supset \neg P(x)$
- Quantified modus ponens holds: $\{\forall x(P(x) \supset Q(x)), P(b)\} \vDash Q(b)$.
- Quantified modus tollens does not hold: $\{\forall x(P(x) \supset Q(x)), \neg Q(b)\} \nvDash \neg P(b)$.

Proof. It is enough to prove, that there is an interpretation and assignment where
$$[\![P(x) \supset Q(x)]\!]_v \neq [\![\neg Q(x) \supset \neg P(x)]\!]_v$$

Let $SBAP$ be a similarity based approximation space such that it has only four base sets, and $\varrho(P) = \langle 1, 0, 0, -1 \rangle, \varrho(Q) = \langle 1, 1, -1, -1 \rangle$. Then
$$[\![P(x) \supset Q(x)]\!]_{v[x:u]} = [\![\neg Q(x) \supset \neg P(x)]\!]_{v[x:u]} \text{ if } u \in B_1 \cup B_2 \cup B_4$$
$$[\![P(x) \supset Q(x)]\!]_{v[x:u]} \neq [\![\neg Q(x) \supset \neg P(x)]\!]_{v[x:u]} \text{ if } u \in B_3$$

Remark 2. $[\![\forall x(P(x) \supset Q(x))]\!]_v = 1$ means only that the positive region of P is a subset of positive region of Q, but it does not mean that the negative region of Q is a subset of negative region of P and so $\forall x(P(x) \supset Q(x)) \not\equiv \forall x(\neg Q(x) \supset \neg P(x))$.

6   Conclusion and Future Work

The main result of the paper is to give a partial first–order three-valued logical system *on* similarity based general approximation spaces. Important advantages of the logical system are the followings:

- its semantics relies on similarity in general (and not on the similarity to a given object);
- its semantics preserves the benefit of the pairwise disjoint system of base sets;
- the semantic values of all formulas with or without quantifiers can be determined by taking into consideration only the values of representatives (i.e. representative objects);
- its semantic treats uncertainty on a precise way;
- logical tools (as for example consequence relation, logical equivalence) can be used in order to make explicit the consequences of embedded knowledge.

The next step is to use the introduced logical system in practice to solve some problems in data mining connected with rough set theory.

Acknowledgements. This work was supported by the construction EFOP–3.6.3–VEKOP–16–2017–00002. The project has been supported by the European Union, co-financed by the European Social Fund.

References

1. Aszalós, L., Mihálydeák, T.: Rough clustering generated by correlation clustering. In: Ciucci, D., Inuiguchi, M., Yao, Y., Ślęzak, D., Wang, G. (eds.) Rough Sets, Fuzzy Sets, Data Mining, and Granular Computing, pp. 315–324. Springer, Heidelberg (2013)
2. Bansal, N., Blum, A., Chawla, S.: Correlation clustering. Mach. Learn. **56**(13), 89–113 (2004). https://doi.org/10.1023/B:MACH.0000033116.57574.95
3. Becker, H.: A survey of correlation clustering. In: Advanced Topics in Computational Learning Theory, pp. 1–10 (2005)
4. Ciucci, D., Mihálydeák, T., Csajbók, Z.E.: On definability and approximations in partial approximation spaces. In: Miao, D., Pedrycz, W., Ślęzak, D., Peters, G., Hu, Q., Wang, R. (eds.) RSKT 2014. LNCS (LNAI), vol. 8818, pp. 15–26. Springer, Cham (2014). https://doi.org/10.1007/978-3-319-11740-9_2
5. Csajbók, Z., Mihálydeák, T.: A general set theoretic approximation framework. In: Greco, S., Bouchon-Meunier, B., Coletti, G., Fedrizzi, M., Matarazzo, B., Yager, R.R. (eds.) Advances on Computational Intelligence, pp. 604–612. Springer, Heidelberg (2012). https://doi.org/10.1007/978-3-642-31709-5_61
6. Csajbók, Z.E., Mihálydeák, T.: From vagueness to rough sets in partial approximation spaces. In: Kryszkiewicz, M., Cornelis, C., Ciucci, D., Medina-Moreno, J., Motoda, H., Raś, Z.W. (eds.) Rough Sets and Intelligent Systems Paradigms, pp. 42–52. Springer, Cham (2014). https://doi.org/10.1007/978-3-319-08729-0_4
7. Golińska-Pilarek, J., Orłowska, E.: Logics of similarity and their dual tableaux a survey. In: Della Riccia, G., Dubois, D., Kruse, R., Lenz, H.J. (eds.) Preferences and Similarities, pp. 129–159. Springer, Vienna (2008). https://doi.org/10.1007/978-3-211-85432-7_5
8. Mani, A.: Choice inclusive general rough semantics. Inf. Sci. **181**(6), 1097–1115 (2011)
9. Mihálydeák, T.: Partial first-order logic with approximative functors based on properties. In: Li, T., Nguyen, H.S., Wang, G., Grzymala-Busse, J., Janicki, R., Hassanien, A.E., Yu, H. (eds.) Rough Sets and Knowledge Technology, pp. 514–523. Springer, Heidelberg (2012). https://doi.org/10.1007/978-3-642-31900-6_63
10. Mihálydeák, T.: Aristotle?s Syllogisms in Logical Semantics Relying on Optimistic, Average and Pessimistic Membership Functions. In: Cornelis, C., Kryszkiewicz, M., Śle?zak, D., Ruiz, E.M., Bello, R., Shang, L. (eds.) RSCTC 2014. LNCS (LNAI), vol. 8536, pp. 59–70. Springer, Cham (2014). https://doi.org/10.1007/978-3-319-08644-6_6
11. Mihálydeák, T.: First-order logic based on set approximation: a partial three-valued approach. In: 2014 IEEE 44th International Symposium on Multiple-Valued Logic, pp. 132–137, May 2014. https://doi.org/10.1109/ISMVL.2014.31
12. Nagy, D., Mihálydeák, T., Aszalós, L.: Similarity based rough sets. In: Polkowski, L., et al. (eds.) Rough Sets, pp. 94–107. Springer, Cham (2017). https://doi.org/10.1007/978-3-319-60840-2_7

13. Pawlak, Z.: Rough sets. Int. J. Parallel Programm. **11**(5), 341–356 (1982)
14. Pawlak, Z., Skowron, A.: Rough sets and Boolean reasoning. Inf. Sci. **177**(1), 41–73 (2007)
15. Pawlak, Z., Skowron, A.: Rudiments of rough sets. Inf. Sci. **177**(1), 3–27 (2007)
16. Pawlak, Z., et al.: Rough Sets: Theoretical Aspects of Reasoning About Data. System Theory, Knowledge Engineering and Problem Solving, vol. 9. Kluwer Academic Publishers, Dordrecht (1991)
17. Skowron, A., Stepaniuk, J.: Tolerance approximation spaces. Fundamenta Informaticae **27**(2), 245–253 (1996)
18. Vakarelov, Dimiter: A modal characterization of indiscernibility and similarity relations in Pawlak's information systems. In: Ślęzak, D., et al. (eds.) RSFDGrC 2005. LNCS (LNAI), vol. 3641, pp. 12–22. Springer, Heidelberg (2005). https://doi.org/10.1007/11548669_2
19. Yao, J., Yao, Y., Ziarko, W.: Probabilistic rough sets: approximations, decision-makings, and applications. Int. J. Approx. Reason. **49**(2), 253–254 (2008)
20. Yao, Y., Yao, B.: Covering based rough set approximations. Inf. Sci. **200**, 91–107 (2012). https://doi.org/10.1016/j.ins.2012.02.065. http://www.sciencedirect.com/science/article/pii/S0020025512001934

Attribute Reduction Algorithms for Relation Systems on Two Universal Sets

Zheng Hua, Qianchen Li, and Guilong Liu[✉]

School of Information Science, Beijing Language and Culture University,
Beijing 100083, China
liuguilong@blcu.edu.cn

Abstract. A relation system on two universal sets is a natural extension of a relation system on a universal set. This paper studies attribute reduction algorithms for relation systems on two universal sets. Based on two new discernibility matrices, we propose two reduction algorithms for relation systems and relation decision systems on two universal sets. As a corollary, we derive respectively the attribute reduction algorithms for relation systems and relation decision systems on one universal set.

Keywords: Attribute reduction · Discernibility matrix
Relation system · Relation decision system

1 Introduction

Attribute reduction is a quite useful technique for preprocessing data. The idea of attribute reduction is selecting a set of attributes which retain the same information for classification purposes as the entire set of attributes. Lots of researchers [1,3,4,8,9] have plunged into the research of attribute reduction and provided varieties of algorithms to obtain reduction set quickly and accurately. Pawlak [10,11] firstly studied attribute reduction for information systems. Skowron and Rauszer [12,13] are the first to propose discernibility matrix based attribute reduction algorithms for information systems. However, their algorithms were designed for dealing with complete and symbolic data sets. We know that lots of data sets are incomplete. In order to explore a better means of dealing with incomplete data sets, many kinds of attribute reductions were presented [15–18]. Jia et al. [2] summarized existing 22 definitions of attribute reductions and compared these definitions through experiments. We [5] proposed an algorithm for general relation decision systems based on a discernibility matrix. Stepaniuk [14] defined the concept of the lth lower approximation reduction for decision tables. We [6,7] considered such a type of reduction and gave the corresponding algorithms based on discernibility matrices.

© Springer Nature Switzerland AG 2018
H. S. Nguyen et al. (Eds.): IJCRS 2018, LNAI 11103, pp. 284–293, 2018.
https://doi.org/10.1007/978-3-319-99368-3_22

Until now, all attribute reduction has focused on one universal set, however, there are lots of the possible two or more different universal sets in the real world. Naturally, we need to consider attribute reduction problems on two universal sets. As for the reduction strategy, we use the discernibility matrix reduction method, different discernibility matrices correspond to different types of reductions, there is no doubt that how to construct discernibility matrix is a key step. In this paper, we will respectively construct a discernibility matrices for a relation system and a relation decision system on two universal sets and give the corresponding attribute reduction algorithms.

The remainder of the paper is organized as follows. In Sect. 2, we briefly retrospect some basic notions and notations of relations and relation decision systems on two universal sets. Section 3 proposes an attribute reduction algorithm for relation systems on two universal sets. In Sect. 4, an attribute reduction algorithm is proposed for relation decision systems on two universal sets. In Sect. 5, as a special case of our proposed algorithms, we give reduction algorithms for a relation system on one universal set. Finally, Sect. 6 concludes the paper.

2 Preliminaries

In this section, we will define some basic knowledge about the notions of relations and relation decision systems on two universal sets. Let $U = \{x_1, x_2, \cdots, x_n\}$ and $V = \{y_1, y_2, \cdots, y_m\}$ be two finite universal sets. Suppose that R is a binary relation from U to V, recall that the left R-relative set of an element y in V is defined as

$$l_R(y) = \{x | x \in U, xRy\}.$$

Similarity, the right R-relative set of an element x in U is defined as

$$r_R(x) = \{y | y \in V, xRy\}.$$

Definition 2.1. Let U and V be two finite universal sets and $A = \{a_1, a_2, ..., a_s\}$ be a family of binary relations from U to V, then (U, V, A) is called a relation system based on two universal sets (a relation system, for short). If $A = C \cup D$, and $C \cap D = \emptyset$, then $(U, V, C \cup D)$ is called a relation decision system based on two universal sets (a relation decision system, for short), where C is called the condition attribute set, and D is called the decision attribute set. For any subset $\emptyset \neq B \subseteq C$, we associate a relation $R_B = \bigcap_{a \in B} a$. The consistent part of the relation decision system $(U, V, C \cup D)$ is defined as $G_{CD} = \{x | r_{R_C}(x) \subseteq r_{R_D}(x), x \in U\}$.

Definition 2.2. Let (U, V, A) be a relation system and Y be an arbitrary subset $Y \subseteq V$, then the lower and upper approximations of Y on two universal sets respected to A are defined respectively as

$$\underline{R_A}(Y) = \{x | x \in U, r_{R_A}(x) \subseteq Y\} \text{ and } \overline{R_A}(Y) = \{x | x \in U, r_{R_A}(x) \cap Y \neq \emptyset\}.$$

Definition 2.3. Let (U, V, A) be a relation system, $B \subseteq A$ and $B \neq \emptyset$. If B satisfies the following two conditions:

(1) $R_B = R_A$.
(2) $R_{B'} \neq R_A$ for any $B' \subset B$.

Then B is called the reduction of (U, V, A).

Definition 2.4. Let $(U, V, C \cup D)$ be a relation decision system, $B \subseteq C$ and $B \neq \emptyset$. If B satisfies the following conditions:

(1) $G_{BD} = G_{CD}$.
(2) $G_{B'D} \neq G_{CD}$ for $\forall B' \subset B$.

Then B is called the reduction of $(U, V, C \cup D)$.

Note that, if $G_{CD} = U$, then $(U, V, C \cup D)$ is called consistent, otherwise it is called inconsistent. Especially, if $G_{CD} = \emptyset$, then $RSDTU$ is called totally inconsistent. In this situation, each singleton set $a(a \in C)$ is a reduction of C. Hence, from now on, we always assume $G_{CD} \neq \emptyset$.

3   An Attribute Reduction Algorithm for Relation Systems

In this section, we propose an attribute reduction algorithm for a relation system on two universal sets. We define the indiscernibility matrix as follows.

Definition 3.1. Let (U, V, A) be a relation system, we define the discernibility matrix $M = (m_{ij})_{n \times m}$ via $m_{ij} = \{a \in A | (x_i, y_j) \notin a\}$.

We will give the reduction algorithm by means of the mathematical proofs.

Theorem 3.1. Let (U, V, A) be a relation system with $\emptyset \neq B \subseteq A$. Then the following conditions are equivalent.

(1) $R_A = R_B$.
(2) If $m_{ij} \neq \emptyset$, then $B \cap m_{ij} \neq \emptyset$.

Proof. (1) $\Rightarrow$ (2): Suppose that $m_{ij} \neq \emptyset$ and $m_{ij} \cap B = \emptyset$, by the definition of the discernibility matrix, we have $(x_i, y_j) \in R_B$. By condition (1), $(x_i, y_j) \in R_A$, so $(x_i, y_j) \in a$ for each $a \in A$. This is in contradiction with $m_{ij} \neq \emptyset$.

(2) $\Rightarrow$ (1): Since $B \subseteq A$, we have $R_A \subseteq R_B$. Now we need to show $R_B \subseteq R_A$.

Suppose that $(x_i, y_j) \notin R_A$, then $\exists a \in A$ satisfies $(x_i, y_j) \notin a$. That means $a \in m_{ij}$. By condition (2), $B \cap m_{ij} \neq \emptyset$, Let $b \in B \cap m_{ij}$, then $(x_i, y_j) \notin b$ and $(x_i, y_j) \notin R_B$. Hence, $R_B \subseteq R_A$ and $R_B = R_A$.   □

Corollary 3.1. Let (U, V, A) be a relation system and $\emptyset \neq B \subseteq A$. Then B is a reduction of A if and only if it is a minimal subset satisfying $m_{ij} \cap B \neq \emptyset$ for any $m_{ij} \neq \emptyset$.

Using Corollary 3.1, we now give a reduction algorithm for a relation system.

Algorithm 1. An attribute reduction algorithm for a relation system

---

Input: A relation system (U, V, A)
Output: All attribute reduction set of (U, V, A)

1 **for** $i = 1$ *to* $i \leq n$ **do**
2 $\quad$ **for** $j = 1$ *to* $j \leq m$ **do**
3 $\quad\quad$ $m_{ij} = \emptyset$;
4 $\quad\quad$ **for** *each* $a \in A$ **do**
5 $\quad\quad\quad$ **if** $(x_i, y_j) \notin a$ **then**
6 $\quad\quad\quad\quad$ $m_{ij} = m_{ij} \cup a$;

7 Transform the discernibility function f from its CNF $f = \prod(\sum m_{ij})$ into a DNF $f = \sum_{t=1}^{s}(\prod B_t), (B_t \subseteq A)$;
8 **return** $reduct(A) = \{B_1, B_2, \cdots, B_s\}$;

Table 1. A relation system

a_1	y_1	y_2	y_3	y_4	y_5	a_2	y_1	y_2	y_3	y_4	y_5	a_3	y_1	y_2	y_3	y_4	y_5	a_4	y_1	y_2	y_3	y_4	y_5
x_1	1	0	0	0	1	x_1	1	1	0	0	0	x_1	1	0	0	0	0	x_1	1	1	0	1	1
x_2	0	1	0	1	0	x_2	1	0	0	0	1	x_2	0	0	1	0	0	x_2	0	1	1	1	1
x_3	1	0	1	1	0	x_3	0	0	0	1	1	x_3	0	0	0	1	0	x_3	1	1	0	1	0
x_4	1	1	1	0	0	x_4	1	1	0	0	0	x_4	1	0	0	1	0	x_4	1	0	1	0	0

Example 3.1. Let $U = \{x_1, x_2, x_3, x_4\}$, $V = \{y_1, y_2, y_3, y_4, y_5\}$ and $A = \{a_1, a_2, a_3, a_4\}$. The relation system (U, V, A) is given by the following table (See Table 1).

(1) Compute the 4×5 discernibility matrix $M = (m_{ij})_{4 \times 5}$ as follows

$$M = \begin{pmatrix} \emptyset & \{a_1, a_3\} & A & \{a_1, a_2, a_3\} & \{a_2, a_3\} \\ \{a_1, a_3, a_4\} & \{a_2, a_3\} & \{a_1, a_2\} & \{a_2, a_3\} & \{a_1, a_3\} \\ \{a_2, a_3\} & \{a_1, a_2, a_3\} & \{a_2, a_3, a_4\} & \emptyset & \{a_1, a_3, a_4\} \\ \emptyset & \{a_3, a_4\} & \{a_2, a_3\} & \{a_1, a_2, a_4\} & A \end{pmatrix}.$$

(2) Transform the discernibility function $f = (a_1 \vee a_2) \wedge (a_1 \vee a_3) \wedge (a_2 \vee a_3) \wedge (a_3 \vee a_4)$ from its CNF into the DNF $f = (a_1 \wedge a_3) \vee (a_2 \wedge a_3) \vee (a_1 \wedge a_2 \wedge a_4)$.
(3) $\{a_1, a_3\}$, $\{a_2, a_3\}$ and $\{a_1, a_2, a_4\}$ are all attribute reduction sets of A.

4 An Attribute Reduction Algorithm for Relation Decision Systems

In this section, we give an attribute reduction algorithm for a relation decision system $(U, V, C \cup D)$. Similar to the previous section, we define the discernibility matrix $M = (m_{ij})_{s \times m}$ as follows.

$$m_{ij} = \begin{cases} \{a | a \in C, (x_i, y_j) \notin a\}, & x_i \in G_{CD}, (x_i, y_j) \notin R_D \\ \emptyset, & \text{otherwise} \end{cases}.$$

where $s = |U_{CD}|$ denotes the cardinality of U_{CD}.

Lemma 4.1. Let $(U, V, C \cup D)$ be a relation decision system, if $x_i \in G_{CD}$ and $(x_i, y_j) \notin R_D$, then $m_{ij} \neq \emptyset$.

Proof. Suppose that $m_{ij} = \emptyset$, then we have $(x_i, y_j) \in a$ for each $a \in C$. That means $(x_i, y_j) \in R_C$. Because of $x_i \in G_{CD}, r_{R_C}(x_i) \subseteq r_{R_D}(x_i)$. So, $y_j \in r_{R_D}(x_i)$. This contradicts $(x_i, y_j) \notin R_D$. $\square$

Theorem 4.1. Let $(U, V, C \cup D)$ be a relation decision system and $\emptyset \neq B \subseteq C$. Then the following conditions are equivalent.

(1) $G_{CD} = G_{BD}$.
(2) If $m_{ij} \neq \emptyset$, then $B \cap m_{ij} \neq \emptyset$.

Proof. (1) $\Rightarrow$ (2): Suppose that $m_{ij} \neq \emptyset$ and $m_{ij} \cap B = \emptyset$. By the definition of the discernibility matrix, we have $(x_i, y_j) \in R_B$, $(x_i, y_j) \notin R_D$ and $x_i \in G_{CD}$. By condition (1), $G_{CD} = G_{BD}$, so $x_i \in G_{BD}$. That means $r_{R_B}(x_i) \subseteq r_{R_D}(x_i)$ and $(x_i, y_j) \in R_D$. This is in contradiction with $(x_i, y_j) \notin R_D$.

(2) $\Rightarrow$ (1): Since $B \subseteq C$, we have $R_C \subseteq R_B$, by definition of G_{CD}, we have $G_{BD} \subseteq G_{CD}$. We now show that $G_{CD} \subseteq G_{BD}$.

Suppose that $x_i \in G_{CD}$, we show $r_{R_B}(x_i) \subseteq r_{R_D}(x_i)$. In fact, if $x_j \notin r_{R_D}(x_i)$, then $(x_i, y_j) \notin R_D$. By Lemma 4.1, $m_{ij} \neq \emptyset$. By condition (2), $B \cap m_{ij} \neq \emptyset$. Let $b \in B \cap m_{ij}$, then $(x_i, y_j) \notin b$ and $(x_i, y_j) \notin R_B$. Hence, $r_{R_B}(x_i) \subseteq r_{R_D}(x_i)$. In other words, $x_i \in G_{BD}$ and $G_{CD} \subseteq G_{BD}$. $\square$

Corollary 4.1. Let $(U, V, C \cup D)$ be a relation decision system and $\emptyset \neq B \subseteq C$. Then B is a reduction of C if and only if it is a minimal subset satisfying $m_{ij} \cap B \neq \emptyset$ for any $m_{ij} \neq \emptyset$.

Example 4.1. Let $U = \{x_1, x_2, x_3, x_4, x_5\}$, $V = \{y_1, y_2, y_3, y_4, y_5, y_6\}$, $C = \{a_1, a_2, a_3, a_4, a_5\}$ and $D = \{d\}$. The relation decision system $(U, V, C \cup D)$ is given by following table (See Table 2). For instance, $(x_1, y_1) \notin a_1$ and $(x_1, y_2) \in a_1$.

According to the Algorithm 2,

(1) Compute the G_{CD} of $(U, V, C \cup D)$, by direct computation, $G_{CD} = \{x_1, x_3, x_4\}$.
(2) Compute the 3×6 discernibility matrix $M = (m_{ij})_{3 \times 6}$ as follows

$$\begin{pmatrix} \{a_1, a_3, a_5\} & \{a_2, a_3, a_5\} & \{a_1, a_2\} & \emptyset & \emptyset & \{a_1, a_2, a_3, a_5\} \\ \emptyset & \{a_3, a_4, a_5\} & \{a_2, a_4, a_5\} & C & \{a_1, a_3, a_4, a_5\} & \emptyset \\ \{a_1, a_2\} & \emptyset & C & \{a_3, a_5\} & \{a_2, a_3, a_5\} & \{a_1, a_4\} \end{pmatrix}.$$

Algorithm 2. Attribute reduction algorithm for a relation decision system

Input: A relation decision system $(U, V, C \cup D)$
Output: All attribute reduction sets of $(U, V, C \cup D)$
1 $G_{CD} = \emptyset$;
2 **for** *each* $x \in U$ **do**
3 $\quad$ **if** $r_{R_C}(x) \subseteq r_{R_D}(x)$ **then**
4 $\quad\quad$ $G_{CD} = G_{CD} \cup x$;

5 **for** $i = 1$ *to* $i \leq n$ **do**
6 $\quad$ **for** $j = 1$ *to* $j \leq m$ **do**
7 $\quad\quad$ $m_{ij} = \emptyset$;
8 $\quad\quad$ **for** *each* $a \in C$ **do**
9 $\quad\quad\quad$ **if** $(x_i, y_j) \notin a$ **then**
10 $\quad\quad\quad\quad$ $m_{ij} = m_{ij} \cup a$;

11 Transform the discernibility function f from its CNF $f = \prod(\sum m_{ij})$ into a
$\quad$ DNF $f = \sum_{t-1}^{s}(\prod B_t), (B_t \subseteq C)$;
12 **return** $reduct(A) = \{B_1, B_2, \cdots, B_s\}$;

---

(3) Transform the discernibility function $f = (a_1 \vee a_2) \wedge (a_1 \vee a_4) \wedge (a_3 \vee a_5) \wedge (a_2 \vee a_4 \vee a_5)$ from its CNF into the DNF $f = (a_1 \wedge a_5) \vee (a_1 \wedge a_2 \wedge a_3) \vee (a_1 \wedge a_3 \wedge a_4) \vee (a_2 \wedge a_3 \wedge a_4) \vee (a_2 \wedge a_4 \wedge a_5)$.
(4) All reduction sets are $\{a_1, a_5\}$, $\{a_1, a_2, a_3\}$, $\{a_1, a_3, a_4\}$, $\{a_2, a_3, a_4\}$ and $\{a_2, a_4, a_5\}$.

5   An Application to Relation Systems on a Universal Set

Since a relation system on one universal set is a special case of a relation system on two universal sets, we can obtain respectively two reduction algorithms for a relation system and a relation decision system on one universal set.

Definition 5.1. Let (U, A) be a relation system and $\emptyset \neq B \subseteq A$, set B is called the attribute reduction of A if B satisfies the following conditions:

(1) $R_A = R_B$;
(2) For any $\emptyset \neq B' \subset B$, $R_A \neq R_{B'}$.

If $U = V$, then (U, V, A) becomes (U, A). The following example illustrates our algorithm.

Example 5.1. Consider the following incomplete information system (U, A) (See Table 3), where $U = \{x_1, x_2, \cdots, x_5\}$ and $A = \{a_1, a_2, a_3, a_4, a_5, a_6, a_7\}$.

Where $*$ denotes missing attribute values (a null or a unknown value). Each $a_k \in A$ can be seen as a relation from U to U via

$$a_k = \{(x_i, x_j) | a_k(x_i) = a(x_j) \text{ or } a_k(x_i) = * \text{ or } a_k(x_j) = *\}.$$

Table 2. A relation decison system

a_1	y_1	y_2	y_3	y_4	y_5	y_6	a_2	y_1	y_2	y_3	y_4	y_5	y_6	a_3	y_1	y_2	y_3	y_4	y_5	y_6
x_1	0	1	0	1	1	0	x_1	1	0	0	1	1	0	x_1	0	0	1	1	1	0
x_2	1	1	1	1	0	1	x_2	0	1	1	1	1	0	x_2	0	1	1	1	0	0
x_3	0	1	0	1	1	0	x_3	0	1	0	1	0	1	x_3	1	1	0	0	0	1
x_4	1	1	1	0	0	1	x_4	1	1	0	0	1	1	x_4	1	0	1	0	0	1
x_5	1	1	0	1	0	1	x_5	1	0	1	0	1	1	x_5	1	1	0	0	1	1

a_4	y_1	y_2	y_3	y_4	y_5	y_6	a_5	y_1	y_2	y_3	y_4	y_5	y_6	d	y_1	y_2	y_3	y_4	y_5	y_6
x_1	1	1	1	1	1	1	x_1	0	0	1	1	1	0	x_1	0	0	0	1	1	0
x_2	0	1	1	1	0	0	x_2	1	1	1	1	1	1	x_2	0	0	1	0	1	0
x_3	1	1	0	1	1	0	x_3	1	1	0	0	0	1	x_3	0	1	0	0	0	1
x_4	1	0	0	0	0	1	x_4	1	0	0	0	0	1	x_4	1	0	0	0	0	1
x_5	1	0	1	0	1	1	x_5	1	1	1	1	1	1	x_5	0	1	0	0	0	1

Table 3. An incomplete information system

U	a_1	a_2	a_3	a_4	a_5	a_6	a_7
x_1	0	0	1	1	1	1	0
x_2	0	0	0	1	0	1	*
x_3	1	1	0	*	0	0	1
x_4	1	1	1	0	0	1	0
x_5	*	0	0	0	0	0	*

For example, $(x_1, x_5) \in a_1$ and $(x_5, x_3) \in a_1$, while $(x_1, x_3) \notin a_1$. Thus (U, A) is a relation system.

According to the Algorithm 1, the lower triangular part of discernibility matrix is as follows:

$$
\begin{pmatrix}
\emptyset \\
\{a_3, a_5\} & \emptyset \\
\{a_1, a_2, a_3, a_5, a_6, a_7\} & \{a_1, a_2, a_6\} & \emptyset \\
\{a_1, a_2, a_4, a_5\} & \{a_1, a_2, a_3, a_4\} & \{a_3, a_6, a_7\} & \emptyset \\
\{a_3, a_4, a_5, a_6\} & \{a_4, a_6\} & \{a_2\} & \{a_2, a_3, a_6\} & \emptyset
\end{pmatrix}.
$$

Transform the discernibility function f from its CNF $f = \prod(\sum m_{ij})$ into a DNF $f = \sum_{t=1}^{s}(\prod B_t), (B_t \subseteq A)$.

Thus $\{a_2, a_3, a_4\}$, $\{a_2, a_3, a_6\}$, $\{a_2, a_5, a_6\}$ and $\{a_2, a_4, a_5, a_7\}$ are the four reduction sets of A.

Definition 5.2 [5]. Let $(U, C \cup D)$ be a relation decision system, then the consistent part is $U_{CD} = \{x | r_{R_C}(x) \subseteq r_d(x)\}$. Let $\emptyset \neq B \subseteq C$, set B is called the attribute reduction of C if B satisfies the following conditions:

(1) $U_{CD} = U_{BD}$;
(2) For any $\emptyset \neq B' \subset B$, $U_{CD} \neq U_{B'D}$.

Similarly, we can derive an attribute reduction algorithm for a relation decision system. The following example illustrates our algorithm.

Example 5.2. Consider the following incomplete decision table shown in Table 4. Where $U = \{x_1, x_2, \cdots, x_5\}$, $C = \{a_1, a_2, a_3, a_4\}$ and $D = \{d\}$. Similarly, $*$ denotes missing attribute values (a null or a unknown value). Each $a_k \in C \cup D$ can be seen as a relation on U via

$$a_k = \{(x_i, x_j) | a_k(x_i) = a(x_j) \text{ or } a_k(x_i) = * \text{ or } a_k(x_j) = *\}.$$

Table 4. An incomplete decision table

U	a_1	a_2	a_3	a_4	d
x_1	0	0	0	1	1
x_2	0	0	1	1	0
x_3	1	*	0	*	1
x_4	1	1	0	0	0
x_5	*	0	1	1	*

According to Algorithm 2, we obtain the $G_{CD} = \{x_1, x_2, x_5\}$. The discernibility matrix M is as follows.

$$(m_{ij})_{3 \times 5} = \begin{pmatrix} \emptyset & \{a_3, a_4\} & \emptyset & \{a_1, a_2, a_4\} & \emptyset \\ \{a_3, a_4\} & \emptyset & \{a_1, a_3\} & \emptyset & \emptyset \\ \emptyset & \emptyset & \emptyset & \emptyset & \emptyset \end{pmatrix}.$$

Transform the discernibility function $f = (a_3 \vee a_4) \wedge (a_1 \vee a_2 \vee a_4) \wedge (a_1 \vee a_3)$ from its CNF into the DNF $f = (a_1 \wedge a_3) \vee (a_1 \wedge a_4) \vee (a_2 \wedge a_3) \vee (a_3 \wedge a_4)$.

Thus $\{a_1, a_3\}$, $\{a_1, a_4\}$, $\{a_2, a_3\}$ and $\{a_3, a_4\}$ are four reduction sets of C.

6   Conclusions

A relation system on two universal sets is an extension of a relation system on one universal set. In this paper, we have introduced the concepts of the attribute reduction for relation systems and relation decision systems on two universal sets. The proposed two algorithms can find all reduction sets for relation systems and relation decision systems, respectively. The corresponding algorithms for one universal set are respectively our special cases of the two algorithms. Now our algorithms are theoretical models, our future work will focus on practical applications of the proposed algorithms.

Acknowledgements. This work is supported by BLCU Scientific Research Ability Cultivation Project for Ph.D Students (Double-First Class Initiative Guiding Fund) (No. 17YPY050) and the Fundamental Research Funds for the Central Universities (the Research Funds of BLCU) (No. 18YCX011).

References

1. Dai, J., Wang, W., Tian, H., Liu, L.: Attribute selection based on a new conditional entropy for incomplete decision systems. Knowl. Based Syst. **39**, 207–213 (2013)
2. Jia, X.Y., Shang, L., Zhou, B., Yao, Y.Y.: Generalized attribute reduct in rough set theory. Knowl. Based Syst. **91**, 204–218 (2016)
3. Liu, G., Li, L., Yang, J., Feng, Y., Zhu, K.: Attribute reduction approaches for general relation decision systems. Pattern Recogn. Lett. **65**, 81–87 (2015)
4. Liu, G., Hua, Z., Zou, J.: A unified reduction algorithm based on invariant matrices for decision tables. Knowl. Based Syst. **109**, 84–89 (2016)
5. Liu, G.L., Hua, Z., Chen, Z.H.: A general reduction algorithm for relation decision systems and its applications. Knowl. Based Syst. **119**, 87–93 (2017)
6. Liu, G.L., Hua, Z., Zou, J.Y.: Local attribute reductions for decision tables. Inf. Sci. **422**, 204–217 (2018)
7. Liu, G.L., Hua, Z.: Partial attribute reduction approaches to relation systems and their applications. Knowl. Based Syst. **139**, 101–107 (2018)
8. Ma, X., Wang, G., Yu, H., Li, T.: Decision region distribution preservation reduction in decision-theoretic rough set model. Inf. Sci. **278**, 614–640 (2014)
9. Mi, J.S., Wu, W.Z., Zhang, W.X.: Approaches to knowledge reduction based on variable precision rough set model. Inf. Sci. **159**, 255–272 (2004)
10. Pawlak, Z.: Rough sets. Int. J. Comput. Inf. Sci. **11**, 341–356 (1982)
11. Pawlak, Z.: Rough Sets: Theoretical Aspects of Reasoning About Data. Kluwer Academic Publishers, Boston (1991)
12. Skowron, A.: Boolean reasoning for decision rules generation. In: 7th International Symposium on Methodologies for Intelligent Systems, pp. 295–305 (1993)
13. Skowron, A., Rauszer, C.: The discernibility matrices and functions in information systems. In: Slowinski, R. (ed.) Intelligent Decision Support, Handbook of Applications and Advances of the Rough Sets Theory, pp. 331–362. Kluwer Academic, Dordrecht (1992)
14. Stepaniuk, J.: Rough sets in knowledge discovery 2: approximation spaces, reducts and representatives. Knowl. Based Syst. **19**, 109–126 (1998)

15. Yu, X., Sun, F.Q., Liu, S.X., Lu, F.Q.: Urban emergency intelligent decision system based on variable precision graded rough set on two universes. In: 2015 27th Chinese IEEE Control and Decision Conference (CCDC), pp. 5202–5205 (2015)
16. Sun, B.Z., Ma, W.M.: An approach to evaluation of emergency plans for unconventional emergency events based on soft fuzzy rough set. Kybernetes **45**, 461–473 (2016)
17. Zhang, C., Li, D.Y., Yan, Y.: A dual hesitant fuzzy multigranulation rough set over two-universe model for medical diagnoses. Comput. Math. Methods Med. **2015**, 1–12 (2015)
18. Zhang, C., Li, D.Y., Mu, Y.M., Song, D.: An interval-valued hesitant fuzzy multi-granulation rough set over two universes model for steam turbine fault diagnosis. Appl. Math. Modell. **42**, 693–704 (2017)

Toward Optimization of Reasoning Using Generalized Fuzzy Petri Nets

Zbigniew Suraj$^{(\boxtimes)}$

Chair of Computer Science, University of Rzeszów, Rzeszów, Poland
zbigniew.suraj@ur.edu.pl

Abstract. Recently, generalized fuzzy Petri nets have been proposed. This paper describes a modified class of generalized fuzzy Petri nets called optimized generalized fuzzy Petri nets. The main difference between the current net model and the previous one is the definition of the operator binding function δ. This function, like in the previous net model, combines transitions with triples of operators (In, Out_1, Out_2) in the form of appropriate triangular norms. The operator In refers to the way in which all input places are connected to a given transition (or more precisely, the statements corresponding to these places) and affects the aggregation power of truth degrees associated with the input places of the transition. However, the operators Out_1 and Out_2 refer to the way in which the new markings of output places of the transition are calculated after firing the transaction. For the operator In, it is assumed that it can belong to one of two classes, i.e., t or s-norms, while the operator Out_1 belongs to the class of t-norms, and the operator Out_2 to the class of s-norms. The meaning of these three operators in the current net model is the same as in the previous one. However, the new net model has been extended to include external knowledge about the partial order between the triangle norms used in the model. In addition, it is assumed that the new net model works in the steps mode. The paper also shows how to use this net model in the fuzzy reasoning algorithm. The tangible benefit of this approach compared to the previous one lies in the fact that the user can now more precisely adapt his model to the real life situation and use it more effectively by choosing the appropriate triples of operators for net transitions. This paper also presents an example of a small rule-based decision support system in the field of control, illustrating the described approach.

1 Introduction

Petri nets (PNs) [13] have broad application areas such as robotic tasks and artificial intelligence. In the past few decades, various types of PNs have been proposed for different applications. Although PN's research and applications have brought a lot of fruit, some flaw remained, namely that they were unable to represent fuzzy data used in knowledge-based systems (KBSs) or a system with uncertainty. To overcome this disadvantage, a novel model of PNs called

© Springer Nature Switzerland AG 2018
H. S. Nguyen et al. (Eds.): IJCRS 2018, LNAI 11103, pp. 294–308, 2018.
https://doi.org/10.1007/978-3-319-99368-3_23

fuzzy Petri net (FPN) was developed in 1984 by Lipp [5]. FPNs are a modification of classical PNs for dealing with imprecise, vague, or fuzzy information in KBSs, which have been extensively used to model fuzzy production rules (FPRs) and formulate fuzzy rule-based reasoning automatically. FPNs support structural organization of information, provide visualization of knowledge reasoning, and facilitate design of efficient fuzzy inference algorithms. All this makes FPNs a potential methodology for knowledge representation and reasoning in KBSs [2,6]. Since the introduction of FPNs for supporting approximate reasoning in a fuzzy rule-based system (FRBS) [7], they have received deal of attention from researches and practitioners in the domain of artificial intelligence. The earlier FPN models, as indicated in the literature on the subject [6], have a number of shortcomings and are not suitable for increasingly complex KBSs. As a result, many alternative models have been proposed in the literature in order to increase FPN power for knowledge representation as well as for a more intelligent implementation of rule-based reasoning [1,2,6,11,16–21].

A few years ago the GFP-nets [16] were proposed for knowledge representation and reasoning in KBSs. This model is a natural extension of classical FPNs [6]. The t-norms and s-norms were introduced to the model as substitutes of min and max operators. The latter ones generalize naturally AND and OR logical operators with the Boolean values 0 and 1. The GFP-net model is not only more comfortable in terms of knowledge representation, but most of all it is more effective in the modeling process of approximate reasoning as in this model the user has the chance to define the input/output operators according to her/his preferences.

This paper describes both the optimized generalized fuzzy Petri nets ($oGFP$-nets for short) and an algorithm for a fuzzy reasoning process. The main difference between this net model and the existing GFP-nets concerns the definition of the operator binding function δ. This function, similarly to GFP-nets, connects transitions with triples of operators (In, Out_1, Out_2) in the form of suitable triangular norms. The meaning of these operators in the $oGFP$-nets is the same as in the case of GFP-nets. However, by building the $oGFP$ model, the external knowledge of the partial order between triangular norms is used. It is also assumed that $oGFP$-nets work in the steps mode. The work also shows the use of this model in the fuzzy reasoning algorithm. Typically, such algorithms are applicable in KBSs to describe fuzzy inference processes in the form of FPRs. For given degrees of truth of some statements from rule promises are determined degrees of truth of other statements which are goal statements. FPRs describe relations between these statements. The speed of a fuzzy reasoning process is very important, especially in real-time decision making systems. The proposed algorithm allows firing of independent FPRs in one reasoning step. In this approach it is assumed that if in a given KBS there are two (or more) FPRs having a common statement in conclusions then operator Out_2 appearing in triples of operators (In, Out_1, Out_2) which are attached to all transitions representing those rules must be the same. Apart from this assumption, you can get different degrees of truth in a joint statement.

Since there exist infinitely numerous triangular norms in the field of fuzzy logic, and the nature of the marking changes variously in given $oGFP$-nets depending on triangular norms used in the net model, it is very difficult to choose the appropriate triangular norms for a specific application without an external knowledge of the relationships between them. However, taking into account some properties of triangular norms described in Proposition 3 in Sect. 2.2, you can build the $oGFP$-net model more efficiently than in the case of the GFP-net one. The choice of suitable operators for the modeled system is very important, especially in control systems or expert systems, which are in many cases described by incomplete, imprecise and/or vague information. Trying to make GFP-nets more useful in practice, in this paper we establish a connection between GFP-nets and the theory of algebraic t-norm properties. This relationship is methodological, demonstrating the possible application of t-norm methodology to transform GFP-nets into a more realistic model.

The rest of this paper is organized in the following way. First, some background knowledge regarding partially ordered sets, triangular norms and their properties are provided in Sect. 2. In Sect. 3, the definition of $oGFP$-net is given. Section 4 describes a reasoning process modelled by means of a given $oGFP$-net. An example illustrating the approach described in this paper is provided in Sect. 5. Finally, Sect. 6 concludes the paper.

2   Preliminaries

2.1   Partially Ordered Sets

Let R be a binary relation on a set A. A relation R on A is said to be a *partial ordering on A* if: (1) it is reflexive, i.e., $(x, x) \in R$ for each $x \in A$, (2) it is transitive, i.e., if $(x, y) \in R$ and $(y, z) \in R$, then $(x, z) \in R$ for any $x, y, z \in A$, (3) it is antisymmetric, i.e., if $(x, y) \in R$ and $(y, x) \in R$, then $x = y$ for any $x, y \in A$. A partial ordering R on A is said to be a *linear ordering on A* if at least one of the following conditions: $(x, y) \in R$, $(y, x) \in R$ or $x = y$ holds for any $x, y \in A$. If R is a partial ordering on A, then the pair $U = (A, R)$ is said to be a *partially ordered set* (abbreviated poset). If R is a linear ordering on A, then the pair $U = (A, R)$ is said to be a *linearly ordered set*.

Let $U = (A, R)$ be a poset, and $X \subseteq A$. The element $a_0 \in A$ is said to be the *upper (lower) bound in U* of a subset $X \subseteq A$ if $(x, a_0) \in R$ $((a_0, x) \in R)$ for all $x \in X$. The upper (lower) bound in U of A is the *greatest (least)* element in U. An element $a \in A$ is said to be *maximal (minimal)* in U if $(a, x) \in U$ (respectively $(x, a) \in R$) implies $x = a$. It is clear that the greatest (least) element is maximal (minimal), and if R is a linear ordering, then the element maximal (minimal) in U is also the greatest (least) in U. It is obvious that if the greatest (least) element in U exists, then all the maximal (minimal) elements are equal. If B is a set of upper bounds in $U = (A, R)$ of a set $A_1 \subseteq A$, then the least element in $(B, R \cap B^2)$ is said to be the *least upper bound in U* of the set A_1 and is denoting $\sup(A_1, U)$. Replacing in the preceding definition "upper" and "least" respectively by "lower" and "greatest" the definition of the *greatest* *lower bound* of A_1 in U is obtained. And this will be denoted as $\inf(A_1, U)$. It is clear that $\sup(A_1, U)$ and $\inf(A_1, U)$ are uniquely determined by A_1 and U if they exist. A poset U is said to be a *lattice* if for any $a, b \in A$ in U there are $\sup(\{a, b\}, U)$ and $\inf(\{a, b\}, U)$.

Detailed information on partially ordered sets is available in [3].

2.2   Triangular Norms

A *triangular* norm (t-norm for short) [4] is a function $T: [0, 1]^2 \to [0, 1]$, such that for all $a, b, c \in [0, 1]$ the following four conditions are satisfied: (1) it has 1 as the unit element; (2) it is monotone; (3) it is commutative; (4) it is associative.

Example 1. We list only a few of basic t-norms known from the literature and used in this paper: (1) $ZtN(a, b) = min(a, b)$ (minimum, Zadeh t-Norm); (2) $HtN(a, b) = 0$ for $a = b = 0$, $HtN(a, b) = ab/(a + b - ab)$ otherwise (Hamaher t-Norm); (3) $GtN(a, b) = ab$ (algebraic product, Goguen t-Norm); (4) $EtN(a, b) = ab/(2 - (a + b - ab))$ (Einstein t-Norm); (5) $LtN(a, b) = max(0, a + b - 1)$ (Lukasiewicz t-Norm); (6) $DtN(a, b) = 0$ for $(a, b) \in [0, 1)^2$, $DtN(a, b) = min(a, b)$ otherwise (drastic product, Drastic t-Norm).

The family of all basic t-norms without the drastic product will be denoted by TN.

The comparison of t-norms is done in the usual way, i.e., pointwise. If, for two t-norms T_1 and T_2, the inequality $T_1(a, b) \leq T_2(a, b)$ holds for all $(a, b) \in [0, 1]^2$, then it is said that T_1 is *weaker* than T_2 and is denoted $T_1 \leq T_2$.

Taking into account the properties of t-norms and the above definitions it is easy to show the following properties:

Proposition 1. (1) For each t-norm T and for each $(a, b) \in [0, 1]^2$ we have: $DtN \leq T \leq ZtN$, i.e., the drastic product DtN is the least, and the minimum ZtN is the greatest t-norm ([4], pages 6–7). (2) Since $LtN \leq EtN \leq GtN \leq HtN$, we get the following linear order for the six basic t-norms: $DtN \leq LtN \leq EtN \leq GtN \leq HtN \leq ZtN$.

An s-norm [4] is a function $S: [0, 1]^2 \to [0, 1]$ such that for all $a, b, c \in [0, 1]$ the following four conditions are satisfied: (1) it has 0 as the unit element, (2) it is monotone, (3) it is commutative, (4) it is associative.

Example 2. We list only a few of basic s-norms corresponding respectively to the basic t-norms presented in Example 1. (1) $ZsN(a, b) = max(a, b)$ (maximum, Zadeh s-Norm); (2) $HsN(a, b) = 1$ for $a = b = 1$, $HsN(a, b) = (a + b - 2ab)/(1 - ab)$ otherwise (Hamaher s-Norm); (3) $GsN(a, b) = a + b - ab$ (probabilistic sum, Goguen s-Norm); (4) $EsN(a, b) = (a + b)/(1 + ab)$ (Einstein s-Norm); (5) $LsN(a, b) = min(1, a + b)$ (bounded sum, Lukasiewicz s-Norm); (6) $DsN(a, b) = 1$ for $(a, b) \in (0, 1]^2$, $DsN(a, b) = max(a, b)$ otherwise (drastic sum, Drastic s-Norm).

The family of all basic s-norms without the drastic sum will be denoted by SN.

As in the case of t-norms, we can also show the following properties for s-norms:

Proposition 2. (1) For each s-norm S and for each $(a, b) \in [0,1]^2$ we have: $ZsN \leq S \leq DsN$, i.e., the maximum ZsN is the least, and the drastic sum DsN is the greatest s-norm ([4], pages 12–13). (2) Since $HsN \leq GsN \leq EsN \leq LsN$, we get the following order for the six basic s-norms: $ZsN \leq HsN \leq GsN \leq EsN \leq LsN \leq DsN$.

Let (x, y, z) and (x', y', z') be two vectors over a non-empty set X. In the following, the comparison of such vectors is done in the usual way, i.e., pointwise. If, for two vectors (x, y, z) and (x', y', z'), the inequalities $x \leq x'$, $y \leq y'$, and $z \leq z'$ hold for all $x, y, z, x', y', z' \in X]$, then we say that the vector (x, y, z) is *less* than the vector (x', y', z') and we write $(x, y, z) \leq (x', y', z')$.

Example 3. Consider two pairs $U = (A, R)$ and $U' = (A, R')$, where the set $A = TN \cup SN$, the relation $R = TN \times TN \times SN$, and the relation $R' = SN \times TN \times SN$ are the sets of all triples over the A. It is easy to show that the pairs $U = (A, R)$ and $U' = (A, R')$ are lattices. The simple proof of this fact is omitted. It is also worth emphasizing that these two lattices are finite, and each of them consists of 125 triples. Due to the large number of nodes in the graphical representation of these lattices, we present only small fragments in the drawings (Figs. 1 and 2). Each lattice contains the least (greatest) element corresponding to the lower (upper) node on the corresponding graph. Moreover, in each graph immediate neighboring vertices to the the lower (upper) node are presented.

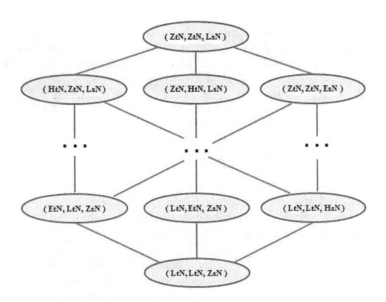

Fig. 1. A fragment of graphical representation of the lattice U (*Case* AND)

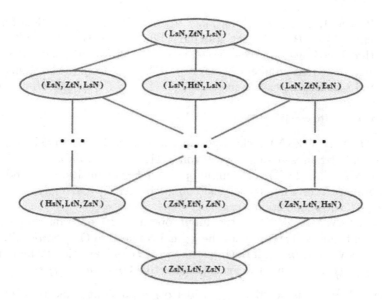

Fig. 2. A fragment of graphical representation of the lattice U' (*Case* OR)

For the lattices $U = (A, R)$ and $U' = (A, R')$ we can show the following properties:

Proposition 3. (1) For each triple (A, B, C), where A, B are any t-norms from TN and C is any s-norm from SN, and for each $(a, b) \in [0, 1]^2$ we have: $(LtN, LtN, ZsN) \leq (A, B, C) \leq (ZtN, ZtN, LsN)$, i.e., (LtN, LtN, ZsN) is the least element in U (*Case* AND, *minimal*), and (ZtN, ZtN, LsN) is the greatest element in U (*Case* AND, *maximal*) (see Fig. 1).

(2) For each triple (D, B, C), where D, C are any s-norms from SN and B is any t-norm from TN, and for each $(a, b) \in [0, 1]^2$ we have: $(ZsN, LtN, ZsN) \leq (D, B, C) \leq (LsN, ZtN, LsN)$, i.e., (ZsN, LtN, ZsN) is the least element in U' (*Case* OR, *minimal*), and (LsN, ZtN, LsN) is the greatest element in U' (*Case* OR, *maximal*) (see Fig. 2).

The properties of triples presented in Proposition 3 will be used in the definition of the new model of fuzzy Petri net presented in the next section.

3   Optimized Generalized Fuzzy Petri Nets

We assume that the reader is familiar with the basic notions of PNs [10, 12].

Let $U = (A, R)$ and $U' = (A, R')$ be the lattices described in Sect. 2. An $oGFP$-net over U and U' is a tuple $N = (P, T, I, O, M_0, S, \alpha, \beta, \gamma, Op, \delta)$, where: (1) $P = \{p_1, p_2, \ldots, p_n\}$ is a finite set of places; (2) $T = \{t_1, t_2, \ldots, t_m\}$ is a finite set of transitions; (3) $I \colon T \to 2^P$ is the input function; (4) $O \colon T \to 2^P$ is the output function, and 2^P denotes a family of all subsets of the set P;

(5) $M_0: P \to [0,1]$ is the initial marking; (6) $S = \{s_1, s_2, \ldots, s_n\}$ is a finite set of statements; (7) $\alpha: P \to S$ is the statement binding function; (8) $\beta: T \to [0,1]$ is the truth degree function; (9) $\gamma: T \to [0,1]$ is the threshold function, and $[0,1]$ denotes the set of real numbers between 0 and 1; (10) Op is the family of all t-norms and s-norms appearing in the set A; (11) $\delta: T \to Op \times Op \times Op$ such that:

(*Case* AND, see Proposition 3)

1. $\delta(t) = (LtN, LtN, ZsN)$, if the input operator In of transition t should belong to t-norms (it represents the logical connective AND, *minimal*),
2. $\delta(t) = (ZtN, ZtN, LsN)$, if the input operator In of transition t should belong to s-norms (it represents the logical connective AND, *maximal*).
   (*Case* OR, see Proposition 3)
3. $\delta(t) = (ZsN, LtN, ZsN)$, if the input operator In of transition t should belong to t-norms (it represents the logical connective OR, *minimal*),
4. $\delta(t) = (LsN, ZtN, LsN)$, if the input operator In of transition t should belong to s-norms (it represents the logical connective OR, *maximal*).

In general case, it is possible to consider other possible connections of triples to the individual transitions of the $oGFP$-net, resulting from the dependencies between the triples illustrated in Figs. 1 and 2. However, we included only these triples of t-norms that are attached to the lowest and highest nodes in the graphs presented in these drawings, because here we are interested in defining the optimized form of our net model.

In the drawing, places are represented as circles and transitions as rectangles. The function I describes the oriented arcs connecting places with transitions, and the function O describes the oriented arcs connecting transitions with places. If $I(t) = \{p\}$ then a place p is called an *input place* of a transition t, and if $O(t) = \{p'\}$, then a place p' is called an *output place* of t. The initial marking M_0 is an initial distribution of real numbers from $[0,1]$ in the places. It can be represented by a vector of dimension n of real numbers over $[0,1]$. For $p \in P$, $M_0(p)$ can be interpreted as a truth value of the statement s bound with a given place p by means of the statement binding function α. In the drawing, the tokens are represented by the appropriate real numbers from $[0,1]$ placed over the circles corresponding to the suitable places. We assume that if $M_0(p) = 0$ then the token does not exist in the place p. The numbers $\beta(t)$ and $\gamma(t)$ are placed in a net picture under the transition t. The first number is interpreted as the truth degree of an implication corresponding to a given transition t. The role of the second one is to limit the possibility of transition firings, i.e., if the input operator In value for all values corresponding to input places of the transition t is less than a threshold value $\gamma(t)$ then this transition cannot be fired (activated). The operator binding function δ connects transitions with triples of operators (In, Out_1, Out_2). The first operator in the triple is called the input operator, and two remaining ones are the output operators. The input operator In concerns the way in which all input places are connected with a given transition t (more precisely, statements corresponding to those places). However, the output operators Out_1 and Out_2 concern the way in which the next marking is computed after firing the transition t. In the case of the input operator we assume that it can belong to one of two classes, i.e., t- or s-norm, whereas the second one belongs to the class of t-norms and the third to the class of s-norms.

It is worth noting that in this definition elements $P, T, I, O, M_0, S, \alpha, \beta, \gamma$ have the same meaning as in the definition of the general fuzzy Petri net introduced in [16]. The main difference between the current net model and the previous one is the definition of the operator binding function δ. This function, like in the previous net model, combines transitions with triples of operators (In, Out_1, Out_2) in the form of appropriate triangular norms. However, this net model has been extended to external knowledge about the partial order between triangle standards (see case AND and OR in the definition). In addition, it is assumed that the new net model operates in the steps mode. This aspect of the net operation will be explained in detail later.

Let N be an $oGFP$-net. A marking of N is a function $M \colon P \to [0,1]$.

The $oGFP$-net dynamics defines how new markings are computed from the current marking when transitions are fired.

There are several ways to increase the usability of Petri nets [14]. They concern different ways of net work. In this paper, we assume that an $oGFP$-net can operate in two modes: *single firings* or *steps*.

Single firings: A transition $t \in T$ is enabled (or ready for firing) for marking M if the number produced by input operator In for all input places of the transition t by M is positive and greater than, or equal to the number being a value of threshold function γ corresponding to the transition t.

Steps are a generalization of nets work in mode of single firings. In the paper, we consider two kinds of steps: *simple* and *generalized*.

Simple steps: A nonempty set U of transitions is called to be *a simple step* by a marking M if and only if there are transitions enabled by M and pairwise structurally independent (concurrent), i.e., these transitions have not joint neither input places nor output places.

Generalized steps: A nonempty set U of transitions is called to be *a generalized step* (for short step) by a marking M if and only if there are transitions enabled by M and fired simultaneously. A step (a simple step) U by a marking M is called to be *maximal*, if there is no any step (simple step) U' by M such that $U' \supset U$.

In the definition of a step we do not demand the structural independency of transitions with a step U, but we demand only the possibility of its simultaneous firing. This means that if the sets of input places and output places for transitions belonging to the step U are not pairwise disjoint, thus simultaneous firing of those transitions will be possible only in *Option* 2 (see in the following). This definition is a natural generalization of the simple step definition.

Only enabled transitions can be fired. We consider two operating options of $oGFP$-nets in the paper.

Option 1. If M is a marking of N enabling a transition t and M' is the marking derived from M by firing t, then for each $p \in P$ a procedure for computing the marking M' is as follows: (1) Tokens from all input places of the transition t are removed. (2) Tokens in all output places of t are modified in the following way: at first the value of input operator In for all input places of t is computed, next the value of output operator Out_1 for the value of In and for the value of truth degree function $\beta(t)$ is determined, and finally, a value corresponding to $M'(p)$ for each $p \in O(p)$ is obtained as a result of output operator Out_2 for the value of Out_1 and the current marking $M(p)$. (3) Tokens in the remaining places of net N are not changed.

Option 2. The main difference in the definition of the marking M' presented above (*Option* 1) concerns input places of the fired transition t. In *Option* 1 tokens from all input places of the fired transition t are removed, whereas in *Option* 2 all tokens from input places of the fired transition t are copied.

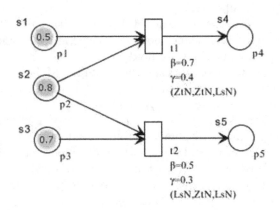

Fig. 3. An *oGFP*-net with the initial marking

Example 4. Consider a *oGFP*-net in Fig. 3. For the net we have: the set of places $P = \{p_1, p_2, p_3, p_4, p_5\}$, the set of transitions $T = \{t_1, t_2\}$, the input function I and the output function O in the form: $I(t_1) = \{p_1, p_2\}$, $I(t_2) = \{p_2, p_3\}$, $O(t_1) = \{p_4\}$, $O(t_2) = \{p_5\}$ and the initial marking $M_0 = (0.5, 0.8, 0.7, 0, 0)$, the set of statements $S = \{s_1, s_2, s_3, s_4, s_5\}$, the statement binding function α: $\alpha(p_1) = s_1$, $\alpha(p_2) = s_2$, $\alpha(p_3) = s_3$, $\alpha(p_4) = s_4$, $\alpha(p_5) = s_5$, the truth degree function β: $\beta(t_1) = 0.7$, $\beta(t_2) = 0.5$, the threshold function γ: $\gamma(t_1) = 0.4$, $\gamma(t_2) = 0.3$, the set of operators $Op = \{ZtN, LsN\}$, the operator binding function δ: $\delta(t_1) = (ZtN, ZtN, LsN)$ (*Case* AND, *maximal*), $\delta(t_2) = (LsN, ZtN, LsN)$ (*Case* OR, *maximal*). The transition t_1 is enabled by the initial marking M_0, since $ZtN(M_0(p_1), M_0(p_2)) = min(0.5, 0.8) = 0.5 \geq 0.4 = \gamma(t_1)$. Firing transition t_1 by the marking M_0 (*Option* 1) transforms M_0 to the marking $M' = (0, 0, 0.7, 0.5, 0)$, because $ZtN(ZtN(M_0(p_1), M_0(p_2)), \beta(t_1)) = ZtN(0.5, 0.7) = 0.5$ and $LsN(M_0(p_3), ZtN(ZtN(M_0(p_1), M_0(p_2)), \beta(t_1))) = LsN(0, 0.5) = min(1, 0 + 0.5) = 0.5$. In a similar way, you can calculate the next marking after firing the transition t_2 by M_0. In this case, the resulting marking will be $M'' = (0.5, 0, 0, 0, 0.5)$. It is easy to see that transitions t_2 by the marking M' and t_1 by M'' are no longer enabled. Let us observe also that a set $U = \{t_1, t_2\}$ is a step by the initial marking M_0 in *Option* 2. The step U can be fired by M_0. However, this set of transitions is not a simple step by M_0, since transitions t_1 and t_2 are not structurally independent. The set U is not also a step by M_0 in *Option* 1, because after firing transition t_1 in this option the transition t_2 is not enabled. A similar situation appears after firing transition t_2 by M_0 in *Option* 1. Whereas maximal step $U = \{t_1, t_2\}$ by marking M_0 in *Option* 2 is enabled. After firing this step by M_0 we obtain the marking $M'' = (0.5, 0.8, 0.7, 0.5, 0.5)$.

In some cases such situations in the net are not accepted, i.e., when, e.g., statements s_4 and s_5 attached to places p_4 and p_5 of the net describe specific decisions in KBS, that is modeled by the net. Then the markings of these places can be interpreted as the true degrees of these statements. Thus, the equality of these values does not allow to unambiguously determine which decision should be chosen. Using the net definition presented above (or more general, using, for example, information about t-norm properties represented in the graphs in Figs. 1 and 2), we can try to find such triples of t-norms attached to net transitions that the problem of ambiguity will be possible to solve. In our example, if we take, for example, the following connections for t_1 and t_2: $\delta(t_1) = (ZtN, ZtN, LsN)$ and $\delta(t_2) = (ZsN, LtN, ZsN)$ (see points 2 and 3 in the definition of $oGFP$-net), then maximal step U is also enabled by marking M_0 and after the step U by M_0 in *Option* 2, we obtain the resulting marking $M''' = (0.5, 0.8, 0.7, 0.5, 0.3)$. Now you can see that places p_4 and p_5 have different markings equal to 0.5 and 0.3, respectively. This means that in this case the problem of ambiguity no longer exists. We omit the detailed description of the relevant calculations illustrating these considerations.

4   Approximate Algorithm

In this section we show how to use the $oGFP$-net model in the fuzzy reasoning algorithm. In order to describe the algorithm, we need earlier two auxiliary concepts.

In some situations we may want to determine the antecedence-consequence relationships between two groups of statements: the starting (given) statements $s_{i1}, \ldots, s_{ik}$, and goal (computed) statements $s_{o1}, \ldots, s_{ol}$. In the Petri net representation, the places associated with the first group of statements are called *starting places*, whereas the places associated with the second one are called *goal places*. Furthermore, if the truth degrees of the starting statements $s_{i1}, \ldots, s_{ik}$ are given, we may want to know what the truth degrees of the goal statements $s_{o1}, \ldots, s_{ol}$ are. These problems can be solved by using an approximate reasoning algorithm based on $oGFP$-nets. We assume that the truth degrees of the starting statements are given by the expert or they are identified by sensors in finite time units. The goal of the reasoning is to determine the truth degrees of the output (goal) statements. In addition, we assume that $oGFP$-net modeling reasoning process works in the step mode (simple or generalized).

In the following section we present an example of this algorithm' use.

Algorithm 1. Reasoning Algorithm Using $oGFP$-net

Input   : A set of the markings of starting places
Output: A set of the markings of goal places
repeat
    Determine the steps ready for firing
    while *Are there any steps ready for firing?* **do**
        Fire a step ready for firing;
        Compute the new markings of places after firing the step;
        Determine the steps ready for firing;
    Read the markings of goal places;
    Reset the markings of all places
until *Is this the end of simulation?*;

---

5   Illustrative Example

Consider the example of KBS, which contains a set of four rules: (r_1) IF s_2 THEN s_4; (r_2) IF s_1 AND s_4 THEN s_5; (r_3) IF s_3 AND s_4 THEN s_6; (r_4) IF s_5 AND s_6 THEN s_7, where the statements' labels have the following meaning: s_1 - 'Plant work is non-stable', s_2 - 'Temperature sensor of plant indicates the temperature over $150\,°C$', s_3 - 'Plant cooling does not work', s_4 - 'Plant temperature is high', s_5 - 'Plant is in failure state', s_6 - 'Plant makes a huge hazard for environment', and s_7 - 'Turn off plant supply'.

At first, using a method for constructing a GFP-net on the base of a given set of rules [16], we present the $oGFP$-net model corresponding to these rules. This net model is shown in Fig. 4. Note that the places $p_1, p_2, p_3, p_4, p_4(copy), p_5, p_6$ and p_7 include the numbers $0.8, 0.7, 0.9, 0, 0, 0, 0, 0$ corresponding to the truth degrees of statements $s_1, s_2, s_3, s_4, s4(copy), s_5, s_6, s_7$, respectively. Moreover, there are: the truth degree function β: $\beta(t_1) = 0.8, \beta(t_2) = 0.9$, $\beta(t_3) = 0.7$, $\beta(t_4) = 1.0$, the threshold function γ: $\gamma(t_1) = \gamma(t_2) = \gamma(t_3) = \gamma(t_4) = 0.1$, the set of operators $Op = \{ZtN, LsN\}$ and the operator binding function δ: $\delta(t_1) = \delta(t_2) = \delta(t_3) = \delta(t_4) = (ZtN, ZtN, LsN)$ (*Case* AND, *maximal*). In addition, it is worth adding that each net transition $t_i, i = 1, 2, ..., 4$) together with the input and output places corresponds exactly to one production rule r_i given above.

Next, we simulate the behavior of the net model shown in Fig. 4(a) using the algorithm 1 in *Option* 1. Assessing the statements attached to the starting places from p_1 to p_3 and choosing the step mode for the net work, we see that only the step $U_1 = \{t_1\}$ is ready for firing by the initial marking $M_0 = (0.8, 0.7, 0.9, 0, 0, 0, 0, 0)$. After firing this step by M_0, we obtain a new marking $M_1 = (0.8, 0, 0.9, 0.7, 0.7, 0, 0, 0)$. Then, we determine steps ready for firing

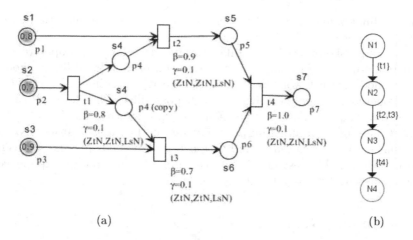

(a)   (b)

Fig. 4. (a) An *oGFP*-net model of the example of KBS constructed by using the method presented in [16], (b) A graph representing all reachable markings of the *oGFP*-net

by M_1. In this case, we also have only one step of the form: $U_2 = \{t_2,t_3\}$. After firing the step U_2 by M_1, we obtain a marking $M_2 = (0,0,0,0,0,0.7,0.7,0)$. Further, we check whether there exist steps ready for firing by M_2. We can see that step $U_3 = \{t_4\}$ is enabled by M_2. After firing step U_3 by M_2 the algorithm 1 stops and the final value, corresponding to the statement s_7 attached to the goal place p_7, equal to 0.7 is obtained. The graphical representation of the algorithm 1 execution is illustrated in Fig. 4(b). We can easily see in this graph a sequence of steps (the reachable path) of the form $\{t_1\}\{t_2, t_3\}\{t_4\}$. The reachable path goes from the initial marking M_0 represented in the graph by the node N_1 to the final marking $M_3 = (0,0,0,0,0,0,0,0.7)$ represented in the graph by the node N_4 (see Table in Fig. 5). Since the marking of place p_7 is the true degree of the statement attached to this place, thus the value 0.7 is the believable degree of final decision in the example of KBS.

N/P	p1	p2	p3	p4	p5	p6	p4 (c...	p7
N1	0.8	0.7	0.9	0.0	0.0	0.0	0.0	0.0
N2	0.8	0.0	0.9	0.7	0.0	0.0	0.7	0.0
N3	0.0	0.0	0.0	0.0	0.7	0.7	0.0	0.0
N4	0.0	0.0	0.0	0.0	0.0	0.0	0.0	0.7

Fig. 5. A table of all nodes in the graph from Fig. 4(b)

It is worth to observe that if we accept for these four transitions the operator binding function δ: $\delta(t_1) = \delta(t_2) = \delta(t_3) = \delta(t_4) = (LtN, LtN, LsN)$ (*Case* AND, *minimal*) and if we choose the same sequences of steps as above, we obtain the final value for the statement s_7 equal to 0. We omit the detailed computations performed in this case.

This example shows clearly that different interpretations of the operator binding function δ may lead to quite different decision results. In addition, choosing the steps mode for net work one can speed up its operation. The $oGFP$-net model proposed in the paper gives us such possibility. Therefore, we can say that this net model is more flexible than the ones known from the subject literature. Choosing a suitable interpretation for the logical operators AND and OR we may apply the mathematical relationships between triangular norms presented in Sect. 2.2. The rest in this case certainly depends on the experience of the model designer to a significant degree.

6   Concluding Remarks

FPNs are one of the most popular and applicable class of PNs in the domain of artificial intelligence, which have been widely studied by researchers and practitioners. In this paper, we have proposed a new approach to fuzzy reasoning process using the $oGFP$-net model. This model uses for the optimized (minimal, maximal in the sense of Proposition 3) operator binding function δ interpretation for the triples (In, Out_1, Out_2) of t-norms. We have shown in the paper by means of the simple example that there exist problems in which, by choosing the appropriate triples (In, Out_1, Out_2) of t-norms, we can force the final conclusion. Of course, it is possible to consider another set of basic t-norms, which is the base for determining the optimized triples (In, Out_1, Out_2) of t-norms. It depends on our preferences. Moreover, thanks to the possibility of firing a set of transitions (steps) at each stage of the net's operation, we speed up the fuzzy reasoning process in the modeled KBS. These two aspects are the main novelty of the presented research work. The algorithm proposed in the paper has been implemented in PNeS [20]. Using an intuitive, realistic example, the practicality and usability of the proposed approach to modeling decision-making systems was demonstrated in the paper. It seems that this paper not only proves that the alternative net model is more suitable than the previous FPNs [6], but it also suggests both practitioners and researchers how to use the FPN more effectively. In addition, this paper can also be seen as a stimulus for further deep analysis of the area and to broaden the knowledge about the FPNs to help practitioners build more effective KBSs for smart decision making.

In this paper, we only considered the extension of AND and OR operators to t-norms in terms of real numbers. It seems useful to study FPNs in the context of the t-norm concept relating to more general mathematical structures (see e.g. [8,9]). In future work, we intend to deal with this problem, focusing in particular on the methodology presented here.

Acknowledgment. This work was partially supported by the Center for Innovation and Transfer of Natural Sciences and Engineering Knowledge at the University of Rzeszów. The author is grateful to the anonymous referees for their helpful comments.

References

1. Bandyopadhyay, S., Suraj, Z., Grochowalski, P.: Modified generalized weighted fuzzy Petri net in intuitionistic fuzzy environment. In: Flores, V., et al. (eds.) IJCRS 2016. LNCS (LNAI), vol. 9920, pp. 342–351. Springer, Cham (2016). https://doi.org/10.1007/978-3-319-47160-0_31

2. Cardoso, J., Camargo, H. (eds.): Fuzziness in Petri Nets. Springer, Heidelberg (1999)

3. Ershov, Y.L., Palyutin, E.A.: Mathematical Logic. MIR Publishers, Moscow (1984)

4. Klement, E.P., Mesiar, R., Pap, E.: Triangular Norms. Springer, Heidelberg (2000). https://doi.org/10.1007/978-94-015-9540-7

5. Lipp, H.P.: Application of a fuzzy Petri net for controlling complex industrial processes. In: Proceedings of IFAC Conference on Fuzzy Information Control, pp. 471–477 (1984)

6. Liu, H.-C., You, J.-X., Li, Z.W., Tian, G.: Fuzzy Petri nets for knowledge representation and reasoning: a literature review. Eng. Appl. Artif. Intell. **60**, 45–56 (2017)

7. Looney, C.G.: Fuzzy Petri nets for rule-based decision-making. IEEE Trans. Syst. Man Cybern. **18**(1), 178–183 (1988)

8. Ma, Z., Wu, W.: Logical operators on complete lattices. Inf. Sci. **55**(97), 77 (1991)

9. Mayor, G., Torrens, J.: On a class of operators for expert systems. Int. J. Intell. Syst. **8**, 771–778 (1993)

10. Murata, T.: Petri nets: properties, analysis and applications. Proc. IEEE **77**(4), 541–580 (1989)

11. Pedrycz, W.: Generalized fuzzy Petri nets as pattern classifiers. Pattern Recog. Lett. **20**(14), 1489–1498 (1999)

12. Peterson, J.L.: Petri Net Theory and the Modeling of Systems. Prentice-Hall Inc., Englewood Cliffs (1981)

13. Petri, C.A.: Kommunikation mit Automaten. Schriften des IIM Nr. 2, Institut für Instrumentelle Mathematik, Bonn (1962)

14. Starke, P.H.: Petri-Netze. In: Grundlagen · Anwendungen · Theorie. VEB Deutscher Verlag der Wissenschaften, Berlin (1980)

15. Suraj, Z.: Knowledge representation and reasoning based on generalised fuzzy Petri nets. In: Proceedings of 12th International Conference on Intelligent Systems Design and Applications, Kochi, India, pp. 101–106. IEEE Press (2012)

16. Suraj, Z.: A new class of fuzzy Petri nets for knowledge representation and reasoning. Fund. Inform. **128**(1–2), 193–207 (2013)

17. Suraj, Z.: Modified generalised fuzzy Petri nets for rule-based systems. In: Yao, Y., Hu, Q., Yu, H., Grzymala-Busse, J.W. (eds.) RSFDGrC 2015. LNCS (LNAI), vol. 9437, pp. 196–206. Springer, Cham (2015). https://doi.org/10.1007/978-3-319-25783-9_18

18. Suraj, Z., Bandyopadhyay, S.: Generalized weighted fuzzy Petri net in intuitionistic fuzzy environment. In: Proceedings of the IEEE World Congress on Computational Intelligence, Vancouver, Canada, pp. 2385–2392. IEEE Press (2016)

19. Suraj, Z., Grochowalski, P., Bandyopadhyay, S.: Flexible generalized fuzzy Petri nets for rule-based systems. In: Martín-Vide, C., Mizuki, T., Vega-Rodríguez, M.A. (eds.) TPNC 2016. LNCS, vol. 10071, pp. 196–207. Springer, Cham (2016). https://doi.org/10.1007/978-3-319-49001-4_16

20. Suraj, Z., Grochowalski, P.: Petri nets and PNeS in modeling and analysis of concurrent systems. In: Proceedings of International Workshop on Concurrency, Specification and Programming, Warsaw, Poland (2017)
21. Zhou, K.-O., Zain, A.M.: Fuzzy Petri nets and industrial applications: a review. Artif. Intell. Rev. **45**, 405–446 (2016)

Sequent Calculi for Varieties of Topological Quasi-Boolean Algebras

Minghui Ma[1], Mihir Kumar Chakraborty[2], and Zhe Lin[1(✉)]

[1] Institute of Logic and Cognition, Sun Yat-sen University, Guangzhou, China
{mamh6,linzhe8}@mail.sysu.edu.cn
[2] School of Cognitive Science, Jadavpur University, Kolkata, India
mihirc4@gmail.com

Abstract. A sequent calculus **wG5** is introduced for the variety of partition topological quasi-Boolean algebras. The sequent calculus **wG5** has the cut elimination property, i.e., every sequent derivable in **wG5** has a cut-free derivation. Furthermore, a sequent calculus **wG4ₜ** is introduced for the variety of topological quasi-Boolean algebras with tense operators, and it is a conservative extension of a sequent calculus **wG4** for the variety of topological quasi-Boolean algebras.

1 Introduction

Rough set theory was systematically established by Pawlak [7]. Indeed there are various different ways to define rough sets, and various algebraic structures were developed to capture them (cf. [1–3]). Rough and pre-rough algebras were defined and Stone-style representation theorems were presented in [3]. Pre-rough and rough algebras are based on quasi-Boolean algebras (also known as De Morgan algebras) and topological quasi-Boolean algebras. Topological Boolean algebras were first investigated by Tarski and McKinsey in [5], and more results on these algebras can be found in Rasiowa [8]. Topological quasi-Boolean algebras were initially defined in [2].

Recently the properties and interrelations between weak pre-rough algebras, particularly the logics of these algebraic structures, have been investigated in [9,10]. Hilbert-style axiomatic systems and Gentzen-style sequent calculi have been established for these logics. However, from proof-theoretic point of view, these sequent calculi in [9,10] do not admit cut elimination. Cut elimination plays central role in proof analysis of various logics, and it allows to obtain various logical properties including subformula property, decidability and interpolation property (cf. e.g. [6]). The aim of the present paper is to make up for such a lack of cut-free sequent calculus for algebras related with rough sets.

M. Ma—The work was supported by the Project Supported by Guangdong Province (China) Pearl River Scholar Funded Scheme (2017–2019).

Z. Lin—The work was supported by Chinese National Funding of Social Sciences (No. 17CZX048).

© Springer Nature Switzerland AG 2018
H. S. Nguyen et al. (Eds.): IJCRS 2018, LNAI 11103, pp. 309–322, 2018.
https://doi.org/10.1007/978-3-319-99368-3_24

The logic of topological Boolean algebras is exactly the modal logic **S4** (cf. [4]). The necessity operator $\Box$ is interpreted as the interior operation in a topological space, and $\Diamond$ is the dual of $\Box$. The characteristic axioms for **S4**, i.e., (T) $\Box p \to p$ and (4) $\Box p \to \Box \Box p$, define the basic properties of the interior operator, and the modal logic **S4** is sound and complete with respect to the class of all topological spaces (cf. [5]). Partition topological spaces are topological spaces defined by the axiom (B) $p \to \Box \Diamond p$, i.e., every closed subset is open. The modal logic **S5** is obtained by extending **S4** with the axiom (B), and it is sound and complete with respect to partition topological spaces. It is worthy to mention that Palwak's approximation spaces are exactly relational frames with equivalence relation, and **S5** is the logic of such approximation space.

If the Boolean basis of topological Boolean algebras is changed into quasi-Boolean algebras, we obtain the class of all topological quasi-Boolean algebras. Similarly we obtain the class of all partition topological quasi-Boolean algebras. The logic of topological quasi-Boolean algebras **tqB4** is a weakening of classical modal logic **S4**, and the logic of partition topological quasi-Boolean algebras **tqB5** is a weakening of classical modal logic **S5**. It is worthy to mention here that there are difficulties in finding cut-free sequent systems that are encountered for quite simple modal systems such as classical modal logic **S5**. Standard Gentzen sequent calculi for classical modal logic fail to be modular and do not satisfy most important properties of sequent calculus (cf. e.g. [6,11]). In the present paper, we shall develop a Gentzen sequent calculus **wG5** for the logic **tqB5** which admits cut elimination. And then we introduce a Gentzen sequent calculus **wG4**$_t$ for the logic of topological quasi-Boolean algebras with tense operators. Finally, by conservativity, we get a Gentzen sequent calculus **wG4** for the logic of topological quasi-Boolean algebras.

2   Partition Topological Quasi-Boolean Algebras

In this section, we shall give the definition of partition topological quasi-Boolean algebras, and a sound and complete consequence system shall be established for these algebras.

Definition 1. A *quasi-Boolean algebra* (qBa) is an algebra $\mathbb{A} = (A, \wedge, \vee, \neg, 0, 1)$ where $(A, \wedge, \vee, 0, 1)$ is a bounded distributive lattice, and $\neg$ is an unary operation on A such that the following conditions hold for all $a, b \in A$:

(DN) $\neg\neg a = a$, (DM) $\neg(a \vee b) = \neg a \wedge \neg b$.

The lattice order $\leq$ on A is defined by: $a \leq b$ if and only if $a \wedge b = a$, or equivalently $a \vee b = b$. A *topological quasi-Boolean algebra* (tqBa) is an algebra $\mathbb{A} = (A, \wedge, \vee, \neg, 0, 1, \Box)$ where $(A, \wedge, \vee, \neg, 0, 1)$ is a quasi-Boolean algebra, and $\Box$ is an unary operation on A such that for all $a, b \in A$:

(K$_\Box$) $\Box(a \wedge b) = \Box a \wedge \Box b$, (N$_\Box$) $\Box \top = \top$.
(T$_\Box$) $\Box a \leq a$, (4$_\Box$) $\Box a \leq \Box \Box a$.

A *partition topological quasi-Boolean algebra* (tqBa5) is a topological quasi-Boolean algebra $\mathbb{A} = (A, \wedge, \vee, \neg, \Box, 0, 1)$ such that for all $a \in A$:

(5) $\Diamond a \leq \Box \Diamond a$,

where $\Diamond$ is an unary operation on A defined by $\Diamond a := \neg \Box \neg a$. The class of all partition topological quasi-Boolean algebra is denoted by **tqBa5**.

Fact 1. *For any tqBa5* $\mathbb{A} = (A, \wedge, \vee, \neg, \Box, 0, 1)$ *and* $a, b \in A$, *the following hold:*

(1) $\neg 0 = 1$ *and* $\neg 1 = 0$.
(2) $\neg(a \wedge b) = \neg a \vee \neg b$.
(3) If $a \leq b$, *then* $\neg b \leq \neg a$.
(4) $\Diamond 0 = 0$ *and* $\Diamond(a \vee b) = \Diamond a \vee \Diamond b$.
(5) $\Box a = \Box \Box a$ *and* $\Diamond a = \Diamond \Diamond a$.
(6) $\Diamond a = \Box \Diamond a$ *and* $\Box a = \Diamond \Box a$.
(7) $\Diamond a \leq b$ *if and only if* $a \leq \Box b$.

Definition 2. Let $\mathbb{X} = \{x_i \mid i < \omega\}$ be the denumerable set of all variables. The set of all *terms* $\mathcal{T}$ is defined inductively by the following rule:

$$\mathcal{T} \ni \varphi ::= x \mid \bot \mid \neg \varphi \mid (\varphi \wedge \varphi) \mid (\varphi \vee \varphi) \mid \Box \varphi, \text{ where } x \in \mathbb{X}.$$

The above definition means that formulas are defined recursively from constant $\bot$, set of proposition variables $\mathbb{X}$ and logical connectives $\neg, \wedge, \vee, \Box$: $\bot$ and $x \in \mathbb{X}$ are formulas; if φ, ψ are a formulas, then $\neg \varphi$, $\varphi \wedge \psi$, $\varphi \vee \psi$ and $\Box \varphi$ are formulas. For convenience, hereafter we frequently use this kind of recursive definition.

We use the abbreviations $\top = \neg \bot$ and $\Diamond \varphi := \neg \Box \neg \varphi$. The *complexity* of a term φ is defined as the number of occurrences of binary connectives or modal operators in φ. The algebra $\mathfrak{T} = (\mathcal{T}, \wedge, \vee, \neg, \Box, \bot, \top)$ is called the *term algebra*.

3   The Sequent Calculus wG5

In this section, we shall introduce the sequent calculus **wG5** for the logic **wS5**. For proof theory of nonclassical logics, we refer to [6]. For this purpose, we introduce two structural operators: the comma for $\wedge$ and the pair of angle brackets $\langle - \rangle$ for $\Diamond$. A *term structure* is an expression Γ defined inductively as follows:

$$\Gamma := \varphi \mid (\Gamma, \Gamma) \mid \langle \Gamma \rangle, \text{ where } \varphi \in \mathcal{T}.$$

Term structures are denoted by Γ, Δ, Σ etc. with or without subscripts. A *context* is a term structure $\Gamma[-]$ with a single position which can be filled with a term structure. Let $\Gamma[\Delta]$ be obtained from the context $\Gamma[-]$ by filling Δ into the single position. We stipulate that a single position $[-]$ itself is a context.

The *complexity* of a term structure Γ (or context $\Gamma[-]$) is the number of occurrences of structural operators in Γ (or $\Gamma[-]$).

A *sequent* is an expression of the form $\Gamma \Rightarrow \varphi$ where Γ is a term structure and φ is a term. A *sequent rule* is a fraction of the form

$$\frac{\Gamma_1 \Rightarrow \varphi_1 \quad \cdots \quad \Gamma_n \Rightarrow \varphi_n}{\Gamma_0 \Rightarrow \varphi_0}(R)$$

where $\Gamma_1 \Rightarrow \varphi_1, \ldots, \Gamma_n \Rightarrow \varphi_n$ are called the *premisses* and $\Gamma_0 \Rightarrow \varphi_0$ is called the *conclusion* of (R).

Definition 3. The Gentzen sequent calculus **wG5** consists of the following axioms and inference rules:

(1) Axioms:

$$(\text{Id}) \; \varphi \Rightarrow \varphi \quad (\bot) \; \Gamma[\bot] \Rightarrow \varphi \quad (\top) \; \Gamma \Rightarrow \top$$

(2) Connective rules:

$$\frac{\Gamma[\varphi, \psi] \Rightarrow \chi}{\Gamma[\varphi \wedge \psi] \Rightarrow \chi}(\wedge\Rightarrow) \qquad \frac{\Gamma \Rightarrow \varphi \quad \Gamma \Rightarrow \psi}{\Gamma \Rightarrow \varphi \wedge \psi}(\Rightarrow\wedge)$$

$$\frac{\Gamma[\varphi] \Rightarrow \chi \quad \Gamma[\psi] \Rightarrow \chi}{\Gamma[\varphi \vee \psi] \Rightarrow \chi}(\vee\Rightarrow) \qquad \frac{\Gamma \Rightarrow \psi_i}{\Gamma \Rightarrow \psi_1 \vee \psi_2}(\Rightarrow\vee)(i = 1, 2)$$

$$\frac{\Gamma[\neg\varphi] \Rightarrow \chi \quad \Gamma[\neg\psi] \Rightarrow \chi}{\Gamma[\neg(\varphi \wedge \psi)] \Rightarrow \chi}(\neg\wedge\Rightarrow) \qquad \frac{\Gamma \Rightarrow \neg\psi_i}{\Gamma \Rightarrow \neg(\psi_1 \wedge \psi_2)}(\Rightarrow\neg\wedge)$$

$$\frac{\Gamma[\neg\varphi, \neg\psi] \Rightarrow \chi}{\Gamma[\neg(\varphi \vee \psi)] \Rightarrow \chi}(\neg\vee\Rightarrow) \qquad \frac{\Gamma \Rightarrow \neg\varphi \quad \Gamma \Rightarrow \neg\psi}{\Gamma \Rightarrow \neg(\varphi \vee \psi)}(\Rightarrow\neg\vee)$$

$$\frac{\Gamma[\varphi] \Rightarrow \chi}{\Gamma[\neg\neg\varphi] \Rightarrow \chi}(\neg\neg\Rightarrow) \qquad \frac{\Gamma \Rightarrow \psi}{\Gamma \Rightarrow \neg\neg\psi}(\neg\neg\Rightarrow)$$

(3) Modal rules:

$$\frac{\Gamma[\langle\varphi\rangle] \Rightarrow \psi}{\Gamma[\Diamond\varphi] \Rightarrow \psi}(\Diamond\Rightarrow) \qquad \frac{\Gamma \Rightarrow \psi}{\langle\Gamma\rangle \Rightarrow \Diamond\psi}(\Rightarrow\Diamond)$$

$$\frac{\Gamma[\neg\varphi] \Rightarrow \psi}{\Gamma[\langle\neg\Diamond\varphi\rangle] \Rightarrow \psi}(\neg\Diamond\Rightarrow) \qquad \frac{\langle\Gamma\rangle \Rightarrow \neg\psi}{\Gamma \Rightarrow \neg\Diamond\psi}(\Rightarrow\neg\Diamond)$$

$$\frac{\Gamma[\varphi] \Rightarrow \psi}{\Gamma[\langle\Box\varphi\rangle] \Rightarrow \psi}(\Box\Rightarrow) \qquad \frac{\langle\Gamma\rangle \Rightarrow \psi}{\Gamma \Rightarrow \Box\psi}(\Rightarrow\Box)$$

$$\frac{\Gamma[\langle\neg\varphi\rangle] \Rightarrow \psi}{\Gamma[\neg\Box\varphi] \Rightarrow \psi}(\neg\Box\Rightarrow) \qquad \frac{\Gamma \Rightarrow \neg\psi}{\langle\Gamma\rangle \Rightarrow \neg\Box\psi}(\Rightarrow\neg\Box)$$

(3) Structural rules and Cut rule:

$$\frac{\Gamma[\Delta] \Rightarrow \psi}{\Gamma[\Delta, \Sigma] \Rightarrow \psi}(\text{Wk}) \qquad \frac{\Gamma[\Delta, \Delta] \Rightarrow \psi}{\Gamma[\Delta] \Rightarrow \psi}(\text{Ctr})$$

$$\frac{\Gamma[\langle\Delta\rangle] \Rightarrow \psi}{\Gamma[\Delta] \Rightarrow \psi}(\text{T}) \qquad \frac{\Gamma[\langle\Delta\rangle] \Rightarrow \psi}{\Gamma[\langle\langle\Delta\rangle\rangle] \Rightarrow \psi}(4) \qquad \frac{\Delta \Rightarrow \varphi \quad \Gamma[\varphi] \Rightarrow \psi}{\Gamma[\Delta] \Rightarrow \psi}(\text{Cut})$$

The term φ in (Cut) is called the *cut term*. A term or term structure in the below sequent of a rule is called *principal* if it is derived by that rule. The notation $\vdash_{\mathbf{wG5}} \Gamma \Rightarrow \psi$ means that $\Gamma \Rightarrow \psi$ is *derivable* in **wG5**. The subscript **wG5** is omitted if no confusion will arise. A sequent rule (R) is *admissible* in **wG5** if the conclusion is derivable whenever the premisses of (R) are derivable in **wG5**.

Lemma 1. *The following rules are admissible in* **wG5***:*

$$\frac{\Gamma[\Delta, \Sigma] \Rightarrow \psi}{\Gamma[\Sigma, \Delta] \Rightarrow \psi}(\text{Ex}) \qquad \frac{\Gamma[\Delta_1, (\Delta_2, \Delta_3)] \Rightarrow \psi}{\Gamma[(\Delta_1, \Delta_2), \Delta_3] \Rightarrow \psi}(\text{As}_1) \qquad \frac{\Gamma[(\Delta_1, \Delta_2), \Delta_3] \Rightarrow \psi}{\Gamma[\Delta_1, (\Delta_2, \Delta_3)] \Rightarrow \psi}(\text{As}_2)$$

Proof. Straightforward by (Wk) and (Ctr). □

For $n \geq 0$, let $\Gamma[-]^n$ be a context with n positions. In particular, if $n = 0$, $\Gamma[-]^0 = \Gamma$. Let $\Gamma[\Delta_1] \ldots [\Delta_n]$ be the term structure obtained from $\Gamma[-]^n$ by filling $\Delta_1, \ldots, \Delta_n$ into the n positions in order. Let $\Gamma[\Delta]^n$ be filling Δ into the n positions. Let **wG5•** be sequent calculus obtained from **wG5** by replacing (Cut) with the following *extended cut rule*:

$$\frac{\Delta \Rightarrow \varphi \quad \Gamma[\varphi]^n \Rightarrow \psi}{\Gamma[\Delta]^n \Rightarrow \psi}(\text{ECut}).$$

Clearly **wG5•** is equivalent to **wG5**, i.e., for any sequent $\Gamma \Rightarrow \psi$, $\vdash_{\mathbf{wG5•}} \Gamma \Rightarrow \psi$ if and only if $\vdash_{\mathbf{wG5}} \Gamma \Rightarrow \psi$. The system **wG5•** is needed in the proof of cut elimination theorem because the system **wG5** contains the contraction rule.

Theorem 2 (Cut Elimination). *If* $\vdash_{\mathbf{wG5}} \Gamma \Rightarrow \psi$, *then there is a derivation of* $\Gamma \Rightarrow \psi$ *in* **wG5** *without using* (Cut).

Proof. Assume that $\vdash_{\mathbf{wG5}} \Gamma \Rightarrow \psi$. Then $\vdash_{\mathbf{wG5•}} \Gamma \Rightarrow \psi$. Let $\mathcal{D}$ be a derivation of $\Gamma \Rightarrow \psi$ in **wG5•**. Take an application of (ECut) in a branch of $\mathcal{D}$ such that there is no application of (ECut) above. We show that such an application can be eliminated and by repeating the process we obtain a cut-free derivation of $\Gamma \Rightarrow \psi$. Consider such an instance of (ECut) with premissed $\Delta \Rightarrow \alpha$ and $\Sigma[\alpha]^n \Rightarrow \beta$ which are derived by (R_1) and (R_2) respectively.

Assume that at least one of (R_1) and (R_2) is an axiom. Then we can derive the conclusion $\Sigma[\Delta] \Rightarrow \beta$ without using (ECut). Similarly, if at least one of (R_1) and (R_2) is a structural rule, (T) or (4), we can easily get the conclusion. For example, let $(\mathbf{R_2})$ be (Ctr). One case is that the derivation

$$\frac{\Delta \Rightarrow \alpha \qquad \dfrac{\dfrac{\Sigma[\alpha, \alpha][\alpha]^{n-1} \Rightarrow \beta}{\Sigma[\alpha][\alpha]^{n-1} \Rightarrow \beta}(\text{Ctr})}{}}{\Sigma[\langle\langle\Delta\rangle\rangle][\Delta]^{n-1} \Rightarrow \beta}(\text{ECut})$$

is transformed into the following derivation:

$$\frac{\dfrac{\Delta \Rightarrow \alpha \quad \Sigma[\alpha, \alpha][\alpha]^{n-1} \Rightarrow \beta}{\Sigma[\Delta, \Delta][\Delta]^{n-1} \Rightarrow \beta}(\text{ECut})}{\Sigma[\Delta]^n \Rightarrow \beta}(\text{Ctr})$$

where (ECut) is applied to sequents with lower height.

Assume that the cut term is not principle in (R_1), we apply (Ecut) to the right premiss of (ECut) and the premisses of (R_1). For example, let (R_1) be $(\Diamond\Rightarrow)$. The derivation

$$\cfrac{\cfrac{\Delta[\langle\varphi\rangle] \Rightarrow \alpha}{\Delta[\Diamond\varphi] \Rightarrow \alpha} (\Diamond\Rightarrow) \qquad \Sigma[\alpha]^n \Rightarrow \beta}{\Sigma[\Delta[\Diamond\varphi]]^n \Rightarrow \beta} (\text{ECut})$$

is transformed into the following derivation:

$$\cfrac{\cfrac{\Delta[\langle\varphi\rangle] \Rightarrow \alpha \quad \Sigma[\alpha]^n \Rightarrow \beta}{\Sigma[\Delta[\langle\varphi\rangle]]^n \Rightarrow \beta} (\text{ECut})}{\Sigma[\Delta[\Diamond\varphi]]^n \Rightarrow \beta} (\Diamond\Rightarrow)^n$$

where $(\Diamond\Rightarrow)^n$ means n times application of $(\Diamond\Rightarrow)$, and (ECut) is applied to sequents with lower height.

Assume that the cut term is not principal in (R_2), we apply cut to the right premiss of (ECut) and the premiss of (R_2). For example, let (R_2) be $(\Box\Rightarrow)$. One case is that the derivation

$$\cfrac{\Delta \Rightarrow \alpha \qquad \cfrac{\Sigma[\beta][\alpha]^n \Rightarrow \beta}{\Sigma[\langle\Box\beta\rangle][\alpha]^n \Rightarrow \beta} (\Box\Rightarrow)}{\Sigma[\langle\Box\beta\rangle][\Delta]^n \Rightarrow \beta} (\text{ECut})$$

is transformed into the following derivation:

$$\cfrac{\cfrac{\Delta \Rightarrow \alpha \quad \Sigma[\beta][\alpha]^n \Rightarrow \beta}{\Sigma[\beta][\Delta]^n \Rightarrow \beta} (\text{ECut})}{\Sigma[\langle\Box\beta\rangle][\Delta]^n \Rightarrow \beta} (\Box\Rightarrow)$$

where (ECut) is applied to sequents with lower height.

Assume that the cut term α is principal in both premisses. The proof proceeds by induction on the complexity of α. Here we show only the following cases and the remaining cases are shown similarly.

(1) $\alpha = \Diamond\alpha'$. The derivation

$$\cfrac{\cfrac{\Delta' \Rightarrow \alpha'}{\langle\Delta'\rangle \Rightarrow \Diamond\alpha'} (\Rightarrow\Diamond) \qquad \cfrac{\Sigma[\langle\alpha'\rangle][\alpha]^{n-1} \Rightarrow \beta}{\Sigma[\Diamond\alpha'][\alpha]^{n-1} \Rightarrow \beta} (\Diamond\Rightarrow)}{\Sigma[\langle\Delta'\rangle]^n \Rightarrow \beta} (\text{ECut})$$

is transformed into the following derivation:

$$\cfrac{\Delta' \Rightarrow \alpha' \qquad \cfrac{\langle\Delta'\rangle \Rightarrow \alpha \quad \Sigma[\langle\alpha'\rangle][\alpha]^{n-1} \Rightarrow \beta}{\Sigma[\langle\alpha'\rangle][\langle\Delta'\rangle]^{n-1} \Rightarrow \beta} (\text{ECut})}{\Sigma[\langle\Delta'\rangle]^n \Rightarrow \beta} (\text{ECut})$$

where (ECut) is applied to sequents with lower height or less complicated term.

(2) $\alpha = \neg\Diamond\alpha'$. The derivation

$$\dfrac{\dfrac{\langle\Delta\rangle \Rightarrow \neg\alpha'}{\Delta \Rightarrow \neg\Diamond\alpha'}\ (\Rightarrow\neg\Diamond) \quad \dfrac{\Sigma[\neg\alpha'][\alpha]^{n-1} \Rightarrow \beta}{\Sigma[\langle\neg\Diamond\alpha'\rangle][\alpha]^{n-1} \Rightarrow \beta}\ (\neg\Diamond\Rightarrow)}{\Sigma[\langle\Delta\rangle][\Delta]^{n-1} \Rightarrow \beta}\ (\text{ECut})$$

is transformed into the following derivation:

$$\dfrac{\langle\Delta\rangle \Rightarrow \neg\alpha' \quad \dfrac{\Delta \Rightarrow \neg\alpha \quad \Sigma[\neg\alpha'][\alpha]^{n-1} \Rightarrow \beta}{\Sigma[\neg\alpha'][\Delta]^{n-1} \Rightarrow \beta}\ (\text{ECut})}{\Sigma[\langle\Delta\rangle][\Delta]^{n-1} \Rightarrow \beta}\ (\text{ECut})$$

where (ECut) is applied to sequents with lower height or less complicated term.

(3) $\alpha = \neg\neg\alpha'$. The derivation

$$\dfrac{\dfrac{\Delta \Rightarrow \alpha'}{\Delta \Rightarrow \neg\neg\alpha'}\ (\Rightarrow\neg\neg) \quad \dfrac{\Sigma[\alpha'][\alpha]^{n-1} \Rightarrow \beta}{\Sigma[\neg\neg\alpha'][\alpha]^{n-1} \Rightarrow \beta}\ (\neg\neg\Rightarrow)}{\Sigma[\Delta]^n \Rightarrow \beta}\ (\text{ECut})$$

is transformed into the following derivation:

$$\dfrac{\Delta \Rightarrow \alpha' \quad \dfrac{\Delta \Rightarrow \alpha \quad \Sigma[\alpha'][\alpha]^{n-1} \Rightarrow \beta}{\Sigma[\alpha'][\Delta]^{n-1} \Rightarrow \beta}\ (\text{ECut})}{\Sigma[\Delta]^n \Rightarrow \beta}\ (\text{ECut})$$

where (ECut) is applied to sequents with lower height or less complicated term. $\square$

Let $\mathbf{wG5}^\circ$ be the sequent calculus obtained from $\mathbf{wG5}$ by dropping (Cut). By the Cut elimination theorem, $\mathbf{wG5}^\circ$ is equivalent to $\mathbf{wG5}$.

Corollary 1. *For any sequent* $\Gamma \Rightarrow \psi$, $\vdash_{\mathbf{wG5}} \Gamma \Rightarrow \psi$ *if and only if* $\vdash_{\mathbf{wG5}^\circ} \Gamma \Rightarrow \psi$.

Hence the cut rule (Cut) is admissible in $\mathbf{wG5}^\circ$. Moreover, by the cut elimination theorem, we have the following property of a derivation which is analogue to the 'subformula property' in proof theory.

Corollary 2. *If* $\vdash_{\mathbf{wG5}} \Gamma \Rightarrow \psi$, *then there is a derivation in* $\mathbf{wG5}$ *in which the complexity of each term in the upper sequent is less or equal to the complexity of a term in the lower sequent.*

4   Soundness and Completeness

In this section, we shall prove that the sequent calculus $\mathbf{wG5}$ is sound and complete with respect to $\mathbf{tqBa5}$. Given a tqBa5 $\mathbb{A} = (A, \wedge, \vee, \neg, \square, 0, 1)$, an *assignment* in $\mathbb{A}$ is a function $\theta : \mathbb{X} \to A$. Every assignment θ can be extended homomorphically to the term algebra $\mathfrak{T}$. Let $\theta(\varphi)$ denote the value of φ under the assignment θ. For any term structure Γ, the term $f(\Gamma)$ associated with Γ is defined inductively by: $f(\varphi) = \varphi$; $f(\Gamma, \Delta) = f(\Gamma) \wedge f(\Delta)$; $f(\langle \Gamma \rangle) = \Diamond f(\Gamma)$. For any tqBa5 $\mathfrak{A}$, a sequent $\Gamma \Rightarrow \psi$ is *valid* in $\mathfrak{A}$, notation $\mathfrak{A} \models \Gamma \Rightarrow \psi$, if $\theta(f(\Gamma)) \leq \theta(\psi)$ for any assignment θ in $\mathfrak{A}$. The notation **tqBa5** $\models \Gamma \Rightarrow \psi$ stands for that $\mathfrak{A} \models \Gamma \Rightarrow \psi$ for all $\mathfrak{A} \in$ **tqBa5**.

Definition 4. *For any sequent* $\Gamma[\varphi] \Rightarrow \psi$, *we obtain* $\varphi \Rightarrow \mathbf{r}(\overline{\Gamma}(\psi))$ *by the following rules:*

$$\frac{\Gamma_1, \Gamma_2 \Rightarrow \psi}{\Gamma_2 \Rightarrow \tau(\Gamma_1) \to \psi}(\mathrm{R}_1) \qquad \frac{\langle \Delta \rangle \Rightarrow \psi}{\Delta \Rightarrow \Box \psi}(\mathrm{R}_2) \qquad \frac{\Gamma[\Delta_1, \Delta_2] \Rightarrow \psi}{\Gamma[\Delta_2, \Delta_1] \Rightarrow \psi}(\mathrm{Ex})$$

We say that φ *is displayed in the sequent* $\varphi \Rightarrow \mathbf{r}(\overline{\Gamma}(\psi))$.

Every formula in the antecedent of a sequent can be displayed. The consequent of $\varphi \Rightarrow \mathbf{r}(\overline{\Gamma}(\psi))$ is the result of displaying φ in $\Gamma[\varphi] \Rightarrow \psi$ and it contains ψ. For example, the formula q in the antecedent of the sequent $\langle p, \langle q, r \rangle \rangle \Rightarrow \Diamond \Diamond p$ can be displayed as follows:

$$\frac{\dfrac{\dfrac{\dfrac{\dfrac{\langle p, \langle q, r \rangle \rangle \Rightarrow \Diamond \Diamond p}{p, \langle q, r \rangle \Rightarrow \Box \Diamond \Diamond p}(\mathrm{R}_2)}{\langle q, r \rangle \Rightarrow p \to \Box \Diamond \Diamond p}(\mathrm{R}_1)}{q, r \Rightarrow \Box(p \to \Box \Diamond \Diamond p)}(\mathrm{R}_2)}{r, q \Rightarrow \Box(p \to \Box \Diamond \Diamond p)}(\mathrm{Ex})}{q \Rightarrow r \to \Box(p \to \Box \Diamond \Diamond p)}(\mathrm{R}_1)$$

Lemma 2. **tqBa5** $\models \Gamma[\varphi] \Rightarrow \psi$ *if and only if* **tqBa5** $\models \varphi \Rightarrow \mathbf{r}(\overline{\Gamma}(\psi))$.

Proof. The rules (R1), (R2) and (Ex) for displaying φ preserve validity in **tqBa5**. The following inverse rules also preserve validity in **tqBa5**:

$$\frac{\Gamma_2 \Rightarrow \tau(\Gamma_1) \to \psi}{\Gamma_1, \Gamma_2 \Rightarrow \psi}(\mathrm{R}_3) \qquad \frac{\Delta \Rightarrow \Box \psi}{\langle \Delta \rangle \Rightarrow \psi}(\mathrm{R}_4)$$

Hence **tqBa5** $\models \Gamma[\varphi] \Rightarrow \psi$ if and only if **tqBa5** $\models \varphi \Rightarrow \mathbf{r}(\overline{\Gamma}(\psi))$.

Theorem 3 (Soundness). *If* $\vdash_{\mathbf{wG5}} \Gamma \Rightarrow \psi$, *then* **tqBa5** $\models \Gamma \Rightarrow \psi$.

Proof. Assume that $\vdash_{\mathbf{wG5}} \Gamma \Rightarrow \psi$. The proof proceeds by induction on the height of a derivation of $\Gamma \Rightarrow \psi$ in **wG5**. It is easy to show that all axioms are valid in **tqBa5**. Note that the axiom ($\bot$) is valid by Lemma 2. It is easy to show that all rules in **wG5** preserve validity in **tqBa5** by Lemma 2. $\qquad \square$

To show the completeness of **wG5**, it suffices to show the completeness of **wG5°**. Henceforth, we use the sequent calculus **wG5°**. Recall that the cut rule (Cut) is admissible in **wG5°**.

Lemma 3. *The following rule of monotonicity is admissible in* **wG5°** *:*

$$\frac{\varphi \Rightarrow \psi}{\Gamma[\varphi] \Rightarrow f(\Gamma[\psi])}(MN).$$

Proof. Assume that $\vdash \varphi \Rightarrow \psi$. The proof proceeds by induction on the complexity of $\Gamma[-]$. The case that $\Gamma[-] = [-]$ is obvious. Suppose that $\Gamma[-] = (\Gamma_1[-], \Gamma_2)$. By induction hypothesis, $\vdash \Gamma_1[\varphi] \Rightarrow f(\Gamma_1[\psi])$. Then it is easy to obtain that $\vdash \Gamma_1[\varphi] \wedge f(\Gamma_2) \Rightarrow f(\Gamma_1[\psi]) \wedge f(\Gamma_2)$. Clearly $\vdash \Gamma_2 \Rightarrow f(\Gamma_2)$. By (Cut), $\vdash \Gamma_1[\varphi], f(\Gamma_2) \Rightarrow f(\Gamma_1[\psi]) \wedge f(\Gamma_2)$. Suppose that $\Gamma[-] = \langle \Delta[-] \rangle$. By induction hypothesis, $\vdash \Delta[\varphi] \Rightarrow f(\Delta[\psi])$. By $(\Rightarrow \Diamond)$, $\vdash \langle \Delta[\varphi] \rangle \Rightarrow \Diamond f(\Delta[\psi])$.   □

Lemma 4. *The following hold in* **wG5°** *:*

(1) $\vdash \neg f(\Gamma[\varphi]) \wedge \neg f(\Gamma[\psi]) \Rightarrow \neg f(\Gamma[\varphi \vee \psi])$.
(2) $\vdash \neg f(\Gamma[\neg\varphi]) \wedge \neg f(\Gamma[\neg\psi]) \Rightarrow \neg f(\Gamma[\neg(\varphi \wedge \psi)])$.
(3) $\vdash \neg f(\Gamma[\neg\varphi, \neg\psi]) \Rightarrow \neg f(\Gamma[\neg(\varphi \vee \psi)])$.
(4) $\vdash \neg f(\Gamma[\varphi]) \Rightarrow \neg f(\Gamma[\neg\neg\varphi])$.
(5) $\vdash \neg f(\Gamma[\neg\varphi]) \Rightarrow \neg f(\Gamma[\langle \neg \Diamond\varphi \rangle])$.
(6) $\vdash \neg f(\Gamma[\varphi]) \Rightarrow \neg f(\Gamma[\langle \Box\varphi \rangle])$.
(7) $\vdash \neg f(\Gamma[\langle \neg\varphi \rangle]) \Rightarrow \neg f(\Gamma[\Box\varphi])$.

Proof. Here we show only (5). The remaining items are shown similarly. The proof proceeds by induction on the complexity of $\Gamma[-]$. Assume that $\Gamma[-] = [-]$. We need to show $\vdash \neg\neg\varphi \Rightarrow \neg\neg\Diamond\varphi$. One derivation is as follows:

$$\frac{\dfrac{\dfrac{\varphi \Rightarrow \varphi}{\langle\varphi\rangle \Rightarrow \Diamond\varphi}(\Diamond\Rightarrow)}{\dfrac{\varphi \Rightarrow \Diamond\varphi}{\neg\neg\varphi \Rightarrow \Diamond\varphi}(\neg\neg\Rightarrow)}(T)}{\neg\neg\varphi \Rightarrow \neg\neg\Diamond\varphi}(\Rightarrow\neg\neg)$$

Assume that $\Gamma[-] = (\Gamma_1[-], \Gamma_2)$. By induction hypothesis, $\vdash \neg f(\Gamma_1[\neg\varphi]) \Rightarrow \neg f(\Gamma_1[\langle\neg\Diamond\varphi\rangle])$. Clearly we have $\neg f(\Gamma_1[\neg\varphi], \Gamma_2) = \neg(f(\Gamma_1[\neg\varphi]) \wedge f(\Gamma_2))$ and $\neg f(\Gamma_1[\langle\neg\Diamond\varphi\rangle], \Gamma_2) = \neg(f(\Gamma_1[\langle\neg\Diamond\varphi\rangle]) \wedge f(\Gamma_2))$. Let $f(\Gamma_1[\neg\varphi]) = \alpha$, $f(\Gamma_1[\langle\neg\Diamond\varphi\rangle]) = \beta$ and $f(\Gamma_2) = \gamma$. One derivation is as follows:

$$\frac{\dfrac{\neg\alpha \Rightarrow \neg\beta}{\neg\alpha \Rightarrow \neg(\beta \wedge \gamma)}(\Rightarrow\neg\wedge) \qquad \dfrac{\neg\gamma \Rightarrow \neg\gamma}{\neg\gamma \Rightarrow \neg(\beta \wedge \gamma)}(\Rightarrow\neg\wedge)}{\neg(\alpha \wedge \gamma) \Rightarrow \neg(\beta \wedge \gamma)}(\neg\wedge\Rightarrow)$$

Assume that $\Gamma[-] = \langle \Sigma[-] \rangle$. Let $\alpha = f(\Sigma[\neg\varphi])$ and $\beta = f(\Sigma[\langle\neg\Diamond\varphi\rangle])$. By induction hypothesis, $\vdash \neg\alpha \Rightarrow \neg\beta$. The derivation of $\neg\Diamond\alpha \Rightarrow \neg\Diamond\beta$ is as follows:

$$\frac{\dfrac{\neg\alpha \Rightarrow \neg\beta}{\langle\neg\Diamond\alpha\rangle \Rightarrow \neg\beta}(\neg\Diamond\Rightarrow)}{\neg\Diamond\alpha \Rightarrow \neg\Diamond\beta}(\Rightarrow\neg\Diamond)$$

This completes the proof.

Lemma 5. *The following contraposition rule is admissible in* **wG5°***:*

$$\frac{\Gamma \Rightarrow \psi}{\neg\psi \Rightarrow \neg f(\Gamma)}(\text{Ctp}).$$

Proof. Assume that $\vdash \Gamma \Rightarrow \psi$. Then there is a derivation $\mathcal{D}$ in **wG5°** for $\Gamma \Rightarrow \psi$. By induction on the height n of $\mathcal{D}$, we prove that $\vdash \neg\psi \Rightarrow \neg f(\Gamma)$. If $n = 0$, then $\Gamma \Rightarrow \psi$ is an axiom. Then it is easy to show that $\vdash \neg\psi \Rightarrow \neg f(\Gamma)$. Note that we can show that $\vdash \neg\psi \Rightarrow \neg f(\Gamma[\bot])$ by induction on the complexity of $\Gamma[-]$. Assume that $n > 0$. Then $\Gamma \Rightarrow \psi$ is obtained by a rule (R). If (R) is a connective rule or modal rule, by (MN) and induction hypothesis, it is easy to show that $\vdash \neg\psi \Rightarrow \neg f(\Gamma)$. For example, let (R) be ($\Box\Rightarrow$) and the derivation end with

$$\frac{\Gamma[\varphi] \Rightarrow \psi}{\Gamma[\langle\Box\varphi\rangle] \Rightarrow \psi}.$$

By induction hypothesis, $\vdash \neg\psi \Rightarrow \neg f(\Gamma[\varphi])$. By Lemma 4 (6), $\vdash \neg f(\Gamma[\varphi]) \Rightarrow \neg f(\Gamma[\langle\Box\varphi\rangle])$. Hence $\vdash \neg\psi \Rightarrow \neg f(\Gamma[\langle\Box\varphi\rangle])$. $\square$

Lemma 6. *For any term structure Γ, the following hold:*

(1) $\vdash_{\textbf{wG5°}} \Gamma \Rightarrow f(\Gamma)$.
(2) if $\vdash_{\textbf{wG5°}} f(\Gamma) \Rightarrow \psi$, *then* $\vdash_{\textbf{wG5°}} \Gamma \Rightarrow \psi$.

Proof. (1) is shown by induction on the complexity of Γ. The case that Γ is a term is trivial. Assume that $\Gamma = (\Gamma_1, \Gamma_2)$. By induction hypothesis, $\vdash_{\textbf{wG5°}} \Gamma_1 \Rightarrow f(\Gamma_1)$ and $\vdash_{\textbf{wG5°}} \Gamma_2 \Rightarrow f(\Gamma_2)$. By (Wk), $\vdash_{\textbf{wG5°}} \Gamma_1, \Gamma_2 \Rightarrow f(\Gamma_1)$ and $\vdash_{\textbf{wG5°}} \Gamma_1, \Gamma_2 \Rightarrow f(\Gamma_2)$. By ($\Rightarrow\wedge$), $\vdash_{\textbf{wG5°}} \Gamma_1, \Gamma_2 \Rightarrow f(\Gamma_1) \wedge f(\Gamma_2)$. Assume that $\Gamma = \langle\Delta\rangle$. By induction hypothesis, $\vdash_{\textbf{wG5°}} \Delta \Rightarrow f(\Delta)$. By ($\Rightarrow\Diamond$), $\vdash_{\textbf{wG5°}} \langle\Delta\rangle \Rightarrow \Diamond f(\Delta)$. For (2), assume that $\vdash_{\textbf{wG5°}} f(\Gamma) \Rightarrow \psi$. By (1) and (Cut), $\vdash_{\textbf{wG5°}} \Gamma \Rightarrow \psi$. $\square$

To show the completeness of **wG5°**, we introduce the Lindenbaum-Tarski algebra. The binary relation $\sim$ on the set of all terms $\mathcal{T}$ as follows:

$$\varphi \sim \psi \text{ if and only if } \vdash_{\textbf{wG5°}} \varphi \Rightarrow \psi \text{ and } \vdash_{\textbf{wG5°}} \psi \Rightarrow \varphi.$$

Clearly $\sim$ is an equivalence relation on $\mathcal{T}$. Let $|\varphi| = \{\psi \in \mathcal{T} \mid \varphi \sim \psi\}$ be the equivalence class of φ under $\sim$. Let $\mathcal{T}/_\sim$ be the set of all such equivalence classes. Moreover, by the rules for $\wedge$ and $\vee$ as well as (Ctp), one can easily show that $\sim$ is a congruence relation on $\mathcal{T}$. Then we define the following operations on $\mathcal{T}/_\sim$:

$$|\varphi| \wedge' |\psi| = |\varphi \wedge \psi| \qquad\qquad |\varphi| \vee' |\psi| = |\varphi \vee \psi|$$
$$\neg'|\varphi| = |\neg\varphi| \qquad\qquad\qquad \Box'|\varphi| = |\Box\varphi|$$
$$0' = |\bot| \qquad\qquad\qquad\qquad 1' = |\top|$$

Let $\mathfrak{T}/_\sim = (\mathcal{T}/_\sim, \wedge', \vee', \neg', \Box', 0', 1')$ be the quotient algebra of the term algebra $\mathfrak{T}$ under $\sim$. One can easily show that $\mathfrak{T}/_\sim$ is a tqBa5.

Lemma 7. *For any terms φ and ψ, if $|\varphi| \leq |\psi|$, then* $\vdash_{\textbf{wG5°}} \varphi \Rightarrow \psi$.

Proof. Assume that $|\varphi| \leq |\psi|$. Then $|\varphi| \wedge' |\psi| = |\varphi \wedge \psi| = |\varphi|$. Hence $\varphi \wedge \psi \sim \varphi$. Then $\vdash_{\mathbf{wG5^\circ}} \varphi \Rightarrow \varphi \wedge \psi$. Clearly $\vdash_{\mathbf{wG5^\circ}} \varphi \wedge \psi \Rightarrow \psi$. By (Cut), $\vdash_{\mathbf{wG5^\circ}} \varphi \Rightarrow \psi$. $\square$

Theorem 4 (Completeness). *If* $\mathbf{tqBa5} \models \Gamma \Rightarrow \psi$, *then* $\vdash_{\mathbf{wG5^\circ}} \Gamma \Rightarrow \psi$.

Proof. Assume that $\not\vdash_{\mathbf{wG5^\circ}} \Gamma \Rightarrow \psi$. By Lemma 6 (2), $\not\vdash_{\mathbf{wG5^\circ}} f(\Gamma) \Rightarrow \psi$. By Lemma 7, $|f(\Gamma)| \not\leq |\psi|$. Let θ be the assignment in $\mathfrak{T}/_\sim$ with $\theta(x) = |x|$ for every variable x. It is easy to show by induction on the complexity of φ that $\theta(\varphi) = |\varphi|$. Hence $\theta(f(\Gamma)) \not\leq \theta(\psi)$. Therefore $\mathbf{tqBa5} \not\models \Gamma \Rightarrow \psi$. $\square$

By the completeness theorem, $\mathbf{wG5}$ is indeed a sequent calculus for partition topological quasi-Boolean algebras.

5   The Sequent Calculus wG4

In this section, we shall introduce a sequent calculus $\mathbf{wG4}$ for the variety of topological quasi-Boolean algebras. We first introduce a sequent calculus $\mathbf{wG4}_t$ for topological quasi-Boolean algebras with tense operators. And then we get $\mathbf{wG4}$ by dropping rules for additional operators, and it is a sequent calculus for $\mathbf{tqBa}$ since $\mathbf{wG4}_t$ is a conservative extension of the logic of $\mathbf{tqBa}$.

Definition 5. A *topological quasi-Boolean algebra with tense operators* (tqBaT) is an algebra $\mathbb{A} = (A, \wedge, \vee, \neg, 0, 1, \blacklozenge, \square)$ where $(A, \wedge, \vee, \neg, 0, 1, \square)$ is a topological quasi-Boolean algebra and $\blacklozenge$ is an unary operation on A such that for all $a, b \in A$:

$$(\mathrm{Adj}_\blacklozenge)\blacklozenge a \leq b \text{ if and only if } a \leq \square b.$$

We define $\blacksquare a := \neg\blacklozenge\neg a$. The class of all topological quasi-Boolean algebras with tense operators is denoted by $\mathbf{tqBaT}$.

Lemma 8. *For any tqBaT* $\mathbb{A}$ *and* $a, b \in A$, *the following hold:*

(1) $\blacklozenge 0 = 0$ *and* $\blacksquare 1 = 1$.
(2) if $a \leq b$, *then* $\blacklozenge a \leq \blacklozenge b$ *and* $\blacksquare a \leq \blacksquare b$.
(3) $\blacklozenge(a \vee b) = \blacklozenge a \vee \blacklozenge b$ *and* $\blacksquare(a \wedge b) = \blacksquare a \wedge \blacksquare b$.
(4) $a \leq \square\blacklozenge a$ *and* $a \leq \blacklozenge a$.
(5) $\lozenge a \leq b$ *if and only if* $a \leq \blacksquare b$.
(6) $a \leq \blacksquare\lozenge a$ *and* $\blacklozenge\blacklozenge a \leq \blacklozenge a$.
(7) $\blacklozenge\blacklozenge a = \blacklozenge a$ *and* $\blacksquare\blacksquare a = \blacksquare a$.

Proof. Here we show only (4) and (5), and the remaining items are shown easily. For (4), by $\blacklozenge a \leq \blacklozenge a$ and (Adj), $a \leq \square\blacklozenge a$. Since $\mathbb{A}$ is a tqBa, we have $\square\blacklozenge a \leq \blacklozenge a$. Then $a \leq \blacklozenge a$. For (5), assume that $\lozenge a \leq b$. Then $\neg b \leq \neg\lozenge a = \square\neg a$. By $(\mathrm{Adj}_\blacklozenge)$, $\blacklozenge\neg b \leq \neg a$. Then $a \leq \blacksquare b$. The other direction is shown similarly.

Definition 6. The set of all *tense terms* $\mathcal{T}_t$ is defined inductively as follows:

$$\mathcal{T}_t \ni \varphi ::= x \mid \perp \mid \neg\varphi \mid (\varphi \wedge \varphi) \mid (\varphi \vee \varphi) \mid \blacklozenge\varphi \mid \square\varphi, \text{ where } x \in \mathbb{X}.$$

Let $\mathfrak{T}_t = (\mathcal{T}, \wedge, \vee, \neg, \perp, \top, \blacklozenge, \square)$ be the tense term algebra.

Now we shall introduce the sequent calculus $\mathbf{wG4}_t$ for the tense logic $\mathbf{wS4}_t$. For this purpose, we introduce three structural operators: (i) the comma for $\wedge$; (ii) $\langle - \rangle^\uparrow$ for $\Diamond$; and (iii) $\langle - \rangle^\downarrow$ for $\blacklozenge$.

Definition 7. A *tense term structure* is an expression Γ defined as follows:

$$\Gamma := \varphi \mid (\Gamma, \Gamma) \mid \langle \Gamma \rangle^\uparrow \mid \langle \Gamma \rangle^\downarrow, \text{ where } \varphi \in T_t.$$

A *sequent* is the form $\Gamma \Rightarrow \varphi$ where Γ is a tense term structure and $\varphi \in T_t$.

Definition 8. The Gentzen sequent calculus $\mathbf{wG4}_t$ consists of axioms and connective rules in $\mathbf{wG5}$ and the following rules:

(1) Modal rules:

$$\frac{\Gamma[\langle\varphi\rangle^\uparrow] \Rightarrow \psi}{\Gamma[\Diamond\varphi] \Rightarrow \psi}(\Diamond\Rightarrow) \qquad \frac{\Gamma \Rightarrow \psi}{\langle\Gamma\rangle^\uparrow \Rightarrow \Diamond\psi}(\Rightarrow\Diamond)$$

$$\frac{\Gamma[\neg\varphi] \Rightarrow \psi}{\Gamma[\langle\neg\Diamond\varphi\rangle^\uparrow] \Rightarrow \psi}(\neg\Diamond\Rightarrow) \qquad \frac{\langle\Gamma\rangle^\uparrow \Rightarrow \neg\psi}{\Gamma \Rightarrow \neg\Diamond\psi}(\Rightarrow\neg\Diamond)$$

$$\frac{\Gamma[\varphi] \Rightarrow \psi}{\Gamma[\langle\Box\varphi\rangle^\uparrow] \Rightarrow \psi}(\Box\Rightarrow) \qquad \frac{\langle\Gamma\rangle^\uparrow \Rightarrow \psi}{\Gamma \Rightarrow \Box\psi}(\Rightarrow\Box)$$

$$\frac{\Gamma[\langle\neg\varphi\rangle^\uparrow] \Rightarrow \psi}{\Gamma[\neg\Box\varphi] \Rightarrow \psi}(\neg\Box\Rightarrow) \qquad \frac{\Gamma \Rightarrow \neg\psi}{\langle\Gamma\rangle^\uparrow \Rightarrow \neg\Box\psi}(\Rightarrow\neg\Box)$$

$$\frac{\Gamma[\langle\varphi\rangle^\downarrow] \Rightarrow \psi}{\Gamma[\blacklozenge\varphi] \Rightarrow \psi}(\blacklozenge\Rightarrow) \qquad \frac{\Gamma \Rightarrow \psi}{\langle\Gamma\rangle^\downarrow \Rightarrow \blacklozenge\psi}(\Rightarrow\blacklozenge)$$

$$\frac{\Gamma[\neg\varphi] \Rightarrow \psi}{\Gamma[\langle\neg\Diamond\varphi\rangle^\downarrow] \Rightarrow \psi}(\neg\blacklozenge\Rightarrow) \qquad \frac{\langle\Gamma\rangle^\downarrow \Rightarrow \neg\psi}{\Gamma \Rightarrow \neg\blacklozenge\psi}(\Rightarrow\neg\blacklozenge)$$

$$\frac{\Gamma[\varphi] \Rightarrow \psi}{\Gamma[\langle\blacksquare\varphi\rangle^\downarrow] \Rightarrow \psi}(\blacksquare\Rightarrow) \qquad \frac{\langle\Gamma\rangle^\downarrow \Rightarrow \psi}{\Gamma \Rightarrow \blacksquare\psi}(\Rightarrow\Box)$$

$$\frac{\Gamma[\langle\neg\varphi\rangle^\downarrow] \Rightarrow \psi}{\Gamma[\neg\blacksquare\varphi] \Rightarrow \psi}(\neg\blacksquare\Rightarrow) \qquad \frac{\Gamma \Rightarrow \neg\psi}{\langle\Gamma\rangle^\downarrow \Rightarrow \neg\blacksquare\psi}(\Rightarrow\neg\blacksquare)$$

(2) Structural rules and Cut rule:

$$\frac{\Gamma[\Delta] \Rightarrow \psi}{\Gamma[\Delta, \Sigma] \Rightarrow \psi}(\text{Wk}) \qquad \frac{\Gamma[\Delta, \Delta] \Rightarrow \psi}{\Gamma[\Delta] \Rightarrow \psi}(\text{Ctr})$$

$$\frac{\Gamma[\langle\Delta\rangle^\uparrow] \Rightarrow \psi}{\Gamma[\Delta] \Rightarrow \psi}(\text{T}_\Diamond) \qquad \frac{\Gamma[\langle\Delta\rangle^\uparrow] \Rightarrow \psi}{\Gamma[\langle\langle\Delta\rangle^\uparrow\rangle^\uparrow] \Rightarrow \psi}(4_\Diamond)$$

$$\frac{\Gamma[\langle\Delta\rangle^\downarrow] \Rightarrow \psi}{\Gamma[\Delta] \Rightarrow \psi}(\text{T}_\blacklozenge) \qquad \frac{\Gamma[\langle\Delta\rangle^\downarrow] \Rightarrow \psi}{\Gamma[\langle\langle\Delta\rangle^\downarrow\rangle^\downarrow] \Rightarrow \psi}(4_\blacklozenge) \qquad \frac{\Delta \Rightarrow \varphi \quad \Gamma[\varphi] \Rightarrow \psi}{\Gamma[\Delta]}(\text{Cut})$$

The notation $\vdash_{\mathbf{wG4}_t} \Gamma \Rightarrow \psi$ means that $\Gamma \Rightarrow \psi$ is *derivable* in $\mathbf{wG4}_t$.

Theorem 5 (Cut Elimination). *If* $\vdash_{\mathbf{wG4}_t} \Gamma \Rightarrow \psi$, *then there is a derivation of* $\Gamma \Rightarrow \psi$ *in* $\mathbf{wG4}_t$ *without using* (Cut).

Proof. The proof proceeds as in the proof of Theorem 2.   □

Let $\mathbf{wG4}_t^{\circ}$ be the sequent calculus obtained from $\mathbf{wG4}_t$ by dropping (Cut). By the Cut elimination theorem, $\mathbf{wG4}_t^{\circ}$ is equivalent to $\mathbf{wG4}_t$.

Lemma 9. *The following contraposition rule is admissible in* $\mathbf{wG4}_t^{\circ}$:

$$\frac{\Gamma \Rightarrow \psi}{\neg\psi \Rightarrow \neg f(\Gamma)}(\text{Ctp}).$$

Proof. The proof is quite similar to the proof of Lemma 5.   □

Theorem 6 (Completeness). *If* $\mathbf{tqBaT} \models \Gamma \Rightarrow \psi$, *then* $\vdash_{\mathbf{wG4}_t^{\circ}} \Gamma \Rightarrow \psi$.

Proof. The soundness part is shown by induction on the height of a derivation. The proof of the completeness part is similar to the proof of Theorem 4.   □

Let $\mathbf{wG4}$ be the sequent calculus obtained from $\mathbf{wG4}_t$ by dropping the modal rules for ♦ and ■. Then $\mathbf{wG4}_t$ is a conservative extension of $\mathbf{wG4}$.

Theorem 7. *For any* $\varphi, \psi \in \mathcal{T}$, $\vdash_{\mathbf{wG4}} \varphi \Rightarrow \psi$ *if and only if* $\vdash_{\mathbf{wG4}_t} \varphi \Rightarrow \psi$.

Proof. The proof proceeds by induction on the height of a derivation of $\varphi \Rightarrow \psi$. Details are omitted here.   □

6   Concluding Remarks

We established a Gentzen sequent calculus $\mathbf{wG5}$ for partition topological quasi-Boolean algebras which admits cut elimination. This is one step in the proof analysis of logics for algebraic structures related with rough sets. Here we conclude with some remarks.

First, if the language is restricted to the language for quasi-Boolean algebras without modal operators and term structures are restricted to finite multisets of terms, the axioms and connective rules of $\mathbf{wG5}$ form a Gentzen sequent calculus for quasi-Boolean algebras. All structural rules and cut rule are admissible in this sequent calculus. And there is also a smooth decision procedure for the derivability of a sequent.

Second, the decidability of $\mathbf{wG5}$ can be proved by showing the finite model property of $\mathbf{wG5}$. This shall be presented in a further paper. Note that the decidability of $\mathbf{wG5}$ does not follow from the cut elimination theorem directly because the system contains the contraction rule.

Third, one can extend the $\mathbf{wG5}$ to sequent calculi for (weak) pre-rough algebras and their non-distributive varieties. It is very likely that we can get cut-free sequent calculi and prove the finite model property and decidability of some logics for (weak) pre-rough algebras and their non-distributive varieties. This provides a proof-theoretic approach to the study of equational theories of these classes of algebraic structures.

References

1. Banerjee, M.: Rough sets and 3-valued łukasiewicz logic. Fundamenta Informatica **31**, 213–220 (1997)
2. Banerjee, M., Chakraborty, M.: Rough algebra. Bull. Pol. Acad. Sci. (Math.) **41**(4), 293–297 (1993)
3. Banerjee, M., Chakraborty, M.: Rough sets through algebraic logic. Fundamenta Informaticae **28**(3–4), 211–221 (1996)
4. van Benthem, J., Bezhanishvili, G.: Modal logic of spaces. In: Aliello, M.I., Pratt-Hartmann, V.B.J. (eds.) Handbook of Spatial Logics, pp. 217–298. Springer, Heidelberg (2007). https://doi.org/10.1007/978-1-4020-5587-4_5
5. McKinsey, J., Tarski, A.: The algebra of topology. Ann. Math. **45**(1), 141–191 (1994)
6. Ono, H.: Proof-theoretic methods in nonclassical logic-an introduction. In: Takahashi, M., Okada, M., Deznai-Ciancaglini, M. (eds.) Theories of Types and Proofs, pp. 207–254. Mathematical Society of Japan, Tokyo (1998)
7. Pawlak, Z.: Rough sets. Int. J. Comput. Inf. Sci. **11**(5), 341–356 (1982)
8. Rasiowa, H.: An Algebraic Approach to Non-Classical Logics. North-Holland Publishing, Amsterdam (1974)
9. Saha, A., Sen, J., Chakraborty, M.: Algebraic structures in the vicinity of pre-rough algebra and their logics. Inf. Sci. **282**, 296–320 (2014)
10. Saha, A., Sen, J., Chakraborty, M.: Algebraic structures in the vicinity of pre-rough algebra and their logics II. Inf. Sci. **333**, 44–60 (2016)
11. Wansing, H.: Sequent systems for modal logics. In: Gabbay, D., Guenther, F. (eds.) Handbook of Philosophical Logic, vol. 8, pp. 61–145. Kluwer Academic Publisher, Dordrecht (2002)

Rule Induction Based on Indiscernible Classes from Rough Sets in Information Tables with Continuous Values

Michinori Nakata[1]([✉]), Hiroshi Sakai[2], and Keitarou Hara[3]

[1] Faculty of Management and Information Science, Josai International University,
1 Gumyo, Togane, Chiba 283-8555, Japan
nakatam@ieee.org
[2] Department of Mathematics and Computer Aided Sciences, Faculty of Engineering,
Kyushu Institute of Technology, Tobata, Kitakyushu 804-8550, Japan
sakai@mns.kyutech.ac.jp
[3] Department of Informatics, Tokyo University of Information Sciences,
4-1 Onaridai, Wakaba-ku, Chiba 265-8501, Japan
hara@rsch.tuis.ac.jp

Abstract. Rule induction based on indiscernible classes from neighborhood rough sets is described in information tables with continuous values. An indiscernible range that a value has in an attribute is determined by a threshold on that attribute. The indiscernible class of every object is derived from using the indiscernible range. First, lower and upper approximations are described in complete information tables by using indiscernible classes. Rules are obtained from the approximations. A rule that an object supports, which is called a single rule, is short of applicability. To improve the applicability of rules, a series of single rules is put into one rule expressed in an interval value, which is called a combined rule. Second, these are addressed in incomplete information tables. Incomplete information is expressed in a set of values or an interval value. Two types of indiscernible classes; namely, certainly and possibly indiscernible ones, are obtained from in an information table. The actual indiscernibility class is between the certainly and possibly indiscernible classes. The family of indiscernible classes of an object has a lattice structure. The minimal element is the certainly indiscernible class while the maximal one is the possibly indiscernible class. By using certainly and possibly indiscernible classes, we obtain four types of approximations: certain lower, certain upper, possible lower, and possible upper approximations. From these approximations we obtain four types of combined rules: certain and consistent, certain and inconsistent, possible and consistent, and possible and inconsistent ones. These combined rules have greater applicability than single rules that individual objects support.

Keywords: Neighborhood rough sets · Rule induction
Incomplete information · Indiscernible classes
Lower and upper approximations · Continuous values

© Springer Nature Switzerland AG 2018
H. S. Nguyen et al. (Eds.): IJCRS 2018, LNAI 11103, pp. 323–336, 2018.
https://doi.org/10.1007/978-3-319-99368-3_25

1   Introduction

Rough sets, constructed by Pawlak [12], are used as an effective method for data mining. The framework is usually applied to information tables with nominal attributes and creates fruitful results in various fields. However, we are frequently faced with attributes taking continuous values, when we describe properties of an object in our daily life. Therefore, we describe rough sets in information tables with continuous values.

Ways how to deal with attributes taking continuous values are broadly classified into two approaches. One is to discretize a continuous domain by dividing it into a collection of disjunctive intervals. Objects included in an interval are regarded as indistinguishable. From this indistinguishability the family of indiscernible classes is derived [1]. Results strongly depend on how discretization is made. Especially, objects that are located in the proximity of the boundary of intervals are strongly affected by discretization. This leads to that results abruptly change by a little alteration of discretization. The other is a way using neighborhood [7]. In this approach when the distance of an object to another one on an attribute is less than or equal to a given threshold, two objects are regarded as indistinguishable on the attribute. Results gradually change as the threshold changes. So, we use the latter approach.

Rules are induced from lower and upper approximations. Concretely speaking, when objects o and o' are included in the approximations, let single rules $a_i = 3.60 \rightarrow a_j = v$ and $a_i = 3.73 \rightarrow a_j = v$ be induced, where objects o and o' are characterized by values 3.60 and 3.73 of attribute a_i and the set approximated is specified by value v of attribute a_j. For example, value 3.66 of attribute a_i is not indiscernible with 3.60 and 3.73 under the threshold 0.05. Therefore, we cannot say anything from these single rules for a rule supported by an object with value 3.66 of attribute a_i. This means that the single rules are short of applicability. To improve such applicability, we consider a combined rule that is derived from a series of single rules supported by individual objects.

In addition, we are frequently confronted with incomplete information in daily life. We cannot sufficiently utilize information obtained from our daily life unless we deal with incomplete information. We express incomplete information in a partial value or an interval value. A missing value that means unknown in an attribute is expressed in all elements over the domain of the attribute. For example, the domain is given in the interval $[1.23, 4.45]$, the missing value is expressed in $[1.23, 4.45]$.

Most of authors fix the indiscernibility of an object with incomplete information with another object [3,16–18], as was done by Kryszkiewicz [4]. However, object o characterized by a value with incomplete information has two possibilities. One possibility is that the object o may have the same value as another one o'; namely, the two objects may be indiscernible. The other possibility is that o may have a different value from o'; namely, the two objects may be discernible. To fix the indiscernibility is to take into account only one of the two possibilities. Therefore, this treatment creates poor results and induces information loss [9,15]. We do not fix the indiscernibility of objects with incomplete information and simultaneously deal with both possibilities. This can be realized by dealing with objects having incomplete information from viewpoints of certainty and possibility [10], as was done by Lipski in the field of incomplete databases [5,6].

We have an approach based on possible world from the viewpoints of certainty and possibility. This way creates possible tables. Unfortunately, infinite possible tables can be derived from an information table with continuous values. Another way uses possible classes of an object, in which the object is possibly indiscernible with anyone [8]. The number of possible classes grows exponentially, as the number of values with incomplete information increases. However, this difficulty can be avoided by using minimum and maximum possible classes in the case of nominal attributes [10]. In this work, we apply this approach to information tables with continuous values.

The paper is organized as follows. In Sect. 2, an approach using indiscernible classes is addressed in complete information tables. In Sect. 3, we develop the approach in incomplete information tables. This is described from two viewpoints of certainty and possibility. In Sect. 4, conclusions are addressed.

2   Rough Sets by Using Indiscernible Classes in Complete Information Systems with Continuous Values

A data set is represented as a two-dimensional table, called an information table. In the information table, each row and each column represent an object and an attribute, respectively. A mathematical model of an information table with complete information is called a complete information system. The complete information system is a triplet expressed by $(U, AT, \{D(a_i) \mid a_i \in AT\})$. U is a non-empty finite set of objects, which is called the universe. AT is a non-empty finite set of attributes such that $a_i : U \rightarrow D(a_i)$ for every $a_i \in AT$ where $D(a_i)$ is the domain of attribute a_i.

Indiscernible class $[o]_{a_i}$ for object o on a_i is:

$$[o]_{a_i} = \{o' \mid |a_i(o) - a_i(o')| \leq \delta_{a_i}\}, \tag{1}$$

where $a_i(o)$ is the value for attribute a_i of object o and δ_{a_i} is a threshold that denotes a range in which $a_i(o)$ is indiscernible with $a_i(o')$. The indiscernible class is a tolerance class. Using the tolerance class, rough sets are generalized [14]. And recently it is used in decision rule induction [13].

Family $\mathcal{F}_{a_i}$ of indiscernible classes on a_i is:

$$\mathcal{F}_{a_i} = \{[o]_{a_i} \mid o \in U\}, \tag{2}$$

where $\cup_i [o]_{a_i} = U$. Using indiscernible classes, lower approximation $\underline{apr}_{a_i}(\mathcal{O})$ and upper approximation $\overline{apr}_{a_i}(\mathcal{O})$ of set $\mathcal{O}$ of objects for a_i are:

$$\underline{apr}_{a_i}(\mathcal{O}) = \{o \mid [o]_{a_i} \subseteq \mathcal{O}\}, \tag{3}$$

$$\overline{apr}_{a_i}(\mathcal{O}) = \{o \mid [o]_{a_i} \cap \mathcal{O} \neq \emptyset\}. \tag{4}$$

Proposition 1. If $\delta 1 \leq \delta 2$, then $\underline{apr}_{a_i}^{\delta 1}(\mathcal{O}) \supseteq \underline{apr}_{a_i}^{\delta 2}(\mathcal{O})$ and $\overline{apr}_{a_i}^{\delta 1}(\mathcal{O}) \subseteq \overline{apr}_{a_i}^{\delta 2}(\mathcal{O})$, where $\underline{apr}_{a_i}^{\delta 1}(\mathcal{O})$ and $\overline{apr}_{a_i}^{\delta 1}(\mathcal{O})$ are lower and upper approximations under threshold $\delta 1$ of attribute a_i and $\underline{apr}_{a_i}^{\delta 2}(\mathcal{O})$ and $\overline{apr}_{a_i}^{\delta 2}(\mathcal{O})$ are lower and upper approximations under threshold $\delta 2$ of attribute a_i.

For object o in the lower approximation of $\mathcal{O}$, all objects with which o is indiscernible are included in $\mathcal{O}$; namely, $[o]_{a_i} \subseteq \mathcal{O}$. On the other hand, for an object o in the upper approximation of $\mathcal{O}$, some objects with which o is indiscernible are in $\mathcal{O}$; namely, $[o]_{a_i} \cap \mathcal{O} \neq \emptyset$. Thus, $\underline{apr}_{a_i}(\mathcal{O}) \subseteq \overline{apr}_{a_i}(\mathcal{O})$.

Rules are induced from lower and upper approximations. Let $\mathcal{O}$ be specified by restriction $a_j = x$. Object $o \in \underline{apr}_{a_i}(\mathcal{O})$ consistently supports a single rule $a_i = a_i(o) \rightarrow a_j = x$. Object $o \in \overline{apr}_{a_i}(\mathcal{O})$ inconsistently supports a single rule $a_i = a_i(o) \rightarrow a_j = x$. The degree of consistency, called accuracy, is $|[o]_{a_i} \cap \mathcal{O}|/|\mathcal{O}|$.

Since attribute a_i has the continuous domain, the antecedent part of single rules that individual objects support is usually different. We obtain lots of single rules, but they have a drawback for applicability. For example, let two values $a_i(o)$ and $a_i(o')$ be 3.65 and 3.75 for objects o and o' in $\underline{apr}_{a_i}(\mathcal{O})$. When $\mathcal{O}$ is specified by restriction $a_j = x$, o and o' support single rules $a_i = 3.65 \rightarrow a_j = x$ and $a_i = 3.75 \rightarrow a_j = x$, respectively. By using these rules, we can say that a object having value 3.68 of a_i, indiscernible with 3.65 under $\delta_{a_i} = 0.03$, supports $a_i = 3.68 \rightarrow a_j = x$. However, we cannot at all say anything for a rule supported by an object with value 3.70 discernible with 3.65 and 3.75. This shows that a single rule is short of applicability.

To improve the applicability of rules, we combine a series of single rules into one rule, which is called a combined rule. Let objects in U be aligned in ascending order of $a_i(o)$ and be attached the serial superscript with 1 to N_U where $|U| = N_U$. $\underline{apr}_{a_i}(\mathcal{O})$ and $\overline{apr}_{a_i}(\mathcal{O})$ consist of collections of objects with serial superscripts. For example, $\underline{apr}_{a_i}(\mathcal{O}) = \{\cdots, o^h, o^{h+1}, \cdots, o^{k-1}, o^k, \cdots\}$ ($h \leq k$). Let o^l in $\underline{apr}_{a_i}(\mathcal{O})$ support a single rule $a_i = a_i(o^l) \rightarrow a_j = x$. Then, single rules derived from collection $(o^h, o^{h+1}, \cdots, o^{k-1}, o^k)$ can be put into one combined rule $a_i = [a_i(o^h), a_i(o^k)] \rightarrow a_j = x$.

Next, when a_j is an attribute with the continuous domain, $\mathcal{O}$ is specified by a restriction with an interval value. The interval value has the lower and the upper bounds that are existing values of attribute. Let the objects be aligned in ascending order of values of a_j and be attached the serial superscript with 1 to N_U. For example, using the ordered objects, $\mathcal{O}$ is specified like $\mathcal{O} = \{o \mid a_j(o) \geq a_j(o^m) \wedge a_j(o) \leq a_j(o^n)\}$ with $m \leq n$; in other words, $\mathcal{O}$ is specified by restriction $a_j = [a_i(o^m), a_i(o^n)]$. In the case, the combined rule, derived from collection $(o^h, o^{h+1}, \cdots, o^{k-1}, o^k)$, is expressed with $a_i = [a_i(o^h), a_i(o^k)] \rightarrow a_j = [a_i(o^m), a_i(o^n)]$. The accuracy of the combined rule is $\min_{h \leq s \leq k} |[o^s]_{a_i} \cap \mathcal{O}|/|\mathcal{O}|$.

Proposition 2. Let $\underline{r}$ and $\overline{r}$ be sets of combined rules obtained from $\underline{apr}_{a_i}(\mathcal{O})$ and $\overline{apr}_{a_i}(\mathcal{O})$, respectively. If $(a_i = [l, u] \to W) \in \underline{r}$, then $\exists l' \leq l, \exists u' \geq u \, (a_i = [l', u'] \to W) \in \overline{r}$, where $\mathcal{O}$ is specified by restriction W.

Example 1. Information tables are depicted in Fig. 1. T0 is the original information table. U is $\{o_1, o_2, \cdots, o_{18}, o_{19}\}$. T1, T2, and T3 are derived from T0, where some attributes are projected and objects are aligned in ascending order of values of attributes a_1, a_2, and a_3, respectively.

T0

U	a_1	a_2	a_3	a_4
1	3.11	2.98	3.02	b
2	2.94	3.65	3.44	b
3	2.33	3.69	3.28	f
4	4.78	2.98	3.52	a
5	3.42	2.35	2.67	b
6	3.03	4.52	4.07	c
7	2.81	2.95	2.91	c
8	4.36	3.11	3.49	a
9	3.22	4.63	4.21	b
10	3.07	3.78	3.57	c
11	2.97	3.98	3.68	b
12	2.63	4.81	4.16	c
13	3.91	3.71	3.77	a
14	3.12	2.78	2.88	b
15	3.05	3.29	3.22	c
16	2.95	3.65	3.44	b
17	2.89	3.51	3.32	c
18	2.45	3.96	3.51	f
19	3.86	3.44	3.57	b

T1

U	a_1	a_4
3	2.33	f
18	2.45	f
12	2.63	c
7	2.81	c
17	2.89	c
2	2.94	b
16	2.95	b
11	2.97	b
6	3.03	c
15	3.05	c
10	3.07	c
1	3.11	b
14	3.12	b
9	3.22	b
5	3.42	b
19	3.86	b
13	3.91	a
8	4.36	a
4	4.78	a

T2

U	a_2	a_3
5	2.35	2.67
14	2.78	2.88
7	2.95	2.91
1	2.98	3.02
4	2.98	3.52
8	3.11	3.49
15	3.29	3.22
19	3.44	3.57
17	3.51	3.32
2	3.65	3.44
16	3.65	3.44
3	3.69	3.28
13	3.71	3.77
10	3.78	3.57
18	3.96	3.51
11	3.98	3.68
6	4.52	4.07
9	4.63	4.21
12	4.81	4.16

T3

U	a_3
5	2.67
14	2.88
7	2.91
1	3.02
15	3.22
3	3.28
17	3.32
2	3.44
16	3.44
8	3.49
18	3.51
4	3.52
19	3.57
10	3.57
11	3.68
13	3.77
6	4.07
12	4.16
9	4.21

Fig. 1. T0 is the original information table. T1, T2, and T3 are derived from T0.

Let threshold δ_{a_1} be 0.05. Indiscernible classes of objects are:

$$[o_1]_{a_1} = \{o_1, o_{10}, o_{14}\}, [o_2]_{a_1} = \{o_2, o_{11}, o_{16}, o_{17}\}, [o_3]_{a_1} = \{o_3\}, [o_4]_{a_1} = \{o_4\},$$
$$[o_5]_{a_1} = \{o_5\}, [o_6]_{a_1} = \{o_6, o_{10}, o_{15}\}, [o_7]_{a_1} = \{o_7\}, [o_8]_{a_1} = \{o_8\}, [o_9]_{a_1} = \{o_9\},$$
$$[o_{10}]_{a_1} = \{o_1, o_6, o_{10}, o_{14}, o_{15}\}, [o_{11}]_{a_1} = \{o_2, o_{11}, o_{16}\}, [o_{12}]_{a_1} = \{o_{12}\},$$
$$[o_{13}]_{a_1} = \{o_{13}, o_{19}\}, [o_{14}]_{a_1} = \{o_1, o_{10}, o_{14}\}, [o_{15}]_{a_1} = \{o_6, o_{10}, o_{15}\},$$
$$[o_{16}]_{a_1} = \{o_2, o_{11}, o_{16}\}, [o_{17}]_{a_1} = \{o_2, o_{17}\}, [o_{18}]_{a_1} = \{o_{18}\}, [o_{19}]_{a_1} = \{o_{13}, o_{19}\}.$$

When $\mathcal{O}$ is specified by restriction $a_4 = b$, $\mathcal{O} = \{o_1, o_2, o_5, o_9, o_{11}, o_{14}, o_{16}, o_{19}\}$. Let $\mathcal{O}$ be approximated by objects on attribute a_1 with continuous values.

Using formulas (3) and (4), lower and upper approximations are:

$$\underline{apr}_{a_1}(\mathcal{O}) = \{o_5, o_9, o_{11}, o_{16}\},$$
$$\overline{apr}_{a_1}(\mathcal{O}) = \{o_1, o_2, o_5, o_9, o_{10}, o_{11}, o_{13}, o_{14}, o_{16}, o_{17}, o_{19}\}.$$

Information table T1 is derived from information table T0, where objects are aligned in ascending order of values of attribute a_1 and are attached the serial superscript from 1 to 19. The above approximations are described using the serial superscript as follows:

$$\underline{apr}_{a_1}(\mathcal{O}) = \{o^7, o^8, o^{14}, o^{15}\},$$
$$\overline{apr}_{a_1}(\mathcal{O}) = \{o^5, o^6, o^7, o^8, o^{11}, o^{12}, o^{13}, o^{14}, o^{15}, o^{16}, o^{17}\},$$

where

$$o^5 = o_{17}, \ o^6 = o_2, \ o^7 = o_{16}, \ o^8 = o_{11}, \ o^{11} = o_{10}, \ o^{12} = o_1,$$
$$o^{13} = o_{14}, \ o^{14} = o_9, \ o^{15} = o_5, \ o^{16} = o_{19}, \ o^{17} = o_{13}.$$

From the lower approximation, consistent combined rules are

$$a_1 = [2.95, 2.97] \rightarrow a_4 = b, \ a_1 = [3.22, 3.42] \rightarrow a_4 = b,$$

from collections $\{o^7, o^8\}$ and $\{o^{14}, o^{15}\}$, respectively, where $a_1(o^7) = 2.95$, $a_1(o^8) = 2.97$, $a_1(o^{14}) = 3.22$, and $a_1(o^{15}) = 3.42$. From the upper approximation, inconsistent combined rules are

$$a_1 = [2.89, 2.97] \rightarrow a_4 = b, \ a_1 = [3.07, 3.91] \rightarrow a_4 = b,$$

from collections $\{o^5, o^6, o^7, o^8\}$ and $\{o^{11}, o^{12}, o^{13}, o^{14}, o^{15}, o^{16}, o^{17}\}$, respectively, where $a_1(o^5) = 2.89$, $a_1(o^{11}) = 3.07$, and $a_1(o^{17}) = 3.91$.

Next, we consider the case where $\mathcal{O}$ is specified by a_3 with the continuous domain. Information table T3 is derived from T0, where the objects are aligned in ascending order of values of a_3 and are attached the serial superscript from 1 to 19. Using lower bound $a_3(o^5) = a_3(o_{15}) = 3.22$ and upper bound $a_3(o^{10}) = a_3(o_8) = 3.49$, $\mathcal{O} = \{o^5, o^6, o^7, o^8, o^9, o^{10}\} = \{o_2, o_3, o_8, o_{15}, o_{16}, o_{17}\}$. We approximate $\mathcal{O}$ by attribute a_2. Information table T2 where the objects are aligned in ascending order of values of a_2 is derived from T0. Let δ_{a_2} be 0.05. Indiscernible classes of objects are:

$$[o_1]_{a_2} = \{o_1, o_4, o_7, o_8\}, [o_2]_{a_2} = \{o_2, o_3, o_{16}\}, [o_3]_{a_2} = \{o_2, o_3, o_{13}, o_{16}\},$$
$$[o_4]_{a_2} = \{o_1, o_4, o_7, o_8\}, [o_5]_{a_2} = \{o_5\}, [o_6]_{a_2} = \{o_6\}, [o_7]_{a_2} = \{o_1, o_4, o_7\},$$
$$[o_8]_{a_2} = \{o_8\}, [o_9]_{a_2} = \{o_9\}, [o_{10}]_{a_2} = \{o_{10}\}, [o_{11}]_{a_2} = \{o_{11}, o_{18}\}, [o_{12}]_{a_2} = \{o_{12}\},$$
$$[o_{13}]_{a_2} = \{o_3, o_{13}\}, [o_{14}]_{a_2} = \{o_{14}\}, [o_{15}]_{a_2} = \{o_{15}\}, [o_{16}]_{a_2} = \{o_2, o_3, o_{16}\},$$
$$[o_{17}]_{a_2} = \{o_{17}\}, [o_{18}]_{a_2} = \{o_{11}, o_{18}\}, [o_{19}]_{a_2} = \{o_{19}\}.$$

Using formulas (3) and (4), lower and upper approximations are:

$$\underline{apr}_{a_2}(\mathcal{O}) = \{o_2, o_8, o_{15}, o_{16}, o_{17}\}, \ \overline{apr}_{a_2}(\mathcal{O}) = \{o_1, o_2, o_3, o_4, o_8, o_{13}, o_{15}, o_{16}, o_{17}\}.$$

Using information table T2 where objects are aligned in ascending order of values of attribute a_2 and are attached the serial superscript from 1 to 19, the above approximations are described as follows:

$$\underline{apr}_{a_2}(\mathcal{O}) = \{o^6, o^7, o^9, o^{10}, o^{11}\}, \quad \overline{apr}_{a_2}(\mathcal{O}) = \{o^4, o^5, o^6, o^7, o^9, o^{10}, o^{11}, o^{12}, o^{13}\},$$

From the lower approximation, consistent combined rules are

$$a_2 = [3.11, 3.29] \rightarrow a_3 = [3.22, 3.49], \quad a_2 = [3.51, 3.65] \rightarrow a_3 = [3.22, 3.49],$$

where $a_2(o^6) = 3.11$, $a_2(o^7) = 3.29$, $a_2(o^9) = 3.51$, and $a_2(o^{11}) = 3.65$. From the upper approximation, inconsistent combined rules are

$$a_2 = [2.98, 3.29] \rightarrow a_3 = [3, 22, 3.49], \quad a_2 = [3.51, 3.71] \rightarrow a_3 = [3.22, 3.49],$$

where $a_2(o^4) = 2.98$ and $a_2(o^{13}) = 3.71$.

This example shows that a combined rule is more applicable than single rules. For example, using the above consistent combined rule $a_2 = [3.11, 3.29] \rightarrow a_3 = [3.22, 3.49]$, we can say that an object with 3.20 for a value of attribute a_2 supports this rule, because 3.20 is included in interval $[3.11, 3.29]$. On the other hand, using single rules $a_2 = 3.11 \rightarrow a_3 = [3.22, 3.49]$ and $a_2 = 3.29 \rightarrow a_3 = [3.22, 3.49]$, we cannot say what rule the object supports under a threshold 0.05.

For formulas on sets A and B of attributes,

$$[o]_A = \cap_{a_i \in A}[o]_{a_i}, \tag{5}$$

$$\underline{apr}_A(\mathcal{O}) = \{o \mid [o]_A \subseteq \mathcal{O}\}, \tag{6}$$

$$\overline{apr}_A(\mathcal{O}) = \{o \mid [o]_A \cap \mathcal{O} \neq \emptyset\}. \tag{7}$$

3   Rough Sets by Indiscernible Classes in Incomplete Information Systems with Continuous Domains

An information table with incomplete information is called an incomplete information system. In incomplete information systems, $a_i : U \rightarrow s_{a_i}$ for every $a_i \in AT$ where s_{a_i} is a set of values over domain $D(a_i)$ of attribute a_i or an interval on $D(a_i)$. Single value v with $v \in a_i(o)$ or $v \subseteq a_i(o)$ is a possible value that may be the actual one as the value of attribute a_i in object o. The possible value is the actual one if $a_i(o)$ is a single value.

In an incomplete information system[1], an indiscernible class is a possible class that may be the actual indiscernible class. We have lots of indiscernible classes. Family $\mathcal{F}[o]_{a_i}$ of indiscernible class is:

$$\mathcal{F}[o]_{a_i} = \{C[o]_{a_i} \cup e \mid e \in \mathcal{P}(P[o]_{a_i} \backslash C[o]_{a_i})\}, \tag{8}$$

---

[1] For the sake of simplicity and space limitation, We describe the case of an attribute, although our approach can be easily extended to the case of more than one attribute.

where $\mathcal{P}(P[o]_{a_i} \backslash C[o]_{a_i})$ is the power set of $P[o]_{a_i} \backslash C[o]_{a_i}$, and certainly indiscernible class $C[o]_{a_i}$ and possibly one $P[o]_{a_i}$ on attribute a_i of object o are:

$$C[o]_{a_i} = \{o' \mid o' = o \vee (\forall u \in a_i(o) \forall v \in a_i(o') | u - v| \le \delta_{a_i})\}, \tag{9}$$

$$P[o]_{a_i} = \{o' \mid o' = o \vee (\exists u \in a_i(o) \exists v \in a_i(o') | u - v| \le \delta_{a_i})\}. \tag{10}$$

The family of indiscernible classes has a lattice structure. The minimal element is the certainly indiscernible class and the maximal one is the possibly indiscernible class. In other words, $C[o]_{a_i}$ is the minimum indiscernible class and $P[o]_{a_i}$ is the maximum indiscernible class. Objects in the certainly indiscernible class of o are certainly indistinguishable with o. Objects in the possibly indiscernible class of o are possibly indistinguishable with o.

We can derive not the actual, but certain and possible approximations from the viewpoint of certainty and possibility, as Lipski obtained in query processing under incomplete information [5,6]. We cannot definitely obtain whether or not an object belongs to the actual approximations, but we can know whether or not the object certainly and/or possibly belongs to approximations. Therefore, we show certain approximations (resp. possible approximations) whose object certainly (resp. possibly) belongs to the actual approximations.

Let $\mathcal{O}$ be a set of objects. Using certainly and possibly indiscernible classes, certain lower approximation $\underline{Capr}_{a_i}(\mathcal{O})$ and possible one $\underline{Papr}_{a_i}(\mathcal{O})$ for a_i are:

$$\underline{Capr}_{a_i}(\mathcal{O}) = \{o \mid P[o]_{a_i} \subseteq \mathcal{O}\}, \tag{11}$$

$$\underline{Papr}_{a_i}(\mathcal{O}) = \{o \mid C[o]_{a_i} \subseteq \mathcal{O}\}. \tag{12}$$

Similarly, Certain upper approximation $\overline{Capr}_{a_i}(\mathcal{O})$ and possible one $\overline{Papr}_{a_i}(\mathcal{O})$ are:

$$\overline{Capr}_{a_i}(\mathcal{O}) = \{o \mid C[o]_{a_i} \cap \mathcal{O} \ne \emptyset\}, \tag{13}$$

$$\overline{Papr}_{a_i}(\mathcal{O}) = \{o \mid P[o]_{a_i} \cap \mathcal{O} \ne \emptyset\}. \tag{14}$$

As with the case of nominal attributes [10], the following proposition holds.

Proposition 3. $\underline{Capr}_{a_i}(\mathcal{O}) \subseteq \underline{Papr}_{a_i}(\mathcal{O}) \subseteq \mathcal{O} \subseteq \overline{Capr}_{a_i}(\mathcal{O}) \subseteq \overline{Papr}_{a_i}(\mathcal{O}).$

Using four approximations denoted by formulae (11)–(14), lower and upper approximations are expressed in interval sets, as is described in [11][2], as follows:

$$\underline{apr}^{\bullet}_{a_i}(\mathcal{O}) = [\underline{Capr}_{a_i}(\mathcal{O}), \underline{Papr}_{a_i}(\mathcal{O})], \tag{15}$$

$$\overline{apr}^{\bullet}_{a_i}(\mathcal{O}) = [\overline{Capr}_{a_i}(\mathcal{O}), \overline{Papr}_{a_i}(\mathcal{O})]. \tag{16}$$

Certain and possible approximations are the lower and upper bounds of the actual approximation. The two approximations $\underline{apr}^{\bullet}_{a_i}(\mathcal{O})$ and $\underline{apr}^{\bullet}_{a_i}(\mathcal{O})$ depend

---

[2] Hu and Yao also say that approximations describes by using an interval set in information tables with incomplete information [2].

on each other; namely, the complementarity property $\underline{apr}^{\bullet}_{a_i}(\mathcal{O}) = U - \overline{apr}^{\bullet}_{a_i}(U - \mathcal{O})$ linked with them holds, as is so in complete information systems.

When objects in $\mathcal{O}$ are specified by attribute a_j with incomplete information, $\mathcal{O}$ is specified by using an element in domain $D(a_j)$. In the case where $\mathcal{O}$ is specified by restriction $a_j = x$ with $x \in D(a_j)$, four approximations: certain lower, possible lower, certain upper, and possible upper ones, are:

$$C\underline{apr}_{a_i}(\mathcal{O}) = \{o \mid P[o]_{a_i} \subseteq CO_{a_j=x}\}, \tag{17}$$

$$P\underline{apr}_{a_i}(\mathcal{O}) = \{o \mid C[o]_{a_i} \subseteq PO_{a_j=x}\}, \tag{18}$$

$$C\overline{apr}_{a_i}(\mathcal{O}) = \{o \mid C[o]_{a_i} \cap CO_{a_j=x} \neq \emptyset\}, \tag{19}$$

$$P\overline{apr}_{a_i}(\mathcal{O}) = \{o \mid P[o]_{a_i} \cap PO_{a_j=x} \neq \emptyset\}, \tag{20}$$

where

$$CO_{a_j=x} = \{o \in \mathcal{O} \mid a_j(o) = x\}, \tag{21}$$

$$PO_{a_j=x} = \{o \in \mathcal{O} \mid a_j(o) \supseteq x\}. \tag{22}$$

For rule induction, we can say as follows:

- $o \in C\underline{apr}_{a_i}(\mathcal{O})$ certainly and consistently supports rule $a_i = a_i(o) \rightarrow a_j(o) = x$.
- $o \in C\overline{apr}_{a_i}(\mathcal{O})$ certainly and inconsistently supports rule $a_i = a_i(o) \rightarrow a_j(o) = x$.
- $o \in P\underline{apr}_{a_i}(\mathcal{O})$ possibly and consistently supports $a_i = a_i(o) \rightarrow a_j(o) = x$.
- $o \in P\overline{apr}_{a_i}(\mathcal{O})$ possibly and inconsistently supports $a_i = a_i(o) \rightarrow a_j(o) = x$.

We create combined rules from them.

Let $U^C_{a_i}$ and $U^I_{a_i}$ be sets of objects having complete information and incomplete information for a_i. $o \in U^C_{a_i}$ is aligned in ascending order of $a_i(o)$ and is attached the serial superscript with 1 to N^C_i where $|U^C_{a_i}| = N^C_i$. Objects $o \in (C\underline{apr}_{a_i}(\mathcal{O}) \cap U^C_{a_i})$, $o \in (C\overline{apr}_{a_i}(\mathcal{O}) \cap U^C_{a_i})$, $o \in (P\underline{apr}_{a_i}(\mathcal{O}) \cap U^C_{a_i})$, and $o \in (P\overline{apr}_{a_i}(\mathcal{O}) \cap U^C_{a_i})$ are aligned in ascending order of $a_i(o)$. And then they are expressed by a sequence of collections of objects with a serial superscript like $\{\cdots, o^h, o^{h+1}, \cdots, o^{k-1}, o^k, \cdots\}$ $(h \leq k)$. From collection $(o^h, o^{h+1}, \cdots, o^{k-1}, o^k)$, four types of combined rules expressed with $a_i = [l, u] \rightarrow a_j = x$ are derived. For a certain and consistent combined rule,

$$l = \min(a_i(o^h), \min_Y e) \text{ and } u = \max(a_i(o^k), \max_Y e),$$

$$Y = \begin{cases} e < a_i(o^{k+1}), & \text{for } h = 1 \wedge k \neq N^C_i \\ a_i(o^{h-1}) < e < a_i(o^{k+1}), & \text{for } h \neq 1 \wedge k \neq N^C_i \\ a_i(o^{h-1}) < e, & \text{for } h \neq 1 \wedge k = N^C_i \end{cases}$$

$$\text{with } e \in a_i(o') \wedge o' \in X, \tag{23}$$

where X is $(C\underline{apr}_{a_i}(\mathcal{O}) \cap U^I_{a_i})$.

For certain and inconsistent, possible and consistent, possible and inconsistent combined rules, X is $(C\overline{apr}_{a_i}(\mathcal{O}) \cap U^I_{a_i})$, $(P\underline{apr}_{a_i}(\mathcal{O}) \cap U^I_{a_i})$, and $(P\overline{apr}_{a_i}(\mathcal{O}) \cap U^I_{a_i})$, respectively.

Proposition 4. Let $C\underline{r}$ and $P\underline{r}$ be sets of combined rules obtained from $\underline{Capr}_{a_i}(\mathcal{O})$ and $\underline{Papr}_{a_i}(\mathcal{O})$, respectively. When $\mathcal{O}$ is specified by restriction W, if $(a_i = [l, u] \rightarrow W) \in C\underline{r}$, then $\exists l' \leq l, \exists u' \geq u \; (a_i = [l', u'] \rightarrow W) \in P\underline{r}$.

Proposition 5. Let $C\overline{r}$ and $P\overline{r}$ be sets of combined rules obtained from $\overline{Capr}_{a_i}(\mathcal{O})$ and $\overline{Papr}_{a_i}(\mathcal{O})$, respectively. When $\mathcal{O}$ is specified by restriction W, if $(a_i = [l, u] \rightarrow W) \in C\overline{r}$, then $\exists l' \leq l, \exists u' \geq u \; (a_i = [l', u'] \rightarrow W) \in P\overline{r}$.

Proposition 6. Let $C\underline{r}$ and $C\overline{r}$ be sets of combined rules obtained from $\underline{Capr}_{a_i}(\mathcal{O})$ and $\overline{Capr}_{a_i}(\mathcal{O})$, respectively. When $\mathcal{O}$ is specified by restriction W, if $(a_i = [l, u] \rightarrow W) \in C\underline{r}$, then $\exists l' \leq l, \exists u' \geq u \; (a_i = [l', u'] \rightarrow W) \in C\overline{r}$.

Proposition 7. Let $P\underline{r}$ and $P\overline{r}$ be sets of combined rules obtained from $\underline{Papr}_{a_i}(\mathcal{O})$ and $\overline{Papr}_{a_i}(\mathcal{O})$, respectively. When $\mathcal{O}$ is specified by restriction W, if $(a_i = [l, u] \rightarrow W) \in P\underline{r}$, then $\exists l' \leq l, \exists u' \geq u \; (a_i = [l', u'] \rightarrow W) \in P\overline{r}$.

Example 2. Let $\mathcal{O}$ be specified by restriction $a_4 = b$ in IT of Fig. 2.

IT

U	a_1	a_2	a_3	a_4
1	$\{3.06, 3.11\}$	2.98	$[3.02, 3.17]$	$\{b, c\}$
2	2.94	$\{3.64, 3.65\}$	3.44	b
3	2.33	$[3.69, 3.72]$	3.28	f
4	4.78	$[2.98, 3.12]$	3.52	a
5	3.42	2.35	2.67	b
6	3.03	4.52	4.07	c
7	2.81	2.95	2.91	c
8	4.36	3.11	3.49	a
9	$\{2.97, 3.22\}$	4.63	4.21	b
10	3.07	3.78	3.57	c
11	$[2.96, 2.97]$	3.98	3.68	b
12	2.63	4.81	4.16	c
13	3.91	3.71	3.77	a
14	3.12	2.78	2.88	b
15	3.05	3.29	3.22	c
16	2.95	$\{3.35, 3.65\}$	3.44	b
17	$[2.89, 2.92]$	3.51	$[3.32, 3.40]$	$\{b, c\}$
18	$[2.45, 2.55]$	3.96	$\{3.49, 3.51\}$	f
19	$[3.86, 3.92]$	3.44	3.57	$\{a, b\}$

Fig. 2. Information table IT with incomplete information

$$CO_{a_4=b} = \{o_2, o_5, o_9, o_{11}, o_{14}, o_{16}\},$$
$$PO_{a_4=b} = \{o_1, o_2, o_5, o_9, o_{11}, o_{14}, o_{16}, o_{17}, o_{19}\}.$$

Each $C[o_i]_{a_1}$ for $i = 1, \ldots, 19$ is, respectively,

$$C[o_1]_{a_1} = \{o_1, o_{10}\}, C[o_2]_{a_1} = \{o_2, o_{11}, o_{16}, o_{17}\}, C[o_3]_{a_1} = \{o_3\},$$
$$C[o_4]_{a_1} = \{o_4\}, C[o_5]_{a_1} = \{o_5\}, C[o_6]_{a_1} = \{o_6, o_{10}, o_{15}\}, C[o_7]_{a_1} = \{o_7\},$$
$$C[o_8]_{a_1} = \{o_8\}, C[o_9]_{a_1} = \{o_9\}, C[o_{10}]_{a_1} = \{o_1, o_6, o_{10}, o_{14}, o_{15}\},$$
$$C[o_{11}]_{a_1} = \{o_2, o_{11}, o_{16}\}, C[o_{12}]_{a_1} = \{o_{12}\}, C[o_{13}]_{a_1} = \{o_{13}, o_{19}\},$$
$$C[o_{14}]_{a_1} = \{o_{10}, o_{14}\}, C[o_{15}]_{a_1} = \{o_6, o_{10}, o_{15}\}, C[o_{16}]_{a_1} = \{o_2, o_{11}, o_{16}\},$$
$$C[o_{17}]_{a_1} = \{o_2, o_{17}\}, C[o_{18}]_{a_1} = \{o_{18}\}, C[o_{19}]_{a_1} = \{o_{13}, o_{19}\}.$$

Each $P[o_i]_{a_1}$ for $i = 1, \ldots, 19$ is, respectively,

$$P[o_1]_{a_1} = \{o_1, o_6, o_{10}, o_{14}, o_{15}\}, P[o_2]_{a_1} = \{o_2, o_9, o_{11}, o_{16}, o_{17}\}, P[o_3]_{a_1} = \{o_3\},$$
$$P[o_4]_{a_1} = \{o_4\}, P[o_5]_{a_1} = \{o_5\}, P[o_6]_{a_1} = \{o_1, o_6, o_{10}, o_{15}\}, P[o_7]_{a_1} = \{o_7\},$$
$$P[o_8]_{a_1} = \{o_8\}, P[o_9]_{a_1} = \{o_2, o_9, o_{11}, o_{16}, o_{17}\}, P[o_{10}]_{a_1} = \{o_1, o_6, o_{10}, o_{14}, o_{15}\},$$
$$P[o_{11}]_{a_1} = \{o_2, o_9, o_{11}, o_{16}, o_{17}\}, P[o_{12}]_{a_1} = \{o_{12}\}, P[o_{13}]_{a_1} = \{o_{13}, o_{19}\},$$
$$P[o_{14}]_{a_1} = \{o_1, o_{10}, o_{14}\}, P[o_{15}]_{a_1} = \{o_1, o_6, o_{10}, o_{15}\},$$
$$P[o_{16}]_{a_1} = \{o_2, o_9, o_{11}, o_{16}, o_{17}\}, P[o_{17}]_{a_1} = \{o_2, o_9, o_{11}, o_{16}, o_{17}\},$$
$$P[o_{18}]_{a_1} = \{o_{18}\}, P[o_{19}]_{a_1} = \{o_{13}, o_{19}\}.$$

Four approximations are:

$$\underline{Capr}_{a_1}(\mathcal{O}) = \{o_5\},$$
$$\underline{Papr}_{a_1}(\mathcal{O}) = \{o_2, o_5, o_9, o_{11}, o_{16}, o_{17}\},$$
$$\overline{Capr}_{a_1}(\mathcal{O}) = \{o_2, o_5, o_9, o_{10}, o_{11}, o_{14}, o_{16}, o_{17}\},$$
$$\overline{Papr}_{a_1}(\mathcal{O}) = \{o_1, o_2, o_5, o_6, o_9, o_{10}, o_{11}, o_{13}, o_{14}, o_{15}, o_{16}, o_{17}, o_{19}\}.$$

$$U_{a1}^C = \{o_2, o_3, o_4, o_5, o_6, o_7, o_8, o_{10}, o_{12}, o_{13}, o_{14}, o_{15}, o_{16}\},$$
$$U_{a1}^I = \{o_1, o_9, o_{11}, o_{17}, o_{18}, o_{19}\}$$

Objects in U_{a1}^C are aligned in ascending order of values of attribute a_1 as follows:

$$o_3, o_{12}, o_7, o_2, o_{16}, o_6, o_{15}, o_{10}, o_{14}, o_5, o_{13}, o_8, o_4$$

A series of superscripts is attached to these objects:

$$o^1, o^2, o^3, o^4, o^5, o^6, o^7, o^8, o^9, o^{10}, o^{11}, o^{12}, o^{13},$$

where $o^1 = o_3, o^2 = o_{12}, \ldots, o^{13} = o_4$. Using objects with the superscript, the four approximations are expressed as follows:

$$\underline{Capr}_{a_1}(\mathcal{O}) = \{o^{10}\},$$
$$\underline{Papr}_{a_1}(\mathcal{O}) = \{o^4, o^5, o^{10}, o_9, o_{11}, o_{17}\},$$
$$\overline{Capr}_{a_1}(\mathcal{O}) = \{o^4, o^5, o^8, o^9, o^{10}, o_9, o_{11}, o_{17}\},$$
$$\overline{Papr}_{a_1}(\mathcal{O}) = \{o^4, o^5, o^6, o^7, o^8, o^9, o^{10}, o^{11}, o_1, o_9, o_{11}, o_{17}, o_{19}\}.$$

where objects with a superscript and with a subscript have complete and incomplete information for attribute a_1, respectively; namely,

$$\underline{Capr}_{a_1}(\mathcal{O}) \cap U_{a_1}^C = \{o^{10}\}, \underline{Capr}_{a_1}(\mathcal{O}) \cap U_{a_1}^I = \emptyset,$$

$$\underline{Papr}_{a_1}(\mathcal{O}) \cap U_{a_1}^C = \{o^4, o^5, o^{10}\}, \underline{Papr}_{a_1}(\mathcal{O}) \cap U_{a_1}^I = \{o_9, o_{11}, o_{17}\},$$

$$\overline{Capr}_{a_1}(\mathcal{O}) \cap U_{a_1}^C = \{o^4, o^5, o^8, o^9, o^{10}\}, \overline{Capr}_{a_1}(\mathcal{O}) \cap U_{a_1}^I = \{o_9, o_{11}, o_{17}\},$$

$$\overline{Papr}_{a_1}(\mathcal{O}) \cap U_{a_1}^C = \{o^4, o^5, o^6, o^7, o^8, o^9, o^{10}, o^{11}\}, \overline{Papr}_{a_1}(\mathcal{O}) \cap U_{a_1}^I = \{o_1, o_9, o_{11}, o_{17}, o_{19}\}.$$

From these expressions, four types combined rules are derived. For certain and consistent rules,

$$a_1 = 3.42 \rightarrow a_4 = b.$$

For possible and consistent rules,

$$a_1 = [2.89, 2.97] \rightarrow a_4 = b, \ a_1 = [3.22, 3.42] \rightarrow a_4 = b.$$

For certain and inconsistent rules,

$$a_1 = [2.89, 2.97] \rightarrow a_4 = b, \ a_1 = [3.07, 3.42] \rightarrow a_4 = b.$$

For possible and inconsistent rules,

$$a_1 = [2.89, 3.92] \rightarrow a_4 = b.$$

Last, we describe the case where $o \in \mathcal{O}$ is specified by numerical attribute a_j with incomplete information. $o \in U_{a_j}^C$ is aligned in ascending order of $a_j(o)$ and is attached with the serial superscript with 1 to N_j^C where $|U_{a_j}^C| = N_j^C$. We specify $\mathcal{O}$ by $a_j(o^m) \in U_{a_j}^C$ and $a_j(o^n) \in U_{a_j}^C$ with $m \leq n$.

$$\underline{Capr}_{a_i}(\mathcal{O}) = \{o \mid P[o]_{a_i} \subseteq CO_{[a_j(o^m), a_j(o^n)]}\}, \tag{24}$$

$$\underline{Papr}_{a_i}(\mathcal{O}) = \{o \mid C[o]_{a_i} \subseteq PO_{[a_j(o^m), a_j(o^n)]}\}, \tag{25}$$

$$\overline{Capr}_{a_i}(\mathcal{O}) = \{o \mid C[o]_{a_i} \cap CO_{[a_j(o^m), a_j(o^n)]} \neq \emptyset\}, \tag{26}$$

$$\overline{Papr}_{a_i}(\mathcal{O}) = \{o \mid P[o]_{a_i} \cap PO_{[a_j(o^m), a_j(o^n)]} \neq \emptyset\}, \tag{27}$$

where

$$CO_{[a_j(o^m), a_j(o^n)]} = \{o \in \mathcal{O} \mid a_j(o) \subseteq [a_j(o^m), a_j(o^n)]\}, \tag{28}$$

$$PO_{[a_j(o^m), a_j(o^n)]} = \{o \in \mathcal{O} \mid a_j(o) \cap [a_j(o^m), a_j(o^n)] \neq \emptyset\}. \tag{29}$$

$o \in U_{a_j}^C$ is aligned in ascending order of $a_j(o)$ and is attached the serial superscript with 1 to N_j^C. Now, $\mathcal{O}$ is specified by attribute values $a_j(o^m)$ and $a_j(o^n)$ with $o^m \in U_{a_j}^C$ and $o^n \in U_{a_j}^C$. $o \in U_{a_i}^C$ is aligned in ascending order of $a_i(o)$ is attached the serial superscript with 1 to N_i^C. Also, four types of combined rules with $a_i = [l, u] \rightarrow a_j = [a_j(o^m), a_j(o^n)]$ are obtained: certain and consistent, certain and inconsistent, possible and consistent, and possible and inconsistent combined rules.

These types of combined rules are obtained in incomplete information table IT in Fig. 2. For example, let $\mathcal{O}$ be specified by numerical attribute a_3 with incomplete information. When $\mathcal{O}$ is approximated on numerical attribute a_2 with incomplete information, the four types of combined rules are derived.

4   Conclusions

We have described rough sets and rule induction from them in information tables with continuous domains. First, we have dealt with complete information tables. Rough sets are obtained from indiscernible classes. Individual objects that belongs to the rough sets support single rules. The single rules are short of applicability. To improve the applicability of rules, we have put a series of single rules derived from the rough sets into one combined rule. The combined rule is expressed by using intervals.

Second, we have dealt with incomplete information tables. Incomplete information is depicted in a disjunctive set of values or an interval of values. We have dealt with it from viewpoints of certainty and possibility, as was introduced by Lipski in the field of incomplete databases. Lots of indiscernible classes are derived. The family of indiscernible classes is expressed by a lattice having the minimal and maximal elements. The number of indiscernible classes increases exponentially as the number of attribute values with incomplete information grows. However, approximations are obtained by using the minimal and the maximal indiscernible classes. Therefore, we have no difficulty of computational complexity. By using the minimal and the maximal indiscernible classes, four types approximations: certain lower, certain upper, possible lower, and possible upper approximations are obtained, as is so in incomplete information tables with nominal attributes. From these approximations, we have derived four types of combined rules that are expressed by using interval values: certain and consistent, certain and inconsistent, possible and consistent, and possible and inconsistent combined rules. The combined rules are more applicable than single ones.

References

1. Grzymala-Busse, J.W.: Mining numerical data – a rough set approach. In: Peters, J.F., Skowron, A. (eds.) Transactions on Rough Sets XI. LNCS, vol. 5946, pp. 1–13. Springer, Heidelberg (2010). https://doi.org/10.1007/978-3-642-11479-3_1
2. Hu, M.J., Yao, Y.Y.: Rough set approximations in an incomplete information table. In: Polkowski, L., et al. (eds.) IJCRS 2017. LNCS (LNAI), vol. 10314, pp. 200–215. Springer, Cham (2017). https://doi.org/10.1007/978-3-319-60840-2_14
3. Jing, S., She, K., Ali, S.: A universal neighborhood rough sets model for knowledge discovering from incomplete hetergeneous data. Expert Syst. **30**(1), 89–96 (2013). https://doi.org/10.1111/j.1468-0394.2012.00633_x
4. Kryszkiewicz, M.: Rules in incomplete information systems. Inf. Sci. **113**, 271–292 (1999)
5. Lipski, W.: On semantics issues connected with incomplete information databases. ACM Trans. Database Syst. **4**, 262–296 (1979)
6. Lipski, W.: On databases with incomplete information. J. ACM **28**, 41–70 (1981)
7. Lin, T.Y.: Neighborhood systems: a qualitative theory for fuzzy and rough sets. In: Wang, P. (ed.) Advances in Machine Intelligence and Soft Computing, vol. IV, pp. 132–155. Duke University (1997)

8. Nakata, M., Sakai, H.: Rough sets handling missing values probabilistically interpreted. In: Ślęzak, D., Wang, G., Szczuka, M., Düntsch, I., Yao, Y. (eds.) RSFDGrC 2005. LNCS (LNAI), vol. 3641, pp. 325–334. Springer, Heidelberg (2005). https://doi.org/10.1007/11548669_34

9. Nakata, M., Sakai, H.: Applying rough sets to information tables containing missing values. In: Proceedings of 39th International Symposium on Multiple-Valued Logic, pp. 286–291. IEEE Press (2009). https://doi.org/10.1109/ISMVL.2009.1

10. Nakata, M., Sakai, H.: Twofold rough approximations under incomplete information. Int. J. Gener. Syst. **42**, 546–571 (2013). https://doi.org/10.1080/17451000.2013.798898

11. Nakata, M., Sakai, H.: Describing rough approximations by indiscernibility relations in information tables with incomplete information. In: Carvalho, J.P., Lesot, M.-J., Kaymak, U., Vieira, S., Bouchon-Meunier, B., Yager, R.R. (eds.) IPMU 2016. CCIS, vol. 611, pp. 355–366. Springer, Cham (2016). https://doi.org/10.1007/978-3-319-40581-0_29

12. Pawlak, Z.: Rough Sets: Theoretical Aspects of Reasoning about Data. Kluwer Academic Publishers, Dordrecht (1991). https://doi.org/10.1007/978-94-011-3534-4

13. Sikora, M.: Decision rule-based data models using TRS and NetTRS – methods and algorithms. In: Peters, J.F., Skowron, A. (eds.) Transactions on Rough Sets XI. LNCS, vol. 5946, pp. 130–160. Springer, Heidelberg (2010). https://doi.org/10.1007/978-3-642-11479-3_8

14. Skowron, A., Stepaniuk, J.: Tolerance approximation spaces. Fundamenta Informaticae **27**, 245–253 (1996)

15. Stefanowski, J., Tsoukiàs, A.: Incomplete information tables and rough classification. Comput. Intell. **17**, 545–566 (2001)

16. Yang, X., Zhang, M., Dou, H., Yang, Y.: Neighborhood systems-based rough sets in incomplete information system. Inf. Sci. **24**, 858–867 (2011). https://doi.org/10.1016/j.knosys.2011.03.007

17. Zenga, A., Lia, T., Liuc, D., Zhanga, J., Chena, H.: A fuzzy rough set approach for incremental feature selection on hybrid information systems. Fuzzy Sets Syst. **258**, 39–60 (2015). https://doi.org/10.1016/j.fss.2014.08.014

18. Zhao, B., Chen, X., Zeng, Q.: Incomplete hybrid attributes reduction based on neighborhood granulation and approximation. In: 2009 International Conference on Mechatronics and Automation, pp. 2066–2071. IEEE Press (2009)

Contextual Probability Estimation from Data Samples – A Generalisation

Hui Wang[1]($\boxtimes$) and Bowen Wang[2]

[1] Ulster University, Jordanstown, UK
h.wang@ulster.ac.uk
[2] Mavern Securities, London, UK
bowenmwang@hotmail.com

Abstract. Contextual probability (G) provides an alternative, efficient way of estimating (primary) probability (P) in a principled way. G is defined in terms of P in a combinatorial way, and they have a simple linear relationship. Consequently, if one is known, the other can be calculated. It turns out G can be estimated based on a set of data samples through a simple process called *neighbourhood counting*. Many results about contextual probability are obtained based on the assumption that the event space is the power set of the sample space. However, the real world is usually not the case. For example, in a multidimensional sample space, the event space is typically the set of hyper tuples which is much smaller than the power set. In this paper, we generalise contextual probability to multidimensional sample space where the attributes may be categorical or numerical. We present results about the normalisation constant, the relationship between G and P and the neighbourhood counting process.

Keywords: Probability estimation · Contextual probability
Neighbourhood counting

1 Introduction

The frequentist view of probability interprets probability as the limit of frequency. Therefore a principled method for probability estimation should be grounded in the notion of frequency and the well known Bayes rule. However a frequency based estimation method is often hindered by the data sparsity dilemma, and a Bayes rule based method is often plagued by the combination explosion problem.

When estimating probability from data samples of a multidimensional space through the notion of frequency, we are usually faced with the problem of data sparsity. In this case, it is not possible to estimate probability via frequency, as if

---

Hui Wang gratefully acknowledges support by EU Horizon 2020 Programme (700381, ASGARD).

H. S. Nguyen et al. (Eds.): IJCRS 2018, LNAI 11103, pp. 337–349, 2018.
https://doi.org/10.1007/978-3-319-99368-3_26

we do so, the probability will be zero for many events. An alternative approach is to break down the problem of estimating probability into simpler sub-problems through the Bayes rule, but then we need to solve an exponential number of sub-problems each corresponding to a simpler event. So this is not feasible if the number of dimensions (attributes, variables) is large.

The contextual probability (G) concept [11] provides an alternative way of estimating (primary) probability (P) in a principled way: we define a secondary probability in terms of the primary probability of interest, establish the relationship between the two probabilities, and then estimate the primary probability via the secondary one. If the secondary probability G can be estimated in a desirable way, the primary probability P can. It has been shown that the secondary probability G can be estimated through neighbourhood counting (see e.g. [11]) for different types of data – multivariate, sequential and graphical. Furthermore, the process of neighbourhood counting can be used for tasks beyond probability estimation (see e.g. [6,9,13]).

When a sample space U is a *structureless set*, the relationship between G and P is linear [11]. However, when U is a structured set (i.e., order structured set, or multidimensional space), their relationship is unknown. In this paper we generalise contextual probability to multidimensional sample spaces when the attributes may be categorical or numerical.

2   Background

In this section we provide some background information on subjects relevant to this paper, in order to make the paper self-contained.

2.1   Notation and Assumption

Let $A = \{a_1, a_2, \cdots, a_n\}$ be a set of attributes. The attributes can be either *categorical* or *numerical*, and all attributes are assumed to be finite. If a_i is categorical, its domain is a finite, un-ordered set $\{x_1, x_2, \cdots, x_{m_i}\}$ and we let $\mathrm{dom}(a_i) \overset{\mathrm{def}}{=} \{x_1, x_2, \cdots, x_{m_i}\}$. If a_i is numerical, its domain is an ordered set $\{x_1, x_2, \cdots, x_{m_i}\}$ and we let $dom(a_i) \overset{\mathrm{def}}{=} \{1, \cdots, m_i\}$. These assumptions about attributes are adopted throughout the rest of the paper. A multivariate sample space defined by A is $U \overset{\mathrm{def}}{=} \prod_{i=1}^{n} dom(a_i)$. A *data set* is $D \subseteq U$ – a set of samples of U.

2.2   Probability

The starting point for probability theory is a set U called the *sample space* whose points are in 1-1 correspondence with the possible outcomes of a random experiment [1]. Any specific subset of these outcomes, which corresponds to a question that can be answered "yes" or "no", is called an *event*. The development of the mathematical theory will be facilitated if we require that the set of events forms a σ-algebra. Thus we may form unions, intersections, and complements of events and be assured that the resulting sets are also events.

Furthermore the basic physical requirement is that the probability $P(E)$ assigned to an event E corresponds to the relative frequency of E in a very large number of independent repetitions of the random experiment. It follows that P should be a nonnegative, additive set function, with $P(U) = 1$.

The above discussion may be summarized as follows. Let U be a set, and $\mathcal{F}$ be a σ-algebra over U. A probability function is a mapping $P : \mathcal{F} \to [0, 1]$ such that the following *axioms of probability* are satisfied:

- $P(E) \geq 0$ for any $E \in \mathcal{F}$;
- $P(U) = 1$;
- For any $E_1, E_2 \in \mathcal{F}$, if $E_1 \cap E_2 = \emptyset$ then $P(E_1 \cup E_2) = P(E_1) + P(E_2)$.

U is the *sample space*, and $\mathcal{F}$ is the *event space* associated with U.

In general any function satisfying the above axioms of probability, however defined, is a probability function [4]. Thus the basic mathematical object of study is a probability space $<U, \mathcal{F}, P>$.

Now we take a closer look at σ-algebra. Let U be a set, and let 2^U be its power set. Then a subset $\mathcal{F} \subseteq U$ is called a σ-*algebra* if it satisfies the following properties[3]:

- $U \in \mathcal{F}$.
- $\mathcal{F}$ is closed under complementation: if $A \in \mathcal{F}$, then so is its complement, $U \setminus A$.
- $\mathcal{F}$ is closed under countable unions: if $A_1, A_2, A_3, \ldots$ are in $\mathcal{F}$, then so is $A = A_1 \cup A_2 \cup A_3 \cup \ldots$

From these properties, it follows that the σ-algebra is also closed under countable intersections (by applying De Morgan's laws). As an example, $\mathcal{F}$ can be the power set of U, i.e., $\mathcal{F} = 2^U$.

2.3   The Contextual Probability

Consider a (mathematical) probability space $<U, \mathcal{F}, P>$, where U is a sample space, $\mathcal{F}$ is a σ-algebra over U and P is a probability function over $\mathcal{F}$. For $X \in \mathcal{F}$ let $f(X)$ be a measure of X. As an example, $f(X)$ can be the counting measure, i.e., $f(X)$ being the number of elements in X.

Definition 1 ([11]). *The* contextual probability *is a mapping from $\mathcal{F}$ to $[0, 1]$ such that, for $X \in \mathcal{F}$,*

$$G(X) = \sum_{E \in \mathcal{F}} P(E)f(X \cap E)/K \tag{1}$$

where $K = \sum_{E \in \mathcal{F}} P(E)f(E)$ and is a constant for a given sample space.

It has been shown that G is a probability function [11]. Since G is defined in terms of P, it is *secondary*. In contrast P is *primary* since the starting point is the probability space $<U, \mathcal{F}, P>$.

This secondary probability is related to works in [5,7,8], but any discussion of these related works is beyond the scope of this paper.

$G(X)$ is defined from all those $E \in \mathcal{F}$ that overlap with X (i.e., $X \cap E \neq \emptyset$). These E's are relevant to X and serve as the *contexts* in which $G(X)$ is defined. Thus $G(X)$ is called the *contextual probability* of X.

Each such E is called a *neighborhood*[1] of X. In other words a neighborhood of X is an element E of $\mathcal{F}$ such that E overlaps X. For simplicity, if E is a singleton set, e.g., $E = \{a\}$, we write $G(a)$ for $G(\{a\})$.

2.4   Relationship Between G and P

Let $U, \mathcal{F}, P, G$ be understood. P is a probability distribution over U, and G is another probability distribution over U which is defined in terms of P. If P is known, G can be calculated by definition. Conversely, if G is known, how can P be calculated? If we can establish the relationship between P and G, we can answer this question.

The following lemma provides a formula to calculate the normalizing factor $K = \sum_{X \in \mathcal{F}} f(X)P(X)$ in the definition of G.

Lemma 1 ([11]). *Assume that U is finite with $N = |U|$, $\mathcal{F} = 2^U$ is the event space associated with U, and $f()$ is a counting measure. Then $K = (N+1)2^{N-2}$.*

The relationship between G and P for elements in U is shown in Theorem 1 below.

Theorem 1 ([11]). *Assume that U is finite with $N = |U|$, $\mathcal{F} = 2^U$ is the event space associated with U, and $f()$ is a counting measure. Then, for $x \in U$, $G(x) = \alpha P(x) + \alpha$, where $\alpha = \frac{1}{N+1}$.*

Since both P and G are probability functions they satisfy the additive axiom. In other words for $E \in \mathcal{F}$, $P(E) = \sum_{x \in E} P(x)$ and $G(E) = \sum_{x \in E} G(x)$. Following Theorem 1 we then have:

Corollary 1 ([11]). *For any $E \in \mathcal{F}$, $G(E) = \alpha P(E) + \alpha |E|$.*

Theorem 1 and Corollary 1 establish the linear relationship between G and P. If we know P we can calculate G, and vice versa.

2.5   Estimation of Contextual Probability

Here we discuss how to estimate G from data. Let $D \subseteq U$ be a given data set. According to Definition 1, G can be calculated from P. Assuming the *principle of indifference*[2], P can be estimated as follows. For any $E \in \mathcal{F}$,

$$\hat{P}(E) = |E^D|/n \tag{2}$$

where $E^D = \{x \in D : x \leq E\}$ is the set of elements in D that are covered by E and $n = |D|$.

---

[1] The concept of neighbourhood is used in different contexts with possibly different definitions. The use of this concept in this paper is defined as such.

[2] This is common in statistics. See, e.g., [3].

Theorem 2 ([11]). *Let U be a finite sample space, $\mathcal{F} = 2^U$ be the event space associated with U, $f()$ be a counting measure, and $D \subseteq U$ be a given data sample with $n = |D|$. Assuming the principle of indifference we have, for any $t \in U$,*

$$\hat{G}(t) = \frac{1}{nK} \sum_{x \in D} c(t, x)$$

where $c(t, x)$ is the number of $E \in \mathcal{F}$ that covers both t and x.

It is shown [10] that $c(t, x)$ is a similarity measure for points t and x, called *neighbourhood counting measure (metric)*, since every $E \in \mathcal{F}$ is a *neighbourhood* of some point. It is the count of all common neighbourhoods for t and x. In fact it is further a similarity metric as it satisfies the similarity axioms [2].

3   Generalisation of Contextual Probability in Multidimensional Sample Space

In this section we seek to generalise contextual probability in a more general setting where the sample space is defined by a set of n attributes, $A = \{a_1, a_2, \cdots, a_n\}$. We assume that the domain of each attribute is a finite set, and we consider two cases: (1) attributes are categorical; and (2) attributes are numerical.

3.1   When All Attributes Are Categorical

Let A be a set of categorical attributes, $A = \{a_1, a_2, \cdots, a_n\}$. The sample space defined by A is denoted by U and is more formally defined as follows,

$$U = \prod_{i=1}^{n} dom(a_i) = \left\{ <v_1, v_2, \ldots, v_n> : v_i \in dom(a_i) \right\}.$$

where $<v_1, v_2, \ldots, v_n>$ is a simple tuple. Thus, every data point is a simple tuple, and vice versa.

As explained earlier, an event is a set of experiment outcomes that corresponds to a question with "yes" or "no" answer. Since the sample space is defined by a set of attributes, a sensible question may be composed in terms of the attributes as follows: a *sub-question* is composed for every attribute, leading to a *sub-event*, and all sub-questions are joined up by the classical logical operators (i.e. conjunction, disjunction and complement) to form an event question. Note that an event is usually a subset of the sample space. In the same spirit, a sub-event can be sensibly a subset of the domain of one attribute. Therefore we sensibly define the event space as a set of arrays of subsets of every attribute domain. More formally,

$$\mathcal{F} = \prod_{i=1}^{n} 2^{dom(a_i)} = \left\{ <s_1, s_2, \ldots, s_n> : s_i \subseteq dom(a_i) \right\}$$

where $<s_1, s_2, \ldots, s_n>$ is a *hyper tuple* [12]. It is clear that this event space is not the same as the power set of U. In fact it is a subset of the power set of U. It can be shown that this $\mathcal{F}$ is a Borel σ-algebra[3]. Therefore it qualifies to be an event space.

Table 1. Sample space defined by three categorical attributes

ID	a_1	a_2	a_3
1	a	α	0
2	a	α	1
3	a	β	0
4	a	β	1
5	a	γ	0
6	a	γ	1
7	b	α	0
8	b	α	1
9	b	β	0
10	b	β	1
11	b	γ	0
12	b	γ	1
13	c	α	0
14	c	α	1
15	c	β	0
16	c	β	1
17	c	γ	0
18	c	γ	1

Example 1 (Data and event space generated by a set of attributes). Consider three categorical attributes $A = \{a_1, a_2, a_3\}$ where

$$dom(a_1) = \{a, b, c\}$$
$$dom(a_2) = \{\alpha, \beta, \gamma\}$$
$$dom(a_3) = \{0, 1\}$$

The (complete) sample space defined by these attributes is shown in Table 1. The event space defined by these attributes is the following,

$$\mathcal{F} = \big\{ <s_1, s_2, s_3>: s_1 \subseteq dom(a_1), s_2 \subseteq dom(a_2), s_3 \subseteq dom(a_3) \big\}.$$

---

[3] https://en.wikipedia.org/wiki/Borel_set.

Table 2. A sample of the event space defined by three categorical attributes

a_1	a_2	a_3
{}	{}	{}
{a}	{α}	{0}
{a}	{α, β}	{0, 1}
{a, b, c}	{α, β, γ}	{0, 1}

There is a total of $2^3 \times 2^3 \times 2^2 = 256$ events. On the other hands, there is a total of $2^{18} = 262144$ subsets of data points. A sample of the event space is shown in Table 2.

The following lemma provides a formula to calculate the normalizing factor K in the definition of G, i.e., $K = \sum_{X \in \mathcal{F}} f(X)P(X)$.

Lemma 2. *Let U be a sample space defined by n categorical attributes a_i, $i = 1, 2, \ldots, n$. Let $M_i = |dom(a_i)|$. Let $K = \sum_{X \in \mathcal{F}} f(X)P(X)$. Then*

$$K = \left((M_1 + 1) \times \cdots \times (M_n + 1)\right) \times \left(2^{M_1 - 2} \times \cdots \times (2^{M_n - 2})\right)$$

Proof.

$$K = \sum_{X \in \mathcal{F}} f(X)P(X) = \sum_{\substack{X \in \mathcal{F} \\ X = <s_1, \ldots, s_n> \\ s_1 \subseteq dom(a_1), \ldots, s_n \subseteq dom(a_n) \\ m_1 = |s_1|, \ldots, m_n = |s_n| \\ m = m_1 \times \ldots \times m_n}} mP(X)$$

$$= \sum_{\substack{s_1 \subseteq dom(a_1) \\ m_1 = |s_1|}} \cdots \sum_{\substack{s_n \subseteq dom(a_n) \\ m_n = |s_n|}} (m_1 \times \ldots \times m_n)P(X)$$

$$= \sum_{\substack{s_1 \subseteq dom(a_1) \\ m_1 = |s_1|}} \cdots \sum_{\substack{s_n \subseteq dom(a_n) \\ m_n = |s_n|}} (m_1 \times \ldots \times m_n) \sum_{x \in X} P(x)$$

$$= \sum_{\substack{x \in U \\ x_1 \in s_1}} \sum_{\substack{s_1 \subseteq dom(a_1) \\ m_1 = |s_1|}} \cdots \sum_{\substack{s_n \subseteq dom(a_n) \\ m_n = |s_n| \\ x_n \in s_n}} (m_1 \times \ldots \times m_n)P(x)$$

$$= \sum_{m_1 = 0}^{M_1 - 1} \cdots \sum_{m_n = 0}^{M_n - 1} \left((m_1 + 1)\binom{M_1 - 1}{m_1} \times \ldots \times (m_n + 1)\binom{M_n - 1}{m_n}\right)$$

$$= \left(\sum_{m_1 = 0}^{M_1 - 1} (m_1 + 1)\binom{M_1 - 1}{m_1}\right) \times \cdots \times \left(\sum_{m_n = 0}^{M_n - 1} (m_n + 1)\binom{M_n - 1}{m_n}\right)$$

$$= \left((M_1 + 1)2^{M_1 - 2}\right) \times \cdots \times \left((M_n + 1)2^{M_n - 2}\right)$$

$$= \left((M_1 + 1) \times \cdots \times (M_n + 1)\right) \times \left(2^{M_1 - 2} \times \cdots \times (2^{M_n - 2})\right)$$

The relationship between G and P for elements in U is shown in Theorem 3.

Theorem 3. *Let U be a sample space defined by n categorical attributes a_i, $i = 1, 2, \ldots, n$. Let $M_i = |dom(a_i)|$. Then, for $x \in U$,*

$$G(x) = \alpha P(x) + \alpha,$$

where $\alpha = \frac{1}{(M_1+1) \times \cdots \times (M_n+1)}$.

Proof.

$$G(x) = \sum_{Y \in \mathcal{F}} \frac{f(x \cap Y) P(Y)}{K} = \sum_{Y \in \mathcal{F}, x \in Y} \frac{P(Y)}{K}$$

$$= \frac{1}{K} \sum_{Y \in \mathcal{F}, x \in Y} P(Y) = \frac{1}{K} \sum_{Y \in \mathcal{F}, x \in Y} \sum_{z \in Y} P(z)$$

$$= \frac{1}{K} \sum_{Y \in \mathcal{F}, x \in Y} \left(\sum_{z \in Y, z \neq x} P(z) + P(x) \right)$$

$$= \frac{1}{K} \left(\sum_{Y \in \mathcal{F}, x \in Y} \sum_{z \in Y, z \neq x} P(z) + \sum_{Y \in \mathcal{F}, x \in Y} P(x) \right)$$

$$= \frac{1}{K} \left(\sum_{z \in U, z \neq x} \sum_{Y \in \mathcal{F}, x \in Y, z \in Y} P(z) + \sum_{Y \in \mathcal{F}, x \in Y} P(x) \right)$$

$$= \frac{1}{K} \left(\left(2^{M_1-2} \times \cdots \times 2^{M_n-2} \right) (1 - P(x)) + \left(2^{M_1-1} \times \cdots \times 2^{M_n-1} \right) P(x) \right)$$

$$= \frac{1}{K} \left(\left(2^{M_1-2} \times \cdots \times 2^{M_n-2} \right) + \left(2^{M_1-2} \times \cdots \times 2^{M_n-2} \right) P(x) \right)$$

$$= \frac{\left(2^{M_1-2} \times \cdots \times 2^{M_n-2} \right)}{K} (1 + P(x)) = \frac{1}{(M_1+1) \times \cdots \times (M_n+1)} (1 + P(x))$$

$$= \alpha(1 + P(x)), \text{ where } \alpha = \frac{1}{(M_1+1) \times \cdots \times (M_n+1)}$$

The claim then follows.

3.2   When Attributes Are Ordinal

Let $A = \{a_1, a_2, \cdots, a_n\}$ be a set of ordinal attributes. For simplicity of presentation we assume that all attributes have finite domains which can be written as $dom(a_i) = \{1, 2, 3, \ldots, m\}$ where $m = |dom(a_i)|$ for attribute a_i. The sample space defined by A is then the following,

$$U = \prod_{i=1}^{n} dom(a_i) = \{<v_1, v_2, \ldots, v_n> : v_i \in dom(a_i)\}.$$

Since there is ordinal relationship between values, the event space is a bit complicated. For one ordinal attribute, we can take the set of all subsets of its domain as the event space, but such a set will lose the ordinal information in the ordinal attribute. We can instead take the set of all intervals of the domain as the event space, but such a set is not a sigma algebra because the complement of one interval is not a single interval. Therefore we need a new definition of event space.

We consider transforming ordinal attributes without losing the ordinal information. There may be different ways of transformation. Here we discuss one way of transformation where every ordinal attribute is replaced by a set of binary attributes.

Consider ordinal attribute a_i where $dom(a_i) = \{1, 2, \ldots, m_i\}$. We construct one binary attribute, $a_{i,j}$, for every ordinal value $j \in dom(a_i)$ and then convert every data instance $<v_1, v_2, \ldots, v_{i-1}, v_i, v_{i+1}, \ldots, v_n>$ into the following:

$$<v_1, v_2, \ldots, v_{i-1}, v_{i,1}, v_{i,2}, \ldots, v_{i,m_i}, v_{i+1}, \ldots, v_n>$$

where

$$v_{i,j} = \begin{cases} 1, & \text{if } v_i \leq j, \\ 0, & \text{otherwise.} \end{cases} \tag{3}$$

which corresponds to a new binary attribute, $a_{i,j}$. Repeating this procedure for all attributes, we will obtain a new binary vector for the original data instance. We thus transform the original sample space U into a binary sample space U_b, which is defined by binary attributes $a_{1,1}, \ldots, a_{1,m_1}, a_{2,1}, \ldots, a_{2,m_2}, \ldots, a_{n,1}, \ldots, a_{n,m_n}$ with domain of $\{0,1\}$ for all. We rename these attributes as $a_1^b, a_2^b, \ldots, a_{n_b}^b$, and we thus have a new binary sample space:

$$U_b = \prod_{i=1}^{n_b} dom(a_i^b) = \prod_{i=1}^{n_b} \{0,1\} = \{<v_1, v_2, \ldots, v_{n_b}> : v_i \in \{0,1\}\}.$$

where $n_b = \sum_{i=1}^{n} |dom(a_i)|$.

Example 2. Table 3 shows a toy data table consisting of 5 data instances from a sample space defined by 3 ordinal attributes. Transforming the attributes in the way as described above, we convert these 5 data instances into binary ones, which are shown in Table 4.

Now that we transform a sample space into a binary one, we can define an event space as follows.

$$\mathcal{F}_b = \prod_{i=1}^{n_b} 2^{\{0,1\}} = \{<s_1, s_2, \ldots, s_{n_b}> : s_i \subseteq \{0,1\}\}$$

This event space is the set of all hyper tuples [12] definable by the set of binary attributes. It is clearly a sigma algebra since the complement of every hyper tuple is another hyper tuple and the union/intersection of any two hyper tuples is another hyper tuple. Probability can thus be rigorously defined on $\mathcal{F}_b$.

On the basis of the above discussions we then have the following corollary from Lemma 2 and Theorem 3.

Table 3. A toy data table with 3 ordinal attributes. The first two attributes have 3 values each in their domain, and the third attribute has 4 values.

ID	a_1	a_2	a_3
1	1	2	3
2	3	1	2
3	2	3	1
4	3	2	4
5	1	3	2

Table 4. A data table with 10 binary attributes, which is transformed from Table 3.

ID	$a_{1,1}$	$a_{1,2}$	$a_{1,3}$	$a_{2,1}$	$a_{2,2}$	$a_{2,3}$	$a_{3,1}$	$a_{3,2}$	$a_{3,3}$	$a_{3,4}$
1	1	1	1	0	1	1	0	0	1	1
2	0	0	1	1	1	1	0	1	1	1
3	0	1	1	0	0	1	1	1	1	1
4	0	0	1	0	1	1	0	0	0	1
5	1	1	1	0	0	1	0	1	1	1

Corollary 2 (All ordinal attributes via binary transformation). *If the sample space U_b is defined by n_b binary attributes, then the normalisation constant is $K = 3^{n_b}$ and $G(x) = aP(x) + a$ for $x \in U_b$ where $a = 1/3^{n_b}$.*

Transforming a single ordinal attribute into a set of binary attributes is the means of working out the relationship between G and P. Now that we have an insight about the transformation, we can work out their relationship without going through the transformation:

Corollary 3 (All ordinal attributes). *If the sample space U is defined by n ordinal attributes with finite domains $\{1, 2, \cdots, m_i\}$ for $i = 1, 2, \cdots, n$, then the normalisation constant is $K = 3^{m_1+m_2+\cdots+m_n}$ and $G(x) = aP(x)+a$ for $x \in U_b$ where $a = 1/3^{m_1+m_2+\cdots+m_n}$.*

Corollary 4 (Mixed attributes). *If the sample space U is defined by a mixture of nominal and ordinal attributes with finite domains. Assume that $a_1, a_2, \cdots, a_h$ are nominal attributes and $a_{h+1}, a_{h+2}, \cdots a_n$ are ordinal attributes. The sizes of their domains are m_i for $i = 1, 2, \cdots, n$. Then the normalisation constant is*

$$K = K_{nom} \times K_{ord}$$

where

$$K_{nom} = \left((M_1 + 1) \times \cdots \times (M_h + 1)\right) \times \left(2^{M_1-2} \times \cdots \times (2^{M_h-2})\right)$$

and

$$K_{ord} = 3^{m_{h+1}+m_{h+2}+\cdots+m_n}$$

and $G(x) = aP(x) + a$ for $x \in U_b$ where $a = 1/(b_{nom} \times b_{ord})$ where

$$b_{nom} = ((m_1 + 1) \times (m_2 + 1) \times \cdots \times (m_h + 1))$$

and

$$b_{ord} = 3^{m_{h+1}+m_{h+2}+\cdots+m_n}$$

4   Estimating Contextual Probability in Multidimensional Sample Space Through Neighbourhood Counting

Following the same line of reasoning as in Theorem 2, we can prove

Theorem 4 (Estimating contextual probability in multidimensional space). *Let U be a multidimensional sample space as discussed above, and $D \subseteq U$ be a set of samples. Then, for any $t \in U$,*

$$\hat{G}(t) = \frac{1}{nK} \sum_{x \in D} c(t, x)$$

where $c(t, x)$ is the number of events (or neighbourhoods) $E \in \mathcal{F}$ that covers both t and x. Therefore, contextual probability $G(t)$ can be estimated through neighbourhood counting.

Next, we follow the same line of reasoning as in [10, Sect. 4.2] to discuss how to count neighbourhoods through a formula. Note U is a multidimensional sample space defined by n attributes $a_1, a_2, \ldots, a_n$ and $m_i = |\operatorname{dom}(a_i)|$. The attributes may be categorical or ordinal. The ordinal attributes are transformed into binary attributes as discussed above. Consider $t, x \in U$ where $t = < t_1, t_2, \ldots, t_n >$ and $x = < x_1, x_2, \ldots, x_n >$, we can count their common neighbourhoods as follows:

$$c(t, x) = \prod_{i}^{n} c_a(t_i, x_i) \tag{4}$$

where

$$c_a(t_i, x_i) = \begin{cases} 2^{m_i - 1}, & \text{if } a_i \text{ is categorical and } x_i = t_i \\ 2^{m_i - 2}, & \text{if } a_i \text{ is categorical and } x_i \neq t_i \\ c_a'(t_i, x_i), & \text{if } a_i \text{ is ordinal} \end{cases}$$

When a_i is ordinal, t_i is transformed into a vector of binary values $<t_{i1}, t_{i2}, \ldots, t_{im_i}>$, and x_i is similarly transformed. We then have

$$c_a'(t_i, x_i) = \prod_{j}^{m_i} c_a''(t_{ij}, x_{ij}) \tag{5}$$

where

$$c_a''(t_{ij}, x_{ij}) = \begin{cases} 2, & \text{if} x_{ij} = t_{ij} \\ 1, & \text{if} x_{ij} \neq t_{ij} \end{cases}$$

Because of the way an ordinal attribute a_i is transformed, we have

$$c_a'(t_i, x_i) = 2^{m_i - |x_i - t_i|} \tag{6}$$

Therefore, in summary, we have

Theorem 5 (Neighbourhood counting). *Let U be a multidimensional sample space defined by n attributes $a_1, a_2, \ldots, a_n$ and let $m_i = |\mathrm{dom}(a_i)|$. The attributes may be categorical or ordinal. The ordinal attributes are transformed into binary attributes as discussed above, resulting in a new sample space U'. For $t, x \in U$ where $t = <t_1, t_2, \ldots, t_n>$ and $x = <x_1, x_2, \ldots, x_n>$, we can count their common neighbourhoods as follows:*

$$c(t, x) = \prod_i^n c(t_i, x_i), \tag{7}$$

where

$$c(t_i, x_i) = \begin{cases} 2^{m_i - 1}, & \text{if } a_i \text{ is categorical and } x_i = t_i \\ 2^{m_i - 2}, & \text{if } a_i \text{ is categorical and } x_i \neq t_i \\ 2^{m_i - |x_i - t_i|}, & \text{if } a_i \text{ is ordinal} \end{cases}$$

5 Conclusion

In this paper we present a generalisation of contextual probability to multidimensional sample space where the attributes are categorical or numerical. We show that under such more realistic conditions, the existing results about contextual probability holds well in a conceptually concise way. One technical challenge is how to handle multidimensional sample space, which is the Cartesian product of multiple sample spaces. The other technical challenge is how to deal with numerical attributes. Both challenges are satisfactorily addressed. In future work, we will apply the generalised contextual probability to real world problems, in particular, financial applications where probability estimation is a key process.

References

1. Ash, R.B., Doléans-Dade, C.: Probability and Measure Theory. Academic Press, San Diego (2000)
2. Chen, S., Ma, B., Zhang, K.: On the similarity and the distance metric. Theoret. Comput. Sci. **410**(24–25), 2365–2376 (2009)

3. Duda, R.O., Hart, P.E.: Pattern Classification and Scene Analysis. Wiley, New York (1973)
4. Feller, W.: An Introduction to Probability Theory and Its Applications. Wiley, New York (1968)
5. Hajek, A.: Probability, logic and probability logic. In: Goble, L. (ed.) Blackwell Companion to Logic, pp. 362–384. Blackwell, Oxford (2000)
6. Lin, Z., Lyu, M., King, I.: Matchsim: a novel similarity measure based on maximum neighborhood matching. Knowl. Inf. Syst. **32**, 141–166 (2012)
7. Mani, A.: Comparing dependencies in probability theory and general rough sets: Part-a. arXiv:1804.02322v1
8. Mani, A.: Probabilities, dependence and rough membership functions. Int. J. Comput. Appl. **39**, 17–35 (2017)
9. TolgaKahraman, H.: A novel and powerful hybrid classifier method: development and testing of heuristic k-nn algorithm with fuzzy distance metric. Data Knowl. Eng. **103**, 44–59 (2016)
10. Wang, H.: Nearest neighbors by neighborhood counting. IEEE Trans. Pattern Anal. Mach. Intell. **28**(6), 942–953 (2006)
11. Wang, H., Düentsch, I., Trindade, L.: Lattice machine classification based on contextual probability. Fundamenta Informaticae **127**(1–4), 241–256 (2013). https://doi.org/10.3233/FI-2013-907
12. Wang, H., Düntsch, I., Gediga, G., Skowron, A.: Hyperrelations in version space. Int. J. Approximate Reasoning **36**(3), 223–241 (2004)
13. Wang, X., Ouyang, J., Chen, G.: Simplifying calculation of graph similarity through matrices. In: Li, D., Li, Z. (eds.) CCTA 2015. IAICT, vol. 479, pp. 417–428. Springer, Cham (2016). https://doi.org/10.1007/978-3-319-48354-2_41

Application of Greedy Heuristics
for Feature Characterisation
and Selection: A Case Study
in Stylometric Domain

Urszula Stańczyk[1], Beata Zielosko[2(✉)], and Krzysztof Żabiński[2]

[1] Institute of Informatics, Silesian University of Technology,
Akademicka 16, 44-100 Gliwice, Poland
urszula.stanczyk@polsl.pl
[2] Institute of Computer Science, University of Silesia in Katowice,
Będzińska 39, 41-200 Sosnowiec, Poland
{beata.zielosko,kzabinski}@us.edu.pl

Abstract. The paper presents research on greedy heuristics used to obtain characteristics of features. The parameters of decision rules induced by heuristics were treated as a source of knowledge about variables. The observations on attributes were exploited for generation of new rules, and for post-processing pruning rule sets, inferred in Classical Rough Set Approach. The proposed framework was applied in stylometric domain.

Keywords: Feature characterisation · Feature selection
Greedy heuristic · Decision rule · Pruning · Stylometry

1 Introduction

Information about roles played by characteristic features in recognition of described concepts is contained not only in the input data, either raw or pre-processed [3]. Knowledge discovered in a data mining process is present in forms constructed by learning algorithms, for example in structures of decision graphs, topologies of artificial neural networks, and in induced rules. These additional representations of knowledge can be used in search of new or optimised solutions.

Association and decision rules are often preferred for description and presentation of information, as due to their transparent structure they enhance understanding of patterns hidden in data. Rule sets can be obtained by many induction algorithms, with the objectives of finding a minimal cover or all rules on examples, providing good generalisation, ensuring high supports, satisfactory classification accuracy, or meeting some other criteria or requirements [13,14].

© Springer Nature Switzerland AG 2018
H. S. Nguyen et al. (Eds.): IJCRS 2018, LNAI 11103, pp. 350–362, 2018.
https://doi.org/10.1007/978-3-319-99368-3_27

Exhaustive algorithms return all rules that can be inferred from input data, which can take time and cause prohibitively high cardinalities of rule sets, yet it gives the widest choice of elements, to be tailored to specific needs in post-processing. Heuristics focused on rule parameters are capable of relatively quickly returning manageable sets of rules, sufficient for the intended purpose. The work of these heuristics on data can be treated as preliminary gathering of information, which is next stored in the inferred rules, ready to be exploited for other ends.

In the research presented in the paper, selected greedy heuristics [1] were employed for induction of decision rules, which were applied as decision algorithms to validation and test sets. The classification results were compared against that of the exhaustive algorithms found for the same learning data in Classical Rough Set Approach [9]. Knowledge represented by rules inferred by greedy heuristics was next used to obtain characterisation of features by the proposed coefficients, which led to construction of attribute rankings based on rule parameters.

Rankings belong with feature selection, a domain dedicated to estimation of importance of variables. Discovering which attributes are essential, redundant, or irrelevant, allows for improvement of predictive models [3]. Techniques of feature selection are typically divided into [5]: filter, wrapper, and embedded methods [10]. Ranking mechanisms can be based on machine learning techniques, statistical measures, information theory, and other approaches [4,12]. They impose an order on variables, assigning to each a specific score. When a scoring function is independent on an inducer used for classification, the ranking performs as a filter, otherwise a wrapper or hybrid solution is obtained.

Knowledge about attributes discovered by greedy heuristics was exploited in two ways: to generate new decision rules, and to prune whole rules from the previously induced exhaustive algorithms. These two processes were governed by the constructed rankings and observations of attributes present as conditions in decision rules inferred by heuristics. Results from the conducted experiments show that with the presented research framework it was possible to discard both some variables and rules without degrading the power of the rule classifier.

Experiments were performed on data sets devoted to two cases of binary authorship attribution [6], with balanced classes and stylometric features. Estimation of performance for rule classifiers was obtained by validation and test sets, and sets discretised with supervised approach described by Kononenko [7].

The paper consists of five sections. Section 2 presents descriptions of greedy heuristics employed in research, and characterisation of attributes by induced rules through defined coefficients. In Sect. 3, the main notions of stylometric processing of texts are explained. Section 4 contains results of experiments and comments to them, while Sect. 5 includes conclusions.

2   Greedy Heuristics

In [1], greedy heuristics were compared from the point of view of optimisation of association rules, relative to length and support. In this paper, an application of four best heuristics (from the point of view of support) in induction of decision rules and feature characterisation is described.

2.1 Main Notions

A *decision table* is defined as $T = (U, A \cup \{d\})$ [9], where $U = \{r_1, \ldots, r_k\}$ is a nonempty, finite set of objects (rows), $A = \{f_1, \ldots, f_n\}$ is a nonempty, finite set of attributes, i.e., $f : U \to V_f$ for any $f \in A$, where V_f is the set of values of an attribute f, called the domain of f. Elements of A are called condition attributes. $d \notin A$ is a distinguishing attribute, called a decision attribute, and a is a value of a decision attribute (called also a decision), $a \in V_d$, where V_d is the domain of d. It is assumed that the decision table is consistent, it does not contain any rows with equal values of condition attributes and different decisions.

The number of rows in the table T is denoted by $N(T)$. For a value a of a decision attribute, $N(T, a)$ is the number of rows r of T with a decision a, and $M(T, a) = N(T) - N(T, a)$. $mcd(T)$ denotes the *most common* decision for T, which is the minimum index of a decision a such that $N(T, a)$ has maximum value. The set of not constant condition attributes on T is denoted by $E(T)$.

A table obtained from T by removal of some rows is called a *subtable* of T. $T(f_{i_1}, a_1), \ldots, (f_{i_m}, a_m)$ denotes a subtable of T that consists of rows which at the intersection with columns $f_{i_1}, \ldots, f_{i_m}$ have values $a_1, \ldots, a_m$.

The expression

$$(f_{i_1} = a_1) \wedge \ldots \wedge (f_{i_m} = a_m) \to d = a \tag{1}$$

is called a *decision rule over* T if $f_{i_1}, \ldots, f_{i_m} \in \{f_1, \ldots, f_n\}$, $a_1, \ldots, a_m$ are values of corresponding attributes, and a is a decision. The rule corresponds to the subtable $T' = T(f_{i_1}, a_1), \ldots, (f_{i_m}, a_m)$ of T. The rule (1) is called *realizable for a row* r if r belongs to T'. This rule is called *true* for T, if each row of T' for which the rule (1) is realizable, has the decision a attached to it. The considered rule is a *rule for T and r*, if this rule is true for T and realizable for r.

The *support* of the rule (1) is the number of rows in T' for which the rule is realizable and which are labeled with the decision a. If the considered rule is a rule for T and r then its support is equal to $N(T')$.

2.2 Description of Heuristics

Algorithm 1 presents a pseudo-code for the greedy heuristic H, for construction of a decision rule for a row r from T with the assigned decision a. At each iteration, an attribute $f_i \in \{f_1, \ldots, f_n\}$ with the minimum index fulfilling heuristic H, is selected. The heuristic H stops when all rows in T' have the same decision. The algorithm is applied sequentially to each row r of T. As a result, for each row of a decision table T, one decision rule for T and r, is obtained.

To describe the work of the heuristic H we denote: $T^{(j+1)} = T^{(j)}(f_i, b_i)$, where j is an index of the subsequent subtable during the execution of H. For

$M(T^{(j+1)}, a) = N(T^{(j+1)}) - N(T^{(j+1)}, a),$
$RM(f_i, r, a) = (N(T^{(j+1)}) - N(T^{(j+1)}, a))/N(T^{(j+1)}),$
$\alpha(f_i, r, a) = N(T^{(j)}, a) - N(T^{(j+1)}, a)$ and $\beta(f_i, r, a) = M(T^{(j)}, a) - M(T^{(j+1)}, a),$

each heuristics H selects the attribute $f_i \in E(T^{(j)})$ in the following manner:

- *Poly* selects an attribute f_i which maximizes the value $\frac{\beta(f_i, r, a)}{\alpha(f_i, r, a) + 1}$,
- *Log* selects an attribute f_i which maximizes the value $\frac{\beta(f_i, r, a)}{\log_2(\alpha(f_i, r, a) + 2)}$,
- *MaxS* selects an attribute f_i which minimizes the value $\alpha(f_i, r, a)$ given that $\beta(f_i, r, a) > 0$,
- *RM* selects an attribute f_i which minimizes the value $RM(f_i, r, a)$.

Algorithm 1. Greedy heuristic H for construction of a decision rule for T and r

---

Require: Decision table T with condition attributes $f_1, \ldots, f_n$, row $r = (b_1, \ldots, b_n)$ with the assigned decision a
Ensure: Decision rule for T, r and a
   begin
      $Q \leftarrow \emptyset$;
      $j \leftarrow 0$;
      $T^{(j)} \leftarrow T$;
      while all rows in $T^{(j)}$ are not assigned with the same decision a **do**
         select $f_i \in \{f_1, \ldots, f_n\}$ with the minimum index fulfilling the heuristic H;
         $T^{(j+1)} \leftarrow T^{(j)}(f_i, b_i)$;
         $Q \leftarrow Q \cup \{f_i\}$;
         $j = j + 1$;
      end while
      $\bigwedge_{f_i \in Q}(f_i = b_i) \rightarrow d = a$, where a is a decision attached to r.
   end

The following example demonstrates calculations executed by all heuristics.

Example 1. The example shows how heuristic H constructs a decision rule for the decision table T_0, row r_1 with the assigned decision A. The decision table T_0 has three condition attributes, so there are considered three subtables:

$$T_1^{(1)} = T_0^{(0)}(f_1, 0), \quad T_2^{(1)} = T_0^{(0)}(f_2, 0) \quad \text{and} \quad T_3^{(1)} = T_0^{(0)}(f_3, 1).$$

$T_0 =$

	f_1	f_2	f_3	d
r_1	0	0	1	A
r_2	2	1	1	B
r_3	2	0	1	A
r_4	2	1	0	B

$T_1^{(1)} =$

	f_1	f_2	f_3	d
r_1	0	0	1	A

$T_2^{(1)} =$

	f_1	f_2	f_3	d
r_1	0	0	1	A
r_3	2	0	1	A

$T_3^{(1)} =$

	f_1	f_2	f_3	d
r_1	0	0	1	A
r_2	2	1	1	B
r_3	2	0	1	A

Heuristic *MaxS*:

$$\alpha(f_1, r_1, A) = 1, \ \beta(f_1, r_1, A) = 2, \ \alpha(f_2, r_1, A) = 0, \ \beta(f_2, r_1, A) = 2,$$
$$\alpha(f_3, r_1, A) = 0, \ \beta(f_3, r_1, A) = 1,$$

so the rule $f_2 = 0 \rightarrow d = A$ is obtained.
Heuristic *Poly*:

$$\frac{\beta(f_1, r_1, A)}{\alpha(f_1, r_1, A) + 1} = \frac{2}{2}, \quad \frac{\beta(f_2, r_1, A)}{\alpha(f_2, r_1, A) + 1} = \frac{2}{1}, \quad \frac{\beta(f_3, r_1, A)}{\alpha(f_3, r_1, A) + 1} = \frac{1}{1},$$

so the rule $f_2 = 0 \rightarrow d = A$ is obtained.

Heuristic *Log*:

$$\frac{\beta(f_1,r_1,A)}{\log_2(\alpha(f_1,r_1,A)+2)} = \frac{2}{\log_2 3}, \frac{\beta(f_2,r_1,A)}{\log_2(\alpha(f_2,r_1,A)+2)} = \frac{2}{\log_2 2}, \frac{\beta(f_3,r_1,A)}{\log_2(\alpha(f_3,r_1,A)+2)} = \frac{1}{\log_2 2},$$

so the rule $f_2 = 0 \rightarrow d = A$ is obtained.
Heuristic *RM*:

$$RM(f_1,r_1,A) = 0,\; RM(f_2,r_1,A) = 0,\; RM(f_3,r_1,A) = \tfrac{1}{3},$$

so the rule $f_1 = 0 \rightarrow d = A$ is obtained.

2.3   Feature Characterisation and Selection

A rule can be characterised by its parameters, such as length corresponding to the number of conditions on attributes, or support indicating for how many training samples the rule is true. When many learning samples support a rule, it means that the rule captures a pattern present in many examples. Greater rule length marks closer, more detailed description of patterns, which runs the risk of overfitting, while shorter rules possess better generalisation properties.

These parameters are often used for formulation of rule quality or interestingness measures [13], which can then be employed in the process of rule selection [11]. On the other hand, the sets of inferred rules can be treated as an additional source of information on features, with the knowledge discovered by the learning algorithm represented in the form of rules.

To mine this new source and exploit it for feature characterisation and selection, to each rule r_i a specific coefficient was assigned, $RuleCoef(r_i)$, equal to the quotient of the rule support divided by length,

$$RuleCoef(r_i) = Support(r_i)/Length(r_i). \tag{2}$$

For an attribute f its coefficient was calculated as a sum of coefficients of all rules that included this attribute among their conditions ($Cond$), divided by the total number of rules ($NrOfRls$)

$$AttrCoef(f) = \sum_{i=1}^{NrOfRls} \frac{RuleCoef(r_i | f \in Cond(r_i))}{NrOfRls}. \tag{3}$$

The cumulative version of attribute coefficient calculated an average of coefficients obtained over various heuristics ($NrOfH$ denotes number of heuristics)

$$CumAC(f) = \frac{\sum_{i=1}^{NrOfH} AttrCoef_i(f)}{NrOfH}. \tag{4}$$

The cumulative coefficient was used as the ranking function applied to all features, with the top positions taken by variables occurring many times in short rules with high supports, and with the attributes included rarely as conditions, in longer rules, with lower support values, at the bottom.

3   Stylometric Analysis of Texts

Authorship attribution is a main task within stylometric analysis of texts [6]. The fundamental notion in this domain comes down to the statement that given a sufficient number of representative samples of writing, any author can be characterised and recognised with a sufficient level of reliability, basing on uniqueness of their style. As authors are to be recognised regardless of what they write about, the subject topics of texts are disregarded, and instead there are considered stylometric features with discriminative properties, specific to authors and their writing styles, habits of expression, linguistic preferences. Thus various sets of attributes are employed in analysis. Techniques applied usually refer to statistic-oriented computations, or to artificial intelligence approaches.

Typical stylometric descriptors are lexical or syntactic. The former specify averages and frequencies of occurrence for words and phrases, while the latter bring information about syntactic aspects of sentence formation, and punctuation marks. Such stylometric features are continuous valued. Mining them for construction of rule classifiers results in transparent description of discovered patterns present in data, which enhances understanding of domain knowledge. However, many rule induction algorithms require nominal values of features, thus discretisation is often implemented as a part of input data pre-processing stage.

When an authorship attribution task is treated as classification, with authors recognised as distinguished classes, to evaluate performance of a constructed classifier it is important to employ independent validation and test samples based on entirely separate source texts. Otherwise (as in case of using cross-validation) the classification results could be overly optimistic [2]. Also, it is documented that authors of the same gender show higher similarity in writing styles [8]. Therefore, texts authored by writers of the opposite gender should not be used in the same input data set as comparing authors without gender distinction falsifies results to such degree as to make them unreliable.

4   Experimental Results

The experiments performed in the research presented in this paper consisted of several steps, as described in the following sections.

4.1   Preparation of Input Data Sets

The pre-processing stage was devoted to the preparation of the input data sets. Firstly, two pairs of authors were chosen, Thomas Hardy and Henry James (denoted as WriterM data set), and Edith Wharton and Mary Johnston (named as WriterF data set). Their works were separated into three groups corresponding to the source texts for learning, validation, and test samples. Each longer text was divided into several smaller pieces of comparable size. For each author the same numbers of samples were selected to ensure balance of data.

Secondly, over all these pieces of texts the frequencies of occurrence were calculated for 25 stylometric descriptors: 18 lexical markers corresponding to common function words used (and, of, in, to, that, for, with, on, this, at, but, from, not, by, as, what, if, without), and 7 syntactic markers referring to employed punctuation marks (exclamation, question, hyphen, colon, semicolon, fullstop, comma). It resulted in the set of continuous condition attributes with all the values in the range [0,1).

Thirdly, for each pair of writers all three sets of samples (learning, validation, and test) were independently discretised with Kononenko's supervised approach [7]. For further considerations there were taken these features for which the number of intervals established in discretisation was greater than one. As discretisation was executed in the limited local context of each set, in discrete WriterF data set there were 19 variables, and 17 for WriterM.

The constructed input data sets were subjected to rule induction algorithms.

4.2 Generation of Decision Rules by Greedy Heuristics

At the second stage of experiments four greedy heuristics, implemented in Java 8 using Spring framework, were applied to WriterM and WriterF training samples, returning four rule sets for both data sets. Heuristics induce one rule for each row of a decision table, regardless of rules inferred for other rows, which means that it is probable that some rules (in particular those with higher supports) are not unique. Thus all generated rules were compared, repeated elements removed, and the numbers listed specify only unique rules.

The rule sets were next employed as decision algorithms to classify samples from validation and test sets (called T1 and T2), using simple majority voting strategy in case of conflicts. In all evaluations of performance constraints on minimal rule support were imposed: there was chosen the highest support that ensured 100% recognition of the training samples. When some rules were discarded the value of support is given with the number of remaining rules. The results are displayed in Table 1.

For *RM* heuristic for WriterF, and *Log* for WriterM data set, from the rule sets some elements were rejected by imposing constraints on rule support, for others all found rules were needed to correctly classify the training data. Classification accuracy for the validation and test sets was not always satisfactory, in fact it was low for *MaxS* and *Poly* for WriterF, and for *MaxS* and *RM* for WriterM. The best results of classification accuracy are shown in bold.

For both data sets exhaustive algorithms in Classical Rough Set Approach (CRSA) were also inferred, with the parameters as listed in Table 2. In the full algorithms (*F-Exh* and *M-Exh* respectively), the numbers of generated rules were two ranks higher than from heuristics. For the minimal algorithms (named as *F-ExhM* and *M-ExhM*), obtained from *Exh* algorithms by rejecting weaker rules with rule supports lower than the listed minimum, the cardinalities of rule sets become manageable, if still higher than those from heuristics. Classification accuracies observed were increased, which is always an advantage.

Table 1. Parameters of rule sets generated by greedy heuristics

	WriterF data set				WriterM data set			
	Log	MaxS	Poly	RM	Log	MaxS	Poly	RM
Number of rules	15	20	17	24	29	65	37	30
Min/Max supp.	32/85	28/81	31/85	1/85	13/62	9/57	18/62	8/54
Average support	68.27	66.10	68.27	36.08	44.70	31.83	41.22	27.60
Min/Max length	2/5	3/8	2/7	1/2	2/5	2/9	2/7	1/4
Average length	2.47	4.85	4.55	1.83	3.45	5.47	4.27	2.37
Class. accuracy for T1 [%]	**94.49**	26.67	26.67	83.33 sup $\geq$ 12 20 rls	73.26 sup $\geq$ 21 28 rls	36.11	**81.48**	75.56
Class. accuracy for T2 [%]	77.63	4.00	12.25	**93.75**	81.47	44.56	**83.13**	50.00

Table 2. Parameters of rule sets generated by exhaustive CRSA algorithm

	WriterF data set		WriterM data set	
	F-Exh	F-ExhM	M-Exh	M-ExhM
Number of rules	2121	98	7291	347
Min/Avg/Max support	1/9.19/85	39/53.16/85	1/5.35/62	24/34.87/62
Min/Avg/Max length	1/3.86/7	2/2.64/5	1/4.90/8	2/3.88/7
Class. accuracy for T1 [%]	94.44	**100.00**	62.22	**90.00**
Class. accuracy for T2 [%]	96.25	**98.75**	85.00	**93.75**

4.3 Characterisation of Features by Induced Rules

In the third stage of experiments for each heuristic rule and attribute coefficients were calculated. The obtained values imposed orderings on attributes, displayed in Tables 3 and 4, which show also ranking based on cumulative attribute coefficients, averaged over all heuristics, and the order based on attribute coefficients calculated for exhaustive algorithms and their minimal forms.

Not all available attributes were always included as conditions in rule sets induced by all tested approaches, which is why some rankings contained fewer positions. The frequency of occurrence of "what" was never used in rules generated by greedy heuristics for WriterF dataset, thus the attribute is separated from others in $CumAC$ ranking. For WriterF heuristics discarded more features from the available set than for WriterM, for which almost always all considered variables were needed, however, the former had more condition attributes to begin with than the latter.

$CumAC$ ranking was next used for inferring new rule sets, and for pruning $ExhM$ rule sets, as described in the following sections of the paper.

4.4   Feature Selection Leading to Induction of New Rule Sets

Generation of new rules governed by a ranking was executed by steps, within which one attribute was added to the considered set, starting at the highest ranking position, and then proceeding down. The process was continued till the list of attributes was exhausted. As for the whole feature sets the algorithms were previously induced (*F-Exh* and *M-Exh*), thus the last induction step corresponded to the one before the lowest ranking position, 18th for WriterF and 16th for WriterM, which is displayed in Table 5.

The induction step indicates the number of attributes involved in generation of each rule set, then the total number of rules inferred in this step is listed, and how, with imposing threshold support given, this number was reduced. The value of support is at the maximal level that still ensures perfect recognition of the learning samples. The tables present only these steps for which inferred rules were capable of 100% recognition of training samples. The initial steps, where some learning examples were incorrectly classified, are omitted.

For WriterF data set 10 highest ranking attributes were sufficient for induction of decision rules correctly classifying all learning samples, yet the threshold

Table 3. Characterisation of available features by rule sets induced through greedy heuristics and exhaustive algorithm (CSRA) for WriterF data set

Ranking position	*Log*	*MaxS*	*Poly*	*RM*	*CumAC*	*F-Exh*	*F-ExhM*
1	comma	comma	comma	comma	comma	on	colon
2	colon	colon	colon	on	colon	but	comma
3	exclam	questi	that	colon	and	as	and
4	semico	without	at	and	at	colon	semico
5	and	that	and	to	not	from	on
6	not	at	to	but	exclam	by	exclam
7	to	and	exclam	not	that	and	to
8	at	not	semico	of	to	to	of
9	on	exclam	not	fullst	semico	of	by
10	by	of	without	as	on	what	as
11	fullst	semico	on	by	without	fullst	from
12	of	to	of	that	of	semico	what
13	from	by	fullst	at	questi	comma	fullst
14	but	on	by		fullst	exclam	at
15		fullst	from		by	not	but
16		from	questi		from	without	not
17		but			as	questi	without
18					but	at	questi
19					*what*	that	that

Table 4. Characterisation of available features by rule sets induced through greedy heuristics and exhaustive algorithm (CSRA) for WriterM data set

Ranking position	Log	MaxS	Poly	RM	CumAC	M-Exh	M-ExhM
1	from	but	but	and	from	at	but
2	for	for	from	from	but	but	from
3	but	semico	for	by	for	with	that
4	by	from	that	that	that	that	at
5	that	that	with	of	by	from	with
6	and	questi	questi	not	and	not	and
7	with	with	by	in	with	if	questi
8	fullst	fullst	fullst	semico	semico	and	for
9	if	hyphen	semico	for	fullst	of	not
10	semico	of	if	but	of	questi	semico
11	not	if	at	what	if	in	if
12	of	by	hyphen	with	questi	by	of
13	at	in	not	questi	not	what	by
14	in	and	of	if	in	semico	hyphen
15	questi	not	what	at	at	hyphen	in
16		at	and	fullst	hyphen	fullst	what
17		what	in		what	for	fullst

value of minimal rule support (7) was lower than for *F-ExhM* (39). Classification accuracy for T1 and T2 sets was slightly decreased, and the number of rules reduced to 59. On the other hand, from all these rule sets only the one studied at the 17th step measured up in performance to *F-ExhM*.

For WriterM the minimal number of variables to be recalled to ensure correct recognition of training examples was 7, but the performance was degraded. Only the 13th step offered the undamaged predictive power for the reduced number of rules, yet again the minimal support (19) was lower than for *M-ExhM* (24).

This part of experiments brought the conclusion that generation of new rule sets driven by characterisation of attributes through heuristics relatively quickly led to induction of rule sets with reduced cardinalities that were capable of perfect classification of the learning samples. Yet obtaining the same predictive power of rule classifiers required more features. Also threshold supports of rules, even though locally maximised, were not necessarily reaching the global maxima.

4.5   Feature Selection Used for Pruning Rule Sets

Greedy heuristics discovered some knowledge with respect to inclusion and exclusion of features from the considered set, while ensuring correct recognition for learning samples. Thus in the research an another approach was tried, relying on

Table 5. Classification results for rule sets generated while following $CumAC$ rankings for WriterF and WriterM data set

WriterF data set				WriterM data set			
Ind. step	Nr of rls (support)/rls	Class.acc. FT1 [%]	Class.acc. FT2 [%]	Ind. step	Nr of rls (support)/rls	Class.acc. MT1 [%]	Class.acc. MT2 [%]
10	132(7)/59	98.89	97.50	7	94(9)/51	78.01	69.13
11	165(22)/41	94.49	91.44	8	139(9)/66	74.72	69.13
12	232(22)/53	94.49	91.44	9	219(9)/80	90.00	76.25
13	255(22)/55	93.40	90.23	10	346(9)/119	75.56	70.00
14	349(26)/59	**100.00**	91.34	11	583(12)/142	87.78	82.50
15	493(27)/73	95.56	97.50	12	857(16)/145	93.33	87.50
16	718(32)/81	97.78	**98.75**	13	1429(19)/177	90.00	**93.75**
17	1104(32)/98	**100.00**	**98.75**	14	2229(23)/184	90.00	91.25
18	1621(39)/86	97.78	**98.75**	15	3434(24)/244	**94.44**	92.58
				16	5004(24)/297	88.89	92.58

pruning rule sets governed by heuristic-based characterisation of features. For each heuristic the set of variables included as conditions in induced rules was composed. Then the rules from $ExhM$ algorithm were pruned by discarding rules referring to variables absent in the considered set. Rule subsets are named after heuristic and the results shown in the upper part of Table 6.

Table 6. Classification results for pruned rule sets for WriterF and WriterM data set

WriterF data set				WriterM data set			
Rule set	Nr of rls	Class.acc. FT1 [%]	Class.acc. FT2 [%]	Rule set	Nr of rls	Class.acc. MT1 [%]	Class.acc. MT2 [%]
$ExhM\_Log$	67	**100.00**	97.50	$ExhM\_Log$	244	94.44	92.58
$ExhM\_MaxS$	75	97.78	**98.75**	$ExhM\_MaxS$	347	90.00	93.75
$ExhM\_Poly$*	68	97.78	**98.75**	$ExhM\_Poly$	347	90.00	93.75
$ExhM\_RM$	46	74.07	95.05	$ExhM\_RM$	291	**96.67**	93.75
$ExhM\_18$	86	97.78	**98.75**	$ExhM\_15$	244	94.44	92.58
$ExhM\_H$	36	72.10	95.05	$ExhM\_16$	297	88.89	92.58
* Only 99% recognition for learning				$ExhM\_RM15$	253	93.33	**93.81**
samples				$ExhM\_H$	214	94.44	92.58

From these subsets all but one maintained the classification for training samples. This was not true for F-$ExhM\_Poly$. For WriterF data set none of the four rule sets offered uncorrupted predictive power for both validation and test sets

T1 and T2. For WriterM $M\text{-}ExhM\_MaxS = M\text{-}ExhM\_Poly = ExhM$. More interesting was $M\text{-}ExhM\_RM$ that included fewer rules (reduction by $56/347 = 16.13\%$ with respect to $M\text{-}ExhM$) and the same or improved performance.

The middle rows (only a single row for WriterF) of Table 6 present selected results from rule set pruning while following a feature ranking. The ranking exploited was the same as previously driving generation of new rule sets, described in Sect. 4.4. The elements from $ExhM$, $ExhM\_Log$, $ExhM\_MaxS$, $ExhM\_Poly$, and $ExhM\_RM$ were pruned, by keeping rules with all attributes included in the subset considered in each step and rejecting others.

The results given are limited to these rule sets that kept the recognition of learning samples intact, and for WriterF it was true only for $F\text{-}ExhM$ for the last possible subset with 18 included variables. For WriterM there were three such cases, two for $M\text{-}ExhM$, and one for $ExhM\_RM$. The numerical index indicates the cardinality of each attribute set. Only $ExhM\_RM15$ rule set challenged results obtained for $M\text{-}ExhM$, with the length reduced by $94/347 = 27.09\%$.

For each data set, for the sets of attributes included in rules induced by each heuristic that perfectly classified the training data, there was executed an intersection and only elements present in this subset were allowed to be used as conditions in rules from $ExhM$, while rules involving other variables were removed. The remaining rule sets $ExhM\_H$, given in the bottom row of the table, had the lowest cardinalities against those studied for rule pruning, but the performance was not impressive, in particular for WriterF.

These experiments showed successful application of feature characterisation by heuristics for pruning rule sets while maintaining the correct classification of training samples, yet without any guarantee of uncorrupted predictive power of rule classifiers. The rules studied in this batch of tests had the advantage of high support values as pruned rule sets were obtained by maximising this parameter.

5 Conclusions

The paper presents research conducted in stylometric domain, dedicated to application of some greedy heuristics for characterisation and selection of features. In the first part of executed experiments selected heuristics were applied to the training data and decision rules were induced by these heuristics. Next, the inferred rule sets and their parameters were treated as an additional source of knowledge on available attributes, which led to construction of feature rankings.

In the second part the rankings were exploited in generation of new rules driven by the ranking, and for pruning rule sets. The results from the two processes were compared to the inferred exhaustive algorithms in their full, and support constrained forms. In both approaches several rule sets were obtained with lowered cardinalities, as well as cases of the same and improved performance for validation and test sets, showing the merit of the proposed methodology.

Acknowledgments. The research described in the paper was performed at the Silesian University of Technology, Gliwice, within the project BK/RAu2/2018, and at the University of Silesia in Katowice, Sosnowiec, within the project "Methods of artificial intelligence in information systems".

References

1. Alsolami, F., Amin, T., Moshkov, M., Zielosko, B.: Comparison of heuristics for optimization of association rules. In: Suraj, Z., Czaja, L. (eds.) Concurrency Specification and Programming. CEUR Workshop Proceedings, vol. 1492, pp. 4–11. CEUR-WS.org (2015)
2. Baron, G.: Comparison of cross-validation and test sets approaches to evaluation of classifiers in authorship attribution domain. In: Czachórski, T., Gelenbe, E., Grochla, K., Lent, R. (eds.) ISCIS 2016. CCIS, vol. 659, pp. 81–89. Springer, Cham (2016). https://doi.org/10.1007/978-3-319-47217-1_9
3. Guyon, I., Gunn, S., Nikravesh, M., Zadeh, L. (eds.): Feature Extraction: Foundations and Applications. Studies in Fuzziness and Soft Computing, vol. 207. Physica-Verlag, Springer, Heidelberg (2006). https://doi.org/10.1007/978-3-540-35488-8
4. Janusz, A., Ślęzak, D.: Rough set methods for attribute clustering and selection. Appl. Artif. Intell. **28**(3), 220–242 (2014)
5. Jensen, R., Shen, Q.: Computational Intelligence and Feature Selection: Rough and Fuzzy Approaches. IEEE Press Series on Computational Intelligence. Wiley-IEEE Press (2008)
6. Jockers, M., Witten, D.: A comparative study of machine learning methods for authorship attribution. Literary Linguist. Comput. **25**(2), 215–223 (2010)
7. Kononenko, I.: On biases in estimating multi-valued attributes. In: 14th International Joint Conference on Articial Intelligence, pp. 1034–1040 (1995)
8. Koppel, M., Argamon, S., Shimoni, A.: Automatically categorizing written texts by author gender. Literary Linguist. Comput. **17**(4), 401–412 (2002)
9. Pawlak, Z.: Rough sets and intelligent data analysis. Inf. Sci. **147**, 1–12 (2002)
10. Stańczyk, U.: Weighting of attributes in an embedded rough approach. In: Gruca, D.A., Czachórski, T., Kozielski, S. (eds.) Man-Machine Interactions 3. AISC, vol. 242, pp. 475–483. Springer, Cham (2014). https://doi.org/10.1007/978-3-319-02309-0_52
11. Stańczyk, U.: Selection of decision rules based on attribute ranking. J. Intell. Fuzzy Syst. **29**(2), 899–915 (2015)
12. Stańczyk, U., Zielosko, B.: On combining discretisation parameters and attribute ranking for selection of decision rules. In: Polkowski, L., et al. (eds.) IJCRS 2017. LNCS (LNAI), vol. 10313, pp. 329–349. Springer, Cham (2017). https://doi.org/10.1007/978-3-319-60837-2_28
13. Wróbel, L., Sikora, M., Michalak, M.: Rule quality measures settings in classification, regression and survival rule induction – an empirical approach. Fundamenta Informaticae **149**, 419–449 (2016)
14. Zielosko, B.: Application of dynamic programming approach to optimization of association rules relative to coverage and length. Fundamenta Informaticae **148**(1–2), 87–105 (2016)

An Optimization View on Intuitionistic Fuzzy Three-Way Decisions

Jiubing Liu[1], Xianzhong Zhou[1,2], Huaxiong Li[1,2(✉)], Bing Huang[3],
Libo Zhang[1], and Xiuyi Jia[4]

[1] School of Management and Engineering, Nanjing University,
Nanjing 210093, People's Republic of China
huaxiongli@nju.edu.cn
[2] Research Center for Novel Technology of Intelligent Equipment,
Nanjing University, Nanjing 210093, People's Republic of China
[3] School of Information Engineering, Nanjing Audit University,
Nanjing 211815, People's Republic of China
[4] School of Computer Science and Engineering,
Nanjing University of Science and Technology,
Nanjing 210094, People's Republic of China

Abstract. From an optimization point of view, we propose a new
method to determine the loss funtion of intuitionistic fuzzy three-way
decisions. First, two linear programming models are constructed to determine a pair of thresholds in three-way decisions based on their practical
semantics. Meanwhile, the validity of the models is verified by KKT conditions. Second, the models are further extended to intuitionistic fuzzy
three-way decisions (IF-3WD) and the corresponding nonlinear models
are established. Third, the uniqueness of solution for models is proven
and a LINGO software is employed to solve the models. We then obtain
both thresholds of IF-3WD and its decision rules. Finally, an example is
given to show the effectiveness of our method.

Keywords: Intuitionistic fuzzy sets · Three-way decisions
Optimization models · KKT conditions

1 Introduction

Three-way decisions (3WD), composed of acceptance, further investigation and
rejection, are initially proposed in 1990 based on the Bayesian decision theory
[1]. Since then, researches on 3WD have received more and more attention and
many related achievements have been achieved [2–5,8], which are widely applied
to various fields such as spam filtering [3], face recognition [4], cognitive concept
learning [6], three-way clustering [7] and multi-attribute decision making [8].

In the studies on three-way decisions, how to determine a pair of thresholds
of three-way decisions has become a crucial step of obtaining three-way decision
rules. Presently many research achievements on such aspect have been obtained

© Springer Nature Switzerland AG 2018
H. S. Nguyen et al. (Eds.): IJCRS 2018, LNAI 11103, pp. 363–377, 2018.
https://doi.org/10.1007/978-3-319-99368-3_28

[1,9–13]. For example, Yao [1] first deduced the analytic solutions to both thresholds in 3WD from decision-theoretic rough sets on the basis of Bayesian decision procedure, which provides a reasonable semantic for a pair of thresholds in probabilistic rough sets. Li et al. [9] derived the mathematical expression of both thresholds in 3WD from decision-theoretic rough sets model with multiple risk preferences. On the basis of optimization models, Jia et al. [10] constructed an optimization model with the minimum of total decision costs, consisting of the positive decision costs, negative decision costs and boundary decision costs in 3WD, based on which an adaptive learning algorithm is designed to solve the model and to determine both thresholds. Also, the similar method is presented in [11]. Based on that, Zhang [12] proposed an approach to the determination for a pair of thresholds of three-way decisions in view of Gini coefficients. In addition, Azam [13] introduced game-theoretic methods to determine these thresholds.

With the above-mentioned literature, it is clearly acknowledged that a pair of thresholds in 3WD are determined based on the loss function assessments with real numbers. In reality, however, decision maker may be difficult to give a crisp evaluation, and more easier to adopt an imprecise or fuzzy evaluation such as interval numbers, linguistic variables and intuitionistic fuzzy sets. Later, researchers explored the determination of both thresholds in 3WD with the loss function expressed by fuzzy evaluation. For example, in light of fuzzy three-way decision models with Bayesian decision procedure, Liang et al. systematically studied the threshold determination of fuzzy three-way decisions based on the loss function given respectively as: interval numbers [15], linguistic variables [21], intuitionistic fuzzy sets [17–19], hesitant fuzzy sets [20] and dual hesitant fuzzy sets [21], and then obtained the corresponding fuzzy three-way decisions.

However, in some cases where the loss function is expressed as fuzzy assessments above (e.g. intuitionistic fuzzy sets), it is usually difficult to determine a pair of thresholds in intuitionistic fuzzy 3WD using existing methods [17–19]. Thus, we only obtain the indirect rules of intuitionistic fuzzy 3WD, which can not facilitate make actual decisions. To overcome these, in this paper a general method for determining these thresholds is proposed based on optimization models, which helps obtain intuitionistic fuzzy 3WD directly.

2   Preliminaries

2.1   Decision-Theoretic Rough Sets

In general, the model of decision-theoretic rough sets is composed of two states and three actions [1,2], denoted by $\Omega = \{X, \neg X\}$, and $A = \{a_P, a_B, a_N\}$, respectively. The loss function regarding three actions under different states is listed by the 3×2 matrix, as shown in Table 1.

For Table 1, $\lambda_{PP}, \lambda_{BP}$ and λ_{NP} denote the risk loss generated by adopting actions of a_P, a_B and a_N, respectively, when an object is in the state of X. Analogously, $\lambda_{PN}, \lambda_{BN}$ and λ_{NN} denote the risk loss for adopting the same actions, respectively, when an object is not in X. Assume $Pr(X|[x])$ is the conditional probability of an object x belonging to X, where x is usually denoted by its

Table 1. The risk loss matrix of actions under different states.

	$X(P)$	$\neg X(N)$
a_P	λ_{PP}	λ_{PN}
a_B	λ_{BP}	λ_{BN}
a_N	λ_{NP}	λ_{NN}

equivalence class $[x]$. Therefore the expected risk loss $R(a_\bullet|[x])(\bullet = P, B, N)$ for each object x is calculated as:

$$R(a_\bullet|[x]) = \lambda_{\bullet P} Pr(X|[x]) + \lambda_{\bullet N} Pr(\neg X|[x]). \tag{1}$$

In light of Bayesian decision procedure, which indicates the following decision rules with the minimum risk losses [1]:

(P) If $R(a_P|[x]) \leq R(a_B|[x])$ and $R(a_P|[x]) \leq R(a_N|[x])$, decide: $x \in POS(X)$;
(B) If $R(a_B|[x]) \leq R(a_P|[x])$ and $R(a_B|[x]) \leq R(a_N|[x])$, decide: $x \in BND(X)$;
(N) If $R(a_N|[x]) \leq R(a_P|[x])$ and $R(a_N|[x]) \leq R(a_B|[x])$, decide: $x \in NEG(X)$.

The above rules (P)–(N) are called three-way decisions. As a matter of fact, these rules can be simplified on the basis of the relation: $Pr(X|[x]) + Pr(\neg X|[x]) = 1$ and the corresponding losses in Table 1. By considering a reasonable case of the loss function with:

$$\lambda_{PP} \leq \lambda_{BP} < \lambda_{NP}, \tag{2}$$
$$\lambda_{NN} \leq \lambda_{BN} < \lambda_{PN}. \tag{3}$$

We can obtain the concise rules (P1)–(N1) as follows:

(P1) If $Pr(X|[x]) \geq \alpha$ and $Pr(X|[x]) \geq \gamma$, decide: $x \in POS(X)$;
(B1) If $Pr(X|[x]) \leq \alpha$ and $Pr(X|[x]) \geq \beta$, decide: $x \in BND(X)$;
(N1) If $Pr(X|[x]) \leq \beta$ and $Pr(X|[x]) \leq \gamma$, decide: $x \in NEG(X)$.

where the thresholds α, β and γ are calculated as:

$$\begin{aligned}
\alpha &= \frac{\lambda_{PN} - \lambda_{BN}}{(\lambda_{PN} - \lambda_{BN}) + (\lambda_{BP} - \lambda_{PP})}, \\
\beta &= \frac{\lambda_{BN} - \lambda_{NN}}{(\lambda_{BN} - \lambda_{NN}) + (\lambda_{NP} - \lambda_{BP})}, \\
\gamma &= \frac{\lambda_{PN} - \lambda_{NN}}{(\lambda_{PN} - \lambda_{NN}) + (\lambda_{NP} - \lambda_{PP})}.
\end{aligned} \tag{4}$$

Based on (2), (3) and (4), it follows that $0 < \alpha \leq 1$, $0 \leq \beta < 1$ and $0 < \gamma < 1$. Additionally, we see from the rule (B1) that there exist two cases: (i) $\alpha > \beta$ and (ii) $\alpha \leq \beta$. Let us first take into account the case: (i) $\alpha > \beta$ which implies:

$$(\lambda_{PN} - \lambda_{BN})(\lambda_{NP} - \lambda_{BP}) > (\lambda_{BP} - \lambda_{PP})(\lambda_{BN} - \lambda_{NN}). \tag{5}$$

(5) induces $\alpha > \gamma > \beta$, by which the rules (P1)–(N1) are further simplified as:

(P2) If $Pr(X|[x]) \geq \alpha$, decide: $x \in POS(X)$;
(B2) If $\beta < Pr(X|[x]) < \alpha$, decide: $x \in BND(X)$;
(N2) If $Pr(X|[x]) \leq \beta$, decide: $x \in NEG(X)$.

(ii) When $\alpha \leq \beta$, we have:

$$(\lambda_{PN} - \lambda_{BN})(\lambda_{NP} - \lambda_{BP}) \leq (\lambda_{BP} - \lambda_{PP})(\lambda_{BN} - \lambda_{NN}), \tag{6}$$

which implies $\alpha \leq \gamma \leq \beta$. Hence the rules (P1)–(N1) are reduced to the two-way decisions below.

(P3) If $Pr(X|[x]) \geq \gamma$, decide: $x \in POS(X)$;
(N3) If $Pr(X|[x]) < \gamma$, decide: $x \in NEG(X)$.

2.2 Intuitionistic Fuzzy Sets

Let $X = \{x_1, x_2, ..., x_n\}$ be a fixed set. An intuitionistic fuzzy set (IFS) E on X is defined as [22]:

$$E = \{(x, \mu_E(x), \nu_E(x))|x \in X\}, \tag{7}$$

where $\mu_E : X \mapsto [0,1]$ and $\nu_E : X \mapsto [0,1]$ denote the membership and non-membership degrees of element x belonging to the IFS E respectively, with $0 \leq \mu_E(x_k) + \nu_E(x_k) \leq 1$ for all $x \in X$. In addition, $\pi_E(x) = 1 - \mu_E(x) - \nu_E(x) \in [0,1]$ is called the hesitation degree of element x belonging to the IFS E. Particularly, if $\pi_E(x) = 0$ for all $x \in X$, then the IFS E reduces to an ordinary fuzzy set.

In light of these results reported in [22], an intuitionistic fuzzy number (IFN) is denoted by $e = (\mu_e, \nu_e)$. Given IFNs $e = (\mu_e, \nu_e)$ and $g = (\mu_g, \nu_g)$, we have:

(1) $e = g$ if and only if $\mu_e = \mu_g$ and $\nu_e = \nu_g$;
(2) $\bar{e} = (\nu_e, \mu_e)$, where $\bar{e}$ is the complement set of e;
(3) $e \oplus g = (\mu_e + \mu_g - \mu_e\mu_g, \nu_e\nu_g)$;
(4) $\lambda e = (1 - (1 - \mu_e)^\lambda, (\nu_e)^\lambda)$, where $\lambda \geq 0$.

To compare IFNs, the ranking function of IFNs based on the risk attitudes of decision maker (DM) is defined in advance.

Definition 1 *[23]. Let $e = (\mu_e, \nu_e)$ be an IFN. Then, the ranking function of e is calculated as:*

$$S_\varpi(e) = (1 - \varpi)\frac{1 - \nu_e}{1 + \pi_e} + \varpi(1 - \frac{1}{2}\pi_e^2), \tag{8}$$

where $\pi_e = 1 - \mu_e - \nu_e$ and $\varpi \in [0,1]$ is the risk coefficient reflecting the risk attitudes of DM. Specially, if $\varpi \in (0.5, 1]$, then the DM is optimistic about decision results; if $\varpi \in [0, 0.5)$, then the DM is pessimistic; otherwise, the DM is neutral.

Based on Definition 1, the rules for ranking IFNs are given as follows:

Definition 2 *[23]. Let $e = (\mu_e, \nu_e)$ and $g = (\mu_g, \nu_g)$ be two IFNs. Then we have:*

(1) If $S_\varpi(e) > S_\varpi(g)$, then e is bigger than g, denoted by $e \succ g$;
(2) If $S_\varpi(e) < S_\varpi(g)$, then e is smaller than g, denoted by $e \prec g$;
(3) If $S_\varpi(e) = S_\varpi(g)$, then e is equal to g, denoted by $e \sim g$.

2.3 KKT Conditions

Considering the model:

$$\min_{x \in R^n} \ f(x)$$

$$\text{s.t.} \begin{cases} g_i(x) \leq 0, i = 1, 2, ..., m, \\ h_j(x) = 0, j = 1, 2, ..., r. \end{cases} \tag{9}$$

It is clearly known from (9) that its feasible region is $D = \{x : g_i(x) \leq 0, i = 1, 2, ..., m; h_j(x) = 0, j = 1, 2, ..., r\}$, which is a closed set. If all locally optimal solutions to (9) are searched, then its globally optimal solution will be found from them. Following this idea, an approach to searching all locally optimal solutions to (9) is given based on the following theorem.

Theorem 1 *[24]. Let x^* be a feasible solution to (9) and $f(x), g_i(x), h_j(x)$ be differentiable functions where $1 \leq i \leq m$ and $1 \leq j \leq r$. If x^* is a locally optimal solution to (9), then there exist multiplier vectors $\Gamma^* = (u_1^*, u_2^*, ..., u_m^*)^T$ and $\Lambda^* = (v_1^*, v_2^*, ..., v_r^*)^T$ of Lagrange such that the following formulas hold.*

$$\begin{cases} \Delta f(x^*) + \sum_{i=1}^{m} u_i^* \Delta g_i(x^*) + \sum_{j=1}^{r} v_j^* \Delta h_j(x^*) = 0, \\ g_i(x^*) \leq 0, u_i^* g_i(x^*) = 0, u_i^* \geq 0, i = 1, 2, ..., m, \\ h_j(x^*) = 0, j = 1, 2, ..., r. \end{cases} \tag{10}$$

Note that Theorem 1 is a necessary KKT condition whether or not (9) exists locally optimal solutions. If there are the locally optimal solutions to (9), then they will be generated among its KKT points. Specially, if it is a convex programming problem, then the KKT points will be its locally optimal solutions. With respect to a linear programming problem, which is regarded as one of convex programming problems, the KKT points are also its globally optimal solutions [24]. Therefore, searching the global optimization solutions to linear programming problems becomes to find their KKT points by KKT conditions, which is the main idea for solving the following α-model and β-model.

3 Constructing Optimization Models to Determine Both Thresholds in 3WD

For readers' convenience, $Pr(X|[x])$ and $Pr(\neg X|[x])$ in (1) are denoted by $s = Pr(X|[x])$ and $t = Pr(\neg X|[x])$ respectively, which can obtain $s + t = 1$. In fact, we see from (P) and (P2) that the value of α should be the minimum one among all conditional probabilities satisfying: $R(a_P|[x]) \leq R(a_B|[x])$ and $R(a_P|[x]) \leq R(a_N|[x])$. Similarly, the value of β should be the maximum one among all conditional probabilities satisfying: $R(a_N|[x]) \leq R(a_P|[x])$ and $R(a_N|[x]) \leq R(a_B|[x])$ from (N) and (N2). These are our motivation that the following two optimization models are established to determine a pair of thresholds α and β in 3WD.

α-model:

$$\alpha = \min\ s$$

$$s.t. \begin{cases} s\lambda_{PP} + t\lambda_{PN} \leq s\lambda_{BP} + t\lambda_{BN}, \\ s\lambda_{PP} + t\lambda_{PN} \leq s\lambda_{NP} + t\lambda_{NN}, \\ s + t = 1. \end{cases}$$

β-model:

$$\beta = \max\ s$$

$$s.t. \begin{cases} s\lambda_{NP} + t\lambda_{NN} \leq s\lambda_{PP} + t\lambda_{PN}, \\ s\lambda_{NP} + t\lambda_{NN} \leq s\lambda_{BP} + t\lambda_{BN}, \\ s + t = 1. \end{cases}$$

In order to verify that the optimal solutions to the above-proposed models are consistent with (4). KKT conditions are used to induce their analytical solutions.

For the α-model, where the KKT conditions are adopted to obtain the following formulas.

$$\begin{cases} 1 - u_1^*(\lambda_{BP} - \lambda_{PP}) - u_2^*(\lambda_{NP} - \lambda_{PP}) + v_1^* = 0, \\ -u_1^*(\lambda_{BN} - \lambda_{PN}) - u_2^*(\lambda_{NN} - \lambda_{PN}) + v_1^* = 0, \\ s^*(\lambda_{BP} - \lambda_{PP}) - t^*(\lambda_{PN} - \lambda_{BN}) \geq 0, \\ s^*(\lambda_{NP} - \lambda_{PP}) - t^*(\lambda_{PN} - \lambda_{NN}) \geq 0, \\ u_1^*[s^*(\lambda_{BP} - \lambda_{PP}) - t^*(\lambda_{PN} - \lambda_{BN})] = 0, u_1^* \geq 0, \\ u_2^*[s^*(\lambda_{NP} - \lambda_{PP}) - t^*(\lambda_{PN} - \lambda_{NN})] = 0, u_2^* \geq 0, \\ s^* + t^* - 1 = 0. \end{cases} \tag{11}$$

In (11), several cases are discussed to obtain the corresponding KKT points.

(1) If $u_1^* = u_2^* = 0$, then $v_1^* = -1$ and $v_1^* = 0$, which are contradictory with each other. Clearly, this case does not hold.

(2) If $u_1^* = 0$ and $u_2^* \neq 0$, then we get:

$$s_1^* = \frac{\lambda_{PN} - \lambda_{NN}}{(\lambda_{PN} - \lambda_{NN}) + (\lambda_{NP} - \lambda_{PP})} \text{ and } u_2^* = \frac{1}{(\lambda_{PN} - \lambda_{NN}) + (\lambda_{NP} - \lambda_{PP})} > 0.$$

(3) If $u_1^* \neq 0$ and $u_2^* = 0$, then it follows:

$$s_2^* = \frac{\lambda_{PN} - \lambda_{BN}}{(\lambda_{PN} - \lambda_{BN}) + (\lambda_{BP} - \lambda_{PP})} \text{ and } u_1^* = \frac{1}{(\lambda_{PN} - \lambda_{BN}) + (\lambda_{BP} - \lambda_{PP})} > 0.$$

(4) If $u_1^* \neq 0$ and $u_2^* \neq 0$, then one has:

$$s_3^* = \frac{\lambda_{BN} - \lambda_{NN}}{(\lambda_{BN} - \lambda_{NN}) + (\lambda_{NP} - \lambda_{BP})},$$

$$u_1^*[(\lambda_{BP} - \lambda_{PP}) + (\lambda_{PN} - \lambda_{BN})] + u_2^*[(\lambda_{NP} - \lambda_{PP}) + (\lambda_{PN} - \lambda_{NN})] = 1.$$

Take $u_1^* = \frac{1}{2[(\lambda_{BP}-\lambda_{PP})+(\lambda_{PN}-\lambda_{BN})]} > 0$ and $u_2^* = \frac{1}{2[(\lambda_{NP}-\lambda_{PP})+(\lambda_{PN}-\lambda_{NN})]} > 0$.

For convenience, let $a = \lambda_{BP} - \lambda_{PP}, b = \lambda_{PN} - \lambda_{BN}, c = \lambda_{NP} - \lambda_{BP}$ and $d = \lambda_{BN} - \lambda_{NN}$. It is clear that $s_1^* = \frac{b+d}{b+d+c+a}, s_2^* = \frac{b}{b+a}$ and $s_3^* = \frac{d}{d+c}$. Nowadays we need to determine whether or not s_1^*, s_2^* and s_3^* are the feasible solutions to α-model, where several cases are discussed as follows:

(1) If $s_2^* > s_3^*$, then it holds that $ad < bc$. In this way,
 (i) When $s_1^* = \frac{b+d}{b+d+c+a}$, we have $t_1^* = \frac{c+a}{b+d+c+a}$, which follows

$$s_1^*(\lambda_{BP} - \lambda_{PP}) - t_1^*(\lambda_{PN} - \lambda_{BN}) = \frac{(b+d)a - (c+a)b}{b+d+c+a} = \frac{ad - bc}{b+d+c+a} < 0.$$

Therefore s_1^* is not a feasible solution to α-model and is not a KKT point.
 (ii) When $s_2^* = \frac{b}{b+a}$, one has $t_2^* = \frac{a}{b+a}$. Thus,

$$s_2^*(\lambda_{BP} - \lambda_{PP}) - t_2^*(\lambda_{PN} - \lambda_{BN}) = \frac{ba-ab}{b+a} = 0 \text{ and}$$

$$s_2^*(\lambda_{NP} - \lambda_{PP}) - t_2^*(\lambda_{PN} - \lambda_{NN}) = \frac{b(c+a)-a(b+d)}{b+a} = \frac{bc-ad}{b+a} > 0,$$

which implies that s_2^* is a feasible solution to α-model and a KKT point.
 (iii) When $s_3^* = \frac{d}{d+c}$, $t_3^* = \frac{c}{d+c}$ holds. That is,

$$s_3^*(\lambda_{BP} - \lambda_{PP}) - t_3^*(\lambda_{PN} - \lambda_{BN}) = \frac{da - cb}{d+c} < 0.$$

Thereby s_3^* is not a feasible solution to α-model and is not a KKT point.
(2) If $s_2^* < s_3^*$, then it follows $ad > bc$. Similarly,
 (i) When $s_1^* = \frac{b+d}{b+d+c+a}$, we have $t_1^* = \frac{c+a}{b+d+c+a}$, which leads to

$$s_1^*(\lambda_{BP} - \lambda_{PP}) - t_1^*(\lambda_{PN} - \lambda_{BN}) = \frac{(b+d)a - (c+a)b}{b+d+c+a} = \frac{ad - bc}{b+d+c+a} > 0,$$

$$\text{and } s_1^*(\lambda_{NP} - \lambda_{PP}) - t_1^*(\lambda_{PN} - \lambda_{NN}) = \frac{(b+d)(c+1) - (c+a)(b+d)}{b+d+c+a} = 0.$$

Thus s_1^* is a feasible solution to α-model and is a KKT point.
 (ii) When $s_2^* = \frac{b}{b+a}$, one has $t_2^* = \frac{a}{b+a}$. Thus,

$$s_2^*(\lambda_{BP} - \lambda_{PP}) - t_2^*(\lambda_{PN} - \lambda_{BN}) = \frac{ba-ab}{b+a} = 0 \text{ and}$$

$$s_2^*(\lambda_{NP} - \lambda_{PP}) - t_2^*(\lambda_{PN} - \lambda_{NN}) = \frac{b(c+a)-a(b+d)}{b+a} = \frac{bc-ad}{b+a} < 0.$$

Obviously, s_2^* is not a feasible solution to α-model and is not a KKT point.

(iii) When $s_3^* = \frac{d}{d+c}$, it yields $t_3^* = \frac{c}{d+c}$. Then we obtain:

$$s_3^*(\lambda_{BP} - \lambda_{PP}) - t_3^*(\lambda_{PN} - \lambda_{BN}) = \frac{da - cb}{d + c} > 0 \text{ and}$$

$$s_3^*(\lambda_{NP} - \lambda_{PP}) - t_3^*(\lambda_{PN} - \lambda_{NN}) = \frac{da - cb}{d + c} > 0,$$

which shows that s_3^* is a feasible solution to α-model. Also it is a KKT point.

(3) If $s_2^* = s_3^*$, then $ad = bc$ is obvious. It is easy to verify that s_1^*, s_2^* and s_3^* are KKT points, and $s_1^* = s_2^* = s_3^*$.

Based on the analysis above, we know when $s_2^* > s_3^*$, s_2^* is a unique KKT point of α-model; when $s_2^* < s_3^*$, s_1^* and s_3^* are both KKT points of α-model. However, it follows $\frac{c}{d} < \frac{a+c}{b+d}$ due to $ad > bc$, thus we get $\frac{1}{1+\frac{c+a}{b+d}} < \frac{1}{1+\frac{c}{d}}$ that is $s_1^* = \frac{b+d}{b+d+c+a} < \frac{d}{d+c} = s_3^*$; when $s_2^* = s_3^*$, s_1^*, s_2^* and s_3^* are its KKT points and $s_1^* = s_2^* = s_3^*$. Considering the α-model is a linear programming model, so the following theorem is obtained:

Theorem 2. *In the α-model, if $s_2^* > s_3^*$, then s_2^* is its unique optimal solution, which is $\alpha = \frac{\lambda_{PN}-\lambda_{BN}}{(\lambda_{PN}-\lambda_{BN})+(\lambda_{BP}-\lambda_{PP})}$; otherwise, s_1^* is its unique optimal solution, that is $\alpha = \frac{\lambda_{PN}-\lambda_{NN}}{(\lambda_{PN}-\lambda_{NN})+(\lambda_{NP}-\lambda_{PP})}$.*

Analogously, we can obtain the similar theorem for the β-model as follows:

Theorem 3. *In the β-model, if $s_2^* > s_3^*$, then s_3^* is its unique optimal solution, which is $\beta = \frac{\lambda_{BN}-\lambda_{NN}}{(\lambda_{BN}-\lambda_{NN})+(\lambda_{NP}-\lambda_{BP})}$; otherwise, s_1^* is its unique optimal solution, that is $\beta = \frac{\lambda_{PN}-\lambda_{NN}}{(\lambda_{PN}-\lambda_{NN})+(\lambda_{NP}-\lambda_{PP})}$.*

Combining Theorem 2 with Theorem 3, We find from (4) that $\alpha = s_2^*, \beta = s_3^*$ and $\gamma = s_1^*$, which is further to deduce the following corollary.

Corollary 1. *In the α-model and β-model, if $\alpha > \beta$, then $\alpha = \frac{\lambda_{PN}-\lambda_{BN}}{(\lambda_{PN}-\lambda_{BN})+(\lambda_{BP}-\lambda_{PP})}$ and $\beta = \frac{\lambda_{BN}-\lambda_{NN}}{(\lambda_{BN}-\lambda_{NN})+(\lambda_{NP}-\lambda_{BP})}$ are their unique optimal solution respectively; otherwise, $\gamma = \frac{\lambda_{PN}-\lambda_{NN}}{(\lambda_{PN}-\lambda_{NN})+(\lambda_{NP}-\lambda_{PP})}$ is their unique optimal solution simultaneously.*

Proof. Theorems 2 and 3 show that $\alpha = \frac{\lambda_{PN}-\lambda_{BN}}{(\lambda_{PN}-\lambda_{BN})+(\lambda_{BP}-\lambda_{PP})}$ and $\beta = \frac{\lambda_{BN}-\lambda_{NN}}{(\lambda_{BN}-\lambda_{NN})+(\lambda_{NP}-\lambda_{BP})}$ are a unique optimal solution of α-model and β-model respectively, when $\alpha > \beta$. We mainly prove the latter part of this corollary, in fact, if $\alpha \leq \beta$, it will hold that $ad \geq bc$. That is $\frac{c}{d} \leq \frac{c+a}{b+d} \leq \frac{a}{b}$, which indicates $\frac{1}{1+\frac{a}{b}} \leq \frac{1}{1+\frac{c+a}{b+d}} \leq \frac{1}{1+\frac{c}{d}}$, namely, $\frac{b}{b+a} \leq \frac{b+d}{b+d+c+a} \leq \frac{d}{d+c}$. Therefore $\alpha \leq \gamma \leq \beta$. However, when $\alpha \leq \beta$, $\alpha = \beta = \frac{\lambda_{PN}-\lambda_{NN}}{(\lambda_{PN}-\lambda_{NN})+(\lambda_{NP}-\lambda_{PP})}$ is an optimal solution of α-model and β-model simultaneously. Thereby γ is also their optimal solution.

Corollary 1 shows that in three-way decisions, a pair of thresholds obtained by the α-model and β-model coincide with the ones derived from the classical method [1,2], which implies that both models-based method for determining thresholds is feasible and effective.

To better understand these models, the semantics of their optimal solutions α and β are presented in Fig. 1. We see from Fig. 1 that if $\alpha > \beta$, the three-way decisions are adopted; otherwise, the two-way decisions are used and here the α-model and β-model converge to the same pont,which is $\alpha = \beta$.

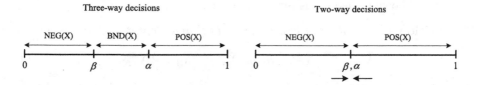

Fig. 1. The semantics of the optimal solutions to these models

It is well known that the loss function in 3WD is usually expressed by precise real numbers. In actual decision process, however, decision maker may be difficult to give a precise assessment on the loss function due to their limited knowledge and tight deadlines. Whence they are much easier to give an imprecise or fuzzy evaluation, such as interval numbers and intuitionistic fuzzy sets (IFSs). In the following, the above method is extended to three-way decision problems where the loss function is expressed by IFSs in Table 2. We then determine both thresholds of intuitionistic fuzzy three-way decisions (for short, IF-3WD), which can overcome these drawbacks that the current methods are difficult to determine both thresholds α and β on IF-3WD in some cases [17–19]. This is the main purpose of proposing an optimization models based IF-3WD method.

Table 2. The IF risk loss matrix of actions under different states.

	$X(P)$	$\neg X(N)$
a_P	$\overline{\lambda}_{PP} = (\mu_{PP}, \nu_{PP})$	$\overline{\lambda}_{PN} = (\mu_{PN}, \nu_{PN})$
a_B	$\overline{\lambda}_{BP} = (\mu_{BP}, \nu_{BP})$	$\overline{\lambda}_{BN} = (\mu_{BN}, \nu_{BN})$
a_N	$\overline{\lambda}_{NP} = (\mu_{NP}, \nu_{NP})$	$\overline{\lambda}_{NN} = (\mu_{NN}, \nu_{NN})$

4   Optimization Models Construction in IF-3WD

In the IF-3WD, there are still two states $\Omega = \{X, \neg X\}$ and three actions $A = \{a_P, a_B, a_N\}$, where the implications of states and actions are the same as the ones in Table 1. The differences in this model are the loss function with intuitionistic fuzzy sets rather than precise real numbers, as shown in Table 2.

By considering a reasonable case of the intuitionistic fuzzy loss function with:

$$\mu_{PP} < \mu_{BP} < \mu_{NP}, \nu_{PP} > \nu_{BP} > \nu_{NP}, \pi_{PP} > \pi_{BP} > \pi_{NP}, \tag{12}$$

$$\mu_{NN} < \mu_{BN} < \mu_{PN}, \nu_{NN} > \nu_{BN} > \nu_{PN}, \pi_{NN} > \pi_{BN} > \pi_{PN}. \tag{13}$$

where $\pi_{\bullet\circ} = 1 - \mu_{\bullet\circ} - \nu_{\bullet\circ} (\bullet = P, B, N; \circ = P, N)$.

Intuitionistic fuzzy sets here are compared on the basis of the Definition 2. Thus, the following theorem is further required:

Proposition 1. *In Table 2, based on (12), (13) and Definition 2, we have:*

$$\overline{\lambda}_{PP} \prec \overline{\lambda}_{BP} \prec \overline{\lambda}_{NP} \ and \overline{\lambda}_{NN} \prec \overline{\lambda}_{BN} \prec \overline{\lambda}_{PN}. \tag{14}$$

Proof. In light of (12) and (13), it is clear that $\frac{1-\nu_{PP}}{1+\pi_{PP}} < \frac{1-\nu_{BP}}{1+\pi_{BP}} < \frac{1-\nu_{NP}}{1+\pi_{NP}}$ and $\frac{1-\nu_{NN}}{1+\pi_{NN}} < \frac{1-\nu_{BN}}{1+\pi_{BN}} < \frac{1-\nu_{PN}}{1+\pi_{PN}}$. Furthermore, $1 - \frac{1}{2}(\pi_{PP})^2 < 1 - \frac{1}{2}(\pi_{BP})^2 < 1 - \frac{1}{2}(\pi_{NP})^2$ and $1 - \frac{1}{2}(\pi_{NN})^2 < 1 - \frac{1}{2}(\pi_{BN})^2 < 1 - \frac{1}{2}(\pi_{PN})^2$ hold as well, which induces $S_{\varpi}(\overline{\lambda}_{PP}) < S_{\varpi}(\overline{\lambda}_{BP}) < S_{\varpi}(\overline{\lambda}_{NP})$ and $S_{\varpi}(\overline{\lambda}_{NN}) < S_{\varpi}(\overline{\lambda}_{BN}) < S_{\varpi}(\overline{\lambda}_{PN})$. Hence, the concludes hold.

It is acknowledged from Proposition 1 that the risk loss for adopting accepted decision is smaller than the one for delayed decision that is smaller than the risk loss for rejected decision, when the object is in the state of X. However, when the object is not in X, the risk losses for taking the same actions are the opposite results. Also, it is the prerequisite of three-way decisions.

According to the operations of IFSs, we can calculate the intuitionistic fuzzy risk loss for taking actions $a_{\bullet} (\bullet = P, B, N)$, denoted by $\overline{R}(a_{\bullet}|[x])$, where

$$\overline{R}(a_{\bullet}|[x]) = s\overline{\lambda}_{\bullet P} \oplus t\overline{\lambda}_{\bullet N} = \left(1 - (1 - \mu_{\bullet P})^s (1 - \mu_{\bullet N})^t, (\nu_{\bullet P})^s (\nu_{\bullet N})^t\right).$$

Based on the Bayesian decision theory, the following decision rules are given:

(P4) If $\overline{R}(a_P|[x]) \preceq \overline{R}(a_B|[x])$ and $\overline{R}(a_P|[x]) \preceq \overline{R}(a_N|[x])$, decide: $x \in POS(X)$;
(B4) If $\overline{R}(a_B|[x]) \preceq \overline{R}(a_P|[x])$ and $\overline{R}(a_B|[x]) \preceq \overline{R}(a_N|[x])$, decide: $x \in BND(X)$;
(N4) If $\overline{R}(a_N|[x]) \preceq \overline{R}(a_P|[x])$ and $\overline{R}(a_N|[x]) \preceq \overline{R}(a_B|[x])$, decide: $x \in NEG(X)$.

As a matter of fact, the rules (P4)–(N4) can be further transformed as these rules based on the ranking function of IFNs as follows:

(P5) If $S_{\varpi}(\overline{R}(a_P|[x])) \leq S_{\varpi}(\overline{R}(a_B|[x]))$ and $S_{\varpi}(\overline{R}(a_P|[x])) \leq S_{\varpi}(\overline{R}(a_N|[x]))$, decide: $x \in POS(X)$;
(B5) If $S_{\varpi}(\overline{R}(a_B|[x])) \leq S_{\varpi}(\overline{R}(a_P|[x]))$ and $S_{\varpi}(\overline{R}(a_B|[x])) \leq S_{\varpi}(\overline{R}(a_N|[x]))$, decide: $x \in BND(X)$;
(N5) If $S_{\varpi}(\overline{R}(a_N|[x])) \leq S_{\varpi}(\overline{R}(a_P|[x]))$ and $S_{\varpi}(\overline{R}(a_N|[x])) \leq S_{\varpi}(\overline{R}(a_B|[x]))$, decide: $x \in NEG(X)$.

At this point, we will extend the optimization models based method in Sect. 3 to IF-3WD and then determine the corresponding threshold values. The similar models are constructed as follows:

$\overline{\alpha}$-model:

$\overline{\alpha} = \min \ s$

$s.t. \begin{cases} S_{\varpi}(\overline{R}(a_P|[x])) \leq S_{\varpi}(\overline{R}(a_B|[x])), \\ S_{\varpi}(\overline{R}(a_P|[x])) \leq S_{\varpi}(\overline{R}(a_N|[x])), \\ s+t=1. \end{cases}$

$\overline{\beta}$-model:

$\overline{\beta} = \max \ s$

$s.t. \begin{cases} S_{\varpi}(\overline{R}(a_N|[x])) \leq S_{\varpi}(\overline{R}(a_P|[x])), \\ S_{\varpi}(\overline{R}(a_N|[x])) \leq S_{\varpi}(\overline{R}(a_B|[x])), \\ s+t=1. \end{cases}$

For the above-constructed models, a pivotal theorem is given as:

Theorem 4. *Under the prerequisites (12) and (13), there is a unique optimal solution to the $\overline{\alpha}$-model and $\overline{\beta}$-model.*

Proof. It is clear from the $\overline{\alpha}$-model that its feasible region is a non-empty and closed set with boundedness, where $(s,t) = (1,0)$ is an actually feasible solution. Thus we conclude that the feasible region is a compact set in $R \times R$ by the Heine-Bore theory. In light of the property of a compact set, that is, a continuous and real function in the compact set is bounded and has minimum and maximum values. Also we note that the objective function is continuous and monotonic in the $\overline{\alpha}$-model, which implies that there is a unique optimal solution to the $\overline{\alpha}$-model such that its objective function reaches a minimum value. Similarly, there is the same conclude for the $\overline{\beta}$-model. Therefore the theorem holds.

Theorem 4 shows that although the $\overline{\alpha}$-model and $\overline{\beta}$-model are nonlinear and their analytic solutions are difficultly induced by KKT conditions, their numerical solutions can be obtained via LINGO solving and IF-3WD are directly acquired. Motivated by this idea, in what follows a general approach to IF-3WD is proposed based on optimization models with LINGO solving.

5 Optimization Models Based Intuitionistic Fuzzy Three-Way Decisions

Based on Theorem 4, we will use LINGO to solve models and then obtain thresholds of IF-3WD and its decision rules, where the detailed steps are as follows:

Step 1: Assume the risk coefficient ϖ and the intuitionistic fuzzy risk loss matrix in Table 2 are given. Thus the $\overline{\alpha}$-model and $\overline{\beta}$-model are constructed.

Step 2: LINGO is used to solve the $\overline{\alpha}$-model and $\overline{\beta}$-model above, and their optimal solutions $\overline{\alpha}$ and $\overline{\beta}$ are obtained. Whence we need to compare the values of $\overline{\alpha}$ and $\overline{\beta}$.

(i) If $\overline{\alpha} > \overline{\beta}$, then the three-way decisions are adopted as follows: (P)When $Pr(X|[x]) \geq \overline{\alpha}$, decide: $x \in POS(X)$; (B)When $\overline{\beta} < Pr(X|[x]) < \overline{\alpha}$, decide: $x \in BND(X)$; (N)When $Pr(X|[x]) \leq \overline{\beta}$, decide: $x \in NEG(X)$.

(ii) If $\overline{\alpha} \leq \overline{\beta}$, here the optimal solution to the $\overline{\alpha}$-model is the same as the one to the $\overline{\beta}$-model (That is $\overline{\gamma} = \overline{\alpha} = \overline{\beta}$). The two-way decisions are used: (P)When $Pr(X|[x]) \geq \overline{\gamma}$, decide: $x \in POS(X)$; (N)When $Pr(X|[x]) < \overline{\gamma}$, decide: $x \in NEG(X)$.

6   An Illustrative Example

To show the feasibility and effectiveness of our method, a numerical example is given [18]. For the selection problem in software plan, suppose there are two states $\Omega = \{X, \neg X\}$, which indicate that the software is good or bad. The set of actions for the new development plan is denoted by $A = \{a_P, a_B, a_N\}$, where a_P, a_B and a_N represent the development, further investigation and not development for the software, respectively. In light of the matrix of the loss function in Table 2, these loss functions are given respectively as follows: $\overline{\lambda}_{PP} = (0.00, 0.60)$, $\overline{\lambda}_{PN} = (0.90, 0.10)$, $\overline{\lambda}_{BP} = (0.40, 0.40)$, $\overline{\lambda}_{BN} = (0.50, 0.40)$, $\overline{\lambda}_{NP} = (0.80, 0.20)$ and $\overline{\lambda}_{NN} = (0.10, 0.50)$.

In this example, the proposed method is implemented to make decision, where the detailed steps are as follows:

Step 1: On the basis of the risk losses above and the risk coefficient of decision maker (assume $\varpi = 0.5$), both models are constructed:

$\overline{\alpha}$-model

$$\overline{\alpha} = \min \ s$$

$$s.t. \begin{cases} \frac{1-0.6^s \times 0.1^t}{2(1+1.0^s \times 0.1^t - 0.6^s \times 0.1^t)} - \frac{1-0.4^s \times 0.4^t}{2(1+0.6^s \times 0.5^t - 0.4^s \times 0.4^t)} \\ \quad \leq \frac{(1.0^s \times 0.1^t - 0.6^s \times 0.1^t)^2 - (0.6^s \times 0.5^t - 0.4^s \times 0.4^t)^2}{4}, \\ \frac{1-0.6^s \times 0.1^t}{2(1+1.0^s \times 0.1^t - 0.6^s \times 0.1^t)} - \frac{1-0.2^s \times 0.5^t}{2(1+0.2^s \times 0.9^t - 0.2^s \times 0.5^t)} \\ \quad \leq \frac{(1.0^s \times 0.1^t - 0.6^s \times 0.1^t)^2 - (0.2^s \times 0.9^t - 0.2^s \times 0.5^t)^2}{4}, \\ s + t = 1. \end{cases}$$

$\overline{\beta}$-model

$$\overline{\beta} = \max \ s$$

$$s.t. \begin{cases} \frac{1-0.2^s \times 0.5^t}{2(1+0.2^s \times 0.9^t - 0.2^s \times 0.5^t)} - \frac{1-0.6^s \times 0.1^t}{2(1+1.0^s \times 0.1^t - 0.6^s \times 0.1^t)} \\ \quad \leq \frac{(0.2^s \times 0.9^t - 0.2^s \times 0.5^t)^2 - (1.0^s \times 0.1^t - 0.6^s \times 0.1^t)^2}{4}, \\ \frac{1-0.2^s \times 0.5^t}{2(1+0.2^s \times 0.9^t - 0.2^s \times 0.5^t)} - \frac{1-0.4^s \times 0.4^t}{2(1+0.6^s \times 0.5^t - 0.4^s \times 0.4^t)} \\ \quad \leq \frac{(0.2^s \times 0.9^t - 0.2^s \times 0.5^t)^2 - (0.6^s \times 0.5^t - 0.4^s \times 0.4^t)^2}{4}, \\ s + t = 1. \end{cases}$$

Step 2: LINGO is employed to solve the $\overline{\alpha}$-model and $\overline{\beta}$-model above, and their optimal solutions $\overline{\alpha}$ and $\overline{\beta}$ are obtained as: $\overline{\alpha} = 0.7617186$ and $\overline{\beta} = 0.3342650$. It is obvious that the three-way decisions are implemented below:

(P) If $Pr(X|[x]) \geq 0.7617186$, then decide: $x \in POS(X)$;
(B) If $0.3342650 < Pr(X|[x]) < 0.7617186$, then decide: $x \in BND(X)$;
(N) If $Pr(X|[x]) \leq 0.3342650$, then decide: $x \in NEG(X)$.

In order to explore the influence of decision maker's risk attitudes on thresholds in IF-3WD, the proposed method is employed to obtain other thresholds, as shown in Table 3.

Table 3. The thresholds change under different DM's risk coefficients.

The thresholds	The risk coefficient of DM ϖ				
	0	0.1	0.2	0.3	0.4
$\overline{\alpha}$	0.7642999	0.7639836	0.7635985	0.7631216	0.7625153
$\overline{\beta}$	0.3217988	0.3233975	0.3253093	0.3276395	0.3305426

0.5	0.6	0.7	0.8	0.9	1
0.7617186	0.7606224	0.7590173	0.7564419	0.7515853	0.7384302
0.3342650	0.3392210	0.3461714	0.3567050	0.3748982	0.4169299

Here, a comparative study is given to illustrate the advantages of the proposed method. For this example, the existing method [17] can determine a pair of thresholds in the special cases of positive and negative viewpoints and then obtain the corresponding decision rules. However, it is difficult to determine these thresholds of intuitionistic fuzzy 3WD in composite situations and thus only obtain the indirect rules of three-way decisions, which are not favourable to make actual decisions, see [17–19] for more details. The proposed method can overcome these limitations and a pair of thresholds determined by our method in composite cases are presented in Fig. 2.

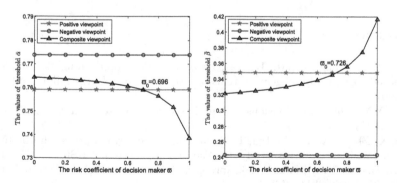

Fig. 2. Comparison of thresholds determined by our method with existing method [17].

From Table 3 and Fig. 2, the following concludes may be obtained:

(1) Our method can effectively determine the thresholds $\overline{\alpha}$ and $\overline{\beta}$ in IF-3WD, which can solve the problem where the current methods are difficult to obtain these thresholds for IF-3WD model in some cases. It is of great significance that this method is extended to the threshold determination of three-way decisions with the loss function expressed by triangular fuzzy numbers, interval numbers, linguistic variables, hesitant fuzzy sets and dual hesitant fuzzy sets respectively.

(2) The thresholds $\overline{\alpha}$ and $\overline{\beta}$ obtained by our method are monotonically decreasing and increasing respectively as ϖ increases. It coincides with human intuition and thus shows the reasonability of our method to some extent.

7   Conclusion

Based on the extended $\overline{\alpha}$-model and $\overline{\beta}$-model, a general method of obtaining IF-3WD is proposed to solve the problem, where the current methods are difficult to determine a pair of thresholds in IF-3WD in some cases. This study provides an idea for deriving three-way decisions from the optimization models, which can enrich the theory of three-way decisions and intuitionistic fuzzy sets. Future researches may focus on the generalization of the proposed optimization models.

Acknowledgments. This work was supported by the Natural Science Foundation of China (Nos. 71671086, 61773208, 61473157, 71732003 and 71201076), the National Key Research and Development Program of China (No.2016YFD0702100), the Fundamental Research Funds for the Central Universities (No. 011814380021), the Central military equipment development of the "13th Five-Year" pre research project (No. 315050202), the Nanjing University Innovation and Creative Program for PhD candidate (No. CXCY17-08) and the pre-research project (No. 3151001**).

References

1. Yao, Y.Y., Wong, S.K.M., Lingras, P.: A decision-theoretic rough set model. In: Methodologies for Intelligent Systems, vol. 5, pp. 17–24. North-Holland, New York (1990)

2. Yao, Y.Y.: Three-way decisions with probabilistic rough sets. Inf. Sci. **180**(3), 341–353 (2010)

3. Jia, X., Zheng, K., Li, W., Liu, T., Shang, L.: Three-way decisions solution to filter spam email: an empirical study. In: Yao, J.T., et al. (eds.) RSCTC 2012. LNCS (LNAI), vol. 7413, pp. 287–296. Springer, Heidelberg (2012). https://doi.org/10.1007/978-3-642-32115-3_34

4. Li, H.X., Zhang, L.B., Huang, B.: Sequential three-way decision and granulation for cost-sensitive face recognition. Knowl. Based Syst. **91**(1), 241–251 (2016)

5. Li, H.X., Zhang, L.B., Zhou, X.Z., et al.: Cost-sensitive sequential three-way decision modeling using a deep neural network. Int. J. Approximate Reasoning **85**, 68–78 (2017)

6. Li, J.H., Huang, C.C., Qi, J.J.: Three-way cognitive concept learning via multi-granularity. Inf. Sci. **378**, 244–263 (2017)
7. Yu, H., Wang, X.C., Wang, G.Y., et al.: An active three-way clustering method via low-rank matrices for multi-view data. Inf. Sci. (2018). https://doi.org/10.1016/j.ins.2018.03.009
8. Sun, B.Z., Ma, W.M., Li, B.J.: Three-way decisions approach to multiple attribute group decision making with linguistic information-based decision-theoretic rough fuzzy set. Int. J. Approximate Reasoning **93**, 424–442 (2018)
9. Li, H.X., Zhou, X.Z.: Risk decision making based on decision-theoretic rough set: a three-way view decision model. Int. J. Comput. Intell. Syst. **4**, 1–11 (2011)
10. Jia, X.Y., Tang, Z.M., Liao, W.H., et al.: On an optimization representation of decision-theoretic rough set model. Int. J. Approximate Reasoning **55**, 156–166 (2014)
11. Deng, X.F., Yao, Y.Y.: A multifaceted analysis of probabilistic three-way decisions. Fundamenta Informaticae **132**, 291–313 (2014)
12. Zhang, Y., Yao, J.T.: Determining three-way decision regions with Gini coefficients. In: Cornelis, C., Kryszkiewicz, M., Ślęzak, D., Ruiz, E.M., Bello, R., Shang, L. (eds.) RSCTC 2014. LNCS (LNAI), vol. 8536, pp. 160–171. Springer, Cham (2014). https://doi.org/10.1007/978-3-319-08644-6_17
13. Azama, N., Zhang, Y., Yao, J.T.: Evaluation functions and decision conditions of three-way decisions with game-theoretic rough sets. Eur. J. Oper. Res. **261**, 704–714 (2017)
14. Qian, Y.H., Zhang, H., Sang, Y.L., et al.: Multigranulation decision-theoretic rough sets. Int. J. Approximate Reasoning **55**, 225–237 (2014)
15. Liang, D.C., Liu, D.: Systematic studies on three-way decisions with interval-valued decision-theoretic rough sets. Inf. Sci. **276**, 186–203 (2014)
16. Liang, D.C., Pedrycz, W., Liu, D., et al.: Three-way decisions based on decision-theoretic rough sets under linguistic assessment with the aid of group decision making. Appl. Soft Comput. **29**, 256–269 (2015)
17. Liang, D.C., Liu, D.: Deriving three-way decisions from intuitionistic fuzzy decision-theoretic rough sets. Inf. Sci. **300**, 28–48 (2015)
18. Liang, D.C., Xu, Z.S., Liu, D.: Three-way decisions with intuitionistic fuzzy decision-theoretic rough sets based on point operators. Inf. Sci. **375**, 183–201 (2017)
19. Liu, J.B., Zhou, X.Z., Huang, B., Li, H.: A three-way decision model based on intuitionistic fuzzy decision systems. In: Polkowski, L. (ed.) IJCRS 2017. LNCS (LNAI), vol. 10314, pp. 249–263. Springer, Cham (2017). https://doi.org/10.1007/978-3-319-60840-2_18
20. Liang, D.C., Liu, D.: A novel risk decision-making based on decision-theoretic rough sets under hesitant fuzzy information. IEEE Trans. Fuzzy Syst. **23**(2), 237–247 (2015)
21. Liang, D.C., Xu, Z.S., Liu, D.: Three-way decisions based on decision-theoretic rough sets with dual hesitant fuzzy information. Inf. Sci. **396**, 127–143 (2017)
22. Atanassov, K.T.: Intuitionistic fuzzy sets. Fuzzy Sets Syst. **20**, 87–96 (1986)
23. Wan, S.P., Wang, F., Dong, J.Y.: A novel risk attitudinal ranking method for intuitionistic fuzzy values and application to MADM. Appl. Soft Comput. **40**, 98–112 (2016)
24. Chen, B.L.: Optimization Theory and Algorithm. Tsinghua University Press, Beijing (2005)

External Indices for Rough Clustering

Matteo Re Depaolini, Davide Ciucci$^{(\boxtimes)}$, Silvia Calegari, and Matteo Dominoni

DISCo, University of Milano-Bicocca, Viale Sarca 336/14, 20126 Milano, Italy
ciucci@disco.unimib.it

Abstract. Clustering external indices are used to compare the clustering result with a given gold standard, represented (in the classical case) by a partition of the dataset. Rough clustering on the other hand splits the dataset in subsets with uncertain boundaries such that different clusters may overlap, i.e., the result is a covering instead of a partition.

The aim of this work is to extend the aforementioned external indices to the rough clustering case, in order to evaluate the results of the clustering with respect to the gold standard. Thus, the comparison of different rough clustering methods among them and with other methods will then be possible.

Keywords: Rough clustering · External indices · Fuzzy clustering

1 Introduction

Clustering is an unsupervised learning technique whose task is to group similar objects together and assign dissimilar objects to different groups. These groups are called *clusters*. In standard clustering methods, clusters have sharp boundaries and objects belong to one and only one cluster. In soft clustering [16], these constraints are relaxed and objects can (partially) belong to more than one cluster. In order to evaluate the performances of a clustering method, two kinds of indices exist: external and internal. The first ones are used when instance labels are available and can be used to partition the universe. In this case, the result of the clustering can be compared with the partition obtained by labels. If this is not the case, the clustering is evaluated by internal indices, on some "good" properties of the clusters' structure.

Here, we are dealing with external indices, the best known one being Rand [17] and its derived ones. These indices have been generalized for some soft techniques, such as fuzzy clustering [6] but both the classical and the fuzzy indices are not applicable to the rough set case (we notice that for internal indices there exist at least one approach based on a decision theoretic rough set approach [12]). Indeed, the classical ones are based on a partition-partition comparison, whereas rough clustering does not generate a partition, as we will see in Sect. 2.2. On the other hand, fuzzy indices suppose the availability of a membership degree of each object to any cluster, and these values are not present in rough clustering. Thus, the aim of the present work is to introduce generalized versions of Rand,

© Springer Nature Switzerland AG 2018
H. S. Nguyen et al. (Eds.): IJCRS 2018, LNAI 11103, pp. 378–391, 2018.
https://doi.org/10.1007/978-3-319-99368-3_29

Jaccard and Fowlkes–Mallows indices suitable for rough clustering and to show their applicability. The importance of these indices is to be able to evaluate and compare rough clustering algorithms with (generalized versions of) standard and well known methods.

In Sect. 2.1, basic notions of clustering, rough clustering and external indices will be recalled. In Sect. 3, the new indices are defined and some properties are given including the relationship with the Frigui index for fuzzy clustering [5]. Some experimental results are shown in Sect. 4 and finally, some remarks and future works are discussed.

2   Clustering

Basic notions of rough clustering and external indices are provided in this section.

2.1   Hard k-means

K-means [8] is the most widely used approach for clustering and rough clustering is mainly based on it. It is a prototype algorithm, that is, based on the idea that each group (cluster) must have a representative called *prototype* or *centroid*. Each instance is grouped in one and only one cluster.

The algorithm executes the following steps.

1. First of all, k instances are elected centroids;
2. Other instances are assigned to their nearest cluster's centroid, so that clusters are built for the first time;
3. Centroids are recomputed averaging the points of their clusters. Centroids will very hardly correspond to dataset instances from now on;
4. Each instance is reassigned to the nearest cluster's centroid;
5. Steps 3 and 4 are repeated until recalculated centroids are closer than a threshold δ to the previous ones.

There are several methods to make the election of centroids described in step 1. In any case, the initial choice of the centroids influences the overall process. Under the assumption that the techniques of election of initial centroids are not deterministic, it is suggested to execute the overall process several times, in order to begin each time with different centroids.

2.2   Rough Clustering

Hard clustering, such as k-means, assigns each instance to just one cluster. This is sometimes questionable since there may exist situations in which we are not able to classify an instance with certainty. Rough clustering, such as rough k-means designed by Lingras [10], exploits rough set theory in order to assign "uncertain" instances to the boundary region of the relative clusters. Indeed, each cluster C_i is made by a *lower* region (or lower approximation) and an uncertain region, named *boundary*. The first one contains the objects that surely belong to the cluster (this region is referred to as $\underline{C_i}$), while the second contains the objects on which we have some evidence that may belong to the cluster, but we are not sure about that. All the points in a cluster C_i, either in the lower or boundary region, fall into the upper approximation, referred as $\overline{C_i}$. Thus, the boundary of each cluster C_i can be obtained as: $\overline{C_i} \setminus \underline{C_i}$. We can consider hard clustering a particular case of rough clustering in which each object of each cluster C_i falls into $\underline{C_i}$.

Remark 1. According to Lingras [10], if an object belongs to a boundary region, then it must belong to at least another one. This requirement is relaxed in [18] under a different interpretation of the boundary region. Here, we do not enter into this discussion, and simply remark that our measures are valid in both cases.

In rough k-means, the assignment of each instance x to a set of clusters is made in the following way. First, the distance between x and all clusters' centroids is computed: $\{d_1^{(x)}, d_2^{(x)}, \ldots, d_n^{(x)}\}$. Then, the minimum distance is taken, let us it $d_{min}^{(x)}$ and each $d_i^{(x)}$ is compared with $d_{min}^{(x)}$. The aim of these comparisons is to determine whether each instance belongs surely to a specific cluster or can be assigned to more clusters and to which clusters can be assigned. Formally, Lingras [10] defines the assignment of each instance x as follows:

1. $\forall$ cluster C_i s.t. $d_{min}^{(x)}/d_i^{(x)} \geq \delta$, x belongs to the *boundary* region of the nearest cluster (whose centroid has distance equal to $d_{min}^{(x)}$ from x) and to the *boundary* region of cluster C_i.
2. Otherwise, x belongs to the *lower* region of the nearest cluster (whose centroid has distance equal to $d_{min}^{(x)}$).

2.3   External Indices

The aim of external indices is to compare a given partition, the "gold standard", with a clustering result. It is expected that the more the clustering result is similar to the partition the more the index is high, and it assumes the maximum value 1 if the two are equal. Vice versa, the more the clustering result and the partition are different, the more the index is low, with 0 as minimum value.

The most famous external indices for hard clustering are Rand [17], Jaccard [7] and Fowlkes-Mallows [4]. They are all based on the following concepts:

- the set of pairs in the same partition and in the same cluster, named a;
- the set of pairs in the same partition and in different clusters, named b;
- the set of pairs in different partitions and in the same cluster, named c;
- the set of pairs in different partitions and in different clusters, named d.

So, it is clear that a and d should be maximized and b and c minimized. The above mentioned indices measure in different ways the ratio of well classified instances with respect to the total number of instances, according to the following formulae:

$$Rand = \frac{|a| + |d|}{|a| + |b| + |c| + |d|} \tag{1}$$

$$Jaccard = \frac{|a|}{|a| + |b| + |c|} \tag{2}$$

$$Fowlkes - Mallows = \sqrt{\frac{|a|}{|a| + |b|} * \frac{|a|}{|a| + |c|}} \tag{3}$$

3   Extending External Indices to Rough Clustering

The aim of this section is to extend the above indices to the rough clustering case. At first, let us discuss why Eqs. (1), (2) and (3) are not applicable to rough k-means and similar algorithms. In hard clustering, every instance is contained in one and only one cluster so every pair of instances is contained in only one of a, b, c or d. For that reason, the following equality holds:

$$\binom{n}{2} = |a| + |b| + |c| + |d| \tag{4}$$

In rough clustering, any pair can be divided in many sets (as shown in Example 1) and can be repeated in the same set (as shown in Example 2).

   Thus, the value of a, b, c, d cannot be computed as previously and we propose to weight each pair according to its number of occurrences.

Example 1. Let us say we have two clusters (C_1, C_2), two partitions (p_1, p_2) and a dataset **D**. Let us consider only two instances $x, y \in$ **D** s.t. $(x \in p_1, y \in p_1)$ and $(x \in C_1, y \in C_1, y \in C_2)$. Then, $a = \{((x, C_1), (y, C_1)) \ldots \}$, $b = \{((x, C_1), (y, C_2)) \ldots \}$, $c = \{\ldots\}$ and $d = \{\ldots\}$.

As we can see in Example 1, in contrast to hard clustering, the same pair appears in two sets (a and b). In this situation it is intuitive to divide the weight of the pair by 2, such that each one will weight $1/2$.

Example 2. Let us suppose again to have two clusters (C_1, C_2), two partitions (p_1, p_2) and a dataset **D**. We consider only two instances $x, y \in$ **D** s.t. $(x \in p_1, y \in p_1)$ and $(x \in C_1, x \in C_2, y \in C_1, y \in C_2)$. Thus, $a = \{((x, C_1), (y, C_1)), ((x, C_2), (y, C_2)) \ldots\}$. Indeed, we have the pair (x, y) with $x \in C_1, y \in C_2$ and the pair (x, y) with $y \in C_1, x \in C_2$. Similarly, $b = \{((x, C_1), (y, C_2)), ((x, C_2), (y, C_1)) \ldots\}$; and $c = \{\ldots\}$, $d = \{\ldots\}$.

As we can see in Example 2, in contrast to hard clustering, the same pair appears in two sets (a and b) and two times in each of these sets. In this case, each pair could be weighted as $1/4$.

   In the following, we formalize this intuition on the weights and give generalized versions of the external indices.

3.1   The New Indices

Our purpose is to validate a rough clustering result comparing it to a given partition. We suppose to have no knowledge on the clustering result and on the mechanism used to obtain it, it could also have been randomly generated. So, it can be stated that given two instances x and y, x belongs to C_i and y belongs to C_j independently. All we know is that, if $x \in \underline{C_i}$, x surely belongs to the cluster C_i and, if $x \in (\overline{C_i} - \underline{C_i})$, x belongs to one or more clusters, but it is not possible to tell the clusters to which x belongs more likely. In the first case, we can set:

$$P(x \in Ci | x \in \underline{Ci}) = 1 \tag{5}$$

Otherwise, the number of boundaries to which x belongs to is denoted as $bn(x)$. Rough clustering does not assert the likelihood of membership of x to the $bn(x)$ clusters. For Laplace's principle of indifference [9], given a set of events, if it is impossible to establish the likelihood of each event, the probability distribution of these events can be considered as uniform. Thus, we can say that

$$P(x \in Ci | x \in (\overline{Ci} - \underline{Ci})) = \frac{1}{bn(x)} \tag{6}$$

The following is straightforward:

$$P(x \in C_i | x \notin C_i) = 0 \tag{7}$$

As stated before, the belonging of an instance x to a cluster is independent from the belonging of an instance y to another (or the same) cluster. Thus, the probability of a pair of instances in any set (a, b, c, d) is as follows:

$$P(x \in C_i, y \in C_j) = P(x \in C_i) * P(y \in C_j) \tag{8}$$

Moreover, we can assert:

$$\sum_{i=1}^{n}(\sum_{j=1}^{n} P(x \in C_i, y \in C_j)) = 1 \tag{9}$$

In hard clustering, every pair has weight equal to one, such that every index presented in Sect. 2.3 exploits the cardinality of a,b,c,d. In rough clustering, the idea is to weight each pair with the value obtained from Eq. (8).

Let $\mathbf{D}$ be the set of instances of the dataset and $\mathbf{C}$ the set of clusters. Taking into account Eqs. (5), (6) and (7), we define $v : \mathbf{D} \times \mathbf{C} \rightarrow \mathbb{R}$:

$$P(x, C_i) = v(x, C_i) = \begin{cases} 0, & \text{if } x \notin \overline{C_i} \\ 1, & \text{if } x \in \underline{C_i} \\ \frac{1}{bn(x)}, & \text{otherwise} \end{cases} \tag{10}$$

Equation 10 can also be applied to hard clustering with the assumption that each object of each cluster C falls into $\underline{C}$.

In order to define a generalized forms of a, b, c, d, we introduce a function $w : \mathcal{P}(\mathbf{D} \times \mathbf{C}) \times \mathcal{P}(\mathbf{D} \times \mathbf{C}) \rightarrow [0, 1]$ that weights each pair $((x, C_i), (y, C_j))$ and it is defined as

$$w((x, C_i), (y, C_j)) = P(x \in C_i, y \in C_j) = v(x, C_i) \cdot v(y, C_j) \qquad (11)$$

Now, let $W : \mathcal{P}(\mathcal{P}(\mathbf{D} \times \mathbf{C}) \times \mathcal{P}(\mathbf{D} \times \mathbf{C})) \rightarrow [0, 1]$, W takes as input a set S of pairs of elements of type $\mathcal{P}(\mathbf{D} \times \mathbf{C})$ and gives as output the weight of S as:

$$W(S) = \sum_{(s,s') \in S} w(s, s') \qquad (12)$$

Using Eq. (12), it is possible to rewrite Rand, Jaccard and Fowlkes–Mallows indices as follows:

$$R - Rand = \frac{W(a) + W(d)}{W(a) + W(b) + W(c) + W(d)} \qquad (13a)$$

$$R - Jaccard = \frac{W(a)}{W(a) + W(b) + W(c)} \qquad (13b)$$

$$R - FowlkesMallows = \sqrt{\frac{W(a)}{W(a) + W(b)} * \frac{W(a)}{W(a) + W(c)}} \qquad (13c)$$

These formulae are clearly an extension of the original ones, since once applied to hard clustering we obtain the indices as previously defined in Eqs. (1), (2) and (3). Moreover, Eq. (4) still holds in this case:

Proposition 1. *The following holds:*

$$\binom{n}{2} = W(a) + W(b) + W(c) + W(d) \qquad (14)$$

Proof. From Eq. (9), it easily follows that all repeated pairs (x, y) sum to 1.

Example 3. Let us suppose to have four instances: e_1, e_2, e_3, e_4, two partitions: P_1, P_2 and two clusters: C_1, C_2. As shown in Fig. 1, $e_1, e_2, e_3 \in P_2$, $e_4 \in P_1$ and as a clustering result we have $e_1, e_2 \in C_1, C_2$, $e_3 \in C_2$, $e_4 \in C_1$. We will omit the cluster in each pair, for simplicity. Thus, we get:

$$a = \{(e_1, e_2), (e_1, e_2), (e_1, e_3), (e_2, e_3)\}$$

$$b = \{(e_1, e_2), (e_2, e_1), (e_3, e_1), (e_3, e_2)\}$$

$$c = \{(e_1, e_4), (e_2, e_4)\}$$

$$d = \{(e_4, e_1), (e_4, e_2), (e_4, e_3)\}$$

$$W(a) = w(e_1, e_2) + w(e_1, e_2) + w(e_1, e_3) + w(e_2, e_3) = \frac{1}{4} + \frac{1}{4} + \frac{1}{2} + \frac{1}{2} = \frac{3}{2}$$

Similarly, it is possible to derive: $W(b) = \frac{3}{2}$, $W(c) = 1$, $W(d) = 2$ and the indices can be computed substituting these values in Eqs. 13a, 13b and 13c with the following results: RAND = 0.583, JACCARD = 0.375, FM = 0.548.

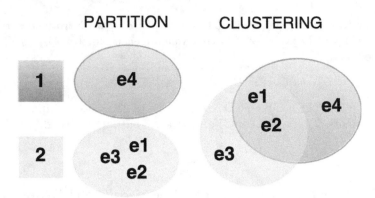

Fig. 1. Example of a partition and a soft clustering

It is possible to infer that, the more the indices grow, the more the clustering looks alike the partitioning (and vice versa). Indeed, $W(S)$ is directly proportional to the number of pairs contained in the set S. Thus, the more $W(a)$ grows, the more "similar" (w.r.t. the partitioning) the instances are clustered together, and vice versa. With the same reasoning, the more $W(d)$ grows, the more "differently" (w.r.t. the partitioning) the instances are clustered apart (and vice versa). On the other hand, the more $W(c)$ and $W(b)$ grow, the more the instances are wrongly clustered w.r.t the partitioning. Considering that $W(a)$ and $W(d)$ are the quantities at the numerator in Eqs. (13a), (13b) and (13c) and that $W(a), W(b), W(c), W(d)$ are the quantities at the denominator, the thesis easily follows. Finally, we have that

Proposition 2. $R - Rand, R - Jaccard, R - FowlkesMallows \in [0, 1]$.

Proof. In the worst case, no pairs are present in a and d, so that the indices are equal to 0. In the best case, no pairs are present in c and b, so that the indices are equal to 1. In all the intermediate cases, of course, the indices are in $(0, 1)$.

3.2   Relationship with Fuzzy Indices

Campello [2] designed a framework to generalize the external indices to fuzzy clustering. His family of indices depends on a t-norm and a t-conorm. Frigui [5] derived from that theoretic framework his indices using multiplication as t-norm and bounded sum as t-conorm. He stated that, in order to compare two partitions P_1 and P_2 generated by two fuzzy algorithms, it is sufficient to compare the respective membership degree matrices[1] D^1 and D^2 by computing the coincidence matrices B^1, B^2 as follows:

$$B^{(i)} = D^{(i)} \cdot D^{(i)T} \tag{15}$$

---

[1] An element d_{ij} of these matrices is a value in $[0, 1]$ and it represents the membership degree of an instance i to a partition P_j.

Once obtained such matrices, it is possible to calculate the generalized versions of a,b,c,d as described below:

$$W_f(a) = \sum_{j=2}^{N} \sum_{k=1}^{j-1} B_{j,k}^{(1)} \cdot B_{j,k}^{(2)} \tag{16a}$$

$$W_f(b) = \sum_{j=2}^{N} \sum_{k=1}^{j-1} B_{j,k}^{(1)} \cdot (1 - B_{j,k}^{(2)}) \tag{16b}$$

$$W_f(c) = \sum_{j=2}^{N} \sum_{k=1}^{j-1} (1 - B_{j,k}^{(1)}) \cdot B_{j,k}^{(2)} \tag{16c}$$

$$W_f(d) = \sum_{j=2}^{N} \sum_{k=1}^{j-1} (1 - B_{j,k}^{(1)}) \cdot (1 - B_{j,k}^{(2)}) \tag{16d}$$

The sense of the indices in the summations of the Eqs. (16a), (16b), (16c) and (16d) is to sum just half resultant matrices, since each $B^{(i)}$ is symmetric: the symmetry is due to unordered pairs. For this reason, if we name L the set of all unordered pairs and let d_k^l be the k-th row of the membership degree matrix D^l, we can rewrite Eqs. (16a)–(16d) as follows:

$$W_f(a) = \sum_{(w_i, w_j) \in L} (d_i^1 \cdot d_j^1) \cdot (d_i^2 \cdot d_j^2) \tag{17a}$$

$$W_f(b) = \sum_{(w_i, w_j) \in L} (d_i^1 \cdot d_j^1) \cdot (1 - (d_i^2 \cdot d_j^2)) \tag{17b}$$

$$W_f(c) = \sum_{(w_i, w_j) \in L} (1 - (d_i^1 \cdot d_j^1)) \cdot (d_i^2 \cdot d_j^2) \tag{17c}$$

$$W_f(d) = \sum_{(w_i, w_j) \in L} (1 - (d_i^1 \cdot d_j^1)) \cdot (1 - (d_i^2 \cdot d_j^2)) \tag{17d}$$

Finally, the Rand, Jaccard and Fowlkes–Mallows indices, generalized to the fuzzy case, are defined as in Eqs. (13a), (13b) and (13c) using $W_f(a)$ in place of $W(a)$.

Now, if we have a fuzzy partition P_1 and a hard partition P_2, and we use the values obtained by Eq. (10) to construct the membership degree matrices we get that $W(x) = W_f(x)$ for $x \in \{a, b, c, d\}$ as formally proved in the following proposition.

Proposition 3. *Let P_1 be the result of a rough clustering and P_2 be a partition. If D_1, D_2 in Eq. (15) are constructed using the values obtained by Eq. (10) then*

$$W(x) = W_f(x) \quad for \quad x \in \{a, b, c, d\}.$$

Proof. Let U be the dataset of N instances: $u_1, u_2, \ldots, u_N$ and suppose to have K clusters. Thus, the membership degree matrix D^1, D^2 have dimension $N \times K$ and by hypothesis they are constructed using Eq. 10.

Now, we interpret each row d_i of D as the set of probabilities $d_{i,k}$ that instance u_i belongs to the cluster C_k, i.e.,

$$P_k(u_i) = d_{i,k} = v(u_i, k) \tag{18}$$

So, given two instances u_i and u_j, $d_i \cdot d_j$ represents the probability of u_i to be in the same cluster together with u_j. Indeed, from Eqs. 8 and 18:

$$d_i \cdot d_j = \sum_{k=1}^{K} P(u_i \in C_k, u_j \in C_k) = \sum_{k=1}^{K} P_k(u_i) \cdot P_k(u_j) = \sum_{f=1}^{K} v(u_i, C_f) \cdot v(u_j, C_f)$$
$$\tag{19}$$

Similarly, the probability that u_i and u_j belong to different clusters can be seen as:

$$\sum_{C_l, C_k, l \neq k} P(u_i \in C_l, u_j \in C_k) = \sum_{C_l, C_k, l \neq k} v(u_i, C_l) \cdot v(u_j, C_k) = 1 - d_i \cdot d_j \tag{20}$$

The below lemma easily follow from the interpretation of $d_i \cdot d_j$ as the probability that u_i and u_j are in the same cluster and the above Eq. (19).

Lemma 1. *Given a hard partition D, two instances u_i and u_j are in the same cluster iff $d_i \cdot d_j = 1$. Vice versa, u_i and u_j are in different clusters iff $d_i \cdot d_j = 0$.*

Now, we prove the main statement. For the sake of space, only the case cases a is shown, being the others proved in a similar way.
$W_f(a) = \sum_{(w_i, w_j) \in L} (d_i^1 \cdot d_j^1) \cdot (d_i^2 \cdot d_j^2)$. For all pairs (u_i, u_j), in which u_i and u_j are not in the same cluster in D^2, $(d_i^1 \cdot d_j^1) \cdot (d_i^2 \cdot d_j^2) = 0$ for Lemma 1. For all pairs (u_i, u_j), in which u_i and u_j are in the same cluster in D^2, $(d_i^1 \cdot d_j^1) \cdot (d_i^2 \cdot d_j^2) = (d_i^1 \cdot d_j^1)$ for Lemma 1. So, with respect to the summation, the only pairs that count are the ones in the same cluster in D^1 and in the same cluster in D^2. So we can conclude that

$$W_f(a) = \sum_{(u_i, u_j) \in a} (d_i^1 \cdot d_j^1) \cdot (d_i^2 \cdot d_j^2) = \sum_{(u_i, u_j) \in a} (d_i^1 \cdot d_j^1).$$

Now, from Eq. 19:

$$\sum_{(u_i, u_j) \in a} (d_i^1 \cdot d_j^1) = \sum_{(u_i, u_j) \in a} \sum_{f=1}^{K} v(u_i, C_f) \cdot v(u_j, C_f)$$

$$= \sum_{((u_i, C_f), (u_j, C_f)) \in a} v(u_i, C_f) \cdot v(u_j, C_f) = W(a)$$

Thus, we can treat rough external indices as a particular case of Frigui indices for fuzzy clustering. We underline, however, that in the rough set case only a comparison soft-hard partition is possible, whereas in the general fuzzy case also a soft-soft partition comparison is possible.

We also notice that Brouwer [1], starting from the contribution of [2,5], asserts that dot multiplication in Eq. (15) is a questionable method for bonding matrices, so that he suggests to normalize the multiplication using cosine similarity, that is: $cos(v, w) = \frac{v \cdot w}{|v| \cdot |w|}$.

Proposition 4. *The computational cost of computing the four indices $W_f(a)$, $W_f(b)$, $W_f(c)$, $W_f(d)$ is $\Theta(K \cdot N^2)$, where N is the number of instances and K is the number of clusters.*

Proof (sketch). Let $D^{(i)}$ be a $N \times K$ matrix represnting the membership matrix. Let M be the $N \times K$ matrix obtained as a results of the rough clustering algorithm. The cost to obtain the corresponding D can be calculated to be $\Theta(N \cdot K)$. This is the substantial overhead introduced by our approach. The rest of the algorithm is identical to Frigui's one, whose cost is $\Theta(K \cdot N^2)$.

4   Experimental Results

At first, we built two simple datasets in order to show that to an evidently better clustering, there corresponds a value closer to 1 of the indices. Then, we test our measures on three well-known datasets.

4.1   On the Relationship Between Clustering's Quality and Indices

We synthesized two 2D datasets that contain the same points and differ only for the labels. They can be graphically seen in Figs. 2 and 3a, whose difference is only in the coloring that represents the instance labeling. Clearly, the first dataset is easy to cluster whereas the second one represents a more challenging task. We applied Lingras and West's rough k-means [11] to both dataset. In the simplest case, the obtained cluster coincides with the original dataset, hence all indices have 1 as a result. The clustering relative to the second dataset is shown in Fig. 3b, where each color represents a different cluster.

The values of all the indices are in this case less than one, thus showing that they correctly measure the performances of the clustering algorithm.

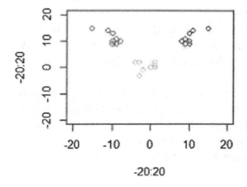

Fig. 2. First dataset, the clustering result is identical (Color figure online)

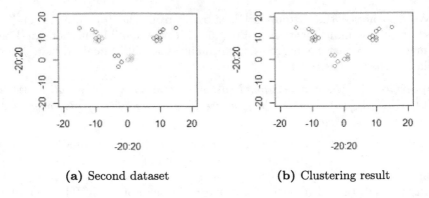

(a) Second dataset   (b) Clustering result

Fig. 3. Original dataset and clustering result. The values of the indices are: Rand = 0.663, Jaccard = 0.290, Fowlkes-Mellows = 0.450. (Color figure online)

4.2 First Application

We have tested our indices on three well-known UCI datasets [3]: Iris, Wine and Glass, whose characteristics are summarized as follows:

- **Iris:** 150 instances, 4 continuous attributes, 3 classes
- **Wine:** 178 instances, 13 numeric attributes, 3 classes
- **Glass:** 214 instances, 10 continuous attributes, 7 classes

We used three versions of rough k-means implemented in the R package *Soft-Clustering* [15]:

- **RoughKMeans_LW**: Lingras & West Rough k-Means [11]
- **RoughKMeans_PI**: PI Rough k-Means [14]
- **RoughKMeans_PE**: Peters Rough k-Means [13]

In all the three algorithms, we set as number of clusters the number of classes of the datasets. We split the dataset into train and test partitions, taking randomly 70% and 30% of the dataset. The same data have been used to learn and test all the algorithms.

In Tables 1, 2 and 3, we report the obtained results of our indices and also of Brouwer "normalized" index: LW stands for results of LW algorithm measured with our indices and LW-Brouwer with Brouwer index, similarly for the other algorithms.

These first results make evident that it is now possible to compare the performances (in presence of a gold standard) of different rough clustering methods. Though it is out of scope of this paper to establish which (and under which conditions) algorithm is better, we can see from these experiments that different patterns exist according to the three datasets:

– LW has better performances in the IRIS case;
– for the wine and glass datasets, all three algorithms have similar performances, with LW slightly better in the wine case w.r.t to Jaccard and Fowlkes-Mallows indices.

Table 1. Clustering results on Iris vs Iris partitions, FM stands for Fowlkes-Mallows.

	a	b	c	d	Rand	Jaccard	FM
LW	239.25	82.75	79.50	588.50	0.84	0.60	0.75
LW-Browuer	245.30	76.70	85.92	582.08	0.84	0.60	0.75
PI	262.83	59.17	255.33	412.67	0.68	0.46	0.64
PI-Browuer	293.62	28.38	269.97	398.03	0.70	0.50	0.69
PE	260.83	61.17	255.33	412.67	0.68	0.45	0.64
PE-Browuer	267.25	54.75	265.18	402.82	0.68	0.46	0.65

Table 2. Clustering results on wine vs wine partitions

	a	b	c	d	Rand	Jaccard	FM
LW	384.00	88.00	315.75	643.25	0.72	0.49	0.67
LW-Browuer	390.21	81.79	321.59	637.41	0.72	0.49	0.67
PI	296.50	175.50	218.50	740.50	0.72	0.43	0.60
PI-Browuer	314.62	157.38	242.66	716.34	0.72	0.44	0.61
PE	291.50	180.50	224.75	734.25	0.72	0.42	0.59
PE-Browuer	312.52	159.48	252.23	706.77	0.71	0.43	0.61

Table 3. Clustering results on glass vs glass partitions

	a	b	c	d	Rand	Jaccard	FM
LW	224.73	244.27	360.93	1 250.08	0.71	0.27	0.43
LW-Browuer	244.18	224.82	424.41	1 186.59	0.69	0.27	0.44
PI	202.25	266.75	334.44	1 276.56	0.71	0.25	0.40
PI-Browuer	223.08	245.92	400.62	1 210.38	0.69	0.26	0.41
PE	198.92	270.08	334.83	1 276.17	0.71	0.25	0.40
PE-Browuer	214.01	254.99	391.17	1 219.83	0.69	0.25	0.40

If we analyze the results of the two families of indices (our vs Brouwer), they are rather similar, that is, the Brouwer normalization factor does not influence the results in the analyzed cases. Of course a deeper investigation, both theoretical and practical, is needed in order to establish the non-influence of this normalization factor.

5   Conclusions

In this work, we extend the classical indices for external clustering evaluation to the case of rough clustering. We showed that:

– the new indices are theoretically sound: the greater value they have, the closer is the clustering to the gold standard;
– they can be seen as a particular case of Frigui indices for fuzzy clustering, thus opening the possibility to have a unique framework to compare rough and fuzzy clustering.
– they can be successfully used in practice to compare different rough clustering algorithms.

As a future work, we plan to perform more experiments in order to test the scalability of the algorithms to compute the indices and to better compare the different rough and also three-way [19] clustering algorithms. Further we will exploit the possibility to use the Frigui indices to compare rough and fuzzy clustering methods. Finally, the indices should be extended to compare rough-rough partitions in order to use them also in case that the gold standard is not a hard partition.

Acknowledgments. The present work has been developed under the Pollicina project, which is supported by the Regional Operational Program of the European Fund for Regional Development 2014–2020 (POR FESR 2014–2020).

References

1. Brouwer, R.K.: Extending the rand, adjusted rand and jaccard indices to fuzzy partitions. J. Intell. Inf. Syst. **32**, 213–235 (2009)
2. Campello, R.: A fuzzy extension of the rand index and other related indexes for clustering and classification assessment. Pattern Recogn. Lett. **28**(7), 833–841 (2007)
3. Dheeru, D., Karra Taniskidou, E.: UCI machine learning repository (2017). http://archive.ics.uci.edu/ml
4. Fowlkes, E.B., Mallows, C.: A method for comparing two hierarchical clusterings. Am. Stat. Assoc. **78**(383), 553–569 (1983)
5. Frigui, H., Hwang, C., Rhee, F.C.H.: Clustering and aggregation of relational data with applications to image database categorization. Pattern Recogn. **40**, 3053–3068 (2007)
6. Hüllermeier, E., Rifqi, M., Henzgen, S., Senge, R.: Comparing fuzzy partitions: a generalization of the rand index and related measures. IEEE Trans. Fuzzy Syst. **20**(3), 546–556 (2012)
7. Jaccard, P.: Novelles recherches sur la distribution florale. Bulletin de la Societe Vaudoise des Sciences Naturelles **44**, 223–270 (1908)
8. Jain, A.K.: Data clustering: 50 years beyond k-means. Pattern Recogn. Lett. **31**(8), 651–666 (2010)
9. Laplace, P.: A Philosophical Essay on Probabilities. Dover Publications, New York (2012)

10. Lingras, P., Peters, G.: Rough clustering. WIREs Data Mining Knowl. Disc. **1**, 65–72 (2011)
11. Lingras, P., West, C.: Interval set clustering of web users with rough k-means. J. Intell. Inf. Syst. **23**, 5–16 (2004)
12. Lingras, P., Chen, M., Miao, D.: Rough cluster quality index based on decision theory. IEEE Trans. Knowl. Data Eng. **21**(7), 1014–1026 (2009)
13. Peters, G.: Some refinements of rough k-means clustering. Pattern Recogn. **39**, 1481–1491 (2006)
14. Peters, G.: Rough clustering utilizing the principle of indifference. Inf. Sci. **277**, 358–374 (2014)
15. Peters, G.: Softclustering: soft clustering algorithms, February 2015. https://cran.r-project.org/web/packages/SoftClustering/index.html
16. Peters, G., Crespo, F.A., Lingras, P., Weber, R.: Soft clustering - fuzzy and rough approaches and their extensions and derivatives. Int. J. Approx. Reasoning **54**(2), 307–322 (2013)
17. Rand, W.M.: Objective criteria for the evaluation of clustering methods. J. Amer. Stat. Assoc. **66**(336), 846–850 (1971)
18. Wang, P., Yang, X., Yao, Y.: C&E re-clustering: Reconstruction of clustering results by three-way strategy. In: Kryszkiewicz, M., Appice, A., Slezak, D., Rybinski, H., Skowron, A., Ras, Z.W. (eds.) Foundations of Intelligent Systems. LNCS, vol. 10352, pp. 540–549. Springer, Cham (2017)
19. Yu, H.: A framework of three-way cluster analysis. In: Polkowski, L., et al. (eds.) IJCRS 2017. LNCS (LNAI), vol. 10314, pp. 300–312. Springer, Cham (2017). https://doi.org/10.1007/978-3-319-60840-2_22

Application of the Pairwise Comparison Matrices into a Dispersed Decision-Making System With Pawlak's Conflict Model

Małgorzata Przybyła-Kasperek[(✉)]

Institute of Computer Science, University of Silesia,
Będzińska 39, 41-200 Sosnowiec, Poland
malgorzata.przybyla-kasperek@us.edu.pl
http://www.us.edu.pl

Abstract. In the article a dispersed system with Pawlak's approach to conflict analysis is used. This system was proposed in a previous work. The novelty that is proposed in this paper is the use of the pairwise comparison method in this system. In the system, at first coalitions of local bases are determined with using Pawlak's approach. Based on an aggregated knowledge, which is defined for a coalition, a pairwise comparison matrix is generated. Then the aggregation of the matrices is realised. Final decisions are made using the row geometric mean method. The proposed approach was tested using two dispersed data sets. Some conclusions are presented in this paper.

Keywords: Dispersed decision-making system · Conflict analysis
Pawlak's model · Pairwise comparison · Geometric mean

1 Introduction

The use of dispersed knowledge that is available from many different sources is considered in this paper. We assume that knowledge is gathered in a set of local decision tables. We do not assume any relations between the sets of objects or the set of attributes of the local tables. The dispersed system with Pawlak's model, which was proposed in the previous work, is considered in this article. In the paper [12] three approaches of using Pawlak's model in a dispersed decision system were discussed, however, it was shown that one of them gives the best results. Therefore, this approach is used in this work. The novelty that is proposed in the study is the use of the pairwise comparison method in this system. The classification process of the proposed model can be described in a few steps. Based on each local table, the classification of object is made. Then, using the conflict analysis method that is based on Pawlak's approach, coalitions of local tables are created. An aggregated decision table is generated for each coalition. Based on the aggregated table, a pairwise comparison matrix is determined.

© Springer Nature Switzerland AG 2018
H. S. Nguyen et al. (Eds.): IJCRS 2018, LNAI 11103, pp. 392–404, 2018.
https://doi.org/10.1007/978-3-319-99368-3_30

Then the matrices obtained for all coalitions are aggregated. Two aggregation methods are considered in this paper – based on the geometric mean and based on the arithmetic mean. Global decisions are made using the aggregated matrix and the row geometric mean method.

The problem of simultaneous use of knowledge that is available in separate data sets is discussed in the context of various computer science problems such as multiple classifier systems [9,10], distributed decision–making [4,15,16], group decision–making [2,7] and data science [8,11]. The model that is considered in this work is not directly related to any of these issues. Of course, the issue of the simultaneous application of knowledge from various data sets is the common denominator of this study and the approaches mentioned above, but these approaches differ in terms of their applications and assumptions. First of all, in the approach that is considered here, the main goal is to use knowledge that is predetermined and given in a dispersed form – the process of knowledge dispersion is not one of the stages of the model building process. Another important difference is its structure. In the system that is considered, the relations that occur between the base classifiers when making decisions for a given object are analysed. A dynamic structure is used – the classifiers are reorganised dynamically – and for each new case a different configuration of classifiers is created. This approach is rather unique and distinguishes the system from the approaches that are known from the literature.

An important concept that is considered in this paper, is the group decision making approach that use geometric mean [5]. It is a technique that is used in pairwise comparison problems, which has very reasonable properties [3]. In this method, the preferences of decision-makers are represented in the form of a numerical answer to the question how much the first alternative is better than the second alternative.

The paper is organised as follows. The second section briefly describes the way Pawlak's model is used in a dispersed system (the approach from the paper [12] that is used in this article). The third section presents the method of generating the pairwise comparison matrices and the technique of their aggregation. The fourth section compares the proposed methods with a fusion method known from the literature. The fifth section describes the experiments that were performed using two data sets from the University of California, Irvine (UCI) repository and presents the results. The article concludes with a short summary in the last section.

2   Pawlak's Model in a Dispersed System

In Pawlak's model, it is assumed that the set Ag is the set of agents that are involved in the conflict. An opinion about the issues being discussed is expressed by each agent by assigning one of three values. -1 means that an agent is against the issue, 0 means it is neutral and 1 means it is for the issue. This knowledge can be written in the form of an information system $S = (U, A)$, where the universe U is the set of agents, A is the set of issues and the set of values of $a \in A$ is equal to $V^a = \{-1, 0, 1\}$. The value $a(x)$, where $x \in U, a \in A$ is the opinion of agent x about issue a.

In the approach that is considered in this paper, it is assumed that a set of local decision tables is available based on which classifiers are created. The classification of a test object is made by such an ensemble of classifiers. In the classification process, the relations between classifiers are analyzed, coalitions are formed and a hierarchical structure of the system is created. In [12], the concepts that were proposed by Pawlak were applied to the analysis of the relations between classifiers. It was assumed that each of the base classifiers made an initial classification that was saved as a vector of ranks. In this vector, one rank was assigned for each decision. More precisely, each classifier is called an agent ag (the concepts classifier and agent are used interchangeably here). It is assumed that for a classified object x and for each classifier ag_i, a vector of ranks $[r_{i,1}(x), \ldots, r_{i,c}(x)]$, where c is the number of decision classes, is generated. For this purpose, the m_1 nearest neighbors' classifier is used. In order to apply Pawlak's model, an information system is generated based on these vectors of ranks. The universe in the information system is equal to the set of classifiers and the set of issues that are being considered by the classifiers is equal to the set of decision classes. The function $a : U \to \{-1, 0, 1\}$ for each $a \in A$ is defined in the following way

$$a(ag) = \begin{cases} 1 & \text{if } r_{ag,a}(x) = 1 \\ 0 & \text{if } r_{ag,a}(x) = 2 \\ -1 & \text{if } r_{ag,a}(x) > 2 \end{cases}$$

This means that agents are favourable only to the decision that received the highest rank – Rank 1. Agents are neutral to the decisions that received Rank 2. For all of the other decision values, the agents are against.

In order to determine the coalitions of agents, the conflict function is used. The conflict function $\rho_B : U \times U \to [0, 1]$ for the set of issues $B \subseteq A$ is defined as follows:

$$\rho_B(x, y) = \frac{card\{\delta_B(x, y)\}}{card\{B\}},$$

where $\delta_B(x, y) = \{a \in B : a(x) \neq a(y)\}$. When we consider the set of all of the attributes A, we write in short $\rho(x, y)$.

We can define the relations between agents by taking into account a set of attributes. A pair $x, y \in U$ is said to be:

– allied $R^+(x, y)$, if $\rho(x, y) < 0.5$,
– in conflict $R^-(x, y)$, if $\rho(x, y) > 0.5$,
– neutral $R^0(x, y)$, if $\rho(x, y) = 0.5$.

Set $X \subseteq U$ is a coalition if for every $x, y \in X$, $R^+(x, y)$ and $x \neq y$.

The classifiers are combined into coalitions as was described above. Then, the common knowledge of classifiers that belongs to one coalition is generated. The method of the elimination of inconsistencies in the knowledge is used for this purpose. One decision table is generated based on relevant objects from all of the decision tables from one coalition. The set of relevant objects is the set of m_2 objects with the greatest similarity to the test object. As was described above, the coalitions of classifiers are generated dynamically. This means that another set of coalitions is determined for each new case. In addition, new aggregated decision tables are generated for each new object. This approach ensures that the aggregated knowledge is relevant to the issue that is currently being considered. For more details, please refer to [13]. Based on each aggregated decision table, a c dimensional vector of values is generated. The m_3 nearest neighbors method is used to do this. The vector's coordinate is equal to the average similarity of m_3 nearest neighbors from a given decision class to a classified object. These vectors are used in order to generate a pairwise comparison matrix, which are described in the following section.

3   Pairwise Comparison Matrices and Row Geometric Mean Method

For j-th coalition of local decision tables, a vector of values

$$\mu_j(x) = [\mu_{j,1}(x), \ldots, \mu_{j,c}(x)]$$

is generated as it was described above. Based on this vector, for each coalition, a comparison matrix is generated.

Pairwise comparison matrix for j-th coalition is a martix $\mathbf{C}^{(j)} = [c_{ik}^{(j)}] \in R_+^{c \times c}$ in which $c_{ik}^{(j)} = \frac{1}{c_{ki}^{(j)}}$ for all $1 \leq i, k \leq c$, where c is the number of decision classes. In this study, it is proposed that, for j-th coalition, the pairwise comparison matrix is calculated according to the formula

$$c_{ik}^{(j)} = \frac{\mu_{j,i}(x)}{\mu_{j,k}(x)} \quad \text{for all } 1 \leq i, k \leq c \text{ and } \mu_{j,i}(x) \neq 0, \mu_{j,k}(x) \neq 0.$$

If any of the values $\mu_{j,i}(x)$ or $\mu_{j,k}(x)$ is equal to zero, then instead of zero we use the value 0.001 in the formula above.

In this way, we get as many pairwise comparison matrices as many coalitions were defined. Then the matrices are aggregated into one matrix. There are two basic ways to aggregate individual preferences into a group preference [6]. Which method should be used depends on whether the group wants to act together as a unit or as separate individuals. In the first case rather the geometric mean should be used, in the second case it is better to use the arithmetic mean. An aggregation method that is equivalent to calculating the geometric mean from all corresponding elements of the matrices is defined next. Let $\mathbf{C}^{(j)}$, where $1 \leq j \leq m$ and m is the number of coalitions, be a set of comparison matrices for all coalitions. An aggregated comparison matrix is equal to

$$\mathbf{C} = \mathbf{C}^{(1)} \odot \ldots \odot \mathbf{C}^{(m)} = \left[\sqrt[m]{c_{ik}^{(1)} \cdots c_{ik}^{(m)}} \right] \in R_+^{c \times c}$$

In [1], it was proved that the aggregated matrix is also a pairwise comparison matrix, i.e. it fulfills the condition $c_{ik} = \frac{1}{c_{ki}}$ for all $1 \leq i, k \leq c$. In addition, multiplying all comparison matrices for coalitions by the same scalar results in an adequate change in the aggregated matrix.

In an aggregation method that is based on the arithmetic mean, an aggregated matrix is equal to

$$\mathbf{C} = \mathbf{C}^{(1)} \oplus \ldots \oplus \mathbf{C}^{(m)} = \left[\frac{c_{ik}^{(1)} + \ldots + c_{ik}^{(m)}}{m} \right] \in R_{+}^{c \times c}$$

This matrix does not have to be a pairwise comparison matrix. However, both methods (the geometric mean and the arithmetic mean) satisfy the Pareto principle, i.e. if $c_{ik}^{(j)} \geq c_{i'k'}^{(j)}$ for all $1 \leq j \leq m$ then $\prod_{j=1}^{m} c_{ik}^{(j)} \geq \prod_{j=1}^{m} c_{i'k'}^{(j)}$ and also $\sum_{j=1}^{m} c_{ik}^{(j)} \geq \sum_{j=1}^{m} c_{i'k'}^{(j)}$.

In the next step, based on the aggregated matrix, a weight vector $\mathbf{w} = [w_i] \in R_{+}^{c}$, $\sum_{i=1}^{c} w_i = 1$, is defined according to the row geometric mean method. The row geometric mean method is the mapping $\mathbf{C} \to \mathbf{w}^{RGM}(\mathbf{C})$ such that the weight vector $\mathbf{w}^{RGM}(\mathbf{C})$ is the unique solution of the optimization problem:

$$\min_{\mathbf{w} \in R^c} \sum_{i=1}^{c} \sum_{k=1}^{c} \left[\log c_{ik} - \log \left(\frac{w_i}{w_k} \right) \right]^2$$

The solution to the above formula is the vector $\mathbf{w}^{RGM} = [w_i^{RGM}]$ defined as follows

$$w_i^{RGM}(\mathbf{C}) = \frac{\prod_{k=1}^{c} c_{ik}^{1/c}}{\sum_{j=1}^{c} \prod_{k=1}^{c} c_{jk}^{1/c}}$$

The weight vector reflect the preferences of all agents. The higher the value w_i^{RGM} is, the more preferred is the i-th decision class for the decision-makers. The global decisions taken by all agents are defined as the decisions with the maximum value of the vector's $\mathbf{w}^{RGM}$ coefficients.

In the next section, it will be justified that in the case considered in the paper – when the pairwise comparison matrices are defined based on the vectors that were generated for the aggregated tables – the use of aggregation based on the geometric mean and the row geometric mean method is equivalent to the fusion method from the measurement level – the product rule. However, the use of aggregation based on the arithmetic mean and the row geometric mean method is not equivalent to any, known from the literature, fusion method and provides an interesting combination of two approaches (sum and product).

4   Comparison of the Aggregation Method Based on the Arithmetic Mean and on the Geometric Mean

In this section, a simple calculations and an example will be presented, which show differences and similarities between the proposed above method that uses the pairwise comparison matrices and aggregation based on the geometric mean or the arithmetic mean and a fusion method known from the literature.

The product rule is well known method for fusion of classifiers' prediction [9]. It belong to the measurement level group and consist in performing simple transformations on vectors generated by the base classifiers. In our case, we use the vectors that were generated based on the aggregated tables

$$\mu_j(x) = [\mu_{j,1}(x), \ldots, \mu_{j,c}(x)], \quad \text{for } j\text{-th coalition.}$$

In the product rule the product of the probability values is determined for each decision class. The set of decisions taken by the dispersed system is the set of classes that have the maximum of these products

$$\arg \max_{i \in \{1, \ldots, c\}} \left\{ \underbrace{\prod}_{j\text{-th coalition}} \mu_{j,i}(x) \right\}.$$

The product rule is very sensitive to the most pessimistic prediction result. To eliminate this drawback, for the probability that is equal to 0, the value 10^{-3} is used instead.

In the first stage, we will justify that the aggregation based on the geometric mean with the row geometric mean, in the considered case, is equivalent to the product rule. Let us assume that

$$w_i^{RGM} \le w_j^{RGM}$$

for certain decision classes i, j. Because we use the row geometric mean it is equivalent to

$$\frac{\prod_{k=1}^{c} c_{ik}^{1/c}}{\sum_{p=1}^{c} \prod_{k=1}^{c} c_{pk}^{1/c}} \le \frac{\prod_{k=1}^{c} c_{jk}^{1/c}}{\sum_{p=1}^{c} \prod_{k=1}^{c} c_{pk}^{1/c}},$$

where c is the number of decision classes. Because we use the aggregation based on the geometric mean it is equivalent to

$$\prod_{k=1}^{c} \sqrt[cm]{c_{ik}^{(1)} \cdots c_{ik}^{(m)}} \le \prod_{k=1}^{c} \sqrt[cm]{c_{jk}^{(1)} \cdots c_{jk}^{(m)}},$$

where m is the number of coalitions. According to the definition of $c_{ik}^{(j)}$ given earlier we have

$$\prod_{k=1}^{c} \frac{\mu_{1,i}(x)}{\mu_{1,k}(x)} \cdots \frac{\mu_{m,i}(x)}{\mu_{m,k}(x)} \le \prod_{k=1}^{c} \frac{\mu_{1,j}(x)}{\mu_{1,k}(x)} \cdots \frac{\mu_{m,j}(x)}{\mu_{m,k}(x)}$$

Thus, this is equivalent to

$$\prod_{p=1}^{m} \mu_{p,i}(x) \le \prod_{p=1}^{m} \mu_{p,j}(x)$$

When we use the aggregation based on the arithmetic mean and the row geometric mean, the j-th decision is preferred over the i-th decision (inequality

$w_i^{RGM} \leq w_j^{RGM}$ is fulfilled, as the row geometric mean is used in the last step) means that

$$\prod_{k=1}^{c} \sqrt[c]{\frac{c_{ik}^{(1)} + \ldots + c_{ik}^{(m)}}{m}} \leq \prod_{k=1}^{c} \sqrt[c]{\frac{c_{jk}^{(1)} + \ldots + c_{jk}^{(m)}}{m}}$$

Thus, we have

$$\prod_{k=1}^{c} \sum_{p=1}^{m} \frac{\mu_{p,i}(x)}{\mu_{p,k}(x)} \leq \prod_{k=1}^{c} \sum_{p=1}^{m} \frac{\mu_{p,j}(x)}{\mu_{p,k}(x)}$$

For example, for two coalitions $m = 2$ and two decision classes $c = 2$, this is equivalent to the following

$$\mu_{1,i}^2(x)\mu_{2,1}(x)\mu_{2,2}(x) + \mu_{2,i}^2(x)\mu_{1,1}(x)\mu_{1,2}(x) + \mu_{1,i}(x)\mu_{2,i}(x)\mu_{1,2}(x)\mu_{2,1}(x)$$
$$+\mu_{1,i}(x)\mu_{2,i}(x)\mu_{1,1}(x)\mu_{2,2}(x) \leq \mu_{1,j}^2(x)\mu_{2,1}(x)\mu_{2,2}(x) + \mu_{2,j}^2(x)\mu_{1,1}(x)\mu_{1,2}(x)$$
$$+\mu_{1,j}(x)\mu_{2,j}(x)\mu_{1,2}(x)\mu_{2,1}(x) + \mu_{1,j}(x)\mu_{2,j}(x)\mu_{1,1}(x)\mu_{2,2}(x)$$

If we put, for example, $i = 1$ and $j = 2$ we have

$$\mu_{1,1}^2(x)\mu_{2,1}(x)\mu_{2,2}(x) + \mu_{2,1}^2(x)\mu_{1,1}(x)\mu_{1,2}(x)$$
$$\leq \mu_{1,2}^2(x)\mu_{2,1}(x)\mu_{2,2}(x) + \mu_{2,2}^2(x)\mu_{1,1}(x)\mu_{1,2}(x)$$

Such formulas can be interpreted as calculating the probability for a given decision class determined by a given coalition in comparison to the probabilities that were designated by the opposite coalition for both decision classes. This method is not equivalent to any of the methods that are known from the literature. In the example below, it will be shown that in some cases it has a certain advantage over and the product rule.

Example 1. Let us assume that two coalitions were created for the set of base classifiers (agents). This means that two aggregated decision tables were created, one for each coalition. Let us assume that the decision attribute that appears in these tables have four decision classes $c = 4$. Based on each aggregated table, a four-dimensional vector is created. The i-th coordinate of such a vector corresponds to the i-th decision value and is equal to the average similarity of m_3 nearest neighbors from a given decision class to a classified object. Due to the limited volume of the article, we will not discuss the entire process of coalitions creation and we will not present the form of decision tables of agents or how the aggregated tables are generated. All this was described in the papers [12,14]. We will only discuss the process of generating one vector based on these aggregated tables.

Let us assume that we have two aggregated tables, with binary conditional attributes, each for one coalition (Table 1). A classified object x is as follows

	a	b	c	e	f	g	h	i	j	k	l	m
x	1	1	1	0	1	0	1	1	0	0	1	0

Table 1. Aggregated tables

U_1	a	b	c	e	f	Decision d
x_1^1	1	1	1	1	0	1
x_2^1	1	0	0	0	1	2
x_3^1	0	0	1	0	1	3
x_4^1	1	0	0	1	0	4

U_2	b	e	f	g	h	i	j	k	l	m	Decision d
x_1^2	1	1	0	1	0	0	1	1	0	0	1
x_2^2	0	1	0	1	0	0	1	1	0	0	2
x_3^2	1	1	0	1	0	0	1	1	0	1	3
x_4^2	1	0	1	0	1	1	1	1	0	1	4

Based on the similarity of the classified object to the objects from the aggregated tables, the coalitions have generated the following two vectors

$$\mu_1(x) = [\mu_{1,1}(x), \mu_{1,2}(x), \mu_{1,3}(x), \mu_{1,4}(x)] = [0.6, 0.6, 0.6, 0.2]$$

$$\mu_2(x) = [\mu_{2,1}(x), \mu_{2,2}(x), \mu_{2,3}(x), \mu_{2,4}(x)] = [0.2, 0.1, 0.1, 0.6]$$

These vectors can be interpreted as follows. The first coalition is the most convinced that the test object should be classified to the first, the second or the third decision classes $\mu_{1,1}(x) = \mu_{1,2}(x) = \mu_{1,3}(x) = 0.6$. Furthermore, the first coalition estimates that the object is the least suited to the fourth decision class. The second coalition believes that the test object should be classified to the fourth decision class with the first decision class on the second place.

For the product rule, the following vector will be generated

$$[0.12, 0.06, 0.06, 0.12]$$

Thus, according to this method (so also for the aggregation based on the geometric mean with the row geometric mean, since these two methods are equivalent) the first and the fourth decisions will be taken.

Now we consider the aggregation based on the arithmetic mean with the row geometric mean. The pairwise comparison matrices, calculated according to the formula $\mathbf{C}^{(j)} = \left[\frac{\mu_{j,i}(x)}{\mu_{j,k}(x)}\right]_{1 \leq i, k \leq c}$, for the first and the second coalitions are as follows

$$\mathbf{C}^{(1)} = \begin{bmatrix} 1 & 1 & 1 & 3 \\ 1 & 1 & 1 & 3 \\ 1 & 1 & 1 & 3 \\ \frac{1}{3} & \frac{1}{3} & \frac{1}{3} & 1 \end{bmatrix} \qquad \mathbf{C}^{(2)} = \begin{bmatrix} 1 & 2 & 2 & \frac{1}{3} \\ \frac{1}{2} & 1 & 1 & \frac{1}{6} \\ \frac{1}{2} & 1 & 1 & \frac{1}{6} \\ 3 & 6 & 6 & 1 \end{bmatrix}$$

The aggregated matrix, calculated according to the aggregation method that is based on the arithmetic mean, is equal to

$$
\mathbf{C} = \begin{bmatrix} 1 & \frac{3}{2} & \frac{3}{2} & \frac{5}{3} \\ \frac{3}{4} & 1 & 1 & \frac{19}{12} \\ \frac{3}{4} & 1 & 1 & \frac{19}{12} \\ \frac{5}{3} & \frac{19}{6} & \frac{19}{6} & 1 \end{bmatrix}
$$

Using the row geometric mean method we have

$$
\mathbf{w}^{RGM} = [0.25, 0.19, 0.19, 0.37]
$$

For example, the value w_1^{RGM} was calculated as follows

$$
w_1^{RGM} = \frac{\sqrt[4]{1 \cdot \frac{3}{2} \cdot \frac{3}{2} \cdot \frac{5}{3}}}{\sqrt[4]{1 \cdot \frac{3}{2} \cdot \frac{3}{2} \cdot \frac{5}{3}} + \sqrt[4]{\frac{3}{4} \cdot 1 \cdot 1 \cdot \frac{19}{12}} + \sqrt[4]{\frac{3}{4} \cdot 1 \cdot 1 \cdot \frac{19}{12}} + \sqrt[4]{\frac{5}{3} \cdot \frac{19}{6} \cdot \frac{19}{6} \cdot 1}}
$$

Thus, the fourth decision is more preferred than the first decision.

When we once again analyze the vectors that were generated by the coalitions $\mu_1(x)$ and $\mu_2(x)$, it can be seen that the first coalition made ambiguous decision. It can therefore be concluded that this coalition was not sure about taken decision. Therefore, perhaps this decision is less important. On the other hand, the second coalition was unambiguous when making decision. Therefore, perhaps the decision of this coalition should be more significant. Such approach was realized only in the method using the aggregation based on the arithmetic mean with the row geometric mean method.

5   Experimental Analysis

In the experimental part, tests on the two data sets that have been dispersed into five different ways are presented. The author does not have access to the dispersed data that are stored in the form of a set of local decision tables, and therefore, some benchmark data that were stored in a single decision table were used. In general, the system with Pawlak's conflict model will be tested in this part (proposed in the paper [12] and described in Sect. 2). The results that were obtained using the system with the aggregation based on the arithmetic mean with the row geometric mean method are compared with the results using the system with the product rule.

Data from the UCI repository were used in the experiments – the Soybean data set and the Vehicle Silhouettes data set. The test set for the Soybean data set was obtained from the repository (specially prepared by the founders of this data set). For the Vehicle Silhouettes data set the test set is not available in the repository. Therefore, it was divided in a random way in the proportion: 70% the training set, 30% the test set.

Table 2. Data set summary

Data set	# Training set	# Test set	# Conditional attributes	# Decision classes
Soybean	307	376	35	19
Vehicle Silhouettes	592	254	18	4

Table 2 presents a numerical summary of the data sets.

Each of the data sets was divided into local decision tables in five different ways. A different number of decision tables were considered in each of these variants – from three local tables to eleven local tables, the number of tables was increased by two. The smallest number of tables is three, because for a smaller number there was no point in studying dependencies and building coalitions of local tables. The largest number of tables is eleven, because with the available set of attributes, the division into a larger number of tables would be impossible.

In the system considered in this paper, there are some parameters that were described in Sect. 2. Their symbols and meaning are repeated below

- m_1 – the parameter that determines the number of relevant objects that are used in the process of generating coalitions;
- m_2 – the parameter of the approximated method of the aggregation of the decision tables;
- m_3 – the parameter that determines the number of relevant objects that are used in the process of generating the vectors of the values that are based on the aggregated tables.

As was mentioned in Sect. 4, the global decisions taken by all agents are defined as the decisions with the maximum value of the vector's $\mathbf{w}^{RGM}$ coefficients. It may happen that many different decisions have the same maximum value in the $\mathbf{w}^{RGM}$ vector. Therefore, the system generates a set of global decisions and special measures are needed to determine the quality of the classification. In order to compare the quality of the classification, the following measures are used:

- estimator of classification error e in which an object is considered to be properly classified if the decision class used for the object belonged to the set of global decisions generated by the system;
- estimator of classification ambiguity error e_{ONE} in which object is considered to be properly classified if only one, correct value of the decision was generated to this object;
- the average size of the global decisions sets $\overline{d}_{WSD_{Ag}^{dyn}}$ generated for a test set.

Obviously, if only one decision is generated for each test object, both e and e_{ONE} measures are equivalent to the error rate. However, if the decisions made by the system are ambiguous, none of these measures is equal to the error rate and each of them defines a completely different value.

Parameters values from the set $m_1, m_2, m_3 \in \{1, \ldots, 10\}$ were tested. Then, the minimum value of the parameters are chosen, which results in the lowest value of the estimator of the classification error to be reached.

The results of experiments with the Soybean data set are presented in Table 3. The results for the Vehicle Silhouettes data set are given in Table 4. These results were obtained for the system with Pawlak's conflict model and the row geometric mean method and the system with Pawlak's conflict model and the product rule. In the tables the following information is given: the number of decision tables (# Local tables); the optimal parameters values m_1, m_2 and ε (Parameters); the measures to determine the quality of the classification: e, e_{ONE} and $\overline{d}_{WSD_{Ag}^{dyn}}$.

Based on the results presented in the tables above, it can be concluded that the use of aggregation based on the arithmetic mean with the row geometric mean method gives certainly not worst quality of inference than the product rule. In four out of ten cases better results were obtained. In the remaining cases, the same results were obtained.

Table 3. Summary of experiments results with the Soybean data set

# Local tables	Aggregation based on the arithmetic mean with the row geometric mean				Product rule (equivalent to aggregation based on the geometric mean with the row geometric mean)			
	$m_1/m_2/m_3$	e	e_{ONE}	$\overline{d}_{WSD_{Ag}}$	$m_1/m_2/m_3$	e	e_{ONE}	$\overline{d}_{WSD_{Ag}}$
3	2/3/6	0.096	0.106	1.011	2/3/6	0.096	0.106	1.011
5	1/2/1	0.114	0.170	1.082	1/2/1	0.114	0.168	1.080
7	1/10/1	0.109	0.218	1.146	1/10/1	0.114	0.218	1.141
9	5/3/1	0.085	0.189	1.122	2/3/1	0.088	0.162	1.093
11	2/4/1	0.120	0.210	1.122	2/5/1	0.128	0.202	1.093

Table 4. Summary of experiments results with the Vehicle Silhouettes data set

# Local tables	Aggregation based on the arithmetic mean with the row geometric mean				Product rule (equivalent to aggregation based on the geometric mean with the row geometric mean)			
	$m_1/m_2/m_3$	e	e_{ONE}	$\overline{d}_{WSD_{Ag}}$	$m_1/m_2/m_3$	e	e_{ONE}	$\overline{d}_{WSD_{Ag}}$
3	9/10/4	0.220	0.220	1	9/10/4	0.224	0.224	1
5	3/4/10	0.303	0.303	1	3/4/10	0.303	0.303	1
7	1/5/6	0.276	0.276	1	1/5/6	0.276	0.276	1
9	1/4/4	0.335	0.335	1	1/3/8	0.335	0.335	1
11	3/2/3	0.280	0.280	1	3/2/3	0.280	0.280	1

Of course, in this paper only preliminary experiments were presented. Further studies are necessary. However, as was shown in the example, the proposed method certainly takes into account a wider aspect when making group decision. Because it considers the value of the vector in relation to the values that were designated for other decisions by other coalitions.

6 Conclusions

In this article, the approach that are known from group decision making (pairwise comparison) was adopted to the system with dispersed knowledge. A method for creating pairwise comparison matrices based on vectors generated by coalitions was proposed. Two approaches to aggregate these matrices (based on the geometric mean and based on the arithmetic mean) were considered. A weight vector is generated based on the aggregated matrix using the row geometric mean method. It was shown that the aggregation based on the geometric mean with the row geometric mean is equivalent to the product rule. It was also justified that in the aggregation based on the arithmetic mean with the row geometric mean, other decisions made by other coalitions are taken into account when making global decisions. Based on the presented experiments it was concluded that the aggregation based on the arithmetic mean provides better results in some cases.

References

1. Aczél, J., Saaty, T.L.: Procedures for synthesizing ratio judgements. J. Math. Psychol. **27**(1), 93–102 (1983)
2. Cabrerizo, F.J., Herrera-Viedma, E., Pedrycz, W.: A method based on PSO and granular computing of linguistic information to solve group decision making problems defined in heterogeneous contexts. Eur. J. Oper. Res. **230**(3), 624–633 (2013)
3. Csató, L.: Eigenvector Method and rank reversal in group decision making revisited. Fundamenta Informaticae **156**(2), 169–178 (2017)
4. Delimata, P., Suraj, Z.: Feature selection algorithm for multiple classifier systems: a hybrid approach. Fundamenta Informaticae **85**(1–4), 97–110 (2008). Amsterdam: IOS Press
5. Dong, Y., Zhang, G., Hong, W.C., Xu, Y.: Consensus models for AHP group decision making under row geometric mean prioritization method. Decis. Support Syst. **49**(3), 281–289 (2010)
6. Forman, E., Peniwati, K.: Aggregating individual judgments and priorities with the analytic hierarchy process. Eur. J. Oper. Res. **108**(1), 165–169 (1998)
7. Greco, S., Matarazzo, B., Słowiński, R.: Rough sets theory for multicriteria decision analysis. Eur. J. Oper. Res. **129**(1), 1–47 (2001)
8. Kanter, J.M., Veeramachaneni, K.: Deep feature synthesis: towards automating data science endeavors. In: IEEE International Conference Data Science and Advanced Analytics (DSAA), pp. 1–10 (2015)
9. Kuncheva, L.: Combining Pattern Classifiers Methods and Algorithms. Wiley, Chichester (2004)

10. Polikar, R.: Ensemble based systems in decision making. IEEE Circuits Syst. Mag. **6**, 21–45 (2006)
11. Provost, F., Fawcett, T.: Data science and its relationship to big data and data-driven decision making. Big Data **1**(1), 51–59 (2013)
12. Przybyła-Kasperek, M.: Methods based on Pawlak's model of conflict analysis - medical applications. In: Polkowski, L., Yao, Y., Artiemjew, P., Ciucci, D., Liu, D., Ślęzak, D., Zielosko, B. (eds.) IJCRS 2017. LNCS (LNAI), vol. 10313, pp. 249–262. Springer, Cham (2017). https://doi.org/10.1007/978-3-319-60837-2_21
13. Przybyła-Kasperek, M., Wakulicz-Deja, A.: The strength of coalition in a dispersed decision support system with negotiations. Eur. J. Oper. Res. **252**, 947–968 (2016)
14. Przybyła-Kasperek, M., Wakulicz-Deja, A.: A dispersed decision-making system - the use of negotiations during the dynamic generation of a systems structure. Inf. Sci. **288**, 194–219 (2014)
15. Schneeweiss, C.: Distributed decision making. Springer, Berlin (2003)
16. Schneeweiss, C.: Distributed decision making - a unified approach. Eur. J. Oper. Res. **150**(2), 237–252 (2003)

Exploring GTRS Based Recommender Systems with Users of Different Rating Patterns

Bingyu Li$^{(\boxtimes)}$ and JingTao Yao

Department of Computer Science, University of Regina, Regina, SK S4S0A2, Canada
{li970,jtyao}@cs.uregina.ca

Abstract. Recommender systems predict a new user's opinion on a collection of items by analyzing preference information of similar users. The Pawlak rough set (PRS) model is one of the effective tools to make personalized recommendations. The game-theoretic rough set (GTRS) model improves the quality of PRS based recommendations by determining a pair of thresholds that could achieve a tradeoff between two prominent recommendation evaluation metrics, accuracy and coverage. It should be noted that the performance of a recommendation algorithm may be affected by the rating patterns of the users in the considered dataset. The aim of this research is to evaluate how the performance of the PRS based and the GTRS based recommendations vary on user groups with different rating patterns. We conducted comparative experiments on five different data samples. The experimental results suggest that compared to the PRS model, the GTRS model could not only obtain an improvement in coverage level, but also achieve an equal accuracy level on each of the considered data samples. In particular, it achieved a bigger advantage over the PRS model on user groups that make a smaller number of rating records. This performance difference indicates that compared to the PRS model, the GTRS model is a better solution to make high quality personalized recommendations on small-scale datasets with fewer rating records stored in the database.

Keywords: Recommender systems · Rough sets
Game-theoretic rough sets

1 Introduction

Recommender systems predict a user's preference among a collection of items by aggregating and analyzing suggestions from similar users [1]. Through the use of data mining techniques, recommender systems help users to find items that they are interested in without searching through the enormous amount of information on the internet [1].

Different approaches are involved in the design phase of a recommender system. Collaborative filtering, content based filtering, knowledge based filtering

© Springer Nature Switzerland AG 2018
H. S. Nguyen et al. (Eds.): IJCRS 2018, LNAI 11103, pp. 405–417, 2018.
https://doi.org/10.1007/978-3-319-99368-3_31

and demographic based filtering are by far the most commonly used techniques in the field of recommender system research [3]. Collaborative filtering (CF) predicts a user's opinion on an item by combining similar users' opinions on this specific item [15]. For the reason that it is easy to implement and highly effective, CF is the most popular one among all these approaches [6].

The methods involved in the implementation phase of a CF recommender system could be further divided into two categories: memory-based methods and model-based methods [19]. Memory-based methods maintain a database to store rating information from all users and make calculations across the whole database whenever a prediction needs to be made [8]. Memory-based methods are widely implemented in e-commerce websites as they take less effort to implement and make moderately accurate recommendations at a low cost [16]. However, the performance of memory-based methods rely highly on the rating density of the database, as the task of finding similarities among different users gets harder when fewer rating records are around [5].

On the other hand, model-based methods transfer the existing information in the database into a preference model through the use of data mining algorithms. When a new user's information is input into the system, the system will approach to the preference model instead of the original database to generate personalized recommendations. It is believed that by using training data to construct a preference model beforehand and making recommendations with the constructed preference model, model-based algorithms are able to overcome the limitation of memory-based algorithms.

Different data mining models are used to predict user preference in model-based CF recommender systems. Some of the well-known models that are commonly used include the Bayesian belief nets model, the clustering model, the latent semantic CF model, and the Pawlak rough set (PRS) model [10].

The PRS model [11] is a powerful mathematical tool to deal with incomplete information. It forms equivalence classes with users that share similar interests on training data, and makes predictions with these formed equivalence classes on test data [17]. One limitation of the PRS based recommendations is that as the model is intolerant to errors, its predictions are only applicable for a limited portion of users. However, this limitation could be eliminated through the use of the game-theoretic rough set (GTRS) model [18].

As a quantitative generalization of the PRS model [13], the GTRS model helps the PRS model with its error-intolerance which further broadens its practical application. It formulates a competitive game between two of the most prominent recommendation evaluation metrics, accuracy and coverage. An optimal threshold pair (α', β') that achieves a tradeoff between the two considered metrics will be returned once the competitive game is completed. The optimal threshold pair is then used to determine the three rough set regions and to carry out rough set analysis.

As the rating pattern of the considered user group may have an impact on the performance of a CF recommendation algorithm [4], we run a comparative study between the PRS and the GTRS model using various data samples with different rating patterns. The two considered models are used to predict user preference on five featured data samples formed by users with different number of rating record respectively. The recommendation quality achieved by the two models are evaluated and compared with each other to address the effect that rating patterns have on the performance of a recommendation algorithm.

The remainder of the paper consists of 5 different parts. Section 2 introduces some important concepts about the PRS model and how it is used to make personalized recommendations. Section 3 gives an insight of how the GTRS model is used to formulate a competitive game between the two recommendation evaluation metrics, accuracy and coverage. In Sect. 4, some data preprocessing and partitioning are performed on the original dataset to form user groups with users that have a certain range of rating records. In Sect. 5, the PRS model and the GTRS model are used to predict user preference on the featured user groups formed in Sect. 4 respectively. Performance evaluations are carried out to compare the recommendation quality of the two models with each other as well as to address the problem of how their performance vary on user groups with different rating patterns. Finally a summary, a conclusion, and limitations of our approach are discussed in Sect. 6.

2   PRS Based Recommendations

The PRS model approximates a set C by a pair of lower and upper approximations, $\underline{apr}(C)$ and $\overline{apr}(C)$ [2]. Let U be a set called universe, and let $[x]$ be an equivalence class formed based on an equivalence relation on U [12]. Set C that is being approximated is normally a subset of set U. The three rough set regions, the positive, the negative and the boundary regions are calculated as follows,

$$POS(C) = \underline{apr}(C) = \{x \in U \mid [x] \subseteq C\} \tag{1}$$

$$NEG(C) = \overline{apr}(C)^c = U - \{x \in U \mid [x] \cap C \neq \emptyset\} \tag{2}$$

$$BND(C) = \overline{apr}(C) - \underline{apr}(C) = \{x \in U \mid [x] \cap C \neq \emptyset\} - \{x \in U \mid [x] \subseteq C\} \tag{3}$$

The example below demonstrates how the PRS model could be used to predict user preference in a CF recommender system. Table 1 is a movie rating table constructed using the rating records in the MovieLens dataset. Let us consider a user set E with a total of 16 users $U_1, U_2, ..., U_{16}$, i.e., $E = \{U_1, U_2, ..., U_8\}$. The considered movie set $M = \{Movie1, Movie2, ..., Movie5\}$ are made up by five different movies that have been rated by all the users in E. Each cell in Table 1 describes a rating record made by a specific user with regard to a specific movie. For instance, the first cell in the first row represents a rating record made by user U_1 with regard to Movie1. For each user, a positive rating to a movie is considered to be a "like" and is transferred into a "+" in the rating table. A negative rating to a movie is considered to be a "dislike" and is transferred into a "−" in the rating table.

Table 1. A movie rating table

	Movie1	Movie2	Movie3	Movie4	Movie5
U_1	−	+	+	+	−
U_2	+	+	−	+	+
U_3	+	−	−	−	+
U_4	−	+	+	+	−
U_5	+	−	−	−	−
U_6	+	−	+	−	+
U_7	+	−	−	−	−
U_8	+	−	+	−	+

The goal of the PRS analysis is to make preference predictions on Movie5. Therefore, in the PRS model, Movie1—Movie4 are defined as the conditional attributes, while Movie5 is defined as the decision attribute. The PRS model identifies the similarities among different users by classifying users with the same conditional attribute values into the same equivalence class, as we are assuming users with the same preference on Movie1—Movie4 might share a similar taste in movies. For instance, user U_1 and U_4 both have a negative rating on Movie1 and positive ratings on Movie2—Movie4. Therefore, they are considered to be similar with each other and are categorized into the same equivalence class X_1. The users in the user set E are classified into four different equivalence classes X_1—X_4 according to the rating records they previously made on Movie1—Movie4.

Table 2. Equivalence classes formed based on Table 1

$X_1 = \{U_1, U_4\}$	$X_2 = \{U_2\}$
$X_3 = \{U_3, U_5, U_7\}$	$X_4 = \{U_6, U_8\}$

When new users enter the system, we first identify the equivalence class they belong to based on the rating records they previously made on Movie1—Movie4. Then we predict their preference on Movie5 according to which rough set region their equivalence classes belong to. For instance, equivalence class X_1 is less likely to like Movie5 and should be classified into the negative region, since both U_1 and U_4 have a negative rating on this movie. On the other hand, equivalence class X_4 is more likely to like Movie5 and should be classified into the positive region, since both U_6 and U_8 have a positive rating on Movie5. However, the preference prediction of the target user could not be specified if the users in the equivalence class do not agree with each other with regard to their opinions on Movie5. For instance, we are unable to tell whether equivalence class X_3 likes Movie5 or not as one of the users in the equivalence class likes the movie while the other two do not. In the PRS model, these equivalence classes are classified into the boundary region, which means the preference of the users that belong to these equivalence classes could not be predicted. For the PRS model, although leaving out the equivalence classes in the boundary region reduces the possibility of it making incorrect recommendations, only being able to make predictions for a limited portion of users is a drawback in its practical application.

On the other hand, different metrics have been proposed to evaluate the performance of PRS based recommendations, and accuracy and coverage are two of the most popular ones among all of them [14]. Accuracy computes how close the recommender system's predictions are to the actual preference of the target user [5]. Coverage measures the portion of users for whom recommendations could be given using only the prediction algorithm [5]. Both accuracy and coverage are the properties we want to pursue in PRS based recommendations, and we may want to optimize them both at the same time. However, this may not be possible in many cases. An attempt to increase accuracy might cause a decrease in coverage, and vice versa [16]. Therefore, instead of trying to optimize both accuracy and coverage simultaneously, we try to realize a tradeoff between the two considered attributes. The problem yet to solve is to what degree is this tradeoff acceptable. As the PRS based recommendations are only applicable for a limited portion of users, how much sacrifice in accuracy level is acceptable in order to improve coverage level requires more tradeoff analysis.

3   GTRS Based Recommendations

The GTRS model [20] provides a near optimal solution to this problem by realizing the tradeoff through a competitive game formulated between accuracy and coverage. There are three major components in the competitive game, a set of players P, a set of strategies S, and a set of payoff functions F [18].

As we are considering a tradeoff between the accuracy and the coverage of the PRS based recommendations, these two attributes are selected as game players, i.e., $P = \{Accuracy, Coverage\}$. The strategies are a set of moves that each game player could choose from [18]. In the GTRS model, strategies are realized by making corresponding adjustments in the thresholds levels [2]. To better compare the GTRS model with the PRS model, the threshold pair values in a GTRS based competitive game is initially configured as $(\alpha, \beta) = (1, 0)$ where accuracy level is at its highest and coverage level is at its lowest. As a result, player accuracy and player coverage have three different types of strategies to choose from, which is to decrease α, to increase β, or to decrease α and increase β at the same time, i.e., $S = \{s_1, s_2, s_3\}$, $s_1 = \alpha \downarrow$, $s_2 = \beta \uparrow$, $s_3 = \alpha \downarrow \beta \uparrow$.

The payoff functions are used to measure the outcome of a game player choosing a specific strategy profile [18]. In the rough set model, the metric of accuracy is defined as the ratio of the number of correctly classified objects in the positive and negative region to the total number of objects in these two regions. The metric of coverage is defined as the ratio of the total number of objects in the positive and negative region to the total number of objects in the universal set. Supposing that the threshold pair is configured as (α, β), the payoffs of the two players $f_A(\alpha, \beta)$ and $f_C(\alpha, \beta)$, i.e., the accuracy and the coverage of the recommendations are calculated using the following equations [2],

$$f_A(\alpha, \beta) = Accuracy_{(\alpha,\beta)} = \frac{\mid (POS_{\alpha,\beta}(C) \cap C) \cup (NEG_{\alpha,\beta}(C) \cap C^c) \mid}{\mid POS_{\alpha,\beta}(C) \cup NEG_{\alpha,\beta}(C) \mid} \tag{4}$$

$$f_C(\alpha, \beta) = Coverage_{(\alpha,\beta)} = \frac{\mid POS_{\alpha,\beta}(C) \cup NEG_{\alpha,\beta}(C) \mid}{\mid U \mid} \tag{5}$$

After the payoffs of all the strategy profiles have been calculated using the corresponding payoff functions, the competitive game is completed. A payoff table like Table 2 will be formed [7]. The rows in Table 3 represent the strategy selection of player accuracy while the columns describe the strategy selection of player coverage. Each cell is assigned with a set of payoffs calculated using the payoff functions with regard to the strategy selections of the two game players.

Table 3. Payoff table for the competitive game between accuracy and coverage

		Coverage		
		$s_1 = \alpha \downarrow$	$s_2 = \beta \uparrow$	$s_3 = \alpha \downarrow \beta \uparrow$
Accuracy	$s_1 = \alpha \downarrow$	$\langle f_A(\alpha \downarrow\downarrow, \beta),$ $f_C(\alpha \downarrow\downarrow, \beta) \rangle$	$\langle f_A(\alpha \downarrow, \beta \uparrow),$ $f_C(\alpha \downarrow, \beta \uparrow) \rangle$	$\langle f_A(\alpha \downarrow\downarrow, \beta \uparrow),$ $f_C(\alpha \downarrow\downarrow, \beta \uparrow) \rangle$
	$s_2 = \beta \uparrow$	$\langle f_A(\alpha \downarrow, \beta \uparrow),$ $f_C(\alpha \downarrow, \beta \uparrow) \rangle$	$\langle f_A(\alpha, \beta \uparrow\uparrow),$ $f_C(\alpha, \beta \uparrow\uparrow) \rangle$	$\langle f_A(\alpha \downarrow, \beta \uparrow\uparrow),$ $f_C(\alpha \downarrow, \beta \uparrow\uparrow) \rangle$
	$s_3 = \alpha \downarrow \beta \uparrow$	$\langle f_A(\alpha \downarrow\downarrow, \beta \uparrow),$ $f_C(\alpha \downarrow\downarrow, \beta \uparrow) \rangle$	$\langle f_A(\alpha \downarrow, \beta \uparrow\uparrow),$ $f_C(\alpha \downarrow, \beta \uparrow\uparrow) \rangle$	$\langle f_A(\alpha \downarrow\downarrow, \beta \uparrow\uparrow),$ $f_C(\alpha \downarrow\downarrow, \beta \uparrow\uparrow) \rangle$

The Nash equilibrium is calculated by going through each cell in the payoff table to check if the following conditions hold [18],

$$\text{for all } k \neq i, \ f_A(s_i, s_j) \geq f_A(s_k, s_j); \tag{6}$$

$$\text{for all } k \neq j, \ f_C(s_i, s_j) \geq f_C(s_i, s_k) \tag{7}$$

The strategy profile (s_i, s_j) that yields the conditions of Nash equilibrium is selected as the solution to the competitive game. With the calculation of the optimal strategy profile (s_i, s_j), the corresponding optimal threshold pair (α', β') could be computed. Similar to the PRS model, when new users enter the system, the GTRS model identifies the appropriate equivalence classes for them based on the rating records they previously made and makes predictions for them based on which rough set region their equivalence classes belong to. An optimal accuracy level and an optimal coverage level could be achieved by determining the three rough set regions with the GTRS optimal threshold pair (α', β').

4   Data Preprocessing and Partitioning

MovieLens is a website that gathers research data to make personalized recommendations. The MovieLens 1M dataset which consists of 1 million 5-star scale ratings on 4,000 different movies provided by 6,000 different users, is used to carry out the comparative evaluation.

Given a dataset, the performance of a CF recommendation algorithm is affected by the rating pattern of the user group which could be represented by the number of rating records each user had in the considered data sample [4]. As the number of rating records directly affects the difficulty of finding users that are similar to the target user, which further affects the performance of a CF recommendation algorithm. Besides, user groups with more rating records and user groups with less rating records have different rating behaviours. A user who tends to make more rating records are more likely to rate items positively, and a user who tends to make fewer rating records are more likely to rate items negatively [9].

We partition the data into 5 groups based on number of ratings as we want to examine model on datasets with different rating patterns. In the MovieLens dataset, users with a rating record number within the range of 1–50, 51–100, 101–150, 151–200, and 201–250 are selected respectively to form five different user groups. The rating records in the original dataset are then partitioned into five data samples according to the formed user groups. The formulation of the featured data samples, $Sample_1 - Sample_5$, is described in Table 4.

Table 4. The featured data samples on MovieLens

Sample	$Sample_1$	$Sample_2$	$Sample_3$	$Sample_4$	$Sample_5$
Total ratings	40,000	40,000	40,000	40,000	40,000
Total users	1,793	556	325	233	182
Range of rating number	1–50	51–100	101–150	151–200	201–250
Average rating number	22	71	123	172	220

After partitioning the original dataset into featured data samples, for each data sample, the non-binary scale ratings in the original dataset is transferred into binary scale ratings. The rating records addressing the top ten most frequently rated movies in each data sample are selected to form equivalence classes, i.e., to discover similarities among different users. For each data sample, we use 80% of it to train and 20% of it to test.

5   Experimental Results and Analysis

Table 5 and Fig. 1 describe the accuracy performance of the two models on data samples $Sample_1 - Sample_5$.

Table 5. Accuracy of the two prediction algorithms on the featured data samples

Model	Sample$_1$	Sample$_2$	Sample$_3$	Sample$_4$	Sample$_5$
GTRS	0.7770	0.6976	0.6160	0.5674	0.5430
PRS	0.7635	0.6898	0.6097	0.5651	0.5414

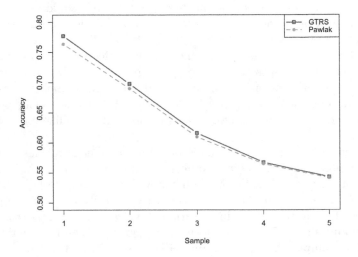

Fig. 1. Accuracy of the two prediction algorithms on the featured data samples

The accuracy level that the PRS model achieved on the five featured user groups ranges from 0.5414 to 0.7635, which is equal to 54.14%—76.35% in percentage. For the GTRS model, the accuracy level it obtained on these data samples ranges from 0.5430 to 0.7770, which is equal to 54.30%—77.70% in percentage. Based on what we can observe from the figure, the accuracy performance of the two algorithms have a tight competition with each other on all five data samples. Since the performance difference between the two models ranges from 0.16% to 1.35%, it is fair to conclude that the two models achieve an equal accuracy level on each of the considered data samples.

In terms of accuracy performance variation on user groups with different rating patterns, the accuracy level of the two considered models both decreases when making recommendations for user groups with a larger number of rating records. For instance, the PRS model achieves an accuracy level of 76.35% on $Sample_1$, while on $Sample_5$ it achieves an accuracy level of 54.14%. For the GTRS model, the accuracy level it obtains on $Sample_1$ is 77.70%, while on $Sample_5$ it obtains an accuracy level of 54.30%.

One reason accounting for this performance decrease is that given a dataset, there are always more users in user groups with a smaller number of rating records than in user groups with a bigger amount of rating records. The equivalence classes formed based on data samples like $Sample_5$ are generally smaller than the ones formed on data samples like $Sample_1$. With less similiar users to learn rating patterns from, the prediction of user preference become less accurate. The other reason accounting for this performance decrease is that the coverage levels on user groups with larger number of rating records are generally higher than user groups with smaller number of rating records. This means on user groups with a large number of rating records, preference predictions are made available for more users. However, both accuracy and coverage have to be considered in association with each other, as an increase in one attribute will result in a decrease in the other. Therefore, as recommendations could be made for more users on user groups with a larger number of rating records, these recommendations are not as accurate as the ones on user groups with a smaller number of rating records.

The coverage performance of the two models on $Sample_1 - Sample_5$ are described in Table 6 and Fig. 2.

Table 6. Coverage of the two prediction algorithms on the featured data samples

Model	$Sample_1$	$Sample_2$	$Sample_3$	$Sample_4$	$Sample_5$
GTRS	0.8761	0.9634	0.9627	0.9665	0.9692
PRS	0.6892	0.8843	0.9154	0.9348	0.9321

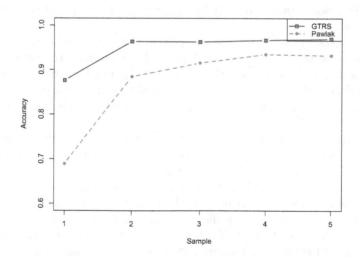

Fig. 2. Coverage of the two prediction algorithms on the featured data samples

Different from what we have discussed in the case of accuracy, the GTRS model achieves a noticeable improvement in coverage level over the PRS model on all five considered user groups $Sample_1 - Sample_5$. This means the GTRS model is able to make preference prediction available for more users no matter how many rating records each user had in the original dataset.

In terms of coverage performance variation on user groups with different rating patterns, the coverage level of both models will increase when making predictions for user groups with more rating records. As the PRS model and the GTRS model achieve their lowest coverage level at 87.61% and 68.92% respectively on $Sample_1$, and obtain their highest coverage level at 96.92% and 93.21% respectively on $Sample_5$. In other words, both models are able to adjust and make recommendations applicable for more users on user groups with a larger number of rating records.

The GTRS model holds a 18.89% advantage in coverage level over the PRS model on $Sample_1$, however, this advantage drops to 3.71% on $Sample_5$. One reason accounting for this performance difference is also that there are a lot more users in user groups with a smaller number of rating records like $Sample_1$, and a lot less users in user groups with a larger number of rating records like $Sample_5$. WAs user groups with smaller number of rating records normally consist of more users, the equivalence classes formed on these user groups are generally larger. The adjustments in the threshold pair values therefore have a bigger impact on data samples with larger equivalence classes compared to the ones with smaller equivalence classes. The GTRS model manipulates the coverage level by adjusting the threshold pair value. Therefore, the increment it brought in coverage level is bigger on user groups with more users like $Sample_1$, and smaller on user groups with less users like $Sample_5$. Moreover, although its advantage in coverage level is not as obvious on $Sample_5$ as it is on $Sample_1$, there is still a noticeable increment in coverage level on each of the considered data samples.

With the accuracy and coverage analysis on the considered data samples, we could summarize that the GTRS model is able to improve the overall quality of PRS based recommendations. Since through the incorporation of the GTRS model, not only could a recommender system make personalized prediction applicable for more users, but also recommend users' preference with an almost equal level of accuracy.

With regard to the performance variation on user groups with different rating patterns, we could conclude that the coverage level of both models will increase while the accuracy level will decrease on user groups with a larger number of rating records. Moreover, although the GTRS model achieves an almost equal accuracy level with the PRS model on all the considered data samples, the advantage it holds in coverage level is bigger on user groups with a smaller number of rating records. Therefore, the overall performance improvement brought by the GTRS model is the bigger on user groups with a smaller number of rating records. This advantage is not as obvious on user groups with a larger number of rating records, as the adjustments in threshold levels have a bigger impact on data samples with larger equivalence classes.

6   Conclusion and Discussion

Recommender systems sift through all available information on the internet to make recommendations for their users. The PRS model is one of the effective techniques to make personalized recommendations in a recommender system. It forms equivalence classes with users that share similar interests on training data, and makes predictions with these formed equivalence classes on test data. The GTRS model improves on the quality of the PRS based recommendations by formulating a competitive game between two of the prominent recommendation evaluation metrics, accuracy and coverage. With the GTRS model, an optimal threshold pair (α, β) will be attained once a tradeoff is achieved between the two considered evaluation metrics. Approximating user preference with the calculated GTRS threshold pair makes the PRS based recommendations applicable for more users, which helps to eliminate the limitation of its practical application.

As the performance of a recommendation algorithm may be affected by the rating patterns of the users in the considered dataset, comparative experiments are carried out on five different data samples to evaluate how the quality of the PRS based and the GTRS based recommendations vary on user groups with different rating patterns. The experimental results suggest that the GTRS model holds an advantage over the PRS model in coverage level, and achieves an equal performance in accuracy level on each of the considered data sample.

Although the GTRS model achieves an overall better performance compared to the PRS model on every considered data sample, the performance improvement on each sample is not the same, as the performance of the two models are affected by the rating pattern of the user group differently. The advantage that the GTRS model holds over the PRS model is bigger on user groups with a smaller number of rating records, and is not as obvious on user groups with a bigger number of rating records.

One reason accounting for this performance difference is that the equivalence classes formed on user groups with a smaller number of rating records are generally larger than those on user groups with a bigger number of rating records. The GTRS model manipulates the accuracy and coverage level by adjusting the thresholds values, and these adjustments have a bigger impact on data samples with larger equivalence classes. More reasonings behind the performance difference could be further addressed by using the GTRS threshold pair attained on one user group to predict user preference on another user group. Conducting these cross recommendations among different data samples in future research will provide us a better insight into the relationship between the rating pattern of the user group and the performance of the recommendation algorithm.

However, not being able to significantly increase its performance as more rating records are added to the database might be a limitation of the GTRS based recommendations. Although it could still achieve an overall better performance compared to the PRS model, the overall performance of the GTRS model is not as competitive as some other data mining models on user groups with a larger number of rating records. Therefore, it is not the best algorithm to predict user preference on large-scale datasets compared to some other model-based techniques such as the latent semantic model and the Bayesian belief nets model. However, some of these techniques require a large number of user rating records in the model building process, and are not able to perform if the provided user rating records are not enough. The GTRS model on the other hand, is able to make moderately accurate recommendations with fewer rating records stored in the database. Therefore, when preference predictions are needed on small-scale datasets with fewer rating records provided, incorporating the GTRS model is an effective solution to make personalized recommendations.

Acknowledgments. This work was partially supported by a Discovery Grant from NSERC Canada.

References

1. Ansari, A., Essegaier, S., Kohli, R.: Internet recommendation systems. J. Mark. Res. **37**(3), 363–375 (2000)
2. Azam, N., Yao, J.T.: Game-theoretic rough sets for recommender systems. Knowl.-Based Syst. **72**, 96–107 (2014)
3. Bobadilla, J., Ortega, F., Hernando, A., Gutiérrez, A.: Recommender systems survey. Knowl.-Based Syst. **46**, 109–132 (2013)
4. Cremonesi, P., Turrin, R., Lentini, E., Matteucci, M.: An evaluation methodology for collaborative recommender systems. In: International Conference on Automated Solutions for Cross Media Content and Multi-channel Distribution, 2008. AXMEDIS 2008, pp. 224–231 (2008)
5. Herlocker, J.L., Konstan, J.A., Terveen, L.G., Riedl, J.T.: Evaluating collaborative filtering recommender systems. ACM Trans. Inf. Syst. **22**(1), 5–53 (2004)
6. Huang, Z., Zeng, D., Chen, H.C.: A comparison of collaborative-filtering recommendation algorithms for e-commerce. IEEE Intell. Syst. **22**(5), 68–78 (2007)
7. Leyton-Brown, K., Shoham, Y.: Essentials of game theory: a concise multidisciplinary introduction. Synthesis Lect. Artif. Intell. Mach. Learn. **2**(1), 1–88 (2008)
8. Liu, F.L., Zhang, B.W., Ciucci, D., Wu, W.Z., Min, F.: A comparison study of similarity measures for covering-based neighborhood classifiers. Inf. Sci. **448**, 1–17 (2018)
9. Middleton, S.E., Roure, D.C.D., Shadbolt, N.R.: Capturing knowledge of user preferences: ontologies in recommender systems. In: The 1st International Conference on Knowledge Capture, pp. 100–107 (2001)
10. Park, D.H., Kim, H.K., Choi, I.Y., Kim, J.K.: A literature review and classification of recommender systems research. Expert Syst. Appl. **39**(11), 10059–10072 (2012)
11. Pawlak, Z.: Rough sets. Int. J. Comput. Inf. Sci. **11**(5), 341–356 (1982)
12. Pawlak, Z.: Rough sets and fuzzy sets. Fuzzy Sets Syst. **17**(1), 99–102 (1985)
13. Qian, Y.H., Zhang, H., Sang, Y.L., Liang, J.Y.: Multigranulation decision-theoretic rough sets. Int. J. Approximate Reasoning **55**(1), 225–237 (2014)
14. Schafer, J.B., Frankowski, D., Herlocker, J., Sen, S.: Collaborative filtering recommender systems. In: The Adaptive Web, pp. 291–324 (2007)
15. Singh, V.K., Mukherjee, M., Mehta, G.K.: Combining collaborative filtering and sentiment classification for improved movie recommendations. In: Multidisciplinary Trends in Artificial Intelligence, pp. 38–50 (2011)

16. Su, X.Y., Khoshgoftaar, T.M.: A survey of collaborative filtering techniques. Adv. Artif. Intell. **2009**, 4–23 (2009)
17. Xu, Y.-Y., Zhang, H.-R., Min, F.: A three-way recommender system for popularity-based costs. In: Polkowski, L., et al. (eds.) IJCRS 2017. LNCS (LNAI), vol. 10314, pp. 278–289. Springer, Cham (2017). https://doi.org/10.1007/978-3-319-60840-2_20
18. Yao, J.T., Herbert, J.P.: A game-theoretic perspective on rough set analysis. J. Chongqing Univ. Posts Telecommun. (Nat. Sci. Edn.) **20**(3), 291–298 (2008)
19. Zhang, H.R., Min, F., Zhang, Z.H., Wang, S.: Efficient collaborative filtering recommendations with multi-channel feature vectors. Int. J. Mach. Learn. Cybernet. 1–8 (2018)
20. Zhang, Y., Yao, J.T.: Multi-criteria based three-way classifications with game-theoretic rough sets. In: International Symposium on Methodologies for Intelligent Systems, pp. 550–559 (2017)

Boundary Region Reduction
for Relation Systems

Guilong Liu$^{(\boxtimes)}$ and Jie Liu

School of Information Science, Beijing Language and Culture University,
Beijing 100083, China
{liuguilong,liujie0829}@blcu.edu.cn

Abstract. Attribute reduction is one of the hottest topics in rough set data analysis. This paper extends the concept of a boundary region to a relation system and studies the boundary region reduction for a given relation system and a fixed set. We present the discernibility matrix and obtain the judgment theorem of such a type of reduction. The discernibility matrix based boundary reduction algorithm for a relation system is established.

Keywords: Attribute reduction · Positive region · Boundary region
Negative region · Rough set · Discernibility matrix

1 Introduction

Attribute reduction in information systems is a fundamental aspect of rough set theory. A reduction is a subset of attributes which reserves the same information for classification purposes as the entire set of attributes. Attribute reduction has been successfully applied in many fields, such as pattern recognition, machine learning and data mining. There are many different types of attribute reductions [1,8,11,12,19], for example, positive region reduction [14], variable precision reduction [18], distribution reduction [10], partial reduction [7], three-way decision based reduction [9] and so on. Jia et al. [2] gave a brief description of twenty-two kinds of existing reduction approaches. Pawlak [13,14] was the first to propose the concept of attribute reduction, Skowron and Rauszer [15,16] proposed discernibility matrix based attribute reduction algorithms for finding all reduction sets in information systems. Recently, Ma and Yao [9] studied class-specific attribute reductions in a decision table from the three-way decision perspective. We [3–7] extended some existing reduction approaches to general relation systems or relation decision systems. For a relation system (U, A) and a fixed non-empty subset $X \subseteq U$, the universal set U is partitioned into the positive, boundary and negative regions via the lower and upper approximations of X. This partition is the theoretical basis of three-way decisions. In fact, we considered the positive and negative region reductions [7] for relation systems. This paper considers the boundary region reduction for a given relation system

© Springer Nature Switzerland AG 2018
H. S. Nguyen et al. (Eds.): IJCRS 2018, LNAI 11103, pp. 418–426, 2018.
https://doi.org/10.1007/978-3-319-99368-3_32

and gives the corresponding reduction algorithm for finding all reduction sets. We also discuss the relationship among positive, boundary and negative region reductions.

The remainder of the paper is organized as follows. In Sect. 2, we briefly recall some basic concepts and properties of binary relations, rough sets and relation systems. In Sect. 3, we present the definition of boundary reduction for a given relation system and a given subset and give a boundary reduction algorithm. Section 4 discusses the relationship among positive, boundary and negative region reductions. Finally, Sect. 5 concludes the paper.

2   Preliminaries

Relationships between numbers, sets and many other entities can be formalized in the idea of a binary relation. This section reviews briefly some basic notations and notions based on binary relations, rough sets and relation systems.

Let $U = \{x_1, x_2, \cdots, x_n\}$ be a finite universal set and $P(U)$ be the power set of U. Suppose that R is an arbitrary binary relation on U. The left and right R-relative sets of an element x in U are defined as

$$l_R(x) = \{y | y \in U, yRx\} \text{ and } r_R(x) = \{y | y \in U, xRy\},$$

respectively. The left and right R-relative sets are a common generalization of equivalence classes. Recall the following terminology: (1) R is reflexive if xRx for each $x \in U$; (2) R is symmetric if $l_R(x) = r_R(x)$ for each $x \in U$; (3) R is transitive if, for each $x, y, z \in U$, $y \in r_R(x)$ and $z \in r_R(y)$ imply $z \in r_R(x)$; and (4) R is an equivalence relation if R is reflexive, symmetric, and transitive. Based on the right R-relative set, for subset $X \subseteq U$, the lower and upper approximations [13, 14, 17] of X are defined as

$$\underline{R}(X) = \{x | x \in U, r_R(x) \subseteq X\} \text{ and } \overline{R}(X) = \{x | x \in U, r_R(x) \cap X \neq \emptyset\},$$

respectively.

Definition 2.1 [5]. Let U be a finite universal set and A be a family of binary relations on U, then (U, A) is called a relation system.

If A consists of equivalence relations on U, then (U, A) is just a usual information system. Thus a relation system is a generalization of an information system. Let (U, A) be a relation system, with respect to a subset $\emptyset \neq B \subseteq A$, we always associate a relation R_B, which is defined as $R_B = \cap_{R \in B} R$.

For a given information system, Pawlak [14] defined the concept of positive, negative and borderline regions of $X \subseteq U$. We extend his definition.

Definition 2.2. Let (U, A) be a relation system and $\emptyset \neq X \subseteq U$, then the positive region $POS_A(X)$, the boundary region $BND_A(X)$ and the negative region $NEG_A(X)$ of X are respectively defined as follows:

$$POS_A(X) = \underline{R_A}(X),$$
$$BND_A(X) = \overline{R_A}(X) - \underline{R_A}(X),$$
$$NEG_A(X) = U - \overline{R_A}(X).$$

This paper studies the boundary region reduction for relation systems. The following proposition gives some basic properties of the boundary region $BND_A(X)$ of X.

Proposition 2.1. Let (U, A) be a relation system, $\emptyset \neq X \subseteq U$ and $\emptyset \neq B \subseteq A$, then the following conditions are equivalent:

(1) $BND_A(X) = BND_B(X)$.
(2) $\overline{R_A}(X) = \overline{R_B}(X)$ and $\underline{R_A}(X) = \underline{R_B}(X)$.
(3) $(\underline{R_A}(X), \underline{R_A}(X^C)) = (\underline{R_B}(X), \underline{R_B}(X^C))$, where $X^C = U - X$ is the complement of X.

Proof. $(2) \Rightarrow (1)$ is clear. By using the negative property $(\overline{R_A}(X))^C = \underline{R_A}(X^C)$, $(2) \Leftrightarrow (3)$ is also clear.

$(1) \Rightarrow (2)$: Since $R_A \subseteq R_B$, we have $\overline{R_A}(X) \subseteq \overline{R_B}(X)$ and $\underline{R_B}(X) \subseteq \underline{R_A}(X)$. $BND_B(X) = \overline{R_B}(X) - \underline{R_B}(X) = \overline{R_A}(X) - \underline{R_A}(X) \subseteq \overline{R_B}(X) - \underline{R_A}(X) \subseteq \overline{R_B}(X) - \underline{R_B}(X)$ implies $\overline{R_A}(X) - \underline{R_A}(X) = \overline{R_B}(X) - \underline{R_A}(X)$, thus $\overline{R_A}(X) = \overline{R_B}(X)$. Similarly, $\underline{R_A}(X) = \underline{R_B}(X)$. $\qquad\square$

3   Boundary Region Reductions

Ma and Yao [9] considered a boundary reduction from the three-way decision perspective on special decision classes for a decision table. Now we extend their definition to a given relation system (U, A) and a given non-empty subset $X \subseteq U$. This section studies such a type of reduction, which keeps $BND_A(X)$ unchanged, we call such a type of reduction a boundary reduction. We first give its definition.

Definition 3.1. Let (U, A) be a relation system and a given subset $\emptyset \neq X \subseteq U$. $\emptyset \neq B \subseteq A$, B is called an X-boundary reduction of (U, A) if B satisfies the following conditions:

(1) $BND_A(X) = BND_B(X)$.
(2) For any $\emptyset \neq B' \subset B$, $BND_A(X) \neq BND_{B'}(X)$.

By Proposition 2.1, an X-boundary reduction of (U, A) keeps both $\underline{R_A}(X)$ and $\overline{R_A}(X)$ unchanged. We [7] considered two types of reductions that keep $\underline{R_A}(X)$ and $\overline{R_A}(X)$ unchanged, respectively. Now, via the strict mathematical proofs, we give an X-boundary reduction algorithm for a given relation system (U, A)

and a given non-empty subset $X \subseteq U$. Suppose that $U = \{x_1, x_2, \cdots, x_n\}$, we define the discernibility matrix $M = (m_{ij})_{n \times n}$ as follows:

$$m_{ij} = \begin{cases} \{a | a \in A, (x_i, x_j) \notin a\}, & \text{if } x_i \in \underline{R_A}(X^C) \text{ and } x_j \in X \\ & \text{or } x_i \in \underline{R_A}(X) \text{ and } x_j \notin X \\ \emptyset, & otherwise \end{cases}.$$

Where X^C denotes the complement of X. We need a technical lemma.

Lemma 3.1. Let (U, A) be a relation system and $\emptyset \neq X \subseteq U$, if x_i and x_j satisfy one of the following conditions:

(1) $x_i \in \underline{R_A}(X^C), x_j \in X.$
(2) $x_i \in \underline{R_A}(X), x_j \notin X.$

Then $m_{ij} \neq \emptyset$.

Proof. Suppose that $x_i \in \underline{R_A}(X^C)$ and $x_j \in X$, if $m_{ij} = \emptyset$, then $x_i R_A x_j$, so $x_j \in r_{R_a}(x_i) \subseteq X^C$, that is, $x_j \notin X$, which contradicts $x_j \in X$. Similarly, if $x_i \in \underline{R_A}(X), x_j \notin X$, then $m_{ij} \neq \emptyset$. $\square$

Theorem 3.1. Let (U, A) be a relation system, $\emptyset \neq X \subseteq U$, and $\emptyset \neq B \subseteq C$. Then the following conditions are equivalent:

(1) $BND_A(X) = BND_B(X).$
(2) If $m_{ij} \neq \emptyset$, then $B \cap m_{ij} \neq \emptyset$.

Proof. $(1) \Rightarrow (2)$: By Proposition 2.1, we have $\underline{R_A}(X^C) = \underline{R_B}(X^C)$ and $\underline{R_A}(X) = \underline{R_B}(X)$. Suppose that $m_{ij} \neq \emptyset$ and $B \cap m_{ij} = \emptyset$, then

(i) $x_i \in \underline{R_A}(X^C)$ and $x_j \in X$ or
(ii) $x_i \in \underline{R_A}(X)$ and $x_j \notin X$.

$B \cap m_{ij} = \emptyset$ implies $x_i R_B x_j$ and $x_j \in R_{R_B}$.

If $x_i \in \underline{R_A}(X^C)$ and $x_j \in X$, by condition (1), $x_i \in \underline{R_B}(X^C)$ and $x_j \in X$, so $x_j \in r_{R_B}(x_i) \subseteq X^C$, which contradicts $x_j \in X$.

If $x_i \in \underline{R_A}(X)$ and $x_j \notin X$, then $x_i \in \underline{R_B}(X)$ and $x_j \notin X$, thus $x_j \in r_{R_B}(x_i) \subseteq X$, which contradicts $x_j \notin X$.

$(2) \Rightarrow (1)$: We first show that $\underline{R_A}(X) = \underline{R_B}(X)$. Note that $\underline{R_B}(X) \subseteq \underline{R_A}(X)$ is clear. If $\underline{R_A}(X) \neq \underline{R_B}(X)$, let $x_i \in \underline{R_A}(X) - \underline{R_B}(X)$, by definition of a lower approximation, we have $r_{R_A}(x_i) \subseteq X$, and $r_{R_B}(x_i) \not\subseteq X$. Let $x_j \in r_{R_B}(x_i)$ and $x_j \notin X$, by Lemma 3.1, $m_{ij} \neq \emptyset$, and from condition (2), $B \cap m_{ij} \neq \emptyset$. Thus $(x_i, x_j) \notin R_B$, which contradicts $x_j \in r_{R_B}$. This shows that $\underline{R_A}(X) = \underline{R_B}(X)$. Similarly, we can show that $\overline{R_A}(X) = \overline{R_B}(X)$. $\square$

From Theorem 3, we have the following corollary.

Corollary 3.1. Let (U, A) be a relation system, $\emptyset \neq X \subseteq U$, and $\emptyset \neq B \subseteq C$, then B is an X-boundary reduction of A if and only if it is a minimal subset satisfying $m_{ij} \cap B \neq \emptyset$ for any $m_{ij} \neq \emptyset$.

According to Corollary 3.1, we propose an X-boundary reduction algorithm for a given relation system (U, A) and a given subset $\emptyset \neq X \subseteq U$ as follows.

Algorithm. An X-boundary reduction for a given relation system.
Input: A given relation system (U, A) and $\emptyset \neq X \subseteq U$.
Output: All X-boundary reduction sets.
(1) Compute a discernibility matrix $M = (m_{ij})_{n \times n}$.
(2) Transform the discernibility function f from its conjunctive normal form (CNF)

$$f = \Pi_{m_{ij} \neq \emptyset, m_{ij} \neq A}(\Sigma m_{ij})$$

into the disjunctive normal form (DNF) $f = \Sigma_{t=1}^{s}(\Pi B_t), (B_t \subseteq A)$.
(3) All reduction sets are $B_1, B_2, \cdots, B_s$ and the core is $\cap_{t=1}^{s} B_t$.
End the algorithm.

We illustrate the algorithm introduced previously with a simple example.

Example 3.1. Let (U, A) be a relation system, where $U = \{1, 2, 3, 4, 5\}$, $A = \{R_1, R_2, R_3, R_4, R_5\}$ and $X = \{1, 3, 5\}$. Each $R_i (i = 1, 2, \cdots, 5)$ is given by its Boolean matrix M_{R_i}.

$$M_{R_1} = \begin{pmatrix} 0 1 1 1 0 \\ 0 1 1 0 1 \\ 1 0 0 0 0 \\ 1 1 0 1 1 \\ 0 1 0 1 0 \end{pmatrix}, M_{R_2} = \begin{pmatrix} 1 1 1 1 0 \\ 0 1 1 0 1 \\ 1 1 0 1 0 \\ 1 1 0 1 0 \\ 0 1 1 1 0 \end{pmatrix}, M_{R_3} = \begin{pmatrix} 1 1 0 1 0 \\ 0 0 1 0 1 \\ 1 0 1 0 0 \\ 1 0 0 1 1 \\ 0 1 1 1 0 \end{pmatrix},$$

$$M_{R_4} = \begin{pmatrix} 0 1 0 1 0 \\ 1 0 1 1 1 \\ 1 0 1 0 1 \\ 1 0 0 1 0 \\ 1 1 1 1 0 \end{pmatrix}, \text{ and } M_{R_5} = \begin{pmatrix} 1 1 1 1 0 \\ 0 1 1 1 1 \\ 1 1 0 1 1 \\ 1 0 0 1 1 \\ 0 1 0 1 1 \end{pmatrix}. \text{ Clearly, } M_{R_A} = \begin{pmatrix} 0 1 0 1 0 \\ 0 0 1 0 1 \\ 1 0 0 0 0 \\ 1 0 0 1 0 \\ 0 1 0 1 0 \end{pmatrix}.$$

By direct computation, $\overline{R_A}(X) = \{2, 3, 4\}$, $\underline{R_A}(X) = \{2, 3\}$ and $BND_A(X) = \overline{R_A}(X) - \underline{R_A}(X) = \{4\}$. The following Table 1 gives the discernibility matrix of the boundary region reduction. Since $1 \in \underline{R_A}(X^C)$ and $1 \in X$, it follows that both R_1 and R_4 are in the entry $(1, 1)$ of Table 1, because $(1, 1) \notin R_1$ and $(1, 1) \notin R_4$. The discernibility function

$$\begin{aligned} f &= (R_1 + R_3)(R_1 + R_4)(R_1 + R_5)(R_3 + R_4) \\ &= (R_1 + R_3 R_4 R_5)(R_3 + R_4) \\ &= R_1 R_3 + R_1 R_4 + R_3 R_4 R_5. \end{aligned}$$

Thus all boundary region reduction sets are $\{R_1, R_3\}$, $\{R_1, R_4\}$, and $\{R_3, R_4, R_5\}$.

4 The Relationship Among Positive, Boundary and Negative Region Reductions

This section will illustrate the relationship among positive, boundary and negative region reductions. Let (U, A) be a relation system and $X \subseteq U$, recall that an X-positive region reduction keeps $\underline{R_C}(X)$ unchanged. Its formal definition is as follows.

Table 1. The discernibility matrix of the reduction

	1	2	3	4	5
1	$\{R_1, R_4\}$	$\emptyset$	$\{R_3, R_4\}$	$\emptyset$	A
2	$\emptyset$	$\{R_3, R_4\}$	$\emptyset$	$\{R_1, R_2, R_3\}$	$\emptyset$
3	$\emptyset$	$\{R_1, R_3, R_4\}$	$\emptyset$	$\{R_1, R_3\}$	$\emptyset$
5	$\{R_1, R_2, R_3, R_5\}$	$\emptyset$	$\{R_1, R_5\}$	$\emptyset$	$\{R_1, R_2, R_3, R_4\}$

Definition 4.1. Let (U, A) be a relation system and a given subset $\emptyset \neq X \subseteq U$. $\emptyset \neq B \subseteq A$, set B is called an X-positive reduction of (U, A) if B satisfies the following conditions:

(1) $POS_A(X) = POS_B(X)$.
(2) For any $\emptyset \neq B' \subset B$, $POS_A(X) \neq POS_{B'}(X)$.

Similarly, an X-negative region reduction keeps $U - \overline{R_C}(X) = \underline{R_C}(X^C)$ unchanged, however, we omit its formal definition. The discernibility matrices $M = (m_{ij})_{s \times (n-t)}$ and $N = (n_{ij})_{u \times t}$ of an X-positive region and X-negative region reduction are given as follows:

$$m_{ij} = \begin{cases} \{a | a \in A, (x_i, x_j) \notin a\}, \ x_i \in \underline{R_A}(X), x_j \notin X \\ \emptyset, \qquad\qquad\qquad\qquad\quad otherwise \end{cases}, \text{ and}$$

$$n_{ij} = \begin{cases} \{a | a \in A, (x_i, x_j) \notin a\}, \ x_i \in \underline{R_A}(X^C), x_j \in X \\ \emptyset, \qquad\qquad\qquad\qquad\quad otherwise \end{cases},$$

respectively. Where $s = |\underline{R_A}(X)|$ denotes the cardinality of $\underline{R_A}(X)$, $t = |X|$ and $u = |\underline{R_C}(X^C)|$.

Using the matrices M and N, we can calculate all positive and negative region reduction sets, respectively. Moreover, we can also derive the boundary region reduction from the positive and negative region reductions. This provides another boundary region reduction algorithm. We use the example below to show the detailed method.

Example 4.1. Let (U, A) and $X \subseteq U$ be as in Example 3.1, the discernibility matrices $M = (m_{ij})_{s \times (n-t)}$ of the X-positive region reduction and $N = (n_{ij})_{u \times t}$ of the X-negative region reduction are shown in Tables 2 and 3:

Table 2. The discernibility matrix of an X-positive reduction

	2	4
2	$\{R_3, R_4\}$	$\{R_1, R_2, R_3\}$
3	$\{R_1, R_3, R_4\}$	$\{R_1, R_3\}$

Table 3. The discernibility matrix of an X-negative reduction

	1	3	5
1	$\{R_1, R_4\}$	$\{R_3, R_4\}$	A
5	$\{R_1, R_2, R_3, R_5\}$	$\{R_1, R_5\}$	$\{R_1, R_2, R_3, R_4\}$

Since the discernibility function of the X-positive region reduction $f_1 = R_3 + R_1 R_4$, so that all the X-positive region reduction sets are $\{R_3\}$ and $\{R_1, R_4\}$, similarly, the discernibility function of the X-negative region reduction $f_2 = R_1 R_3 + R_1 R_4 + R_4 R_5$, so that all the X-negative region reduction sets are $\{R_1, R_3\}$, $\{R_1, R_4\}$ and $\{R_4, R_5\}$. The discernibility function of the X-boundary region reduction is

$$f = f_1 f_2 = (R_3 + R_1 R_4)(R_1 R_3 + R_1 R_4 + R_4 R_5)$$
$$= R_1 R_3 + R_1 R_4 + R_3 R_4 R_5.$$

Thus all boundary region reduction sets are $\{R_1, R_3\}$, $\{R_1, R_4\}$, and $\{R_3, R_4, R_5\}$.

Remark 1. Let B, C and D be respectively X-positive, boundary and negative region reductions of a relation system (U, A), then

(1) $B \cap C$ keeps the negative region unchanged,
(2) $C \cap D$ keeps the positive region unchanged, and
(3) $B \cap D$ keeps the boundary region unchanged.

5 Conclusions

The boundary region consists of hesitation objects. In other words, for these objects, we can neither accept nor reject and, hence, make a non-commitment decision. Naturally, it is an interesting problem to consider the reduction that keeps the boundary region unchanged. Thus we propose the concept of the boundary region reduction for relation systems and obtain a corresponding reduction algorithm for finding all reduction sets. We have also established a relationship among the positive, boundary and negative region reductions. We have provided a way to derive the boundary region reduction sets from the positive and negative region reduction sets. The future work is to apply the reduction model given in this paper to discover knowledge in real life data sets.

Acknowledgements. This work was supported by the National Natural Science Foundation of China (Grant No. 61272031) and supported by Science Foundation of Beijing Language and Culture University (The Fundamental Research Funds for the Central Universities)(Grant No. 18YJ030003).

References

1. Dai, J., Wang, W., Tian, H., Liu, L.: Attribute selection based on a new conditional entropy for incomplete decision systems. Knowl. Based Syst. **39**, 207–213 (2013)
2. Jia, X., Shang, L., Zhou, B., Yao, Y.: Generalized attribute reduct in rough set theory. Knowl. Based Syst. **91**, 204–218 (2016)
3. Liu, G., Li, L., Yang, J., Feng, Y., Zhu, K.: Attribute reduction approaches for general relation decision systems. Pattern Recogn. Lett. **65**, 81–87 (2015)
4. Liu, G., Hua, Z., Zou, J.: A unified reduction algorithm based on invariant matrices for decision tables. Knowl. Based Syst. **109**, 84–89 (2016)
5. Liu, G., Hua, Z., Chen, Z.: A general reduction algorithm for relation decision systems and its applications. Knowl. Based Syst. **119**, 87–93 (2017)
6. Liu, G., Hua, Z., Zou, J.: Local attribute reductions for decision tables. Inf. Sci. **422**, 204–217 (2018)
7. Liu, G., Hua, Z.: Partial attribute reduction approaches to relation systems and their applications. Knowl. Based Syst. **139**, 101–107 (2018)
8. Ma, X., Wang, G., Yu, H., Li, T.: Decision region distribution preservation reduction in decision-theoretic rough set model. Inf. Sci. **278**, 614–640 (2014)
9. Ma, X., Yao, Y.: Three-way decision perspectives on class-specific attribute reducts. Inf. Sci. **450**, 227–245 (2018)
10. Mi, J.S., Wu, W.Z., Zhang, W.X.: Approaches to knowledge reduction based on variable precision rough set model. Inf. Sci. **159**, 255–272 (2004)
11. Mieszkowicz-Rolka, A., Rolka, L.: Variable precision rough rets in analysis of inconsistent decision tables. In: Rutkowski, L., Kacprzyk, J. (eds.) Advances in Soft Computing. Physica-Verlag, Heidelberg (2003)
12. Mieszkowicz-Rolka, A., Rolka, L.: Variable precision fuzzy rough sets. In: Peters, J.F., Skowron, A., Grzymała-Busse, J.W., Kostek, B., Świniarski, R.W., Szczuka, M.S. (eds.) Transactions on Rough Sets I. LNCS, vol. 3100, pp. 144–160. Springer, Heidelberg (2004). https://doi.org/10.1007/978-3-540-27794-1_6
13. Pawlak, Z.: Rough sets. Int. J. Comput. Inf. Sci. **11**, 341–356 (1982)
14. Pawlak, Z.: Rough Sets: Theoretical Aspects of Reasoning About Data. Kluwer Academic Publishers, Boston (1991)
15. Skowron, A., Rauszer, C.: The discernibility matrices and functions in information systems. In: Slowinski, R. (ed.) Intelligent Decision Support-Handbook of Applications and Advances of the Rough Set Theory, pp. 331–362, Springer, Dordrecht (1992)

16. Skowron, A.: Boolean reasoning for decision rules generation. In: Komorowski, J., Raś, Z.W. (eds.) ISMIS 1993. LNCS, vol. 689, pp. 295–305. Springer, Heidelberg (1993). https://doi.org/10.1007/3-540-56804-2_28
17. Yao, Y.: Constructive and algebraic methods of theory of rough Sets. Inf. Sci. **109**, 21–47 (1998)
18. Ziarko, W.: Variable precision rough set model. J. Comput. Syst. Sci. **46**, 39–59 (1993)
19. Zhang, H.Y., Leung, Y., Zhou, L.: Variable-precision-dominance-based rough set approach to interval-valued information systems. Inf. Sci. **244**, 75–91 (2013)

A Method to Determine the Number of Clusters Based on Multi-validity Index

Ning Sun and Hong Yu[✉]

Chongqing Key Laboratory of Computational Intelligence,
Chongqing University of Posts and Telecommunications,
Chongqing 400065, People's Republic of China
sunning_best@foxmail.com, yuhong@cqupt.edu.cn

Abstract. Cluster analysis is a method of unsupervised learning technology which is playing a more and more important role in data mining. However, one basic and difficult question for clustering is how to gain the number of clusters automatically. The traditional solution for the problem is to introduce a single validity index which may lead to failure because the index is bias to some specific condition. On the other hand, most of the existing clustering algorithms are based on hard partitioning which can not reflect the uncertainty of the data in the clustering process. To combat these drawbacks, this paper proposes a method to determine the number of clusters automatically based on three-way decision and multi-validity index which includes three parts: (1) the k-means clustering algorithm is devised to obtain the three-way clustering results; (2) multi-validity indexes are employed to evaluate the results and each evaluated result is weighed according to the mean similarity between the corresponding clustering result and the others based on the idea of the median partition in clustering ensemble; and (3) the comprehensive evaluation results are sorted and the best ranked k value is selected as the optional number of clusters. The experimental results show that the proposed method is better than the single evaluation method used in the fusion at determining the number of clusters automatically.

Keywords: Clustering · Uncertainty · Three-way decisions
Number of clusters · Multi-validity index

1 Introduction

Cluster analysis is a method of unsupervised learning technology, which is playing a more and more important role in data mining. Clustering algorithms aim at categorizing a set of unlabeled objects into clusters so that objects in one cluster are more similar than those in the other clusters [13]. Generally speaking, according to whether there are overlapping regions between clusters, they can be divided into hard clustering and soft clustering. Given the lack of a precise definition of the cluster, one basic and difficult problem for clustering is how to gain the number of clusters automatically [16].

© Springer Nature Switzerland AG 2018
H. S. Nguyen et al. (Eds.): IJCRS 2018, LNAI 11103, pp. 427–439, 2018.
https://doi.org/10.1007/978-3-319-99368-3_33

In general, a good cluster validity index is essential to determine the number of clusters automatically. Yu et al. [16] proposed a hierarchical clustering algorithm which can stop automatically at the perfect number of clusters by extending the decision-theoretic rough set model to clustering. Mok et al. [8] proposed a method which can identify the desired number by integrating the clustering results as a judgment matrix and implementing an iterative graph-partitioning process. Aiming at avoiding the drawback that hard partitioning is still used while constructing the judgment matrix, Chen et al. [2] make full use of the affiliation information in the process of constructing the judgment matrix so that the degree of the sample points belonging to a cluster can be reflected more clearly. To provide a more stable results with less processing time, Azimi et al. [1] introduce the principal component analysis method to the contour coefficient algorithm, run the k-means algorithm with different K values iteratively, evaluate the corresponding results with the modified silhouette algorithm, and select the highest evaluated value as the estimated number of clusters. Based on the idea of particle swarm optimization, Ling et al. [7] proposed a local density model to determine the number of clusters.

Similar to one clustering algorithm can only explore the internal structure of a data set from one certain angle, even if there are so many cluster validity indexes exist, we still cannot find a cluster validity index which is suitable for all clustering evaluations. Each evaluation index has its own features which may lead the index to outperform others or can not compare with others [9]. Therefore, it is difficult for the users to choose a suitable clustering validity index among so many indexes. On the other hand, the traditional methods of determining the number of clusters are mainly based on hard partition which are difficult to reflect the uncertainty of the sample point in the clustering process. But in real production, there exist some three-way phenomenons [15], such as psychology, medical diagnosis, management and so on. Because of the information's inaccuracy or incompleteness, it is difficult for anybody to make an accept or reject judgment directly.

In 2007, Gionis et al. [4] gave a new description of the clustering ensemble: given a set of clustering results, the goal of clustering ensemble is to find a clustering result which is relative to all input clustering results as much as possible. The median partition method is one of the consistency function, the goal of which is to find a clustering result which has a most similarity with the other cluster members [5]. Cristofor et al. [3] obtained an approximate solution under the framework of genetic algorithm. Singhbiostatistics et al. [10] proposed a consistency metric which can be maximized by using 0–1 Semi-definite Program to obtain the center clustering result. Vega-Pons et al. [11] believed the clustering results that are most dissimilar to other cluster members can be removed, which in turn significantly reduces the search space.

Inspired by ensemble learning, one method to overcome the limitation of single index is to utilize multiple validity indexes to construct a multi-index evaluation system. The original intention of the multi-index evaluation system is to enhance the robustness and accuracy of the entire decision system by reducing the inconsistency of different evaluation indicators on the results of clustering results and the probability of selecting poor single model [6]. The idea of this method is similar to the expert committee composed of multiple experts, which integrates each expert's evaluation of a certain problem so that the decision made is more accurate, robust and stable.

In this paper, we firstly apply the idea of the Three-way decision to k-means algorithm and run iteratively the improved k-means algorithm with different values of k. Then, a multi-index evaluation system is constructed. Afterwards, a external validity index is used to measure the similarity between each two clustering results which is used to weight the evaluation result. Then, the weighted evaluation values of different k values clustering results are sorted in each column. Finally, the comprehensive evaluation result of each k value is collected and the best values of k is selected.

The remainder of this paper is organized as follows. Section 2 introduces some basic concepts and theories. Section 3 describes the proposed framework, the three-way k-means clustering algorithm, the weighed method on the evaluated results with different k values and the selecting strategy for the best clustering numbers. Section 4 reports the results of comparative experiments and conclusions are provided in Sect. 5.

2   Preliminaries

In this section, some basic concepts in Three-way clustering and the popular validity indexes are introduced.

2.1   Representation of Three-Way Clustering

The purpose of clustering is to divide the universe $U = \{x_1, x_2, ..., x_n, ..., x_N\}$ into some clusters, here, $x_n = \{x_n^1, \cdots, x_n^m, \cdots, x_n^M\}$, x_n^m is the value of the m-dimensional attribute of the object x_n. If there are K clusters, the family of clusters, $\mathbf{C}$, is represented as $\mathbf{C} = \{C_1, \cdots, C_k, \cdots, C_K\}$. The objects in the set belong to this cluster definitely, the objects not in the set do not belong to this cluster definitely. This is a typical result of two-way decisions. For soft clustering, one object might belong to more than one cluster. However, this representation cannot show which object might belong to this cluster, and it cannot show the degree of the object influence on the form of the cluster intuitively. Thus, the use of three regions to represent a cluster is more appropriate than the use of a crisp set, which also directly leads to three-way decisions based interpretation of clustering.

In contrast to the general crisp representation of a cluster, we represent a three-way cluster C as a pair of sets:

$$C = (Co(C), Fr(C)). \tag{1}$$

Here, $Co(C) \subseteq X$ and $Fr(C) \subseteq X$. Let $Tr(C) = X - Co(C) - Fr(C)$. Then, $Co(C)$, $Fr(C)$ and $Tr(C)$ naturally form the three regions of a cluster as Core

Region, Fringe Region and Trivial Region respectively. That is:

$$CoreRegion(C) = Co(C),$$
$$FringeRegion(C) = Fr(C),$$
$$TrivialRegion(C) = X - Co(C) - Fr(C).$$
(2)

If $x \in CoreRegion(C)$, the object x belongs to the cluster C definitely; if $x \in FringeRegion(C)$, the object x might belong to C; if $x \in TrivialRegion(C)$, the object x does not belong to C definitely.

These subsets have the following properties.

$$X = Co(C) \cup Fr(C) \cup Tr(C),$$
$$Co(C) \cap Fr(C) = \emptyset,$$
$$Fr(C) \cap Tr(C) = \emptyset,$$
$$Tr(C) \cap Co(C) = \emptyset.$$
(3)

If $Fr(C) = \emptyset$, the representation of C in Eq. (1) turns into $C = Co(C)$; it is a single set and $Tr(C) = X - Co(C)$. This is a representation of two-way decisions. In other words, the representation of a single set is a special case of the representation of three-way cluster.

Furthermore, according to Formula (3), we know that it is enough to represent a cluster expediently by the core region and the fringe region.

In another way, we can define a cluster by the following properties:

$$(i) \ Co(C_k) \neq \emptyset, 1 \leq k \leq K;$$
$$(ii) \ \bigcup Co(C_k) \bigcup Fr(C_k) = X, 1 \leq k \leq K.$$
(4)

Property (i) implies that a cluster cannot be empty. This makes sure that a cluster is physically meaningful. Property (ii) states that any object of X must definitely belong to or might belong to a cluster, which ensures that every object is properly clustered.

With respect to the family of clusters, $\mathbf{C}$, we have the following family of clusters formulated by three-way decisions as:

$$\mathbf{C} = \{(Co(C_1), Fr(C_1)), \cdots, (Co(C_k), Fr(C_k)), \cdots, (Co(C_K), Fr(C_K))\}. \ (5)$$

2.2 Review of Validity Indexes

In this section, several popular validity indexes are reviewed as follows.

(1) Dunn

The Dunn index [9] is proposed by Dunn, which is the ratio of the shortest intra-cluster distance to the largest inter-cluster distance. It is defined as:

$$DVI = \frac{\min\limits_{0 < m \neq n < K} \left\{ \min\limits_{\substack{\forall x_i \in \Omega_m \\ \forall x_j \in \Omega_n}} \{\|x_i - x_j\|\} \right\}}{\max\limits_{0 < m < K} \max\limits_{0 < n < K} \{\|x_i - x_j\|\}}.$$
(6)

The larger the DVI, the better the clustering result.

(2) Silhouette coefficient

The Silhouette coefficient [9] is an internal validity index. It is defined as:

$$c(x_i) = \frac{q(x_i) - p(x_i)}{\max\{p(x_i), q(x_i)\}}. \tag{7}$$

The parameters of $p(x_i)$ and $q(x_i)$ are defined as:

$$p(x_i) = \frac{\sum\limits_{x_j \in C_i, x_i \neq x_j} dis(x_i, x_j)}{|C_i| - 1}, q(x_i) = \min_{C_j : 1 \leq j \leq k, j \neq k} \left\{ \frac{\sum\limits_{x_j \in C_i} dis(x_i, x_j)}{|C_j|} \right\}. \tag{8}$$

Here x_i is the object of C_i, $p(x_i)$ is the mean distance between x_i and the other objects in C_i, $q(x_i)$ is the minimal average distance between x_i and all point of other clusters. $c(x_i)$ is the Silhouette coefficient of object x_i. So the Silhouette coefficient of the data set U is defined as:

$$SC = \frac{\sum\limits_{x_i \in U, j=1}^{n} c(x_j)}{n}. \tag{9}$$

The larger the SC, the better the clustering result.

(3) Davies-Bouldin Index

The DB index [9] is proposed by Davies, which is the ratio of compactness to separation. It is defined as:

$$DB = \frac{1}{k} * \sum_{i=1}^{k} \max_{j \neq 1} (\frac{\overline{C_i} + \overline{C_j}}{\|w_i - w_j\|}). \tag{10}$$

where $\overline{C_i}$ and $\overline{C_j}$ are the average within-group distance of the ith and the jth clusters. Respectively, $\|w_i - w_j\|$ is the inter-group distance between these clusters, where $\|\|$ is a norm (e.g. Euclidean).

The smaller the value of DB, the better the clustering result. To achieve a same monotonicity as other, the paper takes a negative operate on the DB index.

(4) Calinski-Harabasz index

The CH index [9] is also the ratio of compactness to separation. It is defined as:

$$CH = \frac{Tr(S_B)/K - 1}{Tr(S_W)/n - K}. \tag{11}$$

The parameters of $Tr(S_B)$ and $Tr(S_W)$ are defined as:

$$Tr(S_B) = \sum_{i=1}^{k} n_i \cdot d(v_i, \bar{v}), Tr(S_w) = \sum_{i=1}^{K} \sum_{j=1}^{n} d(x_j, v_i). \qquad (12)$$

Here n_i is the number of objects assigned to the ith cluster. v_i is the center of the ith cluster, $\bar{v}$ is the center of the whole data set.

The larger the CH is, the better the result is.
(5) XB index

XB index [14] is proposed by Xie et al., which is defined as:

$$XB = \frac{\sum_{i=1}^{K} \sum_{j=1}^{n} \mu_{ij}^{m} d(x_j, v_i)}{n * \min_{i \neq j} d(v_i, v_j)}. \qquad (13)$$

where μ_{ij} represents the probability that object x_j belongs to the ith cluster. m is the ambiguity factor, which is a weight index used to determine the ambiguity of the clustering result.

The smaller the XB index, the better performance of the clustering result. In order to achieve a same monotonicity as other indexes, this paper takes a negative improvement on the XB index.
(6) PBM index

Another criterion, named PBM [9], is defined as:

$$PBM = \left(\frac{1}{K} * \frac{\sum_{i=1}^{n} d(x_i, v_1)}{\sum_{j=1}^{K} \sum_{x_i \in C_j} d(x_i, v_j)} * \max_{i,j=1,2,\cdots,K} d(v_i, v_j) \right)^2. \qquad (14)$$

The larger the PBM is, the better performance of the cluster result is.
(7) Normalized Mutual Information

Normalized Mutual Information (NMI) [12] is used to measure the anastomosing degree of clustering result π_K and π_L. It is defined as:

$$NMI(\pi_K, \pi_L) = \frac{\sum_{k=1}^{K} \sum_{l=1}^{L} n_l^k log(\frac{n n_l^k}{n^k n^l})}{\sqrt{(\sum_{k=1}^{K} n^k log(\frac{n^k}{n})) * (\sum_{l=1}^{L} n^l log(\frac{n^l}{n}))}}. \qquad (15)$$

Here n^k represents the number of objects in the kth cluster generated by the clustering algorithm, n^l represents the number of objects in the lth cluster in the real cluster partition, n_k^l is the number of objects owned by the kth cluster of the generated result and lth cluster of the real partition.

The larger the NMI value, the greater the similarity between the clustering result and the real division.

3   The Proposed Method

In this section, the method to determine the number of cluster and its four components are described.

3.1   The Framework

The proposed automatic method to determine the number of clusters based on multi-validity indexes is shown in Fig. 1, which is consist of four parts: (1) the set of clustering result is generated by different algorithms or some algorithm with different parameters; (2) multiple validity criteria are selected to evaluate those clustering results; (3) the weighting method is introduced to measure the average similarity between the clustering result with some k value and the others; and (4) the selecting strategy of the best clustering numbers.

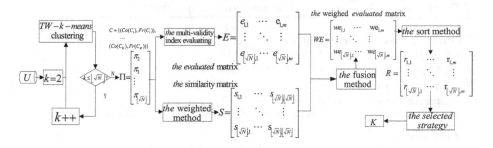

Fig. 1. The framework to determine the number of clusters automatically

3.2   The Three-Way K-Means Clustering Algorithm

The Three-way k-means (TW-k-means, for short) algorithm is improved by the traditional k-means by which Three-way clustering results can be calculated. To determine the thresholds of α and β automatically, a dynamic method is also proposed in this paper. Firstly, the data set is divided by the conventional k-means algorithm. Secondly, the $\alpha[i]$, which is the threshold of the ith cluster, is defined as $d_{ave}[i]$, which is the mean distance of the center of C_i $(1 \leq i \leq K)$ and the other object in C_i. Thirdly, The $\beta[i]$ of the ith cluster is defined as the sum of $d_{ave}[i]$ and $d_{far}[i]$ the farthest distance of the center of C_i and the other objects in C_i. If $dist[i][j](1 \leq i \leq N, 1 \leq j \leq \left\lfloor \sqrt{N} \right\rfloor)$ is the distance of the

ith object and the center of the jth cluster. Naturally, we have the three-way decision rules as follows.

$$if \; dist[i][j] < \alpha[j], \; then \; x_i \; is \; assigned \; to \; Co(C_k);$$
$$if \; \alpha[j] \leq dist[i][j] < \beta[j], \; then \; x_i \; is \; assigned \; to \; Fr(C_k); \qquad (16)$$
$$if \; dist[i][j] \geq \beta[j], \; then \; x_i \; is \; assigned \; to \; Tr(C_k).$$

TW-k-means clustering algorithm is summarized by algorithm 1.

For a data set, the optimal number of clusters does not exceed $\sqrt{N}$ and the traversal ranges of the number of cluster k is $\left[2, \left\lfloor \sqrt{N} \right\rfloor \right]$. Therefore, this paper runs the TW-k-means algorithm with different k from 2 to $\left\lfloor \sqrt{N} \right\rfloor$ and the corresponding results are inserted into Π. The generation is summarised in Algorithm 2.

3.3   Selecting Relative Criteria

The six indexes in Sect. 2 are employed to construct the system. The clustering results with different values of k are evaluated by the six selected indexes and the evaluation matrix $\mathbf{E}_{(\lfloor \sqrt{N} \rfloor -1) \times m}$ is constructed, in which the element E_{ij}

Algorithm 1. The three way k-means clustering algorithm

Input: data set $X = \{x_1, \cdots, x_n, \cdots, x_N\}$, the number of clusters T
Output: the clustering result $\pi_T = \{[\underline{C_1}, \overline{C_1}], \cdots, [\underline{C_t}, \overline{C_t}], \cdots, [\underline{C_T}, \overline{C_T}]\}$
1 **for** *the first T objects* **do**
   $\lfloor$ treat them as the starting cluster centers $v_1, v_2, \cdots, v_T$

2 **for** *each object x_i* **do**
   $\quad$ **for** *each cluster center v_j* **do**
   $\quad\quad \lfloor$ calculate the distance of x_i and v_j, classify the x_i to the nearest cluster

3 **for** *each cluster C_i* **do**
   $\quad$ update the cluster centers $v = \{v^1, v^2, \cdots, v^i, \cdots, v^T\}$
   $\quad$ **if** *the cluster centers don't change* **then**
   $\quad\quad \lfloor$ run to step 4
   $\quad$ **else**
   $\quad\quad \lfloor$ run to step 2

4 **for** *each cluster C_i* **do**
   $\quad$ calculate $d_{ave}[i]$ the distance of all objects to the cluster center and $d_{far}[i]$
   $\quad$ the farthest distance of all objects to the cluster center, based on $d_{ave}[i]$
   $\quad$ and $d_{far}[i]$, calculate $\alpha[i]$ and $\beta[i]$

5 **for** *each object x_i* **do**
   $\quad$ **for** *each cluster center v_j* **do**
   $\quad\quad$ **if** $dist[i][j] < \alpha[j]$, *then x_i is assigned to* $Co(C_k); if \; \alpha[j] \leq dist[i][j] <$
   $\quad\quad \beta[j]$, *then x_i is assigned to* $Fr(C_k); if \; dist[i][j] \geq$
   $\quad\quad \beta[j]$, *then x_i is assigned to* $Tr(C_k)$.

Algorithm 2. The generation of Π

Input: data set $X = \{x_1, \cdots, x_n, \cdots, x_N\}$

Output: the set of clustering results $\Pi = \left\{\pi_2, \cdots, \pi_T, \cdots, \pi_{\lfloor\sqrt{N}\rfloor}\right\}$

for *each k from 2 to* $\left\lfloor\sqrt{N}\right\rfloor$ **do**

    run the TW-k-means algorithm

    insert the result π_k into Π

return $\Pi = \left\{\pi_2, \cdots, \pi_T, \cdots, \pi_{\lfloor\sqrt{N}\rfloor}\right\}$

Algorithm 3. The generation of $\mathbf{E}_{(\lfloor\sqrt{N}\rfloor-1)\times m}$

Input: the set of clustering results $\Pi = \left\{\pi_2, \cdots, \pi_T, \cdots, \pi_{\lfloor\sqrt{N}\rfloor}\right\}$

Output: the evaluated matrix $\mathbf{E}_{(\lfloor\sqrt{N}\rfloor-1)\times m}$

for $i=2$ **to** m **do**

    for $j=2$ **to** $\left\lfloor\sqrt{N}\right\rfloor$ **do**

        $E_{ji}=index_i(\pi_j)$ insert E_{ji} into $\mathbf{E}_{(\lfloor\sqrt{N}\rfloor-1)\times m}$

return $\mathbf{E}_{(\lfloor\sqrt{N}\rfloor-1)\times m}$

---

represents the evaluated result of the $k = i + 2$ under the jth index. Here, $\left\lfloor\sqrt{N}\right\rfloor - 1$ means the number of rows,m means the number of columns. The generation of $\mathbf{E}_{(\lfloor\sqrt{N}\rfloor-1)\times m}$ is summarised in Algorithm 3.

3.4   The Weighed Method Based on Median Partition

In order to select the optimal k from multiple angles, based on the idea of the median partition method, every clustering result is took as the natural result. So the NMI index can be used to measure the similarity between each two clustering results. The mean similarity, stored in the similarity matrix $\mathbf{S}_{\lfloor\sqrt{N}\rfloor-1}(\left\lfloor\sqrt{N}\right\rfloor - 1$ means the number of rows), is took as the weight of the corresponding clustering result and its calculation is shown as follows.

$$S(\pi_i, \Pi) = \frac{\sum_{j=2}^{\sqrt{N}} NMI(\pi_i, \pi_j)}{\left\lfloor\sqrt{N}\right\rfloor - 1} \tag{17}$$

The calculated process of the mean similarity is summarised in Algorithm 4.

Algorithm 4. The generation of $\mathbf{S}_{\lfloor\sqrt{N}\rfloor-1}$

Input: the set of clustering results $\Pi = \left\{\pi_2, \cdots, \pi_T, \cdots, \pi_{\lfloor\sqrt{N}\rfloor}\right\}$
Output: the similarity matrix $\mathbf{S}_{\lfloor\sqrt{N}\rfloor-1}$

for $i=2$ to $\left\lfloor\sqrt{N}\right\rfloor$ **do**

    for $j=2$ to $\left\lfloor\sqrt{N}\right\rfloor$ **do**

        $S_i = S_i + NMI(\pi_i, \pi_j)$

    $S_i = S_i \big/ N^{1}\!/2$ insert the $\mathbf{S}_{\lfloor\sqrt{N}\rfloor-1}$

return $\mathbf{S}_{\lfloor\sqrt{N}\rfloor-1}$

3.5 Choosing the Optimal Number

The evaluation matrix $\mathbf{E}_{(\lfloor\sqrt{N}\rfloor-1)\times m}$ is merged with the similarity matrix $\mathbf{S}_{\lfloor\sqrt{N}\rfloor-1}$. The fusion result is recorded in the weighted evaluation matrix $\mathbf{WE}_{(\lfloor\sqrt{N}\rfloor-1)\times m}$. Here, $\left\lfloor\sqrt{N}\right\rfloor - 1$ means the number of rows, m means the number of columns. The fusion formula is shown as follows:

$$WE_{ij} = E_{ij} * S_i. \tag{18}$$

In order to describe the ranks of the clustering results' quality with different value of k, the function Rank is defined. If the clustering result π_i with k = i in the clustering results set Π is the best clustering result under the u-th weighted evaluation index of the matrix WE, which is written as $R\,(index_u, \pi_m, \Pi) = 1$. Therefore, the weighted evaluation matrix could be transformed to the rank matrix $\mathbf{R}_{(\lfloor\sqrt{N}\rfloor-1)\times m}$. The transformation formula is defined as follows:

$$R_{ij} = R(index_j, \pi_i, \Pi). \tag{19}$$

Based on the rank matrix $\mathbf{R}_{(\lfloor\sqrt{N}\rfloor-1)\times m}$, the acceptance matrix $\mathbf{AC}_{(\lfloor\sqrt{N}\rfloor-1)}$ can be constructed. The transformation formula is defined as follows:

$$AC_i = \sum_{j=1}^{m} R_{ij}. \tag{20}$$

The optional k is corresponding to the clustering result with the highest degree of acceptance. That is, the smallest AC value can be selected as the optional k.

4 Experimental Results

In this section, we have carried out a number of experiments on different real data set to validate the effectiveness of the proposed method.

In order to show the method of multi-validity index (MVI, for short) proposed in this paper is valuable, the six indexes used in the process of constructing the evaluated system, such as the Silhouette index, the DB index and the other indexes, are tested respectively on the real data sets. The performances of the fusion method and the single index in predicting the number of clusters of different data sets are recorded in Table 1. In the table, some information of the data sets is also given, such as the second row of Size means the number of objects, the third row of Clusters denotes the right or natural number of clusters of the corresponding data set. Letter1 (A, B, C) is a subset from the original corresponding UCI data sets.

Table 1. Comparison of experimental results

Datasets	IRIS	Wine	Seeds	Letter(A, B, C)	Ecoli	Vowel
Size	150	178	210	300	336	528
Cluster	3	3	3	3	8	11
Silhouete	9	10	14	16	17	31
Dunn	3	7	12	15	3	2
CH	5	11	13	3	4	26
PBM	6	5	5	2	2	3
DB	9	2	2	6	2	2
XB	2	2	2	3	2	2
MVI	3	4	5	3	7	9

By observing the Table 1, we can find the results of the multi-validity index are much better than the results of the other single indexes. In another words, the performance of the multi-validity index is more accurate and stable to determine the number of clusters of the data sets automatically. To be specific, for a certain validity index, its performance is slightly accurate on some data sets, but on the other data sets it is dissatisfying. For example, the performance of PBM on the data of IRIS, Wine, Seeds, Letter(A, B, C) is more accurate than the data set of Ecoli and Vowel. This phenomenon can prove the starting point of multi-validity index that single validity index is bias to specific condition. However, for some data set, for example the data set of Vowel, the performance of all the single indexes referred in this pater is not so good. On one hand, the performances of single indexes can be divided into 2 classes, the Silhouete's prediction of 31 and the CH's 26 in a class, the other's predictions are range from 2 to 3 and they are in another class. Obviously, the prediction of MVI on the data set of Vowel is more close to the natural clusters of 11. The huge progress of the accuracy on predicting the number of clusters can reflect the rationality of the method proposed in the paper that predict the number of the natural number from multiple aspects.

Since, the time complexity of k-means algorithm is no more than $O(N * k * t)$, here, N is the number of objects, k is the number of clusters. The method proposed in this paper should run iteratively the k-means algorithm $N^{\frac{1}{2}}$ times and the clustering results should also be evaluated by the six single indexes. So the time consumption is somewhat large, but the performance is hard to say perfect, which can prove, on some degree, that the work of determining the number of clusters is really a difficult task and we still have a long way to go.

5   Conclusions

The determination of the number of clusters has always been a question in the research of clustering. Based on the idea of three-way decision, this paper apply the core region, the fringe region and the trivial region to deal with the uncertainty of the objects in the process of partition. The TW-k-means algorithm is run repeatedly with different k from 2 to $\lfloor \sqrt{N} \rfloor$. Aiming at avoiding the bias of single validity index in selecting clustering results, multiple evaluation indexes are employed to construct the multi-validity index, which can evaluate the clustering results with different k values from multiple perspectives. Drawing on the idea of median partition in clustering ensemble, the similarity between each two clustering results is calculated separately, according to which the evaluated results of each k value will be weighted. Finally, the weighted evaluated results will be sorted by the proposed rank function in each column and the optimal k value can be calculated. The experimental results show that the performance of the multi-validity index is better at selecting the optimal k value. Even though, we still need to improve the time consumption of this algorithm and some other aspects so that the method can be used to determine the number of clusters of large-scale data.

Acknowledgments. This work was supported in part by the National Natural Science Foundation of China under Grant Nos. 61533020, 61751312 and 61379114.

References

1. Azimi, R., Ghayekhloo, M., Ghofrani, M., et al.: A novel clustering algorithm based on data transformation approaches. Expert Syst. Appl. Int. J. **76**(C), 59–70 (2017)
2. Chen, H.P., Shen, X.J., Lv, Y.D.: A novel automatic fuzzy clustering algorithm based on soft partition and membership information. Neurocomputing **236**, 104–112 (2016)
3. Cristofor, D., Simovici, D.: Finding median partitions using information-theoretical-based genetic algorithms. J. Univers. Comput. Sci. **8**(2), 153–172 (2002)
4. Gionis, A., Mannila, H., Tsaparas, P.: Clustering aggregation. In: International Conference on Data Engineering, 2005, ICDE 2005. Proceedings. IEEE, pp. 341–352 (2005)
5. Huang, D., Wang, C., Lai, J., et al.: Clustering ensemble by decision weighting. JCAAI Trans. Intell. Syst. **11**(3), 418–424 (2016)

6. Jaskowiak, P.A., Moulavi, D., Furtado, A.C.S.: On strategies for building effective ensembles of relative clustering validity criteria. Knowl. Inf. Syst. **47**(2), 329–354 (2016)
7. Ling, H.L., Wu, J.S., Zhou, Y., et al.: How many clusters? A robust PSO-based local density model. Neurocomputing **207**(C), 264–275 (2016)
8. Mok, P.Y., Huang, H.Q., Kwok, Y.L.: A robust adaptive clustering analysis method for automatic identification of clusters. Pattern Recogn. **45**(8), 3017–3033 (2012)
9. Naldi, M.C., Carvalho, A.C., Campello, R.J.: Cluster ensemble selection based on relative validity indexes. Data Min. Knowl. Discov. **27**(2), 259–289 (2013)
10. Singhbiostatistics, V.: Ensemble clustering using semidefiniteprogramming. Mach. Learn. **79**(1–2), 177–200 (2008)
11. Vega-Pons, S., Avesani, P.: On pruning the search space for clustering ensemble problems. Neurocomputing **150**(1), 481–489 (2015)
12. Yangtao, W., Lihui, C., Jianping, M.: Incremental fuzzy clustering with multiple medoids for large data. IEEE Trans. Fuzzy Syst. **22**(6), 1557–1568 (2014)
13. Wu, X., Kumar, V., Quinlan, J.R.: Top 10 algorithms in data mining. Knowl. Inf. Syst. **14**(1), 1–37 (2007)
14. Xie, X.L., Beni, G.: A validity measure for fuzzy clustering. IEEE Trans. Pami **13**(13), 841–847 (1991)
15. Yao, Y.Y.: Three-way decisions with probabilistic rough sets. Inf. Sci. **180**(3), 341–353 (2010)
16. Yu, H., Liu, Z., Wang, G.: An automatic method to determine the number of clusters using decision-theoretic rough set. Int. J. Approximate Reasoning **55**(1), 101–115 (2014)

Algebras from Semiconcepts
in Rough Set Theory

Prosenjit Howlader and Mohua Banerjee$^{(\boxtimes)}$

Department of Mathematics and Statistics, Indian Institute of Technology,
Kanpur 208016, India
{prosen,mohua}@iitk.ac.in

Abstract. In this article, we propose a notion of a semiconcept in the framework of Yao's object oriented concepts. A study of the algebra of such 'object oriented semiconcepts' is carried out, in the line of the study by Wille for the algebra of semiconcepts in formal concept analysis. Two further unary operators, 'semi-topological' in nature, are introduced on these structures. On abstraction, the properties of these operators lead to the definition of a 'semi-topological double Boolean algebra', of which the algebra of object oriented semiconcepts becomes an instance.

Keywords: Formal concept analysis · Rough sets
Object oriented concepts

1 Introduction

Rough set theory [10] and formal concept analysis (FCA) [4] provide two related methodologies for data analysis. Both investigate the notion of concepts, albeit from different perspectives. Classical rough set theory is developed based on an equivalence relation on a domain of objects. Generalized formulations have been proposed by using a binary relation on two domains, one a set of objects and the other a set of properties – such a binary relation on two domains is called a *formal context* in FCA. Many efforts have been made to compare and combine the two theories [1,3,5,6,9,19].

The central notion in FCA is that of a *concept lattice* on a context $\mathbb{K}$, denoted $\mathfrak{B}(\mathbb{K})$. Düntsch and Gediga, and Yao introduced two kinds of 'rough concept lattices' in rough set theory, based on operators defined in [2]. The former defined *property oriented concept lattices* [1], and Yao proposed *object oriented concept lattices* [17]. Yao also studied the relationship between these two kinds of rough concept lattices and concept lattices of FCA in [17]. It is shown that object oriented concept lattices are dually isomorphic to concept lattices, while property oriented concept lattices are isomorphic to concept lattices. Further algebraic properties of rough concept lattices were investigated in [16].

P. Howlader—This work is supported by the *Council of Scientific and Industrial Research* (CSIR) India - Research Grant No. 09/092(0950)/2016-EMR-I.

H. S. Nguyen et al. (Eds.): IJCRS 2018, LNAI 11103, pp. 440–454, 2018.
https://doi.org/10.1007/978-3-319-99368-3_34

There is also a study of logic in the direction of FCA [13,14]. To formulate what is called *contextual logic*, 'negation of a concept' has to be formalized and Boole's correspondence between negation and set-complement is taken as a basis for the purpose. However, there turns out to be a problem of closure if set-complement is used to define negation of a formal concept. So the latter notion is generalized successively to that of *semiconcept, protoconcept* and *preconcept* [13,15]. Our interest also lies in defining a negation, in the context of rough concepts. This article does so in the framework of Yao's object oriented concepts. We define *object oriented semiconcepts* in Sect. 3, and follow the line of study in [13]. An algebraic structure is developed on the set $\mathfrak{S}(\mathbb{K})$ of all object oriented semiconcepts. We show that it forms a dual of *double Boolean algebra* [13], and contains two special Boolean subalgebras. In Sect. 4, two further unary operators are defined on $\mathfrak{S}(\mathbb{K})$, which turn out to be 'semi-topological' [11] in nature. The properties of these operators lead us to define a *semi-topological double Boolean algebra*, of which $\mathfrak{S}(\mathbb{K})$ becomes an instance.

Considering Boole's correspondence mentioned above, Wille defined another (weak) negation in [13], which can be generated by the negations defined on semiconcepts. This operator gives rise to a 'concept algebra', the abstraction of which is a 'dicomplemented lattice'. In [8], weakly dicomplemented lattices are defined which constitute a superclass of the class of dicomplemented lattices. In Sect. 4.1, we show that weakly dicomplemented lattices are different from the double Boolean algebras considered in this work. Section 5 concludes the article.

In the next section, we give the preliminaries required for the work presented in the rest of the paper.

2   Preliminaries

Definition 1 [4]. *A formal context is a triple* $\mathbb{K} := (G, M, R)$, *where* G, M *are sets of objects and properties respectively, and* $R \subseteq G \times M$. *gRm is interpreted as object g has property m. For $A \subseteq G$ and $B \subseteq M$,*

$$A' := \{m \in M \mid gRm \, for \, all \, g \in A\},$$
$$B' := \{g \in G \mid gRm \, for \, all \, m \in B\}.$$

A concept of $\mathbb{K}$ *is defined to be a pair* (A, B) *where* $A \subseteq G$, $B \subseteq M$, $A' = B$ *and* $B' = A$. *A is called the* extent *and B the* intent *of the concept* (A, B). *The set of all concepts of* $\mathbb{K}$ *is denoted by* $\mathcal{B}(\mathbb{K})$.
For concepts (A_1, B_1) *and* (A_2, B_2) *in* $\mathbb{K}$ *an order is defined as:*

$$(A_1, B_1) \leq (A_2, B_2) \, if \, and \, only \, if \, A_1 \leq A_2.$$

$(\mathfrak{B}(\mathbb{K}), \leq)$ *forms a complete lattice, and is called the* concept lattice *of* $\mathbb{K}$.

Definition 2 [1]. *For a formal context* $\mathbb{K} := (G, M, R)$, $\mathbb{K}^c := (G, M, -R)$, *is called a* complement *of* $\mathbb{K}$, *where* $-R = \{(x, y) \in G \times M : (x, y) \notin R\}$.

Example 1 [15]. The following table gives an example of a formal context. Objects are family members, properties are genders and age variables (Table 1).

Table 1. A formal context

	Male (Ma)	Female (Fe)	Old	Young
Father (Fa)	*		*	
Mother (Mo)		*	*	
Son (So)	*			*
Daughter (Da)		*		*

2.1 Semiconcept Algebra

As mentioned in Sect. 1, there is a problem of closure if set-complement is used to define negation of a formal concept. More explicitly, if (A, B) is a formal concept in a context (G, M, R), the complement $G \setminus A$ $(M \setminus B)$ of the extent (intent) A (B) may not be an extent (intent). The notion of formal concept was then generalized by defining a *semiconcept*.

Definition 3 [12]. *A* semiconcept *of a formal context* $\mathbb{K} := (G, M, R)$ *is defined as a pair* (A, B) *with* $A \subseteq G$ *and* $B \subseteq M$ *such that* $A = B^{'}$ *or* $B = A^{'}$.

The set of all semiconcepts of $\mathbb{K}$ is denoted by $\mathfrak{H}(\mathbb{K})$. The following algebraic operations $\sqcap, \sqcup, \neg, \lrcorner, \bot$ and $\top$ are introduced on $\mathfrak{H}(\mathbb{K})$:

$$(A_1, B_1) \sqcap (A_2, B_2) := (A_1 \cap A_2, (A_1 \cap A_2)^{'})$$
$$(A_1, B_1) \sqcup (A_2, B_2) := ((B_1 \cap B_2)^{'}, B_1 \cap B_2)$$
$$\neg(A, B) := (G \setminus A, (G \setminus A)^{'})$$
$$\lrcorner(A, B) := ((M \setminus B)^{'}, M \setminus B)$$
$$\top := (G, \phi)$$
$$\bot := (\phi, M)$$

$\mathfrak{H}(\mathbb{K})$ with the operations $\sqcap, \sqcup, \neg, \lrcorner, \bot$ and $\top$ is called the *algebra of semiconcepts* of $\mathbb{K}$, and denoted by $\underline{\mathfrak{H}}(\mathbb{K})$. The following sets of idempotent elements are considered, and shown to form Boolean algebras in [12,13]:

$$\mathfrak{H}_{\sqcap} := \{(A, A^{'}) \in \mathfrak{H}(\mathbb{K}) : A \subseteq G\} \text{ and } \mathfrak{H}_{\sqcup} := \{(B^{'}, B) \in \mathfrak{H}(\mathbb{K}) : B \subseteq M\}.$$

2.2 Object Oriented Concept Lattice

Let G and M be two non-empty sets, and $R \subseteq G \times M$ be a relation. For each $x \in G$, the R-range of x is $R(x) := \{y \in M : xRy\}$. The converse R^0 of R is $R^0 := \{(y, x) \in M \times G : xRy\}$.

For a given formal context $\mathbb{K} := (G, M, R)$, $^{\square}, ^{\diamond} : 2^G \to 2^M$ constitute a pair of dual approximation operators defined as:

$$X^{\diamond} := \{y \in M : X \bigcap R^0(y) \neq \emptyset\}, \qquad X^{\square} := \{y \in M : R^0(y) \subseteq X\}.$$

On the other hand, $\square, \lozenge : 2^M \to 2^G$ constitute another pair of dual approximation operators defined as:

$$Y^\lozenge := \{x \in G : Y \cap R(x) \neq \emptyset\}, \qquad Y^\square := \{x \in G : R(x) \subseteq Y\}.$$

$\lozenge$ is called the possibility operator and $\square$ the necessity operator. Note that if we take $G = M$ and R to be an equivalence relation on G then the $\lozenge, \square$ operators coincide respectively with the upper and lower approximation operators (on the approximation space (G, R)) of rough set theory.

Now we list some properties of $\lozenge, \square$. For proof we refer to [1,17,18].

Proposition 1. *Let* $\mathbb{K} := (G, M, R)$ *be a context. For any* $X, X_1, X_2 \subseteq G$ *and* $Y, Y_1, Y_2 \subseteq M$, *the following hold.*

1. $G^\square = M$ *and* $\phi^\lozenge = \phi$.
2. $M^\lozenge = G$ *if and only if* $R(x) \neq \phi$ *for all* $x \in G$.
3. $\phi^\square = \phi$ *if and only if* $R^0(y) \neq \phi$ *for all* $y \in M$.
4. *if* $X_1 \subseteq X_2$ *then* $X_1^\square \subseteq X_2^\square$ *and* $X_1^\lozenge \subseteq X_2^\lozenge$.
5. *if* $Y_1 \subseteq Y_2$ *then* $Y_1^\square \subseteq Y_2^\square$ *and* $Y_1^\lozenge \subseteq Y_2^\lozenge$.
6. $X^{\square\lozenge} \subseteq X \subseteq X^{\lozenge\square}$ *and* $Y^{\square\lozenge} \subseteq Y \subseteq Y^{\lozenge\square}$.
7. $(X)_R^\square = (X^c)'_{-R}$ *and* $(Y)_R^\square = (Y^c)'_{-R}$.
8. $X^{c\square} = X^{\lozenge c}$ *and* $Y^{c\square} = Y^{\lozenge c}$.
9. $X^{\square c} = X^{c\lozenge}$ *and* $Y^{\square c} = Y^{c\lozenge}$.
10. $(X \cap Y)^\square = X^\square \cap Y^\square$.
11. $(X \cup Y)^\lozenge = X^\lozenge \cup Y^\lozenge$.
12. $(X \cap Y)^{\square\lozenge} \subseteq X^{\square\lozenge} \cap Y^{\square\lozenge}$ *and* $X^{\lozenge\square} \cup Y^{\lozenge\square} \subseteq (X \cup Y)^{\lozenge\square}$.
13. $X^{\square\lozenge\square} = X^\square$ *and* $Y^{\lozenge\square\lozenge} = Y^\lozenge$.
14. $X^{\lozenge\square\lozenge} = X^\lozenge$ *and* $Y^{\lozenge\square\lozenge} = Y^\lozenge$.

Proposition 2.

1. $\square\lozenge$ *mapping* X *to* $X^{\lozenge\square}$, *is a closure operator.*
2. $\lozenge\square$ *mapping* X *to* $X^{\square\lozenge}$, *is an interior operator.*

For a given set of objects $A \subseteq G$, the map $\square : 2^G \to 2^M$ assigns to it a set of properties $A^\square$, while the map $\lozenge : 2^M \to 2^G$ assigns to a set of properties $B \subseteq M$, an object set $B^\lozenge$. For special pairs (A, B), we have the following.

Definition 4 [17,18]. *An object oriented concept of the context* $\mathbb{K}$ *is defined as a pair* (A, B) *with* $A \subseteq G$, $B \subseteq M$ *such that* $A^\square = B$ *and* $B^\lozenge = A$. A *is the* extent *and* B *the* intent *of the object oriented concept* (A, B). *The set of all object oriented concepts of* $\mathbb{K}$ *is denoted by* $RO-L(\mathbb{K})$.

With this definition, it is shown in [18] that object oriented concepts are described by disjunctions of properties, whereas formal concepts are described by conjunctions of properties. The two theories together can thus give a more complete picture of data.

An order is defined on the set $RO - L(\mathbb{K})$ of object oriented concepts:

$(A_1, B_1) \leq (A_2, B_2)$ if and only if $A_1 \subseteq A_2$ (which is equivalent to $B_1 \subseteq B_2$). $(RO - L(\mathbb{K}), \leq)$ forms a complete lattice. Moreover, we have

Theorem 1 [16]. $RO-L(\mathbb{K})$ *is dually isomorphic to* $\underline{\mathfrak{B}}(\mathbb{K}^c)$.

3   Object Oriented Semiconcept

We are interested to study the notion of negation in the context of rough concepts. In [13], Wille studied a negation in FCA, by separately negating the extent and intent of a concept, using set-complement. In this work, we consider object oriented concepts and introduce negation using Wille's approach. For a given object oriented concept we also have two negations, one by taking the complement of its extent and the other by taking the complement of its intent.

Example 2. Let us continue with the context given in Example 1. In Tables 2 and 3 below, we list for all $A \subseteq G, B \subseteq M$ respectively, $A^\square, A^{\square\lozenge}$ and $B^\lozenge, B^{\lozenge\square}$.

Table 2. Subsets A of G giving object oriented semiconcepts $(A, A^\square)$

$A \subseteq G$	$A^\square$	$A^{\square\lozenge}$
ϕ	ϕ	ϕ
$\{Fa\}$	ϕ	ϕ
$\{Mo\}$	ϕ	ϕ
$\{so\}$	ϕ	ϕ
$\{Da\}$	ϕ	ϕ
$\{Fa, Mo\}$	$\{old\}$	$\{Fa, Mo\}$
$\{Fa, So\}$	$\{Ma\}$	$\{Fa, So\}$
$\{Fa, Da\}$	ϕ	ϕ
$\{Mo, So\}$	ϕ	ϕ
$\{Mo, Da\}$	$\{Fe\}$	$\{Mo, Da\}$
$\{So, Da\}$	$\{Young\}$	$\{So, Da\}$
$\{Fa, Mo, So\}$	$\{Old, Ma\}$	$\{Fa, Mo, So\}$
$\{Fa, Mo, Da\}$	$\{Fe, Old\}$	$\{Fa, Mo, Da\}$
$\{Mo, So, Da\}$	$\{Young, Fe\}$	$\{Mo, So, Da\}$
$\{Fa, So, Da\}$	$\{Ma, Young\}$	$\{Fa, So, Da\}$
G	M	G

Consider the pair $(\{Mo, So, Da\}, \{Young, Fe\})$. It is clear from Table 2 that it is an object oriented concept. The complement of the extent $A = \{Mo, So, Da\}$ is $\{Fa\}$, which is not the extent of any object oriented concept of $\mathbb{K}$ (cf. Table 3). Now consider the pair $(\{So, Da\}, \{Young\})$, which is also an object oriented concept. The complement of the intent $B = \{Young\}$ is $C = \{Ma, Fe, Old\}$ and C is not the intent of any object oriented concept of $\mathbb{K}$. Analogous to the situation in FCA, simply taking the set-complement of the extent or intent of an object oriented concept, may not result in an object oriented concept. One then relaxes the requirement to consider pairs of the form $(A^c, A^{c\square})$ and $(B^{c\lozenge}, B^c)$

Table 3. Subsets B of M giving Object oriented Semiconcepts $(B^\diamond, B)$

$B \subseteq M$	$B^\diamond$	$B^{\diamond\square}$
ϕ	ϕ	ϕ
$\{old\}$	$\{Fa, Mo\}$	$\{old\}$
$\{Ma\}$	$\{Fa, So\}$	$\{Ma\}$
$\{Fe\}$	$\{Mo, Da\}$	$\{Fe\}$
$\{Young\}$	$\{So, Da\}$	$\{Young\}$
$\{Old, Ma\}$	$\{Fa, Mo, So\}$	$\{Old, Ma\}$
$\{Fe, Old\}$	$\{Fa, Mo, Da\}$	$\{Fe, Old\}$
$\{Young, Fe\}$	$\{Mo, So, Da\}$	$\{Young, Fe\}$
$\{Ma, Young\}$	$\{Fa, So, Da\}$	$\{Ma, Young\}$
$\{Ma, Fe\}$	G	M
$\{Old, Young\}$	G	M
$\{Ma, Fe, Old\}$	G	M
$\{Ma, Old, Young\}$	G	M
$\{Ma, Fe, Young\}$	G	M
$\{Fe, Old, Young\}$	G	M
M	G	M

to define negation, as $A^{c\square}$ collects properties of the objects of A^c only, while $B^{c\diamond}$ contains all objects that have properties belonging to B^c. (Note that these pairs still need not be concepts, as we shall see in an example below). This idea is generalized to give the definition of an *object oriented semiconcept*.

Definition 5. *Let* $\mathbb{K} := (G, M, R)$ *be a formal context. An* object oriented semi-concept *of* $\mathbb{K}$ *is defined as a pair* (A, B) *with* $A \subseteq G, B \subseteq M$ *such that* $A^\square = B$ *or* $B^\diamond = A$. *The set of all object oriented semiconcepts of* $\mathbb{K}$ *is denoted by* $\mathfrak{S}(\mathbb{K})$.

Thus object oriented semiconcepts of $\mathbb{K}$ are pairs of the form $(A, A^\square)$ or $(B^\diamond, B)$. Tables 2 and 3 in Example 2 give us all the object oriented semiconcepts of the context (G, M, R). It may be then observed that an object oriented semiconcept may not always be an object oriented concept: $(\{Fa\}, \phi)$ is an object oriented semiconcept but not an object oriented concept.

Now is there any relation between semiconcepts of FCA and object oriented semiconcepts defined above? The answer is given by

Proposition 3. *For a context* $\mathbb{K}$, $(A, B) \in \mathfrak{S}(\mathbb{K})$ *if and only if* $(A^c, B) \in \mathfrak{H}(\mathbb{K}^c)$, *the set of all semiconcepts of the complement of the context* $\mathbb{K}$.

3.1   Algebra of Object Oriented Semiconcepts

An order $\leq$ and algebraic operations $\sqcap, \sqcup, \neg, \lrcorner, \top$ and $\bot$ are considered on $\mathfrak{S}(\mathbb{K})$:

Definition 6. *For* $(A_1, B_1), (A_2, B_2) \in \mathfrak{S}(\mathbb{K})$,

(a) $(A_1, B_1) \leq (A_2, B_2)$ *if and only if* $A_1 \subseteq A_2$ *and* $B_1 \subseteq B_2$,
(b) $(A_1, B_1) \sqcap (A_2, B_2) := ((B_1 \cap B_2)^{\Diamond}, B_1 \cap B_2)$,
(c) $(A_1, B_1) \sqcup (A_2, B_2) := (A_1 \cup A_2, (A_1 \cup A_2)^{\square})$,
(d) $\neg(A, B) := (G \setminus A, (G \setminus A)^{\square})$,
(e) $\lrcorner(A, B) := ((M \setminus B)^{\Diamond}, M \setminus B)$,
(f) $\top := (G, M)$,
(g) $\bot := (\phi, \phi)$.

The meet ($\sqcap$) and join ($\sqcup$) operations taken in $RO{-}L(\mathbb{K})$ are extended to $\mathfrak{S}(\mathbb{K})$. It is clear from Definition 5 and Proposition 1(1) that $\mathfrak{S}(\mathbb{K})$ is closed with respect to all the operations defined above. The tuple $(\mathfrak{S}(\mathbb{K}), \sqcap, \sqcup, \neg, \lrcorner, \top, \bot)$ is called the *algebra of object oriented semiconcepts* of $\mathbb{K}$ and is denoted by $\underline{\mathfrak{S}}(\mathbb{K})$.

Proposition 4. $(A_1, B_1) \sqcap (A_2, B_2)$ *is a lower bound of* (A_1, B_1) *and* (A_2, B_2), *and* $(A_1, B_1) \sqcup (A_2, B_2)$ *is an upper bound of* (A_1, B_1) *and* (A_2, B_2) *in* $(\mathfrak{S}(\mathbb{K}), \leq)$.

Proof. $(A_1, B_1) \sqcap (A_2, B_2) := ((B_1 \cap B_2)^{\Diamond}, B_1 \cap B_2)$ and $(A_1, B_1) \sqcup (A_2, B_2) := (A_1 \cup A_2, (A_1 \cup A_2)^{\square})$. We have the following cases.

Case I: Suppose $A_1 = B_1^{\Diamond}$ and $A_2 = B_2^{\Diamond}$. Then $(B_1 \cap B_2)^{\Diamond} \subseteq B_1^{\Diamond} = A_1$ and $(B_1 \cap B_2)^{\Diamond} \subseteq B_2^{\Diamond} = A_2$ by (5) of Proposition 1. Now $(A_1 \cup A_2)^{\square} = (B_1^{\Diamond} \cup B_2^{\Diamond})^{\square}$. Using Proposition 1(11) on the rhs, we have $(A_1 \cup A_2)^{\square} = (B_1 \cup B_2)^{\Diamond\square}$ and using Proposition 1(6), we have $B_1, B_2 \subseteq (B_1 \cup B_2)^{\Diamond\square} = (A_1 \cup A_2)^{\square}$.
Case II: $A_1^{\square} = B_1$ and $A_2^{\square} = B_2$. This case is dealt similarly by replacing $\square$ with $\Diamond$ as Case I.
Case III: Now let $A_1^{\square} = B_1$ and $A_2 = B_2^{\Diamond}$. We have $(B_1 \cap B_2)^{\Diamond} \subseteq A_1^{\square\Diamond} \subseteq A_1$ and $(B_1 \cap B_2)^{\Diamond} \subseteq B_2^{\Diamond} = A_2$, using Proposition 1(5) and (6). From Proposition 1(4) and (6), we have $B_1 = A_1^{\square} \subseteq (A_1 \cup A_2)^{\square}$ and $B_2 \subseteq B_2^{\Diamond\square} \subseteq (A_1 \cup A_2)^{\square}$. $\square$

Are these the greatest and least upper bounds? Not necessarily so. In Example 2, consider the two elements $(\{Mo, Da\}, \{Fe\})$ and $(\{So, Da\}, \{Young\})$ in $\mathfrak{S}(\mathbb{K})$. $(\{Mo, Da\}, \{Fe\}) \sqcap (\{So, Da\}, \{Young\}) = (\phi, \phi)$ is a lower bound but is not the greatest lower bound as $(\{Da\}, \phi)$ is also a lower bound of the two object oriented semiconcepts. On the other hand, we can consider $(\{Fa\}, \phi)$ and $(\{Mo, Fa\}, \phi)$, for which $(\{Fa\}, \phi) \sqcup (\{Mo, Fa\}, \phi) = (\{Fa, Mo\}, \{old\})$, which is an upper bound but not least as $(\{Mo, Fa\}, \phi)$ is an upper bound of the two object oriented semiconcepts.

Following the approach of Wille, we now consider the set of idempotent elements in $\underline{\mathfrak{S}}(\mathbb{K})$ with respect to the operations $\sqcup$ and $\sqcap$.

$$\mathfrak{S}(\mathbb{K})_{\sqcup} := \{(A, B) \in \mathfrak{S}(\mathbb{K}) : (A, B) \sqcup (A, B) = (A, B)\}, \text{ and}$$
$$\mathfrak{S}(\mathbb{K})_{\sqcap} := \{(A, B) \in \mathfrak{S}(\mathbb{K}) : (A, B) \sqcap (A, B) = (A, B)\}.$$

It can be easily observed that

$$\mathfrak{S}(\mathbb{K})_{\sqcup} = \{(A, B) \in \mathfrak{S}(\mathbb{K}) : (A, A^{\square}) = (A, B)\} = \{(A, A^{\square}) : A \subseteq G\}, \text{ and}$$
$$\mathfrak{S}(\mathbb{K})_{\sqcap} = \{(A, B) \in \mathfrak{S}(\mathbb{K}) : (B^{\Diamond}, B) = (A, B)\} = \{(B^{\Diamond}, B) : B \subseteq M\}.$$

Note: For two object oriented semiconcepts $(A_1, B_1), (A_2, B_2)$, if the pair with componentwise set-theoretic intersection, viz. $(A_1 \cap A_2, B_1 \cap B_2)$, belongs to $\mathfrak{S}(\mathbb{K})$ then it must be the greatest lower bound of $(A_1, B_1), (A_2, B_2)$. A similar observation holds for $(A_1 \cup A_2, B_1 \cup B_2)$ and least upper bound of $(A_1, B_1), (A_2, B_2)$.

We obtain in a straightforward manner, the following results for any context $\mathbb{K}$.

Proposition 5.

1. $\mathfrak{S}(\mathbb{K})_\sqcap \cap \mathfrak{S}(\mathbb{K})_\sqcup = RO - L(\mathbb{K})$.
2. $\mathfrak{S}(\mathbb{K})_\sqcap \cup \mathfrak{S}(\mathbb{K})_\sqcup = \mathfrak{S}(\mathbb{K})$.
3. $(A_1, B_1) \sqcap (A_2, B_2) = (A_1, B_1) \sqcap (A_1, B_1)$ and $(A_1, B_1) \sqcup (A_2, B_2) = (A_2, B_2) \sqcup (A_2, B_2)$ if and only if $(A_1, B_1) \leq (A_2, B_2)$.

As done for semiconcepts, we define two operations on $\underline{\mathfrak{S}}(\mathbb{K})$:

$$x \vee y := \lrcorner(\lrcorner x \sqcap \lrcorner y), \text{ and } x \wedge y := \neg(\neg x \sqcup \neg y), \text{ for all } x, y \in \underline{\mathfrak{S}}(\mathbb{K}).$$

Theorem 2. *The following equations are valid in $\underline{\mathfrak{S}}(\mathbb{K})$:*

$(1a)$ $(x \sqcap x) \sqcap y = x \sqcap y$ $(1b)$ $(x \sqcup x) \sqcup y = x \sqcup y$

$(2a)$ $x \sqcap y = y \sqcap x$ $(2b)$ $x \sqcup y = y \sqcup x$

$(3a)$ $x \sqcap (y \sqcap z) = (x \sqcap y) \sqcap z$ $(3b)$ $x \sqcup (y \sqcup z) = (x \sqcup y) \sqcup z$

$(4a)$ $\lrcorner(x \sqcap x) = \lrcorner x$ $(4b)$ $\neg(x \sqcup x) = \neg x$

$(5a)$ $x \sqcap (x \sqcup y) = x \sqcap x$ $(5b)$ $x \sqcup (x \sqcap y) = x \sqcup x$

$(6a)$ $x \sqcap (y \vee z) = (x \sqcap y) \vee (x \sqcap z)$ $(6b)$ $x \sqcup (y \wedge z) = (x \sqcup y) \wedge (x \sqcup z)$

$(7a)$ $x \sqcap (x \vee y) = x \sqcap x$ $(7b)$ $x \sqcup (x \wedge y) = x \sqcup x$

$(8a)$ $\lrcorner\lrcorner(x \sqcap y) = x \sqcap y$ $(8b)$ $\neg\neg(x \sqcup y) = x \sqcup y$

$(9a)$ $x \sqcap \lrcorner x = \bot$ $(9b)$ $x \sqcup \neg x = \top$

$(10a)$ $\lrcorner\bot = \top \sqcap \top$ $(10b)$ $\neg\top = \bot \sqcup \bot$

$(11a)$ $\neg\bot = \top$ $(11b)$ $\lrcorner\top = \bot$

(12) $(x \sqcap x) \sqcup (x \sqcap x) = (x \sqcup x) \sqcap (x \sqcup x)$.

Observe that the equations stated in Theorem 2 are dual with respect to $\sqcup$ and $\sqcap$ in the equations defining a double Boolean algebra [13].

In our next result, we prove that $\underline{\mathfrak{S}}(\mathbb{K})$ is dually isomorphic to $\underline{\mathfrak{H}}(\mathbb{K}^c)$. In other words, we show the following for the algebraic structure $\underline{\mathfrak{H}}^\partial(\mathbb{K}^c)$ that is obtained from $\underline{\mathfrak{H}}(\mathbb{K}^c)$ by replacing $\sqcap$ with $\sqcup$ and $\sqcup$ with $\sqcap$.

Theorem 3. *For a context $\mathbb{K}$, $\underline{\mathfrak{S}}(\mathbb{K})$ is isomorphic to $\underline{\mathfrak{H}}^\partial(\mathbb{K}^c)$.*

Proof. We define a map $h : \mathfrak{S}(\mathbb{K}) \to \mathfrak{H}(\mathbb{K}^c)$ such that $h((A, B)) := (A^c, B)$, where $(A, B) \in \mathfrak{S}(\mathbb{K})$. This map is well-defined and onto by Proposition 3. It is trivially one-one. To show h is a homomorphism, we check the case for $\sqcap$.

$$
\begin{aligned}
h((A,B) \sqcap (A_1, B_1)) &= h((B \cap B_1)^\Diamond, B \cap B_1) \\
&= ((B \cap B_1)^{\Diamond c}, B \cap B_1) \\
&= ((B \cap B_1)'_{-R}, B \cap B_1) \quad \text{(by Proposition 1(4) and (5))} \\
&= (A^c, B) \sqcup (A_1^c, B_1) \text{ in } \mathfrak{H}(\mathbb{K}^c) \\
&= (A^c, B) \sqcap (A_1^c, B_1) \text{ in } \mathfrak{H}^\partial(\mathbb{K}^c) \\
&= h((A,B)) \sqcap h((A_1, B_1))
\end{aligned}
$$

$h((G, M)) = (\phi, M) = \bot$, which is the top element of $\mathfrak{H}^\partial(\mathbb{K}^c)$ and $h((\phi, \phi)) = (G, \phi) = \top$, the bottom element of $\mathfrak{H}^\partial(\mathbb{K}^c)$. The case for $\sqcup$ is similar. $\qquad\square$

Recall the algebras of idempotent elements of semiconcepts defined in Sect. 2.1. We get the following relationships.

Corollary 1.

1. $\mathfrak{S}(\mathbb{K})_\sqcap$ is dually isomorphic to $\mathfrak{H}(\mathbb{K}^c)_\sqcup$.
2. $\mathfrak{S}(\mathbb{K})_\sqcup$ is dually isomorphic to $\mathfrak{H}(\mathbb{K}^c)_\sqcap$.

Proof. (1) Let $(A,B) \in \mathfrak{S}(\mathbb{K})$ then $(A^c, B) \in \mathfrak{H}(\mathbb{K}^c)$. Using definitions of $\sqcap, \sqcup$ in algebras of object oriented semiconcepts and semiconcepts respectively, we have
$$(A,B) \sqcap (A,B) = (B^\Diamond, B) \text{ and } (A^c, B) \sqcup (A^c, B) = (B'_{-R}, B).$$
Therefore $(A,B) = (A,B) \sqcap (A,B)$ if and only if $(A,B) = (B^\Diamond, B)$, i.e. if and only if $A^c = B^{\Diamond c}$.

On the other hand, $(A^c, B) = (A^c, B) \sqcup (A^c, B)$ if and only if $(A^c, B) = (B'_{-R}, B)$, i.e. if and only if $A^c = B'_{-R}$. From Proposition 1(7), we have $B^{\Diamond c} = B'_{-R}$ and hence $(A,B) \sqcap (A,B) = (A,B)$ if and only if $(A^c, B) \sqcup (A^c, B) = (A^c, B)$. Similarly one can show that $(A,B) \sqcup (A,B) = (A,B)$ if and only if $(A^c, B) \sqcap (A^c, B) = (A^c, B)$. Therefore image of $\mathfrak{S}(\mathbb{K})_\sqcap$ under h defined in Theorem 3 is equal to $\mathfrak{H}(\mathbb{K}^c)_\sqcup$ and it is also clear that h is an isomorphism from $\mathfrak{S}(\mathbb{K})_\sqcap$ to $\mathfrak{H}^\partial(\mathbb{K}^c)_\sqcup$.
Proof of (2) is similar. $\qquad\square$

4   Semi-topological Operators on $\mathfrak{S}(\mathbb{K})$

Rough concept analysis deals with the necessity and possibility operators $\square$ and $\Diamond$. As mentioned in Proposition 2, $\square\Diamond$ is a closure operator and $\Diamond\square$ is an interior operator. We use this idea and define two unary operators C, I on the set $\mathfrak{S}(\mathbb{K})$ of object oriented semiconcepts. As we shall see, the two operators turn out to have *semi-topological* properties [11].

Definition 7. *For any* $(A, B) \in \mathfrak{S}(\mathbb{K})$,

$$
\begin{aligned}
C((A,B)) &:= (A^{\Diamond\square}, A^{\Diamond\square\square}), \\
I((A,B)) &:= (B^{\square\Diamond\Diamond}, B^{\square\Diamond}).
\end{aligned}
$$

Note. Using the algebraic operations on object oriented semiconcepts, we get for any $x \in \mathfrak{S}(\mathbb{K})$, $C(x) = \neg(\neg x \sqcap \neg x)$, and $I(x) = \lrcorner(\llcorner x \sqcup \lrcorner x)$.

Lemma 1. *Let* $x, y \in \mathfrak{S}(\mathbb{K})$. I *has the following properties.*

1. *If* $x \leq y$ *then* $I(x) \leq I(y)$.
2. $II(x) = I(x)$.
3. $I(x) \sqcap x = I(x) = I(x) \sqcap I(x)$ *and* $x \sqcup I(x) = x \sqcup x$.
4. $I(\top) = \lrcorner(\bot \sqcup \bot)$.
5. $I(x \sqcap y) \leq I(x) \sqcap I(y)$.

Proof. (1) Let $x, y \in \mathfrak{S}(\mathbb{K}) = \mathfrak{S}(\mathbb{K})_\sqcup \cup \mathfrak{S}(\mathbb{K})_\sqcap$ such that $x \leq y$.

Case I: Suppose $x, y \in \mathfrak{S}(\mathbb{K})_\sqcap$. Without loss of generality, we assume that $x = (A^\diamond, A)$ and $y = (B^\diamond, B)$ where $A, B \subseteq M$. Then
$$I(x) = \lrcorner(\lrcorner(A^\diamond, A) \sqcup \llcorner(A^\diamond, A)) = \lrcorner(A^{c\diamond}, A^{c\diamond\square}) = (A^{c\diamond\square c\diamond}, A^{c\diamond\square c}).$$
Similarly we deduce that $I(y) = (B^{c\diamond\square c\diamond}, B^{c\diamond\square c})$. Now $x \leq y$ implies that $A \subseteq B$, which implies that $A^{c\diamond\square c} \subseteq B^{c\diamond\square c}$ and from this we have $A^{c\diamond\square c\diamond} \subseteq B^{c\diamond\square c\diamond}$ and hence $I(x) \leq I(y)$.

Case II: If $x, y \in \mathfrak{S}(\mathbb{K})_\sqcup$ then let $x = (A, A^\square)$ and $y = (B, B^\square)$, $A, B \subseteq G$. Then
$$I(x) = (A^{\square c\diamond\square c\diamond}, A^{\square c\diamond\square c}) \text{ and } I(y) = (B^{\square c\diamond\square c\diamond}, B^{\square c\diamond\square c}).$$
As $x \leq y$, $A^\square \subseteq B^\square$, which implies that $A^{\square c\diamond\square c} \subseteq B^{\square c\diamond\square c}$. So $A^{\square c\diamond\square c\diamond} \subseteq B^{\square c\diamond\square c\diamond}$ and we get $I(x) \leq I(y)$.

Case III: If $x \in \mathfrak{S}(\mathbb{K})_\sqcup$ and $y \in \mathfrak{S}(\mathbb{K})_\sqcap$, we assume that $x = (A, A^\square)$ and $y = (B^\diamond, B)$. Then $I(x) = (A^{\square c\diamond\square c\diamond}, A^{\square c\diamond\square c})$ and $I(y) = (B^{c\diamond\square c\diamond}, B^{c\diamond\square c})$. $x \leq y$ implies that $A^\square \subseteq B$, which gives $A^{\square c\diamond\square c} \subseteq B^{c\diamond\square c}$. From this we have $A^{\square c\diamond\square c\diamond} \subseteq B^{c\diamond\square c\diamond}$ and hence $I(x) \leq I(y)$.

(2) Let $x \in \mathfrak{S}(\mathbb{K})$.

$$
\begin{aligned}
I(I(x)) &= \lrcorner(\lrcorner\lrcorner(\llcorner x \sqcup \lrcorner x) \sqcup \lrcorner\lrcorner(\llcorner x \sqcup \lrcorner x)) \\
&= \lrcorner(((\llcorner x \sqcup \lrcorner x) \sqcap (\llcorner x \sqcup \lrcorner x)) \sqcup ((\llcorner x \sqcup \lrcorner x) \sqcap (\llcorner x \sqcup \lrcorner x))) \\
&= \lrcorner(((\llcorner x \sqcup \lrcorner x) \sqcup (\llcorner x \sqcup \lrcorner x)) \sqcap ((\llcorner x \sqcup \lrcorner x) \sqcup (\llcorner x \sqcup \lrcorner x))) \\
&= \lrcorner((\llcorner x \sqcup \lrcorner x) \sqcap (\llcorner x \sqcup \lrcorner x)) \\
&= \lrcorner(\llcorner x \sqcup \lrcorner x) = I(x).
\end{aligned}
$$

(3) Let $x \in \mathfrak{S}(\mathbb{K}) = \mathfrak{S}(\mathbb{K})_\sqcup \cup \mathfrak{S}(\mathbb{K})_\sqcap$.

Case I: Let $x \in \mathfrak{S}(\mathbb{K})_\sqcup$. Without loss of generality we assume that $x = (A, A^\square)$, for some $A \subseteq G$. $I(x) = (A^{\square c\diamond\square c\diamond}, A^{\square c\diamond\square c})$ and from this we get $I(x) \sqcap x = (A^{\square c\diamond\square c\diamond}, A^{\square c\diamond\square c}) \sqcap (A, A^\square) = ((A^{\square c\diamond\square c} \cap A^\square)^\diamond, A^{\square c\diamond\square c} \cap A^\square)$. Now $A^{\square c} \subseteq A^{\square c\diamond\square}$ for any subset A of G. So $A^{\square c\diamond\square c} \subseteq A^\square$ and hence $I(x) \sqcap x = (A^{\square c\diamond\square c\diamond}, A^{\square c\diamond\square c}) = I(x)$.

Case II: If $x \in \mathfrak{S}(\mathbb{K})_\sqcap$, let $x = (B^\diamond, B)$ for some $B \subseteq M$. Then $I(x) = (B^{c\diamond\square c\diamond}, B^{c\diamond\square c})$ whence $I(x) \sqcap x = (B^{c\diamond\square c\diamond}, B^{c\diamond\square c}) \sqcap (B^\diamond, B) = ((B^{c\diamond\square c} \cap B)^\diamond, (B^{c\diamond\square c} \cap B)) = (B^{c\diamond\square c\diamond}, B^{c\diamond\square c}) = I(x)$, as $B^{c\diamond\square c} \subseteq B$.

Since for any $x \in \mathfrak{S}(\mathbb{K})$ say $x = (A, B)$, $I(x) = (B^{c\diamond\square c\diamond}, B^{c\diamond\square c}) = (D^\diamond, D)$, where $D = B^{c\diamond\square c}$, we have $I(x) \in \mathfrak{S}(\mathbb{K})_\sqcap$ for all $x \in \mathfrak{S}(\mathbb{K})$). Thus $I(x) \sqcap I(x) = I(x)$ and so $I(x) \sqcap x = I(x) = I(x) \sqcap I(x)$.

Now we will show that $x \sqcup I(x) = x \sqcup x$. Let $x \in \mathfrak{S}(\mathbb{K}) = \mathfrak{S}(\mathbb{K})_\sqcup \cup \mathfrak{S}(\mathbb{K})_\sqcap$.

Case I: If $x \in \mathfrak{S}(\mathbb{K})_\sqcup$, say $x = (A, A^\square)$ for some $A \subseteq G$.

$$
\begin{aligned}
(A, A^\square) \sqcup I((A, A^\square)) &= (A, A^\square) \sqcup (A^{\square c \diamond \square c \diamond}, A^{\square c \diamond \square c}) \\
&= (A \cup A^{\square c \diamond \square c \diamond}, (A \cup A^{\square c \diamond \square c \diamond})^\square) \\
&= (A, A^\square) \text{ because } A^{\square c \diamond \square c \diamond} \subseteq A^{\square \diamond} \subseteq A \\
&= (A, A^\square) \sqcup (A, A^\square) = x \sqcup x.
\end{aligned}
$$

Case II: If $x \in \mathfrak{S}(\mathbb{K})_\sqcap$, let $x = (B^\diamond, B)$ for some $B \subseteq M$.

$$
\begin{aligned}
(B^\diamond, B) \sqcup I((B^\diamond, B)) &= (B^\diamond, B) \sqcup (B^{c \diamond \square c \diamond}, B^{c \diamond \square c}) \\
&= (B^\diamond \cup B^{c \diamond \square c \diamond}, (B^\diamond \sqcup B^{c \diamond \square c \diamond})^\square) \\
&= (B^\diamond, B^{\diamond \square}) \text{ because } B^{c \diamond \square c} \subseteq B \\
&= (B^\diamond, B) \sqcup (B^\diamond, B) = x \sqcup x.
\end{aligned}
$$

(4) $I(\top) = \lrcorner(\lrcorner \top \sqcup \lrcorner \top) = \lrcorner(\bot \sqcup \bot)$.

(5) Let $x, y \in \mathfrak{S}(\mathbb{K}) = \mathfrak{S}(\mathbb{K})_\sqcap \cup \mathfrak{S}(\mathbb{K})_\sqcup$.

Case I: Let $x \in \mathfrak{S}(\mathbb{K})_\sqcup$ and $y \in \mathfrak{S}(\mathbb{K})_\sqcap$. Without loss of generality we assume that $x = (A, A^\square)$ and $y = (B^\diamond, B)$, where $A \subseteq G$ and $B \subseteq M$. Then $I(x) = (A^{\square c \diamond \square c \diamond}, A^{\square c \diamond \square c})$ and $I(y) = (B^{c \diamond \square c \diamond}, B^{c \diamond \square c})$.

$I(x) \sqcap I(y) = ((A^{\square c \diamond \square c} \cap B^{c \diamond \square c})^\diamond, A^{\square c \diamond \square c} \cap B^{c \diamond \square c})$ and $I(x \sqcap y) = I((A^\square \cap B)^\diamond, A^\square \cap B) = ((A^\square \cap B)^{c \diamond \square c \diamond}, (A^\square \cap B)^{c \diamond \square c})$. Now $A^{\square c} \subseteq (A^\square \cap B)^c$ and $B^c \subseteq (A^\square \cap B)^c$. From this inequality we have, $A^{\square c \diamond \square} \subseteq (A^\square \cap B)^{c \diamond \square}$ and $B^{c \diamond \square} \subseteq (A^\square \cap B)^{c \diamond \square}$. This implies that $(A^\square \cap B)^{c \diamond \square c} \subseteq A^{\square c \diamond \square c}$ and $(A^\square \cap B)^{c \diamond \square c} \subseteq B^{c \diamond \square c}$. So $(A^\square \cap B)^{c \diamond \square c} \subseteq A^{\square c \diamond \square c} \cap B^{c \diamond \square c}$ and $(A^\square \cap B)^{c \diamond \square c \diamond} \subseteq (A^{\square c \diamond \square c} \cap B^{c \diamond \square c})^\diamond$ and hence $I(x \sqcap y) \leq I(x) \sqcap I(y)$.

Case II: If $x, y \in \mathfrak{S}(\mathbb{K})_\sqcup$, let us assume that $x = (A, A^\square)$ and $y = (B, B^\square)$.

Then $I(x) \sqcap I(y) = ((A^{\square c \diamond \square c \diamond} \cap B^{\square c \diamond \square c \diamond})^\diamond, A^{\square c \diamond \square c} \cap B^{\square c \diamond \square c})$ and $I(x \sqcap y) = I((A \cap B)^{\square \diamond}, (A \cap B)^\square) = ((A \cap B)^{\square c \diamond \square c \diamond}, (A \cap B)^{\square c \diamond \square c})$. Now $(A \cap B)^\square \subseteq A^\square$ and $(A \cap B)^\square \subseteq B^\square$. From this we have,

$$
A^{\square c} \subseteq (A \cap B)^{\square c} \Rightarrow A^{\square c \diamond \square} \subseteq (A \cap B)^{\square c \diamond \square}
$$
$$
\Rightarrow (A \cap B)^{\square c \diamond \square c} \subseteq A^{\square c \diamond \square c}.
$$

Similarly, one can prove that $(A \cap B)^{\square c \diamond \square c} \subseteq B^{\square c \diamond \square c}$. From this inequality we have $(A \cap B)^{\square c \diamond \square c} \subseteq A^{\square c \diamond \square c} \cap B^{\square c \diamond \square c}$ and $(A \cap B)^{\square c \diamond \square c \diamond} \subseteq (A^{\square c \diamond \square c} \cap B^{\square c \diamond \square c})^\diamond$. Hence $I(x \sqcap y) \leq I(x) \sqcap I(y)$.

Case III: If $x, y \in \mathfrak{S}(\mathbb{K})_\sqcap$, the proof is similar to Case II.   □

Dually, one can prove the following for the operator C on $\mathfrak{S}(\mathbb{K})$.

Lemma 2. *For all $x, y \in \mathfrak{S}(\mathbb{K})$,*

1. *If $x \leq y$ then $C(x) \leq C(y)$*
2. *$CC(x) = C(x)$*
3. *$C(x) \sqcup x = C(x) = C(x) \sqcup C(x)$ and $x \sqcap C(x) = x \sqcap x$*
4. *$C(\bot) = \neg(\top \sqcap \top)$*
5. *$C(x) \sqcup C(y) \leq C(x \sqcup y)$.*

4.1   Semi-topological Double Boolean Algebra

Recall our observation after Theorem 2 that $\mathfrak{S}(\mathbb{K})$ satisfies the dual of all equations defining a double Boolean algebra [13]. In this section, we deal with such an abstract 'dual double Boolean algebra', and for the sake of simplicity, retain the name double Boolean algebra for the structure. More precisely, we have the following definition.

Definition 8. *A double Boolean algebra* $(A, \sqcup, \sqcap, \neg, \lrcorner, \top, \bot)$ *is an abstract algebra which satisfies the following properties: For any* $x, y, z \in A$.

(1a) $(x \sqcap x) \sqcap y = x \sqcap y$ *(1b)* $(x \sqcup x) \sqcup y = x \sqcup y$

(2a) $x \sqcap y = y \sqcap x$ *(2b)* $x \sqcup y = y \sqcup x$

(3a) $x \sqcap (y \sqcap z) = (x \sqcap y) \sqcap z$ *(3b)* $x \sqcup (y \sqcup z) = (x \sqcup y) \sqcup z$

(4a) $\lrcorner(x \sqcap x) = \lrcorner x$ *(4b)* $\neg(x \sqcup x) = \neg x$

(5a) $x \sqcap (x \sqcup y) = x \sqcap x$ *(5b)* $x \sqcup (x \sqcap y) = x \sqcup x$

(6a) $x \sqcap (y \vee z) = (x \sqcap y) \vee (x \sqcap z)$ *(6b)* $x \sqcup (y \wedge z) = (x \sqcup y) \wedge (x \sqcup z)$

(7a) $x \sqcap (x \vee y) = x \sqcap x$ *(7b)* $x \sqcup (x \wedge y) = x \sqcup x$

(8a) $\lrcorner\lrcorner(x \sqcap y) = x \sqcap y$ *(8b)* $\neg\neg(x \sqcup y) = x \sqcup y$

(9a) $x \sqcap \lrcorner x = \bot$ *(9b)* $x \sqcup \neg x = \top$

(10a) $\lrcorner\bot = \top \sqcap \top$ *(10b)* $\neg\top = \bot \sqcup \bot$

(11a) $\neg\bot = \top$ *(11b)* $\lrcorner\top = \bot$

(12) $(x \sqcap x) \sqcup (x \sqcap x) = (x \sqcup x) \sqcap (x \sqcup x)$,

where $\vee$ *and* $\wedge$ *are defined as* $x \vee y := \lrcorner(\lrcorner x \sqcap \lrcorner y)$, *and* $x \wedge y := \neg(\neg x \sqcup \neg y)$.
$\neg$ *is called the* negation *and* $\lrcorner$ *the* opposition.

Corollary 2. $\mathfrak{S}(\mathbb{K})$ *is a double Boolean algebra.*

A quasi-order (reflexive and transitive relation) on a double Boolean algebra may be defined [7] for all $x, y \in \mathbf{A}$ as:

$$x \sqsubseteq y \text{ if and only if } x \sqcap y = x \sqcap x \text{ and } x \sqcup y = y \sqcup y.$$

Remark. As we mentioned in Sect. 1, the algebraic structure of a weakly dicomplemented lattice [8,15] also emerged in the context of defining negations in FCA. We now compare this structure with the double Boolean algebra of Definition 8. Note that these are algebras of the same type (2,2,1,1,0,0). However, it can be seen that these are different with respect to the defining axioms. Firstly, in a weakly dicomplemented lattice $(L, \vee, \wedge, ^{\triangle}, ^{\triangledown}, 1, 0)$, the reduct $(L, \vee, \wedge, 1, 0)$ is a *lattice*, while a double Boolean algebra $(A, \sqcup, \sqcap, \neg, \lrcorner, \top, \bot)$ need not be a lattice with respect to the $\sqcup, \sqcap$ operations, as shown in Sect. 3.1. Secondly, to force another comparison, suppose the lattice meet and join in a weakly dicomplemented lattice are relaxed to be lower and upper bound operations $\sqcup, \sqcap$ satisfying the axioms 1a-b, 2a-b, 3a-b, 5a-b and 12 in Definition 8. The remaining defining axioms of the negations $^{\triangle}, ^{\triangledown}$ (cf. [8]) in a weakly dicomplemented lattice are retained. Will a double Boolean algebra then be a special case of such a structure? We find that the negations in the two structures behave differently as well. In particular, it can be shown that the axiom $(x \wedge y) \vee (x \wedge y^{\triangle}) = x$ for $^{\triangle}$ need not hold in a double Boolean algebra, irrespective of whether $^{\triangle}$ is taken as the negation ($\neg$) or opposition ($\lrcorner$) of the double Boolean algebra. Indeed, consider Example 1: take two object oriented semiconcepts $x := (\{Fa, Da\}, \phi)$ and $y := (\phi, \phi)$. $(x \sqcap y) \sqcup (x \sqcap \neg y) = (\phi, \phi) \neq x$. If we take $x := (G, \{Ma, Fe\})$ and $y := (\{Fa\}, \phi)$ then $(x \sqcap y) \sqcup (x \sqcap \lrcorner y) = (G, M) \neq x$. On the other hand, if we force $\sqcap, \sqcup$ in a double Boolean algebra to be infimum and supremum operators respectively, we get the equations $\lrcorner \lrcorner x = x$ and $\neg \neg x = x$ from $(8a)$ and $(8b)$ of Definition 8. However, these do not hold in general for the negations $^{\triangle}, ^{\triangledown}$ in a weakly dicomplemented lattice, so that the latter is not an example of such a special case of a double Boolean algebra either.

Now we define a semi-topological double Boolean algebra.

Definition 9. *A* semi-topological double Boolean algebra *is an abstract algebra* $\boldsymbol{A} := (A, \sqcup, \sqcap, \neg, \lrcorner, \top, \bot, \boldsymbol{I}, \boldsymbol{C})$, *where* $(A, \sqcup, \sqcap, \neg, \lrcorner, \top, \bot)$ *is a double Boolean algebra, and the unary operators* $\boldsymbol{I}$ *and* $\boldsymbol{C}$ *satisfy the following equations for any* $x, y \in A$.

$(sa)^1$ $\boldsymbol{I}(x) \sqcap x = \boldsymbol{I}(x) \sqcap \boldsymbol{I}(x)$ *and*
   $x \sqcup \boldsymbol{I}(x) = x \sqcup x$
$(sa)^2$ $\boldsymbol{I}(x \sqcap y) \sqsubseteq \boldsymbol{I}(x) \sqcap \boldsymbol{I}(y)$
$(sa)^3$ $\boldsymbol{I}(\boldsymbol{I}(x)) = \boldsymbol{I}(x)$

$(sb)^1$ $\boldsymbol{C}(x) \sqcup x = \boldsymbol{C}(x) \sqcup \boldsymbol{C}(x)$ *and*
   $\boldsymbol{C}(x) \sqcap x = x \sqcap x$
$(sb)^2$ $\boldsymbol{C}(x) \sqcup \boldsymbol{C}(y) \sqsubseteq \boldsymbol{C}(x \sqcup y)$
$(sb)^3$ $\boldsymbol{C}(\boldsymbol{C}(x)) = \boldsymbol{C}(x)$

Theorem 4. $\underline{\mathfrak{S}}(\mathbb{K}) := (\mathfrak{S}(\mathbb{K}, \sqcup, \sqcap, \neg, \lrcorner, \top, \bot, I_1, C_1)$ *is a semi-topological double Boolean algebra.*

Proof. Follows from Theorem 2 and Lemmas 1, 2.   □

5   Conclusion

This work introduces the notion of negation in the framework of object oriented concepts in rough concept analysis, and object oriented semiconcepts are defined. The algebra that these semiconcepts form is shown to be (a dual of) double Boolean algebra. Moreover, two unary operators are introduced in this algebra, leading to the definition of a semi-topological double Boolean algebra.

The proposal opens up several directions of further work, including possible applications. The definition of a new algebraic structure warrants some immediate algebraic investigations, such as investigation for representation theorems. Definition of a negation can now facilitate studies in the direction of contextual logic for rough sets. Besides, one can follow up the entire study in the framework of property oriented concepts.

Acknowledgments. We are grateful to the anonymous referees for their suggestions and valuable remarks.

References

1. Düntsch, I., Gediga, G.: Modal-style operators in qualitative data analysis. In: Proceedings of the IEEE International Conference on Data Mining, pp. 155–162 (2002). https://doi.org/10.1109/ICDM.2002.1183898
2. Düntsch, I., Gediga, G.: Approximation operators in qualitative data analysis. In: de Swart, H., Orłowska, E., Schmidt, G., Roubens, M. (eds.) Theory and Applications of Relational Structures as Knowledge Instruments. LNCS, vol. 2929, pp. 214–230. Springer, Heidelberg (2003). https://doi.org/10.1007/978-3-540-24615-2_10
3. Ganter, B., Meschke, C.: A formal concept analysis approach to rough data tables. In: Peters, J.F., et al. (eds.) Transactions on Rough Sets XIV. LNCS, vol. 6600, pp. 37–61. Springer, Heidelberg (2011). https://doi.org/10.1007/978-3-642-21563-6_3
4. Ganter, B., Wille, R.: Formal Concept Analysis: Mathematical Foundations. Springer, Heidelberg (2012)
5. Hu, K., Sui, Y., Lu, Y., Wang, J., Shi, C.: Concept approximation in concept lattice. In: Cheung, D., Williams, G.J., Li, Q. (eds.) PAKDD 2001. LNCS (LNAI), vol. 2035, pp. 167–173. Springer, Heidelberg (2001). https://doi.org/10.1007/3-540-45357-1_21
6. Kent, R.E.: Rough concept analysis. In: Ziarko, W.P. (ed.) Rough Sets, Fuzzy Sets and Knowledge Discovery, pp. 248–255. Springer, London (1994). https://doi.org/10.1007/978-1-4471-3238-7_30
7. Kwuida, L.: Prime ideal theorem for double Boolean algebras. Discussiones Math. Gen. Algebra Appl. **27**(2), 263–275 (2007). https://doi.org/10.7151/dmgaa.1130
8. Kwuida, L., Pech, C., Reppe, H.: Generalizations of Boolean algebras. An attribute exploration. Math. Slovaca **56**(2), 145–165 (2006)
9. Meschke, C.: Approximations in concept lattices. In: Kwuida, L., Sertkaya, B. (eds.) ICFCA 2010. LNCS (LNAI), vol. 5986, pp. 104–123. Springer, Heidelberg (2010). https://doi.org/10.1007/978-3-642-11928-6_8
10. Pawlak, Z.: Rough sets. Int. J. Comput. Inf. Sci. **11**(5), 341–356 (1982). https://doi.org/10.1007/BF01001956
11. Peleg, D.: A generalized closure and complement phenomenon. Discrete Math. **50**, 285–293 (1984). https://doi.org/10.1016/0012-365X(84)90055-4
12. Wille, R.: Concept lattices and conceptual knowledge systems. Comput. Math. Appl. **23**(6), 493–515 (1992). https://doi.org/10.1016/0898-1221(92)90120-7
13. Wille, R.: Boolean concept logic. In: Ganter, B., Mineau, G.W. (eds.) ICCS-ConceptStruct 2000. LNCS (LNAI), vol. 1867, pp. 317–331. Springer, Heidelberg (2000). https://doi.org/10.1007/10722280_22
14. Wille, R.: Boolean judgment logic. In: Delugach, H.S., Stumme, G. (eds.) ICCS-ConceptStruct 2001. LNCS (LNAI), vol. 2120, pp. 115–128. Springer, Heidelberg (2001). https://doi.org/10.1007/3-540-44583-8_9
15. Wille, R.: Preconcept algebras and generalized double boolean algebras. In: Eklund, P. (ed.) ICFCA 2004. LNCS (LNAI), vol. 2961, pp. 1–13. Springer, Heidelberg (2004). https://doi.org/10.1007/978-3-540-24651-0_1
16. Yang, L., Xu, L.: On rough concept lattices. Electron. Notes Theor. Comput. Sci. **257**, 117–133 (2009). https://doi.org/10.1016/j.entcs.2009.11.030
17. Yao, Y.: Concept lattices in rough set theory. Proce. Ann. Meet. North Am. Fuzzy Inf. Process. Soc. **2**, 796–801 (2004). https://doi.org/10.1109/NAFIPS.2004.1337404

18. Yao, Y.: A comparative study of formal concept analysis and rough set theory in data analysis. In: Tsumoto, S., Słowiński, R., Komorowski, J., Grzymała-Busse, J.W. (eds.) RSCTC 2004. LNCS (LNAI), vol. 3066, pp. 59–68. Springer, Heidelberg (2004). https://doi.org/10.1007/978-3-540-25929-9_6
19. Yao, Y., Chen, Y.: Rough set approximations in formal concept analysis. In: Peters, J.F., Skowron, A. (eds.) Transactions on Rough Sets V. LNCS, vol. 4100, pp. 285–305. Springer, Heidelberg (2006). https://doi.org/10.1007/11847465_14

Introducing Dynamic Structures of Rough Sets. The Case of Text Processing: Anaphoric Co-reference in Texts in Natural Language

Wojciech Budzisz[1]([⊠])(iD) and Lech T. Polkowski[2]

[1] Polish-Japanese Academy of IT, Warsaw, Poland
wojciech.budzisz@pja.edu.pl
[2] University of Warmia and Mazury in Olsztyn, Olsztyn, Poland
lech.polkowski@pja.edu.pl

Abstract. Natural language supplies us with a plethora of difficulties in its formal rendering. Common its usage produce cases of uncertain reading one of which is the phenomenon of anaphoric reference. In this notice, we propose to study the anaphoric reference in the framework of distributional models of language according to Dobrushin and Revzin, but with the automatic reading of anaphora as a perspective aim. Anaphora resolution offers us an opportunity to introduce dynamic structures of rough sets. In the studied in this note case, rough sets emerge as primary and secondary anaphoric readings of the text underlying dynamic changes in the process of incremental text deciphering. In a more general perspective, rough set collections are construed as states of dynamic processes with the aim that goals of processes correspond to rough sets in collections representing states becoming exact.

Keywords: Rough sets · Distributive models of language
The dominance relation · Grammatical category · Paradigmatic forms
Anaphora resolution

1 Analytical Models of Natural Language

It is well-known that natural language understanding poses many difficult problems whose part comes from syntax. Many models have been elaborated concerning syntactic problems in natural languages. We focus here on analytical models of language which according to Solomon Marcus [6] attempt at recognizing in a given language, i.e., in a collection of sentences, the structure of sentences, the constitutive elements in them and relations among those elements. Usually we speak of *words* as elementary units, their properly constructed strings called *sentences* and we formalize some relations among words within sentences.

© Springer Nature Switzerland AG 2018
H. S. Nguyen et al. (Eds.): IJCRS 2018, LNAI 11103, pp. 455–463, 2018.
https://doi.org/10.1007/978-3-319-99368-3_35

The inherent ambiguity in such investigations is illustrated by the well-known example from Bar-Hillel and Shamir [1]: the sentence *'they are flying planes'* can admit two distinct readings (1) *'they (are flying) planes'* (2) *they are (flying planes)'*. In language formalization on analytical lines, the language L is considered as a subset of the free semigroup Γ of strings over a given alphabet V and with operation of catenation on words.

In this setting, a *context* is any pair (u, v) of strings. For words x, y, following Dobrushin [3,4] and Sestier [13], we say that y has greater morphological homonymy than x if for each context (u, v) such that $uxv \in L$, we have $uyv \in L$. This is an important example of a relation among words. Another example essential to our purpose is the partition of the set of words into paradigmatic forms.

A paradigmatic form of a word is the set of its flectional forms, e.g., paradigmatic forms of the word *book* are *book, books*, for the pronoun *he*, the paradigmatic class consists of *he, his, him, himself*. We assume that granulating the set of words in a language L into paradigmatic classes results in the partition P on words in the vocabulary V (it may yet happen in general that two non-paradigmatic words have common paradigmatic forms). A triple (the vocabulary of words V, the set of well-formed sequences of words L, the paradigmatic partition P) is a *paradigmatically structured language PL*.

The relation P can be extended over strings, for a string $x_1 x_2 ... x_k$, the corresponding P-sequence is $P_1 P_2 ... P_k$, where P_i is $P(x_i)$ for $i = 1, 2, ..., k$; a P-sequence is *marked* if each P_i is $P(x_i)$ for some string $x_1 x_2 ... x_k \in L$.

Our basic technical notion is that of *domination in the sense of Dobrushin* (op.cit.) We say that $P(x)$ dominates $P(y)$ if for each pair $(P(w), P(v))$ if $P(w)P(x)P(v)$ is marked, then $P(w)P(y)P(v)$ is marked. In particular, a word x dominates a word y if for each context (u, v), if $uxv \in L$ then $uyv \in L$, in symbols $x \to_L y$; we omit the subscript L whenever L is fixed. A word x is *initial* if there is no $y \in L$ such that $y \to x$ and not $x \to y$.

For an initial x, the set of all words dominated by x is the *grammatical category* $C(x)$ of x (cf. Revzin [9]). Within a category $C(x)$, we introduce the notion of *rank* by saying that the rank of y, $rank(y)$ is not greater than rank of z, $rank(z)$ in case $y \to z$.

2   Anaphora. Anaphoric Reference

In linguistics, anaphora means informally that one occurrence of a word in a sentence or a text refers back to some occurrence of another word, a typical example may be *Ann was training hardly before her tennis tournament* in which sentence the *anaphor* is a paradigmatic form *her* of the personal pronoun *she* and the proper name *Ann* is the *antecedent* or *anaphora resolution*.

By an *anaphoric pair*, we will mean the pair (anaphor, antecedent), and we define it formally in a category $C(x)$ with the help of the linear order $\prec$ on occurrences of words in a text. In this setting, we call an *anaphoric C(x)-pair* a pair (u, v) of word occurrences in contexts (a, b) for u and (c, d) for v such that

(i) $u \prec v$;

(ii) $u, v \in C(x)$;

(iii) there exist U', v' with $u' \in P(u)$ and $v' \in P(v)$;

(iv) $rank(v') < rank(u')$;

(v) u' is *paradigmatically accepted* by the context (c, d).

In what follows, we admit as well strings of words as antecedents, and to this aim, we denote by P^* the extension of the paradigmatic partition P over strings.

3 Resolving Anaphoric References. Paradigmatic Sequences of Anaphora

In this section and following ones, we quote from Semeniuk-Polkowska and Polkowski [10–12] concerning basic notions and algorithmic ideas. We consider a text T consisting of separated by space sentences $t_1, t_2, ..., t_n$ where each t_i is a string of word-forms. We consider an initial word-form a along with the set $P^*(C(a))$ of paradigmatic forms of strings in the category $C(a)$ and by $Occ(a, T)$ we denote the set of occurrences in T of strings from $P^*(a)$.

Our first notion aimed at segmenting plausible antecedents of anaphora is that of a *paradigmatic sequence of anaphora* which is a sequence $s = x_1 \prec x_2 \prec ... \prec x_k$ in $P(a)$ such that

(vi) there exists an occurrence of $y \in P^*(C(a))$;

(vii) $y \prec x_1$;

(viii) $rank(y) > rank(a)$;

(ix) there is no occurrence $z \in P^*(C(a))$ with $x_i \prec z \prec x_{i+1}$ and $rank(z) > rank(a)$ (x) the sequence s is maximal with respect to (vi)–(ix).

The meaning of a PSA is that it does collect all anaphora in a block of them not interrupted by any plausible antecedent and separated from other such blocks by occurrences of plausible antecedents. In the process of anaphora resolution, we try to find antecedents for anaphora in that block from the set of antecedents sandwiched between that block and the preceding block. The text in next section supplies the example of partitioning into PSA's.

We denote by $s(x)$ the PSA sequence which contains the occurrence x, and, by $f(s)$, respectively $l(s)$, we denote the first, respectively the last, element of s; $im(s)$, respectively $is(s)$, denote the immediate predecessor PSA, respectively the immediate successor PSA, of s with respect to the order $\prec$.

4 Anaphoric Rough Sets

We now are in the position to indicate rough set structures in the text T. First, we define the set of *strongly plausible antecedents of a sequence* s or, for short *primary readings* as the lower approximation $rough(s)$ which is the set of occurrences y such that $l(im(s)) \prec y \prec f(s)$ and $\overline{(y, x)}$ is an anaphoric pair for each

$x \in s$. Next, we define the *border region* $Bd(rough(s))$ of *secondary readings* as the set of occurrences z such that $z \prec l(s)$ and (z, x) is an anaphoric pair for some $x \in s$. The rough set $rough(s)$ is defined as the union $\underline{rough(s)} \cup Bd(rough(s))$.

Let us give an example of a real text. Our example comes from Chandler [2].

'If *Muriel Chess*[1] impersonated *Crystal Kingsley*[2], *she*[11] murdered *her*[12]. That's elementary. All right, let's look at it. We know *who*[3] *she*[21] was and what kind of *woman*[4] *she*[31] was. *She*[32] had already murdered before *she*[33] met and married *Bill Chess*[1]. *She*[34] had been *Dr Almore's*[2] *office nurse*[5] and *his*[11] *littlepal*[6] and *she*[41] had murdered *Dr Almore's wife*[7] in such a neat way that *Almore*[4] had to cover up for *her*[51].'

In this exemplary text, with double superscripts we singled out paradigmatic sequences for the category of *she* and similarly but with subscripts elements of paradigmatic sequences for the category of *he*, in the former case the first superscript indicates the sentence, the second its consecutive elements. Possible antecedents are marked with superscripts for the category of *she* and subscripts for the category of *he*.

4.1   Exemplary paradigmatic sequences of anaphors (PSA's)

We list those sequences for categories of *she* and *he*. For the category of *she*, we have the following PSA's.

(0) $s^0 = \emptyset$;
(1) $s^1 = < she^{11}, her^{12} >$;
(2) $s^2 = < she^{21} >$;
(3) $s^3 = < she^31, She^{32}, she^{33}, She^{34} >$;
(4) $s^4 = < she^{41} >$;
(5) $s^5 = < her^{51} >$.

PSA's for the category of *he* are
$s_1 = < his_{11} >$.

4.2   Rough Sets Associated with Exemplary PSA's

We list rough sets as indicated above, by defining in each case the lower approximation referring to primary readings of anaphora and the boundary region tied to secondary readings of them.

$For\ s^1 : rough(s^1) = \{Muriel\ Chess^1, Crystal\ Kingsley^2\}$;
  $Bd(\overline{rough(s^1)}) = \emptyset$.
$For\ s^2 : rough(s^2) = \{who^3\}$;
  $Bd(\overline{rough(s^2)}) = \{Muriel\ Chess^1, Crystal\ Kingsley^2\}$.
$For\ s^3 : rough(s^3) = \{woman^4\}$;
  $Bd(\overline{rough(s^3)}) = \{who^3, Muriel\ Chess^1, Crystal\ Kingsley^2\}$.

For s^4 : $\underline{rough(s^4)} = \{office\ nurse^5, little\ pal_3^6\}$;
　　$Bd(\overline{rough(s^4)}) = \{who^3, woman^4, Muriel\ Chess^1, Crystal$
　　$Kingsley^2\}$.

For s^5 : $\underline{rough(s^5)} = \{Dr\ Almore's\ wife^7\}$;
　　$Bd(\overline{rough(s^5)}) = \{little\ pal_3^6, office\ nurse^5, woman^4, who^3,$
　　$Muriel\ Chess^1, Crystal\ Kingsley^2\}$.

For s_1: $\underline{rough(s_1)} = \{Bill\ Chess^1, Dr\ Almore's^2\}$.

4.3　PSA Associated Rough Sets Parsing

Implementation of an exemplary parsing algorithm, that results in a formation of such sets, relies heavily on access to additional syntactic and semantic data, that can be gained during text preprocessing. Initial steps involve sentence segmentation and tokenization followed by part of speech tagging (POS), gender detection and named entity recognition. Tools for such preprocessing are readily available in open source libraries e.g. Stanford CoreNLP Toolkit [5].

Algorithm 1. Rough set parsing algorithm

1: **function** PARSEROUGHSETS(*text, gender*)
2: 　　**let** *anaphoraSequence*　　　　　　　　　▷ a sequence of all anaphora in text
3: 　　**let** *lwApproxSets* ▷ a sequence of lower approx. rough sets for each anaphora
4: 　　**let** *bondarySets* ▷ a sequence of boundary region rough sets for each anaphora
5: 　　**let** *maxBoundarySet*　　　　　　　　▷ a maximal boundary region in text
6: 　　*sentences* ← *splitText(text)*　　　　▷ split text to a sequence of sentences
7: 　　**for** *sentence* ← *sentences* **do**
8: 　　　　*annotatedSentence* ← *annotateWithTags(sentence)*
　　　　　　　　　▷ POS annotation done with external toolkit e.g. Stanford CoreNLP
9: 　　　　*anaphora, segments* ← *splitByPrepositions(annotatedSentence, gender)*
10: 　　　　*anaphoraSequence* ++= *anaphora* ▷ add all anaphora to result sequence
11: 　　　　**for** *segment* ← *segments* **do**
12: 　　　　　　**let** *lowerApprox*　　　　　　　　　▷ an initial empty sequence
13: 　　　　　　**for** *token* ← *segment* **do**
14: 　　　　　　　　**if** (*token* is NamedEntity **and** *token*.gender == *gender*) **or**
(*token*.pos in (NN, NNP, WP) **and** *token* WordNet category person) **then**
　　　　　　　　▷ POS tags as used in Penn Treebank set, WordNet data from external source
15: 　　　　　　　　　　*lowerApprox* += *token.word*
16: 　　　　　　　　**end if**
17: 　　　　　　**end for**
18: 　　　　　　*lwApproxSets* += *lowerApprox*
19: 　　　　　　*bondarySets* += *maxBoundarySet*
20: 　　　　　　*maxBoundarySet* += *lwApproxSets*
21: 　　　　**end for**
22: 　　**end for**
23: 　　**return** *anaphoraSequence, lwApproxSets, bondarySets*
24: **end function**

```
1: function SPLITBYPREPOSITIONS(annotatedSentence, paradigmaticCategory)
                          ▷ In this example paradigmatic category is determined by gender
2:     let segments                    ▷ a sequence of sequences of words from sentence
3:     let subSegment                       ▷ an initial empty sequence of words
4:     let anaphorSequence                 ▷ a sequence of anaphora from sentence
5:     for token ← annotatedSentence do
6:         if token.pos == PRP and token belongs to paradigmaticCategory then
7:             anaphorSequence += token   ▷ preposition (anaphora) was detected
8:             segments += subSegment
9:             subSegment ← empty sequence   ▷ new empty sequence for next prep.
10:        else
11:            subSegment += token
12:        end if
13:    end for
14:    return anaphoraSequence, segments
15: end function
```

The following section presents a pseudocode of an implementation that forms the exact sets as in the example above. Splitting, tokenization and annotation of words was done using the aforementioned NLP toolkit. Additionally WordNet [7] was used to determine the category of nouns that were not marked during the annotation preprocessing as named entities. WordNet unique beginners [8] of these nouns were checked in the database to conform with *person, human being* type. This kind of filtering can also be used for other categories e.g. when resolving for abstract idea or physical object.

5   Semantic Tools. Pruning the Anaphoric Tree of Resolutions

The sequence of anaphora in a text T, $x_1, x_2, ..., x_k$ is resolved by means of an increasing hierarchy of partial functions - partial anaphora resolutions - forming a tree called the *anaphoric tree*. The root of the tree is the empty function $h(0)$ on the empty sequence s^0. Each partial function is defined on an initial segment $x_1, x_2, ..., x_n$, $n <= k$, and the tree order is the Brouwer-Kleene (lexicographic) order. The number of functions in a tree of our example for the category of *she* is 1092 and it is desirable to prune the tree to select most plausible readings. Pruning of the anaphoric tree cannot be done by purely grammatic tools, as understanding of proper readings is intimately related to semantics of the language.

5.1   Identifying Strings, Forbidden Texts

We apply two ideas in order to prune the anaphoric tree. The first is that of *identifying strings* [12]. By an identifying string we mean a string of the form

(is) 'x is (eventually: was, will be, have been, had been, becomes, will become, became, have become, had become, has name of) y', where $x \in P(a)$, a an initial word-form in the category $C(a)$, $y \in P^*(C(a)$, $rank(y) > rank(x)$.

With each identifying string, we associate a *transfer rule*.

Transfer rules. For the string in (is), the transfer rule is:

If y is not of maximal rank in the category $C(a)$, then in each partial anaphora resolution, the occurrence of y as the antecedent to x is replaced by an antecedent in the appropriate rough set $rough(s(x))$ having the maximal rank of a proper name if such antecedent does exist.

In our example, we have the following identifying strings:

(i) 'who^3 she^{21} was';
(ii) 'what kind of $woman^4$ she^{31} was';
(iii) 'She^{34} had been $Dr\ Almore's_2$ $office\ nurse^5$ and his_{11} $little\ pal^6_3$'.

The transfer rules are applicable in each of the three cases. In case (i), we eliminate who^3 as a possible antecedent replacing it with either

$Muriel\ Chess^1$ or $Crystal\ Kingsley^2$. The transfer rule does change the rough set $rough(s^2)$:

$rough(s^2) \leftarrow \{Muriel\ Chess^1, Crystal\ Kingsley^2\}$ and it is an exact set.

In case (ii), we eliminate $woman^4$ replacing it with either $Muriel\ Chess^1$ or $Crystal\ Kingsley^2$.

The rough set $rough(s^3)$ becomes
$rough(s^3) \leftarrow \{Muriel\ Chess^1, Crystal\ Kingsley^2\}$ and it is exact now.

In case (iii), $office\ nurse^5, little\ pal^6$ are replaced each with either $Muriel\ Chess^1$ or $Crystal\ Kingsley^2$. The rough set $rough(s^4)$ becomes:
$rough(s^4) \leftarrow \{Muriel\ Chess^1, Crystal\ Kingsley^2\}$ and it is exact now.

These changes have impact on the rough set $rough(s^5)$ which becomes now:
$rough(s^5) \leftarrow \{Dr\ Almore's\ wife^7\}$;
$Bd(rough(s^5)) = \{Muriel\ Chess^1, Crystal\ Kingsley^2\}$.

Under transfer rules, the reading of our text becomes:
'If Muriel Chess impersonated Crystal Kingsley, Muriel Chess/Crystal Kingsley murdered Crystal Kingsley/ Muriel Chess.

That's elementary. All right, let's look at it.

We know who Muriel Chess/Crystal Kingsley was and what kind of woman Muriel Chess/Crystal Kingsley was.

Muriel Chess/Crystal Kingsley had already murdered before Muriel Chess/Crystal Kingsley met and married Bill Chess.

Muriel Chess/Crystal Kingsley had been Dr Almore's office nurse and his little pal and Muriel Chess/Crystal Kingsley had murdered Dr Almore's wife in such a neat way that Almore had to cover up for Dr Almore's wife/Muriel Chess/Crystal Kingsley.'

Transfer rules reduced the size of the anaphoric tree to 384 possible readings. Which are plausible? We exploit the Revzin idea of *forbidden texts* to further prune the anaphoric tree. The set of forbidden texts we single out consists of:

(a) texts which follow after a phrase x *murdered* y containing any occurrences of y as a possible antecedent.

The transfer rule induced by (a) excludes $Dr\ Almore's\ wife^7$ as the antecedent for her^{51}, changing the rough set $rough(s^5)$ to the exact set:

$\{Muriel\ Chess^1, Crystal\ Kingsley^2\}$

and offering the reading of the form:

'If Muriel Chess impersonated Crystal Kingsley, Muriel Chess/Crystal Kingsley murdered Crystal Kingsley/Muriel Chess.

That's elementary. All right, let's look at it.

We know who Muriel Chess/Crystal Kingsley was and what kind of woman Muriel Chess/Crystal Kingsley was.

Muriel Chess/Crystal Kingsley had already murdered before Muriel Chess/Crystal Kingsley met and married Bill Chess.

Muriel Chess/Crystal Kingsley had been Dr Almore's office nurse and his little pal and Muriel Chess/Crystal Kingsley had murdered Dr Almore's wife in such a neat way that Almore had to cover up for Muriel Chess/Crystal Kingsley.'

(b) in texts following the phrase x *married* y with possible antecedents for $x \in P(a)$ being proper names *name surname*^{1}, ..., *name surname*^{n} and y being a proper name *name surname**, the only possible antecedent for paradigmatic forms of x in the exact rough sets for the sequences containing x is the *name surname*^{k} with *surname*^{k} equiform with *surname**.

(c) if a text contains a necessary antecedent y for a paradigmatic form x of an initial word-form a, then y cannot occur as an object in a preceding y phrase containing verbs of destruction like murdered, destroyed, erased .... .

The transfer rule following (b) excludes $Crystal\ Kingsley^2$ as an antecedent for she^{34} and by virtue of (c) the only possible readings for she^{11} and her^{12} are, respectively $Muriel\ Chess^1$ and $Crystal\ Kingsley^2$. The rough sets resulting from those final transfer rules are exact sets:

$$rough(s^1) = \{Muriel Chess^1, Crystal Kingsley^2\};$$
$$rough(s) = \{Muriel\ Chess^1\}\text{ for } s = s^2, s^3, s^4, s^5\}$$

The final reading of the text becomes:

'If Muriel Chess impersonated Crystal Kingsley, Muriel Chess murdered Crystal Kingsley.

That's elementary. All right, let's look at it.

We know who Muriel Chess was and what kind of woman Muriel Chess was.

Muriel Chess had already murdered before Muriel Chess met and married Bill Chess.

Muriel Chess had been Dr Almore's office nurse and his little pal and Muriel Chess had murdered Dr Almore's wife in such a neat way that Almore had to cover up for Muriel Chess.'

We have observed the dynamic changes in rough sets associated with paradigmatic sequences of anaphora, reducing boundary regions in accordance with advancements in partial readings of the text.

6 Conclusions

We have presented a new venue for rough sets: dynamic structures/collections of them. They are intended as states of dynamic processes, with goals of those processes represented as states at which the collection of rough sets becomes the collection of exact sets. In the case presented here, rough sets are construed as sets of possible antecedents for anaphora in a given grammatical category. We have applied in our schema the idea of R. L. Dobrushin, of contextual domination, which requires for its automatic application the already tagged texts and we intend to use to this end the tools listed in the bibliography. We also intend to extend our analysis to cataphora as for instance the analysis of 'it' must take into account the usage of the pronoun 'it' in phrases like 'it was rain that interrupted the show' where the pair (it, rain) is a cataphoric pair.

One has to be aware that problems of language analysis are difficult and we hope that our approach even restricted to texts not complicated beyond what is typical will bring some tools for automatic reading of anaphora and cataphora resolutions.

References

1. Bar-Hillel, Y., Shamir, E.: Finite state languages: formal representation and adequacy problems. In: Bar-Hillel, Y. (ed.) Language and Information. Selected Essays on Their Theory and Application. Addison-Wesley, Reading, Mass, pp. 87–98 (1964)
2. Chandler, R.: The lady in the lake. In: Chandler. Later Novels and Other Writings. The Library of America, p. 193
3. Dobrushin, R.L.: The elementary grammatical category. Byul. Obiedin. Probl. Mashiinnogo Perevoda No. 5, 19–2 1 (1957). (in Russian)
4. Dobrushin, R.L.: Mathematical methods in linguistics. Applications. Mat. Prosveshchenie **6**, 52–59 (1961). (in Russian)
5. Manning, C. D. et al.: The stanford CoreNLP natural language processing toolkit. In: Proceedings of the 52nd Annual Meeting of the Association for Computational Linguistics: System Demonstrations, pp. 55–60 (2014)
6. Marcus, S.: Algebraic Linguistics; Analytical Models. Academic Press, New York, London (1967)
7. Miller, G.A. et al.: Introduction to WordNet: An Online Lexical Database (1993, preprint)
8. Miller, G.A.: Nouns in WordNet: A Lexical Inheritance System (1993, preprint)
9. Revzin, I.I.: Models of Language. Methuen, London (1962)
10. Semeniuk-Polkowska, M., Polkowski, L.T.: An analytic model of anaphora resolution in algebraic linguistics. Intern. J. Comput. Mathe. **23**, 251–263 (1988)
11. Semeniuk-Polkowska, M., Polkowski, L.T.: Anaphoric trees and an extension of the model of anaphora resolution. Reports Fac. Technical Mathematics and Informatics, no. 89–37, Delft University, Delft, The Netherlands (1989)
12. Semeniuk-Polkowska, M., Polkowski, L.T.: A semantics for anaphora resolution in algebraic linguistics. Intern. J. Comput. Mathe. **32**, 137–147 (1990)
13. Sestier, A.: Contribution á une theorie ensembliste des classifications linguistiques. Premier congres de l'Association francaise de calcul. Grenoble, pp. 293–305 (1960)

Reduct Calculation and Discretization of Numeric Attributes in Entity Attribute Value Model

Wojciech Świeboda[1] and Nguyen Sinh Hoa[2]([✉])

[1] Institute of Computer Science, University of Warsaw,
Banacha 2, 02-097 Warsaw, Poland
[2] Polish-Japanese Academy of Information Technologym,
ul. Koszykowa 86, 02-008 Warsaw, Poland
hoa@pjwstk.edu.pl

Abstract. In this paper we review the problem of short reduct calculation in a sparse decision system. We also address the problem of discretization of numerical attributes in sparse decision systems. We present algorithms that provide an approximate solution to these two problems and analyze the complexity of these algorithms.

1 Introduction

We begin by introducing the necessary notions in Rough Set theory, the problems of reduct calculation and discretization. Afterwards we introduce Maximal Discernibility heuristic. Finally, we discuss an implementation of MD heuristic using contingency tables. We then discuss a version of the algorithm designed for sparse data sets and discuss several theoretical properties of both the algorithm and the minimal reduct problem in the sparse setting.

1.1 Preliminaries

An *information system* is a pair $\mathbb{I} = (\mathbb{U}, \mathbb{A})$ where $\mathbb{U}$ denotes the *universe of objects* and $\mathbb{A}$ is the set of *attributes*. An attribute $a \in \mathbb{A}$ is a mapping $a : U \to V_a$. The codomain V_a of attribute a is often also called the *value set* of attribute a.

A *decision system* is a pair $\mathbb{D} = (\mathbb{U}, \mathbb{A} \cup \{dec\})$ which is an information system with a distinguished attribute $dec : U \to \{1, \ldots, d\}$ called *a decision attribute*. Attributes in $\mathbb{A}$ are called *conditions* or *conditional attributes* and may be either nominal or *numeric*, i.e. with $V_a \subseteq \mathbb{R}$. Throughout this article n will denote the number of objects in a decision system and k will denote the number of conditional attributes.

Table 1 on the left shows a typical decision system with symbolic attributes represented as a table. Attributes *Diploma, Experience, French* and *Reference* are *conditions*, whereas *Decision* is the decision attribute. All conditional attributes in this decision system are nominal.

© Springer Nature Switzerland AG 2018
H. S. Nguyen et al. (Eds.): IJCRS 2018, LNAI 11103, pp. 464–478, 2018.
https://doi.org/10.1007/978-3-319-99368-3_36

Table 1. A typical decision system with symbolic attributes (left) and a decision system in which all conditional attributes are numeric (right)

	Diploma	Experience	French	Reference	Decision		a_1	a_2	a_3	Decision
x_1	MBA	Medium	Yes	Excellent	Accept	x_1	0	1.3	0	F
x_2	MBA	Low	Yes	Neutral	Reject	x_2	3.3	0.9	0	F
x_3	MCE	Low	Yes	Good	Reject	x_3	0	1.5	0	F
x_4	MSc	High	Yes	Neutral	Accept	x_4	0	1.2	2.5	F
x_5	MSc	Medium	Yes	Neutral	Reject	x_5	0	1.3	3.6	F
x_6	MSc	High	Yes	Excellent	Accept	x_6	3.7	2.7	2.4	T
x_7	MBA	High	No	Good	Accept	x_7	4.1	1.0	2.8	T
x_8	MCE	Low	No	Excellent	Reject					

For any $B \subseteq \mathbb{A}$ we define *B-indiscernibility relation* $IND(B) \subseteq \mathbb{U} \times \mathbb{U}$ as follows:

$$IND(B) = \{(x, y) : \forall_{a \in B}\ a(x) = a(y)\}$$

and the *decision-relative B-indiscernibility* relation $IND(B) \subseteq \mathbb{U} \times \mathbb{U}$ by:

$$IND_{dec}(B) = \{(x, y) \in \mathbb{U} \times \mathbb{U} : dec(x) = dec(y) \vee \forall_{a \in B}\ a(x) = a(y)\}$$

The *discernibility relation* $DISC(B)$ and *decision-relative discernibility relation* $DISC_{dec}(B)$ are the complements of $IND(B)$ and $IND_{dec}(B)$ correspondingly:

$$DISC(B) = \mathbb{U} \times \mathbb{U} \smallsetminus IND(B) \qquad DISC_{dec}(B) = \mathbb{U} \times \mathbb{U} \smallsetminus IND_{dec}(B)$$

A decision system $\mathbb{D} = (\mathbb{U}, \mathbb{A} \cup \{d\})$ is *consistent* if

$$\forall_{x,y \in U}\ dec(x) \neq dec(y) \implies \exists_{a \in \mathbb{A}}\ a(x) \neq a(y)$$

Proposition 1. *For arbitrary $B \subseteq \mathbb{A}$, $IND(B)$ is an equivalence relation and thus this relation induces a partitioning of $\mathbb{U}$.*

Definition 1. *A decision-relative reduct or decision reduct of a decision system $\mathbb{D} = (\mathbb{U}, \mathbb{A} \cup \{dec\})$ is a minimal subset of attributes $B \subseteq \mathbb{A}$ such that $IND_{dec}(B) = IND_{dec}(\mathbb{A})$.*

If we loosen the assumption on the minimality of this set, we speak of a *decision superreduct*:

Definition 2. *A decision-relative superreduct or decision superreduct of a decision system $\mathbb{D} = (\mathbb{U}, \mathbb{A} \cup \{dec\})$ is a subset of attributes $B \subseteq \mathbb{A}$ such that $IND_{dec}(B) = IND_{dec}(\mathbb{A})$.*

1.2   Sparse Decision System and Entity Attribute Value Model

In many situations a convenient way to represent the data set is in terms of Entity-Attribute-Value (EAV) Model, which encodes observations in terms of triples. For an information system $I = (\mathbb{U}, \mathbb{A})$, the set of triples is $\{(u, a, v) : a(u) = v\}$. This representation is especially handy for information systems with numerous attributes, missing or default values. Instances with missing and default values are not included in EAV representation, which results in compression of the data set.

In this paper we are only dealing with default values. Their interpretation or semantics is the same as of any other attribute. In practice we store triples corresponding to numeric attributes and to symbolic attributes in two separate tables, and store decisions of objects in a separate vector (Table 2).

Table 2. EAV representation of decision systems in Table 1. The default values for the left table (omitted in this representation) for consecutive attributes are 'MBA', 'Low', 'Yes' and 'Excellent'. The default value for the right table (omitted in this representation) for each attribute is 0.

Entity	Attr.	Value
x_1	a_2	Medium
x_2	a_4	Neutral
x_3	a_1	MCE
x_3	a_4	Good
x_4	a_1	MSc
x_4	a_2	High
x_4	a_4	Neutral
x_5	a_1	MSc
x_5	a_2	Medium
x_5	a_4	Neutral
x_6	a_1	MSc
x_6	a_2	High
x_7	a_2	High
x_7	a_3	No
x_7	a_4	Good
x_8	a_1	MCE
x_8	a_3	No

Entity	Decision
x_1	Accept
x_2	Reject
x_3	Reject
x_4	Accept
x_5	Reject
x_6	Accept
x_7	Accept
x_8	Reject

Entity	Attr.	Value
x_1	a_2	1.3
x_2	a_1	3.3
x_2	a_2	0.9
x_3	a_2	1.5
x_4	a_2	1.2
x_4	a_3	2.5
x_5	a_2	1.3
x_5	a_3	3.6
x_6	a_1	3.7
x_6	a_2	2.7
x_6	a_3	2.4
x_7	a_1	4.1
x_7	a_2	1.0
x_7	a_3	2.8

Entity	Decision
x_1	T
x_2	T
x_3	T
x_4	T
x_5	T
x_6	T
x_7	T

There are various problems related to reduct calculation, e.g. finding all decision reducts or finding the shortest decision reduct [4] in a decision system. In this paper we address the problem of finding a single short decision reduct. The problem of finding the shortest decision reduct is an NP-hard [4], though various heuristics were proposed for this problem, e.g. [1,6]. In this paper we focus on an approximate solution to this problem assuming that the decision system is sparse and is stored in data bases in the EAV Model.

2   Maximal Discernibility Heuristic for Reduct Calculation

A convenient heuristic for the problem of finding a short decision reduct is "maximal discernibility heuristic" (MD-heuristic), presented in Algorithm 1 below.

Definition 3. *A conflict is an unordered pair of objects belonging to different decision classes. For $X \subseteq \mathbb{U}$ we define a function which counts the number of unordered conflicts:*

$$conf(X) = \frac{1}{2}|\{(x, y) \in \mathbb{U} \times \mathbb{U} : dec(x) \neq dec(y)\}|$$

Finally, we define $c : 2^A \to \mathbb{R}$ as: $c(R) = \sum conf([x]_R)$, where the summation is taken over all equivalence classes of the partitioning induced by $IND(R)$ and $[x]_R := [x]_{IND(R)}$.

Definition 4. *For $R \subseteq A, a \in A \setminus R$ we define discernibility measure as follows:*

$$discern(R, a) = c(R) - c(R \cup \{a\})$$

For a fixed $R \subseteq A$, $discern(R, a)$ counts the number of pairs of objects discerned by a, undiscerned by attributes from R alone, and can thus be interpreted as an incremental measure of quality of attribute a.

Algorithm 1. MD-heuristic for superreduct calculation in a consistent decision system.

Data: $\mathbb{D} = (\mathbb{U}, A \cup \{dec\})$: a decision system.
Result: R: a semi-optimal decision superreduct
1 $R \leftarrow \emptyset$;
2 **while** $c(R) \neq 0$ **do**
3 $\quad a \leftarrow argmax_{a \in A \setminus R} \; discern(R, a)$;
4 $\quad R \leftarrow R \cup \{a\}$;
5 **end**

Let k denote the number of attributes and n denote the number of objects in the decision system. The calculation of *discern* in Algorithm 1 may require iterating over each pair of objects. The *argmax* may further require iteration over all attributes. Hence, a naive implementation of the algorithm presented above leads to an algorithm with complexity $O(|R|n^3k)$. In further sections we discuss more efficient implementations of this algorithm and its extension to discretization problem.

3   MD-Heuristic for Discretization Problem

In this section we describe the problem of discretization of numeric attributes. Let $\mathbb{D} = (\mathbb{U}, A \cup \{dec\})$ be a decision system. An attribute a is numeric if $V_a \subseteq \mathbb{R}$. A cut on a numeric attribute $a \in A$ is a pair (a, v) such that $v \in V_a$. We further require that $a(x) \neq v$ for all $x \in \mathbb{U}$ (i.e. we can always tell whether an object is to the left or to the right of a cutpoint).

Let $\mathbb{D} = (\mathbb{U}, A \cup \{dec\})$ be a decision system and (a, c) be a cut. The cut discerns objects $x, y \in \mathbb{U}$ if $a(x) - c$ and $a(y) - c$ are of different signs.

A set of cuts $\mathbb{P}$ is called *consistent with* $\mathbb{D}$ if

$$\forall_{x,y \in \mathbb{U}} \, dec(x) \neq dec(y) \implies \exists_{(a,c) \in \mathbb{P}} \, (a(x) - c)(a(y) - c) < 0$$

and $\mathbb{P}$ is called *optimal* if it is the smallest set of cuts consistent with $\mathbb{D}$.

While there are potentially infinitely many possible cuts on an attribute $a \in \mathbb{A}$, we will only consider cut points that fall in the middle of two consecutive values attained on this attribute. By $M(a)$ we denote the list of middle cuts on attribute $a \in \mathbb{A}$ listed in ascending order. For example the lists of middle cuts for the data set in Table 1 are as follows: $M(a_1) = \langle 1.65, 3.5, 3.9 \rangle$, $M(a_2) = \langle 0.95, 1.1, 1.25, 1.4, 2.1 \rangle$, $M(a_3) = \langle 1.2, 2.45, 2.65, 3.2 \rangle$.

The problem of finding the optimal set of cuts is equivalent to the problem of shortest reduct calculation [2] and thus is NP-hard [2] (Table 3).

Table 3. A discretized version of the decision system presented in Table 1

	a_1	a_2	a_3	Decision
x_1	$(-\infty, +\infty)$	$(1.25, +\infty)$	$(-\infty, 1.2]$	F
x_2	$(-\infty, +\infty)$	$(-\infty, 1.1]$	$(-\infty, 1.2]$	F
x_3	$(-\infty, +\infty)$	$(1.25, +\infty)$	$(-\infty, 1.2]$	F
x_4	$(-\infty, +\infty)$	$(1.1, 1.25]$	$(1.2, +\infty)$	F
x_5	$(-\infty, +\infty)$	$(1.25, +\infty)$	$(1.2, +\infty)$	F
x_6	$(-\infty, +\infty)$	$(1.25, +\infty)$	$(1.2, +\infty)$	T
x_7	$(-\infty, +\infty)$	$(-\infty, 1.1]$	$(1.2, +\infty)$	T

Similarly to c and *discern* for attributes, we can define such functions for cuts. For a set of cuts $\mathbb{P}$ let $c(\mathbb{P}) = \sum conf([x]_{\mathbb{P}})$, where the summation is taken over all equivalence classes of the partitioning induced by the set of cuts $\mathbb{P}$ and equivalence classes of this partitioning are denoted $[x]_{\mathbb{P}}$. For a set of cuts $\mathbb{P}$ and a cut $(a, c) \notin \mathbb{P}$ we define $discern(\mathbb{P}, (a, c)) = c(\mathbb{P}) - c(\mathbb{P} \cup \{(a, c)\})$.

The MD heuristic for optimal discretization problem is presented in Algorithm 2. During the analysis of MD-heuristic for the discretization problem in later sections of this paper it will be convenient to refer to c and *discern* for decision systems with different universes, e.g. $(\mathbb{U}_1, \mathbb{A} \cup \{dec\})$ and $(\mathbb{U}_2, \mathbb{A} \cup \{dec\})$. In order to disambiguate, in such situations we will explicitly write $c_{\mathbb{U}_1}(\mathbb{P})$, $c_{\mathbb{U}_2}(\mathbb{P})$, $discern_{\mathbb{U}_1}(\mathbb{P}, (a, c))$, $discern_{\mathbb{U}_2}(\mathbb{P}, (a, c))$.

Algorithm 2. MD-heuristic for discretization

Result: $\mathbb{P}$: a semi-optimal set of cuts

1 $R \leftarrow \emptyset$;
2 **while** $c(\mathbb{P}) \neq 0$ **do**
3     $(a, c) \leftarrow argmax_{(a,c):a \in \mathbb{A}, c \in M(a)} \, discern(\mathbb{P}, (a, c))$;
4     $\mathbb{P} \leftarrow \mathbb{P} \cup \{(a, c)\}$;
5 **end**

---

4   Contingency Table and Partitioning (CPS)

In this section we introduce a structure which simplifies implementation of several algorithms, including MD heuristic for reduct calculation introduced earlier. We call it Contingency Table and Partitioning, or CPS for short.

CPS is a structure that stores information about (a subset of) objects in the database along with their partition membership. CPS consists of fields:

$\Phi = \{\phi_1, \ldots, \phi_m\}$ a set of labels describing partitions

pid a vector of partition identifiers for objects in the underlying decision system

$C = (a_{ij})$ frequency matrix counting, for each decision value, objects in each partition ϕ_i.

Definition 5. *Let* $\mathbb{D} = (\mathbb{U}, \mathbb{A} \cup \{dec\})$ *be a decision system,* $V_{dec} = \{d_1, \ldots, d_D\}$ *and let* $\mathcal{C} = \{C_1, \ldots, C_m\}$ *be a covering of* $\mathbb{U}$.

A frequency matrix for the pair $(\mathbb{D}, \mathcal{C})$ *is an* $m \times D$ *matrix* (a_{ij}) *such that*

$$a_{ij} = |\{x \in \mathbb{U} : x \in C_i \wedge dec(x) = d_j\}|$$

A contingency table is a frequency matrix in which columns and rows are labeled.

Definition 6. *Let* $\Phi = \langle \phi_1, \ldots, \phi_m \rangle$ *be a list of labels. Let* $\mathbb{D} = (\mathbb{U}, \mathbb{A} \cup \{dec\})$ *be a decision system,* $V_{dec} = \{d_1, \ldots, d_D\}$ *and let* $\mathcal{C} = \{C_1, \ldots, C_m\}$ *be a covering of* $\mathbb{U}$. *A contingency table for the tuple* $(\mathbb{D}, \Phi, \mathcal{C})$ *is a pair* (Φ, C), *where* C *is the frequency matrix for* $(\mathbb{D}, \mathcal{C})$.

We will typically use contingency tables for families $\mathcal{C}$ that form partitionings of $\mathbb{U}$. It is convenient to enumerate partitions and represent the partitioning $\mathcal{P} = \{P_1, \ldots, P_m\}$ by a vector $pid \in \{1, \ldots, m\}^{|\mathbb{U}|}$.

Definition 7. *Let* $\mathbb{D} = (\mathbb{U}, \mathbb{A} \cup \{dec\})$ *be a decision system,* $V_{dec} = \{d_1, \ldots, d_D\}$, $pid \in \{1, \ldots, m\}^{|\mathbb{U}|}$ *and let* $\Phi = \langle \phi_1, \ldots, \phi_m \rangle$ *be a list of partition labels, i.e. label* ϕ_i *corresponds to (or describes) objects* $u \in \mathbb{U}$ *with* $pid[u] = i$.

A CPS (contingency table and partition system) for $(\mathbb{D}, \Phi, pid)$ *is a tuple* (Φ, pid_Φ, C), *where* $pid_\Phi : \mathbb{U} \to \{1, \ldots, m\}$ *and* $pid_\Phi(x)$ *is the partition assigned to object* x, *and where* C *is the frequency matrix for* $(\mathbb{D}, \mathcal{P})$.

Since we discuss the problems of reduct calculation and discretization in this paper, partitions pid and their labels Φ will be of a specific form.

Definition 8. *Let* $\mathbb{D} = (\mathbb{U}, \mathbb{A} \cup \{dec\})$ *be a decision system,* $R = \{a_{i_1}, \ldots, a_{i_l}\} \subseteq \mathbb{A}$. *A CPS (contingency table and partition system) for* $(\mathbb{D}, R)$ *is a tuple* (Φ, pid, C), *where* C *is the contingency table for* $(\mathbb{D}, \Phi)$, *and where* Φ *consists of labels of the form* $(a_{i_1} = v_{i_1}) \wedge \ldots \wedge (a_{i_l} = v_{i_l})$ *and such that the term* $a_{i_j} = v_{i_j}$ *appears in a label of an object* x *iff* $a_{i_j}(x) = v_{i_j}$. *This matches the conventional definition of a contingency table.*

Table 4. Example contingency table for the decision system in Table 1 and $R = \{Diploma, French\}$.

x	$pid_R(x)$
x_1	1
x_2	1
x_3	3
x_4	5
x_5	5
x_6	5
x_7	2
x_8	4

Name	Formulas	$a_{\phi, Reject}$	$a_{\phi, Accept}$
ϕ_1	$a_1 = MBA \wedge a_3 = Yes$	1	1
ϕ_2	$a_1 = MBA \wedge a_3 = No$	0	1
ϕ_3	$a_1 = MCE \wedge a_3 = Yes$	1	0
ϕ_4	$a_1 = MCE \wedge a_3 = No$	1	0
ϕ_5	$a_1 = MSc \wedge a_3 = Yes$	1	2

Table 4 is the illustration of contingency table for the decision system in Table 1 and $R = \{Diploma, French\}$ (a_1 stands for *Diploma* and a_3 for *French*). The first column in this table lists all elements from Φ. Two columns with numbers form a 5×2 contingency matrix.

It turns out that frequency matrices (and therefore contingency tables) provide a sufficient summary of the data for the calculation of functions c and *discern*.

Definition 9. *Let $C = CT(\mathbb{D}, \Phi)$ and let I_Φ denote the equivalence relation on $\mathbb{U} \times \mathbb{U}$ defined as follows: $(x, y) \in I_\Phi \iff \forall_{\phi \in \Phi}(\phi(x) \iff \phi(y))$*

Proposition 1. *(See [2], Proposition 23)*
Let (a_{ij}) $i = 1, \ldots, m; j = 1, \ldots, D$ be the frequency matrix for $(\mathbb{D}, \Phi)$ where $\Phi = \{\phi_1, \ldots, \phi_m\}$.

$$conf([x]_{I_\Phi}) = \frac{1}{2}\left(\left(\sum_{j=1}^{D} a_{ij}\right)^2 - \sum_{j=1}^{D} a_{ij}^2\right)$$

where i is such that $\phi_i(x)$ is satisfied (i.e. $i = pid[x]$) and $[x]_{I_\Phi}$ denotes the equivalence class of x with respect to relation I_Φ.

Let $\mathbb{D} = (\mathbb{U}, \mathbb{A} \cup \{dec\})$. Proposition above shows that the frequency matrix (a_{ij}) is a sufficient summary of the data for the calculation of $c(R)$ or $c(\mathbb{P})$, where $R \subseteq \mathbb{A}$ or $\mathbb{P}$ is a set of cuts in $\mathbb{D}$. In both cases $c(R)$ and $c(\mathbb{P})$ are given by the formula for $conf([x]_{I_\Phi})$.

Finally, we rewrite the MD-heuristic for reduct calculation so that it explicitly calculates c and *discern* using contingency tables. For the algorithm below we define *discern* as a function of contingency tables (further overloading the definition) as follows: $discern(C_1, C_2) = c(C_1) - c(C_2)$.

Finally, we list methods associated with class CPS:

- $Init(\mathbb{U}, dec)$: Initialize the CPS, store contingency table for a trivial partitioning
- $Init(CPS_{old}, npartitions)$: Initialize the CPS given CPS_{old} while allocating extra memory in the underlying contingency table for storing counts of a larger number of partitions
- $remove(object)$: remove the object from any partition it belongs to, decrease appropriate count in the contingency table
- $renumber()$: Reset partition identifiers to $1, \ldots, m$ for some m, i.e. guarantee there are no gaps in their numbering.

and several self-explanatory methods: $getPartition(object)$, $getLabel$ $(partition\_id)$, $setPartition(object, partition\_id)$, $setPartitionLabel$ $(partition\_id, label_{opt})$, $getConflicts()$, $maxPid()$.

Algorithm 3. getBestAttribute

Input: $\mathbb{D} = (\mathbb{U}, \mathbb{A} \cup \{dec\})$: A consistent decision table. For simplicity we assume that $V_a = \{1, \ldots, |V_a|\}$ for each $a \in \mathbb{A}$

Data: CPS: contingency and partition system for $\mathbb{D}$ and R

Data: CPS_t: temporary contingency and partition system

Result: a: $argmax_{a \in \mathbb{A} \setminus R} discern(R, a)$

1   $M \leftarrow max_{a \in \mathbb{A}} |V_a|$;

2   $Init(CPS_t, CPS, M \cdot CPS.maxPid())$;

3   **for** $a \in \mathbb{A} \setminus R$ **do**

4   **for** $x \in \mathbb{U}$ **do**

5       $p \leftarrow getPartition(CPS, x)$;

6       $p' \leftarrow p + a(x) \cdot CPS.maxPid()$;

7       $setPartition(CPS_t, x, p')$;

8   **end**

9     $d_a \leftarrow getConflicts(CPS) - getConflicts(CPS_t)$;

10   // reverse previous setPartition() operations

11   **for** $x \in \mathbb{U}$ **do**

12      $setPartition(CPS_t, x, getPartition(CPS, x))$;

13   **end**

14 **end**

15 $a \leftarrow argmax_{a \in \mathbb{A}} d_a$; // update CPS to reflect inclusion of a in R

16 $CPS_t \leftarrow CPS$;

17 **for** $x \in \mathbb{U}$ **do**

18    $p \leftarrow getPartition(CPS, x)$ $p' \leftarrow p + a(x) \cdot CPS.maxPid()$;

19    $setPartition(CPS_t, x, p')$;

20 **end**

21 $renumber(CPS_t)$;

22 $CPS \leftarrow CPS_t$;

23 return a;

---

5   MD-Heuristic with Contingency Tables

Instead of calculating $argmax_{a \in \mathbb{A} \setminus R} \, discern(R, a)$ directly in Algorithm 1 we now call the function *getBestAttribute* which returns the result while preserving CPS structure helpful for further iterations of the algorithm.

Algorithm 3 is a realization of MD-heuristic for reduct calculation. In this algorithm *discern* is calculated using contingency tables, which in turn are calculated based on *dec* and pid_R assignments. Updating partition identifiers can be done in $O(n)$ time, and determining *discern* can be done in $O(nD)$ time and space.

Lemma 1. *The time complexity of MD-heuristic (Algorithm 1) using procedure getBestAttribute from Algorithm 3 is $O(|R|Dnk)$ and the space complexity is $O(n(k + D))$ (the dependency on $D = |V_{dec}|$ is typically neglected as D is usually small for classification problems).*

Proof. In each step of the algorithm, for each $a \in \mathbb{A} \setminus R$, frequency matrix C can be calculated in $O(nD)$ time. Such a frequency matrix has size at most Dn and further calculation of *discern* is linear in the size of this frequency matrix. All such matrices are calculated $|R|k$ times.

In MD-heuristic for the discretization problem we will use contingency tables with labels Φ describing cuts.

Definition 10. *Let $\mathbb{D} = (\mathbb{U}, \mathbb{A} \cup \{dec\})$ be a decision system, and let $\mathbb{P} = \{(a_{i_1}, c_{i_1}), \ldots, (a_{i_l}, c_{i_l})\}$, with $a_{i_j} \in A$, $c_{i_j} \in \mathbb{R}$, A CPS for $(\mathbb{D}, \mathbb{P})$ is a tuple (Φ, pid, C), where C is the contingency table for $(\mathbb{D}, \Phi)$, and where Φ consists of labels of the form $(b_1(a_{i_1} - c_{i_1}) < 0) \wedge \ldots \wedge (b_l(a_{i_l} - c_{i_l}) < 0)$ for $b_i \in \{-1, 1\}$ $(i = 1, \ldots, l)$ and such that each object $x \in \mathbb{U}$ satisfies exactly one such formula.*

Definition 11. *Let $\mathbb{P} = \{(a_{i_1}, c_{i_1}), \ldots, (a_{i_l}, c_{i_l})\}$, with $a_{i_j} \in A$, $c_{i_j} \in \mathbb{R}$, $V_{a_{i_j}} \subseteq \mathbb{R}$. A contingency table for the pair $(\mathbb{D}, \mathbb{P})$ is a contingency table for $(\mathbb{D}, \Phi)$, where Φ consists of formulas/labels of the form $(b_1(a_{i_1} - c_{i_1}) < 0) \wedge \ldots \wedge (b_l(a_{i_l} - c_{i_l}) < 0)$ for $b_i \in \{-1, 1\}$ $(i = 1, \ldots, l)$ and such that each object $x \in \mathbb{U}$ satisfies exactly one such formula.*

From the requirement that each object satisfies exactly one such formula it follows that for $i = 1, \ldots, l$ and $x \in \mathbb{U}$, $a_{i_j}(x) \neq c_{i_j}$, i.e. cut values never equal values attained by objects in the decision system, and so it is always unambiguous whether an object falls on the left or the right side of a cut point. In practice we only consider middle cuts.

Definition 12. *Let $\mathbb{P} = \{(a_{i_1}, c_{i_1}), \ldots, (a_{i_l}, c_{i_l})\}$, with $a_{i_j} \in A$, $c_{i_j} \in \mathbb{R}$, $V_{a_{i_j}} \subseteq \mathbb{R}$. Let Φ consist of labels/formulas of the form*

$$(b_1(a_{i_1} - c_{i_1}) < 0) \wedge \ldots \wedge (b_l(a_{i_l} - c_{i_l}) < 0)$$

for $b_i \in \{-1, 1\}$ $(i = 1, \ldots, l)$ and such that each object $x \in \mathbb{U}$ satisfies exactly one such formula. We define the partition identifier $pid_{\mathbb{P}} : \mathbb{U} \rightarrow \{1, \ldots, m\}$ as follows: For $x \in \mathbb{U}$ let $pid_{\mathbb{P}}(x)$ denote the index of the formula $\phi_i \in \Phi$ such that $\phi_i(x)$ is satisfied.

Table 5. Example contingency table for the decision system in Table 1 and cuts $\mathbb{P} = \{(a_2, 1.25), (a_3, 1.2)\}$ and an example $pid_\mathbb{P}$ assignment for this set of cuts

x	$pid_\mathbb{P}(x)$
x_1	3
x_2	1
x_3	3
x_4	2
x_5	4
x_6	4
x_7	2

Name	Formula	$a_{\phi,F}$	$a_{\phi,T}$
ϕ_1	$(a_2 < 1.25) \wedge (a_3 < 1.2)$	1	0
ϕ_2	$(a_2 < 1.25) \wedge (a_3 > 1.2)$	1	1
ϕ_3	$(a_2 > 1.25) \wedge (a_3 < 1.2)$	2	0
ϕ_4	$(a_2 > 1.25) \wedge (a_3 > 1.2)$	1	1

Table 5 shows contingency table for the decision system in Table 1 and $\mathbb{P} = \{(a_2, 1.25), (a_3, 1.2)\}$. The first column in this table lists all elements from Φ.

Table 5 also shows $pid_\mathbb{P}$ assignment for the decision system in Table 1 and $\mathbb{P}$. The table on the right lists formulae in Φ. In practice the formulae do not need to be stored as only partition identifiers (formulae indices) are necessary for calculations.

The algorithm below keeps two contingency tables: for objects $\mathbb{U}_L$ on the left and for objects $\mathbb{U}_R$ on the right side of a (variable) cut, and iterates over cuts in $M(a)$ (the list of middle cuts) in increasing order.

Lemma 2. *Suppose that we are given a set $U_0 \subseteq \mathbb{U}$ and a set of cuts $\mathbb{P}$ on U_0. If a cut (a, c) partitions U_0 into two disjoint subsets U_1 and U_2, then*

$$discern_{U_0}(\mathbb{P}, (a, c)) = c_{U_0}(\mathbb{P}) - c_{U_1}(\mathbb{P}) - c_{U_2}(\mathbb{P})$$

Proof. Since any objects $x_1 \in U_1, x_2 \in U_2$ are discerned by (a, c), we have $c_{U_0}(\mathbb{P} \cup \{(a, c)\}) = c_{U_1}(\mathbb{P}) + c_{U_2}(\mathbb{P})$.

Definition 13. *Let frequency matrices for $(U_0, \mathbb{P})$, $(U_1, \mathbb{P})$ and $(U_2, \mathbb{P})$ be C_0, C_1 and C_2. For the discretization problem we may thus define discern as a function of contingency tables as: $discern(C_0, C_1, C_2) = c_{U_0}(\mathbb{P}) - c_{U_1}(\mathbb{P}) - c_{U_2}(\mathbb{P})$.*

Notice that by iterating over cut points in increasing order, C_L and C_R can be updated with minimal effort (only one entry needs to be changed in each of these tables). Furthermore, *discern* does not need to be explicitly recalculated in each of the iterations in the innermost loop, as it can be sequentially updated, accessing only a few elements of the involved contingency tables as follows.

Suppose that the object x is counted in the i-th row in C_R and is moved from the right partition to the left partition, with $U_R' = U_R \setminus \{x\}$ and $U_L' = U_L \cup \{x\}$. The following holds:

$$discern(C^*, C_L', C_R') = c_\mathbb{U}(\mathbb{P}) - c_{U_L'}(\mathbb{P}) - c_{U_R'}(\mathbb{P})$$
$$= c_\mathbb{U}(\mathbb{P}) - (c_{U_L}(\mathbb{P}) + |\{y \in [x]_\mathbb{P} \cap U_L : dec(x) \neq dec(y)\}|)$$
$$- (c_{U_R}(\mathbb{P}) - |\{y \in [x]_\mathbb{P} \cap U_R : dec(x) \neq dec(y)\}|)$$
$$= discern(C^*, C_L, C_R) + \sum_{j=1}^{D} C_L[i, j] I(dec(x) \neq d_j) - \sum_{j=1}^{D} C_R[i, j] I(dec(x) \neq d_j)$$

Algorithm 4. getBestCut

Data: $\mathbb{D} = (\mathbb{U}, \mathbb{A} \cup \{dec\})$: A consistent decision table.
Data: C: a set of cuts
Data: CPS: a contingency and partition system for $\mathbb{D}$ and a set of cuts $\mathbb{P}$
Data: CPS_L: a temporary contingency and partition system for the set of cuts $\mathbb{P}$ with all counts equal 0
Data: CPS_R: a temporary contingency and partition system for $\mathbb{D}$ and a set of cuts $\mathbb{P}$
Input: $a \in \mathbb{A}$: attribute under consideration
Result: v: best cut value.
Result: d_v: $disc(\mathbb{P}, (a, c))$

1 **for** $x \in \mathbb{U}$ ordered by values of attribute a **do**
2 | $p \leftarrow getPartition(CPS_R, x)$;
3 | $setPartition(CPS_L, x, p)$;
4 | $remove(CPS_R, x)$;
5 | $v \leftarrow a(x)$;
6 | $d_v \leftarrow discern(C^*, C'_L, C'_R)$;
7 **end**
8 Return v with maximal d_v;

Only the row describing indiscernibility class containing x needs to be read in order to update *discern* in this step. Moreover, if in addition to contingency tables C_L and C_R we also store vectors with row totals, only four entries need to be accessed at each step: one in C_L, C_R and one in each of the corresponding totals.

Theorem 1. *(See [2], Theorem 22)*
The time complexity of MD-heuristic for discretization (Algorithm 2) using procedure getBestCut from Algorithm 4 is $O(kn(|\mathbb{P}|D + \log n))$ and the space complexity is $O(n(k + D))$. The dependency on $D = |V_{dec}|$ is typically neglected as D is usually small for classification problems, and thus the time complexity is $O(kn(|\mathbb{P}| + \log n))$.

6 MD-Heuristic for Sparse Decision Systems

We will now discuss the MD heuristic for superreduct calculation and for discretization problems for datasets in EAV format. In what follows, $E(i), A(i), V(i)$ will denote the entity, attribute and value of the i-th object, respectively. We will assume that there are N EAV triples in the database. We focus on scenarios in which N is much smaller than $n \times k$.

In the algorithm for (super)reduct caculation, the contingency table was recalculated from scratch each time an attribute was considered for addition to the (super)reduct set. Thus, *pid* assignment had to be accessed for each of the n objects.

If we use EAV representation, it suffices to update assigned *pid* identifiers only for objects that attain non-missing values on an attribute a. In other words, suppose we consider $a \in \mathbb{A}$ and we are given pid_R. We define $pid_{R \cup \{a\}}$ as follows: We can assume that an object x_i retains its *pid* identifier if it has default value on attribute a (so that the corresponding row is missing in the EAV database), otherwise it gets a new *pid* assigned. We set the new *pid* identifier of x_i to $pid_R[i] + j \cdot \max_{i'} pid_R[i']$, where j is the index of the attained attribute value on the list of V_a elements: $v_0 = *, v_1, \ldots, v_j, \ldots, v_l$.

There are at most $\max_{a \in \mathbb{A} \setminus R} |V_a| \max_{i'} pid_R[i']$ values of new $pid_{R \cup \{a\}}$.

In order to simplify the notation, we will set $m = \max_{i'} pid_R[i'] = |\Phi|$, where $|\Phi|$ is the header in contingency table for $(\mathbb{D}, R)$.

In the version of the algorithm for sparse datasets, when we consider an attribute $a \in \mathbb{A}$ for inclusion in R, we do not store the new (temporary) $pid_{R \cup \{a\}}$ unless we include a in the result.

For each $a \in \mathbb{A} \setminus R$ we calculate the frequency matrix C of contingency table $CT(\mathbb{D}, R \cup \{a\})$ and calculate $discern(R, a)$ by using:

$$c(R) = \frac{1}{2} \sum_{i=1}^{m} \left(\left(\sum_{d=1}^{D} C_{i,d}^* \right)^2 - \sum_{d=1}^{D} (C_{i,d}^*)^2 \right)$$

$$c(R \cup \{a\}) = \frac{1}{2} \sum_{i=0}^{m} \sum_{j=0}^{|V_a|} \left(\left(\sum_{d=1}^{D} C_{i+jm,d} \right)^2 - \sum_{d=1}^{D} (C_{i+jm,d})^2 \right)$$

where C^* is the frequency matrix for $(\mathbb{D}, R)$, m is the number of rows in C^*, i.e. $m = |\Phi_R|$ and D is the number of decision classes. The temporary *pid* does not need to be stored anywhere to perform these calculations.

Entries corresponding to $j = 0$ in the equation defining $c(R \cup \{a\})$ count objects with missing value on attribute a.

In *getBestAttr* algorithm for sparse decision systems we construct frequency matrix C for an attribute a_j while simultaneously updating *discern* calculation for this frequency matrix. Similarly to *discern* calculation for discretization, only the row describing indiscernibility class containing x needs to be read in order to update *discern* in this step. Moreover, if in addition to C we also store the vector with row totals (T_i), $T_i = \sum_{d=1}^{D} C_{i,d}$, only four entries need to be accessed at each step: source row in C and T and destination row in C and T, where source and destination describe the initial and final *pid* reassigned to the object.

Theorem 2. *For sparse decision systems with N rows in EAV database the time complexity of MD-heuristic (Algorithm 1) using procedure getBestAttribute from Algorithm 3 is $O(N \log N + |R|(Dn + N))$ and the space complexity is $O(N + nD)$ (the dependency on $D = |V_{dec}|$ is typically neglected as D is usually small for classification problems).*

Proof. Storing pid_R requires $O(n)$ space, storing frequency matrices requires at most $O(nD)$ space (and storing headers is optional). EAV database may further need to be sorted, hence $O(N)$ space and $O(N \log N)$ time complexity.

Consider function *getBestAttr*. We initialize the frequency matrix C in $O(Dn)$ time. We iterate over the EAV database, sorted by attributes, and keep track of frequency matrix C counting, for a given attribute a_j, all objects with default value on a_j as well as all objects whose a_j value was visited till that point. We simulaneously update *discern* as described in the text preceeding this lemma. Resetting matrix C (when a new attribute is found) imposes no additional cost as it corresponds to reversing previous operations. The summary cost of these operations for all attributes is $O(N)$. This step is repeated $|R|$ times, hence the time complexity bound.

Function *getBestCut* for discretization in sparse decision systems takes an additional input parameter v_{miss} which denotes the default value for attribute a. The algorithm iterates over permitted cut points in $M(a)$ and updates contingency tables C_L, C_R, vectors with totals T_L, T_R and corresponding *discern* (or conflicts) moving points from U_R to U_L.

7   Experimental Results

The usefulness of decision reducts and discretization of numerical attributes was illustrated in practice in numerous applications [2]. A natural question is whether these algorithms remain useful for sparse decision systems.

Our first experiment focused on the study of select papers from the PubMed Central Open Access Subset [3] repository. Each document was assigned several medical headings-subheading pairs (MeSH) [5], i.e. medical terms from a fixed ontology that describe documents. In our study we neglected MeSH headings and focused on subheadings. The input files consited of NXML files which contained either the full text, abstract or merely the metadata without the abstract. We have only used input files with abstracts and/or full text papers in our experiments. We focused on documents that were assigned either subheading **"drug-effects"** (14202 documents) or **"toxicity"** (3928 documents), with 2175 documents assigned both of these subheadings. Our goal is to discern documents pertaining to these two subheadings.

The number of words in each document has different characteristics for the two subsets of document corresponding to different subheadings and is summarized on histograms on Fig. 1. The number of words in the two subsets of documents has slightly different characteristics: while both are bimodal with similar peaks, the average (and the median) of the number of words in "drug-effects" subset is larger than in "toxicity" subset This is due to the fact that the underlying mixtures represent abstract-only and full-paper documents. Documents in "toxicity" subset have a much larger fraction of papers which are only represented by abstracts.

In our first experiment we tried to assess whether the attributes obtained during discretization are informative. We changed all letters to lowercase and

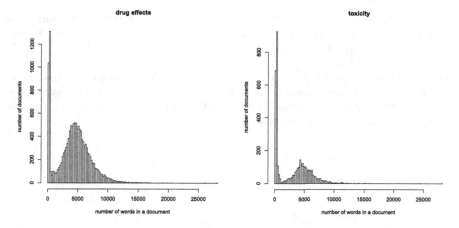

Fig. 1. The number of words for subsets "drug-effects" and "toxicity".

removed all non-alphanumeric characters. For the sake of simplicity, in this experiment we removed duplicated words from each document. No additional preprocessing (like stemming or stop words removal) was performed in the first experiment. The consecutive attributes are:

ml, exposure, many, dose, toxicity, evidence, animals, activity, images, effects, dna, development, these, health, compounds, from, studied, blood, this, caused, induced, less, or, acid, various, on, clinical, are, assessed, as, examined, have, human, lower, all, . . .

We obtained similar results when stemming was performed before applying the discretization algorithm. Most of the words which were picked in the first steps of the algorithm are informative on their own.

8   Conclusions

In this article we introduced sparse decision system versions of classical algorithms for semi-optimal decision reduct calculation and for discretization of numerical attributes. We analyzed the computational complexity of these algorithms for sparse decision systems and their application to dimensionality reduction in text mining.

References

1. Hoa, N.S., Son, N.H.: Some efficient algorithms for rough set methods. In: Proceedings IPMU 1996 Granada, Spain, pp. 1541–1457 (1996)
2. Nguyen, H.: Discretization of real value attributes, boolean reasoning approach. Ph.D. thesis, Warsaw University (1997)
3. Roberts, R.J.: PubMed central: the GenBank of the published literature. Proce. Nat. Acad. Sci. U.S.A **98**(2), 381–382 (2001)

4. Skowron, A., Rauszer, C.: The discernibility matrices and functions in information systems. In: Słowiński, R. (ed.) Intelligent Decision Support. Theory and Decision Library, vol. 11, pp. 331–362. Springer, Dordrecht (1992). https://doi.org/10.1007/978-94-015-7975-9_21
5. United States National Library of Medicine. Introduction to MeSH - 2011 (2011)
6. Wróblewski, J.: Finding minimal reducts using genetic algorithm (extended version). In: Proceedings of Second Joint Annual Conference on Information Sciences, Wrightsville Beach, North Carolina (1995)

Medical Diagnosis from Images
with Intuitionistic Fuzzy Distance
Measures

Roan Thi Ngan[1,4(✉)], Bui Cong Cuong[2(✉)], Tran Manh Tuan[3(✉)],
and Le Hoang Son[4(✉)]

[1] Hanoi University of Natural Resources and Environment, Hanoi, Vietnam
rtngan@hunre.edu.vn
[2] Institute of Mathematics, Hanoi, Vietnam
bccuong@math.ac.vn
[3] Faculty of Computer Science and Engineering, ThuyLoi University, Hanoi, Vietnam
tmtuan@tlu.edu.vn
[4] VNU University of Science, Vietnam National University, Hanoi, Vietnam
sonlh@vnu.edu.vn

Abstract. Medical diagnosis from images supports clinicians in their profession. In practical dentistry, diseases are found mainly on experience of dentists regarding dental structures and explicit symptoms of patients. In this paper, in order to reduce errors in medical diagnosis problem from images, we introduce a new diagnostic model based on intuitionistic fuzzy distance measures with parameter learning. A new intuitionistic fuzzy distance measure named Modified H-max is proposed to calculate similarity degree between an input image and all patterns of corresponding disease patterns. Parameters of the proposed measure are trained to optimize performance. Hence, the new diagnosis model has the advantages of using the cross-evaluation degree of H-max measure and weight optimization. The proposed algorithm is experimentally validated on real datasets of Hanoi Medical University, Vietnam against related methods.

Keywords: Distance measures · Dental features · X-ray images
Intuitionistic fuzzy sets · Similarity measures

1 Introduction

Fuzzy set (FS) of Zadeh was introduced to handle uncertainty [21]. It is characterized by a membership function whose range is within the unit interval. In 1986, intuitionistic fuzzy set (IFS) proposed by Atanassov [1] generalizes FS

This research is supported by the Vietnam National Foundation for Science and Technology Development (NAFOSTED) under grant number 102.01-2017.02. The author (R. T. Ngan) would like to thank the Project 911 of VNU University of Science, Vietnam National University for supporting her work.

© Springer Nature Switzerland AG 2018
H. S. Nguyen et al. (Eds.): IJCRS 2018, LNAI 11103, pp. 479–490, 2018.
https://doi.org/10.1007/978-3-319-99368-3_37

by adding a non-membership function. It overcomes limitations of FSs in handling conflicting information concerning memberships of objects. Intuitionistic fuzzy distance measure [20] which is an important content in IFS theory was used to calculate similarity degree between intuitionistic fuzzy information. It is researched and applied in many different fields such as pattern recognition [3], decision-making, medical diagnosis [8], etc.

Predicting dental diseases plays a significant role for treatment of patients, especially in their early stage, as well as for studying the diseases in nature. It is performed from examination of a dental X-ray image through its structures namely bones, soft tissues, and teeth [4–6,9,13–19]. There are several machine learning methods which have been recently used in supporting dental diagnosis. The fuzzy inference system (FIS) [10], for instance, is a common diagnosis model which uses fuzzy rules. The fuzzy k-nearest neighbor method (FKNN) [2], is used in different problems of handling dental images. A hybrid approach combining decision making, classification, and segmentation methods named Dental Diagnosis System (DDS) [12] was recently introduced. Some other methods such as the kruskal spanning tree (GCK), the prim spanning tree (GCP), and the affinity propagation clustering (APC) [18]. We have tested the previous methods on the same dataset and have received the not low error except that of DDS. Moreover, these methods are almost complex.

In this paper, in order to obtain lower error than those of the previous methods for dental diagnosis, we propose a new method denoted by DIMHM for medical diagnosis from images based on intuitionistic fuzzy distance measures. Here, instead of building fuzzy rules which require experts' experience or using a complex combination of many different algorithms, a new intuitionistic fuzzy distance measure named Modified H-max is proposed to calculate similarity degree between input and patterns of corresponding disease patterns. The largest similarity degree implies diagnosis results. The Modified H-max measure adds weights to the component functions in the H-max measure which is introduced [7]. Based on the cross-evaluation degree, H-max is more effective than other existing distance measures such as the intuitionistic Hamming, Euclidean and Hausdorff measure, etc., in decision making.

Moreover, parameters of the proposed measure are trained to optimize the mean absolute error of the DIMHM method. Hence, the new diagnosis model has the advantages of using the cross-evaluation degree of H-max measure and weight optimization. Besides, it can be seen that the approach of DIMHM is not too complex. This algorithm is implemented and experimentally validated against the related algorithms on the real dataset [12].

In what follows, Sect. 2 is the preliminary. Medical diagnosis method from images with intuitionistic fuzzy distance measures is showed in Sect. 3. The experiment results and performance comparison are presented in Sect. 4. Section 5 highlights the conclusions.

2   Preliminary

Let $FS(U)$ and $IFS(U)$ denote the sets of all FSs and IFSs in U, respectively. Here, U is a space of points.

Definition 1 [21]. *In a universal set U, a function μ_I named membership function determines a fuzzy set I which is given as follows:*

$$I = \{ (x, \mu_I(x)) \mid x \in U, \mu_I \in [0,1] \}. \tag{1}$$

Definition 2 [20]. *A function $d : FS(U) \times FS(U) \to R$ is a distance measure on FS(U) if it satisfies the following axioms:*

1. $d(I_1, I_2) \geq 0$,
2. $d(I_1, I_2) = d(I_2, I_1)$,
3. $d(I_1, I_2) = 0 \Leftrightarrow I_1 = I_2$,
4. *If $I_0 \subseteq I_1 \subseteq I_2$ then $d(I_0, I_2) \geq d(I_0, I_1)$ and $d(I_0, I_2) \geq d(I_1, I_2)$,*

where I_0, I_1 and I_2 are in IF(U).

Definition 3 [1]. *In a universal set U, two functions μ_I and ν_I named membership function and non-membership function, respectively, determine an IFS I which is given as follows:*

$$I = \{ \langle x, \mu_I(x), \nu_I(x) \rangle \mid x \in U; \mu_I, \nu_I, \text{ and } \mu_I + \nu_I \in [0,1] \}. \tag{2}$$

Definition 4 [20]. *A function $d : IFS(U) \times IFS(U) \to R$ is a distance measure on IFS(U) if it satisfies the following axioms:*

1. $d(I_1, I_2) \geq 0$,
2. $d(I_1, I_2) = d(I_2, I_1)$,
3. $d(I_1, I_2) = 0 \Leftrightarrow I_1 = I_2$,
4. *If $I_0 \subseteq I_1 \subseteq I_2$ then $d(I_0, I_2) \geq d(I_0, I_1)$ and $d(I_0, I_2) \geq d(I_1, I_2)$,*

where I_0, I_1 and I_2 are in IFS(U).

3   Proposed Method

3.1   Problem Statement

Given a dental X-ray image, let us predict the disease that can occur on this image. The disease set consists of missing teeth, resorption of periodontal bone, incluse teeth, decay, and root fracture. The dataset taken from Hanoi Medical University Hospital includes 56 images of intraoral and panoramic images (Fig. 1) [12].

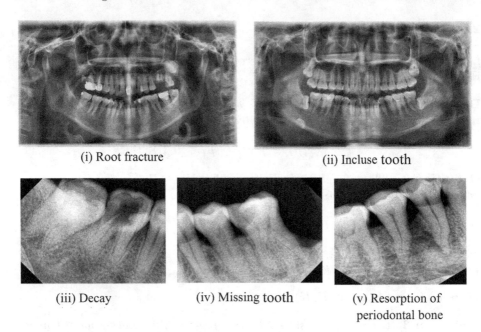

(i) Root fracture   (ii) Incluse tooth

(iii) Decay   (iv) Missing tooth   (v) Resorption of
                                                  periodontal bone

Fig. 1. The dental X-ray images with the corresponding diseases.

3.2 Extracted Dental Features

In this research, we extract and analyze five basis features of X-Ray images, which are Entropy, edge-value and intensity (EEI); Gradient feature (GRA); Local Patterns Binary feature (LBP); Patch level feature (Pat); and Red-Green-Blue (RGB) [11].

GRA: The various tiny parts of teeth which are the enamel, gum, root canal, and cementum are identified by the GRA. Firstly, the background noises of the dental image is reduced by applying the Gaussian filter. Secondly, the gradient of the image in 2D space is calculated by using Difference of Gaussian filter. Lastly, each pixel is determined by a normalized gradient vector.

EEI: This feature plays a role in the simulation of the structure of the dental image includes the background, teeth areas, and dental structure. In a certain range, the achieved information has the randomness level which is measured by Entropy in EEI. Besides, in a domain, in order to calculate the numbers of value changes of pixels, we use Edge-value and intensity.

LBP: In the dental image, we use the LBP feature to effectively distinguish clusters. In a given domain, the density order of pixels is ensured by LBP. For any light intensity transformation, this order is considered to be unchanged.

RGB: Three types of color of the dental image which are Red, Green, and Blue are measured by the RGB features.

Pat: In a patch of pixels, all gradient vectors are calculated by this feature.

3.3   Proposed Measure

The novelty of the proposed method is the introduction of the weights to the H-max measure which is trained and validated using cross-validation approach.

Definition 5. *Let A and B be in IFS($U = \{x_1, x_2, \ldots, x_m\}$) defined by the membership and non-membership degrees μ_1, ν_1 and μ_2, ν_2 respectively. The* **H-max measure** *is defined as,*

$$d(A, B) = \frac{1}{3m} \sum_{i=1}^{m} (d_\mu(x_i) + d_\nu(x_i) + d_{\mu\nu}(x_i)). \tag{3}$$

The **modified H-max measure** *is:*

$$d(A, B) = \sum_{i=1}^{m} w_i (u_1.d_\mu(x_i) + u_2.d_\nu(x_i) + u_3.d_{\mu\nu}(x_i)), \tag{4}$$

where

$$d_\mu(x_i) = |\mu_1(x_i) - \mu_2(x_i)|, \tag{5}$$
$$d_\nu(x_i) = |\nu_1(x_i) - \nu_2(x_i)|, \tag{6}$$

$$d_{\mu\nu}(x_i) = |\max\{\mu_1(x_i), \nu_2(x_i)\} - \max\{\mu_2(x_i), \nu_1(x_i)\}|, \tag{7}$$

and

$$\sum_{i=1}^{m} w_i = 1; \ \sum_{s=1}^{3} u_s = 1; \ w_{i(i=1,2,\ldots,m)} \geq 0; \ u_{s(s=1,2,3)} \geq 0. \tag{8}$$

In (3), (4), and (7), $d_{\mu\nu}$ called the cross-evaluation function is a characteristic of the H-max and modified H-max measures. The difference between A and B is fully evaluated through this cross-evaluation. By adding the weights of x_i, d_μ, d_ν and $d_{\mu\nu}$, the modified measure provides a more flexible assessment than H-max measure.

3.4   DIMHM Algorithm

The basic idea of DIMHM is to use the intuitionistic fuzzy distance measure in Sect. 3.3 to calculate the similarity degrees between an input image and all patterns of corresponding disease patterns. The largest similarity degree implies diagnosis result for the input image.

Suppose we have m images $\{I_1, I_2, \ldots, I_m\}$ and h diseases in numeric labels $\{D_1, D_2, \ldots, D_h\}$. The Hold-out or K-Fold cross-validation method is used to divide the initial images dataset into two subdatasets, which are the training and testing datasets. Here, the chosen values of K are 4, 5 and 6. The training dataset is divided into the validation dataset and the Basic Medical Knowledge by the Hold-out method.

Let $\{I_1, I_2, \ldots, I_t\}$ be the Basic Medical Knowledge, $\{I_{t+1}, \ldots, I_g\}$ be the validation dataset, and $\{I_{g+1}, \ldots, I_m\}$ be the testing dataset. The proposed diagnosis method involves some basic steps:

1. Feature extraction: The images I_i $(i = 1, 2, \ldots, m)$ are digitized by n extracted dental features denoted by F_{il} $(l = 1, 2, \ldots, n)$ (see Sect. 3.2).

2. Fuzzification: The values F_{il} $(l = 1, 2, \ldots, n)$ of the images $I_i (i = 1, 2, \ldots, m)$ are fuzzified in the form of $(\mu_{F_{il}}, \nu_{F_{il}})$:

$$\mu_{F_{il}} = \frac{\mu_{F_{il}} - \mu_{F_l \min}}{\mu_{F_l \max} - \mu_{F_l \min}}, \text{ and } \nu_{F_{il}} = \frac{1 - \mu_{F_{il}}}{1 + \lambda \mu_{F_{il}}}, \tag{9}$$

where

$$\mu_{F_l \min} = \min_i (\mu_{F_{il}}), \ \mu_{F_l \max} = \max_i (\mu_{F_{il}}), \text{ and } \lambda \in [0, 1].$$

The values $\mu_{F_{il}}$ and $\nu_{F_{il}}$ are the degrees of membership and non-membership of the image I_i in the features F_l $(l = 1, 2, \ldots, n)$, respectively. Table 1 illustrates the fuzzified dataset, where $y_{i(i=1,2,\ldots,m)} \in \{D_1, D_2, \ldots, D_h\}$.

Table 1. The fuzzified dataset

	F_1	...	F_l	...	F_n	Class Y
I_1	(μ_{11}, ν_{11})	...	(μ_{1l}, ν_{1l})	...	(μ_{1n}, ν_{1n})	y_1
...	...	... ...	...	... ...	...	...
I_i	(μ_{i1}, ν_{i1})	...	(μ_{il}, ν_{il})	...	(μ_{in}, ν_{in})	y_i
...	...	... ...	...	... ...	...	...
I_m	(μ_{m1}, ν_{m1})	...	(μ_{ml}, ν_{ml})	...	(μ_{mn}, ν_{mn})	y_m

3. Disease identification: the diagnosis of image I_i is identified based on the calculating the modified H-max distance measures between the feature values of the image I_i and those of all the images I_j $(j = 1, 2, \ldots, t)$ in the Basic Medical Knowledge.

$$d_{ij} = d(I_i, I_j) = \sum_{l=1}^{n} w_l \cdot (u_1 \cdot d_\mu (F_l) + u_2 \cdot d_\nu (F_l) + u_3 \cdot d_{\mu\nu} (F_l)), \tag{10}$$

where

$$d_\mu (F_l) = |\mu_{F_{il}} - \mu_{F_{jl}}|, \tag{11}$$

$$d_\nu (F_l) = |\nu_{F_{il}} - \nu_{F_{jl}}|, \tag{12}$$

$$d_{\mu\nu} (F_l) = |\max \{\mu_{F_{il}}, \nu_{F_{jl}}\} - \max \{\mu_{F_{jl}}, \nu_{F_{il}}\}|, \tag{13}$$

and u_1, u_2, u_3 are the parameters of the measure, w_l are the weights of the features F_l, which satisfy

$$\sum_{l=1}^{n} w_l = 1; \quad \sum_{s=1}^{3} u_s = 1; \quad w_{l(l=1,2,\ldots,n)} \geq 0; \quad u_{s(s=1,2,3)} \geq 0. \tag{14}$$

Let the measure value between feature of the image I_i and those of the image I_{j_0} be the smallest, i.e.,

$$d_{ij_0} = d\left(I_i, I_{j_0}\right) = \min_{j}\left(d_{ij}\right), \tag{15}$$

and image I_{j_0} belongs to D_{h_0} disease group. Indeed, D_{h_0} is diagnosis result of image I_i.

3.5   Training

Training Weights: Weights w_l of features F_l, where $l = 1, 2, \ldots, n$, are calculated based on the Pearson correlation coefficient function between F_l and Y on the Basic Medical Knowledge:

$$w_l = \frac{W_l}{\sum\limits_{l=1}^{n} W_l}, \tag{16}$$

where

$$W_l = \frac{|E[\mu_{F_l} Y] - E[\mu_{F_l}]E[Y]|}{\sqrt{E[\mu_{F_l}^2] - E[\mu_{F_l}]^2}.\sqrt{E[Y^2] - E[Y]^2}}. \tag{17}$$

Training Parameters: The training of parameters $u = \{u_1, u_2, u_3\}$ of the modified H-max distance (Eq. 10) on the validation dataset is defined as an optimization problem as follows:

$$F(u) = MAE(u) = \frac{1}{k} \sum_{i=1}^{k} |\hat{y}_i(u) - y_i| \rightarrow \min, \tag{18}$$

where

$$u = [u_1, u_2, u_3]^T; \quad u_1, u_2, u_3 \in [0, 1], \quad \sum_{i=1}^{3} u_i = 1, \tag{19}$$

and the objective function, $F(u)$, is the mean absolute error (MAE) function [12]. Here, k is the number of elements of the validation dataset, $\hat{y}_i(u)$ is the prediction result of the image I_i, and y_i is the observed result of I_i.

   In fact, we usually evaluate $u_1 = u_2$, therefore $u_1 = u_2 = 1 - u_3 = t \in [0, 1]$. We use the proposed diagnosis method with trained weights of features to diagnose for all the images in the validation dataset. For each set of parameters $u = \{u_1, u_2, u_3\}$ of the measure (Eq. 10), we determine the set which gives the best MAE value of the proposed algorithm. The obtained u is corresponding to the minimum MAE value on the validation dataset.

Testing: Finally, the proposed algorithm with the trained weights and parameters is used to diagnose for all the images in the testing dataset.

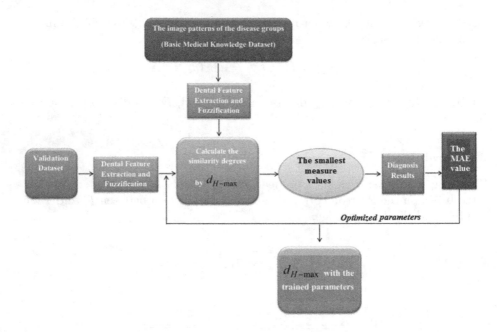

Fig. 2. Training parameters

Figures 2 and 3 illustrate the proposed medical diagnosis system from images with the training process. In Fig. 2, the parameters in the modified H-max distance measure are trained on the validation dataset to obtain the best of MAE value. Figure 3 presents the complete model, which uses the optimized parameters. In this model, the input is a medical image and the output is the diagnosis result of the input image.

Obviously, the proposed model which uses the modified H-max measure (DIMHM) is better than that uses the H-max measure (DIHM) because the parameters in the measure are trained to the optimal value.

4   Experiments

4.1   Experimental Environments

Database, Tools and Evaluation: Based on the same real dataset [12], the proposed method is validated by MAE and MSE against the related methods such as DIHM, FIS [10], FKNN [2], GCP, GCK, APC [18], and DDS [12] in

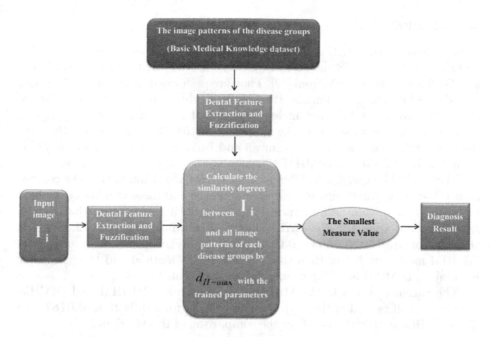

Fig. 3. The model of DIMHM.

Matlab 2015a and R languages. In details, the dataset includes 56 dental X-ray images with 5 labels which are Decay, Root fracture, Missing teeth, Resorption of periodontal bone, and Incluse teeth; and 5 extracted features which are GRA, EEI, LBP, RGB, and Pat. The link in Appendix provides the codes and the used datasets of this paper.

Parameters: In the DIMHM algorithm, the K-fold cross-validation method is used to divide the initial images dataset with $K = 4, 5, 6$. The chosen value of the parameter λ in the Fuzzification step is 0.8.

Validity Indices: Two indices used to validate the methods are MSE (Mean Squared Error) and MAE (Mean Absolute Error) as follows

$$MAE = \frac{1}{q} \sum_{i=1}^{q} |\hat{y}_i - y_i|, \tag{20}$$

$$MSE = \frac{1}{q} \sum_{i=1}^{q} (\hat{y}_i - y_i)^2, \tag{21}$$

where $\hat{y}_i$ and y_i is the prediction result and observed result of the image I_i, respectively, and q is the number of elements of the test dataset.

4.2 Results

Table 2 presents the MAE and MSE values of the proposed method (DIMHM) and the others. The MSE and MAE results of the GCP, GCK, APC, FIS, FKNN, and DDS methods are cited from [12]. They are performed on the original dataset including 87 dental X-ray images [12]. Currently, due to objective conditions, we just only have the subdataset includes 56 dental X-ray images of the original dataset. Hence, in this paper, the proposed method is validated on the subdataset. Besides, we calculated the mean and variance values of MSE and MAE from running DIHM and DIMHM 10 times on the dataset of 56 images.

The DIMHM algorithm is different from DIHM algorithm in that the component measures in the H-max measure are weighted and these weights are trained to optimize algorithm performance. From Table 2, it can be seen that for the same Hold-out approach, the MAE value of DIHM, **0.0701 ± 0.0122**, is higher than that of DIMHM, **0.0605 ± 0.0134**. That means the diagnostic results of the DIMHM model are better than those of the DIHM method and the parameter learning in DIMHM is really meaningful and efficient.

The variance values in the MAE and MSE results of DIHM and DIMHM are quite small ranged in the narrow value domain from **0.0122** to **0.0187** (see Table 2). Hence, they do not affect the comparison of the MAE and MSE results of the algorithms, i.e., it just needs to pay attention to the mean MAE and MSE values. For instance, Table 2 presents the MSE values of FKNN and DIMHM based on the Hold-out approach, which are **0.2863** and **0.0605 ± 0.0134**, respectively. It is obvious that the DIMHM diagnostic algorithm is more efficient than FKNN.

Table 2. The performance of 7 methods.

	Cross-validation	MSE	MAE
DDS	Hold-out	0.0804	0.0804
FKNN	Hold-out	0.2863	0.2346
FIS	Hold-out	0.2098	0.1982
APC	Hold-out	0.845	0.805
GCK	Hold-out	1.908	1.007
GCP	Hold-out	1.908	1.002
DIHM (using H-max)	Hold-out	0.0701 ± 0.0122	0.0701 ± 0.0122
DIMHM (using the modified H-max)	Hold-out	0.0605 ± 0.0134	0.0605 ± 0.0134
DIMHM (using the modified H-max)	4-fold	0.0558 ± 0.0175	0.0558 ± 0.0175
DIMHM (using the modified H-max)	5-fold	$\mathbf{0.0469 \pm 0.013}$	$\mathbf{0.0469 \pm 0.013}$
DIMHM (using the modified H-max)	6-fold	0.0477 ± 0.0187	0.0477 ± 0.0187

Obviously, on the same cross-validation method (Hold-out), the MSE and MAE values of the proposed method are smaller than those of the other methods. Specifically, in Table 2, the MAE value of DIMHM with Hold-out is

0.0605 ± 0.0134 while those of DIHM, GCP, GCK, APC, FIS, FKNN, and DDS are 0.0701 ± 0.0122, 1.002, 1.007, 0.805, 0.1982, 0.2346, and 0.0804, respectively. In all the cases of cross-validation (Hold-out, 4-fold, 5-fold, and 6-fold), DIMHM has the best result in the 5-fold cross-validation case with the MSE and MAE values are both 0.0469 ± 0.013. From Table 2, we also can see that the higher the algorithmic performance is, the greater the equal ability of MAE and MSE. The GCP method has the highest error, i.e., it is the worst algorithm on the used dataset. For details, the MSE value of GCP is 1.908 and the MAE value of GCP is 1.002.

In summary, the Modified H-max measure and parameter learning in the proposed method (DIMHM) are efficient tools in decision making. On the same dataset DIMHM has the best diagnostic error in the all considered methods for the medical diagnosis problem from images.

5   Conclusions

Concerning Medical Diagnosis from Images, this paper proposed a new diagnostic algorithm named DIMHM based on Intuitionistic Fuzzy Distance Measures. It uses the H-max measure with trained weights. Hence, the new diagnosis model has the advantages of using the cross-evaluation degree of H-max measure and parameter optimization. DIMHM has the best performance when comparing to 7 other methods namely DIHM, GCP, GCK, APC, FIS, FKNN, and DDS on the real medical datasets.

In this paper, the proposed method is also considered as the single nearest neighbor (1-NN) method. In fact, the k-NN method can be used on the 56-image dataset. The number of considered neighbors allow to capture the nature of a processed set. However, it will be more complicated than choosing $k = 1$. In the near future, we will research the appropriate k value on the bigger image dataset.

Validating performance of the proposed method on larger datasets will be performed in the future work. We will improve DIMHM by replacing fuzzification functions for dental features by other functions.

Appendix

The link https://source-forge.net/projects/DIMHM/ provides the code and dataset of this paper.

References

1. Atanassov, K.-T.: Intuitionistic fuzzy sets. Fuzzy Sets Syst. **20**(1), 87–96 (1986)
2. Castillo, E.-O.-R., Soria, J.: Hybrid system for cardiac arrhythmia classification with fuzzy K-Nearest neighbors and neural networks combined by a fuzzy inference system. In: Melin, P., Kacprzyk, J., Pedrycz, W. (eds.) Soft Computing for Recognition Based on Biometrics, pp. 37–55. Springer, Berlin (2010). https://doi. org/10.1007/978-3-642-15111-8_3

3. Chen, S.-M., Cheng, S.-H., Lan, T.-C.: A novel similarity measure between intuitionistic fuzzy sets based on the centroid points of transformed fuzzy numbers with applications to pattern recognition. Inf. Sci. **343–344**, 15–40 (2016)
4. Madoz, L.-V., Giuliodori, M.-J., Migliorisi, A.-L., Jaureguiberry, M., De la Sota, R.-L.: Endometrial cytology, biopsy, and bacteriology for the diagnosis of subclinical endometritis in grazing dairy cows. J. Dairy Sci. **97**(1), 195–201 (2014)
5. Meurer, M.-I., Caffery, L.-J., Bradford, N.-K., Smith, A.-C.: Accuracy of dental images for the diagnosis of dental caries and enamel defects in children and adolescents: a systematic review. J. Telemed. Telecare **21**(8), 449–458 (2015)
6. Nelson, S.-J.: Wheeler's Dental Anatomy, Physiology and Occlusion-E-Book. Elsevier Health Sciences, St Louis (2014)
7. Ngan, R.-T., Son, L.-H., Cuong, B.-C., Ali, M.: H-max distance measure of intuitionistic fuzzy sets in decision making. Appl. Soft Comput. **69**, 393–425 (2018)
8. Ngan, R.-T., Ali, M., Son, L.-H.: δ-equality of intuitionistic fuzzy sets: a new proximity measure and applications in medical diagnosis. Appl. Intell. **48**(2), 499–525 (2018)
9. Ngan, T.-T., Tuan, T.-M., Son, L.-H., Minh, N.-H., Dey, N.: Decision making based on fuzzy aggregation operators for medical diagnosis from dental X-ray images. J. Med. Syst. **40**(12), 280 (2016). 1–7
10. Oad, K.-K., DeZhi, X., Butt, P.-K.: A fuzzy rule based approach to predict risk level of heart disease. Glob. J. Comput. Sci. Technol **14**(3), 16–22 (2014)
11. Said, E., Fahmy, G.-F., Nassar, D., Ammar, H.: Dental X-ray image segmentation. In: Defense and Security. International Society for Optics and Photonics, pp. 409–417 (2004)
12. Son, L.-H., Tuan, T.-M., Fujita, H., Dey, N., Ashour, A.-S., Ngoc, V.-T.-N., Chu, D.-T.: Dental diagnosis from X-ray images: an expert system based on fuzzy computing. Biomed. Sig. Process. Control **39**, 64–73 (2018)
13. Son, L.-H., Tuan, T.-M.: A cooperative semi-supervised fuzzy clustering framework for dental X-ray image segmentation. Expert Syst. Appl. **46**, 380–93 (2016)
14. Son, L.-H., Tuan, T.-M.: Dental segmentation from X-ray images using semi-supervised fuzzy clustering with spatial constraints. Eng. Appl. Artif. Intell. **59**, 186–195 (2017)
15. Tuan, T.-M., Duc, N.-T., Hai, P.-V., Son, L.-H.: Dental diagnosis from X-ray images using fuzzy rule-based systems. Int. J. Fuzzy Syst. Appl. **6**(1), 1–16 (2017)
16. Tuan, T.-M., Ngan, T.-T., Son, L.-H.: A novel semi-supervised fuzzy clustering method based on interactive fuzzy satisficing for dental X-ray image segmentation. Appl. Intell. **45**(2), 402–428 (2016)
17. Tuan, T.-M., Son, L.-H., Dung, L.-B.: Dynamic semi-supervised fuzzy clustering for dental X-ray image segmentation: an analysis on the additional function. J. Comput. Sci. Cybern. **31**(4), 323–339 (2015)
18. Tuan, T.-M., Son, L.-H.: A novel framework using graph-based clustering for dental X-ray image search in medical diagnosis. Int. J. Eng. Technol. **8**(6), 422–427 (2016)
19. Tuan, T.-M., Son, L.-H.: A novel framework using graph-based clustering for dental X-ray image search in medical diagnosis. Int. J. Eng. Technol. **8**(6), 428–433 (2016)
20. Wang, W., Xin, X.: Distance measure between intuitionistic fuzzy sets. Pattern Recogn. Lett. **26**(13), 2063–2069 (2005)
21. Zadeh, L.-A.: Fuzzy sets. Inf. Control **8**(3), 338–353 (1965)

Rough Set Approach
to Sufficient Statistics

Huynh Bao Tuyen[1], Ta Thi Thu Phuong[1,2(✉)], and Dang Phuoc Huy[1]

[1] Department of Mathematics and Informatics, Dalat University, Dalat, Vietnam
{tuyenhb,phuongttt,huydp}@dlu.edu.vn
[2] University of Science, VNU-HCM, Ho Chi Minh City, Vietnam

Abstract. In the paper, the approach of using rough sets to verifying sufficiency of a statistic is presented. The notions of the rough set approximation operators on statistics, consistency between statistics and its properties are introduced. Then, based on these materials, the results on the sufficiency of a statistic are given.

Keywords: Rough set approximations · Consistency
Sufficient statistics

1 Introduction

In the statistical inference problems of the parametric statistical structures (or parametric statistical models), the sufficient statistics are used to replace the entire data of a random sample because they exhaust all the information that a sample has about the parameter. In some sense, because all the available information about the parameter is contained in the observations (i.e. in a random sample), using the sufficient statistics can be thought of as reducing the original observation data or data compression without loss of information about the parameter (see [4,5,12]). The formal definition of sufficiency is as follows: for a given random sample $X = (X_1, \cdots, X_n)$ taking values in the statistical structure $(\mathcal{X}, \mathcal{A}, \mathcal{P})$, distributed according to a distribution from the family $\mathcal{P} = \{P_\theta : \theta \in \Theta\}$ (where $(\mathcal{X}, \mathcal{A})$ is the sample space of X and Θ is a parameter space), a statistic $T = T(X)$ is sufficient for θ (or $\mathcal{P}$) if the conditional distribution of X given T does not depend on θ. This concept was introduced by R. A. Fisher in 1922. It plays an important role in statistical methods because the sufficient statistics preserve the Fisher information about parameters as in a sample. Many topics on such statistics have been widely investigated by many scholars (see [1,3,6–8]).

A method for finding the sufficient statistics was developed by R. A. Fisher in 1922, J. Neyman in 1935, and P. R. Halmos and L. J. Savage in 1949 which is known as the Neyman–Fisher factorization Theorem (see, e.g. [6]). Lehmann and Scheffé proposed a method to find the minimal sufficient statistic [6], and minimal sufficiency in statistics emerges from the observed likelihood functions under weak conditions is established by Fraser in [2].

© Springer Nature Switzerland AG 2018
H. S. Nguyen et al. (Eds.): IJCRS 2018, LNAI 11103, pp. 491–501, 2018.
https://doi.org/10.1007/978-3-319-99368-3_38

Notice that the Neyman–Fisher factorization Theorem only gives us a convenient way of finding sufficient statistics. In general, it is not easy to use this factorization criterion to show that a given statistic is not sufficient. The theory of rough sets, proposed by Pawlak [9], can be used as a tool for solving this problem. In this paper, we consider the discrete version of statistical structures, and introduce the notions of the rough set approximation operators on statistics. Then, the concept of the consistency between statistics is defined and its properties are also considered. The results on the sufficiency of a statistic are given by using rough sets.

2   Preliminaries

In this section, we briefly recall some basic concepts in mathematical statistics and rough set theory that are used in the next sections.

Concepts in Statistics:

A measurable space is a pair $(\Omega, \mathcal{F})$, where Ω is a non-empty set of elements and $\mathcal{F}$ is a σ-field (or σ-algebra) on Ω, i.e., a collection of subsets of Ω satisfying the conditions: (i) $\emptyset \in \mathcal{F}$; (ii) $A \in \mathcal{F}$ implies $A^c = \Omega \setminus A \in \mathcal{F}$; (iii) $\bigcup_n A_n \in \mathcal{F}$ for any countable family of subsets A_n belonging to $\mathcal{F}$. If $(\Omega, \mathcal{F})$ is a measurable space modeling an experiment, then the set Ω represents all possible outcomes of the experiment, $\mathcal{F}$ contains all events of conceivable interest to the experimenter.

Let $(\Omega, \mathcal{F})$ and $(\mathcal{X}, \mathcal{A})$ be two measurable spaces and let the mapping $X : \Omega \to \mathcal{X}$ be $(\mathcal{F}, \mathcal{A})$-measurable, i.e. $X^{-1}(B) \in \mathcal{F}$ for all $B \in \mathcal{A}$. Then, X is called a random element with values in $\mathcal{X}$ (or a $\mathcal{X}$-valued random variable). If $(\mathcal{X}, \mathcal{A}) = (\mathbb{R}, \mathcal{B}(\mathbb{R}))$ (where $\mathcal{B}(\mathbb{R})$ is the Borel σ-field on $\mathbb{R}$), then X is called a random variable.

A random sample of size n from a population is a set of n independent and identically distributed observable random variables $X_1, \ldots, X_n$. We shall denote the random sample by $X = (X_1, \ldots, X_n)$. Note that the random sample X is a $\mathbb{R}^n$-valued random variable (i.e. $(\mathcal{X}, \mathcal{A}) = (\mathbb{R}^n, \mathcal{B}(\mathbb{R}^n))$, where $\mathcal{B}(\mathbb{R}^n)$ is the Borel σ-field on $\mathbb{R}^n$, and this measurable space is called the sample space of X).

In this work, we only consider parametric statistical model as follows: let $(\mathcal{X}, \mathcal{A})$ be a measurable space and let $\mathcal{P} = \{P_\theta : \theta \in \Theta\}$ be a family of parametrized probability distributions on $(\mathcal{X}, \mathcal{A})$ with the property that every distribution P_θ is known, and where only the parameter θ is unknown and belongs to a parameter space Θ of finite dimension. Then the triplet $(\mathcal{X}, \mathcal{A}, \mathcal{P})$ is called a statistical structure. We say that a random sample X takes values in the statistical structure $(\mathcal{X}, \mathcal{A}, \mathcal{P})$ if the sample space of X is $(\mathcal{X}, \mathcal{A})$ and $\mathcal{P}$ is family of distributions of X.

Let X be a random sample taking values in the statistical structure $(\mathcal{X}, \mathcal{A}, \mathcal{P})$ and $(\mathcal{T}, \mathcal{B})$ a measurable space. If the mapping $T : \mathcal{X} \to \mathcal{T}$ is $(\mathcal{A}, \mathcal{B})$-measurable and does not depend on any unknown parameter, then T is called a statistic (of X). The statistic T of X is also sometimes written as $T : (\mathcal{X}, \mathcal{A}) \to (\mathcal{T}, \mathcal{B})$. The measurable space $(\mathcal{T}, \mathcal{B})$ is called a range space of statistic T.

Given any family $\mathcal{U}$ of subsets of a set $\mathcal{S}$ there is a smallest σ-field containing $\mathcal{U}$, which is denoted by $\sigma(\mathcal{U})$. We call $\sigma(\mathcal{U})$ the σ-field generated by $\mathcal{U}$. In particular, if $T : (\mathcal{X}, \mathcal{A}) \rightarrow (\mathcal{T}, \mathcal{B})$ is a statistic of the random sample X then the smallest σ-field containing all the sets $T^{-1}(B)$ $(B \in \mathcal{B})$ is called the σ-field generated by T and is denoted by $\sigma(T)$,

$$\sigma(T) = \sigma(\{T^{-1}(B) : B \in \mathcal{B}\}).$$

Note that we have $\sigma(T) = \{T^{-1}(B) : B \in \mathcal{B}\}$, i.e. the smallest σ-field such that T is measurable.

Definition 1 *(Sufficient statistic). Let X be a random sample taking values in the statistical structure $(\mathcal{X}, \mathcal{A}, \mathcal{P})$ and let T be a statistic of X with range space $(\mathcal{T}, \mathcal{B})$. Then, the statistic $T = T(X)$ is said to be sufficient for θ (or $\mathcal{P}$) if the conditional distribution of X given $T = t$ does not depend on θ for any value of $t \in \mathcal{T}$.*

Definition 2 *(Minimal sufficient statistic). Let X be a random sample taking values in the statistical structure $(\mathcal{X}, \mathcal{A}, \mathcal{P})$ and T a sufficient statistic for θ (of X). Then T is called a minimal sufficient statistic if for any sufficient statistic S of X there is a measurable function g such that $T = g(S)$.*

More detailed descriptions can be found in [4,11].

Concepts in Rough Set Theory:

Let U be a non-empty finite set of objects and R an equivalence relation on U. The family of all equivalence classes of R is denoted by U/R. A set of objects is characterized by a pair of definable concepts- called the *lower* and the *upper approximations*. Formally, each subset X of U is associated with two subsets

$$\underline{R}(X) = \cup\{Y \in U/R \mid Y \subseteq X\}$$
$$\overline{R}(X) = \cup\{Y \in U/R \mid Y \cap X \neq \emptyset\},$$

which are called the R-lower and R-upper approximations of X respectively.

The set $X (\subseteq U)$ is said to be a definable (precise) set (with respect to R) if $X = \underline{R}(X)$. Otherwise the set is undefinable (rough).

Let R and S be equivalence relations on U. The R-positive region $POS_R(\cdot)$ of S is defined by

$$POS_R(S) = \bigcup_{X \in U/S} \underline{R}(X).$$

For more details of rough sets can be found, e.g., in [9,10].

3   Rough Set Approximation Operators on Statistics

In this section, we introduce the notions of the rough set approximation operators on statistics.

Consider the statistical structures in a "usual" discrete setting: let X be a random sample taking values in the statistical structure $(\mathcal{X}, \mathcal{A}, \mathcal{P})$, where $\mathcal{X}$ is a non-empty discrete set, $\mathcal{A}$ is a σ-field on $\mathcal{X}$ (so, the measurable space $(\mathcal{X}, \mathcal{A})$ is a sample space); X is distributed according to a distribution from the family of probability measures $\mathcal{P} = \{P_\theta : \theta \in \Theta\}$ (on $(\mathcal{X}, \mathcal{A})$) that indexed by a parameter space Θ. The statistics of X satisfy condition as usual that all singleton sets are measurable.

Rough Set Approximation Operators on Statistics

Definition 3 *(Basic granule of a statistic). Let S be a statistic of the random sample X. For each element x of $\mathcal{X}$, the basic granule of S containing x, denoted by $[x]_S$, is defined by*

$$[x]_S = \{y \in \mathcal{X} : S(y) = S(x)\}.$$

The set $[S] = \{[x]_S : x \in \mathcal{X}\}$ is called the set of all basic granules of S.

Definition 4. *Let S, T be two statistics of the random sample X. For any $x \in \mathcal{X}$, the approximations of a basic granule $[x]_S$ (in T) are defined as:*

– the lower approximation of $[x]_S$ in T:

$$\underline{app}_T([x]_S) = \bigcup\{Z \in [T] : Z \subseteq [x]_S\}$$
$$= \{z \in \mathcal{X} : [z]_T \subseteq [x]_S\},$$

– the upper approximation of $[x]_S$ in T:

$$\overline{app}_T([x]_S) = \bigcup\{Z \in [T] : Z \cap [x]_S \neq \emptyset\}$$
$$= \{z \in \mathcal{X} : [z]_T \cap [x]_S \neq \emptyset\}.$$

From the above, the positive region of S in T is defined by:

$$POS_T(S) = \bigcup_{x \in \mathcal{X}} \underline{app}_T([x]_S).$$

Notation 1. *From the properties of the Pawlak's rough sets (see [9]), we have*

$$\forall x \in \mathcal{X} : \underline{app}_T([x]_S) \subseteq [x]_S \subseteq \overline{app}_T([x]_S),$$

and the basic granule $[x]_S$ is called a definable (precise) set in T iff $[x]_S = \underline{app}_T([x]_S)$, otherwise the set is undefinable (rough) in T.

Remark 1. Let $S : (\mathcal{X}, \mathcal{A}) \longrightarrow (\mathcal{J}, \mathcal{C})$ be a statistic of X. Recall that $\sigma(S)$ is the σ-field generated by S (i.e. the smallest σ-field such that S is measurable)

$$\sigma(S) = S^{-1}(\mathcal{C}) = \{S^{-1}(C) : C \in \mathcal{C}\}.$$

Using the fact that all singleton sets are measurable, we get by taking $\mathcal{J} = S(\mathcal{X})$ and $\mathcal{C} = \mathcal{P}(\mathcal{J})$ (where $\mathcal{P}(\mathcal{J})$ is the power set of $\mathcal{J}$)

$$S^{-1}(C) = S^{-1}(\bigcup_{s \in C} \{s\}) = \bigcup_{s \in C} S^{-1}(\{s\}) \in \mathcal{A} \quad \text{for all } C \subseteq \mathcal{J}.$$

This implies that

$$\sigma(S) = \sigma(\{S^{-1}(s) : s \in \mathcal{J}\}) = \sigma(\{[x]_S : x \in \mathcal{X}\}) = \sigma([S]),$$

i.e. the σ-field generated by the set of all basic granules of S.

4 Consistency Between Statistics

In this section we present the concept of the consistency between statistics and give its properties.

Definition 5. *Let S, T be two statistics of the random sample X. T is called a consistent statistic with respect to S if $[x]_T \subseteq [x]_S$ for any $x \in \mathcal{X}$.*

The following proposition gives the equivalent conditions for the consistency of statistics.

Proposition 1. *Let S, T be two statistics of the random sample X. Then, the following conditions are equivalent:*

(i) T is consistent with respect to S;
(ii) $\underline{app}_T([x]_S) = [x]_S, \forall x \in \mathcal{X}$;
(iii) $\overline{POS}_T(S) = \mathcal{X}$.

Proof. $[(i) \iff (ii)]$ Assume that condition (i) is satisfied. By Notation 1, we have that $\underline{app}_T([x]_S) \subseteq [x]_S$ for any $x \in \mathcal{X}$. Now use (i) to see that, for any $z \in [x]_S$, we have $[z]_T \subseteq [z]_S = [x]_S$. This implies that $[x]_S \subseteq \underline{app}_T([x]_S)$. So condition (ii) is satisfied.

Conversely, assume that condition (ii) is satisfied. Then (i) follows immediately from the definition of $\underline{app}_T([x]_S)$ and the fact that $x \in [x]_S$ (for any $x \in \mathcal{X}$).

$[(ii) \iff (iii)]$ The implication (iii) $\Rightarrow$ (ii) is trivial. To prove that (ii) implies (iii), we see that if condition (ii) is satisfied, then

$$POS_T(S) = \bigcup_{x \in \mathcal{X}} \underline{app}_T([x]_S) = \bigcup_{x \in \mathcal{X}} [x]_S = \mathcal{X},$$

so condition (iii) is satisfied. This completes the proof. □

We illustrate this proposition with an example.

Example 1. Let $X = (X_1, X_2, X_3)$ be a random sample from a Bernoulli distribution with probability of success $\theta(\theta \in (0;1))$ [11]. Notice that $\mathcal{X} = \{0;1\}^3$ and $\mathcal{A} = \mathcal{P}(\mathcal{X})$. We consider the following two statistics of the random sample X

$$S = (X_1 + X_2, X_3) \text{ and } T = \begin{cases} (\frac{1}{3}, 0), & \text{if } \min\{X_1, X_3\} = 1, X_2 = 0, \\ (\frac{2}{3}, 0), & \text{if } \min\{X_2, X_3\} = 1, X_1 = 0, \\ S, & \text{otherwise.} \end{cases}$$

The values of X, S and T are presented in Table 1.

Table 1. Values of X, S and T

X	S	T
$(0,0,0)$	$(0,0)$	$(0,0)$
$(1,0,0)$	$(1,0)$	$(1,0)$
$(0,1,0)$	$(1,0)$	$(1,0)$
$(0,0,1)$	$(0,1)$	$(0,1)$
$(1,1,0)$	$(2,0)$	$(2,0)$
$(1,0,1)$	$(1,1)$	$(\frac{1}{3},0)$
$(0,1,1)$	$(1,1)$	$(\frac{2}{3},0)$
$(1,1,1)$	$(2,1)$	$(2,1)$

From this we have the sets of all basic granules of S, T as follows:

$$[S] = \{\{(0,0,0)\}, \{(1,0,0), (0,1,0)\}, \{(0,0,1)\}, \{(1,1,0)\},$$
$$\{(1,0,1), (0,1,1)\}, \{(1,1,1)\}\}$$
$$[T] = \{\{(0,0,0)\}, \{(1,0,0), (0,1,0)\}, \{(0,0,1)\}, \{(1,1,0)\},$$
$$\{(1,0,1)\}, \{(0,1,1)\}, \{(1,1,1)\}\}.$$

Therefore the cardinalities of the lower approximations of all basic granules of S in T are given in the following Table 2 (where $|A|$ denotes the cardinality of the set A). Hence

$$|POS_T(S)| = \sum_{x \in \mathcal{X}} |\underline{app}_T([x]_S)| = 8 = |\mathcal{X}|$$

i.e.,

$$POS_T(S) = \mathcal{X}.$$

So by Proposition 1 it follows that T is consistent with respect to S.

Table 2. Cardinalities of the lower approximations of all basic granules of S in T

$[S]$	$|app_T([x]_S)|$
$\{(0,0,0)\}$	1
$\{(1,0,0),(0,1,0)\}$	2
$\{(0,0,1)\}$	1
$\{(1,1,0)\}$	1
$\{(1,0,1),(0,1,1)\}$	2
$\{(1,1,1)\}$	1

5   Rough Set Approach to Sufficient Statistics

This section presents the results on the sufficiency of a statistic by using rough sets.

First we need the following lemma.

Lemma 1. *Let X be a random sample taking values in the statistical structure $(\mathcal{X}, \mathcal{A}, \mathcal{P})$ and S, T the statistics of X. Then*

$$T \text{ is consistent with respect to } S \iff \forall x \in \mathcal{X} : [x]_S = \bigcup_{z \in [x]_S} [z]_T.$$

Proof. Assume that T is consistent with respect to S. Then, for any $x \in \mathcal{X}$ and $z \in [x]_S$, by $S(z) = S(x)$, we have that

$$[z]_T \subseteq [z]_S = [x]_S.$$

Therefore

$$\bigcup_{z \in [x]_S} [z]_T \subseteq [x]_S \subseteq \bigcup_{z \in [x]_S} [z]_T$$

and hence $[x]_S = \bigcup_{z \in [x]_S} [z]_T$.

Conversely, assume that $[x]_S = \bigcup_{z \in [x]_S} [z]_T$ for any $x \in \mathcal{X}$. Then for each $x \in \mathcal{X}$, since $x \in [x]_S$, we have

$$[x]_T \subseteq \bigcup_{z \in [x]_S} [z]_T = [x]_S.$$

So T is consistent with respect to S.   □

Theorem 1. *Let X be a random sample taking values in the statistical structure $(\mathcal{X}, \mathcal{A}, \mathcal{P})$ and S a sufficient statistic for θ of X. Let T be a statistic of X. If T is consistent with respect to S, then T is also a sufficient statistic for θ of X.*

Proof. Assume that T is consistent with respect to S. Then by Remark 1 and Lemma 1 we have

$$\sigma(S) = \sigma([S]) = \sigma(\{[x]_S : x \in \mathcal{X}\}) = \sigma\left(\left\{\bigcup_{z \in [x]_S} [z]_T : x \in \mathcal{X}\right\}\right)$$

$$\subseteq \sigma(\{[x]_T : x \in \mathcal{X}\}) = \sigma([T]) = \sigma(T), \tag{1}$$

so S is $\sigma(T)$-measurable.

For notational convenience, we shall assume that T and S are statistics with range spaces $(\mathcal{T}, \mathcal{B})$ and $(\mathcal{J}, \mathcal{C})$, respectively. Then by a classical result of Doob-Dynkin (see, e.g. [4], Lemma 2.3.1) there exists a $\mathcal{B}$-measurable function f such that $S(x) = f(T(x))$ for all $x \in \mathcal{X}$.

Now from the sufficiency of statistic S for θ and by the Neyman–Fisher factorization Theorem (see, e.g. [4,6]), there exist nonnegative $\mathcal{C}$-measurable functions g_θ and a nonnegative $\mathcal{A}$-measurable function h such that $P_\theta(x) = g_\theta[S(x)]h(x)$ for all $x \in \mathcal{X}$ and $\theta \in \Theta$. Hence we have

$$P_\theta(x) = g_\theta[S(x)]h(x) = g_\theta\big[f(T(x))\big]h(x)$$
$$= g_\theta \circ f(T(x))h(x) = u_\theta(T(x))h(x),$$

where u_θ denotes the composite function $g_\theta \circ f$.

From this we get, again, by the Neyman–Fisher factorization Theorem, that T is also a sufficient statistic for θ. $\qquad\square$

Example 2. We return to Example 1. Since X is a random sample from a Bernoulli distribution $B(1; \theta)$ with probability of success θ ($\theta \in (0; 1)$), we have the probability mass function $P_\theta(x)$ of X as follows [11]:

$$P_\theta(x) = \theta^{\sum_{k=1}^{3} x_k}(1 - \theta)^{3 - \sum_{k=1}^{3} x_k}$$

for each $x = (x_1, x_2, x_3) \in \mathcal{X}$.

Recall that we consider the following two statistics of the random sample X

$$S = (X_1 + X_2, X_3) \text{ and}$$

$$T = \begin{cases} (\frac{1}{3}, 0), & \text{if } \min\{X_1, X_3\} = 1, X_2 = 0, \\ (\frac{2}{3}, 0), & \text{if } \min\{X_2, X_3\} = 1, X_1 = 0, \\ S, & \text{otherwise.} \end{cases}$$

Notice that the statistics $S_1 = X_1 + X_2$ and $S_2 = X_3$ have the Binomial distributions $B(2; \theta)$ and $B(1; \theta)$, respectively. So we obtain the probability mass function $P_\theta^S(s) = P_\theta(S = s)$ of S by using that S_1 is independent of S_2

$$P_\theta^S(s) = C_2^{t_1}\theta^{s_1+s_2}(1 - \theta)^{3-(s_1+s_2)} \quad \text{for } s = (s_1, s_2) \in \{0; 1; 2\} \times \{0; 1\}.$$

Hence we may rewrite $P_\theta(x)$ as

$$P_\theta(x) = C_2^{S_1(x)}\theta^{S_1(x)+S_2(x)}(1 - \theta)^{3-(S_1(x)+S_2(x))}\frac{1}{C_2^{S_1(x)}} \quad \text{for } x \in \mathcal{X}.$$

From this we get, by applying the Neyman–Fisher factorization Theorem, that S is a sufficient statistic for θ. Then, since T is consistent with respect to S (see Example 1), by Theorem 1 we conclude that T is a sufficient statistic for θ.

The converse assertion of theorem above does not hold generally. To see this, we consider another statistic of X as $H = (X_1, X_2 + X_3)$. Using the same argument as with the statistic S, we obtain that H is also a sufficient statistic for θ. Now consider the set of all basic granules of H

$$[H] = \{\{(0,0,0)\}, \{(1,0,0)\}, \{(0,1,0),(0,0,1)\}, \{(1,1,0),(1,0,1)\},$$
$$\{(0,1,1)\}, \{(1,1,1)\}\}.$$

Then the cardinalities of the lower approximations of all basic granules of S in H are given in the following Table 3. We have

$$|POS_H(S)| = \sum_{x \in \mathcal{X}} |\underline{app}_H([x]_S)| = 3 \neq 8 = |\mathcal{X}|$$

i.e., $POS_H(S) \neq \mathcal{X}$. So by Proposition 1 it follows that H is inconsistent with respect to S.

Table 3. Cardinalities of the lower approximations of all basic granules of S in H

$[S]$	$|\underline{app}_H([x]_S)|$
$\{(0,0,0)\}$	1
$\{(1,0,0),(0,1,0)\}$	1
$\{(0,0,1)\}$	0
$\{(1,1,0)\}$	0
$\{(1,0,1),(0,1,1)\}$	0
$\{(1,1,1)\}$	1

Theorem 2. *Let X be a random sample taking values in the statistical structure $(\mathcal{X}, \mathcal{A}, \mathcal{P})$ and S a minimal sufficient statistic for θ of X. Let T be a statistic of X. Then*

$$T \text{ is sufficient for } \theta \iff T \text{ is consistent with respect to } S.$$

Proof. Recall that a statistic is said to be minimal sufficient for θ if and only if it is sufficient for θ and is a measurable function of all other sufficient statistics for θ. Now, assume that T is sufficient for θ. As mentioned above we assume that T and S are statistics with range spaces $(\mathcal{T}, \mathcal{B})$ and $(\mathcal{J}, \mathcal{C})$, respectively. Then, since S is minimal sufficient for θ, again using the Doob-Dynkin Lemma leads to that there exists a $\mathcal{B}$-measurable function f such that $S(x) = f(T(x))$ for all $x \in \mathcal{X}$. Hence, for each $x \in \mathcal{X}$ we have

$$z \in [x]_T \Rightarrow T(z) = T(x) \Rightarrow f(T(z)) = f(T(x))$$
$$\Rightarrow S(z) = S(x) \Rightarrow z \in [x]_S.$$

This implies that $[x]_T \subseteq [x]_S$. So T is consistent with respect to S.

Conversely, assume that T is consistent with respect to S. Then the sufficiency of statistic T for θ is immediate consequence of Theorem 1.   □

Example 3. Let X be a random sample from a Bernoulli distribution with probability of success θ and $P_\theta(x)$ the probability mass function of X as in Example 2. We define two other statistics of the random sample X by putting $S = X_1 + X_2 + X_3$ and $T = X_1 + X_2 - X_3$. Notice that the statistic S has the Binomial distributions $B(3; \theta)$ and since ratio

$$\frac{P_\theta(x)}{P_\theta(y)} = \frac{\theta^{S(x)}(1-\theta)^{3-S(x)}}{\theta^{S(y)}(1-\theta)^{3-S(y)}} = \left(\frac{\theta}{1-\theta}\right)^{S(x)-S(y)}$$

is independent of θ if and only if $S(x) = S(y)$ (for all $x, y \in \mathcal{X}$), we obtain, by the Lehmann-Scheffé Theorem for minimal sufficient statistics (see, e.g. [6]), that S is a minimal sufficient statistic for θ. Table 4 gives the values of X, S and T, respectively.

Table 4. Values of X, S and T

X	S	T
$(0,0,0)$	0	0
$(1,0,0)$	1	1
$(0,1,0)$	1	1
$(0,0,1)$	1	-1
$(1,1,0)$	2	2
$(1,0,1)$	2	0
$(0,1,1)$	2	0
$(1,1,1)$	3	1

We have the following sets of all basic granules of S, T

$$[S] = \{\{(0,0,0)\}, \{(1,0,0), (0,1,0), (0,0,1)\}, \{(1,1,0), (1,0,1), (0,1,1)\}, \{(1,1,1)\}\}$$
$$[T] = \{\{(0,0,1)\}, \{(0,0,0), (1,0,1), (0,1,1)\}, \{(1,0,0), (0,1,0), (1,1,1)\}, \{(1,1,0)\}\},$$

and the cardinalities of the lower approximations of all basic granules of S in T are given in Table 5.

From this we have

$$|POS_T(S)| = \sum_{x \in \mathcal{X}} |\underline{app}_T([x]_S)| = 2 \neq 8 = |\mathcal{X}|,$$

so by Proposition 1 it follows that T is inconsistent with respect to S. Hence, by Theorem 2 we conclude that T is not a sufficient statistic for θ.

Table 5. Cardinalities of the lower approximations of all basic granules of S in T

$[S]$	$\lvert \underline{app}_T([x]_S) \rvert$
$\{(0,0,0)\}$	0
$\{(1,0,0),(0,1,0),(0,0,1)\}$	1
$\{(1,1,0),(1,0,1),(0,1,1)\}$	1
$\{(1,1,1)\}$	0

6   Conclusion

In this paper, we have studied the sufficiency of a statistic by using rough sets. We introduced the concept of consistency between statistics, and based on this concept, the results on the sufficiency of a statistic were given.

Acknowledgments. The authors would like to thank all the anonymous reviewers for their comments to improve the quality of the paper.

References

1. Benavoli, A., de Campos, C.P.: Statistical tests for joint analysis of performance measures. In: Suzuki, J., Ueno, M. (eds.) AMBN 2015. LNCS (LNAI), vol. 9505, pp. 76–92. Springer, Cham (2015). https://doi.org/10.1007/978-3-319-28379-1_6
2. Fraser, D.A.S., Naderi, A.: Minimal sufficient statistics emerge from the observed likelihood functions. Int. J. Stat. Sci. 5(Special Issue) (2006)
3. Lehmann, E.L.: An interpretation of completeness and Basu's theorem. J. Am. Stat. Assoc. **76**(374), 335–340 (1981)
4. Lehmann, E.L., Romano, J.P.: Testing Statistical Hypotheses, 3rd edn. Springer Science+Business Media Inc, New York (2005)
5. Ly, A., Marsman, M., Verhagen, J., Grasman, R.P.P.P., Wagenmakers, E.-J.: A tutorial on Fisher information. J. Math. Psychol. **80**, 40–55 (2017)
6. Martin, R.: Exponential Families, Sufficiency & Information. Stat 511. Lecture Notes II (2014)
7. Mukhopadhyay, N., Banerjee, S.: Fisher information, sufficiency, and ancillary: some clarifications. In: METRON, vol. 71, pp. 33–38 (2013). https://doi.org/10.1007/s40300-013-0005-0
8. Park, S., Ng, H.K.T., Chan, P.S.: On the Fisher information and design of a flexible progressive censored experiment. Stat. Probab. Lett. **97**, 142–149 (2015)
9. Pawlak, Z.: Rough Sets: Theoretical Aspects of Reasoning About Data. Kluwer Academic Publishers, Dordrecht (1991)
10. Pawlak, Z., Skowron, A.: Rudiments of rough sets. Inf. Sci. **177**, 3–27 (2007)
11. Ramachandran, K.M., Tsokos, C.P.: Mathematical Statistics with Applications. Elsevier Academic Press (2009)
12. Stein, M.S., Nossek, J.A., Barbé, K.: Fisher information lower bounds with applications in hardware-aware nonlinear signal processing. arXiv Preprint arXiv:1512.03473v2 [cs.IT] 27 May 2018

A Formal Study of a Generalized Rough Set Model Based on Relative Approximations

Md. Aquil Khan$^{(\boxtimes)}$ and Vineeta Singh Patel$^{(\boxtimes)}$

Discipline of Mathematics, Indian Institute of Technology Indore,
Indore 453552, India
aquilk@iiti.ac.in, vineetasingh1994@gmail.com

Abstract. We propose a generalization of the rough set model where approximation operators are defined relative to a given collection of subsets of the domain of objects. A modal logic with semantics based on relative accessibility relations is also proposed, that can be used to reason about the proposed approximations.

1 Introduction

Rough set theory, introduced by Pawlak in the early 1980s [13] offers an approach to deal with the uncertainty inherent in real-life problems, more specifically that stemming from inconsistency or vagueness in data. Pawlak's rough set model is based on the simple notion of approximation space (W, R), where R is an equivalence relation on the domain W. Objects being in the same equivalence class of R are indiscernible using knowledge provided by R. In general, a concept $X \subseteq W$ may not be precisely describable in terms of information provided by the equivalence relation R. It is then approximated from 'within' and 'outside', by its *lower* and *upper* approximations $\underline{X}_R$ and $\overline{X}_R$, respectively, where

$$\underline{X}_R := \{x \in W : R(x) \subseteq X\} \text{ and } \overline{X}_R := \{x \in W : R(x) \cap X \neq \emptyset\}. \quad (1)$$

Here, $R(x)$ denotes the set $\{y \in W : (x, y) \in R\}$.

With time, many generalizations of Pawlak's rough set model have been proposed in the literature (e.g. [5,11,14–17]). A useful natural generalization is the one where the distinguishability relation R is not necessarily an equivalence. For instance, in [8,15], a *tolerance approximation space* is considered, where R is a tolerance (i.e., reflexive and symmetric) relation. The notions of lower and upper approximations of a set in these generalized approximation spaces are then defined naturally using (1).

Another natural generalization of Pawlak's rough set model is one where we consider a number of relations instead of just one. For instance, we have the following notion of *tolerance information structure*.

V. S. Patel—This work has been supported by the Council of Scientific and Industrial Research (CSIR) India, Research Grant No. 09/1022(0028)/2016-EMR-I.

H. S. Nguyen et al. (Eds.): IJCRS 2018, LNAI 11103, pp. 502–510, 2018.
https://doi.org/10.1007/978-3-319-99368-3_39

Definition 1. *A* tolerance information structure *is defined as a tuple* $(W, \{R_B\}_{B \subseteq \mathcal{A}})$, *where* $\mathcal{A}$ *is a non-empty set of attributes, and for each* $B \subseteq \mathcal{A}$, R_B *is a tolerance relation on the* W *satisfying (i)* $R_\emptyset := W \times W$ *and (ii)* $R_B := \bigcap_{a \in B} R_{\{a\}}$.

The relations R_B are intended to represent the similarity relations relative to attribute set B obtained from the incomplete information systems (cf. [9,10]). We note that in the original definition of *information structure* proposed in [12], the relations R_B were taken as equivalence relations as they were intended to represent indiscernibility relations. But, in this article, our study will be based on similarity relation and hence, accordingly, we made the necessary changes.

   Let us return to the notions of approximations once again and note that the definitions of the same given by (1) are defined relative to the whole *domain* W of the (generalized) approximation space. But in some situation it may be useful to consider *a subset of the domain instead of the whole domain*. Thus, we consider the following notion of relative approximations.

Definition 2. *Let* (W, R) *be a generalized approximation space and* $Y \subseteq W$. *The lower and upper approximations of a set* $X \subseteq W$ *relative to* Y, *denoted as* $\underline{X}_{R,Y}$ *and* $\overline{X}_{R,Y}$, *respectively, are defined as follows.*

$$\underline{X}_{R,Y} := \{x \in Y : R(x) \cap Y \subseteq X\} \text{ and } \overline{X}_{R,Y} := \{x \in Y : R(x) \cap Y \cap X \neq \emptyset\}.$$

Observe that $\underline{X}_{R,W}$ and $\overline{X}_{R,W}$ are the standard lower and upper approximations defined on generalized approximation space.

We now propose the following generalization of the notion of information structure.

Definition 3 (Tolerance Subset Information Structure). *A* tolerance subset information structure, *in brief TSIS,* $(W, \sigma, \{R_B\}_{B \subseteq \mathcal{A}})$ *consists of a tolerance information structure* $(W, \{R_B\}_{B \subseteq \mathcal{A}})$ *along with a non-empty collection* σ *of subsets of* W.

Here, $\sigma \subseteq \wp(W)$ gives the collection of subsets of W, called *the sets of interest*, with respect to which we are interested to calculate the relative approximations (cf. Definition 2).

   Let us try to explain the above concept with the help of an example. Recall the notion of incomplete information system (in brief, IIS) and similarity relation defined on it. Consider a situation where there is a spread of an unknown disease, and we aim to study its symptoms. Suppose the IIS $\mathcal{K}$ (cf. Table 1) provides information gathered from a hospital, and we need to make decisions based on this information. $\mathcal{K}$ contains four attributes a_1, a_2, a_3, d representing three symptoms a_1, a_2, a_3 and the presence/absence of the disease, respectively. Let X be the concept 'infected with the disease'. Based on the information provided by $\mathcal{K}$, we obtain $X := \{P_1, P_6\}$. Let $B := \{a_1, a_2, a_3\}$. Note that P_1 and P_2 belong to the undecidable region $\overline{X}_{Sim_B^S} \setminus \underline{X}_{Sim_B^S}$ of the concept X. At this point, one may like not to take into account the patients P_4 and P_5 as for these patients we do

Table 1. IIS $\mathcal{K}$

Patient	a_1	a_2	a_3	d	Patient	a_1	a_2	a_3	d
P_1	+	+	+	Yes	P_4	+	*	*	No
P_2	+	+	−	No	P_5	*	*	+	No
P_3	+	−	+	No	P_6	+	+	*	Yes

not have enough information. Therefore, one may wish to consider the relative approximations $\underline{X}_{Sim_B^S,Y}$ and $\overline{X}_{Sim_B^S,Y}$, where $Y := \{P_1, P_2, P_3, P_6\}$. Observe that with respect to these approximations, P_1 does not remain undecidable and moves to the region $\underline{X}_{Sim_B^S,Y}$.

Similarly, one may be interested in the approximations relative to the set $Z := \{P_1, P_2, P_3\}$, the set of patients about whom we have complete information regarding the attributes. Thus, under the above circumstances, we may be interested on the TSIS $(W, \sigma, \{Sim_B^{\mathcal{K}}\}_{B \subseteq \{a_1,a_2,a_3\}})$, where $W := \{P_1, \ldots, P_6\}$ and $\sigma := \{\{P_1, \ldots, P_4\}, \{P_1, P_2, P_3\}\}$.

In this article, we aim to study the behaviour of rough sets, more specifically, relative approximations, under the framework of TSIS. In such a study many natural questions arise. For example, which objects are 'definitely' (not) elements of a concept relative to all the sets of interest? Or, which objects are definitely elements of a concept relative to some sets of interest? Accordingly, we will propose notions of approximations based on TSIS in Sect. 2 and some ensuing properties will be discussed.

There have been extensive studies on the logics that can be used to reason about the approximations of concepts. For a detailed survey on rough set logics, we refer to [3,4]. In literature one can find several proposals of logics with semantics based on relative accessibility relations where we have a family of relations indexed with attribute sets (cf. e.g. [1,2,6,12]). These relations are intended to capture the distinguishability relations (like indiscernibility, similarity etc.) relative to different attribute sets. These proposals, as required, are multi-modal logics with a modal operator $[P]$ for each subset P for the attribute set. Modal operators $[P]$ are intended to capture the approximations of concepts with respect to distinguishability relations relative to attribute set P. It should be mentioned here that Orłowska [12] cited the axiomatization of a logic with semantics based on information structures as an open problem. Later, Balbiani gave a complete axiomatization of the set of wffs valid in every information structure. In fact, in [2], complete axiomatizations of logics with semantics based on various types of structures with relative accessibility relations are presented. One of these is a logic for information structures (cf. [1]).

At this point, it is pertinent to mention that we have not come across any proposals of rough set logics that can capture the approximations of concepts relative to different subsets of the domain of the underlined (generalized) approximation space. Hence we are not aware of a logic that can be used to reason about the approximations of concepts proposed in this article (cf. Definitions 2 and 4).

In Sect. 3, we will introduce such logic for TSISs, and it will be shown in Sect. 4 how the language can be used for this purpose. Section 5 concludes the article.

2 Relative Approximations and Tolerance Subset Information Structure

Let us first recall the notion of relative approximations (cf. Definition 2) and note the following properties.

Proposition 1. *Let* (W, R) *be a tolerance approximation space and* $X, Y, V \subseteq W$. *Then the following hold.*

- $\underline{\emptyset}_{R,V} = \overline{\emptyset}_{R,V} = \emptyset$ *and* $\underline{X}_{R,V} = \overline{X}_{R,V} = V$ *for all* $X \supseteq V$.
- $\underline{X}_{R,V} \subseteq V$ *and* $\overline{X}_{R,V} \subseteq V$.
- $\underline{X}_{R,V} \subseteq X$.
- $X \subseteq \overline{X}_{R,V}$ *if and only if* $X \subseteq V$.
- $\left(\underline{X^{cv}}_{R,V}\right)^{cv} = \overline{X}_{R,V}$, *where, for* $Y \subseteq W$, Y^{cv} *denotes the set* $V \setminus Y$.
- $\underline{X \cap Y}_{R,V} = \underline{X}_{R,V} \cap \underline{Y}_{R,V}$.
- $\underline{X}_{R,V} \cup \underline{Y}_{R,V} \subseteq \underline{X \cup Y}_{R,V}$.
- $\overline{X \cup Y}_{R,V} = \overline{X}_{R,V} \cup \overline{Y}_{R,V}$.
- $\overline{X \cap Y}_{R,V} \subseteq \overline{X}_{R,V} \cap \overline{Y}_{R,V}$.
- *If* $X \subseteq Y$, *then* $\underline{X}_{R,V} \subseteq \underline{Y}_{R,V}$ *and* $\overline{X}_{R,V} \subseteq \overline{Y}_{R,V}$.
- $X \subseteq \overline{\underline{X}_{R,V}}_{R,V}$ *holds if* $X \subseteq V$.
- $\underline{X}_{R,V} \subseteq \underline{\overline{X}_{R,V}}_{R,V}$.

Next, we propose the following notions of approximations based on TSIS. Let $\mathfrak{F} := (W, \sigma, \{R_B\}_{B \subseteq A})$ be a TSIS, and $X \subseteq W$.

Definition 4. *The necessity lower approximation* $L^n_{R_B}(X)$, *possibility lower approximation* $L^p_{R_B}(X)$, *necessity upper approximation* $U^n_{R_B}(X)$, *and possibility upper approximation* $U^p_{R_B}(X)$ *with respect to the relation* R_B, *respectively, are defined as follows.*

$$L^n_{R_B}(X) := \bigcap_{V \in \sigma} \underline{X}_{R_B,V}; \qquad L^p_{R_B}(X) := \bigcup_{V \in \sigma} \underline{X}_{R_B,V};$$

$$U^n_{R_B}(X) := \bigcap_{V \in \sigma} \overline{X}_{R_B,V}; \qquad U^p_{R_B}(X) := \bigcup_{V \in \sigma} \overline{X}_{R_B,V}.$$

Thus, $L^p_{R_B}(X)$ $(L^n_{R_B}(X))$ consists of objects that are in the lower approximation of the concept X with respect to R_B, relative to some (respectively, all) sets from σ. Similarly, $U^p_{R_B}(X)$ $(U^n_{R_B}(X))$ consists of objects that are in the upper approximation of the concept X with respect to R_B, relative to some (respectively, all) sets from σ. At this point, it is important to note that the above-defined approximations are very different and based on the entirely different structure and ideas from the possibility and necessity approximations considered in [7], although we have used the same name.

The obvious relationship between the defined approximations are:

$$L_{R_B}^n(X) \subseteq L_{R_B}^p(X), \ U_{R_B}^n(X) \subseteq U_{R_B}^p(X) \text{ and}$$
$$L_{R_B}^p(X) \subseteq U_{R_B}^n(X) \text{ if } X \subseteq \bigcap_{V \in \sigma} V.$$

It is not difficult to see that a tolerance information structure $(W, \{R_B\}_{B \subseteq \mathcal{A}})$ can be viewed as the TSIS $(W, \sigma, \{R_B\}_{B \subseteq \mathcal{A}})$, where $\sigma := \{W\}$. Moreover, in such a TSIS, we obtain

$$L_{R_B}^n(X) := \underline{X}_{R_B} = L_{R_B}^p(X) \text{ and } U_{R_B}^n(X) := \overline{X}_{R_B} = U_{R_B}^p(X).$$

Next proposition lists a few properties of the proposed approximations.

Proposition 2. *1.* $M_{R_B}(X) \subseteq X$ *for* $M \in \{L^n, L^p\}$.
2. $X \subseteq U_{R_B}^p(X)$ *if and only if for all* $x \in X$, *there exists a* $V \in \sigma$ *such that* $x \in V$.
3. $X \subseteq U_{R_B}^n(X)$ *if and only if* $X \subseteq \bigcap_{V \in \sigma} V$.
4. $L_{R_B}^n(X \cap Y) = L_{R_B}^n(X) \cap L_{R_B}^n(Y)$.
5. $M_{R_B}(X \cap Y) \subseteq M_{R_B}(X) \cap M_{R_B}(Y)$ *for* $M \in \{L^p, U^p, U^n\}$.
6. $M_{R_B}(X) \cup M_{R_B}(Y) \subseteq M_{R_B}(X \cup Y)$ *for* $M \in \{L^n, L^p, U^n\}$.
7. $U_{R_B}^p(X \cup Y) = U_{R_B}^p(X) \cup U_{R_B}^p(Y)$.
8. $L_{R_B}^p(W) = W$ *if and only if for all* x, *there exists a* $V \in \sigma$ *such that* $x \in V$.
9. $M_{R_B}(W) = W$ *if and only if* $\sigma := \{W\}$ *for* $M \in \{L^n, U^n\}$.
10. $M_{R_B}(\emptyset) = \emptyset$ *for* $M \in \{L^n, L^p, U^p, U^n\}$.
11. $L_{R_B}^p(X^c) \subseteq (U_{R_B}^n(X))^c$, *where* $X^c := W \setminus X$.
12. $\bigcap_{V \in \sigma} \cap L_{R_B}^p(X^c) \supseteq \bigcap_{V \in \sigma} \cap (U_{R_B}^n(X))^c$.
13. $L_{R_B}^n(X^c) \subseteq (U_{R_B}^p(X))^c$.
14. $\bigcap_{V \in \sigma} \cap L_{R_B}^n(X^c) \supseteq \bigcap_{V \in \sigma} \cap (U_{R_B}^p(X))^c$.
15. If $X \subseteq Y$, *then* $M_{R_B}(X) \subseteq M_{R_B}(Y)$ *for all* $M \in \{L^n, L^p, U^n, U^p\}$.

3 Proposal of a Logic with Semantics Based on the Relative Accessibility Relations

In this section, we shall propose a logic that can be used to reason about relative approximations defined in Definition 2 with respect to similarity relations corresponding to different set of attributes. The semantics of the logic will be based on TSISs.

3.1 Syntax

The alphabet of the language $\mathcal{L}$ contains (i) a non-empty countable set PV of propositional variables, (ii) a non-empty empty set $\mathcal{A}$ of attribute constants, and (iii) the propositional constants $\top, \bot$. The propositional variables $p \in PV$ and propositional constants $\top, \bot$ constitute the set of atomic well-formed formulae. Using atomic well-formed formulae, the standard Boolean logical connectives $\neg$

(negation) and $\wedge$ (conjunction), the modal connectives $\square, \square_C$ where $C \subseteq \mathcal{A}$, the well-formed formulae (in brief, wffs) of $\mathcal{L}$ is then defined recursively as:

$$p \mid \top \mid \bot \mid \neg\alpha \mid \alpha \wedge \beta \mid \square\alpha \mid \square_C\alpha,$$

where $p \in PV$ and α, β are wffs. Apart from the usual derived connectives $\vee$, $\rightarrow$, $\leftrightarrow$, we have the connectives $\Diamond$, and $\Diamond_C$ defined as follows:

$$\Diamond_C\alpha := \neg\square_C\neg\alpha, \text{ and } \Diamond\alpha := \neg\square\neg\alpha.$$

We will make use of the same symbol $\mathcal{L}$ to denote the set of all wffs of the language $\mathcal{L}$.

3.2 Semantics

We have the following definition of model.

Definition 5. *A model of $\mathcal{L}$ is a tuple $\mathfrak{M} := (\mathfrak{F}, V)$, where*

- *$\mathfrak{F} := (W, \sigma, \{R_B\}_{B \subseteq \mathcal{A}})$ is a TSIS,*
- *$V : PV \rightarrow \wp(W)$ is a* valuation function.

The satisfiability of a wff α in a model $\mathfrak{M} := (\mathfrak{F}, V)$, where $\mathfrak{F} := (W, \sigma, \{R_B\}_{B \subseteq \mathcal{A}})$, at (x, U) with $x \in U \in \sigma$, denoted as $\mathfrak{M}, x, U \models \alpha$, is defined inductively as follows. We omit the cases of propositional constants and Boolean connectives.

Definition 6.

$$\mathfrak{M}, x, U \models \square_B\alpha \iff \text{ for all } y \in U \text{ with } xR_By, \ \mathfrak{M}, y, U \models \alpha.$$
$$\mathfrak{M}, x, U \models \square\alpha \iff \text{ for all } V \in \sigma \text{ with } x \in V, \ \mathfrak{M}, x, V \models \alpha.$$

The satisfiability conditions of the derived connectives are then obtained as follows.

Proposition 3.

$$\mathfrak{M}, x, U \models \Diamond_B\alpha \iff \quad \text{there exists a } y \in U \text{ with } xR_By \text{ such that } \mathfrak{M}, y, U \models \alpha.$$
$$\mathfrak{M}, x, U \models \Diamond\alpha \iff \quad \text{there exists a } V \in \sigma \text{ with } x \in V \text{ such that } \mathfrak{M}, x, V \models \alpha.$$

For any wff α, model $\mathfrak{M}$ and $U \in \sigma$, let

$$[\![\alpha]\!]_{\mathfrak{M}, U} := \{x \in W : \mathfrak{M}, x, U \models \alpha\}.$$
$$[\![\alpha]\!]_{\mathfrak{M}} := \{(x, U) \in W \times \sigma : x \in U \ \& \ \mathfrak{M}, x, U \models \alpha\}.$$

Let us use $\mathcal{L}^*$ to denote the set of all wffs α such that for all models $\mathfrak{M}$, object x and $U \in \sigma$, we have,

$$\mathfrak{M}, x, U \models \alpha \iff \mathfrak{M}, x, V \models \alpha \text{ for all } V \in \sigma \text{ with } x \in V.$$

That is, the satisfiability of wffs from $\mathcal{L}^*$ do not depend on the elements from σ. Therefore, for $\alpha \in \mathcal{L}^*$, we will use $[\![\alpha]\!]_{\mathfrak{M}}^*$ to denote the set

$$\{x \in W : \mathfrak{M}, x, U \models \alpha \text{ for some } U \in \sigma\}.$$

Observe that wffs that do not involve modal operators $\square_C$ and $\lozenge_C$ belong to the set $\mathcal{L}^*$.

A wff α is said to be *valid in* $\mathfrak{M}$, notation: $\mathfrak{M} \models \alpha$, if $[\![\alpha]\!]_{\mathfrak{M}} = \{(x, U) : W \times \sigma : x \in U\}$. α is said to be valid if $\mathfrak{M} \models \alpha$ for all $\mathfrak{M}$.

4   Rough Set Interpretation

Let us consider a model $\mathfrak{M} := (\mathfrak{F}, V)$, where $\mathfrak{F} := (W, \sigma, \{R_B\}_{B \subseteq \mathcal{A}})$. Then, we have the following.

Proposition 4. *For a model $\mathfrak{M}$, $\alpha \in \mathcal{L}$ and $\beta \in \mathcal{L}^*$, we have the following.*

1. $[\![\square_B \alpha]\!]_{\mathfrak{M},U} = \underline{[\![\alpha]\!]_{\mathfrak{M},U}}_{R_B,U}$, $[\![\lozenge_B \alpha]\!]_{\mathfrak{M},U} = \overline{[\![\alpha]\!]_{\mathfrak{M},U}}_{R_B,U}$;

2. $[\![\square_B \beta]\!]_{\mathfrak{M},U} = \underline{[\![\beta]\!]_{\mathfrak{M}}^*}_{R_B,U}$, $[\![\lozenge_B \beta]\!]_{\mathfrak{M},U} = \overline{[\![\alpha]\!]_{\mathfrak{M}}^*}_{R_B,U}$;

3. $[\![\square\alpha]\!]_{\mathfrak{M},U} = \bigcap_{V \in \sigma} [\![\alpha]\!]_{\mathfrak{M},V}$, $[\![\lozenge\alpha]\!]_{\mathfrak{M},U} = \bigcup_{V \in \sigma} [\![\alpha]\!]_{\mathfrak{M},U}$.

From Items 1 and 2, it is evident that the operators $\square_B$ and $\lozenge_B$ capture lower and upper approximations, respectively, with respect to the relation R_B relative to the set at which the wffs are evaluated. The operator $\square_B$ can be combined with the operator $\square$ to capture necessity and possibility approximations, as shown by the following proposition. Let us define the following connectives for each $B \subseteq \mathcal{A}$.

$$\triangle_B^n \alpha := \square\square_B \alpha, \qquad\qquad \triangle_B^p \alpha := \lozenge\square_B \alpha,$$
$$\triangledown_B^p \alpha := \neg\triangle_B^n \neg\alpha, \qquad\qquad \triangledown_B^n \alpha := \neg\triangle_B^p \neg\alpha.$$

Proposition 5. *For a model $\mathfrak{M}$ and $\beta \in \mathcal{L}^*$, we have the following.*

1. $[\![\triangle_B^n \beta]\!]_{\mathfrak{M},U} = L_{R_B}^n([\![\beta]\!]_{\mathfrak{M}}^*)$, $[\![\triangledown_B^p \beta]\!]_{\mathfrak{M},U} = U_{R_B}^p([\![\beta]\!]_{\mathfrak{M}}^*)$;
2. $[\![\triangle_B^p \beta]\!]_{\mathfrak{M},U} = L_{R_B}^p([\![\beta]\!]_{\mathfrak{M}}^*)$, $[\![\triangledown_B^n \beta]\!]_{\mathfrak{M},U} = U_{R_B}^n([\![\beta]\!]_{\mathfrak{M}}^*)$.

It follows from Proposition 5 that if $\beta \in \mathcal{L}^*$, then we also have $\triangle_B^n \beta$, $\triangle_B^p \beta$, $\triangledown_B^n \beta$, $\triangledown_B^p \beta \in \mathcal{L}^*$. The properties listed in Propositions 1 and 2 translate into valid wffs of the language $\mathcal{L}$. We end this section with the following proposition that lists a few such valid wffs.

Proposition 6. *The following wffs are valid in the model $\mathfrak{M} := (\mathfrak{F}, V)$, where $\mathfrak{F} := (W, \sigma, \{R_B\}_{B \subseteq \mathcal{A}})$.*

- $\square_B \alpha \to \alpha$.
- $\alpha \to \lozenge_B \alpha$.
- $\square_B(\alpha \wedge \beta) \leftrightarrow \square_B \alpha \wedge \square_B \beta$.

- $\Box_B \alpha \vee \Box_B \beta \rightarrow \Box_B(\alpha \vee \beta)$.
- $\alpha \rightarrow \Box_B \Diamond_B \alpha$.
- $\triangle_B^n \alpha \rightarrow \alpha$.
- $\triangle_B^p \top \leftrightarrow \top$.
- $\triangle_B^p \alpha \rightarrow \alpha$ *if* $\alpha \in \mathcal{L}^*$.
- $\alpha \rightarrow \nabla_B^p \alpha$.
- $\triangle_B^n(\alpha \wedge \beta) \leftrightarrow \triangle_B^n \alpha \wedge \triangle_B^n \beta$.
- $\triangle_B^p(\alpha \wedge \beta) \rightarrow \triangle_B^p \alpha \wedge \triangle_B^p \beta$.
- $\nabla_B^p(\alpha \vee \beta) \leftrightarrow \nabla_B^p \alpha \vee \nabla_B^p \beta$.
- $\nabla_B^n \alpha \vee \nabla_B^n \beta \rightarrow \nabla_B^n(\alpha \vee \beta)$.

5 Conclusions

In this article, we proposed a generalization of the rough set model where approximation operators are defined relative to a given collection of subsets of the domain of objects. A few properties of the proposed approximations are studied, but a detailed study on the proposed generalization covering the standard notions like definability, membership function, dependency etc. needs to be done. Similarly, the axiomatization and decidability problems of the proposed logic also need to be answered.

References

1. Balbiani, P.: Axiomatization of logics based on Kripke models with relative accessibility relations. In: Orłowska, E. (ed.) Incomplete Information: Rough Set Analysis, pp. 553–578. Physica Verlag, Heidelberg, New York (1998)
2. Balbiani, P., Orłowska, E.: A hierarchy of modal logics with relative accessibility relations. J. Appl. Non-Class. Log. **9**(2–3), 303–328 (1999)
3. Banerjee, M., Khan, M.A.: Propositional logics from rough set theory. In: Peters, J.F., Skowron, A., Düntsch, I., Grzymała-Busse, J., Orłowska, E., Polkowski, L. (eds.) Transactions on Rough Sets VI. LNCS, vol. 4374, pp. 1–25. Springer, Heidelberg (2007). https://doi.org/10.1007/978-3-540-71200-8_1
4. Demri, S., Orłowska, E.: Incomplete Information: Structure, Inference, Complexity. Springer, Heidelberg (2002). https://doi.org/10.1007/978-3-662-04997-6
5. Dubois, D., Prade, H.: Rough fuzzy sets and fuzzy rough sets. Int. J. Gen. Syst. **17**, 191–209 (1990)
6. Farinas Del Cerro, L., Orłowska, E.: *DAL* - a logic for data analysis. Theor. Comput. Sci. **36**, 251–264 (1985)
7. Khan, M.A.: A probabilistic approach to rough set theory with modal logic perspective. Inf. Sci. **406–407**, 170–184 (2017)
8. Komorowski, J., Pawlak, Z., Polkowski, L., Skowron, A.: Rough sets: a tutorial. In: Pal, S.K., Skowron, A. (eds.) Rough Fuzzy Hybridization: A New Trend in Decision-Making, pp. 3–98. Springer, Singapore (1999)
9. Kryszkiewicz, M.: Rough set approach to incomplete information systems. Inf. Sci. **112**, 39–49 (1998)
10. Kryszkiewicz, M.: Rules in incomplete information systems. Inf. Sci. **113**, 271–292 (1999)

11. Lin T.Y., Yao, Y.Y.: Neighborhoods system: measure, probability and belief functions. In: Proceedings of the 4th International Workshop on Rough Sets and Fuzzy Sets and Machine Discovery, pp. 202–208, November 1996
12. Orłowska, E.: Kripke semantics for knowledge representation logics. Studia Logica **49**, 255–272 (1990)
13. Pawlak, Z.: Rough sets. Int. J. Comput. Inf. Sci. **11**(5), 341–356 (1982)
14. J. A. Pomykała. Approximation, similarity and rough constructions. ILLC pre-publication series for computation and complexity theory CT-93-07, University of Amsterdam (1993)
15. Skowron, A., Stepaniuk, J.: Tolerance approximation spaces. Fundam. Inform. **27**, 245–253 (1996)
16. Ślęzak, D., Ziarko, W.: The investigation of the Bayesian rough set model. Int. J. Approx. Reason. **40**, 81–91 (2005)
17. Ziarko, W.: Variable precision rough set model. J. Comput. Syst. Sci. **46**, 39–59 (1993)

Decidability in Pre-rough Algebras: Extended Abstract

Zhe Lin[1], Mihir Kumar Chakraborty[2], and Minghui Ma[1(✉)]

[1] Institute of Logic and Cognition, Sun Yat-sen University, Guangzhou, China
{linzhe8,mamh6}@mail.sysu.edu.cn
[2] School of Cognitive Science, Jadavpur University, Kolkata, India
mihir4@gmail.com

Abstract. Some classes of topological quasi-Boolean algebras, including algebraic structures related with rough sets, are enriched with residuated and adjoint pairs. The strong finite model property for these classes of algebraic structures is established. The decidability of equational theories of these classes of algebras is derived from the finite model property.

Keywords: Pre-rough algebra · Finite model property · Decidability

1 Introduction

Rough set theory was introduced by Z. Pawlak in 1982 [8]. It was immediately observed that rough set models have both algebraic and topological components. Investigations on both these aspects exist abundantly. In this paper we shall deal with only the algebraic aspect. We wish to mention that Pomykala's work [9] probably was the beginning of this research direction. There are also many other researches who make contributions to this area [1,3,11,12].

Classical rough set theory starts with approximation spaces which are pairs of the form $\langle X, R \rangle$, where X is a non-empty set, and R is an equivalence relation on X that gives a partition. In the literature (cf. e.g. [8]), for any subset A of X, the pair $\langle \underline{A}, \overline{A} \rangle$ is called a *rough set* in the approximation space $\langle X, R \rangle$. The sets $\underline{A}$ and $\overline{A}$ are called *lower* and *upper* approximations of A respectively, and they are formally defined as follows:

$$\underline{A} = \{x \in X \mid [x]_R \subseteq A\} \text{ and } \overline{A} = \{x \in X \mid [x]_R \cap A \neq \emptyset\}$$

Z. Lin—The work was supported by Chinese National Funding of Social Sciences (No. 17CZX048).

M. Ma—The work was supported by Guangdong Province (China) Pearl River Scholar Funded Scheme (2017–2019).

© Springer Nature Switzerland AG 2018
H. S. Nguyen et al. (Eds.): IJCRS 2018, LNAI 11103, pp. 511–521, 2018.
https://doi.org/10.1007/978-3-319-99368-3_40

where $[x]_R$ is the equivalence class of $x \in X$ with respect to R. Now we define meet ($\sqcap$), join ($\sqcup$) and complementation ($\neg$) on rough sets as follows:

$$\langle \underline{A}, \overline{A} \rangle \sqcap \langle \underline{B}, \overline{B} \rangle = \langle \underline{A} \cap \underline{B}, \overline{A} \cap \overline{B} \rangle.$$

$$\langle \underline{A}, \overline{A} \rangle \sqcup \langle \underline{B}, \overline{B} \rangle = \langle \underline{A} \cup \underline{B}, \overline{A} \cup \overline{B} \rangle.$$

$$\neg \langle \underline{A}, \overline{A} \rangle = \langle \underline{A}^c, \overline{A}^c \rangle.$$

One can observe from [2,3] that the algebraic structure of rough sets forms a quasi-Boolean algebra (qBa), the formal definition of which will be given in the next section. If R discretizes X totally, that is, if each equivalence class is a singleton, the qBa becomes a Boolean algebra viz. the power set algebra $\mathcal{P}(X)$. In a slightly modified definition [10], rough sets form a topological quasi-Boolean algebra (tqBa) with respect to a topological operator. In the present work, we shall use this modified definition (cf. [3]).

From the perspective of tqBa, a rough set is a pair $\langle D_1, D_2 \rangle$ such that $D_1 \subseteq D_2 \subseteq X$ where D_1 and D_2 are unions of equivalence classes with respect to R. Such unions are called *definable sets*. Let $\mathcal{D} = \{\langle D_1, D_2 \rangle \mid D_1 \subseteq D_2\}$. The structure $\mathfrak{D} = \langle \mathcal{D}, \sqcap, \sqcup, \neg, \langle \emptyset, \emptyset \rangle, \langle X, X \rangle \rangle$ is a qBa. It is also observed that the approximations $\underline{A}$ and $\overline{A}$ are definable sets and $\underline{A} \subseteq \overline{A}$. Furthermore, we define the unary operator $\Diamond$ on $\mathcal{D}$ by $\Diamond \langle D_1, D_2 \rangle = \langle D_2, D_1 \rangle$. Then $\langle \mathfrak{D}, \Diamond \rangle$ forms a tqBa.

The aim of this work is to enhance some algebraic structures in [11] with two additional binary operators: the product ($\bullet$) and implication/residual ($\rightarrow$). In such a way, we can obtain more logical properties of algebraic structures related with rough sets using tools from partially ordered residuated algebras (cf. e.g. [4]). We shall explore the finite model property (FMP) of these enriched algebraic structures. Although the study of FMP of algebraic structures has a long tradition, in order to make this work as self contained as possible, we shall give the definition of FMP, strong finite model property (SFMP) and related concepts in the next section.

It should be mentioned that current investigations on rough set theory have already traversed a long way form the original starting point of a set X with an equivalence relation. First, in place of an equivalence relation, any arbitrary relation has been taken and lower/upper approximations of a set are defined. This gives immediate connection with Kripke frames and modal logic (cf. e.g. [5,13,14,16]). Second, in place of a partition of X due to the equivalence relation, a general covering is taken and this has emerged a wide branch called *covering-based rough sets* (cf. e.g. [6,15]). Abstract algebraic studies in the former case have been carried out [2,3,11,12]. There have been category-theoretic studies as well [7]. But to our best knowledge, there are no attempts to enrich algebraic structures with residuation pairs. We hope that a new branch of logical-algebraic studies in abstract rough sets will emerge out of the present research.

2   Pre-rough Algebras

As mentioned in the introduction, rough algebra are algebraic structures based on quasi-Boolean algebras.

Definition 1. A *quasi-Boolean algebra* (qBa) is an algebra $\mathbb{A} = (A, \wedge, \vee, \neg, \perp, \top)$ where $(A, \wedge, \vee, \perp, \top)$ is a bounded distributive lattice, and $\neg$ is an unary operation on A such that the following conditions hold for all $a, b \in A$:

(DN) $\neg\neg a = a$.
(DM) $\neg(a \vee b) = \neg a \wedge \neg b$.

The lattice order $\leq$ on A is defined by: $a \leq b$ if and only if $a \wedge b = a$, or equivalently $a \vee b = b$.

Definition 2. A *topological quasi-Boolean algebra* (tqBa) is an algebra $\mathbb{A} = (A, \wedge, \vee, \neg, \perp, \top, \square)$ where $(A, \wedge, \vee, \neg, \perp, \top)$ is a qBa, and $\square$ is an unary operation on A such that the following conditions hold for all $a, b \in A$:

(N$_\square$) $\square\top = \top$.
(T$_\square$) $\square a \leq a$.
(K$_\square$) $\square(a \wedge b) = \square a \wedge \square b$.
(4$_\square$) $\square a \leq \square\square a$.

A *topological quasi-Boolean 5 algebra* (tqBa5) is a tqBa $\mathbb{A}$ such that the following condition holds for all $a \in A$:

(B) $\Diamond\square a = a$,

where $\Diamond$ is the unary operation on A defined by $\Diamond a := \neg\square\neg a$. We use **qBa**, **tqBa** and **tqBa5** to denote classes of quasi-Boolean algebras, topological quasi-Boolean algebras and topological quasi-Boolean 5 algebras, respectively.

Definition 3. An *intermediate algebra of type 1* (IA1) is a tqBa5 $\mathbb{A}$ satisfying the following condition for all $a \in A$:

(IA1) $\neg\square a \vee \square a = 1$.

An *intermediate algebra of type 2* (IA2) is a tqBa5 $\mathbb{A}$ satisfying the following condition for all $a, b \in A$:

(IA2) $\square(a \vee b) = \square a \vee \square b$.

An *intermediate algebra of type 3* (IA3) is a tqBa5 $\mathbb{A}$ satisfying the following condition for all $a, b \in A$:

(IA3) if $\square a \leq \square b$ and $\Diamond a \leq \Diamond b$, then $a \leq b$.

A *pre-rough algebra* (Pra) is an IA1, IA2 or IA3. We use **IA1**, **IA2**, **IA3** and **Pra** to denote classes of intermediate algebras of type 1, intermediate algebras of type 2, intermediate algebras of type 3 and pre-rough algebras, respectively.

Now we shall introduce equational logics of topological quasi-Boolean algebras. Let $\mathbb{X}$ be a denumerable set of variables.

Definition 4. The set $T(\mathbb{X})$ of all *terms* for tqBa is defined as follows:

$$T(\mathbb{X}) \ni \varphi ::= x \mid (\varphi \wedge \varphi) \mid (\varphi \vee \varphi) \mid \neg\varphi \mid I\varphi, \text{ where } x \in \mathbb{X}.$$

Terms are denoted by φ, ψ, χ etc. We define $C\varphi := \neg I\neg\varphi$. An *equation* is an expression of the form $\varphi \approx \psi$ where $\varphi, \psi \in T(\mathbb{X})$. Equations are denoted by s, t etc. with or without subscripts. A *quasi-equation* is an expression of the form $(s_1 \mathbin{\&} \cdots \mathbin{\&} s_n) \supset s_{n+1}$.

An *assignment* in a tqBa $\mathbb{A}$ is a function $\sigma : \mathbb{X} \to A$. An assignment σ is extended homomorphically to all terms, and $\sigma(\varphi)$ is the value of φ. An equation $\varphi \approx \psi$ is *valid* in **A**, if $\sigma(\varphi) = \sigma(\psi)$ for any assignment σ in **A**. A quasi-equation $(\varphi_1 \approx \psi_1 \mathbin{\&} \ldots \mathbin{\&} \varphi_n \approx \psi_n) \supset \varphi_0 \approx \psi_0$ is *valid* in **A**, if for any assignment σ in $\mathbb{A}$, $\sigma(\varphi_i) = \sigma(\psi_i)$ for all $1 \le i \le n$ imply $\sigma(\varphi_0) = \sigma(\varphi_0)$.

An equation or quasi-equation is valid in a class of algebras **K** if it is valid in all algebras in **K**. Let **K** be any class of algebras. The *equational theory* of **K** is defined as the set $Eq(\mathbf{K})$ of all equations which are valid in **K**. For any set of equations or quasi-equations Σ, let $\mathbf{Alg}(\Sigma)$ be the class of all algebras which validate all equations in Σ. A class of algebras **K** is called a *variety* if there is a set of equations Σ such that $\mathbf{K} = \mathbf{Alg}(\Sigma)$. A class of algebras **K** is called a *quasi-variety* if there is a set of quasi-equations Θ such that $\mathbf{K} = \mathbf{Alg}(\Theta)$.

It is obvious that **qBa**, **tqBa**, **tqBa5**, **IA1** and **IA2** are varieties since they are defined by equations. **IA3** and **Pra** are quasi-varieties since they are defined by quasi-equations. Now, given a variety or quasi-variety **K**, a natural question is the decidability of its equational theory $Eq(\mathbf{K})$.

We shall prove some decidability results in terms of finite model property. A class of algebras **K** has the *finite model property* (FMP), if any equation which is not valid in **K** is refuted by a finite member of **K**. The FMP of **K** yields the decidability of the equational theory $Eq(\mathbf{K})$.

Given a set of equations Φ and an equation $\varphi \approx \psi$, a more general question is whether $\varphi \approx \psi$ is valid in **K** if all equations in Φ are valid in **K**. A positive answer to this question follows from the *strong finite model property* of the *Horn theory* of **K**. A *Horn sentence* is a universal sentence of the form $\forall x_1 \ldots x_m(s_1 \mathbin{\&} \cdots \mathbin{\&} s_n \supset s_{n+1})$ where $n, m \ge 0$ and each s_i $(1 \le i \le n+1)$ is an equation. The *Horn theory* of **K**, denoted by $Horn(\mathbf{K})$, is the set of all Horn sentences that are valid in **K**. We say that a quasi-variety **K** has the *strong finite model property* (SFMP), if any Horn sentence not valid in **K** is refuted in a finite member of **K**. If **K** has the SFMP, $Horn(\mathbf{K})$ is decidable.

3 Residuated Pre-rough Algebras

Definition 5. A *bounded commutative residuated groupoid* (crg) is a partially ordered algebraic structure $\mathbb{G} = (G, \cdot, \to, \top.\bot, \le)$ where $(G, \le)$ is poset, $\bot$ and $\top$ are the least and greatest elements in G, and $\cdot$ and $\to$ are binary operations on G satisfying the following conditions for all $a, b, c \in G$:

(COM) $a \cdot b = b \cdot a$.

(RES) $a \cdot b \leq c$ if and only if $a \leq b \to c$.

Let **crg** be the class of all bounded commutative residuated groupoids.

Definition 6. A quasi-Boolean commutative residuated groupoid (qBacrg) is an algebra $\mathbb{G} = (G, \cdot, \to, \bot, \top, \wedge, \vee, \neg)$ where (i) $(G, \wedge, \vee, \bot, \top)$ is an bounded distributive lattice, and (ii) the following double negation law holds:

$$(\text{DNE}) \; \neg\neg a \leq a$$

where $\neg$ is the unary operation on G defined by $\neg a = a \to \bot$ for all $a \in G$, and (iii) $(G, \cdot, \to, \bot, \top, \leq)$ is a crg where $\leq$ is the lattice order. Let **qBacrg** be the class of all quasi-Boolean commutative residuated groupoid.

Example 1. Let $\mathfrak{B} = (B, \wedge, \vee, \bot, \top, \neg, \cdot, \leq_B)$ be a Boolean algebra and $\leq_B$ be the lattice order on B. Let $\cdot$ be a binary operator on B such that for all $a, b \in B$:

(1) if $a \leq_B b$, then $c \cdot a \leq_B c \cdot b$ and $a \cdot c \leq_B b \cdot c$.
(2) $a \cdot \bot = \bot$.

Let $A = \{\langle a, b \rangle \in B \times B \mid a \leq_B b\}$. We define the binary relation $\leq$ and the operations $\sqcap, \sqcup, \sim$ and $\odot$ on A as follows:

$$\langle a, b \rangle \leq \langle a', b' \rangle \text{ iff } a \leq a' \text{ and } b \leq b'.$$

$$\langle a, b \rangle \sqcap \langle a'b' \rangle = \langle a \wedge a', b \wedge b' \rangle.$$

$$\langle a, b \rangle \sqcup \langle a'b' \rangle = \langle a \vee a', b \vee b' \rangle.$$

$$\sim \langle a, b \rangle = \langle \neg b, \neg a \rangle.$$

$$\langle a, b \rangle \odot \langle a', b' \rangle = \langle a \cdot a', b \cdot b' \rangle.$$

Then $Q(\mathfrak{B}) = (A, \leq, \sqcup, \sqcap, \sim, \langle 0, 0 \rangle, \langle 1, 1 \rangle, \odot)$ is a partially ordered quasi-Boolean algebra with the binary operator $\odot$ satisfying the following conditions:

(1) if $\langle a, b \rangle \leq \langle a', b' \rangle$, then $(c, c') \odot \langle a, b \rangle \leq (c, c') \odot \langle a', b' \rangle$ and $\langle a, b \rangle \odot (c, c') \leq \langle a', b' \rangle \odot (c, c')$.
(2) $\langle a, b \rangle \odot \langle 0, 0 \rangle = \langle 0, 0 \rangle$.
(3) $\langle a, b \rangle \odot \langle c, d \rangle = \langle c, d \rangle \odot \langle a, b \rangle$

Now we define an implication operation $\to$ on A as follows:

$$\langle a, b \rangle \to \langle b, c \rangle = \bigvee \{ \langle a'', b'' \rangle \in A \mid \langle a, b \rangle \odot \langle a'', b'' \rangle \leq \langle a', b' \rangle \}.$$

Since $\langle a, b \rangle \odot \langle 0, 0 \rangle = \langle 0, 0 \rangle$, the supermum $\bigvee \{ \langle a'', b'' \rangle \in A \mid \langle a, b \rangle \odot \langle a'', b'' \rangle \leq \langle a', b' \rangle \}$ exists. It is easy to show that

$$\langle a, b \rangle \odot \langle a', b' \rangle \leq \langle a'', b'' \rangle \text{ if and only if } \langle a, b \rangle \leq \langle a', b' \rangle \to \langle a'', b'' \rangle.$$

Then $(A, \leq, \sqcup, \sqcap, \sim, \langle 0, 0 \rangle, \langle 1, 1 \rangle, \odot, \to)$ is qBacrg.

Example 2. Let $\mathfrak{G} = (G, \cdot)$ be a commutative groupoid. We define the binary operation $\odot$ on the powerset $\mathcal{P}(G)$ as follows:

$$X \bullet Y = \{a \cdot b \mid a \in X \text{ and } b \in Y\}.$$

Consider the algebraic structure $\mathfrak{B} = (\mathcal{P}(G), \bullet, \cup, \cap, {}^c, \emptyset, \mathcal{P}(G), \subseteq)$ where the reduct $(\mathcal{P}(G), \cup, \cap, {}^c, \emptyset, \mathcal{P}(G))$ is the powerset Boolean algebra. Clearly $\bullet$ satisfies the following conditions for all $X, Y, Z \in \mathcal{P}(G)$:

(1) if $X \subseteq Y$, then $Z \bullet X \subseteq Z \bullet Y$ and $X \bullet Z \subseteq Y \bullet Z$.
(2) $X \bullet \emptyset = \emptyset$.

Let $A = \langle X, Y \rangle \in \mathcal{P}(G) \times \mathcal{P}(G) \mid X \subseteq Y\}$. Using the construction in Example 1, we obtain a qBacrg on A.

Definition 7. A *topological quasi-Boolean commutative residuated groupoid* (tqBacrg) is an algebra $\mathbb{G} = (G, \cdot, \rightarrow, \Diamond, \Box^{\downarrow}, \bot, \top, \wedge, \vee)$ where $(G, \cdot, \rightarrow, \bot, \top, \wedge, \vee)$ is a qBacrg, and $\Diamond$ and $\Box^{\downarrow}$ are unary operations on G satisfying the following conditions for all $a, b \in G$,

(Adj) $\Diamond a \leq b$ if and only if $a \leq \Box^{\downarrow} b$.
($4_\Diamond$) $\Diamond \Diamond a \leq \Diamond a$.
($T_\Diamond$) $a \leq \Diamond a$.

The condition (Adj) is called the *adjointness* law for the pair $(\Diamond, \Box^{\downarrow})$. Let **tqBacrg** be the class of all topological quasi-Boolean commutative residuated groupoids. (Note that the algebra $(G, \wedge, \vee, \bot, \top, \neg, \Box)$ is a tqBa.)

A *topological quasi-Boolean 5 commutative residuated groupoid* (tqBacrg5) is a tqBacrg $\mathbb{G}$ satisfying the following condition for all $a \in G$:

($5_{\Diamond\Box}$) $\Diamond a \leq \Box \Diamond a$.

Let **tqBacrg5** be the class of all topological quasi-Boolean 5 commutative residuated groupoids.

Definition 8. A *pre-rough algebra with commutative residuated groupoid* (Pracrg) is a tqBacrg5 $\mathbb{G}$ satisfying the following conditions for all $a, b \in G$:

(IA1$\Diamond$) $\Diamond a \wedge \neg \Diamond a \leq \bot$.
(IA2$\Diamond$) $\Diamond a \wedge \Diamond b \leq \Diamond (a \wedge b)$.
(IA3$\Diamond$) if $\Box a \leq \Box b$ and $\Diamond a \leq \Diamond b$, then $a \leq b$.

Let **Pracrg** be the class of all pre-rough algebra with commutative residuated groupoids. Intermediate algebras of type 1 (IA1crg), type 2 (IA2crg), type 3 (IA3crg), and their combinations IA12crg and IA23crg, are defined naturally.

We consider all algebras between **tqBacrg** and **Pracrg** (including **tqBacrg** and **Pracrg**) defined above. These classes of algebras are quasi-varieties. The algebras presented in Sect. 2 can be expanded to corresponding algebras defined above. An algebra $\mathbf{A}'$ is called an *expansion* of $\mathbf{A}$, if $\mathbf{A}'$ is obtained from $\mathbf{A}$ by adding new operations such that $\mathbf{A}$ is a reduct of $\mathbf{A}'$.

Lemma 1. *The following hold:*

(1) *Every* qBa *is expanded to a* qBacrg.
(2) *Every* tqBa *is expanded to a* tqBacrg.
(3) *Every* tqBa5 *is expanded to a* tqBacrg5.
(4) *Every* IA1 *is expanded to a* IA1crg.
(5) *Every* IA2 *is expanded to a* IA2crg.
(6) *Every* IA3 *is expanded to a* IA3crg.
(7) *Every* IA12 *is expanded to a* IA12crg.
(8) *Every* IA23 *is expanded to a* IA23crg.
(9) *Every* Pra *is expanded to a* Pracrg.

We present a construction of powerset algebra from a residuated algebra defined above, which will be essentially used in the proof of SFMP.

Definition 9. Let $\mathbb{G} = (G, \cdot, \dagger)$ be a commutative groupoid with an unary operation $\dagger$ on G. We define the following operations on the powerset $\mathcal{P}(G)$:

$$U \odot V = \{a \cdot b \in \mathbb{G} : a \in U, b \in V\},$$
$$\Diamond U = \{\dagger a \in \mathbb{G} : a \in U\},$$
$$U \to V = \{a \in \mathbb{G} : U \odot \{a\} \subseteq V\},$$
$$\square^{\downarrow} U = \{a \in \mathbb{G} : \dagger a \in U\},$$
$$U \vee V = U \cup V,$$
$$U \wedge V = U \cap V.$$

where $U, V \subseteq G$. Let $\mathfrak{P}(\mathbb{G}) = (\mathcal{P}(G), \odot, \Diamond, \to, \square^{\downarrow}, \vee, \wedge, \emptyset, G)$.

Definition 10. Let $\mathbf{G} = (G, \cdot, \dagger)$ be a commutative groupoid with an unary operation $\dagger$ on G. An operation $C : \mathcal{P}(G) \to \mathcal{P}(G)$ is called a *closure operator* on $\mathcal{P}(G)$, if the following conditions are satisfied:

(C1) $U \subseteq C(U)$.
(C2) if $U \subseteq V$, then $C(U) \subseteq C(V)$.
(C3) $C(C(U)) \subseteq C(U)$.
(C4) $C(U) \odot C(V) \subseteq C(U \odot V)$.
(C5) $\Diamond C(U) \subseteq C(\Diamond U)$.

A subset $U \subseteq G$ is called *C-closed*, if $U = C(U)$. The set of all C-closed subsets of G is denote by $C(G)$. The operations $\otimes, \blacklozenge, \vee_C$ on $C(G)$ are defined as follows:

$$U \otimes V = C(U \odot V), \quad \blacklozenge U = C(\Diamond U), \quad U \vee_C V = C(U \vee V).$$

Clearly $C(G)$ is closed under $\otimes, \blacklozenge$ and $\vee_C$.

Let $\mathbf{C}(\mathbf{G}) = (C(G), \otimes, \to, \wedge, \vee_C, \blacklozenge, \square^{\downarrow}, C(\emptyset), C(G))$ where the operations $\to$ and $\square^{\downarrow}$ are defined as in Definition 9. One can prove that $C(G)$ is closed under $\to$ and $\square^{\downarrow}$. Moreover, $\mathbf{C}(\mathbf{G})$ is a lattice with a residuated pair $(\otimes, \to)$ and an adjoint pair $(\blacklozenge, \square^{\downarrow})$. We define $\neg U := U \to C(\emptyset)$. If $\mathbf{C}(\mathbf{G})$ is distributive and $\neg\neg U \subseteq U$ for all $U \in C(G)$, then $\mathbf{C}(\mathbf{G})$ is a qBacrg. $\mathbf{C}(\mathbf{G})$ can be any algebra between **tqBacrg** and **Pracrg** if $(\blacklozenge, \square^{\downarrow})$ satisfies corresponding conditions.

4   Sequent Calculi

In this section, we shall introduce sequent calculi for residuated algebras. The language is defined inductively as follows:

$$\varphi ::= p \mid \bot \mid \top \mid (\varphi \bullet \varphi) \mid (\varphi \rightarrow \varphi) \mid (\varphi \wedge \varphi) \mid (\varphi \vee \varphi) \mid \Diamond\varphi \mid \Box^{\downarrow}\varphi,$$

where $p \in \mathbf{Prop}$ is a propositional variable. Formula trees are defined inductively as follows:

$$\Gamma ::= \varphi \mid (\Gamma \circ \Gamma) \mid \langle \Gamma \rangle$$

where φ is a formula. The binary operation $\circ$ and unary operation $\langle \rangle$ corresponded to connectives $\bullet$ and $\Diamond$ respectively.

A *context* is a formula tree containing one occurrence of special atom $-$ (a place for substitution). If $\Gamma[-]$ is a context, then $\Gamma[\Delta]$ is the formula tree obtained from $\Gamma[-]$ by substituting Δ for $-$. A sequent is an expression of the form $\Gamma \Rightarrow \varphi$ where Γ is a formula tree and φ is a formula.

Definition 11. The sequent calculus for tqBa, denoted by **StqBacrg**, consists of the following axioms and rules:

– Axioms:

$$(\mathrm{Id})\ \varphi \Rightarrow \varphi \quad (\bot)\ \Gamma[\bot] \Rightarrow \varphi \quad (\top)\ \Gamma \Rightarrow \top$$

$$(\mathrm{DN1})\ \neg\neg\varphi \Rightarrow \varphi \quad (\mathrm{D})\ \varphi \wedge (\psi \vee \chi) \Rightarrow (\varphi \wedge \psi) \vee (\varphi \wedge \chi)$$

– Inference rules:

$$(\rightarrow\mathrm{L})\ \frac{\Delta \Rightarrow \varphi \quad \Gamma[\psi] \Rightarrow \chi}{\Gamma[(\Delta \circ \varphi \rightarrow \psi)] \Rightarrow \chi} \quad (\rightarrow\mathrm{R})\ \frac{(\varphi \circ \Gamma) \Rightarrow \psi}{\Gamma \Rightarrow \varphi \rightarrow \psi}$$

$$(\bullet\mathrm{L})\ \frac{\Gamma[(\varphi \circ \psi)] \Rightarrow \chi}{\Gamma[\varphi \bullet \psi] \Rightarrow \chi} \quad (\bullet\mathrm{R})\ \frac{\Gamma \Rightarrow \varphi \quad \Delta \Rightarrow \psi}{(\Gamma \circ \Delta) \Rightarrow \varphi \bullet \psi}$$

$$(\Diamond\mathrm{L})\ \frac{\Gamma[\langle\varphi\rangle] \Rightarrow \psi}{\Gamma[\Diamond\varphi] \Rightarrow \psi} \quad (\Diamond\mathrm{R})\ \frac{\Gamma \Rightarrow \varphi}{\langle\Gamma\rangle \Rightarrow \Diamond\varphi}$$

$$(\Box^{\downarrow}\mathrm{L})\ \frac{\Gamma[\varphi] \Rightarrow \psi}{\Gamma[\langle\Box^{\downarrow}\varphi\rangle] \Rightarrow \psi} \quad (\Box^{\downarrow}\mathrm{R})\ \frac{\langle\Gamma\rangle \Rightarrow \varphi}{\Gamma \Rightarrow \Box^{\downarrow}\varphi}$$

$$(\wedge\mathrm{L})\ \frac{\Gamma[\varphi_i] \Rightarrow \psi}{\Gamma[\varphi_1 \wedge \varphi_2] \Rightarrow \psi} \quad (\wedge\mathrm{R})\ \frac{\Gamma \Rightarrow \varphi \quad \Gamma \Rightarrow \psi}{\Gamma \Rightarrow \varphi \wedge \psi}$$

$$(\vee\mathrm{L})\ \frac{\Gamma[\varphi_1] \Rightarrow \psi \quad \Gamma[\varphi_2] \Rightarrow \psi}{\Gamma[\varphi_1 \vee \varphi_2] \Rightarrow \psi} \quad (\vee\mathrm{R})\ \frac{\Gamma \Rightarrow \varphi_i}{\Gamma \Rightarrow \varphi_1 \vee \varphi_2}$$

In $(\wedge\mathrm{L})$ and $(\vee\mathrm{R})$, the subscript i equals 1 or 2.
– Structural rules:

$$(\mathrm{Com})\ \frac{\Gamma[(\Delta_1 \circ \Delta_2)] \Rightarrow \varphi}{\Gamma[(\Delta_2 \circ \Delta_1)] \Rightarrow \varphi} \quad (\mathrm{S4})\ \frac{\Gamma[\langle\Delta\rangle] \Rightarrow \varphi}{\Gamma[\langle\langle\Delta\rangle\rangle] \Rightarrow \varphi} \quad (\mathrm{T})\ \frac{\Gamma[\langle\Delta\rangle] \Rightarrow \varphi}{\Gamma[\Delta] \Rightarrow \varphi}$$

– Cut rule:

$$(\text{Cut}) \ \frac{\Delta \Rightarrow \varphi \quad \Gamma[\varphi] \Rightarrow \psi}{\Gamma[\Delta] \Rightarrow \psi}$$

StqBa5crg is obtained from **StqBacrg** by adding the following rule:

$$(\Diamond\Box^{\downarrow}) \frac{(\langle\Gamma\rangle_1 \circ \Gamma_2) \Rightarrow \bot}{(\Gamma_1 \circ \langle\Gamma_2\rangle) \Rightarrow \bot}$$

Sequent calculi **SIA1crg**, **SIA2crg**, **SIA3crg**, and **Spracrg** are obtained from **StqBa5crg** by adding the following corresponding axioms and rules:

$$(\text{IA1}\Diamond) \ \Diamond\varphi \wedge \neg\Diamond\varphi \Rightarrow \bot \quad (\text{IA2}\Diamond) \ \Diamond\varphi \wedge \Diamond\psi \Rightarrow \Diamond(\varphi \wedge \psi)$$

$$(\text{IA3}\Diamond) \ \frac{\Diamond\varphi \Rightarrow \Diamond\psi \quad \Box^{\downarrow}\varphi \Rightarrow \Box^{\downarrow}\psi}{\varphi \Rightarrow \psi}$$

The cut elimination does not hold for all these sequent calculi. We first show an interpolation property for all these sequent calculi. And then in Sect. 4, by interpolation property and model-theoretic method, we obtain the SFMP.

Henceforth, let **S** be one of sequent calculi **StqBacrg**, **StqBa5crg**, **SIA1crg**, **SIA2crg**, **SIA12crg**, **SIA3crg**, **SIA23crg** and **Spracrg**. Let T be a set of formulas. A sequent $\Gamma \Rightarrow \varphi$ is call a T-*sequent* if all formulas appearing in it belong to T. A derivation of a T-sequent $\Gamma \Rightarrow \varphi$ is called a T-*derivation* if all sequents appearing in the derivation are T-sequents. The notation $\vdash_{\mathbf{S}} \Gamma \Rightarrow_T \varphi$ means that $\Gamma \Rightarrow \varphi$ has a T-derivation in **S**. In the following lemma, we assume that T contains $\bot$, $\top$ and is closed under taking subformulas as well as operations $\vee$, $\wedge$ and $\neg$. Let Φ be any finite set of sequents of the form $\varphi \Rightarrow \psi$.

Lemma 2 (Interpolation). *If $\Phi \vdash_{\mathbf{S}} \Gamma[\Delta] \Rightarrow_T \varphi$, then there exists $\chi \in T$ such that $\Phi \vdash_{\mathbf{S}} \Delta \Rightarrow_T \chi$ and $\Phi \vdash_{\mathbf{S}} \Gamma[\chi] \Rightarrow_T \varphi$.*

5 Strong Finite Model Property

Let $\mathbf{Alg(S)}$ be the class of algebras corresponding to **S**. We show the SFMP of $\mathbf{Alg(S)}$. Let T be a nonempty set of formulas. By T^* we denote the set of all formula trees built from formulas in T. Let $T^*[-]$ be the set of all contexts in which all formulas belong to T. Then $\mathbf{G(T^*)} = (T^*, (- \circ -), \langle-\rangle)$ is a groupoid with a unary operation $\langle-\rangle$. Let $\Gamma[-] \in T^*[-]$ and $\varphi \in T$. We define

$$[\Gamma[-], \varphi] = \{\Delta \mid \Delta \in T^* \text{ and } \Phi \vdash_{\mathbf{S}} \Gamma[\Delta] \Rightarrow_T \varphi\},$$

$$[\varphi] = \{\Gamma \mid \Gamma \in T^* \text{ and } \Phi \vdash_{\mathbf{S}} \Gamma \Rightarrow_T \varphi\}.$$

Let $B(T)$ be the family of all sets of the form $[\Gamma[-], \varphi]$ defined above. We define the function $C_T \colon \wp(T^*) \to \wp(T^*)$ on the powerset of T^* as follows:

$$C_T(U) = \bigcap\{[\Gamma[-], \varphi] \in B(T) \mid U \subseteq [\Gamma[-], \varphi]\}.$$

Proposition 1. C_T *is a closure operator.*

Then $\mathbf{C_T}(\mathbf{G}(\mathbf{T}^*))$ is a lattice with residuated pair $(\otimes, \rightarrow)$ and adjoint pair $(\blacklozenge, \square^\downarrow)$. We define $\neg U := U \rightarrow C(\emptyset)$. Then $U \subseteq \neg\neg U$. One can easily show that $\blacklozenge\blacklozenge U \subseteq \blacklozenge U$ and $U \subseteq \blacklozenge U$ in $\mathbf{C_T}(\mathbf{G}(\mathbf{T}^*))$. Moreover, if $\mathbf{S}$ is not $\mathbf{StqBacrg}$, then $\blacklozenge\square^\downarrow U \subseteq U$. The following equations hold in $\mathbf{C_T}(\mathbf{G}(\mathbf{T}^*))$ provided that all formulas appearing in them belong to T:

$$[\varphi] \otimes [\psi] = [\varphi \bullet \psi] \qquad\qquad [\varphi] \rightarrow [\psi] = [\varphi \rightarrow \psi]$$
$$\blacklozenge[\varphi] = [\Diamond\varphi] \qquad\qquad \square^\downarrow[\varphi] = [\square^\downarrow\varphi]$$
$$[\varphi] \wedge [\psi] = [\varphi \wedge \psi] \qquad\qquad [\varphi] \vee_C [\psi] = [\varphi \vee \psi].$$

Let T be a finite nonempty set of formulas such that $\top, \bot \in T$. Let $\overline{T}$ be the smallest set of formulas containing all formulas in T and is closed under taking subformulas and $\wedge$, $\vee$, $\neg$, $\Diamond$ and $\square^\downarrow$. For any $\varphi, \psi \in \overline{T}$, we say that φ and ψ are $\overline{T}$-*equivalent with respect to* $\mathbf{S}$, notation $\varphi \sim_{\mathbf{S}} \psi$, if $\vdash_{\mathbf{S}} \varphi \Rightarrow \psi$ and $\vdash_{\mathbf{S}} \psi \Rightarrow \varphi$.

Lemma 3. $\overline{T}$ *is finite up to the equivalence relation* $\sim_{\mathbf{S}}$.

Let $r(\overline{T})$ be the set of all representatives in the quotient of $\overline{T}$ with respect to $\sim_{\mathbf{S}}$. Clearly $r(\overline{T})$ is a nonempty finite subset of $\overline{T}$.

Lemma 4. *For any set* $U \in \mathrm{C}_T(\overline{T}^*)$, *there exists* $\varphi \in r(\overline{T})$ *with* $U = [\varphi]$.

By Lemma 4, we can show that $\mathbf{C_T}(\mathbf{G}(\mathbf{T}^*))$ is a qBa, and that it satisfies the defining conditions of $\mathbf{Alg}(\mathbf{S})$.

Lemma 5. *The algebra* $\mathbf{C_T}(\mathbf{G}(\overline{\mathbf{T}}^*))$ *is finite and belongs to* $\mathbf{Alg}(\mathbf{S})$.

Lemma 6. *Let* T *be the set of all formulas appearing in* $\Gamma \Rightarrow A$ *or* Φ. *If* $\Phi \nvdash_{\mathbf{S}} \Gamma \Rightarrow_{\overline{T}} A$, *then* $\mathbf{C_T}(\mathbf{G}(\overline{\mathbf{T}}^*)) \nvDash \Gamma \Rightarrow A$.

Theorem 1. $\mathbf{Alg}(\mathbf{S})$ *has the SFMP.*

Let $\mathbf{Alg}^*(\mathbf{S})$ be the class of algebras obtained from $\mathbf{Alg}(\mathbf{S})$ by deleting operators $\bullet$ and $\rightarrow$.

Theorem 2. $\mathbf{Alg}^*(\mathbf{S})$ *has the SFMP.*

Theorem 3. $\mathbf{Alg}(\mathbf{S})$ *and* $\mathbf{Alg}^*(\mathbf{S})$ *are decidable.*

6   Conclusion

In this extended abstract, we describe the model-theoretic approach to show the strong finite model property of residuated algebras related with rough sets from which the decidability of equational theories of some classes of rough algebras follows. In a forthcoming full paper, we shall construct decision algorithm for these sequent calculi. Furthermore, the approach given in the present paper can be extended to more general algebraic structures. For example, we can introduce non-distributive topological quasi-Boolean algebras, and obtain results on the strong finite model property and decidability.

References

1. Banerjee, M.: Rough sets and 3-valued łukasiewicz logic. Fundamenta Informatica **31**, 213–220 (1997)
2. Banerjee, M., Chakraborty, M.: Rough algebra. Bull. Pol. Acad. Sci. (Math.) **41**(4), 293–297 (1993)
3. Banerjee, M., Chakraborty, M.: Rough sets through algebraic logic. Fundamenta Informaticae **28**(3–4), 211–221 (1996)
4. Buszkowski, W.: Interpolation and FEP for logics of residuated algebras. Log. J. IGPL **19**, 437–454 (2011)
5. Liu, G.L., Zhu, W.: The algebraic structures of generalized rough set theory. Inf. Sci. **178**(21), 4105–4133 (2008)
6. Ma, M., Chakraborty, M.K.: Covering-based rough sets and modal logics. Part I. Int. J. Approx. Reason. **77**, 55–65 (2016)
7. Ma, M., Chakraborty, M.K.: Covering-based rough sets and modal logics. Part II. Int. J. Approx. Reason. **95**, 113–123 (2018)
8. Pawlak, Z.: Rough sets. Int. J. Comput. Inf. Sci. **11**(5), 341–356 (1982)
9. Pomykala, J., Pomykala, J.A.: The stone algebra of rough sets. Bull. Pol. Acad. Sci. Math. **36**, 498–508 (1988)
10. Rasiowa, H.: An Algebraic Approach to Non-Classical Logics. North-Holland Publishing, Amsterdam (1974)
11. Saha, A., Sen, J., Chakraborty, M.K.: Algebraic structures in the vicinity of pre-rough algebra and their logics. Inf. Sci. **282**, 296–320 (2014)
12. Saha, A., Sen, J., Chakraborty, M.K.: Algebraic structures in the vicinity of pre-rough algebra and their logics II. Inf. Sci. **333**, 44–60 (2016)
13. Yao, Y.Y.: On generalizing Pawlak approximation operators. In: Polkowski, L., Skowron, A. (eds.) RSCTC 1998. LNCS (LNAI), vol. 1424, pp. 298–307. Springer, Heidelberg (1998). https://doi.org/10.1007/3-540-69115-4_41
14. Yao, Y.: Relational interpretations of neighborhood operators and rough set approximation operators. Inf. Sci. **111**(1–4), 239–259 (1998)
15. Yao, Y., Yao, B.: Covering based rough set approximations. Inf. Sci. **200**, 91–107 (2012)
16. Zhu, W.: Generalized rough sets based on relations. Inf. Sci. **177**(22), 4997–5011 (2007)

A Conflict Analysis Model Based on Three-Way Decisions

Yan Fan[1], Jianjun Qi[2], and Ling Wei[1(✉)]

[1] School of Mathematics, Northwest University,
Xi'an 710127, People's Republic of China
fyan0411@163.com, wl@nwu.edu.cn
[2] School of Computer Science and Technology, Xidian University,
Xi'an 710071, People's Republic of China
qijj@mail.xidian.edu.cn

Abstract. In decision-making, three-way decisions play an essential role and have been widely used in many fields and disciplines. In this paper, we propose a conflict analysis model based on three-way decisions, so as to explore the inter structure of conflict situation. Firstly, by adopting including degree, two pairs of evaluation functions are defined specifically based on the conflict situation. After that, with restricting the evaluations, three regions of agent set and issue set can be obtained. Comparing with existing conflict analysis models, this trisection model is more efficient, practical and pragmatical. Finally, the trisection of agent set and issue set could be used to ascertain sub-optimal feasible consensus strategies, and determine the scope of the kernel issues in conflict situation, respectively.

Keywords: Three-way decisions · Conflict analysis · Including degree

1 Introduction

Conflict, as an essential characteristic of human life, exists in a wide variety of social problems. To make proper decisions in conflict situations, conflict study is of significance both in theory and practice. Conflict analysis, purposed to explore the structure of conflict, has attracted enormous attention [1–13]. For example, Pawlak initially proposed discernibility matrix and distance functions based on rough set [2,3], then presented an approach dividing the agent set into several coalitions. Deja [4,5] subsequently extended Pawlak conflict analysis model through adding three basic questions:

(1) What are the intrinsic reasons for the conflict?
(2) How can a feasible consensus strategy be found?
(3) Is it possible to satisfy all the agents?

To tackle the problems mentioned by Deja, Sun et al. [6,7] developed a rough set-based conflict analysis model. However, there are still many problems should

© Springer Nature Switzerland AG 2018
H. S. Nguyen et al. (Eds.): IJCRS 2018, LNAI 11103, pp. 522–532, 2018.
https://doi.org/10.1007/978-3-319-99368-3_41

be studied further, such as the more feasible strategy. Ali et al. [8] provided a new conflict analysis model based on soft preference relation and soft dominance relation, disclosing the information more efficiently. Nevertheless, this model paid more attention to domination relations between agents, so that the relations between issues and agents were ignored. That would end up with missing more benefit strategies.

In conflict situations, the main problem is how find an efficient way to model uncertainty in conflict situations [4,5]. For a feasible consensus strategy, the way of model uncertainty is to ascertain the agents' attitudes towards any strategy: agreed, opposed or neutral.

The notion of three-way decisions was proposed and used to interpret three regions in rough set. More specifically, positive, negative and boundary region are viewed respectively as acceptance, rejection, and non-commitment in a ternary classification [14–17]. The intrinsic ideas of three-way decisions has been widely applied to many fields, for instance, medical decision-making [18], management sciences [19], and peering review process [20].

The essential ideas of three-way decisions are described in terms of a ternary classification according to the evaluations of a set of criteria [17]. This kind of classification is, to some extent, consensus with the trisection of agent set based on every agent's attitude to a specific strategy, and the trisection of issue set based on agent group's whole attitude to every single issue. Therefore, our main research are as follows. On the one hand, we define a pair of evaluation functions to estimate the extent to which agent u accepts or opposes a strategy Y. Then, the three regions of agents could be determined through restricting the value of the evaluation function subsequently; On the other hand, another pair of evaluation functions is also defined to estimate the extent to which an issue a is accepted or opposed by the whole agent group X. Then, three regions of issues could be determined as well. Finally, we can find that this model is more appropriate than existing conflict analysis models.

Basic notions of Sun's conflict analysis model and three-way decisions are recalled in Sect. 2. Then, the conflict analysis model based on three-way decisions is proposed in Sect. 3. Finally, we conclude our researches and give further research directions in Sect. 4.

2   Preliminaries

Conflict situation consists of agents and their attitudes to some issues. In Pawlak's model, conflict situation can be presented as a pair (U, V), where $U = \{u_1, ..., u_m\}$ is the universe of agents, and $V = \{a_1, ..., a_n\}$ is the universe of issues. The attitude of agent u to an issue a can be interpreted as a function $a : U \to V_a$, where $V_a = \{+, -, 0\}$. $a(u) = +$ represents agent u agrees with issue a, $a(u) = -$ means agent u objects to issue a, and $a(u) = 0$ means agent u is neutral towards issue a. An example of conflict situation is presented in Table 1. The relationship of each agent u_i to a specific issue a_j could be clearly shown in this table.

Table 1. The conflict situation of the Middle East conflict.

	a_1	a_2	a_3	a_4	a_5
u_1	−	+	+	+	+
u_2	+	0	−	−	−
u_3	+	−	−	−	0
u_4	0	−	−	0	−
u_5	+	−	−	−	−
u_6	0	+	−	0	+

Sun et al. [7] focused on the first two questions proposed by Deja [4,5], "What are the intrinsic conflict reasons" and "How can a feasible consensus strategy be found". Inspired by Pawlak's model, they tried to introduce a new analysing method of conflict situation based on rough set theory over two universes.

According to [7], for any subset $Y \subseteq V$, Y is called a strategy. Subsequently, Y is called a feasible consensus strategy if it satisfies all agents. A sub-optimal feasible consensus strategy Y satisfies the agents as many as possible. The feasible consensus strategy does not exist usually since there are different opinions for every issue. Thus, it is more meaningful to determine sub-optimal feasible consensus strategies. In order to find a sub-optimal feasible consensus strategy, the most important thing is to determine the attitudes of all agents to every strategy. On the basis of Pawlak rough set, Sun et al. [7] described an agent's attitude in the conflict situation as follows:

Let $f = \{f^+, f^-\}$ be the set valued mappings from U to $P(V)$, where

$$f^+ : U \rightarrow P(V), \ f^+(u) = \{a \in V | a(u) = +\}, \forall u \in U,$$
$$f^- : U \rightarrow P(V), \ f^-(u) = \{a \in V | a(u) = -\}, \forall u \in U.$$

The image of f^+ represents the subset of issue universe V which satisfy agent u. The image of f^- represents the subset of issue universe V which are opposed by agent u.

For any strategy $Y \subseteq V$, the lower and upper approximations are:

$$\underline{apr}_f^+(Y) = \{u \in U | f^+(u) \subseteq Y\}, \ \overline{apr}_f^+(Y) = \{u \in U | f^+(u) \cap Y \neq \varnothing\};$$
$$\underline{apr}_f^-(Y) = \{u \in U | f^-(u) \subseteq Y\}, \ \overline{apr}_f^-(Y) = \{u \in U | f^-(u) \cap Y \neq \varnothing\}.$$

Then the agreement subset, disagreement subset, neutral subset for the strategy Y are denoted as follows:

Agreement subset: $R_f^+(Y) = \underline{apr}_f^+(Y) - \underline{apr}_f^-(Y)$;

Disagreement subset: $R_f^-(Y) = \underline{apr}_f^-(Y) - \underline{apr}_f^+(Y)$;

Neutral subset: $R_f^0(Y) = U - R_f^+(Y) \cup R_f^-(Y)$.

Thus, a sub-optimal feasible consensus strategy Y can be found through selecting the maximum cardinality of the agreement subset $R_f^+(Y)$.

Example 1. We consider the Middle East conflict in Table 1. Given strategy $Y = \{a_2, a_3, a_5\} \subseteq V$, and then $\underline{apr}_f^+(Y) = \{u_6\}$, $\overline{apr}_f^+(Y) = \{u_1, u_6\}$, $\underline{apr}_f^-(Y) = \{u_4, u_6\}$, $\overline{apr}_f^-(Y) = \{u_2, u_3, u_4, u_5, u_6\}$. According to Sun et al. [7], there is no agent agrees with the strategy Y, since $R_f^+(Y) = \varnothing$. Additionally, the agents in $R_f^-(Y) = \{u_4\}$ oppose the strategy Y, and all the agents in $R_f^0(Y) = \{u_1, u_2, u_3, u_5, u_6\}$ hold neutral attitude.

The following facts can be observed: (1) For agents u_4 and u_5, they agree on strategy Y, but they are grouped into different coalitions. (2) In Table 1, the issues in Y are all agreed by agent u_1, but according to the above method, agent u_1 is considered neutral about strategy Y. Both of the two aspects are not very suitable for assuring the agents' attitude to a specific strategy in practice. Moreover, more feasible strategy may be missed. Actually, the reason for these confusions is the inconformity between the approximation in rough set based on two universes and semantics of the three subsets of agents for a strategy. Therefore, we need more efficient conflict analysis model to determine the structure in conflict situation.

The theory of three-way decisions can be used to interpret the regions of acceptance, rejection, and non-commitment in a ternary classification. This theory is applicable to divide agent universe U into three subsets according to their attitude to a strategy. Three kinds of evaluation-based three-way decisions are proposed in [17], and then the corresponding three-way decision models are introduced and studied. Among these three kinds of models, the first one as follows is more consensus to the semantics of determining the three subsets in conflict analysis.

Definition 1 [17]. *Suppose U is a finite nonempty set and $(L_a, \preceq_a)$, $(L_r, \preceq_r)$ are two posets. A pair of functions $v_a : U \to L_a$ and $v_r : U \to L_r$ is called an acceptance evaluation and a rejection evaluation, respectively. For $u \in U$, $v_a(u)$ and $v_r(u)$ are called the acceptance and rejection values of u, respectively.*

In conflict situation (U, V), the acceptance value $v_a(u)$ and rejection value $v_r(u)$ can be constructed by evaluating the extent to which agent u agrees with or disagrees with strategy Y, respectively. What's more, if the agent u_1 accepts strategy Y, $v_a(u_1)$ must be in a certain subset of L_a representing the acceptance region of L_a. Similarly, $v_a(u_2)$ included in the rejection region of L_r means agent u_2 reject strategy Y to a large extent. Therefore, L_a and L_r should be defined. These values are called designated values for acceptance and designated values for rejection, respectively. Based on the two sets of designated values, one can easily obtain three regions for three-way decisions.

Definition 2 [17]. *Let $\varnothing \neq L_a^+ \subseteq L_a$ be a subset of L_a called the designated values for acceptance, and $\varnothing \neq L_r^- \subseteq L_r$ be a subset of L_r called the designated* *values for rejection. The positive, negative, and boundary regions of three-way decisions induced by (v_a, v_r) are defined by:*

$$POS_{(L_a^+, L_r^-)}(v_a; v_r) = \{u \in U | v_a(u) \in L_a^+ \wedge v_r(u) \notin L_r^-\},$$

$$NEG_{(L_a^+, L_r^-)}(v_a; v_r) = \{u \in U | v_a(u) \notin L_a^+ \wedge v_r(u) \in L_r^-\},$$

$$BND_{(L_a^+, L_r^-)}(v_a; v_r) = (POS_{(L_a^+, L_r^-)}(v_a, v_r) \cup NEG_{(L_a^+, L_r^-)}(v_a, v_r))^c$$
$$= \{u \in U | (v_a(u) \notin L_a^+ \wedge v_r(u) \notin L_r^-) \vee (v_a(u) \in L_a^+ \wedge v_r(u) \in L_r^-)\}.$$

From the above analysis, we know that there are two essential problems. One is how to evaluate the extent to which agent u agrees and disagrees with a certain strategy Y, and the other is how to define the designated values for acceptance and rejection.

3   Conflict Analysis Model Based on Three-Way Decisions

This section mainly introduces an conflict analysis model on the basis of three-way decisions, which is considered from two perspectives. Based on three-way decisions, Sect. 3.1 shows how to obtain three subsets of agents subjecting to each agent's attitude to a specific strategy, which helps to determine the suboptimal feasible consensus strategy. Similarly, Sect. 3.2 proposes an approach to get trisection of the issue set related to the unitary attitude of an agent group to a specific strategy. The outcome helps to determine the scope of the core issues causing conflict. Furthermore, compared with Sun's conflict analysis model, the superiorities of this model are showed as well.

3.1   Trisection of Agent Set Based on Each Agent's Attitude to a Specific Strategy

To trisect the agent set, we just have to tackle the problems in the last paragraph of Sect. 2. That is how to evaluate the extent to which agent u agrees and disagrees with strategy Y, and how to define the designated values for acceptance and rejection. Including degree can be adopted to estimate the extent to which agent u accepts or opposes strategy Y. Then the designated values can be determined through restricting the including degree.

Definition 3 [21]. *Let $(L, \leq)$ be a partially ordered set. If for any $X, Y \subseteq L$, there is a real number $D(Y/X)$ with the following properties:*

(1) $0 \leq D(Y/X) \leq 1$
(2) $X \subseteq Y$ implies $D(Y/X) = 1$
(3) $X \subseteq Y \subseteq Z$ implies $D(X/Z) \leq D(X/Y)$

then D is called an including degree on L.

The including degree $D(Y/X)$ represents the extent to which set Y contains the set X. It is obvious that $D(Y/X) = \frac{|X \cap Y|}{|X|}$ is an including degree.

Definition 4. *Let (U, V) be a conflict situation. $([0, 1], \leq)$ a totally ordered set. $Y \subseteq V$, Y is a strategy. A pair of evaluation functions v_a and v_r are defined as:*

$$v_a : U \times P(V) \to [0, 1], v_a(u, Y) = D(f^+(u)|Y),$$

$$v_r : U \times P(V) \to [0, 1], v_r(u, Y) = D(f^-(u)|Y).$$

v_a is called agent acceptance evaluation function, and $v_a(u, Y)$ evaluates the extent to which agent u accepts strategy Y; v_r is called agent rejection evaluation function, and $v_r(u, Y)$ evaluates the extent to which agent u rejects strategy Y, where, $D(f^+(u)|Y)$ and $D(f^-(u)|Y)$ are defined as

$$D(f^+(u)|Y) = \frac{|f^+(u) \cap Y|}{|Y|}, \ D(f^-(u)|Y) = \frac{|f^-(u) \cap Y|}{|Y|}.$$

Property 1. Let (U, V) be a conflict situation. $\forall u \in U, Y \subseteq V$, we have $v_a(u, Y) + v_r(u, Y) \leq 1$.

Proof. It is obvious that $f^+(u) \cap f^-(u) = \varnothing$. Then $(f^+(u) \cap Y) \cap (f^-(u) \cap Y) = \varnothing$, so $|f^+(u) \cap Y| + |f^-(u) \cap Y| \leq |Y|$. Therefore, $\frac{|f^+(u) \cap Y|}{|Y|} + \frac{|f^-(u) \cap Y|}{|Y|} \leq 1$. That is, $v_a(u, Y) + v_r(u, Y) \leq 1$.

Example 2. Consider the Middle East conflict presented in Table 1. For strategy $Y = \{a_2, a_3, a_5\} \subseteq V$, we obtain the following results:

Table 2. Evaluations for the Middle East conflict.

U	u_1	u_2	u_3	u_4	u_5	u_6
$v_a(u_i, Y)$	1	0	0	0	0	$\frac{2}{3}$
$v_r(u_i, Y)$	0	$\frac{2}{3}$	$\frac{2}{3}$	1	1	$\frac{1}{3}$

From Table 2, we know that the extent to which agent u_6 accepts strategy Y is $\frac{2}{3}$, and the extent to which agent u_6 opposes strategy Y is $\frac{1}{3}$ and so on.

Let $\alpha \geq 0.5$, $\beta \geq 0.5$, and then $(\alpha, 1]$ represent the designated values for acceptance, which are used to restrict the extent to which an agent accepts strategy Y in the agreement subset. $(\beta, 1]$ represent the designated values for rejection, which are used to restrict the extent to which an agent rejects the strategy Y in the disagreement subset. On the basis of two sets of designated values, we can easily obtain three regions of agents based on their attitudes to strategy Y.

Definition 5. *Let (U, A) be a conflict situation, $(\alpha, 1]$ the designated values for acceptance, $(\beta, 1]$ the designated values for rejection, $Y \subseteq V$ a strategy, $v_a(u, Y) = D(f^+(u)|Y)$ and $v_r(u, Y) = D(f^-(u)|Y)$. Then, we denote:*

$$AS_{\alpha,\beta}(Y) = \{u \in U | v_a(u, Y) \in (\alpha, 1] \wedge v_r(u, Y) \notin (\beta, 1]\},$$

$$DS_{\alpha,\beta}(Y) = \{u \in U | v_a(u, Y) \notin (\alpha, 1] \wedge v_r(u, Y) \in (\beta, 1]\},$$

$$NS_{\alpha,\beta}(Y) = U - AS_{\alpha,\beta}(Y) \cup DS_{\alpha,\beta}(Y).$$

We call $AS_{\alpha,\beta}(Y)$ the $(\alpha, \beta)-$agreement subset of strategy Y, $DS_{\alpha,\beta}(Y)$ the $(\alpha, \beta)-$disagreement subset of strategy Y, and $NS_{\alpha,\beta}(Y)$ the $(\alpha, \beta)-$neutral subset of strategy Y.

Remark. It should be noted that when $\alpha \geq 0.5$ and $\beta \geq 0.5$, we have $v_a(u, Y) \in (\alpha, 1] \iff v_r(u, Y) \notin (\beta, 1]$, and $v_a(u, Y) \notin (\alpha, 1] \iff v_r(u, Y) \in (\beta, 1]$. It can be proved easily through Property 1, $v_a(u, Y) + v_r(u, Y) \leq 1$. Therefore, the definition of $AS_{\alpha,\beta}(Y)$ and $DS_{\alpha,\beta}(Y)$ can be simplified as

$$AS_\alpha(Y) = \{u \in U | v_a(u) \in (\alpha, 1]\},$$

$$DS_\beta(Y) = \{u \in U | v_r(u) \in (\beta, 1]\}.$$

Similarly, $AS_\alpha(Y)$ is named the $\alpha-$agreement subset of strategy Y, and $DS_\beta(Y)$ is called the $\beta-$disagreement subset of strategy Y. Therefore, the agents in $AS_\alpha(Y)$ agree with strategy Y to designated value α, the agents in $DS_\beta(Y)$ object to strategy Y to designated value β, and the agents in $NS_{\alpha,\beta}(Y)$ have neutral attitude for strategy Y to designated values (α, β).

Proposition 1. *Let (U, A) be a conflict situation, $Y \subseteq V$ a strategy. $\alpha \geq 0.5$, and $\beta \geq 0.5$. The following relations hold: $AS_\alpha(Y) \cap DS_\beta(Y) = \varnothing$, $AS_\alpha(Y) \cap NS_{\alpha,\beta}(Y) = \varnothing$, and $DS_\beta(Y) \cap NS_{\alpha,\beta}(Y) = \varnothing$.*

Proof. For any $u \in AS_\alpha(Y)$, we have $v_a(u, Y) > \alpha \geq 0.5$. Since $v_a(u, Y) + v_r(u, Y) \leq 1$, then $v_r(u, Y) \leq 1 - v_a(u, Y) < 1 - \alpha \leq 0.5$, so $v_r(u, Y) \not> 0.5$, which means $u \notin DS_\beta(Y)$. Thus, we obtain $AS_\alpha(Y) \cap DS_\beta(Y) = \varnothing$. According to the definition of neutral subset $NS_{\alpha,\beta}(Y)$, we have $AS_\alpha(Y) \cap NS_{\alpha,\beta}(Y) = \varnothing$ and $DS_\beta(Y) \cap NS_{\alpha,\beta}(Y) = \varnothing$. Therefore, the three regions are pair-wise disjoint.

For simplicity, we denote $I_1 = AS_\alpha(Y)$, $I_2 = DS_\beta(Y)$, and $I_3 = NS_{\alpha,\beta}(Y)$.

Proposition 2. *Let (U, A) be a conflict situation, $Y \subseteq V$ a strategy. $\forall u_1, u_2 \in U$, if $f^+(u_1) \cap Y = f^+(u_2) \cap Y$ and $f^-(u_1) \cap Y = f^-(u_2) \cap Y$, then $u_1 \in I_t \iff u_2 \in I_t$, $t = \{1, 2, 3\}$.*

Proof. If $f^+(u_1) \cap Y = f^+(u_2) \cap Y$, and $f^-(u_1) \cap Y = f^-(u_2) \cap Y$, then $v_a(u_1, Y) = v_a(u_2, Y)$ and $v_r(u_1, Y) = v_r(u_2, Y)$. Furthermore, we have that

$$u_1 \in I_1 \iff v_a(u_1, Y) > \alpha \iff v_a(u_2, Y) > \alpha \iff u_2 \in I_1;$$

$$u_1 \in I_2 \iff v_r(u_1, Y) > \beta \iff v_r(u_2, Y) > \beta \iff u_2 \in I_2;$$

$$u_1 \in I_3 \Longleftrightarrow v_a(u_1, Y) < \alpha \& v_r(u_1, Y) < \beta$$
$$\Longleftrightarrow v_a(u_2, Y) < \alpha \& v_r(u_2, Y) < \beta \Longleftrightarrow u_2 \in I_3.$$

The proposition is proved.

This proposition shows that if two agents of universe U have the same attitude to strategy Y, they will be grouped together. That is to say, in the terms of determining agreement subset, disagreement subset and neutral subset for strategy Y, the model proposed in this paper improves the first inconformity in Sun's model, which is presented in Example 1.

Proposition 3. *Let (U, A) be a conflict situation, $Y \subseteq V$ a strategy. $\forall u_1, u_2 \in U$, if $v_a(u_1, Y) \geq v_a(u_2, Y)$, and $u_2 \in AS_\alpha(Y)$, then we have $u_1 \in AS_\alpha(Y)$; Similarly, if $v_r(u_1, Y) \geq v_r(u_2, Y)$, and $u_2 \in DS_\beta(Y)$, then we have $u_1 \in DS_\beta(Y)$.*

Proof. If $v_a(u_1, Y) \geq v_a(u_2, Y)$ and $u_2 \in AS_\alpha(Y)$, then we have $v_a(u_1, Y) > \alpha$, which means $u_1 \in AS_\alpha(Y)$. Similarly, If $v_r(u_1, Y) \geq v_r(u_2, Y)$ and $u_2 \in DS_\beta(Y)$, then we conclude $v_r(u_1, Y) > \beta$, which means $u_1 \in DS_\beta(Y)$.

From above we can know that if agent u agrees with all issues of strategy Y, then u would be grouped into the $\alpha-$agreement subset. This conclusion is tenable for any $\alpha \in [0.5, 1]$. Similarly, the model proposed in this paper improves the second inconformity in Sun's model, which is presented in Example 1. Therefore, compared with the outcomes of Sun's conflict analysis model in Sect. 2, the approach to determine the three regions of agent set proposed in this paper is more appropriate.

Example 3 (continued from Example 2). Consider the Middle East conflict presented in Table 1. For strategy $Y = \{a_2, a_3, a_5\}$, let $\alpha = 0.6$, $\beta = 0.6$, and we obtain the following results: $AS_{0.6}(Y) = \{u_1, u_6\}$, $DS_{0.6}(Y) = \{u_2, u_3, u_4, u_5\}$ and $NS_{0.6,0.6}(Y) = \varnothing$.

Therefore, the agents in $AS_{0.6}(Y) = \{u_1, u_6\}$ agree with strategy Y to designated value 0.6, the agents in $DS_{0.6}(Y) = \{u_2, u_3, u_4, u_5\}$ object to strategy Y to designated value 0.6, and no agent has neutral attitude for strategy Y to designated values (0.6,0.6). Furthermore, the agents u_4 and u_5 are grouped together, besides, u_1 is assigned to the 0.6-agreement subset because of its full agreements with the issues in Y.

In this section, we proposed an effective approach to determine three regions of agents for any strategy Y. The result can be used to resolve some problems, such as finding the sub-optimal feasible consensus strategy by selecting the maximum cardinality of the $\alpha-$agreement subset [7].

3.2 Trisection of Issue Set Based on the Whole Attitude of Agent Group to Every Issue

We call $X \subseteq U$ an agent group. This subsection defines two evaluation functions to estimate the extent to which the issue a is accepted or opposed by the whole agent group X. Then three regions of issues: α-agreement strategy, β-disagreement strategy and (α, β)-noncommittal strategy are determined as well. Since the theories in this section are dual to that in Sect. 3.1. We omit the proofs of theories in this section.

Let $g = \{g^+, g^-\}$ be the set valued mappings from V to $P(U)$, where

$$g^+ : V \to P(U), g^+(a) = \{u \in U | a(u) = +\}, \forall a \in V,$$
$$g^- : V \to P(U), g^-(a) = \{u \in U | a(u) = -\}, \forall a \in V.$$

Definition 6. *Let (U, A) be a conflict situation, $([0, 1], \leq)$ a totally ordered set, $X \subseteq U$ an agent group. A pair of evaluation functions w_a and w_r are defined as:*

$$w_a : V \times P(U) \to [0, 1], w_a(a, X) = D(g^+(a)|X),$$
$$w_r : V \times P(U) \to [0, 1], w_r(a, X) = D(g^-(a)|X).$$

w_a is called issue acceptance evaluation function, and $w_a(a, X)$ evaluates the extent to which agent group X accepts issue a; w_r is called issue rejection evaluation function, and $w_r(a, X)$ evaluates the extent to which agent group X rejects issue a, where $D(g^+(a)|X)$ and $D(g^-(a)|X)$ are defined as

$$D(g^+(a)|X) = \frac{|g^+(a) \cap X|}{|X|}, \quad D(g^-(a)|X) = \frac{|g^-(a) \cap X|}{|X|}.$$

Property 2. Let (U, A) be a conflict situation. $\forall a \in V$, $X \subseteq U$, we have $w_a(a, X) + w_r(a, X) \leq 1$.

The designated values for acceptance and rejection of issue set are identical to that in Sect. 3.1 numerically. Therefore, the three regions of issues can be determined similarly.

Definition 7. *Let (U, A) be a conflict situation, $(\alpha, 1]$ the designated values for acceptance, $(\beta, 1]$ the designated values for rejection, $X \subseteq U$ an agent group. $w_a(a, X) = D(g^+(a)|X)$ and $w_r(a, X) = D(g^-(a)|X)$, then we denote:*

$$AT_\alpha(X) = \{a \in V | w_a(a, X) \in (\alpha, 1]\},$$

$$DT_\beta(X) = \{a \in V | w_r(a, X) \in (\beta, 1]\},$$

$$NT_{\alpha,\beta}(X) = U - AT_\alpha(X) \cup DT_\beta(X).$$

We name $AT_\alpha(X)$ the α-agreement strategy of agent group X, which represents the issues agreed by agent group X to designated value α; $DT_\beta(X)$ is called the β-disagreement strategy of agent group X, which represents the issues disagreed by agent group X to designated value β; $NT_{\alpha,\beta}(X)$ is called the (α, β)-noncommittal strategy of agent group X, which represents the noncommittal issues to designated values (α, β).

From Definition 7, the (α, β)-noncommittal strategy contains issues with $w_a(a, X) \leq \alpha$ and $w_r(a, X) \leq \beta$. Thus, the attitude of the whole agent group X to issue a would be not inclined to agree or disagree greatly. Consequently, and the issues in $NT_{\alpha,\beta}(X)$ could be essential points causing the conflict.

Proposition 4. *Let (U, A) be a conflict situation, $X \subseteq U$ an agent group. $\alpha > 0.5, \beta > 0.5$. The following relations hold: $AT_\alpha(X) \cap DT_\beta(X) = \varnothing$, $AT_\alpha(X) \cap NT_{\alpha,\beta}(X) = \varnothing$, $DT_\beta(X) \cap NT_{\alpha,\beta}(X) = \varnothing$.*

For simplicity, we denote $F_1 = AT_\alpha(X)$, $F_2 = DT_\beta(X)$, and $F_2 = NT_{\alpha,\beta}(X)$.

Proposition 5. *Let (U, A) be a conflict situation, $X \subseteq U$ an agent group. $\forall a_1, a_2 \in V$, if $g^+(a_1) \cap X = g^+(a_2) \cap X$ and $g^-(a_1) \cap X = g^-(a_2) \cap X$, then $a_1 \in F_t \iff a_2 \in F_t$, $t = \{1, 2, 3\}$.*

This proposition shows that if the agents in group X have the same attitude to issues a_1 and a_2, then the two issues will be assigned to identical strategy.

Proposition 6. *Let (U, A) be a conflict situation, $X \subseteq U$ an agent group. $\forall a_1, a_2 \in V$, if $w_a(a_1, X) \geq w_a(a_2, X)$, and $a_2 \in AT_\alpha(X)$, then we have $a_1 \in AT_\alpha(X)$; Similarly, if $w_r(a_1, X) \geq w_r(a_2, X)$, and $a_2 \in DT_\beta(X)$, then we have $a_1 \in DT_\beta(X)$.*

4   Conclusion

A new conflict analysis model based on three-way decisions is proposed in this paper. This model analyzes the structure of conflict situation from two aspects.

On the one hand, we define a pair of evaluation functions, through including degree, to estimate the extent to which agent u accepts or opposes a strategy Y, and then trisect the agent set into three regions. Those ideas are all based on the theory of three-way decisions. Subsequently, the better strategy can be acquired.

On the other hand, another pair of evaluation functions are defined to estimate the extent to which issue a is accepted or opposed by an agent group X, and trisection of issue set is confirmed as well. Then the core conflict issues of agent group would be contained in (α, β)−noncommittal strategy. Moreover, we conclude that this model is more suitable to our cognizance than the existing models.

Open problems remaining for future research include: the algorithm of finding the sub-optimal feasible consensus strategy should be acquired; the determination of core conflict issues need to be studied explicitly further.

Acknowledgments. The authors gratefully acknowledge the support of the Natural Science Foundation of China (No. 61772021).

References

1. Pawlak, Z.: Analysis of conflicts. In: Proceedings of the 1997 Joint Conference on Information Sciences, pp. 350–352 (1997)
2. Pawlak, Z.: An inquiry into anatomy of conflicts. J. Inf. Sci. **109**, 65–68 (1998)
3. Pawlak, Z.: Some remarks on conflict analysis. Eur. J. Oper. Res. **166**, 649–654 (2005)

4. Deja, R., Ślęak, D.: Rough set theory in conflict analysis. In: Terano, T., Ohsawa, Y., Nishida, T., Namatame, A., Tsumoto, S., Washio, T. (eds.) JSAI 2001. LNCS (LNAI), vol. 2253, pp. 349–353. Springer, Heidelberg (2001). https://doi.org/10.1007/3-540-45548-5_44

5. Deja, R.: Conflict analysis. Int. J. Intell. Syst. **17**, 235–253 (2002)

6. Sun, B.Z., Ma, W.M.: Rough approximation of a preference relation by multi-decision dominance for a multi-agent conflict analysis problem. Inf. Sci. **315**, 39–53 (2015)

7. Sun, B.Z., Ma, W.M., Zhao, H.Y.: Rough set-based conflict analysis model and method over two universes. Inf. Sci. **372**, 111–125 (2016)

8. Ali, A., Ali, M.I., Rehmana, N.: A more efficient conflict analysis based on soft preference relation. J. Intell. Fuzzy Syst. **34**, 283–293 (2018)

9. Lang, G.M., Miao, D.Q., Cai, M.J.: Three-way decision approaches to conflict analysis using decision-theoretic rough set theory. Inf. Sci. **406–407**, 185–207 (2017)

10. Liu, Y., Lin, Y.: Intuitionistic fuzzy rough set model based on conflict distance and applications. Appl. Soft Comput. **31**, 266–273 (2015)

11. Silva, L.G.D.O., Almeida-Filho, A.T.D.: A multicriteria approach for analysis of conflicts in evidence theory. Inf. Sci. **346–347**, 275–285 (2016)

12. Yang, J.P., Huang, H.Z., Miao, Q., Sun, R.: A novel information fusion method based on Dempster-Shafer evidence theory for conflict resolution. Intell. Data Anal. **15**, 399–411 (2011)

13. Yu, C., Yang, J., Yang, D., Ma, X., Min, H.: An improved conflicting evidence combination approach based on a new supporting probability distance. Expert Syst. Appl. **42**, 5139–5149 (2015)

14. Yao, Y.Y.: Three-way decision: an interpretation of rules in rough set theory. In: Wen, P., Li, Y., Polkowski, L., Yao, Y., Tsumoto, S., Wang, G. (eds.) RSKT 2009. LNCS (LNAI), vol. 5589, pp. 642–649. Springer, Heidelberg (2009). https://doi.org/10.1007/978-3-642-02962-2_81

15. Yao, Y.Y.: Three-way decisions with probabilistic rough sets. Inf. Sci. **180**, 341–353 (2010)

16. Yao, Y.Y.: The superiority of three-way decisions in probabilistic rough set models. Inf. Sci. **181**, 1080–1096 (2011)

17. Yao, Y.Y.: An outline of a theory of three-way decisions. In: Yao, J., Yang, Y., Slowinski, R., Greco, S., Li, H., Mitra, S., Polkowski, L. (eds.) RSCTC 2012. LNCS (LNAI), vol. 7413, pp. 1–17. Springer, Heidelberg (2012). https://doi.org/10.1007/978-3-642-32115-3_1

18. Lurie, J.D., Sox, H.C.: Principles of medical decision making. Spine **24**, 493–498 (1999)

19. Goudey, R.: Do statistical inferences allowing three alternative decision give better feedback for environmentally precautionary decision-making. J. Environ. Manag. **85**, 338–344 (2007)

20. Weller, A.C.: Editorial Peer Review: Its Strengths and Weaknesses. Information Today Inc., Medford (2001)

21. Zhang, W.X., Leung, Y.: Theory of including degrees and its applications to uncertainty. In: Soft Computing in Intelligent Systems and Information Processing: Proceeding of the 1996 Asian Fuzzy Systems Symposium, Kenting, Taiwan, 11–14 December 1996, pp. 496–501 (1996)

Tolerance Relations and Rough Approximations in Incomplete Contexts

Tong-Jun Li[1,2(✉)], Wei-Zhi Wu[1,2], and Xiao-Ping Yang[1,2]

[1] School of Mathematics, Physics and Information Science,
Zhejiang Ocean University, Zhoushan 316022, Zhejiang, China
{litj,wuwz}@zjou.edu.cn
[2] Key Laboratory of Oceanographic Big Data Mining and Application of Zhejiang
Province, Zhejiang Ocean University, Zhoushan 316022, Zhejiang, China
yxpzyp@sina.com

Abstract. The rough approximation operations are induced to incomplete contexts, two binary relation from the object set to the attribute set of an incomplete context are defined, by means of the rough approximation operators based on which, four pairs of rough approximation operators are constructed. The relationships and equivalence among them are discussed in detail.

Keywords: Incomplete contexts · Rough approximations
Tolerance relations · Formal contexts

1 Introduction

Rough set theory [1], proposed by Pawlak in 1982, is an effective mathematic approach, which can be used to deal with vague and uncertain information, therein unknown concepts are approximated by two known concepts called lower and upper approximations respectively. In the classical rough sets, equivalence relations are used to depict the known concepts. In order to generalize the rough set theory, various approaches for concept description are introduced, for example, relation-based rough sets [2], probabilistic rough sets [3], covering rough sets [4], etc.

The traditional rough set theory is usually used for knowledge discovery in complete information systems. Making decision with partial information is ultimately inevitable [5], so it is very important that the rough set technique are taken to deal with incomplete information systems. An incomplete information systems means a system with unknown values, the unknown values have two explanations [6,7]: all unknown values are "do not care" condition, or lost. With incomplete information systems, some important results on rough set have been obtained [8–10]. Recently, Du and Hu [11] investigate dominance-based rough sets in incomplete ordered information systems. Liu et al. [12] introduce

© Springer Nature Switzerland AG 2018
H. S. Nguyen et al. (Eds.): IJCRS 2018, LNAI 11103, pp. 533–545, 2018.
https://doi.org/10.1007/978-3-319-99368-3_42

three-way decision analysis in incomplete information systems. Dai et al. [13] examine the uncertainty measurements of rough approximations based on α-weak similarity in incomplete interval-valued information systems.

Formal context is a primary notion in formal concept analysis. In this framework, Wille [14] first establishes the formal concept analysis. Yao [15] and Duntsch [16] introduce rough approximation operations in formal contexts, so the object oriented concept lattices and the attribute oriented concept lattices are defined, and Shao et al. [17] explore the attribute reduction of the two concept lattices. Kent [18] and Pagliani [19] introduce the approaches of rough sets into concept lattices, so that the concept approximations of formal concepts are put forward. Li et al. [20] define four pairs of rough approximation operators in formal contexts, and compare them. Being analogous to incomplete information systems, incomplete contexts have unknown relation values for many objects and attributes [21]. The unknown values in incomplete contexts are generally considered as being lost, which exist or can not be determined on the current condition.

Many results have been gained for concept analysis and its application in incomplete contexts [22–25]. Li et al. [23] propose one kind of definitions of formal concepts in incomplete contexts, and explore the rule extraction and the attribute reduction. Li and Wang [24] construct approximate concepts based on the theory of three-way decision in incomplete contexts, and present the attribute reduction approaches. Yao [25] introduce interval sets in formal concept analysis of incomplete contexts, so some existing studies on concept analysis are interpreted and extended better. As well known, the theory of rough sets and the formal concept analysis are closely related in formal contexts. However, most studies focus on formal concept analysis in incomplete contexts, and there are few studies on rough set theory. The objective of this paper is to introduce rough set approaches for knowledge discovery in incomplete contexts, and our focus is on the construction of rough set models with a novel approach, and the models proposed are mostly related to tolerance relations.

The rest of this paper is organized as follows. We briefly review in the next section some basic notions and knowledge related to the work. In Sect. 3 we define two binary relations in an incomplete context, by means of the rough approximation operators based on the relations, construct some new rough approximation operators, and investigate the properties of these operators. The paper is then concluded with a brief summary.

2   Preliminaries

In this section, a lot of basic knowledge about rough approximations in formal contexts are reviewed briefly.

2.1   Rough Approximations Based on Binary Relations

Let U be a finite and nonempty set called the universe of discourse. The family of all subsets of U will be denoted by $\mathcal{P}(X)$. The complement of a subset A in U will be denoted by $\sim A$, that is, $\sim A = \{x \in U | x \notin A\}$.

Let U and W be two finite and nonempty universes of discourse, and R a binary relation from U to W, that is, $R \subseteq U \times W$. The *inverse relation* of R, denoted by R^{-1}, is defined as $R^{-1} = \{(x, y) \in W \times U | (y, x) \in R\}$. For any $x \in U$ the *successor neighborhood* of x is $R(x) = \{y \in W | (x, y) \in R\}$. For any $y \in W$ the *predecessor neighborhood* of y is $R^{-1}(y) = \{x \in U | (x, y) \in R\}$.

When $W = U$, the relation R is said to be *reflexive* if $x \in R(x), \forall x \in U$; R is said to be *symmetric* if $y \in R(x) \Rightarrow x \in R(y), \forall x, y \in U$. If R is reflexive and symmetric, then R is said to be a *tolerance relation* on U.

Let R be a binary relation from U to W. The triple (U, W, R) is called a *generalized approximation space* in [26]. For $X \in \mathcal{P}(W)$, *the generalized lower and upper rough approximations* of X with respect to (w.r.t.) (U, W, R), denoted by $\underline{R}(X)$ and $\overline{R}(X)$ respectively, are defined by

$$\underline{R}(X) = \{x \in U | R(x) \subseteq X\}, \ \ \overline{R}(X) = \{x \in U | R(x) \cap X \neq \emptyset\}. \tag{1}$$

The basic properties of the rough approximation operators, $\underline{R}$ and $\overline{R}$, are enumerated as follows: $\forall X, Y \in \mathcal{F}(W)$,

(L1) $\underline{R}(X) = \sim (\overline{R}(\sim X))$,   (U1) $\overline{R}(X) = \sim (\underline{R}(\sim X))$;
(L2) $\underline{R}(W) = U$,   (U2) $\overline{R}(\emptyset) = \emptyset$;
(L3) $\underline{R}(X \cap Y) = \underline{R}(X) \cap \underline{R}(Y)$, (U3) $\overline{R}(X \cup Y) = \overline{R}(X) \cup \overline{R}(X)$;
(L4) $X \subseteq Y \Rightarrow \underline{R}(X) \subseteq \underline{R}(Y)$,   (U4) $X \subseteq Y \Rightarrow \overline{R}(X) \subseteq \overline{R}(Y)$.

Properties (L1) and (U1) show that $\underline{R}$ and $\overline{R}$ are dual to each other. The rough approximation operators based on a variety of binary relations have different properties, conversely some kinds of binary relations can be characterized by corresponding rough approximation operators [26,27].

2.2   Rough Approximations Induced in Formal Contexts

Definition 1. *A formal context is a triple (U, A, I), where U is a nonempty and finite set of objects, A is a nonempty and finite set of attributes, and I is a binary relation from U to A with $(x, a) \in I$ indicating that the object x has the attribute a and $(x, a) \notin I$ indicating the opposite.*

A formal context (U, A, I) can be represented by a two-dimensional table filled with, for example, 1 and 0 numbers, where $I(x, a) = 1$ indicates the object x has the attribute a and $I(x, a) = 0$ indicates the opposite. For convenience, $I(x, a) = 1$ and $I(x, a) = 0$ can also denoted as $a(x) = 1$ and $a(x) = 0$, respectively.

Example 1. Table 1 shows a formal context $T = (U, A, I)$, where

$$U = \{x_1, x_2, x_3, x_4, x_5, x_6\}, \quad A = \{a_1, a_2, a_3, a_4, a_5, a_6\}.$$

In this table, for example, the object x_4 has the properties a_1, a_3 and a_6, and does not have a_2, a_4 and a_5.

Table 1. A formal context $T = (U, A, I)$

U	a_1	a_2	a_3	a_4	a_5	a_6
x_1	1	0	0	1	0	0
x_2	0	1	0	1	1	0
x_3	0	1	1	0	0	1
x_4	1	0	1	0	0	1
x_5	0	1	0	0	1	0
x_6	1	0	0	1	0	1

A formal context (U, A, I) can be viewed as a generalized approximation space, for $B \subseteq A$, the lower approximation $\underline{I}(B)$ and the upper approximation $\overline{I}(B)$ are subsets of U. In [20], the four types of rough approximation operators are defined on (U, A, I), that is, $(\underline{apri}, \overline{apri})$, $(\underline{aprii}, \overline{aprii})$, $(\underline{apriii}, \overline{apriii})$, and $(\underline{apr}, \overline{apr})$ from $\mathcal{P}(A)$ to $\mathcal{P}(U)$, It should be noted that for any $B \subseteq A$, the approximation subsets, $\underline{apri}(B)$, $\overline{apri}(B)$, $\underline{aprii}(B)$, $\overline{aprii}(B)$, $\underline{apriii}(B)$, $\overline{apriii}(B)$, $\underline{apr}(B)$, and $\overline{apr}(B)$, are included in the another universe U.

Considering the relevance to this work, $\underline{aprii}$ and $\overline{aprii}$, re-denoted as $\underline{S_I}$ and $\overline{S_I}$ respectively, are reviewed as follows:

A tolerance relation S_I on U defined on (U, A, I) is

$$S_I = \{(x, y) \in U \times U | I(x) \cap I(y) \neq \emptyset\}.$$

For any $X \subseteq U$, $\underline{S_I}(X)$ and $\overline{S_I}(X)$ are represented as

$$\underline{S_I}(X) = \{x \in U | S_I(x) \subseteq X\}, \quad \overline{S_I}(X) = \{x \in U | S_I(x) \cap X \neq \emptyset\}.$$

By generalized rough approximation operators $\underline{S_I}(X)$ and $\overline{S_I}(X)$ can be expressed as

$$\underline{S_I}(X) = \underline{I}(I^{-1}(X)), \quad \overline{S_I}(X) = \overline{I}(I^{-1}(X)), \quad \forall X \subseteq U. \tag{2}$$

Equation (2) shows that $\underline{S_I}$ and $\overline{S_I}$ are the compositions of the generalized lower and upper approximation operators respectively, and the internal and external operators are respectively based on I and I^{-1}.

The next proposition will be used in the following.

Proposition 1. *Let (U, R_1) and (U, R_2) be two approximation spaces. Then*

(1) $\underline{R_2}(X) \subseteq \underline{R_1}(X)$ or $\overline{R_1}(X) \subseteq \overline{R_2}(X), \forall X \subseteq U$ if and only if $R_1 \subseteq R_2$;
(2) $\forall X \subseteq U$, $\underline{R_1 \cup R_2}(X) = \underline{R_1}(X) \cap \underline{R_2}(X), \overline{R_1 \cup R_2}(X) = \overline{R_1}(X) \cup \overline{R_2}(X)$.

Proof. (1) If $R_1 \subseteq R_2$, then $\forall x \in U$, $R_1^{-1}(x) \subseteq R_2^{-1}(x)$. For any $X \subseteq U$ we have

$$\overline{R_1}(X) = \bigcup_{x \in X} \overline{R_1}(\{x\}) = \bigcup_{x \in X} R_1^{-1}(x) \subseteq \bigcup_{x \in X} R_2^{-1}(x) = \bigcup_{x \in X} \overline{R_2}(\{x\}) = \overline{R_2}(X).$$

From the duality it follows that $\underline{R_2}(X) \subseteq \underline{R_1}(X), \forall X \subseteq U$.

Conversely, if $\overline{R_1}(X) \subseteq \overline{R_2}(X), \forall X \subseteq U$, then $\overline{R_1}(\{x\}) \subseteq \overline{R_2}(\{x\}), \forall x \in U$, thus $R_1^{-1}(x) \subseteq R_2^{-1}(x), \forall x \in U$, which implies that $R_1 \subseteq R_2$.

(2) For any $X \subseteq U$, we have

$$\begin{aligned}
\underline{R_1 \cup R_2}(X) &= \{x \in U | (R_1 \cup R_2)(x) \subseteq X\} \\
&= \{x \in U | R_1(x) \cup R_2(x) \subseteq X\} \\
&= \{x \in U | R_1(x) \subseteq X\} \cap \{x \in U | R_2(x) \subseteq X\} \\
&= \underline{R_1}(X) \cap \underline{R_2}(X).
\end{aligned}$$

By the duality we have $\overline{R_1 \cup R_2}(X) = \overline{R_1}(X) \cup \overline{R_2}(X), \forall X \subseteq U$.

3   Rough Sets in Incomplete Contexts

In this section, we investigate some binary relations induced from incomplete contexts, and explore properties of the rough approximation operators based on them.

3.1   Incomplete Contexts

Definition 2. *An incomplete context is a quadruple $(U, A, \{1, *, 0\}, I)$ where U and A are sets of objects and attributes respectively, $\{1, *, 0\}$ is the set of values, I is a mapping from $U \times A$ to $\{1, *, 0\}$ such that*

$I(x, a) = 1$ *or* $a(x) = 1$ *means the object x has the attribute a,*
$I(x, a) = 0$ *or* $a(x) = 0$ *means the object x does not have the attribute a,*
$I(x, a) = *$ *or* $a(x) = *$ *means it is unknown whether or not the object x has the attribute a.*

Example 2. Table 2 provides an exemplary incomplete context $(U, A, \{1, *, 0\}, I)$ in which $U = \{x_1, x_2, x_3, x_4, x_5, x_6\}$ and $A = \{a_1, a_2, a_3, a_4, a_5, a_6\}$. In this table, for example, the two asterisks in line 3 means that it is unknown whether or not the object x_3 has the attribute a_2 or a_4.

Table 2. An incomplete context $(U, A, \{1, *, 0\}, I)$

U	a_1	a_2	a_3	a_4	a_5	a_6
x_1	1	0	0	1	0	0
x_2	0	1	0	*	1	0
x_3	0	*	1	*	0	1
x_4	*	0	1	0	0	1
x_5	*	1	0	0	1	0
x_6	1	0	0	1	0	*

Let $(U, A, \{1, *, 0\}, I)$ be an incomplete context. Four neighborhood operators can be induced as follows: $\forall x \in U, \forall a \in A$,

$$f(x) = \{a \in A | a(x) = 1\}, \quad f^*(x) = \{a \in A | a(x) = 1 \, or \, a(x) = *\};$$

$$g(a) = \{x \in U | a(x) = 1\}, \quad g^*(a) = \{x \in U | a(x) = 1 \, or \, a(x) = *\}.$$

Then f and f^* correspond to two binary relation from U to A, meanwhile g and g^* correspond to two binary relation from A to U. It is obvious that $\forall x \in U, \forall a \in A$, $a \in f(x)$ and $x \in g(a)$, and $a \in f^*(x)$ and $x \in g^*(a)$ are equivalent, respectively. Furthermore, $f(x) \subseteq f^*(x)$, $g(a) \subseteq g^*(a)$.

Based on the above four neighborhood operators and Eq. (1), four pairs of rough approximation operators can be constructed as follows: $\forall X \subseteq U, \forall B \subseteq A$,

$$
\begin{aligned}
\underline{f}(B) &= \{x \in U | f(x) \subseteq B\}, & \overline{f}(B) &= \{x \in U | f(x) \cap B \neq \emptyset\}; \\
\underline{f^*}(B) &= \{x \in U | f^*(x) \subseteq B\}, & \overline{f^*}(B) &= \{x \in U | f^*(x) \cap B \neq \emptyset\}; \\
\underline{g}(X) &= \{a \in A | g(a) \subseteq X\}, & \overline{g}(X) &= \{a \in A | g(a) \cap X \neq \emptyset\}; \\
\underline{g^*}(X) &= \{a \in A | g^*(a) \subseteq X\}, & \overline{g^*}(X) &= \{a \in A | g^*(a) \cap X \neq \emptyset\}.
\end{aligned}
$$

Formal contexts defined in Definition 1 are called complete contexts w.r.t. the incomplete contexts defined in Definition 2. Complete contexts and incomplete contexts are all called contexts.

A complete context (U, A, I') is called a completion of the incomplete context $(U, A, \{1, *, 0\}, I)$ if $\forall x \in U, \forall a \in A$, $I(x, a) \neq *$ implies $I'(x, a) = I(x, a)$.

An incomplete context $(U, A, \{1, *, 0\}, I)$ is called regular [23] if it satisfies the following conditions:

(1) $\forall x \in U$, $\exists a, b \in A$ such that $a(x) = 1$, $b(x) = 0$,
(2) $\forall a \in A$, $\exists x, y \in U$ such that $a(x) = 1$, $a(y) = 0$.

In this paper, we assume that all incomplete contexts are regular.

3.2   Tolerance Relations and Rough Approximations in Incomplete Contexts

Let $(U, A, \{1, *, 0\}, I)$ be an incomplete context. It can be seen that the operators $\underline{f}, \overline{f}, \underline{f^*}$, and $\overline{f^*}$ are from $\mathcal{P}(A)$ to $\mathcal{P}(U)$, and $\underline{g}, \overline{g}, \underline{g^*}$, and $\overline{g^*}$ are all from $\mathcal{P}(U)$

to $\mathcal{P}(A)$. Imitating the right sides of Eq. (2) and using the compositions of the generalized lower and upper approximation operators based on f, g, f^*, and g^*, we can establish four pairs of operators as follows: $\forall X \subseteq U$,

(I) $\underline{I_1}(X) = \underline{f}(g(X))$, $\overline{I_1}(X) = \overline{f}(\overline{g}(X))$;
(II) $\underline{I_2}(X) = \underline{f}(g^*(X))$, $\overline{I_2}(X) = \overline{f}(\overline{g^*}(X))$;
(III) $\underline{I_3}(X) = \underline{f^*}(g(X))$, $\overline{I_3}(X) = \overline{f^*}(\overline{g}(X))$;
(V) $\underline{I_5}(X) = \underline{f^*}(g^*(X))$, $\overline{I_5}(X) = \overline{f^*}(\overline{g^*}(X))$.

Theorem 1. *Let $(U, A, \{1, *, 0\}, I)$ be an incomplete context, and*

$$\begin{aligned}
S_1 &= \{(x,y) \in U \times U | f(x) \cap f(y) \neq \emptyset\}, \\
S_2 &= \{(x,y) \in U \times U | f(x) \cap f^*(y) \neq \emptyset\}, \\
S_3 &= \{(x,y) \in U \times U | f^*(x) \cap f(y) \neq \emptyset\}, \\
S_5 &= \{(x,y) \in U \times U | f^*(x) \cap f^*(y) \neq \emptyset\},
\end{aligned}$$

then $\forall X \subseteq U$,

$$\begin{aligned}
\underline{I_1}(X) &= \underline{S_1}(X),\ \overline{I_1}(X) = \overline{S_1}(X), \\
\underline{I_2}(X) &= \underline{S_2}(X),\ \overline{I_2}(X) = \overline{S_2}(X), \\
\underline{I_3}(X) &= \underline{S_3}(X),\ \overline{I_3}(X) = \overline{S_3}(X), \\
\underline{I_5}(X) &= \underline{S_5}(X),\ \overline{I_5}(X) = \overline{S_5}(X).
\end{aligned}$$

Proof. Since the proofs are similar, as an example, we only give the proof for S_1 as follows.

For any $X \subseteq U$, by (U3) it can be proved easily that $\overline{I_1}(X) = \bigcup_{x \in X} \overline{I_1}(\{x\})$ and $\overline{S_1}(X) = \bigcup_{x \in X} \overline{S_1}(\{x\})$. For any $x \in U$, we have

$$\begin{aligned}
\overline{I_1}(\{x\}) &= \{y \in U | f(y) \cap \overline{g}(\{x\}) \neq \emptyset\} \\
&= \{y \in U | \exists a \in A(a \in f(y), a \in \overline{g}(\{x\}))\} \\
&= \{y \in U | \exists a \in A(a \in f(y), x \in g(a))\} \\
&= \{y \in U | \exists a \in A(a \in f(y), a \in f(x))\} \\
&= \{y \in U | (y, x) \in S_1\} \\
&= S_1^{-1}(x) \\
&= \overline{S_1}(\{x\}).
\end{aligned}$$

Hence $\overline{I_1}(X) = \overline{S_1}(X), \forall X \subseteq U$, by the duality we get $\underline{I_1}(X) = \underline{S_1}(X)$.

Theorem 1 indicates that the four pairs of operator, (I), (II), (III) and (V), are all rough approximation operators based on binary relation on U. It can be verified that S_2 and S_3 are inverse to each other, S_1 and S_5 are two tolerance relations on U. In fact, S_1 and S_5 can be induced from two completions of $(U, A, \{1, *, 0\}, I)$. Let (U, A, I^0) be the completion of $(U, A, \{1, *, 0\}, I)$ by replacing the relation values $*$ in $(U, A, \{1, *, 0\}, I)$ with 0, and (U, A, I^1) the completion of $(U, A, \{1, *, 0\}, I)$ by substituting 1 for $*$ in $(U, A, \{1, *, 0\}, I)$, then $S_1 = S_{I^0}$ and $S_5 = S_{I^1}$. However, S_2 and S_3 may be not tolerance relations.

Example 3. For the incomplete context in Example 2, the successor neighborhoods of all elements of U for the binary relations S_2 and S_3 are listed as follows:

$$
\begin{aligned}
&S_2(x_1) = U, &&S_2(x_2) = \{x_2, x_3, x_5\}, &&S_2(x_3) = \{x_3, x_4, x_6\}, \\
&S_2(x_4) = \{x_3, x_4, x_6\}, &&S_2(x_5) = \{x_2, x_3, x_5\}, &&S_2(x_6) = U; \\
&S_3(x_1) = \{x_1, x_6\}, &&S_3(x_2) = \{x_1, x_2, x_5, x_6\}, &&S_3(x_3) = U, \\
&S_3(x_4) = \{x_1, x_3, x_4, x_6\}, &&S_3(x_5) = \{x_1, x_2, x_5, x_6\}, &&S_3(x_6) = \{x_1, x_3, x_4, x_6\}.
\end{aligned}
$$

Then $x_2 \in S_2(x_1)$ and $x_1 \in S_3(x_2)$, but $x_1 \notin S_2(x_2)$ and $x_2 \notin S_3(x_1)$. Hence, S_2 and S_3 are not tolerance relations.

3.3   Comparison Among Rough Approximations

Let $(U, A, \{1, *, 0\}, I)$ be an incomplete context. For the relations S_1, S_2, S_3 and S_5 defined in Theorem 1, we have that

$$S_1 \subseteq S_2 \cap S_3 \subseteq S_2(\text{or } S_3) \subseteq S_2 \cup S_3 \subseteq S_5. \tag{3}$$

From $S_2 = S_3^{-1}$, or equivalently $S_3 = S_2^{-1}$, we know that $S_2 \cap S_3$ and $S_2 \cup S_3$ are two tolerance relations. But S_1 and $S_2 \cap S_3$, $S_2 \cap S_3$ and $S_2 \cup S_3$, and $S_2 \cup S_3$ and S_5 may not be equal, respectively.

Example 4. For the incomplete context shown in Table 2, referring to Example 3 we can get the successor neighborhoods of all elements of U for $S_2 \cap S_3$ and $S_2 \cup S_3$, and list them as follows:

$$
\begin{aligned}
&(S_2 \cap S_3)(x_1) = \{x_1, x_6\}, &&(S_2 \cap S_3)(x_2) = \{x_2, x_5\}, \\
&(S_2 \cap S_3)(x_3) = \{x_3, x_4, x_6\}, &&(S_2 \cap S_3)(x_4) = \{x_3, x_4, x_6\}, \\
&(S_2 \cap S_3)(x_5) = \{x_2, x_5\}, &&(S_2 \cap S_3)(x_6) = \{x_1, x_3, x_4, x_6\}; \\
&(S_2 \cup S_3)(x_1) = U, &&(S_2 \cup S_3)(x_2) = \{x_1, x_2, x_3, x_5, x_6\}, \\
&(S_2 \cup S_3)(x_3) = U, &&(S_2 \cup S_3)(x_4) = \{x_1, x_3, x_4, x_6\}, \\
&(S_2 \cup S_3)(x_5) = \{x_1, x_2, x_3, x_5, x_6\}, &&(S_2 \cup S_3)(x_6) = U.
\end{aligned}
$$

It can be seen that $S_2 \cap S_3$ des not equal $S_2 \cup S_3$. In order to compare $S_2 \cap S_3$ and S_1, and $S_2 \cup S_3$ and S_5, the successor neighborhoods of all elements of U for S_1 and S_5 are wrote as follows:

$$
\begin{aligned}
&S_1(x_1) = \{x_1, x_6\}, &&S_1(x_2) = \{x_2, x_5\}, &&S_1(x_3) = \{x_3, x_4\}, \\
&S_1(x_4) = \{x_3, x_4\}, &&S_1(x_5) = \{x_2, x_5\}, &&S_1(x_6) = \{x_1, x_6\}; \\
&S_5(x_1) = U, &&S_5(x_2) = \{x_1, x_2, x_3, x_5, x_6\}, &&S_5(x_3) = U, \\
&S_5(x_4) = \{x_1, x_3, x_4, x_5, x_6\}, &&S_5(x_5) = U, &&S_5(x_6) = U.
\end{aligned}
$$

From $(S_2 \cap S_3)(x_3) = \{x_3, x_4, x_6\}$ and $S_1(x_3) = \{x_3, x_4\}$, and $(S_2 \cup S_3)(x_5) = \{x_1, x_2, x_3, x_5, x_6\}$ and $S_5(x_5) = U$, we know that $S_2 \cap S_3$ is not equal to S_1, and $S_2 \cup S_3$ is not equal to S_5.

Denote $S_2 \cup S_3$ as S_4, and $\underline{S_4}$ and $\overline{S_4}$ as $\underline{I_4}$ and $\overline{I_4}$ respectively, we have

Proposition 2. Let $(U, A, \{1, *, 0\}, I)$ be an incomplete context. Then $\forall X \subseteq U$,

$$\underline{I_4}(X) = \underline{I_2}(X) \cap \underline{I_3}(X), \quad \overline{I_4}(X) = \overline{I_2}(X) \cup \overline{I_3}(X).$$

Proof. It directly follows from Proposition 1 and $S_4 = S_2 \cup S_3$.

With the rough approximation operators w.r.t $S_i, i = 1, 2, 3, 4, 5$, we have

Theorem 2. *Let* $(U, A, \{1, *, 0\}, I)$ *be an incomplete context. Then* $\forall X \subseteq U$,

$$\underline{I_5}(X) \subseteq \underline{I_4}(X) \subseteq \underline{I_2}(X)(\text{or } \underline{I_3}(X)) \subseteq \underline{I_1}(X);$$

$$\overline{I_1}(X) \subseteq \overline{I_2}(X)(\text{or } \overline{I_3}(X)) \subseteq \overline{I_4}(X) \subseteq \overline{I_5}(X).$$

Proof. It directly follows from Proposition 1 and Inequation (3).

Example 5. With the incomplete context shown in Table 2, if we take $X = \{x_1, x_2, x_3, x_5, x_6\}$, then $\underline{I_5}(X) = \{x_2\}$, $\underline{I_4}(X) = \{x_2, x_5\}$, $\underline{I_3}(X) = \{x_1, x_2, x_5\}$, $\underline{I_2}(X) = \{x_2, x_5\}$, $\underline{I_1}(X) = \{x_1, x_2, x_5, x_6\}$. Thus

$$\underline{I_5}(X) \subset \underline{I_4}(X) = \underline{I_2}(X) \subset \underline{I_3}(X) \subset \underline{I_1}(X).$$

Choosing $Y = \{x_2, x_5\}$, we have $\overline{I_1}(Y) = \{x_2, x_5\}$, $\overline{I_2}(Y) = \{x_1, x_2, x_5, x_6\}$, $\overline{I_3}(Y) = \{x_2, , x_3, x_5\}$, $\overline{I_4}(Y) = \{x_1, x_2, x_3, x_5, x_6\}$, $\overline{I_5}(Y) = U$. Hence

$$\overline{I_1}(Y) \subset \overline{I_2}(Y) \subset \overline{I_4}(Y) \subset \overline{I_5}(Y),$$

$$\overline{I_1}(Y) \subset \overline{I_3}(Y) \subset \overline{I_4}(Y) \subset \overline{I_5}(Y).$$

But $\overline{I_2}(Y) \not\subseteq \overline{I_3}(Y)$ and $\overline{I_3}(Y) \not\subseteq \overline{I_2}(Y)$.

In the following, we examine the equivalence among the five pairs of rough approximation operators.

Firstly, with the equivalence between $\underline{I_1}$ and $\underline{I_2}$, or $\underline{I_3}$, or $\overline{I_1}$ and $\overline{I_2}$, or $\overline{I_3}$, The following conclusions hold.

Theorem 3. *Let* $(U, A, \{1, *, 0\}, I)$ *be an incomplete context. Then the following statements are equivalent:*

(1) $S_1 = S_2$,
(2) $S_1 = S_3$,
(3) (c1) $\forall x, y \in U$, if $f(x) \cap f(y) = \emptyset$ then $f(x) \cap f^*(y) = \emptyset$.

Proof. (1) $\Rightarrow$ (2) If $S_1 = S_2$, then $S_3 = S_2^{-1} = S_1^{-1}$, since S_1 is symmetric, so $S_1 = S_3$.

(2) $\Rightarrow$ (1) It can be proved similarly.

(1) $\Rightarrow$ (3) Assume that $S_1 = S_2$, then $S_2 \subseteq S_1$, that is, $\forall x, y \in U$, if $(x, y) \in S_2$ then $(x, y) \in S_1$. In terms of the definition S_1 and S_2, we have that $\forall x, y \in U$, if $f(x) \cap f^*(y) \neq \emptyset$ then $f(x) \cap f(y) \neq \emptyset$. Equivalently, $\forall x, y \in U$, if $f(x) \cap f(y) = \emptyset$ then $f(x) \cap f^*(y) = \emptyset$, that is, the condition (c) holds.

(3) $\Rightarrow$ (1) It can be proved similarly.

From Theorem 3 it follows immediately that $S_1 = S_4$ is equivalent to the condition (c1), and the following corollary hold.

Corollary 1. *Let* $(U, A, \{1, *, 0\}, I)$ *be an incomplete context. Then the following statements are equivalent:*

(1) $\underline{I_1}(X) = \underline{I_2}(X)$, *or* $\overline{I_1}(X) = \overline{I_2}(X)$, $\forall X \subseteq U$,
(2) $\underline{I_1}(X) = \underline{I_3}(X)$, *or* $\overline{I_1}(X) = \overline{I_3}(X)$, $\forall X \subseteq U$,
(3) *the condition* (c1) *holds.*

Secondly, the following conclusions show the equivalence between $\underline{I_4}$ and $\underline{I_2}$, or $\underline{I_3}$, or $\overline{I_4}$ and $\overline{I_2}$, or $\overline{I_3}$.

Theorem 4. *Let* $(U, A, \{1, *, 0\}, I)$ *be an incomplete context. Then the following statements are equivalent:*

(1) $S_2 = S_4$,
(2) $S_3 = S_4$,
(3) (c2) $\forall x, y \in U$, *if* $f(x) \cap f^*(y) = \emptyset$ *then* $f^*(x) \cap f(y) = \emptyset$.

Proof. (1) $\Rightarrow$ (2) If $S_2 = S_4$, then by $S_4 = S_2 \cup S_3$ we have $S_3 \subseteq S_2$. By the definitions of S_2 and S_3, we have that $\forall x, y \in U$, if $f^*(x) \cap f(y) \neq \emptyset$ then $f(x) \cap f^*(y) \neq \emptyset$. That is to say, $\forall x, y \in U$, if $(x, y) \in S_3$ then $(y, x) \in S_3$. Thus, S_3 is symmetric, of course S_2 is symmetric, so $S_2 = S_3$. Clearly $S_3 = S_4$.
　　(2) $\Rightarrow$ (1) Similarly it can be proved.
　　(1) $\Rightarrow$ (3) If $S_2 = S_4$, according to the above proof we have that $\forall x, y \in U$, if $f^*(x) \cap f(y) \neq \emptyset$ then $f(x) \cap f^*(y) \neq \emptyset$. Equivalently, $\forall x, y \in U$, if $f(x) \cap f^*(y) = \emptyset$ then $f^*(x) \cap f(y) = \emptyset$, that is, the condition (c2) holds.
　　(3) $\Rightarrow$ (1) Similarly it can be proved.

From Theorem 4 we can see that S_2 or S_3 is symmetric if and only if the condition (c2) holds, and the following corollary follows.

Corollary 2. *Let* $(U, A, \{1, *, 0\}, I)$ *be an incomplete context. Then the following statements are equivalent:*

(1) $\underline{I_2}(X) = \underline{I_4}(X)$, *or* $\overline{I_2}(X) = \overline{I_4}(X)$, $\forall X \subseteq U$,
(2) $\underline{I_3}(X) = \underline{I_4}(X)$, *or* $\overline{I_3}(X) = \overline{I_4}(X)$, $\forall X \subseteq U$,
(3) *the condition* (c2) *holds.*

For the conditions (c1) and (c2) we have the following conclusion.

Proposition 3. *Let* $(U, A, \{1, *, 0\}, I)$ *be an incomplete context. Then the condition* (c1) *implies the condition* (c2).

Furthermore, the below conclusions depict the equivalence between $\underline{I_4}$ and $\underline{I_5}$, or $\overline{I_4}$ and $\overline{I_5}$.

Theorem 5. *Let* $(U, A, \{1, *, 0\}, I)$ *be an incomplete context. Then* $S_4 = S_5$ *if and only if* (c3) $\forall x, y \in U$, $f^*(x) \cap f(y) = \emptyset$ *and* $f(x) \cap f^*(y) = \emptyset$ *implies* $f^*(x) \cap f^*(y) = \emptyset$.

Proof. If $S_4 = S_5$, that is, $S_4 \supseteq S_5$, then by the definitions of S_4 and S_5 we have that $\forall x, y \in U$, if $f^*(x) \cap f^*(y) \neq \emptyset$ then $f^*(x) \cap f(y) \neq \emptyset$ or $f(x) \cap f^*(y) \neq \emptyset$. Equivalently, the condition (c3) holds.

Conversely, if the condition (c3) holds, then it can be proved similarly that $S_4 = S_5$.

Similarly the next corollary can be gotten.

Corollary 3. *Let* $(U, A, \{1, *, 0\}, I)$ *be an incomplete context. Then* $\underline{I_4}(X) = \underline{I_5}(X)$, *or* $\overline{I_4}(X) = \overline{I_5}(X)$, $\forall X \subseteq U$ *if and only if the condition* (c3) *holds.*

With respect to the equivalence of the five pairs of rough approximation operators, the next conclusions can be proved similarly.

Theorem 6. *Let* $(U, A, \{1, *, 0\}, I)$ *be an incomplete context. Then* $S_1 = S_5$ *if and only if* (c4) $\forall x, y \in U$, $f(x) \cap f(y) = \emptyset$ *implies* $f^*(x) \cap f^*(y) = \emptyset$.

Corollary 4. *Let* $(U, A, \{1, *, 0\}, I)$ *be an incomplete context. Then* $\underline{I_1}(X) = \underline{I_5}(X)$, *or* $\overline{I_1}(X) = \overline{I_5}(X)$, $\forall X \subseteq U$, *if and only if the condition* (c4) *holds.*

4 Summaries

Much attention has been paid on formal concept analysis in incomplete contexts, however little study on data analysis in incomplete contexts by rough set approaches has been made, thus it is significant to find suitable way to exploit the knowledge hide in incomplete contexts. In this paper, two binary relations in an incomplete context are induced, by the lower and upper rough approximation operators based on the relations, four pairs of rough approximation operators are constructed via compound operation. The derived rough approximation operators are all relation-based rough approximation operators, two pairs of them are based on tolerance relations, and the other two are based on reflexive relations. Furthermore, the comparison among the operators are made, so an ordered relation among them are gained, and the equivalence among them is also characterized by different ways.

It is well known that attribute reduction is a key issue in rough set theory, as for the rough approximation operators proposed in the paper, we will study attribute reduction of incomplete contexts in the future.

Acknowledgements. This work was supported by grants from the National Natural Science Foundation of China (Nos. 61773349, 61075120, 61272021, 61202206).

References

1. Pawlak, Z.: Rough sets. Int. J. Comput. Inf. Sci. **11**, 341–356 (1982)
2. Slowinski, R., Vanderpooten, D.: A generalized definition of rough approximations based on similarity. IEEE Trans. Knowl. Data Eng. **12**, 331–336 (2000)
3. Yao, Y.Y.: The superiority of three-way decisions in probabilistic rough set models. Inf. Sci. **181**, 1080–1096 (2011)
4. Bonikowski, Z., Bryniarski, E., Wybraniec, U.: Extensions and intentions in the rough set theory. Inf. Sci. **107**, 149–167 (1998)
5. Ebenbach, D.H., Moore, C.F.: Incomplete information, inferences, and individual differences: the case of environmental judgments. Organ. Behav. Hum. Decis. Process. **2000**(81), 1–27 (2000)
6. Grzymała-Busse, J.W.: Characteristic relations for incomplete data: a generalization of the indiscernibility relation. In: Tsumoto, S., Słowiński, R., Komorowski, J., Grzymała-Busse, J.W. (eds.) RSCTC 2004. LNCS (LNAI), vol. 3066, pp. 244–253. Springer, Heidelberg (2004). https://doi.org/10.1007/978-3-540-25929-9_29
7. Wang, G.Y., Guan, J.Y., Hu, F.: Rough set extensions in incomplete information system. Front. Electr. Electron. Eng. China **3**, 399–405 (2008)
8. Qian, Y.H., Liang, J.Y., Li, D.Y., Wang, F., Ma, N.N.: Approximation reduction in inconsistent incomplete tables. Knowl. Based Syst. **21**, 427–433 (2010)
9. Leung, Y., Wu, W.Z., Zhang, W.X.: Knowledge acquisition in incomplete information systems: a rough set approach. Eur. J. Oper. Res. **168**, 164–180 (2006)
10. Yang, X.B., Yu, D.Y., Yang, J.Y., Song, X.N.: Difference relation based rough sets and negative rules in incomplete information system. Int. J. Uncertainty Fuzziness Knowl. Based Syst. **17**, 649–665 (2009)
11. Du, W.S., Hu, B.Q.: Dominance-based rough set approach to incomplete ordered information systems. Inf. Sci. **346–347**, 106–129 (2016)
12. Liu, D., Liang, D., Wang, C.: A novel three-way decision model based on incomplete information system. Knowl. Based Syst. **91**, 32–45 (2016)
13. Dai, J., Wei, B., Zhang, X., Zhang, Q.: Uncertainty measurement for incomplete interval-valued information systems based on α-weak similarity. Knowl. Based Syst. **136**, 159–171 (2017)
14. Wille, R.: Restructuring lattice theory: an approach based on hierarchies of concepts. In: Rival, I. (ed.) Ordered Sets. NATO Science Series, vol. 83, pp. 445–470. Reidel, Dordrecht (1982). https://doi.org/10.1007/978-94-009-7798-3_15
15. Yao, Y.Y.: Concept lattices in rough set theory. In: Proceedings of 23rd International Meeting of the North American Fuzzy Information Processing Society, pp. 796–801 (2004)
16. Gediga, G., Duntsch, I.: Modal-style operators in qualitative data analysis, In: Proceedings of the 2002 IEEE International Conference on Data Mining, pp. 155–162 (2002)
17. Shao, M.-W., Zhang, W.-X.: Approximation in formal concept analysis. In: Ślęzak, D., Wang, G., Szczuka, M., Düntsch, I., Yao, Y. (eds.) RSFDGrC 2005. LNCS (LNAI), vol. 3641, pp. 43–53. Springer, Heidelberg (2005). https://doi.org/10.1007/11548669_5
18. Kent, R.E.: Rough concept analysis. In: Ziarko, W.P. (ed.) Rough Sets, Fuzzy Sets and Knowledge Discovery. Workshops in Computing, pp. 248–255. Springer, London (1994). https://doi.org/10.1007/978-1-4471-3238-7_30
19. Pagliani, P.: From concept lattices to approximation spaces: algebraic structures of some spaces of partial objects. Fundamenta Informaticae **18**(1), 1–25 (1993)

20. Li, T.J., Zhang, W.X.: Rough approximations in formal contexts. In: Proceedings of the Fourth International Conference on Machine Learning and Cybernetics, ICMLC 2005, Guangzhou, pp. 18–21 (2005)
21. Burmeister, P., Holzer, R.: On the treatment of incomplete knowledge in formal concept analysis. In: Ganter, B., Mineau, G.W. (eds.) ICCS-ConceptStruct 2000. LNCS (LNAI), vol. 1867, pp. 385–398. Springer, Heidelberg (2000). https://doi.org/10.1007/10722280_27
22. Holzer, R.: Knowledge acquisition under incomplete knowledge using methods from formal concept analysis: Parts I and II. Fundamenta Informaticae **63**(1), 17–39 (2004)
23. Li, J.H., Mei, C.L., Lv, Y.: Incomplete decision contexts: approximation concept construction, rule acquisition and knowledge reduction. Int. J. Approx. Reason. **54**, 149–165 (2013)
24. Li, M., Wang, G.: Approximate concept construction with three-way decisions and attribute reduction in incomplete contexts. Knowl. Based Syst. **91**, 165–178 (2016)
25. Yao, Y.Y.: Interval sets and three-way concept analysis in incomplete contexts. Int. J. Mach. Learn. Cybern. **8**, 3–20 (2017)
26. Wu, W.Z., Zhang, W.X.: Constructive and axiomatic approaches of fuzzy approximation operators. Inf. Sci. **159**, 233–254 (2004)
27. Yao, Y.Y.: Constructive and algebraic methods of the theory of rough sets. Inf. Sci. **109**, 21–47 (1998)

On Granular Rough Computing: Epsilon Homogenous Granulation

Krzysztof Ropiak[(✉)] and Piotr Artiemjew

Faculty of Mathematics and Computer Science,
University of Warmia and Mazury in Olsztyn, Olsztyn, Poland
{kropiak,artem}@matman.uwm.edu.pl

Abstract. In this work we have proposed a new technique of granulation in the family of methods inspired by Polkowski standard granulation algorithm. The new method is called epsilon homogenous granulation. The idea is to create the epsilon granules around the training objects lowering the r-indiscernibility ratio until the group of objects is homogenous in the sense of their belongingness to decision class of central object. We use epsilon granules, which means that during granulation process of numerical data we consider indiscernibility ratio of descriptors. The main advantage of this method in addition to reduction in the number of training objects is that there is no need to estimate the optimal granulation radii. The process of granulation is run only once, and the radii for particular objects are formed in automatic way - dependent on indiscernibility ratio of data and their homogeneity in decision concepts. Next step is to cover the original decision system with formed granules and get the final granular decision system by ε-majority voting method. We have performed preliminary experiments with use of multiple cross validation methods. We have used selected data sets from University of California, Irvine machine learning repository for our research. To verify the quality of approximation we used k-NN classifier designed for our granulation method. The method seems to be comparable with the ones of previous algorithms, with satisfying effectiveness in classification and significant reduction in number of training data.

Keywords: Epsilon homogenous granulation · Rough sets
Decision systems · Classification

1 Introduction

Data approximation methods play a crucial role in big data analysis. One of the most important paradigm, in which researchers consider the problem of data approximation, is granular rough computing. In the granular rough computing we deal with granules in terms of rough sets theory [4]. The term 'granule' was initially used by Zadeh [27] to define the group of objects put together with respect to a similarity relation.

© Springer Nature Switzerland AG 2018
H. S. Nguyen et al. (Eds.): IJCRS 2018, LNAI 11103, pp. 546–558, 2018.
https://doi.org/10.1007/978-3-319-99368-3_43

One of the approximation techniques family, in the frame of rough set theory, was proposed by Polkowski in [10, 11]; it was a brilliant, simple idea of data approximation using rough inclusions. The main idea was to create granules of r-indiscernible objects and to cover the original training data using a selected strategy, where finally the granular reflections of granules are formed by majority voting. This process was named as standard granulation.

This idea was the source of many new techniques and their applications ([1–3], Polkowski [9–14], and Polkowski and Artiemjew [17–24]. Recent years showed use of such granulation among other applications of data approximation process, classification and missing values absorption - see [16].

In this family of methods, were developed new techniques such as concept-dependent and layered granulation variant, also the variants with descriptors indiscernibility ratio based on weak rough inclusions. The methods were extensively checked in experiments and turned out to be effective in data reduction with maintenance of internal knowledge in terms of classification effectiveness.

In this particular work we have proposed a new technique of data granulation called epsilon homogenous granulation. Detailed description is to be found in the next sections.

The motivation to conduct this research was to consider the idea in which we are lowering the r-indiscernibility ratio during granulation until the granule is homogenous in the sense of their decision class.

The new method turned out to be different from previously proposed techniques - where the r-indiscernibility ratio for objects is set in automatic way. The optimal radius estimation is not needed.

The approximation level is up to 50% of the original training size - and the effectiveness in terms of classification suggests that internal knowledge, in comparison with original training set, is preserved.

The rest of the paper has the following content. In Sect. 1 we introduce the theoretical introduction to granular rough computing. In Sect. 2 we detail the description of our new granulation method. In Sect. 3 we present the classifier used in experimental part. In Sect. 4 we show the results of the experiments, and we conclude the paper in Sect. 5.

The granulation process consists of three basic steps, the granules are formed around the training objects, the covering of universe of training objects is chosen, and finally granular reflection from covering granules is obtained by majority voting procedure. We begin with the basic notions of rough inclusions to introduce the first step.

1.1 Theoretical Background - Granular Rough Inclusions

The models for rough mereology which give us methods by which the rough inclusions are defined are presented in Polkowski [6–10]; a detailed discussion may be found in Polkowski [15].

For a rough inclusion μ on the universe U of a decision system $D = (U, A, d)$. We introduce the parameter r_{gran}, the *granulation radius* with values $0, \frac{1}{|A|}, \frac{2}{|A|}, ..., 1$. For each object $u \in U$, and $r = r_{gran}$, the *standard granule*

$g(u, r, \mu)$, *of radius r about u*, is defined as

$$g(u, r, \mu) \text{ is } \{v \in U : \mu(v, u, r)\}. \tag{1}$$

The standard rough inclusion is defined as

$$\mu(v, u, r) \Leftrightarrow \frac{|Ind(u, v)|}{|A|} \geq r \tag{2}$$

where

$$IND(u, v) = \{a \in A : a(u) = a(v)\}, \tag{3}$$

It follows that this rough inclusion extends the indiscernibility relation to a degree of r.

1.2   ε–modification of the Standard Rough Inclusion

Given a parameter ε valued in the unit interval $[0, 1]$, we define the set

$$Ind_\varepsilon(u, v) = \{a \in A : dist(a(u), a(v)) \leq \varepsilon\}, \tag{4}$$

and, we set

$$\mu_\varepsilon(v, u, r) \Leftrightarrow \frac{|Ind_\varepsilon(u, v)|}{|A|} \geq r \tag{5}$$

The rough inclusion extends the indiscernibility relation to a degree of r.

1.3   Covering of Decision System

In this step the universe of training objects should be covered by computed granules using a selected strategy. One of the most effective methods among the studied ones (see [24]) is simple random choice and thus this method is selected for our experiments. In the next section there is a description of the last step of the granulation process.

1.4   Granular Reflections

Once the granular covering is selected, the idea is to represent granules by single objects. The strategy for obtaining it can be the *majority voting MV*, so for each granule $g \in COV(U, \mu, r)$, the final representation is formed as follows

$$\{MV(\{a(u) : u \in g\}) : a \in A \cup \{d\}\} \tag{6}$$

where for numerical data we treat the descriptors as indiscernible in case $\frac{|a_i(u) - a_j(u)|}{max_a - min_a} \leq \varepsilon$, i, j are the numbers of objects in granule.

The granular reflection of the decision system $D = (U, A, d)$, (where U is the universe of objects, A the set of conditional attributes and d is decision attribute), $(COV(U, \mu, r))$ is formed from granules.

$$v \in g_r^{cd}(u) \text{ if and only if } \mu(v, u, r) \text{ and } (d(u) = d(v)) \tag{7}$$

for a given rough (weak) inclusion μ.

In the next section we introduce our new method of granulation.

2   Epsilon Homogenous Granulation

The method is defined in the following way,

$$g_{r_u}^{\varepsilon, homogenous} = \{v \in U : |g_{r_u}^{\varepsilon - cd}| - |g_{r_u}^{\varepsilon}| == 0, \ for \ minimal \ r_u \ fulfills \ the \ equation\}$$

where

$$g_{r_u}^{\varepsilon, cd}(u) = \{v \in U : \frac{IND_\varepsilon(u, v)}{|A|} \le r_u \ AND \ d(u) == d(v)\}$$

and

$$g_{r_u}^{\varepsilon}(u) = \{v \in U : \frac{IND_\varepsilon(u,v)}{|A|} \le r_u\}$$
$$r_u = \{\frac{0}{|A|}, \frac{1}{|A|}, ..., \frac{|A|}{|A|}\}$$
$$IND_\varepsilon(u, v) = \{a \in A : \frac{|a(u) - a(v)|}{max_a - min_a} \le \varepsilon\}$$

where max_a, min_a are the maximal and minimal attribute values for $a \in A$ in the original data set (Table 1).

2.1   Metrics for Granulation and Classification

The Hamming metric - for symbolic data is defined as

$$d_H(u, v) = |\{a \in A : a(u) \ne a(v)\}|. \tag{8}$$

ε-**normalized Hamming metric** is a modification for numerical, for given ε, is defined as

$$d_{H,\varepsilon}(u, v) = |\{a \in A : \frac{|a(u) - a(v)|}{max_a - min_a} > \varepsilon\}|. \tag{9}$$

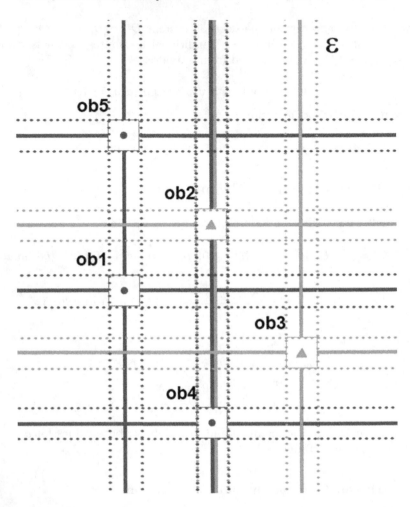

Fig. 1. Exemplary toy demonstration for objects represented as pairs of attributes. We have two decision concepts circles and rectangles. Epsilon homogenous granules can be $g^\varepsilon_{0.5}(ob1) = \{ob1, ob5\}$, $g^\varepsilon_1(ob2) = \{ob2\}$, $g^\varepsilon_{0.5}(ob3) = \{ob3\}$, $g^\varepsilon_1(ob4) = \{ob4\}$, $g^\varepsilon_{0.5}(ob1) = \{ob5, ob1\}$. The set of possible radii is $\{\frac{0}{2}, \frac{1}{2}, \frac{2}{2}\}$. The descriptors can be shifted in the range determined by ε and still were treated as indiscernible.

Table 1. Training data system (U_{trn}, A, d), (a sample from australian credit data set), for $varepsilon = 0.05$

	a_1	a_2	a_3	a_4	a_5	a_6	a_7	a_8	a_9	a_{10}	a_{11}	a_{12}	a_{13}	a_{14}	d
u_1	1	20.17	8.17	2	6	4	1.96	1	1	14	0	2	60	159	1
u_2	1	34.92	5	2	14	8	7.5	1	1	6	1	2	0	1001	1
u_3	1	58.58	2.71	2	8	4	2.415	0	0	0	1	2	320	1	0
u_4	1	29.58	4.5	2	9	4	7.5	1	1	2	1	2	330	1	1
u_5	0	19.17	0.58	1	6	4	0.585	1	0	0	1	2	160	1	0
u_6	1	23.08	2.5	2	8	4	1.085	1	1	11	1	2	60	2185	1
u_7	0	21.67	11.5	1	5	3	0	1	1	11	1	2	00	1	1
u_8	1	27.83	1	1	2	8	3	0	0	0	0	2	176	538	0
u_9	1	41.17	1.33	2	2	4	0.165	0	0	0	0	2	168	1	0
u_{10}	1	41.58	1.75	2	4	4	0.21	1	0	0	0	2	160	1	0
u_{11}	1	22.5	0.12	1	4	4	0.125	0	0	0	0	2	200	71	0
u_{12}	1	33.17	3.04	1	8	8	2.04	1	1	1	1	2	180	18028	1
u_{13}	1.234	22.08	11.46	2	4	4	1.585	0	0	0	1	2	100	1213	0
u_{14}	0	58.67	4.46	2	11	8	3.04	1	1	6	0	2	43	561	1
u_{15}	1	33.5	1.75	2	14	8	4.5	1	1	4	1	2	253	858	1
u_{16}	0	18.92	9	2	6	4	0.75	1	1	2	0	2	88	592	1
u_{17}	1	20	1.25	1	4	4	0.125	0	0	0	0	2	140	5	0
u_{18}	1	19.5	9.58	2	6	4	0.79	0	0	0	0	2	80	351	0
u_{19}	0	22.67	3.8	2	8	4	0.165	0	0	0	0	2	160	1	0
u_{20}	1	17.42	6.5	2	3	4	0.125	0	0	0	0	2	60	101	0
u_{21}	1	41.42	5	2	11	8	5	1	1	6	1	2	470	1	1
u_{22}	1	20.67	1.25	1	8	8	1.375	1	1	3	1	2	140	211	0
u_{23}	1	48.08	6.04	2	4	4	0.04	0	0	0	0	2	0	2691	1
u_{24}	0	28.17	0.58	2	6	4	0.04	0	0	0	0	2	260	1005	0

2.2   Toy Example of Epsilon Homogenous Granulation

Considering training decision system

Epsilon Homogenous granules for all training objects:

$g_0.571429(u_1) = (u_1),$
$g_0.5(u_2) = (u_2, u_4, u_{15}, u_{21}),$
$g_0.571429(u_3) = (u_3, u_9, u_{19}, u_{20}),$
$g_0.5(u_4) = (u_1, u_2, u_4, u_6, u_{21}),$
$g_0.5(u_5) = (u_5, u_{10}, u_{19}, u_{24}),$
$g_0.5(u_6) = (u_1, u_4, u_6),$
$g_0.5(u_7) = (u_7),$
$g_0.5(u_8) = (u_8, u_9, u_{11}, u_{17}),$

$g_0.642857(u_9) = (u_9, u_{10}, u_{11}, u_{17}, u_{19}, u_{20})$,
$g_0.642857(u_{10}) = (u_9, u_{10}, u_{19})$,
$g_0.642857(u_{11}) = (u_9, u_{11}, u_{17}, u_{19}, u_{20})$,
$g_0.642857(u_{12}) = (u_{12})$,
$g_0.571429(u_{13}) = (u_{13})$,
$g_0.428571(u_{14}) = (u_2, u_{14}, u_{16}, u_{21})$,
$g_0.5(u_{15}) = (u_2, u_{12}, u_{15}, u_{21})$,
$g_0.5(u_{16}) = (u_1, u_{14}, u_{16})$,
$g_0.642857(u_{17}) = (u_9, u_{11}, u_{17}, u_{20})$,
$g_0.642857(u_{18}) = (u_{18})$,
$g_0.571429(u_{19}) = (u_3, u_9, u_{10}, u_{11}, u_{17}, u_{19}, u_{20}, u_{24})$,
$g_0.642857(u_{20}) = (u_9, u_{11}, u_{17}, u_{19}, u_{20})$,
$g_0.5(u_{21}) = (u_2, u_4, u_{14}, u_{15}, u_{21})$,
$g_0.642857(u_{22}) = (u_{22})$,
$g_0.642857(u_{23}) = (u_{23})$,
$g_0.642857(u_{24}) = (u_{24})$,

Granules covering training system by random choice:

Covering granules: $g_0.5(u_2) = (u_2, u_4, u_{15}, u_{21})$,
$g_0.571429(u_3) = (u_3, u_9, u_{19}, u_{20})$,
$g_0.5(u_5) = (u_5, u_{10}, u_{19}, u_{24})$,
$g_0.5(u_6) = (u_1, u_4, u_6)$,
$g_0.5(u_7) = (u_7)$,
$g_0.5(u_8) = (u_8, u_9, u_{11}, u_{17})$,
$g_0.642857(u_{12}) = (u_{12})$,
$g_0.571429(u_{13}) = (u_{13})$,
$g_0.5(u_{16}) = (u_1, u_{14}, u_{16})$,
$g_0.642857(u_{18}) = (u_{18})$,
$g_0.642857(u_{20}) = (u_9, u_{11}, u_{17}, u_{19}, u_{20})$,
$g_0.5(u_{21}) = (u_2, u_4, u_{14}, u_{15}, u_{21})$,
$g_0.642857(u_{22}) = (u_{22})$,
$g_0.642857(u_{23}) = (u_{23})$,

Granular decision system from above granules is as follows (Table 2):

Table 2. Granular decision system formed from Covering granules

	a_1	a_2	a_3	a_4	a_5	a_6	a_7	a_8	a_9	a_{10}	a_{11}	a_{12}	a_{13}	a_{14}	d
$g_0.5(u_2)$	1	34.92	5	2	14	8	7.5	1	1	6	1	2	0	1001	1
$g_0.571429(u_3)$	1	58.58	2.71	2	8	4	0.165	0	0	0	0	2	320	1	0
$g_0.5(u_5)$	0	19.17	0.58	2	6	4	0.21	1	0	0	0	2	160	1	0
$g_0.5(u_6)$	1	20.17	8.17	2	6	4	1.96	1	1	14	1	2	60	159	1
$g_0.5(u_7)$	0	21.67	11.5	1	5	3	0	1	1	11	1	2	0	1	1
$g_0.5(u_8)$	1	27.83	1.33	1	2	4	0.165	0	0	0	0	2	176	1	0
$g_0.642857(u_{12})$	1	33.17	3.04	1	8	8	2.04	1	1	1	1	2	180	18028	1
$g_0.571429(u_{13})$	1.234	22.08	11.46	2	4	4	1.585	0	0	0	1	2	100	1213	0
$g_0.5(u_{16})$	0	20.17	8.17	2	6	4	1.96	1	1	14	0	2	60	561	1
$g_0.642857(u_{18})$	1	19.5	9.58	2	6	4	0.79	0	0	0	0	2	80	351	0
$g_0.642857(u_{20})$	1	22.5	1.33	2	4	4	0.165	0	0	0	0	2	168	1	0
$g_0.5(u_{21})$	1	34.92	5	2	14	8	7.5	1	1	6	1	2	0	1001	1
$g_0.642857(u_{22})$	1	20.67	1.25	1	8	8	1.375	1	1	3	1	2	140	211	0
$g_0.642857(u_{23})$	1	48.08	6.04	2	4	4	0.04	0	0	0	0	2	0	2691	1

In the Fig. 1 we have added a simple visualization of granulation process.

3 k-NN Method for Evaluation of Epsilon Homogenous Granulation

The k-NN classifier use modified epsilon Hamming metric, where the descriptors are treated as indiscernible in case $\frac{|a(u)-a(v)|}{max_a-min_a} \leq \varepsilon$. The similar form of this classification was proposed in [24].

Procedure

Step 1. Granulated training data set $(G^{trn}_{r_{gran}}, A, d)$ and the test decision set (U_{tst}, A, d) have been chosen, where A is a set of conditional attributes, d the decision attribute, and, r_{gran} a granulation radius.

Step 2. Classification of test objects by means of granules of training objects is performed as follows.

For all conditional attributes $a \in A$, training objects $v \in G^{trn}$, and test objects $u \in U_{tst}$, we compute weights $w(u, v)$ based on the ε-normalized Hamming metric.

In the voting procedure of the kNN classifier, we use optimal k estimated by CV5 (Cross Validation with 5 folds citeboosting), details of the procedure are highlighted in next section.

If the cardinality of the smallest training decision class is less than k, we apply the value for $k = |the\ smallest\ training\ decision\ class|$.

The test object u is classified by means of weights computed for all training objects v. Weights are sorted in increasing order as,

$$w_1^{c_1}(u, v_1^{c_1}) \leq w_2^{c_1}(u, v_2^{c_1}) \leq \ldots \leq w_{|C_1|}^{c_1}(u, v_{|C_1|}^{c_1});$$
$$w_1^{c_2}(u, v_1^{c_2}) \leq w_2^{c_2}(u, v_2^{c_2}) \leq \ldots \leq w_{|C_2|}^{c_2}(u, v_{|C_2|}^{c_2});$$
$$\ldots$$
$$w_1^{c_m}(u, v_1^{c_m}) \leq w_2^{c_m}(u, v_2^{c_m}) \leq \ldots \leq w_{|C_m|}^{c_m}(u, v_{|C_m|}^{c_m}),$$

where $C_1, C_2, ..., C_m$ are all decision classes in the training set.

Based on computed and sorted weights, training decision classes vote by means of the following parameter, where c runs over decision classes in the training set,

$$Concept\_weight_c(u) = \sum_{i=1}^{k} w_i^c(u, v_i^c). \tag{10}$$

Finally, the test object u is classified into the class c with a minimal value of $Concept\_weight_c(u)$.

After all test objects u are classified, the quality parameter of *accuracy, acc* is computed, according to the formula

$$acc = \frac{number\ of\ correctly\ classified\ objects}{number\ of\ classified\ objects}.$$

3.1   Parameter Estimation in kNN Classifier

The parameter for experiments were estimated in [24]. The optimal k is presented in Table 3.

Table 3. Estimated parameters for kNN based on $5 \times$ CV5

Name	Optimal k
Australian-credit	5
German-credit	18
Heartdisease	19
Hepatitis	3

4   Experimental Session

To verify effectiveness and to obtain first sight on behaviour of epsilon homogenous granulation we have performed a series of experiments with data from UCI Repository [26] - see Table 4. We have implemented the tests in C++.

Table 4. Data sets description

Name	Attr type	Attr no.	Obj no.	Class no.
Australian-credit	categorical, integer, real	15	690	2
German-credit	categorical, integer	21	1000	2
Heartdisease	categorical, real	14	270	2
Hepatitis	categorical, integer, real	20	155	2

Table 5. The result for homogenous granulation (HG) and for epsilon homogenous granulation ($\varepsilon - HGS$) - 5 times CV5 method; HG_acc = average accuracy for HG, $\varepsilon - HG\_acc$ average accuracy for $\varepsilon - HGS$, HGS_size = HG decision system size, $\varepsilon - HGS\_size$ = $\varepsilon - HGS$ decision system size, TRN_size = training set size, $HG_T RN\_red$ = reduction in object number in training set for HG, $\varepsilon - HGS\_size$ = reduction in object number in training set for $\varepsilon - HGS$, HG_r_range = spectrum of radii for HG, $\varepsilon - HG\_r\_range$ = spectrum of radii for $\varepsilon - HGS$

Results	Australian-credit	German-credit	Heartdisease	Hepatitis
HG_acc	0.835	0.725	0.833	0.88
$\varepsilon - HG\_acc$	0.842	0.725	0.831	0.87
HGS_size	286.52	513.3	120.5	46.16
$\varepsilon - HGS\_size$	274.52	503	109.4	46.2
TRN_size	552	800	216	124
$HG_T RN\_red$	48.1%	35.8%	44.2%	62.8%
$\varepsilon HG_T RN\_red$	50.3%	37.1%	49.4%	62.7%
HG_r_range	$r_u \geq 0.5$	$r_u \geq 0.6$	$r_u \geq 0.461$	$r_u \geq 0.579$
$\varepsilon - HG\_r\_range$	$r_u \geq 0.571$	$r_u \geq 0.65$	$r_u \geq 0.615$	$r_u \geq 0.579$

The model we used is multiple cross validation 5. The main classifier used to verify the protection of internal knowledge in the process of granulation was k-NN with modified epsilon hamming metric. The optimal values of k that were used in this research where the ones identified in [24] and presented in Table 3. Seeing the results for considered data in [24] we used $\varepsilon = 0.05$ in granulation and classification.

The result of experiments is presented in Table 5. The approximation quality seems to be comparable with our best previous methods. To show the difference we published the result for concept dependent granulation in Table 6. In Table 5 we can see also the result for homogenous granulation dedicated to symbolic data. We observed a slight lowering of granular decision system size for *varepsilon*-homogenous granulation in comparison with homogenous granulation with similar result of classification. Due to the lack of space we have shown only exemplary results.

Table 6. Summary of results, k-NN vs Naive Bayes Classifier, granular and non granular case, acc = accuracy of classification, red = percentage reduction in object number, r = granulation radius, $method$ = variant of Naive Bayes classifier

Name	$k - NN(acc, red, r)$	$k - NN.nil(acc)$
Australian-credit	$0.851, 71.86, 0.571$	0.855
Car Evaluation	$0.865, 73.23, 0.833$	0.944
Diabetes	$0.616, 74.74, 0.25$	0.631
German-credit	$0.724, 59.85, 0.65$	0.73
Heartdisease	$0.83, 67.69, 0.538$	0.837
Hepatitis	$0.884, 60, 0.632$	0.89
Nursery	$0.696, 77.09, 0.875$	0.578
SPECTF Heart	$0.802, 60.3, 0.114$	0.779

5   Conclusions

The paper contains theoretical introduction and experimental effectiveness verification of a new granulation technique called epsilon homogenous granulation. This is a method from the family of techniques proposed by Polkowski in [10, 11]. This new method is based on granules created as r-indiscernible group of objects by lowering the granulation radius until the granules are homogenous in the sense of their decision class. There is no need to estimate any optimal granulation radii in this method, this being its main advantage. Another positive conclusion which came out after experimental verification of classification effectiveness using granule data set was reduction of training dataset size by up to ca. 50% while retaining internal knowledge in a high degree. The radii for ε homogenous granulation are in many cases larger than for homogenous granulation, because the granules move faster to homogenous form. The result of classification for both methods are comparable, but in case of ε homogenous granulation we obtained better reduction in training set size. In the future works we have a plan to check the effectiveness of the new method in the process of missing values absorbtion. Another direction of research is to find the most effective classifier and to check the boosting effect in the Ensemble models for homogenous granulation. As a further research one could provide a tolerance level in acceptance of objects from the other classes to check the influence on the internal knowledge preservation.

Acknowledgements. The research has been supported by grant 23:610:007-300 from Ministry of Science and Higher Education of the Republic of Poland.

References

1. Artiemjew, P.: Classifiers from granulated data sets: concept dependent and layered granulation. In: Proceedings of RSKD 2007, the Workshops at ECML/PKDD 2007, pp. 1–9. Warsaw University Press, Warsaw (2007)
2. Artiemjew, P.: Natural versus granular computing: classifiers from granular structures. In: Proceedings of 6th International Conference on Rough Sets and Current Trends in Computing RSCTC 2008, Akron OH, USA (2008)
3. Artiemjew, P.: A review of the knowledge granulation methods: discrete vs. continuous algorithms. In: Skowron, A., Suraj, Z. (eds.) Rough Sets and Intelligent Systems - Professor Zdzisław Pawlak in Memoriam. Intelligent Systems Reference Library, vol. 43, pp. 41–59. Springer, Heidelberg (2013). https://doi.org/10.1007/978-3-642-30341-8_4
4. Pawlak, Z.: Rough sets. Int. J. Comput. Inf. Sci. **11**, 341–356 (1982)
5. Polap, D., Wozniak, M., Wei, W., Damasevicius, R.: Multi-threaded Learning Control Mechanism for Neural Networks. Future Generation Computer Systems. Elsevier (2018)
6. Polkowski, L.: Rough Sets. Mathematical Foundations. Physica Verlag, Heidelberg (2002)
7. Polkowski, L.: A rough set paradigm for unifying rough set theory and fuzzy set theory. In: Wang, G., Liu, Q., Yao, Y., Skowron, A. (eds.) RSFDGrC 2003. LNCS (LNAI), vol. 2639, pp. 70–77. Springer, Heidelberg (2003). https://doi.org/10.1007/3-540-39205-X_9
8. Polkowski, L.: Toward rough set foundations. Mereological approach. In: Tsumoto, S., Słowiński, R., Komorowski, J., Grzymała-Busse, J.W. (eds.) RSCTC 2004. LNCS (LNAI), vol. 3066, pp. 8–25. Springer, Heidelberg (2004). https://doi.org/10.1007/978-3-540-25929-9_2
9. Polkowski, L.: Granulation of knowledge in decision systems: the approach based on rough inclusions. The method and its applications. In: Kryszkiewicz, M., Peters, J.F., Rybinski, H., Skowron, A. (eds.) RSEISP 2007. LNCS (LNAI), vol. 4585, pp. 69–79. Springer, Heidelberg (2007). https://doi.org/10.1007/978-3-540-73451-2_9
10. Polkowski, L.: Formal granular calculi based on rough inclusions. In: Proceedings of IEEE 2005 Conference on Granular Computing GrC05, Beijing, China, pp. 57–62. IEEE Press (2005)
11. Polkowski, L.: A model of granular computing with applications. In: Proceedings of IEEE 2006 Conference on Granular Computing GrC06, Atlanta, USA, pp. 9–16. IEEE Press (2006)
12. Polkowski, L.: The paradigm of granular rough computing. In: Proceedings ICCI 2007, Lake Tahoe NV, pp. 145–163. IEEE Computer Society, Los Alamitos (2007)
13. Polkowski, L.: A unified approach to granulation of knowledge and granular computing based on rough mereology: a survey. In: Pedrycz, W., Skowron, A., Kreinovich, V. (eds.) Handbook of Granular Computing, pp. 375–401. Wiley, New York (2008)
14. Polkowski, L.: Granulation of knowledge: similarity based approach in information and decision systems. In: Meyers, R.A. (ed.) Encyclopedia of Complexity and System Sciences. Springer, Heidelberg (2009). https://doi.org/10.1007/978-1-4614-1800-9_94. article 00788
15. Polkowski, L.: Approximate Reasoning by Parts. An Introduction to Rough Mereology. Springer, Heidelberg (2011). https://doi.org/10.1007/978-3-642-22279-5

16. Polkowski, L., Artiemjew, P.: On granular rough computing with missing values. In: Kryszkiewicz, M., Peters, J.F., Rybinski, H., Skowron, A. (eds.) RSEISP 2007. LNCS (LNAI), vol. 4585, pp. 271–279. Springer, Heidelberg (2007). https://doi.org/10.1007/978-3-540-73451-2_29

17. Polkowski, L., Artiemjew, P.: On granular rough computing: factoring classifiers through granulated decision systems. In: Kryszkiewicz, M., Peters, J.F., Rybinski, H., Skowron, A. (eds.) RSEISP 2007. LNCS (LNAI), vol. 4585, pp. 280–289. Springer, Heidelberg (2007). https://doi.org/10.1007/978-3-540-73451-2_30

18. Polkowski, L., Artiemjew, P.: Towards granular computing: classifiers induced from granular structures. In: Proceedings RSKD 2007, the Workshops at ECML/PKDD 2007, pp. 43–53. Warsaw University Press, Warsaw (2007)

19. Polkowski, L., Artiemjew, P.: Classifiers based on granular structures from rough inclusions. In: Proceedings of 12th International Conference on Information Processing and Management of Uncertainty in Knowledge-Based Systems IPMU 2008, Torremolinos, Malaga, Spain, pp. 1786–1794 (2008)

20. Polkowski, L., Artiemjew, P.: Rough sets in data analysis: foundations and applications. In: Smoliński, T.G., Milanova, M., Hassanien, A.-E. (eds.) Applications of Computational Intelligence in Biology: Current Trends and open Problems, SCI, vol. 122, pp. 33–54. Springer, Heidelberg (2008)

21. Polkowski, L., Artiemjew, P.: Rough mereology in classification of data: voting by means of residual rough inclusions. In: Chan, C.-C., Grzymala-Busse, J.W., Ziarko, W.P. (eds.) RSCTC 2008. LNCS (LNAI), vol. 5306, pp. 113–120. Springer, Heidelberg (2008). https://doi.org/10.1007/978-3-540-88425-5_12

22. Polkowski, L., Artiemjew, P.: A study in granular computing: on classifiers induced from granular reflections of data. In: Peters, J.F., Skowron, A., Rybiński, H. (eds.) Transactions on Rough Sets IX. LNCS, vol. 5390, pp. 230–263. Springer, Heidelberg (2008). https://doi.org/10.1007/978-3-540-89876-4_14

23. Polkowski, L., Artiemjew, P.: On classifying mappings induced by granular structures. In: Peters, J.F., Skowron, A., Rybiński, H. (eds.) Transactions on Rough Sets IX. LNCS, vol. 5390, pp. 264–286. Springer, Heidelberg (2008). https://doi.org/10.1007/978-3-540-89876-4_15

24. Polkowski, L., Artiemjew, P.: Granular computing in decision approximation - an application of rough mereology. In: Intelligent Systems Reference Library, vol. 77, pp. 1–422. Springer, Heidelberg (2015). ISBN 978-3-319-12879-5. https://doi.org/10.1007/978-3-319-12880-1

25. Ohno-Machado, L.: Cross-validation and bootstrap ensembles, bagging, boosting, Harvard-MIT division of health sciences and technology (2005). http://ocw.mit.edu/courses/health-sciences-and-technology/hst-951j-medical-decision-support-fall-2005/lecture-notes/hst951_6.pdf HST.951J: Medical Decision Support, Fall

26. University of California, Irvine Machine Learning Repository. https://archive.ics.uci.edu/ml/index.php

27. Zadeh, L.A.: Fuzzy sets and information granularity. In: Gupta, M., Ragade, R., Yager, R.R. (eds.) Advances in Fuzzy Set Theory and Applications, North-Holland, Amsterdam, pp. 3–18 (1979)

Fuzzy Bisimulations in Fuzzy Description Logics Under the Gödel Semantics

Quang-Thuy Ha[1], Linh Anh Nguyen[2,3(✉)], Thi Hong Khanh Nguyen[4], and Thanh-Luong Tran[5]

[1] Faculty of Information Technology, VNU University of Engineering and Technology, 144 Xuan Thuy, Hanoi, Vietnam
thuyhq@vnu.edu.vn
[2] Division of Knowledge and System Engineering for ICT, Faculty of Information Technology, Ton Duc Thang University, Ho Chi Minh City, Vietnam
nguyenanhlinh@tdt.edu.vn
[3] Institute of Informatics, University of Warsaw, Banacha 2, 02-097 Warsaw, Poland
nguyen@mimuw.edu.pl
[4] Faculty of Information Technology, Electricity Power University, 235 Hoang Quoc Viet, Hanoi, Vietnam
khanhnth@epu.edu.vn
[5] Department of Information Technology, University of Sciences, Hue University, 77 Nguyen Hue, Hue, Vietnam
ttluong@hueuni.edu.vn

Abstract. Description logics (DLs) are a suitable formalism for representing knowledge about domains in which objects are described not only by attributes but also by binary relations between objects. Fuzzy DLs can be used for such domains when data and knowledge about them are vague. One of the possible ways to specify classes of objects in such domains is to use concepts in fuzzy DLs. As DLs are variants of modal logics, indiscernibility in DLs is characterized by bisimilarity. The bisimilarity relation of an interpretation is the largest auto-bisimulation of that interpretation. In (fuzzy) DLs, it can be used for concept learning. In this paper, for the first time, we define fuzzy bisimulation and (crisp) bisimilarity for fuzzy DLs under the Gödel semantics. The considered logics are fuzzy extensions of the DL $\mathcal{ALC}_{reg}$ with additional features among inverse roles, nominals, qualified number restrictions, the universal role and local reflexivity of a role. We give results on invariance of concepts as well as conditional invariance of TBoxes and ABoxes for bisimilarity in fuzzy DLs under the Gödel semantics. We also provide a theorem on the Hennessy-Milner property for fuzzy bisimulations in fuzzy DLs under the Gödel semantics.

© Springer Nature Switzerland AG 2018
H. S. Nguyen et al. (Eds.): IJCRS 2018, LNAI 11103, pp. 559–571, 2018.
https://doi.org/10.1007/978-3-319-99368-3_44

1 Introduction

In traditional machine learning, objects are usually described by attributes, and a class of objects can be specified, among others, by a logical formula using attributes. Decision trees and rule-based classifiers are variants of classifiers based on logical formulas. To construct a classifier, one can restrict to using a sublanguage that allows only essential attributes and certain forms of formulas. If two objects are indiscernible w.r.t. that sublanguage, then they belong to the same decision class. Indiscernibility is an equivalence relation that partitions the domain into equivalence classes, and each decision class is the union of some of those equivalence classes.

There are domains in which objects are described not only by attributes but also by binary relations between objects. Examples include social networks and linked data. For such domains, description logics (DLs) are a suitable formalism for representing knowledge about objects. Basic elements of DLs are concepts, roles and individuals (objects). A concept name is a unary predicate, a role name is binary predicate. A concept is interpreted as a set of objects. It can be built from atomic concepts, atomic roles and individual names (as nominals) by using constructors. As DLs are variants of modal logics, indiscernibility in DLs is characterized by bisimilarity. The bisimilarity relation of an interpretation $\mathcal{I}$ w.r.t. a logic language is the largest auto-bisimulation of $\mathcal{I}$ w.r.t. that language. It has been exploited for concept learning in DLs [6,10,15,17,18].

In practical applications, data and knowledge may be imprecise and vague, and fuzzy logics can be used to deal with them. Fuzzy DLs have attracted researchers for two decades (see [1,3] for overviews and surveys). If objects are described by attributes and binary relations, and data about them are vague, then one of the possible ways to specify classes of objects is to use concepts in fuzzy DLs. Bisimilarity in fuzzy DLs can be used for learning such concepts. Thus, bisimilarity and bisimulation in fuzzy DLs are worth studying.

There are different families of fuzzy operators. The Gödel, Łukasiewicz, Product and Zadeh families are the most popular ones. The first three of them use t-norms for defining implication. The Gödel and Zadeh families define conjunction and disjunction of truth values as infimum and supremum, respectively. Each family of fuzzy operators represents a semantics, which is extended to fuzzy DLs appropriately (see, e.g., [2]).

The objective of this paper is to introduce and study bisimulations in fuzzy DLs under the Gödel semantics. Apart from the works [7,12,14] on bisimulation/bisimilarity in traditional or paraconsistent DLs and the earlier mentioned works on using bisimilarity for concept learning in traditional DLs, other notable related works are [5,8,9]. In [8] Eleftheriou et al. presented (weak) bisimulation and bisimilarity in Heyting-valued modal logics and proved the Hennessy-Milner property for those notions. A Heyting-valued modal logic uses a Heyting algebra as the space of truth values. There is a close relationship between Heyting-valued modal logics and fuzzy modal logics under the Gödel semantics, as every linear Heyting algebra is a Gödel algebra [8] and every Gödel algebra is a Heyting algebra with the Dummett condition [4]. In [5] Ćirić et al. introduced bisimulations for fuzzy automata. Such a bisimulation is a fuzzy relation between the sets of states of the two considered automata. One of the results of [5] states that there is a uniform forward bisimulation between fuzzy automata $\mathcal{A}$ and $\mathcal{B}$ iff there is a special isomorphism between the factor fuzzy automata of them w.r.t. their greatest forward bisimulation fuzzy equivalence relations. It is a kind of the Hennessy-Milner property. In [9] Fan introduced fuzzy bisimulations for some Gödel modal logics, which are fuzzy modal logics using the Gödel semantics. The considered logics include the fuzzy monomodal logic K and its extensions with converse and/or involutive negation. She proved that fuzzy bisimulations in those logics have the Hennessy-Milner property. The work [9] follows the app-roach of [5] in defining bisimulation as a fuzzy relation and expressing conditions of bisimulation by using relational composition. As discussed in [9], there is a relationship between fuzzy bisimulations in Gödel modal logics and weak bisim-ulations in Heyting-valued modal logics [8], especially for the case when the underlying Heyting algebra is linear.

In this paper, we define fuzzy bisimulation and (crisp) bisimilarity for fuzzy DLs under the Gödel semantics. The considered logics are fuzzy extensions of the DL $\mathcal{ALC}_{reg}$ with additional features among inverse roles, nominals, qualified number restrictions, the universal role and local reflexivity of a role. The DL $\mathcal{ALC}_{reg}$ is a variant of Propositional Dynamic Logic (PDL) [16]. It extends the basic DL $\mathcal{ALC}$ with role constructors like program constructors of PDL. We give results on invariance of concepts as well as conditional invariance of TBoxes and ABoxes for bisimilarity in fuzzy DLs under the Gödel semantics. Moreover, we provide a theorem on the Hennessy-Milner property for fuzzy bisimulations in fuzzy DLs under the Gödel semantics. Roughly speaking, it states that, if fuzzy interpretations $\mathcal{I}$ and $\mathcal{I}'$ are witnessed and modally saturated, then $Z : \Delta^{\mathcal{I}} \times \Delta^{\mathcal{I}'} \to [0,1]$ is the greatest fuzzy bisimulation between $\mathcal{I}$ and $\mathcal{I}'$ iff $Z(x,x') = \inf\{C^{\mathcal{I}}(x) \Leftrightarrow C^{\mathcal{I}'}(x) \mid C$ is a concept$\}$ for all $x \in \Delta^{\mathcal{I}}$ and $x' \in \Delta^{\mathcal{I}'}$, where $\Leftrightarrow$ denotes the Gödel equivalence.

The motivations of our work are as follows:

– (Fuzzy) bisimulation has potential applications to concept learning in fuzzy DLs, i.e., for machine learning in information systems based on fuzzy DLs. It was not studied for fuzzy DLs under the Gödel semantics.
– The class of fuzzy DLs studied in this paper is large. In comparison with [9], not only are they variants of multimodal (instead of monomodal) logics, but they also allow PDL-like role constructors, qualified number restrictions, nom-inals, the universal role and the concept constructor that represents local reflexivity of a role.
– To deal with qualified number restrictions, the approach of using relational composition for defining conditions of (fuzzy) bisimulation in [5,9] is not suit-able, and we have to use "elementary" conditions for defining bisimulation. Consequently, when restricting to the fuzzy monomodal logic K, our notion of fuzzy bisimulation is different in nature from the one introduced by Fan [9] (see Remark 3), although the *greatest* fuzzy bisimulation relations specified by these two different approaches coincide. This means that our study on fuzzy

bisimulations in fuzzy DLs under the Gödel semantics is not a simple extension of Fan's work [9] on fuzzy bisimulations in Gödel monomodal logics. Due to the mentioned difference, proofs of our results are more complicated.

- This paper serves as a starting point for studying bisimulation and bisimilarity in fuzzy DLs under other t-norm based semantics (e.g., Łukasiewicz and Product).

The remainder of this paper is structured as follows. In Sect. 2, we formally specify the considered fuzzy DLs and their Gödel semantics. In Sect. 3, we define fuzzy bisimulations. In Sect. 4, we present our results on invariance of concepts, TBoxes and ABoxes for bisimilarity in fuzzy DLs under the Gödel semantics. Section 5 contains our results on the Hennessy-Milner property of fuzzy bisimulations. Concluding remarks are given in Sect. 6. Due to the lack of space, all proofs of our results are omitted. They will be made available online or published in an extended version of the paper.

2 Preliminaries

In this section, we recall the Gödel fuzzy operators, fuzzy DLs under the Gödel semantics and define related notions that are needed for this paper.

2.1 The Gödel Fuzzy Operators

The family of Gödel fuzzy operators are defined as follows, where $p, q \in [0, 1]$:

$$p \otimes q = \min\{p, q\}$$
$$p \oplus q = \max\{p, q\}$$
$$\ominus p = (\text{if } p = 0 \text{ then } 1 \text{ else } 0)$$
$$(p \Rightarrow q) = (\text{if } p \le q \text{ then } 1 \text{ else } q)$$
$$(p \Leftrightarrow q) = (p \Rightarrow q) \otimes (q \Rightarrow p).$$

Note that $(p \Leftrightarrow q) = 1$ if $p = q$, and $(p \Leftrightarrow q) = \min\{p, q\}$ otherwise.

For a set Γ of values in $[0, 1]$, we define $\otimes\Gamma = \inf \Gamma$ and $\oplus\Gamma = \sup \Gamma$, where the extrema are taken in the complete lattice $[0, 1]$.

Given $R, S : \Delta \times \Delta' \to [0, 1]$, if $R(x, y) \le S(x, y)$ for all $\langle x, y \rangle \in \Delta \times \Delta'$, then we write $R \le S$ and say that S is *greater than or equal to* R. We write $R \oplus S$ to denote the function of type $\Delta \times \Delta' \to [0, 1]$ defined as follows:

$$(R \oplus S)(x, y) = R(x, y) \oplus S(x, y).$$

If $\mathcal{Z}$ is a set of functions of type $\Delta \times \Delta' \to [0, 1]$, then by $\oplus\mathcal{Z}$ we denote the function of the same type defined as follows:

$$(\oplus\mathcal{Z})(x, y) = \oplus\{Z(x, y) \mid Z \in \mathcal{Z}\}.$$

Given $R : \Delta \times \Delta' \to [0, 1]$ and $S : \Delta' \times \Delta'' \to [0, 1]$, the composition $R \circ S$ is a function of type $\Delta \times \Delta'' \to [0, 1]$ defined as follows:

$$(R \circ S)(x, y) = \oplus\{R(x, z) \otimes S(z, y) \mid z \in \Delta'\}.$$

2.2 Fuzzy Description Logics Under the Gödel Semantics

By Φ we denote a set of features among I, O, Q, U and $\mathtt{Self}$, which stand for inverse roles, nominals, qualified number restrictions, the universal role and local reflexivity of a role, respectively. In this subsection, we first define the syntax of roles and concepts in the fuzzy DL $\mathcal{L}_\Phi$, where $\mathcal{L}$ extends the DL $\mathcal{ALC}_{reg}$ with fuzzy (truth) values and $\mathcal{L}_\Phi$ extends $\mathcal{L}$ with the features from Φ. We then define fuzzy interpretations and the Gödel semantics of $\mathcal{L}_\Phi$.

Our logic language uses a set $\mathbf{C}$ of *concept names*, a set $\mathbf{R}$ of role names, and a set $\mathbf{I}$ of individual names. A *basic role* of $\mathcal{L}_\Phi$ is either a role name or the inverse r^- of a role name r (when $I \in \Phi$).

Roles and *concepts* of $\mathcal{L}_\Phi$ are defined as follows:

- if $r \in \mathbf{R}$, then r is a role of $\mathcal{L}_\Phi$,
- if R, S are roles of $\mathcal{L}_\Phi$ and C is a concept of $\mathcal{L}_\Phi$,
  then $R \circ S$, $R \sqcup S$, R^* and $C?$ are roles of $\mathcal{L}_\Phi$,
- if $I \in \Phi$ and R is a role of $\mathcal{L}_\Phi$, then R^- is a role of $\mathcal{L}_\Phi$,
- if $U \in \Phi$, then U is a role of $\mathcal{L}_\Phi$, called the *universal role*
  (we assume that $U \notin \mathbf{R}$),
- if $p \in [0,1]$, then p is a concept of $\mathcal{L}_\Phi$,
- if $A \in \mathbf{C}$, then A is a concept of $\mathcal{L}_\Phi$,
- if C, D are concepts of $\mathcal{L}_\Phi$ and R is a role of $\mathcal{L}_\Phi$, then:
  - $C \sqcap D$, $C \to D$, $\neg C$, $C \sqcup D$, $\forall R.C$, $\exists R.C$ are concepts of $\mathcal{L}_\Phi$,
  - if $O \in \Phi$ and $a \in \mathbf{I}$, then $\{a\}$ is a concept of $\mathcal{L}_\Phi$,
  - if $Q \in \Phi$, R is a basic role of $\mathcal{L}_\Phi$ and $n \in \mathbb{N}$,
    then $\geq n\,R.C$ and $\leq n\,R.C$ are concepts of $\mathcal{L}_\Phi$,
  - if $\mathtt{Self} \in \Phi$ and $r \in \mathbf{R}$, then $\exists r.\mathtt{Self}$ is a concept of $\mathcal{L}_\Phi$.

The concept 0 stands for $\bot$, and 1 for $\top$.

By $\mathcal{L}_\Phi^0$ we denote the largest sublanguage of $\mathcal{L}_\Phi$ that disallows the role constructors $R \circ S$, $R \sqcup S$, R^*, $C?$ and the concept constructors $\neg C$, $C \sqcup D$, $\forall R.C$, $\leq n\,R.C$.

We use letters A and B to denote *atomic concepts* (which are concept names), C and D to denote arbitrary concepts, r and s to denote *atomic roles* (which are role names), R and S to denote arbitrary roles, a and b to denote individual names.

Given a finite set $\Gamma = \{C_1, \ldots, C_n\}$ of concepts, by $\bigsqcap \Gamma$ we denote $C_1 \sqcap \ldots \sqcap C_n$, and by $\bigsqcup \Gamma$ we denote $C_1 \sqcup \ldots \sqcup C_n$. If $\Gamma = \emptyset$, then $\bigsqcap \Gamma = 1$ and $\bigsqcup \Gamma = 0$.

Definition 1. A *(fuzzy) interpretation* is a pair $\mathcal{I} = \langle \Delta^\mathcal{I}, \cdot^\mathcal{I} \rangle$, where $\Delta^\mathcal{I}$ is a non-empty set, called the *domain*, and $\cdot^\mathcal{I}$ is the *interpretation function*, which maps every individual name a to an element $a^\mathcal{I} \in \Delta^\mathcal{I}$, every concept name A to a function $A^\mathcal{I} : \Delta^\mathcal{I} \to [0,1]$, and every role name r to a function $r^\mathcal{I} : \Delta^\mathcal{I} \times \Delta^\mathcal{I} \to [0,1]$. The function $\cdot^\mathcal{I}$ is extended to complex roles and concepts as follows (cf. [2]):

$$U^{\mathcal{I}}(x,y) = 1$$
$$(r^-)^{\mathcal{I}}(x,y) = r^{\mathcal{I}}(y,x)$$
$$(C?)^{\mathcal{I}}(x,y) = (\text{if } x = y \text{ then } C^{\mathcal{I}}(x) \text{ else } 0)$$
$$(R \circ S)^{\mathcal{I}}(x,y) = \oplus\{R^{\mathcal{I}}(x,z) \otimes S^{\mathcal{I}}(z,y) \mid z \in \Delta^{\mathcal{I}}\}$$
$$(R \sqcup S)^{\mathcal{I}}(x,y) = R^{\mathcal{I}}(x,y) \oplus S^{\mathcal{I}}(x,y)$$
$$(R^*)^{\mathcal{I}}(x,y) = \oplus\{\otimes\{R^{\mathcal{I}}(x_i, x_{i+1}) \mid 0 \leq i < n\} \mid$$
$$n \geq 0,\ x_0, \ldots, x_n \in \Delta^{\mathcal{I}},\ x_0 = x,\ x_n = y\}$$
$$p^{\mathcal{I}}(x) = p$$
$$\{a\}^{\mathcal{I}}(x) = (\text{if } x = a^{\mathcal{I}} \text{ then } 1 \text{ else } 0)$$
$$(\neg C)^{\mathcal{I}}(x) = \ominus C^{\mathcal{I}}(x)$$
$$(C \sqcap D)^{\mathcal{I}}(x) = C^{\mathcal{I}}(x) \otimes D^{\mathcal{I}}(x)$$
$$(C \sqcup D)^{\mathcal{I}}(x) = C^{\mathcal{I}}(x) \oplus D^{\mathcal{I}}(x)$$
$$(C \to D)^{\mathcal{I}}(x) = (C^{\mathcal{I}}(x) \Rightarrow D^{\mathcal{I}}(x))$$
$$(\exists r.\mathbf{Self})^{\mathcal{I}}(x) = r^{\mathcal{I}}(x,x)$$
$$(\exists R.C)^{\mathcal{I}}(x) = \oplus\{R^{\mathcal{I}}(x,y) \otimes C^{\mathcal{I}}(y) \mid y \in \Delta^{\mathcal{I}}\}$$
$$(\forall R.C)^{\mathcal{I}}(x) = \otimes\{R^{\mathcal{I}}(x,y) \Rightarrow C^{\mathcal{I}}(y) \mid y \in \Delta^{\mathcal{I}}\}$$
$$(\geq n\,R.C)^{\mathcal{I}}(x) = \oplus\{\otimes\{R^{\mathcal{I}}(x,y_i) \otimes C^{\mathcal{I}}(y_i) \mid 1 \leq i \leq n\} \mid$$
$$y_1, \ldots, y_n \in \Delta^{\mathcal{I}},\ y_i \neq y_j \text{ if } i \neq j\}$$
$$(\leq n\,R.C)^{\mathcal{I}}(x) = \otimes\{(\otimes\{R^{\mathcal{I}}(x,y_i) \otimes C^{\mathcal{I}}(y_i) \mid 1 \leq i \leq n+1\} \Rightarrow$$
$$\oplus\{y_j \neq y_k \mid 1 \leq j < k \leq n+1\}) \mid y_1, \ldots, y_{n+1} \in \Delta^{\mathcal{I}}\}. \ \blacksquare$$

For definitions of the Zadeh, Łukasiewicz and Product semantics for fuzzy DLs, we refer the reader to [2].

Remark 1. Observe that $(\leq nR.C)^{\mathcal{I}}(x)$ is either 1 or 0. Namely, $(\leq n\,R.C)^{\mathcal{I}}(x) = 1$ if, for every set $\{y_1, \ldots, y_{n+1}\}$ of $n+1$ pairwise distinct elements of $\Delta^{\mathcal{I}}$, there exists $1 \leq i \leq n+1$ such that $R^{\mathcal{I}}(x,y_i) \otimes C^{\mathcal{I}}(y_i) = 0$. Otherwise, $(\leq n\,R.C)^{\mathcal{I}}(x) = 0$. $\blacksquare$

Example 1. Let $\mathbf{R} = \{r\}$, $\mathbf{C} = \{A\}$ and $\mathbf{I} = \emptyset$. Consider the fuzzy interpretation $\mathcal{I}$ illustrated and specified below:

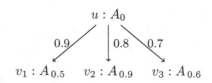

- $\Delta^{\mathcal{I}} = \{u, v_1, v_2, v_3\}$,
- $A^{\mathcal{I}}(u) = 0$, $A^{\mathcal{I}}(v_1) = 0.5$, $A^{\mathcal{I}}(v_2) = 0.9$, $A^{\mathcal{I}}(v_3) = 0.6$,
- $r^{\mathcal{I}}(u, v_1) = 0.9$, $r^{\mathcal{I}}(u, v_2) = 0.8$, $r^{\mathcal{I}}(u, v_3) = 0.7$,
  and $r^{\mathcal{I}}(x, y) = 0$ for other pairs $\langle x, y \rangle$.

We have that:

- $(\forall r.A)^{\mathcal{I}}(a) = 0.5$, $(\exists r.A)^{\mathcal{I}}(a) = 0.8$, $(\leq 1\, r.A)^{\mathcal{I}}(a) = 0$, $(\geq 2\, r.A)^{\mathcal{I}}(a) = 0.6$,
- for $C = \forall (r \sqcup r^-)^*.A$ and $1 \leq i \leq 3$: $C^{\mathcal{I}}(v_i) = 0$,
- for $C = \exists (r \sqcup r^-)^*.A$: $C^{\mathcal{I}}(v_1) = 0.8$, $C^{\mathcal{I}}(v_2) = 0.9$ and $C^{\mathcal{I}}(v_3) = 0.7$.   ∎

A fuzzy interpretation $\mathcal{I}$ is *witnessed* (w.r.t. $\mathcal{L}_{\Phi}$) [11] if any infinite set under the prefix operator $\otimes$ (resp. $\oplus$) in Definition 1 has the smallest (resp. biggest) element. The notion of being "*witnessed w.r.t. $\mathcal{L}_{\Phi}^0$*" is defined similarly under the assumption that only roles and concepts of $\mathcal{L}_{\Phi}^0$ are allowed. A fuzzy interpretation $\mathcal{I}$ is *finite* if $\Delta^{\mathcal{I}}$, **C**, **R** and **I** are finite, and is *image-finite* w.r.t. Φ if, for every $x \in \Delta^{\mathcal{I}}$ and every basic role R of $\mathcal{L}_{\Phi}$, $\{y \in \Delta^{\mathcal{I}} \mid R^{\mathcal{I}}(x, y) > 0\}$ is finite. Observe that every finite fuzzy interpretation is witnessed and every image-finite fuzzy interpretation w.r.t. Φ is witnessed w.r.t. $\mathcal{L}_{\Phi}^0$.

A *fuzzy assertion* in $\mathcal{L}_{\Phi}$ is an expression of the form $a \doteq b$, $a \not\doteq b$, $C(a) \bowtie p$ or $R(a, b) \bowtie p$, where C is a concept of $\mathcal{L}_{\Phi}$, R is a role of $\mathcal{L}_{\Phi}$, $\bowtie \in \{\geq, >, \leq, <\}$ and $p \in [0, 1]$. A *fuzzy ABox* in $\mathcal{L}_{\Phi}$ is a finite set of fuzzy assertions in $\mathcal{L}_{\Phi}$.

A *fuzzy GCI* (general concept inclusion) in $\mathcal{L}_{\Phi}$ is an expression of the form $(C \sqsubseteq D) \rhd p$, where C and D are concepts of $\mathcal{L}_{\Phi}$, $\rhd \in \{\geq, >\}$ and $p \in (0, 1]$. A *fuzzy TBox* in $\mathcal{L}_{\Phi}$ is a finite set of fuzzy GCIs in $\mathcal{L}_{\Phi}$.

Given a fuzzy interpretation $\mathcal{I}$ and a fuzzy assertion or GCI φ, we say that $\mathcal{I}$ *validates* φ, denoted by $\mathcal{I} \models \varphi$, if:

- case $\varphi = (a \doteq b)$: $a^{\mathcal{I}} = b^{\mathcal{I}}$,
- case $\varphi = (a \not\doteq b)$: $a^{\mathcal{I}} \neq b^{\mathcal{I}}$,
- case $\varphi = (C(a) \bowtie p)$: $C^{\mathcal{I}}(a^{\mathcal{I}}) \bowtie p$,
- case $\varphi = (R(a, b) \bowtie p)$: $R^{\mathcal{I}}(a^{\mathcal{I}}, b^{\mathcal{I}}) \bowtie p$,
- case $\varphi = (C \sqsubseteq D) \rhd p$: $(C \to D)^{\mathcal{I}}(x) \rhd p$ for all $x \in \Delta^{\mathcal{I}}$.

A fuzzy interpretation $\mathcal{I}$ is a *model* of a fuzzy ABox $\mathcal{A}$, denoted by $\mathcal{I} \models \mathcal{A}$, if $\mathcal{I} \models \varphi$ for all $\varphi \in \mathcal{A}$. Similarly, $\mathcal{I}$ is a model of a fuzzy TBox $\mathcal{T}$, denoted by $\mathcal{I} \models \mathcal{T}$, if $\mathcal{I} \models \varphi$ for all $\varphi \in \mathcal{T}$.

Two concepts C and D are *equivalent*, denoted by $C \equiv D$, if $C^{\mathcal{I}} = D^{\mathcal{I}}$ for every fuzzy interpretation $\mathcal{I}$. Two roles R and S are *equivalent*, denoted by $R \equiv S$, if $R^{\mathcal{I}} = S^{\mathcal{I}}$ for every fuzzy interpretation $\mathcal{I}$.

We say that a role R is in *inverse normal form* if inverse constructor is applied in R only to role names. In this paper, we assume that roles are presented in inverse normal form because every role can be translated to an equivalent role in inverse normal form using the following rules:

$$
\begin{array}{ll}
U^- \equiv U & (R \circ S)^- \equiv S^- \circ R^- \\
(R^-)^- \equiv R & (R \sqcup S)^- \equiv R^- \sqcup S^- \\
(C?)^- \equiv C? & (R^*)^- \equiv (R^-)^*.
\end{array}
$$

Remark 2. The concept constructors $\neg C$ and $C \sqcup D$ can be excluded from $\mathcal{L}_\Phi$ and $\mathcal{L}_\Phi^0$ because

$$\neg C \equiv (C \to 0)$$
$$C \sqcup D \equiv ((C \to D) \to D) \sqcap ((D \to C) \to C).$$ ∎

3   Fuzzy Bisimulations

Let $\Phi \subseteq \{I, O, Q, U, \mathtt{Self}\}$ be a set of features and $\mathcal{I}$, $\mathcal{I}'$ fuzzy interpretations. A function $Z : \Delta^\mathcal{I} \times \Delta^{\mathcal{I}'} \to [0,1]$ is called a *fuzzy $\mathcal{L}_\Phi$-bisimulation* (under the Gödel semantics) between $\mathcal{I}$ and $\mathcal{I}'$ if the following conditions hold for every $x \in \Delta^\mathcal{I}$, $x' \in \Delta^{\mathcal{I}'}$, $A \in \mathbf{C}$, $a \in \mathbf{I}$, $r \in \mathbf{R}$ and every basic role R of $\mathcal{L}_\Phi$:

$$Z(x,x') \leq (A^\mathcal{I}(x) \Leftrightarrow A^{\mathcal{I}'}(x')) \tag{1}$$
$$\forall y \in \Delta^\mathcal{I} \, \exists y' \in \Delta^{\mathcal{I}'} \; Z(x,x') \otimes R^\mathcal{I}(x,y) \leq Z(y,y') \otimes R^{\mathcal{I}'}(x',y') \tag{2}$$
$$\forall y' \in \Delta^{\mathcal{I}'} \, \exists y \in \Delta^\mathcal{I} \; Z(x,x') \otimes R^{\mathcal{I}'}(x',y') \leq Z(y,y') \otimes R^\mathcal{I}(x,y); \tag{3}$$

if $O \in \Phi$, then

$$Z(x,x') \leq (x = a^\mathcal{I} \Leftrightarrow x' = a^{\mathcal{I}'}); \tag{4}$$

if $Q \in \Phi$, then, for any $n \geq 1$,

if $Z(x,x') > 0$ and $y_1, \ldots, y_n$ are pairwise distinct elements of $\Delta^\mathcal{I}$ such that $R^\mathcal{I}(x,y_j) > 0$ for all $1 \leq j \leq n$, then there exist pairwise distinct elements $y_1', \ldots, y_n'$ of $\Delta^{\mathcal{I}'}$ such that, for every $1 \leq i \leq n$, there exists $\quad$ (5) $1 \leq j \leq n$ such that $Z(x,x') \otimes R^\mathcal{I}(x,y_j) \leq Z(y_j, y_i') \otimes R^{\mathcal{I}'}(x', y_i')$,

if $Z(x,x') > 0$ and $y_1', \ldots, y_n'$ are pairwise distinct elements of $\Delta^{\mathcal{I}'}$ such that $R^{\mathcal{I}'}(x', y_j') > 0$ for all $1 \leq j \leq n$, then there exist pairwise distinct elements $y_1, \ldots, y_n$ of $\Delta^\mathcal{I}$ such that, for every $1 \leq i \leq n$, there $\quad$ (6) exists $1 \leq j \leq n$ such that $Z(x,x') \otimes R^{\mathcal{I}'}(x', y_j') \leq Z(y_i, y_j') \otimes R^\mathcal{I}(x, y_i);$

if $U \in \Phi$, then

$$\forall y \in \Delta^\mathcal{I} \, \exists y' \in \Delta^{\mathcal{I}'} \; Z(x,x') \leq Z(y,y') \tag{7}$$
$$\forall y' \in \Delta^{\mathcal{I}'} \, \exists y \in \Delta^\mathcal{I} \; Z(x,x') \leq Z(y,y'); \tag{8}$$

if $\mathtt{Self} \in \Phi$, then

$$Z(x,x') \leq (r^\mathcal{I}(x,x) \Leftrightarrow r^{\mathcal{I}'}(x',x')). \tag{9}$$

For example, if $\Phi = \{I, Q\}$, then only Conditions (1)–(3), (5) and (6) are essential. By definition, the function $\lambda\langle x, x' \rangle \in \Delta^\mathcal{I} \times \Delta^{\mathcal{I}'}.0$ is a fuzzy $\mathcal{L}_\Phi$-bisimulation between $\mathcal{I}$ and $\mathcal{I}'$.

Remark 3. Observe that Condition (2) (resp. (3)) together with the qualification over x and x' implies $Z^{-1} \circ R^{\mathcal{I}} \leq R^{\mathcal{I}'} \circ Z^{-1}$ (resp. $Z \circ R^{\mathcal{I}'} \leq R^{\mathcal{I}} \circ Z$). However, in general, the converse does not hold.   ∎

Example 2. Let $\mathbf{R} = \{r\}$, $\mathbf{C} = \{A\}$, $\mathbf{I} = \emptyset$ and $\Phi = \emptyset$. Consider the fuzzy interpretations $\mathcal{I}$ and $\mathcal{I}'$ illustrated below (and specified similarly as in Example 1).

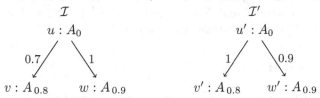

If Z is a fuzzy $\mathcal{L}_{\Phi}$-bisimulation between $\mathcal{I}$ and $\mathcal{I}'$, then:

- $Z(v, w') \leq 0.8$ and $Z(w, v') \leq 0.8$ due to (1),
- $Z(u, u') \leq 0.8$ due to (3) for $x = u$, $x' = u'$ and $y' = v'$,
- $Z(u, v') = Z(u, w') = Z(v, u') = Z(w, u') = 0$ due to (1).

It can be check that the function $Z : \Delta^{\mathcal{I}} \times \Delta^{\mathcal{I}'} \to [0, 1]$ specified by

- $Z(v, v') = Z(w, w') = 1$,
- $Z(v, w') = Z(w, v') = Z(u, u') = 0.8$,
- $Z(u, v') = Z(u, w') = Z(v, u') = Z(w, u') = 0$

is a fuzzy $\mathcal{L}_{\Phi}$-bisimulation between $\mathcal{I}$ and $\mathcal{I}'$, and hence is the greatest fuzzy $\mathcal{L}_{\Phi}$-bisimulation between $\mathcal{I}$ and $\mathcal{I}'$.   ∎

Proposition 1. *Let $\mathcal{I}$, $\mathcal{I}'$ and $\mathcal{I}''$ be fuzzy interpretations.*

1. *The function $Z : \Delta^{\mathcal{I}} \times \Delta^{\mathcal{I}} \to [0, 1]$ specified by*

$$Z(x, x') = (if \ x = x' \ then \ 1 \ else \ 0)$$

   is a fuzzy $\mathcal{L}_{\Phi}$-bisimulation between $\mathcal{I}$ and itself.
2. *If Z is a fuzzy $\mathcal{L}_{\Phi}$-bisimulation between $\mathcal{I}$ and $\mathcal{I}'$, then Z^{-1} is a fuzzy $\mathcal{L}_{\Phi}$-bisimulation between $\mathcal{I}'$ and $\mathcal{I}$.*
3. *If Z_1 is a fuzzy $\mathcal{L}_{\Phi}$-bisimulation between $\mathcal{I}$ and $\mathcal{I}'$, and Z_2 is a fuzzy $\mathcal{L}_{\Phi}$-bisimulation between $\mathcal{I}'$ and $\mathcal{I}''$, then $Z_1 \circ Z_2$ is a fuzzy $\mathcal{L}_{\Phi}$-bisimulation between $\mathcal{I}$ and $\mathcal{I}''$.*
4. *If $\mathcal{Z}$ is a finite set of fuzzy $\mathcal{L}_{\Phi}$-bisimulations between $\mathcal{I}$ and $\mathcal{I}'$, then $\oplus \mathcal{Z}$ is also a fuzzy $\mathcal{L}_{\Phi}$-bisimulation between $\mathcal{I}$ and $\mathcal{I}'$.*

The proof of this proposition is straightforward.

Remark 4. It seems that the assertion 4 of Proposition 1 cannot be strengthened for infinite $\mathcal{Z}$. So, the greatest fuzzy $\mathcal{L}_{\Phi}$-bisimulation between $\mathcal{I}$ and $\mathcal{I}'$ may not exist. As stated later by Theorem 4, if $\mathcal{I}$ and $\mathcal{I}'$ are witnessed w.r.t. $\mathcal{L}_{\Phi}^{0}$ and modally saturated w.r.t. $\mathcal{L}_{\Phi}^{0}$ (see Definition 2), then the greatest fuzzy $\mathcal{L}_{\Phi}$-bisimulation between $\mathcal{I}$ and $\mathcal{I}'$ exists.   ∎

Let $\mathcal{I}$ and $\mathcal{I}'$ be fuzzy interpretations. For $x \in \Delta^{\mathcal{I}}$ and $x' \in \Delta^{\mathcal{I}'}$, we write $x \sim_\Phi x'$ to denote that there exists a fuzzy $\mathcal{L}_\Phi$-bisimulation Z between $\mathcal{I}$ and $\mathcal{I}'$ such that $Z(x, x') = 1$. If $x \sim_\Phi x'$, then we say that x and x' are $\mathcal{L}_\Phi$-bisimilar to each other. Let $\sim_{\Phi, \mathcal{I}}$ be the binary relation on $\Delta^{\mathcal{I}}$ such that, for $x, x' \in \Delta^{\mathcal{I}}$, $x \sim_{\Phi, \mathcal{I}} x'$ iff $x \sim_\Phi x'$. By Proposition 1, $\sim_{\Phi, \mathcal{I}}$ is an equivalence relation. We call it the $\mathcal{L}_\Phi$-bisimilarity relation of $\mathcal{I}$. If $\mathbf{I} \neq \emptyset$ and there exists a fuzzy $\mathcal{L}_\Phi$-bisimulation Z between $\mathcal{I}$ and $\mathcal{I}'$ such that $Z(a^{\mathcal{I}}, a^{\mathcal{I}'}) = 1$ for all $a \in \mathbf{I}$, then we say that $\mathcal{I}$ and $\mathcal{I}'$ are $\mathcal{L}_\Phi$-bisimilar to each other and write $\mathcal{I} \sim_\Phi \mathcal{I}'$.

4   Invariance Results

A concept C of $\mathcal{L}_\Phi$ is said to be *invariant for $\mathcal{L}_\Phi$-bisimilarity between witnessed interpretations* if, for any witnessed interpretations $\mathcal{I}$, $\mathcal{I}'$ and any $x \in \Delta^{\mathcal{I}}$ and $x' \in \Delta^{\mathcal{I}'}$, if $x \sim_\Phi x'$, then $C^{\mathcal{I}}(x) = C^{\mathcal{I}'}(x')$.

Theorem 1. *All concepts of $\mathcal{L}_\Phi$ are invariant for $\mathcal{L}_\Phi$-bisimilarity between witnessed interpretations.*

This theorem is a corollary of the following stronger result.

Lemma 1. *Let $\mathcal{I}$ and $\mathcal{I}'$ be witnessed interpretations and Z a fuzzy $\mathcal{L}_\Phi$-bisimulation between $\mathcal{I}$ and $\mathcal{I}'$. Then, the following properties hold for every concept C of $\mathcal{L}_\Phi$, every role R of $\mathcal{L}_\Phi$, every $x \in \Delta^{\mathcal{I}}$ and every $x' \in \Delta^{\mathcal{I}'}$:*

$$Z(x, x') \leq (C^{\mathcal{I}}(x) \Leftrightarrow C^{\mathcal{I}'}(x')) \tag{10}$$

$$\forall y \in \Delta^{\mathcal{I}} \; \exists y' \in \Delta^{\mathcal{I}'} \; Z(x, x') \otimes R^{\mathcal{I}}(x, y) \leq Z(y, y') \otimes R^{\mathcal{I}'}(x', y') \tag{11}$$

$$\forall y' \in \Delta^{\mathcal{I}'} \; \exists y \in \Delta^{\mathcal{I}} \; Z(x, x') \otimes R^{\mathcal{I}'}(x', y') \leq Z(y, y') \otimes R^{\mathcal{I}}(x, y). \tag{12}$$

The following lemma differs from Lemma 1 in that $\mathcal{L}_\Phi^0$ is used instead of $\mathcal{L}_\Phi$. Its proof is a shortened version the one of Lemma 1, as (11) and (12) are the same as (2) and (3) when R is a role of $\mathcal{L}_\Phi^0$, respectively, and we can ignore the cases when C is $\forall R.D$ or $\leq n\,R.D$.

Lemma 2. *Let $\mathcal{I}$ and $\mathcal{I}'$ be witnessed interpretations w.r.t. $\mathcal{L}_\Phi^0$ and Z a fuzzy $\mathcal{L}_\Phi$-bisimulation between $\mathcal{I}$ and $\mathcal{I}'$. Then, for every concept C of $\mathcal{L}_\Phi^0$, every $x \in \Delta^{\mathcal{I}}$ and every $x' \in \Delta^{\mathcal{I}'}$, $Z(x, x') \leq (C^{\mathcal{I}}(x) \Leftrightarrow C^{\mathcal{I}'}(x'))$.*

A fuzzy TBox $\mathcal{T}$ is said to be *invariant for $\mathcal{L}_\Phi$-bisimilarity between witnessed interpretations* if, for every witnessed interpretations $\mathcal{I}$ and $\mathcal{I}'$ that are $\mathcal{L}_\Phi$-bisimilar to each other, $\mathcal{I} \models \mathcal{T}$ iff $\mathcal{I}' \models \mathcal{T}$. The notion of invariance of fuzzy ABoxes for $\mathcal{L}_\Phi$-bisimilarity between witnessed interpretations is defined similarly.

Theorem 2. *If $U \in \Phi$ and $\mathbf{I} \neq \emptyset$, then all fuzzy TBoxes in $\mathcal{L}_\Phi$ are invariant for $\mathcal{L}_\Phi$-bisimilarity between witnessed interpretations.*

Theorem 3. *Let $\mathcal{A}$ be a fuzzy ABox in $\mathcal{L}_\Phi$. If $O \in \Phi$ or $\mathcal{A}$ consists of only fuzzy assertions of the form $C(a) \bowtie p$, then $\mathcal{A}$ is invariant for $\mathcal{L}_\Phi$-bisimilarity between witnessed interpretations.*

5   The Hennessy-Milner Property

Definition 2. A fuzzy interpretation $\mathcal{I}$ is said to be *modally saturated* w.r.t. $\mathcal{L}_{\Phi}^0$ (and the Gödel semantics) if the following conditions hold:

- for every $p \in (0,1]$, every $x \in \Delta^{\mathcal{I}}$, every basic role R of $\mathcal{L}_{\Phi}$ and every infinite set Γ of concepts in $\mathcal{L}_{\Phi}^0$, if for every finite subset Λ of Γ there exists $y \in \Delta^{\mathcal{I}}$ such that $R^{\mathcal{I}}(x,y) \otimes C^{\mathcal{I}}(y) \geq p$ for all $C \in \Lambda$, then there exists $y \in \Delta^{\mathcal{I}}$ such that $R^{\mathcal{I}}(x,y) \otimes C^{\mathcal{I}}(y) \geq p$ for all $C \in \Gamma$;
- if $Q \in \Phi$, then for every $p \in (0,1]$, every $x \in \Delta^{\mathcal{I}}$, every basic role R of $\mathcal{L}_{\Phi}$, every infinite set Γ of concepts in $\mathcal{L}_{\Phi}^0$ and every $n \in \mathbb{N}$, if for every finite subset Λ of Γ there exist n pairwise distinct $y_1, \ldots, y_n \in \Delta^{\mathcal{I}}$ such that $R^{\mathcal{I}}(x,y_i) \otimes C^{\mathcal{I}}(y_i) \geq p$ for all $1 \leq i \leq n$ and $C \in \Lambda$, then there exist n pairwise distinct $y_1, \ldots, y_n \in \Delta^{\mathcal{I}}$ such that $R^{\mathcal{I}}(x,y_i) \otimes C^{\mathcal{I}}(y_i) \geq p$ for all $1 \leq i \leq n$ and $C \in \Gamma$;
- if $U \in \Phi$, then for every $p \in (0,1]$ and every infinite set Γ of concepts in $\mathcal{L}_{\Phi}^0$, if for every finite subset Λ of Γ there exists $y \in \Delta^{\mathcal{I}}$ such that $C^{\mathcal{I}}(y) \geq p$ for all $C \in \Lambda$, then there exists $y \in \Delta^{\mathcal{I}}$ such that $C^{\mathcal{I}}(y) \geq p$ for all $C \in \Gamma$. ∎

Clearly, every finite fuzzy interpretation is modally saturated w.r.t. $\mathcal{L}_{\Phi}^0$ for any Φ. If $U \notin \Phi$, then every image-finite fuzzy interpretation w.r.t. Φ is modally saturated w.r.t. $\mathcal{L}_{\Phi}^0$.

Theorem 4. *Let $\mathcal{I}$ and $\mathcal{I}'$ be fuzzy interpretations that are witnessed w.r.t. $\mathcal{L}_{\Phi}^0$ and modally saturated w.r.t. $\mathcal{L}_{\Phi}^0$. Let $Z : \Delta^{\mathcal{I}} \times \Delta^{\mathcal{I}'} \to [0,1]$ be specified by*

$$Z(x,x') = \otimes \{C^{\mathcal{I}}(x) \Leftrightarrow C^{\mathcal{I}'}(x) \mid C \text{ is a concept of } \mathcal{L}_{\Phi}^0\}.$$

Then, Z is the greatest fuzzy $\mathcal{L}_{\Phi}$-bisimulation between $\mathcal{I}$ and $\mathcal{I}'$.

Given fuzzy interpretations $\mathcal{I}$, $\mathcal{I}'$ and $x \in \Delta^{\mathcal{I}}$, $x' \in \Delta^{\mathcal{I}'}$, we write $x \equiv_{\Phi} x'$ to denote that $C^{\mathcal{I}}(x) = C^{\mathcal{I}'}(x')$ for every concept C of $\mathcal{L}_{\Phi}$. Similarly, we write $x \equiv_{\Phi}^0 x'$ to denote that $C^{\mathcal{I}}(x) = C^{\mathcal{I}'}(x')$ for every concept C of $\mathcal{L}_{\Phi}^0$.

Corollary 1. *Let $\mathcal{I}$, $\mathcal{I}'$ be fuzzy interpretations and let $x \in \Delta^{\mathcal{I}}$, $x' \in \Delta^{\mathcal{I}'}$.*

1. *If $\mathcal{I}$ and $\mathcal{I}'$ are witnessed w.r.t. $\mathcal{L}_{\Phi}^0$ and modally saturated w.r.t. $\mathcal{L}_{\Phi}^0$, then*

$$x \sim_{\Phi} x' \quad \text{iff} \quad x \equiv_{\Phi}^0 x'.$$

2. *If $\mathcal{I}$ and $\mathcal{I}'$ are image-finite fuzzy interpretations w.r.t. Φ, then*

$$x \sim_{\Phi} x' \quad \text{iff} \quad x \equiv_{\Phi}^0 x'.$$

3. *If $\mathcal{I}$ and $\mathcal{I}'$ are witnessed w.r.t. $\mathcal{L}_{\Phi}$ and modally saturated w.r.t. $\mathcal{L}_{\Phi}^0$, then*

$$x \equiv_{\Phi} x' \quad \text{iff} \quad x \sim_{\Phi} x' \quad \text{iff} \quad x \equiv_{\Phi}^0 x'.$$

The assertion 1 (resp. 3) directly follows from Theorem 4 and Lemma 2 (resp. 1). The assertion 2 directly follows from the assertion 1. The following corollary directly follows from Theorem 4 and Lemma 1.

Corollary 2. *Let $\mathcal{I}$ and $\mathcal{I}'$ be fuzzy interpretations that are witnessed w.r.t. $\mathcal{L}_{\Phi}$ and modally saturated w.r.t. $\mathcal{L}_{\Phi}^0$. Then, $\mathcal{I}$ and $\mathcal{I}'$ are $\mathcal{L}_{\Phi}$-bisimilar iff $a^{\mathcal{I}} \equiv_{\Phi}^0 a^{\mathcal{I}'}$ for all $a \in \mathbf{I}$.*

6 Concluding Remarks

We have defined fuzzy bisimulations and (crisp) bisimilarity relations for a large class of fuzzy DLs under the Gödel semantics. We have provided results on invariance of concepts as well as conditional invariance of TBoxes and ABoxes for such bisimilarity. We have also provided results on the Hennessy-Milner property for such bisimulations. As far as we know, this is the first time fuzzy bisimulations are defined and studied for fuzzy DLs under the Gödel semantics.

As mentioned in the Introduction, we use "elementary" Conditions (2), (3) and (5)–(8) instead of the ones based on relational composition for defining bisimulations. Consequently, our notion of fuzzy bisimulation is different in nature from the one introduced by Fan [9], although the greatest fuzzy bisimulation relations specified by these two different approaches coincide when restricting to the fuzzy modal logics without involutive negation considered in [9]. Furthermore, in comparison with [9], not only is the class of logics considered by us much larger, we also study invariance of TBoxes and ABoxes for bisimilarity, and our theorem on the Hennessy-Milner property is formulated for witnessed and modally saturated interpretations, which are more general than image-finite interpretations.

Like the relationship between [9] and [8], our notion of fuzzy bisimulation is also related to the notion of weak bisimulation introduced by Eleftheriou et al. [8] for Heyting-valued modal logics, especially for the case when the considered logic is K and the underlying Heyting algebra is the complete lattice $\langle [0, 1], \leq \rangle$. In this case, the latter notion can be treated as a cut-based variant of our notion (see [9] for a more detailed discussion). The differences are that the considered classes of logics are essentially different and our approach uses fuzzy relations as in [5,9], while the approach of [8] uses families of crisp relations, where each of the families is specified by a cut-value. Following [9], we use the term "fuzzy bisimulation" instead of "bisimulation" to emphasize its fuzziness.

Our notions and results have potential applications to concept learning in fuzzy DLs. As future work, apart from such applications, it is also worth studying bisimulation and bisimilarity in fuzzy DLs under other t-norm based semantics (e.g., Łukasiewicz and Product). Recently, Nguyen [13] studied bisimilarity in fuzzy DLs under the Zadeh semantics, which does not use t-norms for defining implication. His approach is essentially different, as it uses (crisp) simulation instead of (fuzzy) bisimulation because the latter notion does not seem to be definable for fuzzy DLs under the Zadeh semantics.

Acknowlegements. This paper was partially supported by VNU-UET and VNU.

References

1. Bobillo, F., Cerami, M., Esteva, F., García-Cerdaña, Á., Peñaloza, R., Straccia, U.: Fuzzy description logics. In: Handbook of Mathematical Fuzzy Logic, Volume 58 of Studies in Logic, Mathematical Logic and Foundations, vol. 3, pp. 1105–1181. College Publications (2015)
2. Bobillo, F., Delgado, M., Gómez-Romero, J., Straccia, U.: Fuzzy description logics under Gödel semantics. Int. J. Approximate Reasoning **50**(3), 494–514 (2009)
3. Borgwardt, S., Peñaloza, R.: Fuzzy description logics – a survey. In: Moral, S., Pivert, O., Sánchez, D., Marín, N. (eds.) SUM 2017. LNCS (LNAI), vol. 10564, pp. 31–45. Springer, Cham (2017). https://doi.org/10.1007/978-3-319-67582-4_3
4. Cattaneo, G., Ciucci, D., Giuntini, R., Konig, M.: Algebraic structures related to many valued logical systems. Part I: Heyting Wajsberg algebras. Fundamenta Informaticae **63**(4), 331–355 (2004)
5. Ćirić, M., Ignjatović, J., Damljanović, N., Bašic, M.: Bisimulations for fuzzy automata. Fuzzy Sets Syst. **186**(1), 100–139 (2012)
6. Divroodi, A.R., Ha, Q.-T., Nguyen, L.A., Nguyen, H.S.: On the possibility of correct concept learning in description logics. Vietnam J. Comput. Sci. **5**(1), 3–14 (2018)
7. Divroodi, A.R., Nguyen, L.A.: On bisimulations for description logics. Inf. Sci. **295**, 465–493 (2015)
8. Eleftheriou, P.E., Koutras, C.D., Nomikos, C.: Notions of bisimulation for Heyting-valued modal languages. J. Log. Comput. **22**(2), 213–235 (2012)
9. Fan, T.-F.: Fuzzy bisimulation for Gödel modal logic. IEEE Trans. Fuzzy Syst. **23**(6), 2387–2396 (2015)
10. Ha, Q.-T., Hoang, T.-L.-G., Nguyen, L.A., Nguyen, H.S., Szałas, A., Tran, T.-L.: A bisimulation-based method of concept learning for knowledge bases in description logics. In: Proceedings of SoICT 2012, pp. 241–249. ACM (2012)
11. Hájek, P.: Making fuzzy description logic more general. Fuzzy Sets Syst. **154**(1), 1–15 (2005)
12. Lutz, C., Piro, R., Wolter, F.: Description logic TBoxes: model-theoretic characterizations and rewritability. In: Walsh, T. (ed.) Proceedings of IJCAI 2011, pp. 983–988 (2011)
13. Nguyen, L.A.: Bisimilarity in fuzzy description logics under the Zadeh semantics, submitted
14. Nguyen, L.A., Nguyen, T.H.K., Nguyen, N.-T., Ha, Q.-T.: Bisimilarity for paraconsistent description logics. J. Intell. Fuzzy Syst. **32**(2), 1203–1215 (2017)
15. Nguyen, L.A., Szałas, A.: Logic-based roughification. In: Skowron, A., Suraj, Z. (eds.) Rough Sets and Intelligent Systems (To the Memory of Professor Zdzisław Pawlak), vol. 1, pp. 517–543. Springer, Heidelberg (2012). https://doi.org/10.1007/978-3-642-30344-9_19
16. Schild, K.: A correspondence theory for terminological logics: preliminary report. In: Proceedings of IJCAI 1991, pp. 466–471. Morgan Kaufmann (1991)
17. Tran, T.-L., Ha, Q.-T., Hoang, T.-L.-G., Nguyen, L.A., Nguyen, H.S.: Bisimulation-based concept learning in description logics. Fundamenta Informaticae **133**(2–3), 287–303 (2014)
18. Tran, T.-L., Nguyen, L.A., Hoang, T.-L.-G.: Bisimulation-based concept learning for information systems in description logics. Vietnam J. Comput. Sci. **2**(3), 149–167 (2015)

An Efficient Method for Mining Clickstream Patterns

Bang V. Bui[1], Bay Vo[2], Huy M. Huynh[3], Tu-Anh Nguyen-Hoang[1], and Bao Huynh[4($\boxtimes$)]

[1] University of Information Technology, Vietnam National University HCMC,
Ho Chi Minh City, Vietnam
vanbang0208@gmail.com, anhnht@uit.edu.vn
[2] Faculty of Information Technology, Ho Chi Minh City University
of Technology (HUTECH), Ho Chi Minh City, Vietnam
vd.bay@hutech.edu.vn
[3] Faculty of Electrical Engineering and Computer Science, Technical University
of Ostrava (VŠB), Ostrava-Poruba, Czech Republic
huy.minh.huynh.st@vsb.cz
[4] Faculty of Information Technology, Ton Duc Thang University,
Ho Chi Minh City, Vietnam
huynhquocbao@tdt.edu.vn

Abstract. Recently, hybrid approaches, which combine an FP-tree-like data structure with an interaction-based approach, are efficient approaches for mining frequent itemsets. However, applying those approaches for sequential pattern mining arose some challenges. In this paper, we introduce a hybrid approach for a specific version of sequential pattern mining, clickstream pattern mining, with our proposed B-List structure and SMUB algorithm. The SMUB algorithm exploited the B-List structure that is generated from the SPPC tree and the B-List intersection are used to discover all sequential patterns in the given sequence database. Via our experiments on various databases, SMUB has been shown to be more efficient than the current state-of-the-art algorithm, CM-Spade, in terms of runtime, and scalability, especially on huge databases with very small thresholds.

Keywords: Data mining · Clickstream pattern · Sequence pattern

1 Introduction

The problem of sequential pattern mining was first brought up by Srikant and Agrawal in 1995 [2]. Since then, there have been quite a lot of approaches and algorithms proposed to solve this problem. However, finding an effective method is still challenging. Recently, hybrid approaches using DiffNodeSets [10], N-List [9] data structures are reported as very efficient for mining frequent itemsets. But can those approaches be applied for mining pattern with a sequential order? To the best of our knowledge, there have not any work that was based on the hybrid approaches using

© Springer Nature Switzerland AG 2018
H. S. Nguyen et al. (Eds.): IJCRS 2018, LNAI 11103, pp. 572–583, 2018.
https://doi.org/10.1007/978-3-319-99368-3_45

those data structures. Itemset patterns are easier to deal with because each item only appears once at most in each transaction of the database, and the order of items in the itemsets can be assigned by users. On the other hand, sequential patterns consist of multiple transactions in sequential or timely order. Thus, each item can appear more than one in a sequence, in various transactions, and in an order that users cannot predict.

In this paper, we propose the SMUB algorithm to tackle a part of sequential pattern mining problem by solving clickstream pattern mining, a special version of sequential pattern mining. SMUB is a hybrid-based approach algorithm, based on B-List, an extension of N-List data structure. B-Lists are generated from an SPPC tree. Via our experiments on various datasets have shown that SMUB was more efficient than the recent state-of-the-art algorithm, CM-Spade [11], with respect to runtime, especially on huge datasets with low minimum support thresholds.

We organized this paper as follows. In Sect. 2, we describe the basic concepts. In Sect. 3, we introduce related work. In Sect. 4, we introduce SPPC tree and definitions. In Sect. 5, we present our B-List and SMUB algorithm for clickstream pattern mining. In Sect. 6, we present our experiments. In Sect. 7, we conclude our study and present our future work.

2 Basic Concepts

Let $I = \{i_1, i_2, \ldots, i_j\}$ be a set of distinct elements, each element is called an item. A sequence is a list of items that are ordered. A clickstream sequence S is denoted as $\langle s_1, s_2, \ldots, s_q \rangle$, where $s_p \in S (1 \leq p \leq q)$ is an item. The number of items in clickstream is called the size or length of the clickstream. A clickstream sequence having length k is denoted as k-sequence. A clickstream sequence $S_a = \langle a_1, a_2, \ldots, a_n \rangle$ is a subsequence of another clickstream sequence $S_b = \langle b_1, b_2, \ldots, b_m \rangle$, denoted by $S_a \subseteq S_b$, if there exist integers $x_1 < x_2 < \cdots < x_y$ that $a_t = b_{x_y}$ with all of a_t. In other words, S_b is called a super sequence of S_a.

A clickstream sequence database SDB is a collection of clickstream and each sequence has a unique id (called sid). Support of a clickstream pattern P is defined as the number of clickstreams in SDB that are the super sequences of P. Given a threshold, a clickstream sequence is a frequent clickstream pattern if its support is more than or equal to the given threshold. The clickstream pattern mining task is discovering all frequent clickstream patterns in SDB.

Table 1. A clickstream sequence database

SID	Clickstream
100	<2,5,1>
200	<2,5,1,5,1>
300	<2,3>
400	<1,5,1,5,1>

3 Related Work

Several algorithms have been proposed for sequential pattern mining such as AproriAll [2], GSP [3] and SPADE [5]. All of them find all sequential patterns by using "generate and test candidate" approach which consumes a lot of time and memory. PrefixSpan [6], FUSP [7] and Sequential Pattern Tree [8] does not generate any candidate sequences, but the structure of the tree is complex; thus, they create lots of projected databases and in order to find new sequential patterns, they need to completely scan the projected databases.

SPADE algorithm [5] identifies all frequent items (viz., 1-sequences) at the beginning, converts the database to the vertical database format and identify the rest of sequential pattern by BFS or DFS based on lattice decomposition concept. Though experiments, it is more efficient than the GSP algorithm. However, SPADE needs to convert database from horizontal to the vertical format, so the memory usage for storing the databases increased and it is even bigger than the original databases.

In 2008, Lin et al. proposed FUSP-tree [7] data structure and its maintenance algorithm for mining sequential patterns in incremental databases. FUSP-tree consists of one root node and a set of prefix subtrees as the children branches of the root. Each node in the prefix subtrees contains three values: *item − name* represents the node contains that item, *count* is the number of sequences represented by the section of the path reaching the node and *node − link* links to the next node of the same item in another branch of the FUSP-tree. The FUSP-tree contains a Header-Table which stores frequent items, their count and the link to the first occurrence in the tree corresponding to the item. This table assists on finding appropriate items or sequences in the tree.

Fournier-Viger et al. proposed CM-Spade in 2014 [11]. In their work, they proposed the CMAP data structure to store co-occurrence information of items and used the CMAP to produce a candidate pruning mechanism. Basically, CM-Spade integrates CMAP data structure into the SPADE algorithm. It was reported to have better performance than previous algorithms, SPADE and SPAM. But CM-Spade still suffers from spending much time evaluating candidates that do not exist in the sequence database.

There have been quite a few several efficient algorithms recently for mining frequent itemset from transaction databases [1] such as FP-growth [4], N-List [9] and DiffNodeSets [10].

In 2012, Deng proposed PrePost [9] algorithms. PrePost was based on the N-List structure that was generated from PPC-tree, which was a new structure for representing transaction databases. This data structure saves all information of itemsets. By combining the approach of candidate-generation-and-test and the approach of mining sequence itemset directly without candidate generation, PrePost was reported as an efficient algorithm for mining frequent itemsets. PPC-tree structure includes a root node and a set of children nodes, the structure of each node includes five properties: *item-name, count, children-list, pre-order,* and *post-order. Item-name* registers which item this node represents, *count* registers the number of transactions presented by the portion of the path reaching this node, *children-list* registers all children of the node, *pre-order* is the pre-order rank of the node and *post-order* is the post-order rank of the node. PPC-tree structure is like an FP-tree [4].

4   SPPC-Tree Structure

Definition 1. SPPC-tree is a tree data structure. The tree consists of a root and a set of item prefix subtrees as the children of the root. Each node of the tree consists of eight fields: *item-name, count, first-child, first-father, right-sibling, label-sibling, pre-order, post-order. Item-name* is the item that the current node represents. *Count* is the number of sequences that have the same path reaching to the current node. *First-child* is a list that contains the first children of the node. *First-father* is the first previous node that is reached from the root node. *Right-sibling* is the first sibling node of the current node. *Label-sibling* is a list of nodes that have the same *item-name* even they may be in different branches of the tree. *Pre-order* is a list of pre-order ranks that were generated by pre-order traversal of the tree. *Post-order* is a list of post-order ranks that were generated by post-order traversal of the tree. SPPC-tree is derived from PPC-tree [9]. However, there are two differences between SPPC-tree and PPC-tree:

1. The support of frequent item is not the sum of all counts of nodes with same item name on SPPC-tree.
2. The *item-name* of an item can appear more than in one node in the same branch of SPPC tree.

Based on Definition 1, an SPPC-tree can be built by the following algorithm.

Algorithm 1 (Building an SPPC-tree)
Input: A sequence database *SDB* and a minimum support ξ.
Output: An SPPC-tree and the set of frequent items *F1*.
Procedure: *Construct-SPPC-tree* (*SDB*, ξ)
[Finding frequent items in the database]
1:   Scan *SDB* once to find *F1*, the set of frequent items, with their supports $\geq \xi$.
[Start building SPPC-tree]
2:   Create an SPPC node, called S_r, and assign it as a root node.
3:   **for** each sequence *Seq* in *SDB* **do**
4:       Remove the infrequent items from *Seq* and let *p* be the remnants of *Seq*. Thus, *p* is a sequence that only contains frequent items.
[Start inserting the sequence into the tree]
5:       **for** each item in *p* **do**
6:           **if** S_r has a child *N* such that *N.item-name* = *p.item-name* **then**
7:               *N*.count++;
8:           **else**
9:               create a new node *N* with the default value;
10:          **if** *N.right-sibling* == *null* **then**
11:              add new node *N* to *first-child* list of S_r;
12:          **else**
13:              add new node *N* to *right-sibling* list of S_r;
14:          **end if**
15:      **end for**
16: **end for**
[Adding Pre-Post code after building the tree]
17: Traverse the SPPC-tree with pre-order and post-order traversals to generate the *pre-order* and the *post-order* values for each node.

For example, assuming that we use an example sequence database *SDB* in Table 1 with minimum support threshold $\xi = 0.5$. First, we convert the value of minimum support from a double value to an integer value: $4 * 0.5 = 2$. Then, we scan *SDB* to find the frequent items with their support count greater than or equal to ξ. The final set is *SP1* = {<1>,<2>,<5>} with their support counts. With all infrequent items eliminated, we have a newly transformed sequence database as in Table 2.

Table 2. The new sequence database with infrequent items already removed

SID	Clickstream
100	<2,5,1>
200	<2,5,1,5,1>
300	<2>
400	<1,5,1,5,1>

Based on the newly transformed database, we build an SPPC-tree by the following steps. First, we create an empty node and assign it as a root node, then we add sequence 100 to the tree. The adding process starts at the root node. From there, each item in the sequence will have a node created and appended to the tree in a sequential order. The first item of the sequence will be appended to the root, the second will be appended to the first node and so on. The tree will look like in Fig. 1a the sequence 100 is added. After which, we add sequence 200 to the tree. Because the subsequence <2,5,1> was previously added into the tree during adding sequence 100, so we increase the count of each same node, the process for the rest of the items the same as adding sequence 100. After the sequence 200 is added, the tree will be like Fig. 1b and so on. However, the sequence 400 does not start with the same start item with other previous sequences. Thus, we create a new branch and add each item in this sequence into the tree like what we did to 100. The tree then will be like in Fig. 1d. Considering the node 2:2, it means that this is the node of item 2 and its support count is 2.

After adding all sequences in *SDB* in Table 2 into the tree, we travel the tree using depth-first search (DFS) algorithm to add *pre-order* and *post-order* for each node. The tree looks like in Fig. 1e, which depicts the final result tree from *SDB* in Table 2 after executing the Algorithm 1. The node (0,4)2:3 mean this is the node of item 2, the *count* is 3, and the *pre-order* and *post-order* of the node is 0 and 4 respectively.

5   Sequential Pattern Mining Using B-Lists

In this section, we describe the idea and step by step of our proposed SMUB algorithm (sequential clickstream mining using B-List). SMUB is a hybrid approach for mining frequent sequences. Main steps of SMUB algorithm include: (1) build SPPC-tree and identify all frequent 1-sequences (2) based on SPPC-tree, conduct the B-List for each frequent 1-sequence (3) mine the remaining frequent k-sequences ($k > 1$). The details of the algorithm are presented in Sect. 5.2.

Definition 2 (SPP-code). Given an SPPC-tree S_{tr} and a node $N \in S_{tr}$, an SPP-code of N is an element represented in the form of (*N.pre-order*, *N.post-order*):*count*.

Definition 3 (B-List of a frequent item; viz., frequent 1-sequence). Given an SPPC-tree, the B-List of a specified frequent item is an ordered set of all the SPP-codes of nodes having the same *item-name* with respect to the frequent item. The SPP-codes are sorted in an ascending order based on their *pre-order* values and the B-List is represented in the form of $(x_1, y_1) : z_1 \rightarrow \cdots \rightarrow (x_n, y_n) : z_n$. For each SPP-code in a B-List, there should always be a node in SPPC-tree that is registered with the SPP-code.

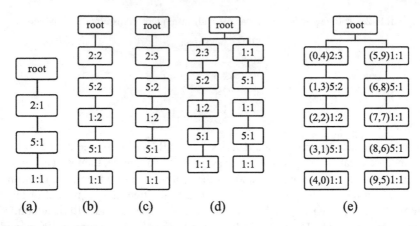

Fig. 1. Step by step SPPC-Tree construction: (a) after adding sequence 100 (b) after adding sequence 200 (c) after adding sequence 300 (d) after adding sequence 400 (e) after adding *pre-order* and *post-order*

Definition 4 (Support count of a B-List). Given a B-List $BL = (x_1, y_1) : z_1 \rightarrow \cdots \rightarrow (x_n, y_n) : z_n$, and $BL_m = BL \backslash \{(x, y) : z \in BL | \exists (x_i, y_i) : z_i \in BL : x \rangle x_i \wedge y < y_i\}$. The support of BL can be calculated via BL_m by the sum of all z_k with $(x_k, y_k) : z_k \in BL_m$. For example, consider the B-List of the frequent 1-sequence <1> in Table 3, its BL_m is $(2,2):2 \rightarrow (5,9):1$. So the support count would be 3.

Table 3. The B-Lists of frequent 1-sequences

Frequent 1-sequence	B-List
1	$(2,2):2 \rightarrow (4,0):1 \rightarrow (5,9):1 \rightarrow (7,7):1 \rightarrow (9,5):1$
2	$(0,4):3$
5	$(1,3):2 \rightarrow (3,1):1 \rightarrow (6,8):1 \rightarrow (8,6):1$

5.1   B-List Generation for *k*-Sequences

Let $BL1$ and $BL2$ be the B-Lists of two k-frequent sequences $P_1 = \langle i_1, i_2, \ldots, i_{k-1}, x \rangle$ and $P_2 = \langle i_1, i_2, \ldots, i_{k-1}, y \rangle$, P_1 and P_2 share the same $(k-1)$ prefix, the B-List of $(k+1)$-sequence $P_3 = \langle i_1, \ldots, i_{k-1}, x, y \rangle$ is formed by following the procedure in Algorithm 2. In other words, BL_intersection only works between two frequent k-patterns that share $(k-1)$ prefix. A special case is that frequent 1-sequences are considered sharing an empty prefix.

Algorithm 2 (BL_intersection)
Input: $BL1 = (x_{11}, y_{11}): z_{11} \to \cdots \to (x_{1m}, y_{1m}): z_{1m}$ and
$BL2 = (x_{21}, y_{21}): z_{21} \to \cdots \to (x_{2n}, y_{2n}): z_{2n}$.
Output: $BL3$, the B-List of $P3$.
Procedure: BL_intersection $(BL1, BL2)$

```
1:   i ← 1; j ← 1;
2:   while i ≤ m && j ≤ n do
3:       if (x₁ᵢ < x₂ⱼ) then
4:           if (y₁ᵢ > y₂ⱼ) then
5:               Insert (x₂ⱼ,y₂ⱼ):z₂ⱼ into BL3; j++;
6:           else
7:               j++;
8:           end if
9:       else
10:          i++;
11:      end if
12: end while
```

For example, assuming that we have frequent 1-sequence <5> and we want to generate the B-List of 2-sequence <5,5> . As shown in Table 3, the B-List of <5> is $(1,3):2 \to (3,1):1 \to (6,8):1 \to (8,6):1$. The generation of the B-List of <5,5> is done by combining the B-List of <5> with itself. First, we check $(1,3):2$ with every element in the B-List of itself. However, the *pre-order* of the SPP-code $(1,3):2$ is 1, which is not greater than the *pre-order* of $(1,3):2$ itself. So we move to $(3,1):1$. The *pre-order* of $(3,1):1$ is 3, which is higher than *pre-order* of $(1,3):2$. The *post-order* of $(3,1):1$ is 1, which is less than *post-order* of $(1,3):2$. So $(3,1):1$ is added to the B-List of <5,5> . Finishing the BL_intersection, we have the B-List of <5,5> , which is $(3,1):1 \to (8,6):1$.

5.2   Mining Clickstream Sequential Patterns

Based on previous definitions, Algorithm 3 illustrates the process of SMUB with high-level pseudocodes.

Algorithm 3 (Mining frequent clickstream patterns)
Input: the minimum support ξ, the sequential patterns 1-sequences SP_1 and set of all frequent 1-sequence B-List BL_1.
Output: The set of all sequential patterns SP.
1: Initialize SP and assign $SP = SP_1$
2: Call mining_L(SP_1, BL_1)
Procedure: mining_L$(SP_k, \ BL_k)$
3: Initialize $SP_{k+1} = \emptyset$ and $BL_{k+1} = \emptyset$
4: **for** each pattern P_a in SP_k **do**
5:      **for** each pattern P_b in SP_k that share $(k-1)$ prefix with P_a **do**
6:          Assuming $P_a = <i_1, i_2, ..., i_{k-1}, x>$ and $P_b = <i_1, i_2, ..., i_{k-1}, y>$, create $P_c = <i_1, i_2, ..., i_{k-1}, x, y>$
7:          Create B-List of P_c by calling BL_intersection for B-Lists of P_a and P_b
8:              **if** support count of B-List of $P_c \geq \xi$ **then**
9:                  Put P_c into SP and SP_{k+1}
10:                 Put B-List of P_c into BL_{k+1}
11:             **end if**
12:         **end for**
13: **end for**
14: Call mining_L(SP_{k+1}, BL_{k+1})

For example, considering the minimum support $\xi = 3$, we have SP_1 as the set of frequent 1-sequences <5>, <2> and <1> mined from the example database SDB and their respective B-List set BL_1. Running 3, we first join <5> with <5>, <2> and <1> to form 2-sequence candidates <5,5> , <5,1> and <5,2> . By generating B-Lists for aforementioned candidates, we can use them to check for support count of each candidate. Only <5,5> and <5,2> have their support counts higher than ξ, so they are frequent 2-sequences and are added into the set of frequent 2-patterns SP_2. In the same way, <2> is joined with <5>, <2> and <1>, and <1> is joined with <5>, <2> and <1>. The resultant frequent 2-patterns are added into SP_2 and their respective B-Lists are added into BL_2. Recursively, we re-run mining_L procedure with SP_2 and BL_2 and so on, until no candidate can be generated. Figure 2 illustrates the full set of frequent clickstream patterns.

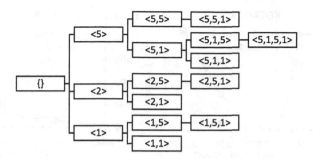

Fig. 2. The tree of frequent clickstream patterns

6 Experimental Evaluation

In this section, we performed experiments to assess the performance of the proposed algorithm. We performed experiments on a computer running Intel Core i7 2.2 GHz CPU, 16 GB memory, and macOS Sierra 10.12.6 operating system. We configured JVM with the flags of -Xmx10G -Xms10G (viz., the maximum memory allowed was 10 GB). The state-of-art algorithm, CM-Spade, for sequential pattern mining that was proved more efficient than previous algorithms, which were GSP, PrefixSpan and FUSP in [11]. So, in this paper we just compared the proposed algorithm, SMUB, with CM-Spade. We use Kosarak, FIFA, MSNBC, and BMS2 datasets (Table 4) for testing performance. We implemented the SMUB in Java 8. The experiments are conducted on each database by decreasing the minimum support thresholds until algorithm took too long time to execute (more than 2000s) or ran out of memory. The running time is the total execution time of the algorithm.

Table 4. Database description

Database	Sequences	Unique items	Average sequence length
Kosarak	990,002	41,270	8.1
FIFA	20,450	2,990	34.74
MSNBC	989,818	17	4.75
BMS2	77,512	3,340	4.62

Figure 3 shows the running time of SMUB and CM-Spade on Kosarak, FIFA, MSNBC, and BMS2 correspondingly. Generally, SMUB ran faster than CM-Spade and the gap kept getting bigger at smaller minimum support. Thus, we can see that SMUB is more efficient than CM-Spade at low minimum support threshold.

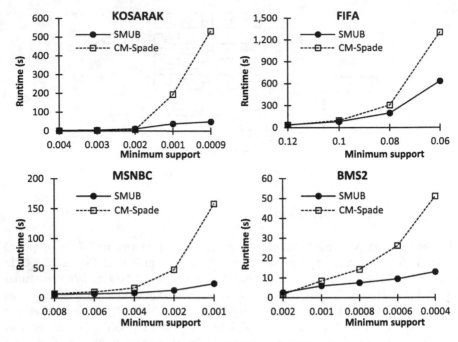

Fig. 3. Runtime of SMUB and CM-Spade

7   Conclusions and Future Work

In this paper, we proposed a novel data structure, B-List, for compressing and storing information for clickstream patterns. Based on B-Lists, we developed an algorithm, SMUB, for fast mining clickstream patterns in clickstream databases. The advantages of the SMUB algorithm compared to other previous algorithms are as follow: First, it uses a compact data structure, B-List, which is usually substantially smaller than the original databases, and thus avoids costly database scans in the subsequent mining processes. Second, counting the support of sequence is transformed into the intersection of B-Lists and it employs an efficient strategy with the complexity of $O(m + n)$ for intersecting two B-Lists, where m and n are the cardinalities of the two B-Lists respectively. We have implemented the SMUB algorithm and studied its performance in comparison with CM-Spade, a well-known sequential pattern mining algorithm, on a variety of real and synthetic datasets. Our performance study shows that the SMUB algorithm is more efficient than CM-Spade.

In future work, we will further explore our method to fully work with sequential pattern mining problem (viz., there is more than one element in itemsets). We also consider using the parallel approach for SMUB so that it can work even bigger databases.

References

1. Agrawal, R., Imieliński, T., Swami, A.: Mining association rules between sets of items in large databases. ACM Sigmod Rec. **22**(2), 207–216 (1993)
2. Agrawal, R., Srikant, R.: Mining sequential patterns. In: The Eleventh International Conference on Data Engineering, pp. 3–14. IEEE (1995)
3. Srikant, R., Agrawal, R.: Mining sequential patterns: generalizations and performance improvements. In: Apers, P., Bouzeghoub, M., Gardarin, G. (eds.) EDBT 1996. LNCS, vol. 1057, pp. 1–17. Springer, Heidelberg (1996). https://doi.org/10.1007/BFb0014140
4. Han, J., Pei, J., Yin, Y.: Mining frequent patterns without candidate generation. ACM Sigmod Rec. **29**(2), 1–2 (2000)
5. Zaki, M.J.: SPADE: an efficient algorithm for mining frequent sequences. Mach. Learn. **42** (1–2), 31–60 (2001)
6. Han, J., et al.: PrefixSpan: mining sequential patterns efficiently by prefix-projected pattern growth. In: The 17th International Conference on Data Engineering, pp. 215–224 (2001)
7. Lin, C.-W., et al.: An incremental FUSP-tree maintenance algorithm. In: Proceedings of 2008 Eighth International Conference on Intelligent Systems Design and Applications, vol. 1, pp. 445–449. IEEE (2008)
8. Bithi, A.A., Ferdaus, A.A.: Sequential pattern tree mining. IOSR J. Comput. Eng. **5**(5), 79–89 (2013)
9. Deng, Z.-H., Wang, Z., Jiang, J.: A new algorithm for fast mining frequent itemsets using N-Lists. Sci. China Inf. Sci. **55**(9), 2008–2030 (2012)
10. Deng, Z.-H.: DiffNodesets: an efficient structure for fast mining frequent itemsets. Appl. Soft Comput. **41**, 214–223 (2016)
11. Fournier-Viger, P., Gomariz, A., Campos, M., Thomas, R.: Fast vertical mining of sequential patterns using co-occurrence information. In: Tseng, V.S., Ho, T.B., Zhou, Z.-H., Chen, A.L. P., Kao, H.-Y. (eds.) PAKDD 2014. LNCS (LNAI), vol. 8443, pp. 40–52. Springer, Cham (2014). https://doi.org/10.1007/978-3-319-06608-0_4

Transformation Semigroups
for Rough Sets

Anuj Kumar More$^{(\boxtimes)}$ and Mohua Banerjee

Department of Mathematics and Statistics, Indian Institute of Technology, Kanpur,
Kanpur 208016, India
{anujmore,mohua}@iitk.ac.in

Abstract. In this article we define transformation semigroups for rough sets. Basic constructions such as closures, products, coverings and partitions for transformation semigroups are defined. A decomposition theorem for reset transformation semigroups is given. A connection with automata is also presented by defining a semiautomaton for rough sets.

Keywords: Transformation semigroups · Rough sets · Automata

1 Introduction

Rough set theory [13] has been studied extensively over the years, from applicational as well as foundational points of view. One of the directions of work on foundational aspects, is the study of categories of rough sets and generalizations (cf. [12]). An instance of the generalizations is found to be the special class of categories $RSC(\textbf{M-Set})$ for monoids $\textbf{M}$, which yields the definition of *monoid actions on rough sets* [12]. Monoid or semigroup actions have direct connection with 'transformation semigroups' and automata theory [5]. We follow this line of study in the present article to explore semigroup actions on rough sets.

An important class of semigroups is the collection $PF(Q)$ of all partial functions from a finite set Q to itself, representing *transformations of Q*. The binary operation involved is function composition, and in fact, results in a monoid structure, with the identity function on Q as the identity element. Any subset S of this collection that is closed under function composition is a subsemigroup of $PF(Q)$. The pair (Q, S) for such S, is called a *transformation semigroup (ts)* [4,5,7]. We observe that the objects of the category $RSC(\textbf{M-Set})$ mentioned above, may be interpreted as transformations for rough sets. By taking the more general structure of semigroups instead of monoids, we obtain here a natural definition of a *transformation semigroup for a rough set*. The algebra of these transformation semigroups is developed in this article, by defining basic constructions of *ts* theory such as resets, coverings, products and admissible partitions for the

This work has been supported by the *Council of Scientific and Industrial Research* (CSIR) India, Research Grant No. 09/092(0875)/2013-EMR-I.

H. S. Nguyen et al. (Eds.): IJCRS 2018, LNAI 11103, pp. 584–598, 2018.
https://doi.org/10.1007/978-3-319-99368-3_46

structures (cf. Sects. 3 and 4). Our main goal is to look for a Krohn-Rhodes style decomposition result (cf. [1]) for these semigroups. Here, we present the first step towards that direction by obtaining a decomposition theorem for the special case of reset transformation semigroups (cf. Sect. 5).

One of the reasons to study transformation semigroups has been a natural connection with automata theory [7]. We shall also study this connection in case of rough sets, by defining a *semiautomaton for a rough set* (cf. Sect. 6). Rough sets have been connected with automata theory and transformation semigroups earlier, by Basu and Tiwari [2,15]. Our approach differs from theirs, and a comparison is presented in Sect. 6. We conclude in Sect. 7.

In the next section, we present preliminaries of transformation semigroups that are required for this work. We shall follow the notations and terminologies of [7] throughout the paper.

2   Transformation Semigroups

Semigroup *actions* give an alternative and equivalent way of viewing transformation semigroups [4]. Recall that an action of a semigroup S on the set Q is a function $\delta : Q \times S \to Q$ satisfying $\delta(\delta(q, s_1), s_2) = \delta(q, s_1 s_2)$, for all $q \in Q$ and $s_1, s_2 \in S.$, where $s_1 s_2$ denotes the application of the binary operation of S on s_1 and s_2. If the function δ is partial, δ is called a *partial semigroup action* of S on the set Q. Then we have the following definition.

Definition 1 (Transformation semigroups) [5]. *A transformation semigroup is a pair $\mathcal{A} := (Q, S)$ consisting of a finite set Q, a finite semigroup S, along with a partial semigroup action δ of S on Q that satisfies:*

$$\text{for any } s_1, s_2 \in S, \text{if } \delta(q, s_1) = \delta(q, s_2) \text{ for all } q \in Q, \text{ then } s_1 = s_2. \quad (1)$$

Observation 1. Condition (1) is termed the *faithfulness* of the action δ. For a fixed $s \in S$, the partial function $\delta_s := \delta(-, s) : Q \to Q$ can be viewed as a transformation of the set Q, and δ can also be interpreted as a set $\{\delta_s\}_{s \in S}$ of transformations of Q. Faithfulness of δ ensures a bijection between S and $\{\delta_s\}_{s \in S}$. Thus both the definitions of transformation semigroup are equivalent.

Notation 1. Hereafter, $\delta(q, s)$ shall be denoted by 'qs' and $Qs := \{qs \mid q \in Q\}$.

Constant functions motivate the definition of a special kind of ts: a ts $\mathcal{A} := (Q, S)$ is called *reset* if $|Qs| \leq 1$ for any $s \in S$. Given a ts $\mathcal{A} := (Q, S)$, the *closure* $\overline{\mathcal{A}}$ of $\mathcal{A}$ is the subsemigroup of $PF(Q)$ that is generated by the set $S \cup \{\overline{q} \mid q \in Q\}$, where $\overline{q}$ represents a constant function on Q mapping any element of Q to q. In notation, $\overline{\mathcal{A}} := (Q, \langle S \cup \{\overline{q} \mid q \in Q\}\rangle)$.

Example 1. A trivial example of a reset *ts* is the pair $(Q, \emptyset)$. If $|Q| = n$, the *ts* $(Q, \emptyset)$ is denoted as $\boldsymbol{n}$. Then $\overline{\boldsymbol{n}} := (Q, \{\overline{q} \mid q \in Q\})$, which is again a reset *ts*.

Observation 2

1. Let Q be a finite set, S a finite semigroup and δ a partial semigroup action of Q on S. Does (Q, S) form a ts? Not necessarily, as δ may not be faithful. One then defines a relation $\sim$ on S by $s \sim s' \Leftrightarrow qs = qs'$ for all $q \in Q$. $\sim$ is a congruence relation on S and $S/\sim$ is a quotient semigroup of S. The pair $(Q, S/\sim)$ forms a ts with the action defined by $q[s] := qs$, for all $q \in Q$, $[s] \in S/\sim$. If $Q = \emptyset$ then $S/\sim$ is the singleton $\{S\}$.

2. Two different definitions of *restriction* of a given ts are found in literature [5]. Consider a ts $\mathcal{A} := (Q, S)$, $P \subseteq Q$ and the inclusion function $i : P \rightarrow Q$.

   (a) Define a subsemigroup $T := \{s \mid s \in S \text{ and } Ps \subseteq P\}$. Using part (1) of this observation, $\mathcal{A}_P := (P, T/\sim)$ forms a ts.

   (b) Define a partial function i^{-1} from Q to P given by $i^{-1}(q) = q$ for all $q \in P$ and not defined for $q \notin P$. Let S' be the semigroup generated by the partial functions $s' = isi^{-1} : P \rightarrow P$ for all $s \in S$. Then $\mathcal{A}|P := (P, S')$ is also a ts.

In some cases, these definitions coincide [5]:

Proposition 1 *For a ts $\mathcal{A} := (Q, S)$ and $P \subseteq Q$, if $Ps \subseteq P$ for all $s \in S$, then $\mathcal{A}|P = \mathcal{A}_P$.*

2.1   Algebra on Transformation Semigroups

Consider two ts $\mathcal{A} := (Q, S)$ and $\mathcal{B} := (P, T)$. Let $\alpha : Q \rightarrow P$ be a set function and $\beta : S \rightarrow T$ a semigroup homomorphism such that $\alpha(qs) = \alpha(q)\beta(s)$, whenever qs is defined for any $q \in Q$ and $s \in S$. The pair (α, β) is called a *transformation semigroup homomorphism* from $\mathcal{A}$ to $\mathcal{B}$. If both α and β are bijective maps then $\mathcal{A}$ is said to be equivalent to $\mathcal{B}$ and this is denoted by $\mathcal{A} \cong \mathcal{B}$.

Therefore, one can easily see that ts constitute a category.

Definition 2 (The category TS of transformation semigroups). *Objects of TS are ts and morphisms of TS are ts homomorphisms.*

Note that the object class of **TS** is different from the class **TS** defined in [5].

For ts $\mathcal{A} := (Q, S)$ and $\mathcal{B} := (P, T)$, $\mathcal{A} \times \mathcal{B} := (Q \times P, S \times T)$ is also a ts, called the *direct product* of $\mathcal{A}$ and $\mathcal{B}$. The semigroup operation/action involved is defined componentwise.

Consider a ts $\mathcal{A} := (Q, S)$ and $\pi := \{H_i\}_{i \in I}$ a set of non-empty subsets of Q. π is called a *partition* of Q if $\bigcup_{i \in I} H_i = Q$ and $H_i \cap H_j = \emptyset$ for any $i, j \in I$. π is called *admissible* if for every $H_i \in \pi$ and $s \in S$, if $H_i s$ is non-empty then there exists $H_j \in \pi$ such that $H_i s \subseteq H_j$. Note that such a choice for H_j would be unique for the H_i. Then a partial semigroup action $*$ of S on π can be defined as follows: For any $H_i \in \pi$, $s \in S$,

(1) $H_i * s := H_j$ if $H_i s \subseteq H_j$; (2) $H_i * s$ is not defined if $H_i s = \emptyset$.

This action may not be faithful. However as discussed in Observation 2(1), one can obtain the *quotient ts* $\mathcal{A}/\langle \pi \rangle := (\pi, S/\sim)$, using the congruence relation $\sim$. When $|Q| > 2$, π is said to be *non-trivial* if $1 < |H_i| < |Q|$ for some $i \in I$.

For a non-trivial admissible partition $\pi := \{H_i\}_{i \in I}$ on ts (Q, S), if there exists another non-trivial admissible partition $\tau := \{K_j\}_{j \in J}$ such that $|H_i \cap K_j| \le 1$ for all $i \in I$ and $j \in J$, then π is called an *orthogonal* partition on Q [7]. The condition '$|H_i \cap K_j| \le 1$ for all $i \in I$ and $j \in J$' is denoted as '$\pi \cap \tau = 1_Q$'.

Definition 3 (Coverings) [7]. *A ts $\mathcal{B} := (P, T)$ covers the ts $\mathcal{A} := (Q, S)$, written as $\mathcal{A} \preccurlyeq \mathcal{B}$, if there exists a partial surjective function $\eta : P \to Q$ such that for each $s \in S$, there is $t_s \in T$ satisfying $\eta(p)s = \eta(pt_s)$ whenever $\eta(p)s$ is defined for any $p \in P$. η is called a* covering of $\mathcal{A}$ by $\mathcal{B}$, *or $\mathcal{B}$ is said to* cover $\mathcal{A}$ *by η. t_s is said to* cover s.

Using the definition and the fact that any element in the semigroup $\langle S \rangle$ generated by S can be written as a finite product of elements of S, one gets

Proposition 2. *For ts $(Q, \langle S \rangle)$ and (P, T), the following are equivalent.*

(a) $(Q, \langle S \rangle) \preccurlyeq (P, T)$.
(b) There exists a partial surjective function $\eta : P \to Q$ satisfying the following property: for each $s \in S$ there exists a $t_s \in T$ such that for any $p \in P$, if $\eta(p)s$ is defined then $\eta(p)s = \eta(pt_s)$.

3   Transformation Semigroup for Rough Sets

Iwiński [8] gave a generalized interpretation of rough sets based on a Boolean algebra. A pair (A_1, A_2) is called an *I-rough set* of the *rough universe* $(U, \mathbf{B})$, where U is the domain, $\mathbf{B}$ is a subalgebra of the power set Boolean algebra $\mathcal{P}(U)$ and A_1, A_2 in $\mathbf{B}$ are such that $A_1 \subseteq A_2$. Observe that any pair of sets (Q_1, Q_2), where $Q_1 \subseteq Q_2 \subseteq C$ for some set C, can then be interpreted as an *I*-rough set of the rough universe $(C, \mathcal{P}(C))$. This approach was followed in defining the category *RSC* [9] of *I*-rough sets, which was shown to be equivalent to the category *ROUGH* defined earlier (cf. [12]). *I*-rough sets are referred to simply as rough sets. A generalization of *RSC* leads to the class of categories *RSC*(**M-Set**) for monoids **M**; the properties of objects and morphisms therein yield the definition of *monoid actions on rough sets* [12]. We apply the definition to the more general structure of semigroups, and to rough sets (Q_1, Q_2) with Q_2 finite.

Definition 4 (Semigroup action on rough sets). *A semigroup action on a rough set (Q_1, Q_2) with Q_2 finite, is a triple (Q_1, Q_2, δ) where $\delta : Q_2 \times S \to Q_2$ is an action of a semigroup S on Q_2 such that the restriction $\delta|_{Q_1} : Q_1 \times S \to Q_1$ is an action of S on Q_1. Note that $\delta|_{Q_1}((q, s)) := \delta((q, s))$, for all $q \in Q_1, s \in S$.*

Using Observation 2(1), we get a ts $(Q_2, S/\sim_2)$ for δ, where $s_1 \sim_2 s_2$ if and only if $qs_1 = qs_2$ for all $q \in Q_2$. As Q_2 is finite, $S/\sim_2$ is also finite. Now consider the action $\delta' : Q_2 \times S/\sim_2 \to Q_2$ associated with the ts $(Q_2, S/\sim_2)$. By Observation 1, we can identify $\{\delta'_{[s]_{\sim_2}}\}_{s\in S}$ with $S/\sim_2$ which is a semigroup of transformations of Q_2. These transformations also restrict Q_1 to Q_1. Thus, $(Q_1, S/\sim_1)$ forms another ts, where $s_1 \sim_1 s_2$ if and only if $qs_1 = qs_2$ for all $q \in Q_1$. We now arrive at the definition of a ts for a rough set. Henceforth, when the contexts are clear, we shall drop suffixes and simply write $\sim$.

Definition 5 (Transformation semigroups for rough sets). *A transformation semigroup for a rough set* (Q_1, Q_2) *is a triple* $\mathcal{A} := (Q_1, Q_2, S)$, *where* (Q_2, S) *is a ts and* $Q_1 \subseteq Q_2$ *such that* $Q_1 s \subseteq Q_1$ *for all* $s \in S$. (Q_2, S) *is called the* upper ts *and* $(Q_1, S/\sim)$ *the* lower ts *for* $\mathcal{A}$, *where* $s_1 \sim s_2$ *if and only if* $qs_1 = qs_2$ *for all* $q \in Q_1$.

Observation 3

1. Relating Definitions 4 and 5: Given a semigroup action (Q_1, Q_2, δ) on rough set (Q_1, Q_2), we can obtain a ts $(Q_1, Q_2, S/\sim_2)$ for the same rough set (Q_1, Q_2). Conversely, a ts (Q_1, Q_2, S) for a rough set (Q_1, Q_2) gives a semigroup action (Q_1, Q_2, δ) on (Q_1, Q_2), where δ is the action associated with ts (Q_2, S) (cf. Definition 1).
2. Relating Definitions 1 and 5: For a ts $\mathcal{A} := (Q, S)$, (Q, Q, S) is a ts for rough set for which, trivially, $\mathcal{A}$ is the upper ts, and also the lower ts up to isomorphism. (Q, Q, S) shall also be denoted as (Q, S), by abuse of notation. $\mathcal{A}$ is also the upper ts for the ts $(\emptyset, Q, S)$ for rough set $(\emptyset, Q)$. The lower ts of $(\emptyset, Q, S)$ is $(\emptyset, S/\sim)$, where $S/\sim$ is a 1-element semigroup.
3. For ts (Q_1, Q_2, S), $(Q_1, S/\sim) = (Q_2, S)_{Q_1} = (Q_2, S)|Q_1$. Indeed, recall Observation 2(2) and Proposition 1. Since $Q_1 s \subseteq Q_1$ for all $s \in S$, we have $(Q_2, S)_{Q_1} = (Q_2, S)|Q_1$. Moreover by definition, $(Q_2, S)_{Q_1}$ is just $(Q_1, S/\sim)$.

Example 2. Consider the ts (Q_2, S) from [10] where $Q_2 := \{1, 2, 3, 4, 5, 6, 7\}$ and $S := \{s_i \mid 1 \leq i \leq 7\}$ is the semigroup with $s_1 := 1_{Q_2}$ and

$$s_2 := \begin{pmatrix} 1\,2\,3\,4\,5\,6\,7 \\ 1\,2\,4\,3\,6\,5\,7 \end{pmatrix} \quad s_3 := \begin{pmatrix} 1\,2\,3\,4\,5\,6\,7 \\ 1\,5\,6\,7\,1\,1\,1 \end{pmatrix} \quad s_4 := \begin{pmatrix} 1\,2\,3\,4\,5\,6\,7 \\ 1\,5\,7\,6\,1\,1\,1 \end{pmatrix}$$

$$s_5 := \begin{pmatrix} 1\,2\,3\,4\,5\,6\,7 \\ 1\,6\,5\,7\,1\,1\,1 \end{pmatrix} \quad s_6 := \begin{pmatrix} 1\,2\,3\,4\,5\,6\,7 \\ 1\,6\,7\,5\,1\,1\,1 \end{pmatrix} \quad s_7 := \begin{pmatrix} 1\,2\,3\,4\,5\,6\,7 \\ 1\,1\,1\,1\,1\,1\,1 \end{pmatrix}$$

Take $Q_1 := \{1, 5, 6, 7\}$. Since $Q_1 s_i \subseteq Q_1$ for all $s_i \in S$, the triple (Q_1, Q_2, S) forms a ts for rough set (Q_1, Q_2). The upper ts is (Q_2, S) and the lower ts is $(Q_1, S/\sim)$, where $S/\sim = \{\{s_1\}, \{s_2\}, \{s_3, s_4, s_5, s_6, s_7\}\}$.

We should remark here that a notion of *rough transformation semigroup* was defined in [15], and was motivated by the *rough finite semi-automaton* defined by Basu [2]. We shall make a comparison of all the structures in Sect. 6.

3.1 Resets and Closures

What could be an appropriate definition for a reset ts here?

Definition 6 (Reset ts for rough sets). (Q_1, Q_2, S) *is a reset if the upper ts* (Q_2, S) *is a reset, that is,* $|Q_2 s| \leq 1$ *for all* $s \in S$.

It is then easy to observe that the lower ts $(Q_1, S/\sim)$ is also a reset.

Example 3. Recall Example 1. A trivial reset ts for a rough set (Q_1, Q_2) is the triple $\mathcal{A} := (Q_1, Q_2, \emptyset)$. The reset ts $(Q_1, \emptyset)$ and $(Q_2, \emptyset)$ are respectively the lower and upper ts. If $|Q_1| = m \leq n = |Q_2|$, $\mathcal{A}$ shall be denoted by $(m, n, \emptyset)$.

For defining closure of a ts $\mathcal{A} := (Q_1, Q_2, S)$, we note that for $q \in Q_2 \setminus Q_1$, the constant functions $\bar{q}$ on Q_2 do not restrict Q_1 into Q_1. We have the following.

Definition 7 (Closure of transformation semigroup for a rough set). *The closure of* $\mathcal{A} := (Q_1, Q_2, S)$ *is defined as*

$$\overline{\mathcal{A}} := (Q_1, Q_2, S') \text{ with } S' := \langle S \cup \{\bar{q} \mid q \in Q_1\} \cup \{\tilde{q} \mid q \in Q_2 \setminus Q_1\}\rangle, \text{ where}$$

$\bar{q}$ *is the constant function on* Q_2 *mapping any element of* Q_2 *to* q,
$\tilde{q}$ *is the partial constant function on* Q_2 *mapping the elements of* $Q_2 \setminus Q_1$ *to* q *and not defined otherwise.*

Observation 4

1. If $Q_1 = \emptyset$ or $Q_1 = Q_2$ then the semigroup S' in $\overline{\mathcal{A}}$ is just the semigroup in $\overline{(Q_2, S)}$, as expected.
2. Let $\emptyset_{Q_2}$ denote the empty partial function, i.e. it is not defined for any $q \in Q_2$. If $Q_1 \neq \emptyset$ and $Q_1 \neq Q_2$, then S' contains the following:
   (a) all $\tilde{q}$ for $q \in Q_2$, since if $q \in Q_1$ then $\tilde{q} = \bar{q'}\tilde{q} \in S'$ for any $q' \in Q_2 \setminus Q_1$,
   (b) $\emptyset_{Q_2}$, because $\emptyset_{Q_2} = \overline{q'}\tilde{q} \in S'$ for any $q \in Q_2 \setminus Q_1$ and $q' \in Q_1$.
3. Closure is idempotent, i.e. $\overline{\mathcal{A}} = \overline{(\overline{\mathcal{A}})}$.

Example 4. Consider the reset ts $(m, n, \emptyset)$ (Example 3). For $m \neq n$ and $m \neq 0$,

$$\overline{(m, n, \emptyset)} := (m, n, S'), \text{ where } S' = \{\bar{q} \mid q \in Q_1\} \cup \{\tilde{q} \mid q \in Q_2\} \cup \{\emptyset_{Q_2}\}.$$

In particular, for $Q_2 := \{0, 1\}$ and $Q_1 := \{0\}$, $S' := \{\bar{0}, \tilde{1}, \tilde{0}, \emptyset_{Q_2}\}$. Diagrammatically, the upper ts (Q_2, S') is the following.

Note that the upper ts (Q_2, S') of the closure $\overline{(1, 2, \emptyset)}$ of ts $(1, 2, \emptyset)$ is not isomorphic to the closure $\overline{\mathbf{2}}$ of the upper reset ts $\mathbf{2}$ of $(1, 2, \emptyset)$ (cf. Example 1). The lower ts $(Q_1, \{\{\bar{0}\}, \{\emptyset_{Q_2}, \tilde{0}, \tilde{1}\}\})$ of the closure $\overline{(1, 2, \emptyset)}$ of $(1, 2, \emptyset)$ is also not isomorphic to the closure $\overline{\mathbf{1}}$ of the lower reset ts $\mathbf{1}$ of $(1, 2, \emptyset)$. However, the following holds.

Proposition 3. *For ts* $\mathcal{A} := (Q_1, Q_2, S)$,

(a) the closure of the upper ts of $\mathcal{A}$ *covers the upper ts of the closure of* $\mathcal{A}$,
(b) the closure of the lower ts of $\mathcal{A}$ *covers the lower ts of the closure of* $\mathcal{A}$.

Proof. We refer to $\overline{\mathcal{A}}$ as in Definition 7. The coverings η (cf. Definition 3) are the maps 1_{Q_2} and 1_{Q_1} for cases (a) and (b) respectively. It is then easy to find covers for elements of S' and $S'/\sim$ in the two cases, using Proposition 2. $\quad\square$

4   Algebra on Transformation Semigroups for Rough Sets

Consider two *ts* $\mathcal{A} := (Q_1, Q_2, S)$ and $\mathcal{B} := (P_1, P_2, T)$ for rough sets (Q_1, Q_2) and (P_1, P_2) respectively, and the *ts* homomorphism (α, β) from the upper *ts* (Q_2, S) to the upper *ts* (P_2, T) (cf. Sect. 2.1) satisfying the condition $\alpha(Q_1) \subseteq P_1$. Would this imply that the pair $(\alpha|_{Q_1}, \widehat{\beta})$ is a *ts* homomorphism from $(Q_1, S/\sim)$ to $(P_1, T/\sim')$, where $\widehat{\beta} : S/\sim \to T/\sim'$ is defined as $\widehat{\beta}([s]_\sim) := [\beta(s)]_{\sim'}$, $s \in S$? The answer is no, as $\widehat{\beta}$ may not be well-defined: consider the *ts* (Q_1, Q_2, S) from Example 2. By Observation 3(2), (Q_2, Q_2, S) is also a *ts* for rough set with lower *ts* identifiable with (Q_2, S). $(1_{Q_2}, 1_S) : (Q_2, S) \to (Q_2, S)$ is a *ts* homomorphism and $1_{Q_2}(Q_1) \subseteq Q_2$. $\widehat{1_S} : S/\sim \to S$ is such that $\widehat{1_S}([s_i]_\sim) := s_i$, $s_i \in S$; however, $[s_3]_\sim = [s_7]_\sim$ but $s_3 \neq s_7$. So we have the following.

Definition 8 (Homomorphisms). (α, β) *is a* ts *homomorphism from* $\mathcal{A} := (Q_1, Q_2, S)$ *to* $\mathcal{B} := (P_1, P_2, T)$, *provided*

(a) (α, β) *is a* ts *homomorphism between the upper* ts (Q_2, S) *and* (P_2, T),
(b) $\alpha(Q_1) \subseteq P_1$, *and*
(c) for any $s, s' \in S$,

$$\text{if } qs = qs' \text{ for all } q \in Q_1 \text{ then } p\beta(s) = p\beta(s') \text{ for all } p \in P_1. \quad (2)$$

Observation 5

1. Condition (2) ensures that the pair $(\alpha|_{Q_1}, \widehat{\beta})$ is a *ts* homomorphism between the lower *ts* $(Q_1, S/\sim)$ and $(P_1, T/\sim)$.
2. If $\alpha|_{Q_1}$ is a bijection, (2) is always true.

How are *ts* for rough sets and *ts* for sets related? A direct relationship may be observed using category theory. Recall Definition 2 of the category **TS** of transformation semigroups for sets.

Definition 9 (The category RTS of transformation semigroups for rough sets). *Objects are ts for rough sets and morphisms are homomorphisms of ts for rough sets.*

Using Observation 3(2) and Definition 8 of homomorphisms, we easily obtain

Theorem 1. *The category* **TS** *is isomorphic to each of the following categories:*

(a) the full subcategory of **RTS** *with objects of the type* (Q, Q, S), *and*
(b) the full subcategory of **RTS** *with objects of the type* $(\emptyset, Q, S)$.

Let us now define the direct product of *ts* for rough sets.

Definition 10 (Direct products). *The* direct product $\mathcal{A} \times \mathcal{B}$ *of ts* $\mathcal{A} :=$ (Q_1, Q_2, S) *and* $\mathcal{B} := (P_1, P_2, T)$ *is defined as the ts* $(Q_1 \times P_1, Q_2 \times P_2, S \times T)$.

Note that the direct product is indeed a *ts*, as for any $(q, p) \in Q_1 \times P_1$, we have $(q, p)(s, t) = (qs, pt) \in Q_1 \times P_1$.

The relation between the upper (lower) *ts* of the direct product and the direct product of the upper (lower) *ts* is given by the following.

Proposition 4

(a) $(Q_2 \times P_2, S \times T) = (Q_2, S) \times (P_2, T)$.
(b) $(Q_1 \times P_1, (S \times T)/{\sim}) \cong (Q_1, S/{\sim}) \times (P_1, T/{\sim})$.

We next move to admissible partitions and quotients.

Definition 11 (Admissible partitions and quotients). *Let* $\mathcal{A} := (Q_1, Q_2, S)$ *be a ts for rough set* (Q_1, Q_2) *and* $\pi_2 := \{H_i\}_{i \in I}$ *be an admissible partition on* Q_2 *in ts* (Q_2, S). *Consider the quotient ts* $(\pi_2, S/{\sim})$, *and let* $\pi_1 := \{H_i \in \pi_2 \mid H_i \cap Q_1 \neq \emptyset\}$. *If* π_1 *satisfies the condition:*

$$\pi_1 * [s] \subseteq \pi_1 \text{ for all } [s] \in S/{\sim}, \tag{3}$$

then $\pi := (\pi_1, \pi_2)$ *is termed an* admissible partition *on rough set* (Q_1, Q_2) *in* $\mathcal{A}$, *and the* quotient *of* $\mathcal{A}$ *with respect to* π *is the ts* $\mathcal{A}/\langle \pi \rangle := (\pi_1, \pi_2, S/{\sim})$. *An admissible partition* π *in* $\mathcal{A}$ *is non-trivial, if* π_2 *is non-trivial on* Q_2 *in ts* (Q_2, S).

Does $(\pi_1, \pi_2, S/{\sim})$ form a *ts*, if condition (3) is not satisfied? No: let us consider the reset *ts* $(2, 4, \emptyset)$ (Example 4). $\overline{(2, 4, \emptyset)} := (Q_1, Q_2, S)$ where $Q_1 := \{q_1, q_2\}$, $Q_2 := \{q_1, q_2, q_3, q_4\}$ and $S := \{\overline{q}_1, \overline{q}_2, \widetilde{q}_1, \widetilde{q}_2, \widetilde{q}_3, \widetilde{q}_4, \emptyset_{Q_2}\}$. Consider the admissible partition $\pi_2 := \{\{q_1, q_3\}, \{q_2\}, \{q_4\}\}$ on Q_2 in the *ts* (Q_2, S). Then $\pi_1 = \{\{q_1, q_3\}, \{q_2\}\}$. For $\{q_1, q_3\} \in \pi_1$ and $[\widetilde{q}_4] \in S/{\sim}$, $\{q_1, q_3\} * [\widetilde{q}_4] = \{q_4\} \notin \pi_1$, i.e. $\pi_1 * [s] \not\subseteq \pi_1$ for some $[s] \in S/{\sim}$. Therefore $(\pi_1, \pi_2, S/{\sim})$ is not a *ts*.

Definition 12 (Orthogonal partitions). *For a ts* $\mathcal{A} := (Q_1, Q_2, S)$, *a non-trivial admissible partition* $\pi := (\pi_1, \pi_2)$ *on rough set* (Q_1, Q_2) *in* $\mathcal{A}$ *is called* orthogonal *if there exists a non-trivial admissible partition* $\tau := (\tau_1, \tau_2)$ *on rough set* (Q_1, Q_2) *in* $\mathcal{A}$ *such that* $\pi_2 \cap \tau_2 = 1_{Q_2}$ *and* $\pi_1 \cap \tau_1 = 1_{Q_1}$.

It is clear that τ is also orthogonal.

Example 5. Consider the *ts* $\mathcal{A} := (Q_1, Q_2, S)$ of Example 2. Define a partition π_2 on Q_2 as $\pi_2 := \{\{1\}, \{2, 3, 4\}, \{5, 6, 7\}\}$. π_2 is an admissible partition on Q_2 in the *ts* (Q_2, S). The semigroup $S/\sim = \{\{s_1, s_2\}, \{s_3, s_4, s_5, s_6\}, \{s_7\}\}$ and $\pi_1 = \{\{1\}, \{5, 6, 7\}\}$. For $[s_1] \in S/\sim$, $\pi_1 * [s_1] = \pi_1$, while $\pi_1 * [s_i] = \{\{1\}\} \subseteq \pi_1$ for $i = 3, 7$. Therefore $\pi := (\pi_1, \pi_2)$ is an admissible partition on (Q_1, Q_2) in $\mathcal{A}$. Another admissible partition on (Q_1, Q_2) in $\mathcal{A}$ is $\tau := (\tau_1, \tau_2)$, where $\tau_1 = \{\{1, 7\}, \{5\}, \{6\}\}$ and $\tau_2 := \{\{1, 7\}, \{2\}, \{3\}, \{4\}, \{5\}, \{6\}\}$. Then $\pi_2 \cap \tau_2 = 1_{Q_2}$ and $\pi_1 \cap \tau_1 = 1_{Q_1}$. Therefore π is an orthogonal partition on (Q_1, Q_2) in $\mathcal{A}$.

We now come to the last definition in this work. If $\mathcal{A}$ and $\mathcal{B}$ are *ts* for rough sets, a covering of $\mathcal{A}$ by $\mathcal{B}$ should result in two coverings (cf. Definition 3): one of upper *ts* of $\mathcal{A}$ by upper *ts* of $\mathcal{B}$ and another of lower *ts* of $\mathcal{A}$ by lower *ts* of $\mathcal{B}$.

Definition 13 (Coverings). *A ts* $\mathcal{A} := (Q_1, Q_2, S)$ *is covered by ts* $\mathcal{B} := (P_1, P_2, T)$, *written as* $\mathcal{A} \preccurlyeq \mathcal{B}$, *if there exists a surjective partial morphism* $\eta : P_2 \to Q_2$ *such that*

(a) η *restricts* P_1 *onto* Q_1, *that is* $\eta(P_1) = Q_1$, *and*
(b) η *is a covering of* (Q_2, S) *by* (P_2, T).

It is then straightforward to show that

Proposition 5. *If* η *is a covering of* $\mathcal{A} := (Q_1, Q_2, S)$ *by* $\mathcal{B} := (P_1, P_2, T)$, $\eta|_{P_1}$ *is a covering of the lower ts* $(Q_1, S/\sim)$ *of* $\mathcal{A}$ *by the lower ts* $(P_1, T/\sim)$ *of* $\mathcal{B}$.

The following results on coverings can be obtained, and will be helpful in the study of decomposition theorems of *ts* for rough sets. We omit the proofs, as the required coverings are not difficult to obtain.

Proposition 6. *Let* $\mathcal{A}, \mathcal{B}, \mathcal{C}, \mathcal{D}$ *be ts for rough sets.*

(a) $\mathcal{A} \preccurlyeq \overline{\mathcal{A}} = \overline{(\underline{\mathcal{A}})}$.
(b) *If* $\mathcal{A} \preccurlyeq \mathcal{B}$ *then* $\overline{\mathcal{A}} \preccurlyeq \overline{\mathcal{B}}$.
(c) *If* $\mathcal{A} \preccurlyeq \mathcal{C}$ *and* $\mathcal{B} \preccurlyeq \mathcal{D}$ *then* $\mathcal{A} \times \mathcal{B} \preccurlyeq \mathcal{C} \times \mathcal{D}$.
(d) *If* $\mathcal{A} \preccurlyeq \mathcal{B}$ *and* $\mathcal{B} \preccurlyeq \mathcal{C}$ *then* $\mathcal{A} \preccurlyeq \mathcal{C}$.

Proposition 7. *A reset ts* $\mathcal{A} := (Q_1, Q_2, S)$ *is covered by the reset ts* $\overline{(Q_1, Q_2, \emptyset)}$.

Proof. The covering η will be 1_{Q_2}, and then we argue for the two cases obtained by Observation 4: (1) $Q_1 = \emptyset$ or $Q_1 = Q_2$, and (2) $Q_1 \neq \emptyset$, $Q_1 \neq Q_2$.   □

5   Decomposition Theorems

In *ts* theory, a ('useful') 'decomposition' of a *ts* $\mathcal{A}$ is a covering of $\mathcal{A}$ by products of some $\overline{\mathcal{A}_i}$'s where each $\mathcal{A}_i$ is 'smaller' than $\mathcal{A}$ – in terms of cardinality of components in the pairs constituting the *ts*. Products involved in the decomposition may not always be direct products; there are other products defined on *ts*, e.g. wreath or cascade products. Our goal is to study decomposition results of the above kind in the case of a *ts* $\mathcal{A} := (Q_1, Q_2, S)$ for rough sets. So we shall look for coverings of $\mathcal{A}$ by products of the closure of non-decomposable and smaller *ts* $\mathcal{A}_i := (Q_{1i}, Q_{2i}, S_i)$, $i \in I$. We present our first result in this direction here, for the special case of reset *ts*. In *ts* theory, the decomposition result obtained for reset *ts* is the following. $\prod^k \mathcal{A}_i$ denotes the direct product of $\mathcal{A}_i$, $i = 1, \ldots, k$.

Proposition 8 [7]. *Any reset ts can be covered by* $\prod^k \overline{\mathbf{2}}$.

In the case of reset *ts* for rough sets, we prove

Theorem 2. *For a reset ts* $\mathcal{A} := (Q_1, Q_2, S)$ *for rough set* (Q_1, Q_2) *with* $|Q_2| = n$, $|Q_1| = m \geq 2$ *and* $Q_1 \neq Q_2$, *we have*

$$(Q_1, Q_2, S) \preccurlyeq \overline{(1, n - m + 1, \emptyset)} \times \prod^{m-1} \overline{(2, 2, \emptyset)}.$$

Proof. Since $|Q_2| = n$ is finite, let us enumerate the elements $\{q_i\}_{i=1}^n$ of Q_2 such that the first m elements belong to $Q_1 = \{q_i\}_{i=1}^m$.

Using Proposition 7, $\mathcal{A} \preccurlyeq \mathcal{B} := (Q_1, Q_2, \emptyset)$. Therefore we shall focus on the reset $\mathcal{B}$. By Example 4, $\mathcal{B} := (Q_1, Q_2, S')$ where $S' = \{\overline{q}_i \mid q_i \in Q_1\} \cup \{\widetilde{q}_i \mid q_i \in Q_2\} \cup \{\emptyset_{Q_2}\}$.

Case 1: $|Q_1| = 2$ and $|Q_2| > 2$. We have $Q_1 = \{q_1, q_2\}$. Define the following partitions on Q_2.

$$\pi_2 := \{ \{q_1\}, Q_2 \setminus \{q_1\} \}$$
$$\tau_2 := \{ \{q_1, q_2\}, \{q_3\}, \{q_4\}, \{q_5\}, \ldots, \{q_n\} \}$$

Then $\pi := (\pi_1, \pi_2)$, where $\pi_1 = \pi_2 = \{H_i \in \pi_2 \mid H_i \cap Q_1 \neq \emptyset\}$, is an admissible partition on (Q_1, Q_2). The semigroup in $\mathcal{B}/\langle \pi \rangle := (\pi_1, \pi_2, S'/\sim_{\pi_2})$ is

$$S'/\sim_{\pi_2} = \{ \{\overline{q}_1\}, \{\overline{q}_2\}, \{\widetilde{q}_1\}, \{\widetilde{q}_i \mid 2 \leq i \leq n\}, \{\emptyset_{Q_2}\} \},$$

and the reset $(\pi_1, \pi_2, S'/\sim_{\pi_2}) \preccurlyeq \overline{(2, 2, \emptyset)}$, using Proposition 7. The semigroup in the quotient $\mathcal{B}/\langle \tau \rangle := (\tau_1, \tau_2, S'/\sim_{\tau_2})$ is

$$S'/\sim_{\tau_2} = \{ \{\overline{q}_1, \overline{q}_2\}, \{\widetilde{q}_1, \widetilde{q}_2\}, \{\widetilde{q}_i\}_{i=3}^n, \{\emptyset_{Q_2}\} \}.$$

Further, $\tau_1 = \{K_i \in \tau_2 \mid K_i \cap Q_1 \neq \emptyset\} = \{\{q_1, q_2\}\}$ and $\tau_1 * [s] \subseteq \tau_1$ for all $[s] \in S'/\sim_{\tau_2}$. Thus, $\tau := (\tau_1, \tau_2)$ is also an admissible partition on (Q_1, Q_2). In fact π and τ are orthogonal admissible partitions on (Q_1, Q_2) because $\pi_2 \cap \tau_2 = 1_{Q_2}$, and $\pi_1 \cap \tau_1 = 1_{Q_1}$.

We claim that $\mathcal{B} \preccurlyeq \mathcal{B}/\langle \pi \rangle \times \mathcal{B}/\langle \tau \rangle$. Define the map $\eta : \pi_2 \times \tau_2 \rightarrow Q_2$ as follows: For $H_i \in \pi_2$ and $K_j \in \tau_2$,

$$\eta(H_i, K_j) := q_k \text{ if } H_i \cap K_j = \{q_k\}, \text{ and}$$
$$\eta(H_i, K_j) \text{ is not defined if } H_i \cap K_j = \emptyset.$$

- η is well-defined and onto because τ_2 is orthogonal to π_2.
- η is a covering of (Q_2, S') by $(\pi_2, S'/\sim_{\pi_2}) \times (\tau_2, S'/\sim_{\tau_2})$, where
  - $\bar{q}_1 \in S'$ is covered by $([\bar{q}_1]_{\pi_2}, [\bar{q}_1]_{\tau_2}) \in S'/\sim_{\pi_2} \times S'/\sim_{\tau_2}$,
  - $\bar{q}_2 \in S'$ is covered by $([\bar{q}_2]_{\pi_2}, [\bar{q}_1]_{\tau_2}) \in S'/\sim_{\pi_2} \times S'/\sim_{\tau_2}$, and
  - $\tilde{q}_i \in S'$ is covered by $([\bar{q}_2]_{\pi_2}, [\tilde{q}_i]_{\tau_2}) \in S'/\sim_{\pi_2} \times S'/\sim_{\tau_2}$ for all $3 \le i \le n$.
- η restricts $\pi_1 \times \tau_1$ to Q_1.

Therefore η is a covering of $\mathcal{B}$ by $\mathcal{B}/\langle\pi\rangle \times \mathcal{B}/\langle\tau\rangle$. Observe the following for the quotient ts $\mathcal{B}/\langle\tau\rangle$ for rough set (τ_1, τ_2).

1. $|\tau_1| = 1$, $|\tau_2| = |Q_2| - 1$, and
2. $\mathcal{B}/\langle\tau\rangle$ is again a reset ts for rough set, and can be covered by $\overline{(1, n-1, \emptyset)}$.

Thus we have, using Proposition 6(c) and (d),

$$\mathcal{A} \preccurlyeq \mathcal{B} \preccurlyeq \mathcal{B}/\langle\pi\rangle \times \mathcal{B}/\langle\tau\rangle \preccurlyeq \overline{(2,2,\emptyset)} \times \overline{(1, n-1, \emptyset)}.$$

<u>Case 2</u>: $|Q_1| > 2$. $\{q_1, q_2, q_3\} \subseteq Q_1$. We consider the following partitions on Q_2.

$$\pi_2 := \{ \{q_1, q_2\}, Q_2 \setminus \{q_1, q_2\} \}$$
$$\tau_2 := \{ \{q_1, q_3\}, \{q_2\}, \{q_4\}, \{q_5\}, \ldots, \{q_n\} \}$$

This results in orthogonal admissible partitions $\pi := (\pi_1, \pi_2)$ and $\tau := (\tau_1, \tau_2)$ on (Q_1, Q_2) and $\mathcal{B}/\langle\pi\rangle := (\pi_1, \pi_2, S'/\sim_{\pi_2}) \preccurlyeq \overline{(2, 2, \emptyset)}$ by Proposition 7. It can be shown as in Case 1 that $\mathcal{B} \preccurlyeq \mathcal{B}/\langle\pi\rangle \times \mathcal{B}/\langle\tau\rangle$. Moreover, $|\tau_1| = |Q_1| - 1$, $|\tau_2| = |Q_2| - 1$, and $\mathcal{B}/\langle\tau\rangle$ is a reset ts covered by $\overline{(m-1, n-1, \emptyset)}$. Thus as in Case 1, using Proposition 6(c) and (d), we get

$$\mathcal{A} \preccurlyeq \mathcal{B} \preccurlyeq \mathcal{B}/\langle\pi\rangle \times \mathcal{B}/\langle\tau\rangle \preccurlyeq \overline{(2,2,\emptyset)} \times \overline{(m-1, n-1, \emptyset)}.$$

By repeating the above process $m - 2$ times, we obtain the following decomposition:

$$\mathcal{A} \preccurlyeq \prod^{m-2} \overline{(2,2,\emptyset)} \times \overline{(2, n-m+2, \emptyset)}$$

Applying Case 1 on $\overline{(2, n-m+2, \emptyset)}$, we have

$$\mathcal{A} \preccurlyeq \prod^{m-1} \overline{(2,2,\emptyset)} \times \overline{(1, n-m+1, \emptyset)} \qquad \square$$

What about other cases for reset ts – when $Q_1 = Q_2$, or $|Q_1| = 0$, or $|Q_1| = 1$?

<u>Case 3</u>: $|Q_1| = |Q_2| = 2$. $(Q_1, Q_2, S) \preccurlyeq \overline{(2, 2, \emptyset)}$, by Proposition 7.

<u>Case 4</u>: $|Q_1| = |Q_2| \ne 2$. Consider the reset ts $\overline{(n, n, \emptyset)}$ with $n > 2$. The proof of its decomposition is similar to the proof of Case 2 in Theorem 2: define the sets π_2 and τ_2 as in the proof. We have $\tau_1 = \tau_2$, $\pi_1 = \pi_2$ and the partition

$\pi = (\pi_1, \pi_2)$ is admissible and orthogonal. Proceeding similarly as above and repeating the process $n - 2$ times, we obtain

$$\overline{(n, n, \emptyset)} \preccurlyeq \prod^{n-1} \overline{(2, 2, \emptyset)}$$

<u>Case 5</u>: $|Q_1| = 0$. For the reset ts $\overline{(0, n, \emptyset)}$ where $n > 2$, the proof of decomposition is again similar as that of Case 2 in Theorem 2, with the change that $\pi_1 = \tau_1 = \emptyset$. We get in this case

$$\overline{(0, n, \emptyset)} \preccurlyeq \prod^{n-1} \overline{(0, 2, \emptyset)}$$

Combining all the cases, we have the following.

Corollary 1. *Any reset ts $\mathcal{A} := (Q_1, Q_2, S)$ of the rough set (Q_1, Q_2) with $Q_2 \neq \emptyset$, can be covered by the direct product of the resets $\overline{(0, 2, \emptyset)}$, $\overline{(2, 2, \emptyset)}$ and $\overline{(1, n, \emptyset)}$.*

6 Rough Sets and Automata Theory

We now focus on connections of transformation semigroups with *semiautomata*, and how these could apply to the study here in the context of rough sets. Semiautomata are automata without outputs, defined in the following way [7]. Note that in literature, a semiautomaton is sometimes referred to as an 'automaton' or as a 'state machine'. Here, we shall use the term 'semiautomaton' only.

Definition 14 (Semiautomaton). *A semiautomaton is a triple $\mathcal{M} := (Q, \Sigma, \delta)$, where Q and Σ are finite sets, and $\delta : Q \times \Sigma \to Q$ is a partial function.*

A semiautomaton $\mathcal{M}$ can be associated with the free semigroup Σ^*, and the partial function δ can be extended to define a semigroup action of Σ^* on the set of states Q. So Σ^* can be seen as a collection of transformations of Q. The relation between semiautomata and transformation semigroups of finite sets is given as follows. Given any semiautomaton $\mathcal{M} := (Q, \Sigma, \delta)$, one can obtain a ts by forcing the action of the free semigroup Σ^* on Q to be faithful, as done in Observation 2(1) by defining a congruence relation $\sim$ on Σ^*. The pair $TS(\mathcal{M}) := (Q, \Sigma^*/\sim)$ forms a ts. Conversely, given a ts $\mathcal{A} := (Q, S)$, the triple $SM(\mathcal{A}) := (Q, S, \delta)$ is a semiautomaton, where δ is the semigroup action associated with the ts $\mathcal{A}$.

Definition 15 [7]. *For semiautomata (Q, Σ, δ) and (P, Λ, γ), consider the functions $\alpha : Q \to P$ and $\beta : \Sigma \to \Lambda$ such that if $\alpha(\delta(q, s))$ is defined then*

$$\alpha(\delta(q, s)) = \gamma(\alpha(q), \beta(s)) \text{ for any } q \in Q \text{ and } s \in \Sigma.$$

The pair (α, β) is called a semiautomaton homomorphism.

It can be shown that for a ts $\mathcal{A}$ and a semiautomaton $\mathcal{M}$, $TS(SM(\mathcal{A}))$ is isomorphic to $\mathcal{A}$, while there is a homomorphism from $\mathcal{M}$ to $SM(TS(\mathcal{M}))$.

6.1   Semiautomata for Rough Sets

As mentioned earlier, our aim is to see how the *ts* for rough sets that we study in this work, are related to an appropriate notion of semiautomaton that may be defined in the context of rough sets. We must use the concept of a 'subautomaton' for the purpose. Substructures of an automaton were first defined by Ginsburg [6], and studied extensively by others – the literature contains various definitions of subautomata depending on the applications. A discussion can be found in [11]. We consider the following.

Definition 16 (Subautomaton) [3]. $\mathcal{M}' := (Q', \Sigma, \nu)$ *is a* subautomaton *of the semiautomaton* $\mathcal{M} := (Q, \Sigma, \delta)$ *if* $Q' \subseteq Q$ *and* $\nu = \delta$ *on* $Q' \times \Sigma$.

This definition suits us here, as a natural relation with *ts* for rough sets in the lines described above (for *ts* and semiautomata) may be arrived at, if the input states constitute a rough set and Σ is fixed.

Definition 17 (Semiautomaton for a rough set).   *A* semiautomaton for a rough set (Q_1, Q_2) *is a quadruple* $\mathcal{M} := (Q_1, Q_2, \Sigma, \delta)$, *where* (Q_2, Σ, δ) *is a semiautomaton and* $Q_1 \subseteq Q_2$ *such that* $(Q_1, \Sigma, \delta \mid_{Q_1})$ *is a subautomaton of* (Q_2, Σ, δ).

Remark. Let us compare semiautomaton for rough sets with Basu's definition [2,14] of *rough semi-automaton*, and also compare *rough transformation semigroups* defined in [15] with the *ts* for rough sets considered in this work.

1. A rough semi-automaton generalizes the concept of a non-deterministic automaton, in which the transition function maps an input state to a set of input states. For the definition in [2], the set Q of input states has a partition R yielding an approximation space on Q. For a given state and an input symbol, the transition function gives an output that is a rough set on the approximation space (Q, R).

    In our case also, there is an underlying partition R of a set Q of states; for some subset X of Q in the approximation space (Q, R), Q_1, Q_2 may be taken respectively as the set of equivalence classes contained in X and the set of equivalence classes properly intersecting X. On any given input symbol, the transition function from (Q_1, Q_2) to (Q_1, Q_2) maps each equivalence class in Q_2 to an equivalence class in Q_2 such that classes in Q_1 remain in Q_1.
2. Rough transformation semigroups are derived from rough semi-automata [2], and thus involve transformations of the set Q into the collection of rough sets on the approximation space on Q. In contrast, if we consider the interpretation given above for *ts* for rough sets defined here, these structures are semigroups of transformations of the set Q_2 of equivalence classes to itself that also preserve the set Q_1.

Now to get the exact connection with *ts* for rough sets, we define homomorphisms. Recall Definition 15.

Definition 18 (Homomorphisms). *Let* $\mathcal{M} := (Q_1, Q_2, \Sigma, \delta)$ *and* $\mathcal{N} :=$ $(P_1, P_2, \Lambda, \gamma)$ *be two semiautomata for rough sets* (Q_1, Q_2) *and* (P_1, P_2) *respectively. The semiautomaton homomorphism* (α, β) *from* (Q_2, Σ, δ) *to* (P_2, Λ, γ) *such that* $\alpha(Q_1) \subseteq P_1$ *is called a* semiautomaton homomorphism for rough sets.

Let us now consider a *ts* $\mathcal{A} := (Q_1, Q_2, S)$ for a rough set (Q_1, Q_2). The tuple (Q_1, Q_2, S, δ) is a semiautomaton for rough set, denoted by $RSM(\mathcal{A})$, where δ is the partial semigroup action associated with the upper *ts* of $\mathcal{A}$. On the other hand, starting from a semiautomaton $\mathcal{M} := (Q_1, Q_2, \Sigma, \delta)$ for a rough set (Q_1, Q_2), we have $TS(Q_2, \Sigma, \delta) := (Q_2, \Sigma^*/\sim)$ as a *ts*. As $Q_1 \subseteq Q_2$ and $q[s] \in Q_1$ for all $q \in Q_1$, $s \in \Sigma^*$, $(Q_1, Q_2, \Sigma^*/\sim)$ forms a *ts* for rough set – it is denoted as $RTS(\mathcal{M})$.

Theorem 3. *Consider a ts* $\mathcal{A} := (Q_1, Q_2, S)$ *for a rough set* (Q_1, Q_2) *and a semiautomaton* $\mathcal{M} := (P_1, P_2, \Sigma, \delta)$ *for a rough set* (P_1, P_2). *The following results hold.*

(a) $RTS(RSM(\mathcal{A})) \cong \mathcal{A}$, *and*
(b) there exists a homomorphism from $\mathcal{M}$ *to* $RSM(RTS(\mathcal{M}))$.

Proof. *(a)* $RTS(RSM(\mathcal{A})) = RTS(Q_1, Q_2, S, \delta) = (Q_1, Q_2, S^*/\sim)$. The semigroup $S^*/\sim$ is isomorphic to S, because $S^* \cong S$ and the congruence relation $\sim$ is the identity. Thus $RTS(RSM(\mathcal{A})) \cong \mathcal{A}$.

(b) $RTS(\mathcal{M}) = (P_1, P_2, \Sigma^*/\sim)$ and $RSM(RTS(\mathcal{M})) = (P_1, P_2, \Sigma^*/\sim, \widetilde{\delta})$. Define the semiautomaton homomorphism $(1_{P_2}, \beta)$ from (P_1, Σ, δ) to $(P_2, \Sigma^*/\sim, \widetilde{\delta})$, where $\beta : \Sigma \to \Sigma^*/\sim$ maps s to $[s]$ for all $s \in \Sigma$. Since $1_{P_2}(P_1) \subseteq P_1$, we have the required semiautomaton homomorphism from $\mathcal{M}$ to $RSM(RTS(\mathcal{M}))$. $\square$

Due to Theorem 3, studying any one of semiautomata for rough sets or transformation semigroups for rough sets is enough to get similar results for the other. In particular, all the concepts defined in our work on *ts* for rough sets can be carried over to semiautomata for rough sets.

7   Conclusion

The theory of transformation semigroups has two strong motivations – one from semigroup theory, and other from automata theory. This work marks the beginning of a study of transformation semigroups for rough sets, that is distinct from the notion of rough transformation semigroups defined earlier by [15]. A goal is to obtain decomposition results; the work introduces some basic notions for the purpose, culminating in a decomposition theorem for reset *ts* for rough sets.

There are various other concepts such as wreath products, heights, admissible subset systems that can be the subject of further investigation, and one can try for a Krohn-Rhodes style decomposition or holonomy decomposition result.

The decomposition theorem for reset *ts* for rough sets presented here, differs from that for reset *ts* in that the basic entities in the decomposition are not just transformations of the 2-element set, rather n-element reset ts of type $\overline{(1, n, \emptyset)}$. We expect that these may not be further decomposable. If true, this will be a major deviation of the theory of ts for rough sets from ts theory.

In this work we have mainly focused on the algebraic side of the transformation semigroups or automata theory. However, an important goal of studying automata theory is to understand real world models. Rough sets have applications in various fields. It would be interesting to find some applications of automata theory, where the set of states are taken as rough sets. One particular application which seems promising is in cellular automata.

Acknowledgments. We are grateful to the anonymous referees for their suggestions and valuable remarks.

References

1. Arbib, M.A., Krohn, K., Rhodes, J.L.: Algebraic Theory of Machines, Languages, and Semigroups. Academic Press, London (1968)
2. Basu, S.: Rough finite-state automata. Cybern. Syst. **36**(2), 107–124 (2005). https://doi.org/10.1080/01969720590887324
3. Bavel, Z.: The source as a tool in automata. Inf. Control **18**(2), 140–155 (1971). https://doi.org/10.1016/S0019-9958(71)90324-X
4. Clifford, A.H., Preston, G.B.: The Algebraic Theory of Semigroups. Volume II. American Mathematical Society (1961)
5. Eilenberg, S., Tilson, B.: Automata, Languages, and Machines. Volume B. Pure & Applied Mathematics, vol. B. Academic Press, New York (1976)
6. Ginsburg, S.: Some remarks on abstract machines. Trans. Am. Math. Soc. **96**(3), 400–444 (1960). https://doi.org/10.1090/S0002-9947-60-99988-8
7. Holcombe, W.M.L.: Algebraic Automata Theory. Cambridge University Press, New York (1982)
8. Iwiński, T.B.: Algebraic approach to rough sets. Bull. Polish Acad. Sci. Math. **35**, 673–683 (1987)
9. Li, X.S., Yuan, X.H.: The category RSC of I-rough sets. In: Fifth International Conference on Fuzzy Systems and Knowledge Discovery, vol. 1, pp. 448–452, October 2008. https://doi.org/10.1109/FSKD.2008.106
10. Linton, S.A., Pfeiffer, G., Robertson, E.F., Ruškuc, N.: Groups and actions in transformation semigroups. Math. Z. **228**(3), 435–450 (1998). https://doi.org/10.1007/PL00004628
11. Mikolajczak, B.: Algebraic and Structural Automata Theory. Annals of Discrete Mathematics. North-Holland, Amsterdam (1991)
12. More, A.K., Banerjee, M.: Categories and algebras from rough sets: new facets. Fundam. Inf. **148**(1–2), 173–190 (2016). https://doi.org/10.3233/FI-2016-1429
13. Pawlak, Z.: Rough sets. Int. J. Comput. Inform. Sci. **11**(5), 341–356 (1982). https://doi.org/10.1007/BF01001956
14. Sharan, S., Srivastava, A.K., Tiwari, S.P.: Characterizations of rough finite state automata. Int. J. Mach. Learn. Cybernet. **8**(3), 721–730 (2017). https://doi.org/10.1007/s13042-015-0372-3
15. Tiwari, S.P., Sharan, S., Singh, A.K.: On coverings of products of rough transformation semigroups. Int. J. Found. Comput. Sci. **24**(03), 375–391 (2013). https://doi.org/10.1142/S0129054113500093

A Sequential Three-Way Approach to Constructing a Co-association Matrix in Consensus Clustering

Mengjun Hu$^{(\boxtimes)}$, Xiaofei Deng, and Yiyu Yao

Department of Computer Science, University of Regina,
Regina, SK S4S 0A2, Canada
{hu258,deng200x,yyao}@cs.uregina.ca

Abstract. The main task in consensus clustering is to produce an optimal output clustering based on a set of input clusterings. The co-association matrix based consensus clustering methods are easy to understand and implement. However, they usually have high computational cost with big datasets, which restricts their applications. We propose a sequential three-way approach to constructing the co-association matrix progressively in multiple stages. In each stage, based on a set of input clusterings, we evaluate how likely two data points are associated and accordingly, divide a set of data-point pairs into three disjoint positive, negative and boundary regions. A data-point pair in the positive region is associated with a definite decision of clustering the two data points together. A pair in the negative region is associated with a definite decision of separating the two data points into different clusters. For a pair in the boundary region, we do not have sufficient information to make a definite decision. The decision on such a pair is deferred into the next stage where more input clusterings will be involved. By making quick decisions on early stages, the overall computational cost of constructing the matrix and the consensus clustering may be reduced.

Keywords: Sequential three-way decision · Consensus clustering
Co-association matrix

1 Introduction

Given a set of data points described by a set of attributes or features, the main task of clustering is to divide these data points into groups such that the data points in the same group are as similar as possible and those in different groups are as dissimilar as possible. Each group is called a cluster, and the family of all groups is called a clustering. The results of some popular clustering methods [2,4,5,8,16] depend on their initial configurations that involve a priori parameters such as a given number of clusters. In order to improve the robustness and accuracy, these methods are usually repeatedly applied with different

This work is partially supported by a Discovery Grant from NSERC, Canada.

H. S. Nguyen et al. (Eds.): IJCRS 2018, LNAI 11103, pp. 599–613, 2018.
https://doi.org/10.1007/978-3-319-99368-3_47

initial configurations. The family of produced clusterings are then combined into a single clustering via consensus clustering. This is one of the main motivations for consensus clustering that produces a final clustering by synthesizing a set of input clusterings.

The consensus clustering methods based on co-association matrix [6,7,12, 13,21,22] are very popular and well studied in the literature. The first step in the main procedure is to synthesize the set of input clusterings into an $n \times n$ co-association matrix where n is the total number of data points. The values in the matrix reflect how likely the corresponding two data points are clustered together in the input clusterings. The second step is to obtain the final clustering by applying a basic clustering method to the matrix. These consensus clustering methods are easy to understand and implement. However, since they focus on all data-point pairs when constructing the matrix, they usually have high computational cost when applied to large datasets, which restricts their applications.

The consensus clustering can be viewed as a decision making process. In the co-association matrix based methods, we make decisions of whether to cluster two data points together or not based on the information provided by input clusterings. The theory of three-way decisions [23] offers a framework of decision making by dividing a set of objects into three disjoint decision regions according to some criterion. Each region is associated with a specific decision. Generally, the three regions include the positive, negative and boundary regions. The objects in the positive region are associated with an acceptance decision, that is, we accept that these objects satisfy the criterion. The objects in the negative region are associated with a rejection decision, that is, we decide that these objects do not satisfy the criterion. Those in the boundary region cannot be definitely determined to satisfy the criterion or not. They are associated with a third non-commitment decision due to the uncertainty. The theory of three-way decisions has been applied to basic clustering methods by researchers [27–30].

The sequential three-way decision model [26] iteratively applies the three-way decision model to refine the boundary region and reduce the uncertainty. Definite decisions (i.e., acceptance and rejection) are made on objects in each stage if sufficient information is available. Otherwise, the decision on the objects will be postponed into the next stage where more detailed and sufficient information will be involved. It has been applied to many real-world applications such as face recognition in [14,15]. Four modes of sequential three-way decisions are examined in [26], including multiple levels of granularity, probabilistic rough set theory, multiple models of classification, and ensemble classifications. Our presented approach in this paper follows a similar mode as ensemble classifications.

The presented approach integrates the sequential three-way decision model into the construction of a co-association matrix. In each stage, based on a set of input clusterings, we put a data-point pair into a positive region if the corresponding value in the matrix is high enough or into a negative region if the value is low enough. The corresponding entry in the matrix is then updated with the largest value 1 or the smallest value 0, respectively. Otherwise, the pair is put into a third boundary region and the corresponding entry is to be determined in the next stage that involves more input clusterings. In this way, we determine the entries in the matrix and correspondingly, make quick decisions on the clustering of some data points in early stages. As a result, we may be able to reduce the overall computational cost of constructing the matrix.

The remaining part of this paper is arranged as follows. Section 2 reviews consensus clustering methods based on co-association matrix. The sequential three-way approach to constructing the matrix is presented in Sect. 3. Section 4 shows the experimental results. Section 5 concludes the paper and discusses possible directions for the future work.

2   A Review of Co-association Matrix Based Consensus Clustering Methods

The main task of consensus clustering is to combine different clusterings of a dataset into one single clustering, usually without referring to the original features or attributes of the data points. A general framework of consensus clustering includes two steps [20], namely, the Generation and Consensus steps. The Generation step generates the set of input clusterings for a given dataset. They can be produced by different basic clustering methods or multiple applications of the same method with different parameters. The Consensus step combines the input clusterings into a final consensus clustering according to a particular consensus function.

A co-association matrix based method includes two steps in the main procedure. The first step is to synthesize the input clusterings into an intermediate representation called a co-association matrix. Each entry in the matrix measures how many times the two corresponding data points are associated or clustered together in the input clusterings. The second step is to get the final consensus clustering by applying a basic clustering method to the matrix.

Suppose $X = \{x_1, x_2, \cdots, x_n\}$ is a given dataset and $\mathbb{C}^{in} = \{\mathcal{C}_1, \mathcal{C}_2, \cdots, \mathcal{C}_m\}$ is a set of input clusterings on X. In a co-association matrix based method, an input clustering $\mathcal{C}_k (1 \leq k \leq m)$ is commonly represented by an $n \times n$ matrix. Moreover, the input clusterings are widely assumed to be hard clusterings where a data point belongs to exactly one cluster. Thus, the entries in a matrix $\mathcal{C}_k (1 \leq k \leq m)$ are formally defined as: for $1 \leq i \leq n$ and $1 \leq j \leq n$,

$$\mathcal{C}_k(i, j) = \begin{cases} 1, & \text{if } x_i \text{ and } x_j \text{ are clustered together,} \\ 0, & \text{otherwise.} \end{cases} \tag{1}$$

Based on the set $\mathbb{C}^{in}$, a simple way to construct the co-association matrix $M_{n \times n}$ is to use the proportion of input clusterings where the two corresponding data points are associated, which is the evidence accumulation framework proposed in [7]. Accordingly, M is constructed as: for $1 \leq i \leq n$ and $1 \leq j \leq n$,

$$M(i,j) = \frac{1}{m} \sum_{k=1}^{m} \mathcal{C}_k(i,j). \tag{2}$$

More complex measures are proposed to construct the matrix by taking into account more information. The Connected-Triple based Similarity (CTS) and SimRank based Similarity (SRS) [12] consider the transitivity property of clustering data points. A Weighted Co-Association Matrix is presented in [21] which takes into consideration the size of the clusters containing the two data points and the total number of clusters in the corresponding input clustering. The Probability Accumulation Matrix [22] considers the size of the clusters containing the two data points and the number of attributes used to describe the data points.

To cluster the data points based on the co-association matrix, two hierarchical clustering methods are proposed in [7,13]. A graph based method proposed in [19] generates a similarity graph based on the matrix and obtains the final clustering by partitioning the graph. Two threshold based methods are presented in [6,7].

The co-association matrix based methods are advantageous in several aspects. They use the co-association idea to avoid the labeling correspondence problem which is a common difficulty in some popular categories of current consensus clustering methods. For instance, in the relabeling and voting based methods [20], the first step is to relabel the input clusters in all the input clusterings where the labeling correspondence problem needs to be solved in order to find the correspondence between clusters in different clusterings. The labeling correspondence problem can only be solved, with certain accuracy, when the input clusterings have the same number of clusters, which is a very restrictive condition in these methods. Besides, the co-association matrix based methods are easy to understand and implement since the constructions of the matrix and the basic clustering methods are usually quite intuitive. However, since they need to compute the value for each data-point pair to construct the co-association matrix, they usually have high computational cost with big datasets, which restricts their applications.

3   A Sequential Three-Way Approach to Constructing a Co-association Matrix

Based on a general framework of sequential three-way decisions proposed in [26], we present a sequential three-way approach to progressively constructing a co-association matrix in multiple stages.

3.1   An (α, β)-cut of a Co-association Matrix

The values in a co-association matrix quantitatively evaluate how likely two data points are clustered together. In order to decide whether two data points should be clustered together in the final clustering, it may be sufficient to qualitatively know whether they are likely enough to be associated, that is, whether the corresponding value in the matrix is large enough. Similarly, to decide whether they should be separated into different clusters, a qualitatively small enough value may be sufficient. Based on this idea, we can use a pair of thresholds to cut the values and divide the data-point pairs into three decision regions. The matrix is then updated by assigning different values to the pairs in different regions.

Suppose (α, β) is a pair of thresholds with $0 \leq \beta < \alpha \leq 1$ and $eval : X \times X \rightarrow [0, 1]$ is a measure to evaluate how likely two data points are associated based on a set of input clusterings (e.g., Eq. (2)). By using the pair (α, β) to cut the evaluation values, the set of data-point pairs $\mathbb{X} = X \times X$ is divided into three disjoint positive POS, negative NEG and boundary BND regions:

$$\text{POS}(\mathbb{X}) = \{(x_i, x_j) \in \mathbb{X} \mid eval(x_i, x_j) \geq \alpha\},$$
$$\text{NEG}(\mathbb{X}) = \{(x_i, x_j) \in \mathbb{X} \mid eval(x_i, x_j) \leq \beta\},$$
$$\text{BND}(\mathbb{X}) = \{(x_i, x_j) \in \mathbb{X} \mid \beta < eval(x_i, x_j) < \alpha\}. \tag{3}$$

The entries in the co-association matrix $M_{n \times n}$ are accordingly determined as:

(M^P) If $(x_i, x_j) \in \text{POS}(\mathbb{X})$, then $M(i, j) = 1$,
(M^N) If $(x_i, x_j) \in \text{NEG}(\mathbb{X})$, then $M(i, j) = 0$,
(M^B) If $(x_i, x_j) \in \text{BND}(\mathbb{X})$, then $M(i, j) = eval(x_i, x_j)$ or a constant value $v \in (0, 1)$.

As a result, for two data points x_i and x_j, if their evaluation value $eval(x_i, x_j)$ is high enough to indicate that they are associated (i.e., $eval(x_i, x_j) \geq \alpha$), then we cluster them together by assigning the largest evaluation value 1 to the entry $M(i, j)$. If the evaluation value is low enough to indicate that they are not associated (i.e., $eval(x_i, x_j) \leq \beta$), then we separate them into different clusters by assigning the smallest evaluation value 0 to the entry $M(i, j)$. Otherwise, we cannot make a definite decision due to insufficient information. The entry $M(i, j)$ may take the original evaluation value or a default constant value $v \in (0, 1)$ such as 0.5.

3.2   An l-stage Sequential Three-Way Approach to Constructing a Co-association Matrix

In the (α, β)-cut discussed in the previous subsection, a definite decision cannot be made on the data-point pairs in the boundary region due to insufficient information provided by the input clusterings. By involving more input clusterings, we may be able to refine the boundary region, which results in a sequential three-way approach to constructing a co-association matrix.

Suppose we have the following sequence of sets of input clusterings:

$$\mathbb{C}_1^{in} \subsetneq \mathbb{C}_2^{in} \subsetneq \cdots \subsetneq \mathbb{C}_l^{in}. \tag{4}$$

The proper subset relationship $\mathbb{C}_k^{in} \subsetneq \mathbb{C}_{k+1}^{in} (1 \leq k < l)$ ensures that $\mathbb{C}_{k+1}^{in}$ contains at least one more input clustering than $\mathbb{C}_k^{in}$, which gives more information about the clustering of data points. By using these sets one by one, we can obtain an l-stage sequential three-way approach to constructing the co-association matrix. Suppose X is the given dataset and $\mathbb{X}_k$ is the set of data-point pairs considered in the kth stage. The three regions in the kth stage are constructed as: let $\mathbb{X}_1 = X \times X$ and $\mathbb{X}_k = \text{BND}_{k-1}(\mathbb{X}_{k-1})(1 < k \leq l)$,

$$\text{POS}_k(\mathbb{X}_k) = \{(x_i, x_j) \in \mathbb{X}_k \mid eval(x_i, x_j|\mathbb{C}_k^{in}) \geq \alpha_k\},$$
$$\text{NEG}_k(\mathbb{X}_k) = \{(x_i, x_j) \in \mathbb{X}_k \mid eval(x_i, x_j|\mathbb{C}_k^{in}) \leq \beta_k\},$$
$$\text{BND}_k(\mathbb{X}_k) = \{(x_i, x_j) \in \mathbb{X}_k \mid \beta_k < eval(x_i, x_j|\mathbb{C}_k^{in}) < \alpha_k\}, \tag{5}$$

where $eval(x_i, x_j|\mathbb{C}_k^{in})$ is the evaluation value of x_i and x_j calculated based on the set $\mathbb{C}_k^{in}$, and the thresholds satisfy the condition $0 \leq \beta_k < \alpha_k \leq 1$. Accordingly, the entries in the co-association matrix $M_{n \times n}$ are determined as follows:

(M_k^{P}) If $(x_i, x_j) \in \text{POS}_k(\mathbb{X}_k)$, then $M(i, j) = 1$,
(M_k^{N}) If $(x_i, x_j) \in \text{NEG}_k(\mathbb{X}_k)$, then $M(i, j) = 0$,
(M_k^{B}) If $(x_i, x_j) \in \text{BND}_k(\mathbb{X}_k)$, then $M(i, j) = eval(x_i, x_j|\mathbb{C}_k^{in})$.

One may take special actions to deal with a nonempty final boundary region $\text{BND}_l(\mathbb{X}_l)$ instead of using the original evaluation values. For example, one may use a two-way process with a threshold r (e.g., 0.5) to clean up the boundary region or use a fixed value (e.g., 0.5) to replace the original evaluation values.

There are several assumptions in the above sequential three-way approach. Firstly, it is assumed that we are more biased towards putting the data-point pairs into the boundary region in an early stage where limited information is available. It leads to the relationships of all the thresholds [25]: $0 \leq \beta_1 \leq \beta_2 \leq \cdots \leq \beta_l < \alpha_l \leq \alpha_{l-1} \leq \cdots \leq \alpha_1 \leq 1$. By using a more restrictive pair of thresholds in an early stage, a data-point pair is more likely to be put into the boundary region, which indicates a more conservative opinion due to limited information. A third assumption is that we do not go back to update the positive and negative regions constructed in earlier stages. In other words, the definite decisions associated with these regions are not updated although they might be inappropriate when more input clusterings are available in some stage later on. Consequently, in each stage, we only focus on refining the boundary region constructed in the previous stage.

Example 1. We illustrate the construction of a co-association matrix by the presented approach. Suppose the data set is $X = \{o_1, o_2, o_3, o_4, o_5, o_6, o_7, o_8, o_9, o_{10}\}$. The set $\mathbb{C}^{in}$ of all input clusterings on X includes the following ten clusterings:

$$\mathcal{C}_1 = \{\{o_1, o_2, o_8\}, \{o_3, o_9, o_{10}\}, \{o_4, o_6, o_7\}, \{o_5\}\},$$
$$\mathcal{C}_2 = \{\{o_1, o_4, o_6\}, \{o_2, o_5, o_8\}, \{o_3, o_7, o_9, o_{10}\}\},$$
$$\mathcal{C}_3 = \{\{o_1, o_4, o_6\}, \{o_2, o_8\}, \{o_3, o_5, o_9, o_{10}\}, \{o_7\}\},$$
$$\mathcal{C}_4 = \{\{o_1, o_2, o_7, o_8\}, \{o_3, o_5, o_9, o_{10}\}, \{o_4, o_6\}\},$$
$$\mathcal{C}_5 = \{\{o_1, o_2, o_7, o_8\}, \{o_3, o_9, o_{10}\}, \{o_4, o_5, o_6\}\},$$
$$\mathcal{C}_6 = \{\{o_1, o_4, o_6\}, \{o_2, o_3, o_5, o_9\}, \{o_7\}, \{o_8, o_{10}\}\},$$
$$\mathcal{C}_7 = \{\{o_1, o_4, o_6, o_7\}, \{o_2, o_3, o_8\}, \{o_5, o_9, o_{10}\}\},$$
$$\mathcal{C}_8 = \{\{o_1, o_3, o_7, o_9, o_{10}\}, \{o_2, o_8\}, \{o_4, o_6\}, \{o_5\}\},$$
$$\mathcal{C}_9 = \{\{o_1, o_2, o_4\}, \{o_3, o_5, o_9, o_{10}\}, \{o_6, o_7, o_8\}\},$$
$$\mathcal{C}_{10} = \{\{o_1, o_4, o_6\}, \{o_2, o_7, o_8\}, \{o_3, o_5, o_9, o_{10}\}\}.$$

We use Eq. (2) to calculate the evaluation values, which is a symmetric measure. Thus, we need to compute the entries in the top right half of the matrix, not including the diagonal line. Suppose $\mathbb{C}_1^{in} = \{\mathcal{C}_1, \mathcal{C}_2, \mathcal{C}_3, \mathcal{C}_4, \mathcal{C}_5, \mathcal{C}_6\}$. The evaluation values are given in Table 1(a). By using thresholds $(1, 0)$, the entries with grey background are in the boundary region and the remaining entries are in either the positive or negative region. In stage 2, $\mathbb{C}_2^{in} = \mathbb{C}_1^{in} \cup \{\mathcal{C}_7\}$. The evaluation values for the previous boundary region are modified and given in Table 1(b). By using thresholds $(0.9, 0.1)$, the previous boundary region stays the same. In stage 3, $\mathbb{C}_3^{in} = \mathbb{C}_2^{in} \cup \{\mathcal{C}_8\}$ and the evaluation values are given in Table 1(c). By using thresholds $(0.8, 0.2)$, some entries in the previous boundary region are moved to either the positive or negative region and the corresponding values in the matrix are changed to either 1 or 0. This process goes on with stage 4 using $\mathbb{C}_4^{in} = \mathbb{C}_3^{in} \cup \{\mathcal{C}_9\}$ and thresholds $(0.7, 0.3)$ and stage 5 using $\mathbb{C}_5^{in} = \mathbb{C}^{in}$ and thresholds $(0.6, 0.4)$. If we do not allow overlap between clusters (i.e., we consider the hard clusterings) and assume that two data points are clustered together if they are both clustered together with a third data point, then the nonempty boundary region in stage 5 can be cleaned up and the final consensus clustering is $\{\{o_1, o_4, o_6\}, \{o_2, o_8\}, \{o_3, o_5, o_9, o_{10}\}\}$.

3.3   Two Issues in the Presented Approach

The first issue in the presented sequential three-way approach is to avoid an easy agreement on a definite decision in early stages where we have limited input clusterings. In other words, the data-point pairs should be less likely to be put into the positive and negative regions in early stages. There are at least two possible solutions to this issue. One solution is to use very restrictive thresholds in early stages, such as $(1, 0)$ in the first few stages. Another solution is to carefully select the input clusterings used in an early stage so that it is not easy

Table 1. The construction of a co-association matrix in Example 1

	o1	o2	o3	o4	o5	o6	o7	o8	o9	o10
o1	3/6	0	3/6	0	3/6	2/6	3/6	0	0	0
o2		1/6	0	2/6	0	2/6	5/6	1/6	0	
o3			0	3/6	0	1/6	0	1		5/6
o4				1/6	1	1/6	0	0	0	0
o5					1/6	0	1/6	3/6	2/6	
o6						1/6	0	0	0	
o7								2/6	1/6	1/6
o8								0		1/6
o9										5/6
o10										

(a) Stage 1: $\mathbb{C}^{in}_1 = \{\mathcal{C}_1, \mathcal{C}_2, \mathcal{C}_3, \mathcal{C}_4, \mathcal{C}_5, \mathcal{C}_6\}$

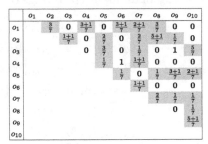

	o1	o2	o3	o4	o5	o6	o7	o8	o9	o10
o1	3/7	0	(3+1)/7	0	(3+1)/7	(2+1)/7	3/7	0	0	0
o2		(1+1)/7	0	2/7	0	2/7	(5+1)/7	1/7	0	
o3			0	3/7	0	1/7	0	1		5/7
o4				1/7	1					
o5					1/7	0	1/7	(3+1)/7	(2+1)/7	
o6						(1+1)/7	0	0	0	
o7								2/7	1/7	1/7
o8								0		1/7
o9										(5+1)/7
o10										

(b) Stage 2: $\mathbb{C}^{in}_2 = \mathbb{C}^{in}_1 \cup \{\mathcal{C}_7\}$

	o1	o2	o3	o4	o5	o6	o7	o8	o9	o10
o1		3/8	0	4/8	0	4/8	(3+1)/8	3/8	0	0
o2			2/8	0	2/8	0	(6+1)/8 ⇒1	1/8 ⇒0	0	
o3				0	3/8	0	(1+1)/8	0	1	(5+1)/8
o4					1/8 ⇒0	1	2/8	0	0	0
o5						1/8 ⇒0	0	1/8 ⇒0	4/8	3/8
o6							2/8	0	0	0
o7								2/8	(1+1)/8	(1+1)/8
o8									0	1/8 ⇒0
o9										(6+1)/8 ⇒1
o10										

(c) Stage 3: $\mathbb{C}^{in}_3 = \mathbb{C}^{in}_2 \cup \{\mathcal{C}_8\}$

	o1	o2	o3	o4	o5	o6	o7	o8	o9	o10
o1		(3+1)/9	0	(4+1)/9	0	4/9	2/9 ⇒0	3/9	0	0
o2			2/9 ⇒0	0	2/9 ⇒0	0	2/9 ⇒0	1	0	0
o3				0	(3+1)/9	0	2/9 ⇒0	0	1	(6+1)/9 ⇒1
o4					0	1	2/9 ⇒0	0	0	0
o5						0	0	0	(4+1)/9	(3+1)/9
o6							(2+1)/9	0	0	0
o7								(2+1)/9	2/9 ⇒0	2/9 ⇒0
o8									0	0
o9										1
o10										

(d) Stage 4: $\mathbb{C}^{in}_4 = \mathbb{C}^{in}_3 \cup \{\mathcal{C}_9\}$

	o1	o2	o3	o4	o5	o6	o7	o8	o9	o10
o1		4/10 ⇒0	0	(5+1)/10 ⇒1	0	(4+1)/10	4/10 ⇒0	3/10 ⇒0	0	0
o2			0	0	0	0	0	1	0	0
o3				0	(4+1)/10	0	0	0	1	1
o4					0	1	0	0	0	0
o5						0	0	0	(5+1)/10 ⇒1	(4+1)/10
o6							3/10 ⇒0	0	0	0
o7								(3+1)/10 ⇒0	0	0
o8									0	0
o9										1
o10										

(e) Stage 5: $\mathbb{C}^{in}_5 = \mathbb{C}^{in}$

for them to agree on a definite decision. This involves the determination of a proper total number of input clusterings and the selection of the basic clustering methods to generate the input clusterings. Intuitively, the group of input clusterings should be large enough since a small group is more likely to agree on a definite decision. The basic clustering methods that are used to generate the input clusterings should be as various as possible so that we can capture different views of clustering the data points. Repeated applications of the same method, such as k-means, are likely to produce similar clusterings although they start with different initial configurations. We should involve basic clustering methods in various categories, such as density-based clustering methods [5] that model clusters as areas with high density and EM algorithms [2] that model clusters as probability distributions.

The second issue is the determination of thresholds. The computation and interpretation of thresholds have been studied with respect to one-step three-way decisions, such as a probabilistic approach proposed in [24], a game-theoretic approach proposed in [9], and a decision-theoretic approach proposed in [3]. In order to apply these studies in the presented approach, we need to generalize the current methods with respect to the sequential case and the specific topic of consensus clustering.

These two issues can also be empirically solved by tuning related parameters in the experiments. For instance, one may use a fixed decreasing step and a fixed increasing step to update α and β in each stage. The two step lengths can be tuned though experiments to find the optimal lengths.

4   Experiments

The experiments are implemented using R Studio (IDE) based on Microsoft R Open 3.4.2. The implemented algorithm, which is called a Sequential THREE-Way algorithm to Consensus Clustering based on Co-Association Matrix (S3WCC-CAM), constructs a co-association matrix based on a set of input matrices representing the input clusterings and applies a hierarchical clustering method to generate the final clustering. The main procedure in S3WCC-CAM is given as follows.

Input:
- A set $\mathbb{C}^{in}$ of $n \times n$ matrices where n is the number of data points in the dataset. The values in these matrices are in the unit interval [0,1].
- A number m of input matrices to be used in the first iteration.
- A number $r(r \geq 1)$ used to refine the thresholds.

Output: A hierarchical final clustering $\mathcal{HC}$ of the dataset.

Step 1: Construct the co-association matrix $M_{n\times n}$.
   (1) Generate a sequence Seq of thresholds refined by r.
   (2) Initialize all the entries in the co-association matrix $M_{n\times n}$ to be N/A (i.e., not available) and the subset $\mathbb{C}_{it}^{in}$ of input matrices used in the next iteration to be empty. As a result, $\mathbb{C}_{it}^{in}$ is the set of visited input matrices in $\mathbb{C}^{in}$ and $(\mathbb{C}^{in} - \mathbb{C}_{it}^{in})$ is the set of non-visited input matrices.

(3) Perform the following steps iteratively until either the boundary region or the set $(\mathbb{C}^{in} - \mathbb{C}_{it}^{in})$ is empty:
- Get the next pair of thresholds (α, β) from the sequence Seq.
- If it is the first iteration, select a set of m matrices from $(\mathbb{C}^{in} - \mathbb{C}_{it}^{in})$ and add them to $\mathbb{C}_{it}^{in}$. Otherwise, select one matrix from $(\mathbb{C}^{in} - \mathbb{C}_{it}^{in})$ and add it to $\mathbb{C}_{it}^{in}$.
- Based on the set $\mathbb{C}_{it}^{in}$, update the evaluation values of all data-point pairs in the current boundary region, divide these pairs into three regions and update the entries in M accordingly.

(4) If the boundary region is not empty, update all entries in the boundary region with 0.5.

Step 2: Generate the hierarchical clustering $\mathcal{HC}$ by applying a hierarchical clustering method to M.

The input matrices in $\mathbb{C}^{in}$ are produced by applying basic clustering algorithms to a dataset. These basic clustering algorithms include 12 algorithms implemented in the package diceR [1], namely, AP, BLOCK, CMEANS, GMM, SC, SOM, DIANA_Euclidean, HC_Euclidean, HDBSCAN, KM_Euclidean, NMF_Scd (or NMF_Lee), and PAM_Euclidean. Every clustering algorithm can be repeatedly applied with different sets of tuning parameters, such as a given number of clusters and a distance measure. In the current implementation, we only consider Euclidean distance and run each algorithm three times with the number of clusters as 3, 4, and 5, respectively. In total, they produce 36 clusterings represented by 36 $n \times n$ matrices that comprise the input set $\mathbb{C}^{in}$.

The sequence Seq of thresholds starts from the most restrictive pair $(1, 0)$. The other pairs are generated according to two step lengths, one δ_α for decreasing α and another δ_β for increasing β. In the current implementation, we consider a simple case where $\delta_\alpha = \delta_\beta = \delta$. The step length δ is calculated as:

$$\delta = \frac{1}{2 * (|\mathbb{C}^{in}| - m + 1) - 1} \cdot \frac{1}{r}, \tag{6}$$

where the number $|\mathbb{C}^{in}| - m + 1$ is the maximum number of iterations.

Each iteration in (3) of Step 1 represents a stage in the presented sequential three-way approach. In order to use as various input clusterings as possible, when selecting matrices from $(\mathbb{C}^{in} - \mathbb{C}_{it}^{in})$, we prefer the matrices produced by non-visited clustering algorithms, that is, these algorithms do not produce any matrix in $\mathbb{C}_{it}^{in}$ that is the set of visited matrices. If there are more candidate matrices than required, we randomly select a required number of matrices from them. To deal with a nonempty boundary region after the iterations, we update all the entries in the boundary region with a value 0.5. The hierarchical clustering method used in Step 2 adopts an agglomerative strategy using the average linkage (UPGMA) [18] to find and merge similar clusters, which is implemented in the package diceR [1].

The algorithm S3WCC-CAM is applied to two datasets, that is, iris[1] from UCI and hgsc[2] from the diceR package. The dataset iris includes 150 data points described by 4 attributes. A fifth attribute of class labels is ignored in the clustering process and used as an external reference in the evaluations. The dataset hgsc includes 489 data points described by 321 attributes without an attribute of class labels. Due to the limitation of our experimental environments, the algorithm is not applied with large datasets in the current experiments. This might be a direction of our future work. The evaluation value of a data-point pair is computed as the proportion of times that the two data points are clustered together out of the times that they are chosen in the bootstrap resampling [1], which is implemented in the package diceR. Table 2 lists the configurations of m and r considered in our experiments.

Table 2. Configurations of m and r in the experiments

id	m	r	id	m	r	id	m	r	id	m	r
c1	3	1	c5	6	1	c9	9	1	c13	12	1
c2	3	3	c6	6	3	c10	9	3	c14	12	3
c3	3	6	c7	6	6	c11	9	6	c15	12	6
c4	3	9	c8	6	9	c12	9	9	c16	12	9

The results of S3WCC-CAM are compared with Cluster-based Similarity Partitioning Algorithm (CSPA) [19] and Link-based Cluster Ensemble method (LCE) [11]. The clustering results are measured by both internal and external indices implemented in the package diceR [1]. The internal indices include avg_within that measures the average distance within clusters, avg_between that measures the average distance between clusters and avg_silwidth that measures the average distance between clusters based on Silhouette width. Thus, a smaller avg_within, a bigger avg_between and a bigger avg_silwidth indicate a better clustering. The external indices measure the similarity of two clusterings by using the class labels as an external reference. The two external indices used in our experiments are the corrected Rand index (corrected_rand) [10] and Meila's variation index (vi) [17]. The corrected Rand index ranges from -1 to 1 with -1 indicating no agreement and 1 indicating perfect agreement. The Meila's variation index measures the variation of information for two clusterings based on mutual information. It has an upper bound $\log n$ where n is the number of data points in the dataset. A smaller Meila's variation index indicates a better clustering. Table 3 summarizes the results of all the above indices. Besides, Table 3 also shows the run time (run_time) and the percentage of boundary region when the iterations stop (BND_perc) in S3WCC-CAM. Since the dataset hgsc does not contain the class labels, only internal indices are evaluated.

---

[1] https://archive.ics.uci.edu/ml/datasets/Iris.
[2] https://www.rdocumentation.org/packages/diceR/versions/0.3.2/topics/hgsc.

Table 3. A summary of the experiment results

		internal indices			external indices		run_time(s)	BND_perc(%)
		avg_within	avg_between	avg_silwidth	corrected_rand	vi		
S3WCC-CAM	c1	0.132±0.096	0.991±0.007	0.877±0.089	0.739±0.078	0.468±0.075	0.227±0.017	1.701±0.714
	c2	0.093±0.043	0.993±0.003	0.916±0.036	0.747±0.020	0.461±0.030	0.720±0.024	0.457±0.180
	c3	0.161±0.042	0.969±0.017	0.830±0.043	0.750±0.024	0.453±0.033	0.835±0.026	13.610±2.325
	c4	0.210±0.044	0.943±0.021	0.750±0.050	0.751±0.025	0.446±0.043	0.837±0.028	20.914±3.404
	c5	0.126±0.074	0.990±0.008	0.882±0.067	0.746±0.013	0.467±0.028	0.158±0.014	3.014±1.113
	c6	0.102±0.039	0.993±0.003	0.908±0.034	0.747±0.007	0.470±0.025	0.497±0.024	1.225±0.353
	c7	0.130±0.003	0.989±0.000	0.879±0.003	0.745±0.001	0.483±0.002	0.780±0.028	8.480±0.172
	c8	0.186±0.015	0.959±0.004	0.793±0.016	0.749±0.007	0.464±0.029	0.783±0.028	18.013±1.180
	c9	0.126±0.062	0.987±0.011	0.881±0.056	0.748±0.008	0.466±0.027	0.098±0.010	5.817±1.667
	c10	0.112±0.058	0.991±0.005	0.896±0.053	0.746±0.009	0.469±0.026	0.275±0.015	1.566±0.501
	c11	0.098±0.031	0.993±0.002	0.912±0.028	0.747±0.007	0.471±0.026	0.546±0.023	1.643±0.357
	c12	0.106±0.000	0.992±0.000	0.903±0.000	0.745±0.000	0.483±0.000	0.727±0.030	5.241±0.015
	c13	0.482±0.028	0.872±0.058	0.194±0.045	0.477±0.053	0.790±0.128	0.031±0.000	61.272±5.289
	c14	0.162±0.051	0.972±0.018	0.834±0.049	0.748±0.008	0.463±0.028	0.053±0.001	12.911±2.732
	c15	0.150±0.046	0.976±0.017	0.849±0.044	0.749±0.007	0.461±0.029	0.075±0.002	11.956±2.699
	c16	0.116±0.055	0.990±0.007	0.892±0.049	0.747±0.007	0.470±0.026	0.124±0.005	4.330±1.145
CSPA		0.687±0	8.464±0	0.898±0	0.745±0	0.483±0	0.615±0.020	N/A
LCE		0.339±0	6.133±0	0.906±0	0.759±0	0.422±0	1.438±0.034	N/A

(a) iris

		internal indices			run_time(s)	BND_perc(%)
		avg_within	avg_between	avg_silwidth		
S3WCC-CAM	c1	0.272±0.091	0.932±0.025	0.695±0.085	0.349±0.021	3.323±0.654
	c2	0.254±0.042	0.932±0.017	0.714±0.038	1.069±0.043	0.899±0.120
	c3	0.293±0.062	0.910±0.022	0.642±0.060	1.291±0.064	26.236±4.778
	c4	0.334±0.067	0.888±0.036	0.570±0.081	1.308±0.069	36.903±6.814
	c5	0.252±0.080	0.929±0.021	0.710±0.071	0.268±0.020	6.470±0.785
	c6	0.251±0.058	0.934±0.015	0.723±0.042	0.770±0.028	2.514±0.191
	c7	0.307±0.015	0.922±0.003	0.655±0.012	1.239±0.043	17.703±0.408
	c8	0.324±0.031	0.885±0.017	0.569±0.027	1.274±0.046	36.643±1.549
	c9	0.275±0.075	0.933±0.022	0.689±0.063	0.182±0.017	11.348±1.292
	c10	0.258±0.058	0.932±0.019	0.703±0.056	0.451±0.029	3.335±0.290
	c11	0.259±0.047	0.935±0.015	0.709±0.044	0.854±0.039	3.298±0.194
	c12	0.275±0.004	0.918±0.002	0.678±0.004	1.151±0.047	11.190±0.033
	c13	0.484±0.009	0.685±0.046	0.110±0.048	0.084±0.001	91.439±8.501
	c14	0.281±0.054	0.909±0.024	0.649±0.048	0.125±0.022	26.261±2.971
	c15	0.286±0.064	0.910±0.028	0.650±0.050	0.163±0.022	25.775±2.708
	c16	0.259±0.065	0.933±0.019	0.703±0.054	0.238±0.027	8.938±0.836
CSPA		2.485±0	11.752±0	0.710±0	20.341±0.099	N/A
LCE		2.401±0	10.508±0	0.677±0	11.261±0.070	N/A

(b) hgsc

As shown in Table 3, S3WCC-CAM generally produces as good clustering results as CSPA and LCE based on the internal and external indices. In terms of the run time, S3WCC-CAM outperforms LCE with all the configurations and CSPA with most configurations, especially on the dataset hgsc. Different configurations of m and r in S3WCC-CAM have a significant influence on run_time and BND_perc. A further study, either experimental or theoretical, on the optimal configuration is necessary and might be a direction for future work.

5   Conclusions and Future Work

We present a sequential three-way approach to progressively constructing a co-association matrix in multiple stages. In each stage, we calculate the evaluation values based on a set of input clusterings. A pair of thresholds is then used to cut the evaluation values, and accordingly, the data-point pairs are divided into three disjoint positive, negative and boundary regions. The entries in the co-association matrix corresponding to the positive and negative regions are updated with the highest evaluation value 1 and the lowest evaluation value 0, respectively. Accordingly, a definite decision of either clustering two data points together or separating them is associated. By gradually involving more input clusterings, we are able to refine the evaluation values in the boundary regions and make a definite decision if possible. By determining some entries to be 1 or 0 once sufficient information can be obtained from the input clusterings, the presented approach makes quick definite decisions on the clustering of some data points in early stages. In this way, we may reduce the overall computational cost of constructing the co-association matrix and obtaining the final clustering.

One direction of the future work is to solve the two issues in the presented approach as mentioned. A second direction is to generalize the presented sequential approach with respect to other consensus clustering methods that do not use co-association matrix. A third direction is a further experimental study, including the optimal configuration of S3WCC-CAM as well as its applications on larger datasets.

References

1. Chiu, D.S., Talhouk, A.: diceR: an R package for class discovery using an ensemble driven approach. BMC Bioinform. **19**, 11–18 (2018)
2. Dempster, A., Laird, N., Rubin, D.: Maximum likelihood from incomplete data via the EM algorithm (with discussion). J. Royal Stat. Soc. Ser. B **39**, 1–38 (1977)
3. Deng, X.F., Yao, Y.Y.: An information-theoretic interpretation of thresholds in probabilistic rough sets. In: Li, T., et al. (eds.) RSKT 2012. LNCS (LNAI), vol. 7414, pp. 369–378. Springer, Heidelberg (2012). https://doi.org/10.1007/978-3-642-31900-6_46
4. Donath, W.E., Hoffman, A.J.: Algorithms for partitioning of graphs and computer logic based on eigenvectors of connection matrices. IBM Tech. Discl. Bull. **15**, 938–944 (1972)
5. Ester, M., Kriegel, H.P., Sander, J., Xu, X.W.: A density-based algorithm for discovering clusters in large spatial databases with noise. In: Simoudis, E., et al. (eds.) KDD 1996, pp. 226–231. AAAI Press (1996)
6. Fred, A.: Finding consistent clusters in data partitions. In: Kittler, J., Roli, F. (eds.) MCS 2001. LNCS, vol. 2096, pp. 309–318. Springer, Heidelberg (2001). https://doi.org/10.1007/3-540-48219-9_31
7. Fred, A., Jain, A.K.: Combining multiple clustering using evidence accumulation. IEEE Trans. Pattern Anal. Mach. Intell. **27**, 835–850 (2005)
8. Hastie, T., Tibshirani, R., Friedman, J.: The Elements of Statistical Learning: Data Mining, Inference, and Prediction, 2nd edn. Springer, New York (2009). https://doi.org/10.1007/978-0-387-84858-7

9. Herbert, J.P., Yao, J.T.: Game-theoretic rough sets. Fundamenta Informaticae **108**, 267–286 (2011)
10. Hubert, L., Arabie, P.: Comparing partitions. J. Classif. **2**, 193–218 (1985)
11. Iam-on, N., Boongoen, T., Garrett, S.: LCE: a link-based cluster ensemble method for improved gene expression data analysis. Bioinformatics **26**, 1513–1519 (2010)
12. Iam-on, N., Boongoen, T., Garrett, S.: Refining pairwise similarity matrix for cluster ensemble problem with cluster relations. In: Jean-Fran, J.-F., Berthold, M.R., Horváth, T. (eds.) DS 2008. LNCS, vol. 5255, pp. 222–233. Springer, Heidelberg (2008). https://doi.org/10.1007/978-3-540-88411-8_22
13. Li, Y., Yu, J., Hao, P., Li, Z.: Clustering ensembles based on normalized edges. In: Zhou, Z.-H., Li, H., Yang, Q. (eds.) PAKDD 2007. LNCS, vol. 4426, pp. 664–671. Springer, Heidelberg (2007). https://doi.org/10.1007/978-3-540-71701-0_71
14. Li, H.X., Zhang, L.B., Huang, B., Zhou, X.Z.: Sequential three-way decision and granulation for cost-sensitive face recognition. Knowl. Based Syst. **91**, 241–251 (2016)
15. Li, H.X., Zhang, L.B., Zhou, X.Z., Huang, B.: Cost-sensitive sequential three-way decision modeling using a deep neural network. Int. J. Approx. Reason. **85**, 68–78 (2017)
16. MacQueen, J.B.: Some methods for classification and analysis of multivariate observations. In: Proceedings of the 5th Berkeley Symposium on Mathematical Statistics and Probability, pp. 281–297. University of California Press (1967)
17. Meila, M.: Comparing clusterings - an information based distance. J. Multivar. Anal. **98**, 873–895 (2007)
18. Sokal, R., Michener, C.: A statistical method for evaluating systematic relationships. Univ. Kansas Sci. Bull. **38**, 1409–1438 (1958)
19. Strehl, A., Ghosh, J.: Cluster ensembles - a knowledge reuse framework for combining multiple partitions. J. Mach. Learn. Res. **3**, 583–617 (2002)
20. Vega-Pons, S., Ruiz-Shulcloper, J.: A survey of clustering ensemble algorithms. Int. J. Pattern Recogn. Artif. Intell. **25**, 337–372 (2011)
21. Vega-Pons, S., Ruiz-Shulcloper, J.: Clustering ensemble method for heterogeneous partitions. In: Bayro-Corrochano, E., Eklundh, J.-O. (eds.) CIARP 2009. LNCS, vol. 5856, pp. 481–488. Springer, Heidelberg (2009). https://doi.org/10.1007/978-3-642-10268-4_56
22. Wang, X., Yang, C., Zhou, J.: Clustering aggregation by probability accumulation. Pattern Recogn. **42**, 668–675 (2009)
23. Yao, Y.Y.: An outline of a theory of three-way decisions. In: Yao, J.T., et al. (eds.) RSCTC 2012. LNCS, vol. 7413, pp. 1–17. Springer, Heidelberg (2012). https://doi.org/10.1007/978-3-642-32115-3_1
24. Yao, Y.Y.: Probabilistic rough set approximations. Int. J. Approx. Reason. **49**, 255–271 (2008)
25. Yao, Y.Y., Deng, X.F.: Sequential three-way decisions with probabilistic rough sets. In: Wang, Y., et al. (eds.) ICCI-CC 2011, pp. 120–125 (2011)
26. Yao, Y.Y., Hu, M., Deng, X.F.: Modes of sequential three-way classifications. In: Medina, J., Ojeda-Aciego, M., Verdegay, J.L., Pelta, D.A., Cabrera, I.P., Bouchon-Meunier, B., Yager, R.R. (eds.) IPMU 2018. CCIS, vol. 854, pp. 724–735. Springer, Cham (2018). https://doi.org/10.1007/978-3-319-91476-3_59
27. Yao, Y.Y., Lingras, P., Wang, R., Miao, D.: Interval set cluster analysis: a reformulation. In: Sakai, H., Chakraborty, M.K., Hassanien, A.E., Ślęzak, D., Zhu, W. (eds.) RSFDGrC 2009. LNCS, vol. 5908, pp. 398–405. Springer, Heidelberg (2009). https://doi.org/10.1007/978-3-642-10646-0_48

28. Yu, H.: A framework of three-way cluster analysis. In: Polkowski, L., et al. (eds.) IJCRS 2017. LNCS, vol. 10314, pp. 300–312. Springer, Cham (2017). https://doi.org/10.1007/978-3-319-60840-2_22
29. Yu, H., Wang, X., Wang, G.: A semi-supervised three-way clustering framework for multi-view data. In: Polkowski, L., et al. (eds.) IJCRS 2017. LNCS, vol. 10314, pp. 313–325. Springer, Cham (2017). https://doi.org/10.1007/978-3-319-60840-2_23
30. Yu, H., Zhang, H.: A three-way decision clustering approach for high dimensional data. In: Flores, V., et al. (eds.) IJCRS 2016. LNCS, vol. 9920, pp. 229–239. Springer, Cham (2016). https://doi.org/10.1007/978-3-319-47160-0_21

Fuzzy Partition Distance Based Attribute Reduction in Decision Tables

Van Thien Nguyen[1], Long Giang Nguyen[2($\boxtimes$)], and Nhu Son Nguyen[2]

[1] Hanoi University of Industry, Hanoi, Vietnam
nguyenthien@haui.edu.vn
[2] Institute of Information Technology, VAST, Hanoi, Vietnam
{nlgiang,nnson}@ioit.ac.vn

Abstract. In recent years, researchers have proposed fuzzy rough set based attribute reduction methods direct on original decision tables to improve the accuracy of the classification model. Most of the previously proposed methods are filter methods, which means that the classification accuracy is evaluated after finding reduct. Therefore, the obtained reduct is not optimal both in terms of number of attributes and classification accuracy. In this paper, we propose a fuzzy partitioning distance and a fuzzy partitioning distance based algorithm to find approximate reduct according to filter-wrapper approach. Experimental results on some data sets show that the classification accuracy on reduct of proposed algorithm is more efficient than that of traditional filter algorithms. Furthermore, by using distance measurements, the execution time of the proposed algorithm is more efficient than the execution time of entropy based filter-wrapper algorithms.

Keywords: Fuzzy rough set · Fuzzy equivalence relation
Fuzzy distance · Decision tables · Attribute reduction · reduct

1 Introduction

Attribute reduction is an important problem in the preprocessing step. The objective of attribute reduction is to eliminate redundant attributes to increase the efficiency of data mining algorithms. Rough set theory proposed by Pawlak [25] is considered to be an effective tool for solving attribute reduction problem. According to rough set approach, the researchers have proposed different measures based on the cardinality of equivalence classes, typically positive region, discernibility function, information entropy, information granule, distance measure. Using these measures, the researchers have proposed attribute reduction algorithms in decision tables. In the proposed measures, distance is considered to be an effective measure to solve attribute reduction problem [7–9,22]. However, rough set based attribute reduction algorithms are implemented on tables with discrete value domain. It is clear that discrete methods do not preserve the

---

Supported by Institute of Information Technology, VAST, Vietnam.

H. S. Nguyen et al. (Eds.): IJCRS 2018, LNAI 11103, pp. 614–627, 2018.
https://doi.org/10.1007/978-3-319-99368-3_48

original differences between objects in the original data. Therefore, the classification accuracy on the obtained reduct is reduced. To improve the classification accuracy, researchers have proposed a fuzzy rough set approach.

Fuzzy rough set proposed by Dubois et al. [2] is considered as an effective tool for solving attribute reduction problem direct on original decision tables, without data preprocessing step. In fuzzy rough set, a fuzzy equivalence relation is defined on the attribute value domain. Based on the fuzzy equivalence, the concepts in traditional set theory are redefined as: fuzzy lower approximation, fuzzy upper approximation, fuzzy domain region some measures are rebuilt as fuzzy discernibility matrix, fuzzy entropy and some fuzzy rough set based attribute reduction methods are proposed. In recent years, many researches have proposed fuzzy rough set based attribute reduction methods, typically fuzzy domain region based methods [11,13,17–21,24], fuzzy discernibility matrix based methods [3,4], fuzzy entropy based methods [6,12,13,23] and fuzzy distance methods [1,10]. For fuzzy domain region based methods, Jensen et al. [17–19] proposed the QUICKREDUCT algorithm to find reduct. Bhatt et al. [21] improved the QUICKREDUCT algorithm to improve the execution time. Jensen et al. [20] proposed three improved directions of QUICKREDUCT to optimize the obtained reduct. Hu et al. [13] proposed the FAR-VPFRS algorithm to find a reduct on hybrid decision tables. Qian et al. [24] proposed improved versions of approximations and proposed the FA-FPR algorithm to minimize the execution time. Authors in [11] proposed an algorithm for finding reduct using fuzzy dependency function on real-valued decision tables. Chen et al. [3,4] proposed algorithms to find reduct based on fuzzy discernibility matrix. For fuzzy entropy based methods, Hu et al. [12,13] constructed fuzzy entropies and proposed some attribute reduction algorithms using fuzzy entropies. Dai et al. [6] constructed a fuzzy gain ratio and developed the GAIN-RATION-AS-FRS algorithm to find a reduct. Using fuzzy distance, authors in [1,10] constructed a fuzzy Jaccard distance and proposed the algorithm F-DBAR to find reduct. The experimental results in the above publications show that fuzzy rough set based attribute reduction methods has a higher classification accuracy than traditional rough set based methods. Furthermore, fuzzy distance based methods are more effective than other methods on both classification accuracy and execution time. However, most of the above attribute reduction methods are filter approach, which means that the classification accuracy is evaluated after obtaining reduct. Therefore, the reduct of the above methods has not optimized both the cardinality of reduct and the classification accuracy, which means that the obtained reduct does not have the best classification accuracy.

In order to improve the classification accuracy on the obtained reduct, Zhang et al. [23] proposed a filter-wrapper algorithm using -fuzzy entropy. With this approach, the filter phase finds candidates for reduct, called the approximate reduct, the wrapper phase finds the reduct with the highest classification accuracy. The experimental results on some data sets show that the filter-wrapper algorithm reduced significantly the cardinality of reduct and increase significantly the classification accuracy. The execution time of the algorithm is higher than traditional filter algorithms due to the time cost of computational the classification accuracy in the wrapper phase. However, the filter-wrapper algorithm in [23] computed fuzzy positive region to compute and computed logarithm expressions in the fuzzy entropy formula. Therefore, the execution time increases compared with the fuzzy distance formula calculated in [1,10].

In this paper, we propose a filter-wrapper algorithm to find approximate reduct using a fuzzy partition distance measure. First of all, we construct a new fuzzy partition distance, which is to improve the fuzzy Jaccard distance in [1,10]. Using the fuzzy partition distances, we propose a filter-wrapper algorithm to find the approximate reduct with the best classification accuracy. Experimental results on some data sets show that the proposed algorithm reduced the execution time compared with the filter-wrapper algorithm using -fuzzy entropy in [23]. Furthermore, the proposed algorithm has a higher classification accuracy than filter algorithms using fuzzy distance in [1,10]. The structure of paper is as follows. Section 2 presents some basic concepts. Section 3 shows the method to construct a fuzzy partition distance between two sets of attributes. Section 4 proposes fuzzy partition distance based attribute reduction method. In Sect. 5, we present experimental results on some data sets. Finally, the conclusion and further research directions.

2   Some Basic Concepts

A decision tables is a pair $DS = (U, C \cup D)$ in which U is a finite set of non-empty objects; C is a conditional attribute set, D is a decision attribute set where $C \cap D = \emptyset$.

Pawlak's rough set theory [25] uses the equivalence relation to approximate the set. Consider the decision table $DS = (U, C \cup D)$, each attribute subset $P \subseteq C$ defines an equivalence relation on the attribute value domain, denoted by R_P.

$$R_P = \{(x, y) \in U \times U \,|\, \forall a \in P,\, a(x) = a(y)\}$$

where $a(x)$ is the value of the attribute a of the object x. The relation R_P determines a partition on U, denoted by $K(P) = U/R_p = \{[x]_P \,|\, x \in U\}$ where $[x]_P$ is the equivalence class contains the object x, $[x]_P = \{y \in U \,|\, (x, y) \in R_P\}$. For $X \subseteq U$, the lower approximation and the upper approximation of X are $\underline{P}X = \{x \in U \,|\, [x]_P \subseteq X\}$, $\overline{P}X = \{x \in U \,|\, [x]_P \cap X \neq \emptyset\}$ respectively. The pair $\langle \underline{P}X, \overline{P}X \rangle$ is called rough set of X with respect to R_P.

Fuzzy rough set proposed by Dubois et al. [2] uses a fuzzy equivalence to approximate fuzzy sets. Let us consider the decision table $DS = (U, C \cup D)$, a relation $\widetilde{R}$ defined on the attribute value domain is called a fuzzy equivalence relation if it satisfies the following conditions:

(1) Reflectivity: $\widetilde{R}(x, x) = 1$;
(2) Symmetry: $\widetilde{R}(x, y) = \widetilde{R}(y, x)$;
(3) Max-min transitive: $\widetilde{R}(x, z) \geq \min\left\{\widetilde{R}(x, y), \widetilde{R}(y, z)\right\}$ for any $x, y, z \in U$;

Given two fuzzy equivalence relations $\tilde{R}_P$, $\tilde{R}_Q$ defined on $P, Q \subseteq C$, then for any $x, y \in U$ we have [14]:

(1) $\tilde{R}_P = \tilde{R}_Q \Leftrightarrow \tilde{R}_P(x, y) = \tilde{R}_Q(x, y)$

(2) $\tilde{R} = \tilde{R}_P \cup \tilde{R}_Q \Leftrightarrow \tilde{R}(x, y) = \max\left\{\tilde{R}_P(x, y), \tilde{R}_Q(x, y)\right\}$

(3) $\tilde{R} = \tilde{R}_P \cap \tilde{R}_Q \Leftrightarrow \tilde{R}(x, y) = \min\left\{\tilde{R}_P(x, y), \tilde{R}_Q(x, y)\right\}$

(4) $\tilde{R}_P \subseteq \tilde{R}_Q \Leftrightarrow \tilde{R}_P(x, y) \leq \tilde{R}_Q(x, y)$

The relation $\tilde{R}_P$ is represented by the fuzzy equivalent matrix $M\left(\tilde{R}_P\right) = [p_{ij}]_{n \times n}$ as follows:

$$M(\tilde{R}_P) = \begin{bmatrix} p_{11} & p_{12} & \cdots & p_{1n} \\ p_{21} & p_{22} & \cdots & p_{2n} \\ \cdots & \cdots & \cdots & \cdots \\ p_{n1} & p_{n2} & \cdots & p_{nn} \end{bmatrix}$$

where $p_{ij} = \tilde{R}_P(x_i, x_j)$ is the value of the relationship between two objects x_i and x_j on the attribute set P, $p_{ij} \in [0, 1]$.

For $P, Q \subseteq C$, as indicated in [14], we have $\tilde{R}_P = \cap_{a \in P} \tilde{R}_a$ and $\tilde{R}_{P \cup Q} = \tilde{R}_P \cap \tilde{R}_Q$, that is for any $x, y \in U$, $\tilde{R}_{P \cup Q}(x, y) = \min\left\{\tilde{R}_P(x, y), \tilde{R}_Q(x, y)\right\}$. Assume that $M\left(\tilde{R}_P\right) = [p_{ij}]_{n \times n}$ and $M(\tilde{R}_Q) = [q_{ij}]_{n \times n}$ are the fuzzy equivalent matrices of $\tilde{R}_P$, $\tilde{R}_Q$, then fuzzy equivalent matrix on the attribute set $S = P \cup Q$ is:

$$M(\tilde{R}_S) = M\left(\tilde{R}_{P \cup Q}\right) = [s_{ij}]_{n \times n} \text{ where } s_{ij} = \min\{p_{ij}, q_{ij}\}$$

For $P \subseteq C$, $U = \{x_1, x_2, ..., x_n\}$, fuzzy equivalence relation $\tilde{R}_P$ determines a fuzzy partition $\pi\left(\tilde{R}_P\right) = U/\tilde{R}_P$ on U

$$\pi\left(\tilde{R}_P\right) = U/\tilde{R}_P = \left\{[x_i]_{\tilde{P}}\right\}_{i=1}^{n} = \left\{[x_1]_{\tilde{P}}, ..., [x_n]_{\tilde{P}}\right\}$$

where $[x_i]_{\tilde{P}} = p_{i1}/x_1 + p_{i2}/x_2 + ... + p_{in}/x_n$ is a fuzzy set as a fuzzy equivalence class of object x_i. Membership functions of objects is determined by $\mu_{[x_i]_{\tilde{P}}}(x_j) = \mu_{\tilde{R}_P}(x_i, x_j) = \tilde{R}_P(x_i, x_j) = p_{ij}$ for any $x_j \in U$. Then, the cardinality of fuzzy equivalence class $[x_i]_{\tilde{R}_P}$ is calculated by $\left|[x_i]_{\tilde{P}}\right| = \sum_{j=1}^{n} p_{ij}$.

Assume that $\mathcal{P}$ is the set of all fuzzy partition on U defined by fuzzy equivalence relations on attribute sets, then $\mathcal{P}$ is called a fuzzy partition space on U. Let us consider fuzzy partition $\pi\left(\tilde{R}_P\right) = \left\{[x_i]_{\tilde{P}}\right\}_{i=1}^{n}$. Specially, if $p_{ij} = 0$ for $1 \leq i, j \leq n$ then $\left|[x_i]_{\tilde{P}}\right| = 0$ for $i \leq n$, $\pi\left(\tilde{R}_P\right)$ is called finest, denoted as $\pi(\tilde{\omega})$. If $p_{ij} = 1$ for $1 \leq i, j \leq n$ then $\left|[x_i]_{\tilde{P}}\right| = |U|$ for $i \leq n$, $\pi\left(\tilde{R}_P\right)$ is called coarseness, denoted as $\pi\left(\tilde{\delta}\right)$.

For $\pi\left(\widetilde{R}_P\right)$, $\pi\left(\widetilde{R}_Q\right) \in \mathcal{P}$, a partial order relation $\preceq$ is defined as [15]: $\pi\left(\widetilde{R}_P\right)$ $\preceq\pi\left(\widetilde{R}_Q\right) \Leftrightarrow [x_i]_{\widetilde{R}_P} \subseteq [x_i]_{\widetilde{R}_Q}$, $i \leq n \Leftrightarrow p_{ij} \leq q_{ij}$, $i,j \leq n$, $\widetilde{R}_P\preceq\widetilde{R}_Q$ for short. Equality $\pi\left(\widetilde{R}_P\right) = \pi\left(\widetilde{R}_Q\right) \Leftrightarrow [x_i]_{\widetilde{R}_P} = [x_i]_{\widetilde{R}_Q}$, $i \leq n \Leftrightarrow p_{ij} = q_{ij}$, $i,j \leq n$, $\widetilde{R}_P = \widetilde{R}_Q$ for short. $\pi\left(\widetilde{R}_P\right) \prec \pi\left(\widetilde{R}_Q\right) \Leftrightarrow \pi\left(\widetilde{R}_P\right)\preceq\pi\left(\widetilde{R}_Q\right)$ and $\pi\left(\widetilde{R}_P\right) \neq \pi\left(\widetilde{R}_Q\right)$, $\widetilde{R}_P \prec \widetilde{R}_Q$ for short.

3   Fuzzy Partition Distance and Its Properties

Given a decision table $DS = (U, C \cup D)$ where $U = \{x_1, x_2, ..., x_n\}$, $P,Q \subseteq C$ and $K(P) = \{[x_i]_P | x_i \in U\}$, $K(Q) = \left\{[x_i]_Q | x_i \in U\right\}$ are two crisp partitions on P and Q. Liang et al. [7] indicated

$$D(K(P), K(Q)) = \frac{1}{|U|} \sum_{i=1}^{|U|} \left(\frac{|[x_i]_P| \oplus |[x_i]_Q|}{|U|} \right)$$

where $|[x_i]_P| \oplus |[x_i]_Q| = |[x_i]_P \cup [x_i]_Q| - |[x_i]_P \cap [x_i]_Q|$ is the distance between partitions $K(P)$ and $K(Q)$. Based on above partition distance, in this section we construct a fuzzy partition distance according to fuzzy rough set approach.

3.1   Fuzzy Distance Between Two Fuzzy Sets

First of all, in this section we construct a distance measure between two fuzzy sets, called fuzzy distance.

Lemma 1 [10]. Given three fuzzy sets $\widetilde{A}, \widetilde{B}, \widetilde{C}$ on the object set U. Then we have $\left|\widetilde{A}\right| - \left|\widetilde{A} \cap \widetilde{B}\right| + \left|\widetilde{C}\right| - \left|\widetilde{C} \cap \widetilde{A}\right| \geq \left|\widetilde{C}\right| - \left|\widetilde{C} \cap \widetilde{B}\right|$.

Proposition 1. Given two fuzzy sets $\widetilde{A}, \widetilde{B}$ on U. Then $d\left(\widetilde{A}, \widetilde{B}\right) = \left|\widetilde{A} \cup \widetilde{B}\right| - \left|\widetilde{A} \cap \widetilde{B}\right|$ is a distance between fuzzy sets $\widetilde{A}$ and $\widetilde{B}$.

Proof. It is clear that $\left|\widetilde{A} \cup \widetilde{B}\right| \geq \left|\widetilde{A} \cap \widetilde{B}\right|$, so $d\left(\widetilde{A}, \widetilde{B}\right) \geq 0$. Furthermore, $d\left(\widetilde{A}, \widetilde{B}\right) = d\left(\widetilde{B}, \widetilde{A}\right)$. Next, we have to prove triangle inequality $d\left(\widetilde{A}, \widetilde{B}\right) + d\left(\widetilde{A}, \widetilde{C}\right) \geq d\left(\widetilde{B}, \widetilde{C}\right)$. According to Lemma 1 we have:

$$\left|\widetilde{A}\right| - \left|\widetilde{A} \cap \widetilde{B}\right| + \left|\widetilde{C}\right| - \left|\widetilde{C} \cap \widetilde{A}\right| \geq \left|\widetilde{C}\right| - \left|\widetilde{C} \cap \widetilde{B}\right| \tag{1}$$

$$\left|\widetilde{A}\right| - \left|\widetilde{A} \cap \widetilde{C}\right| + \left|\widetilde{B}\right| - \left|\widetilde{B} \cap \widetilde{A}\right| \geq \left|\widetilde{B}\right| - \left|\widetilde{B} \cap \widetilde{C}\right| \tag{2}$$

Adding (1) to (2), we have

$$\left(\left|\tilde{A}\right| + \left|\tilde{B}\right| - 2\left|\tilde{A} \cap \tilde{B}\right|\right) + \left(\left|\tilde{A}\right| + \left|\tilde{C}\right| - 2\left|\tilde{A} \cap \tilde{C}\right|\right) \geq \left|\tilde{B}\right| + \left|\tilde{C}\right| - 2\left|\tilde{B} \cap \tilde{C}\right| \tag{3}$$

On the other hand, for any two real numbers a, b we have $\max(a, b) = a + b - \min(a, b)$. So we have for any $x_i \in U$, $\max\left(\mu_{\tilde{A}}(x_i), \mu_{\tilde{B}}(x_i)\right) = \mu_{\tilde{A}}(x_i) + \mu_{\tilde{B}}(x_i) - \min\left(\mu_{\tilde{A}}(x_i), \mu_{\tilde{B}}(x_i)\right)$, that is $\left|\tilde{A} \cup \tilde{B}\right| = \left|\tilde{A}\right| + \left|\tilde{B}\right| - \left|\tilde{A} \cap \tilde{B}\right|$. From (3) we obtain $\left(\left|\tilde{A} \cup \tilde{B}\right| - \left|\tilde{A} \cap \tilde{B}\right|\right) + \left(\left|\tilde{A} \cup \tilde{C}\right| - \left|\tilde{A} \cap \tilde{C}\right|\right) \geq \left|\tilde{B} \cup \tilde{C}\right| - \left|\tilde{B} \cap \tilde{C}\right|$ or $d\left(\tilde{A}, \tilde{B}\right) + d\left(\tilde{A}, \tilde{C}\right) \geq d\left(\tilde{B}, \tilde{C}\right)$. Finally, $d\left(\tilde{A}, \tilde{B}\right)$ is a fuzzy distance between two fuzzy sets $\tilde{A}$, $\tilde{B}$. Based on this fuzzy distance, we construct a fuzzy partition distance in next section.

3.2   Fuzzy Partition Distance and Its Properties

Proposition 2. Given $DS = (U, C \cup D)$ where $U = \{x_1, x_2, ..., x_n\}$ and $\pi\left(\tilde{R}_P\right)$, $\pi\left(\tilde{R}_Q\right)$ is two fuzzy partitions induced by $P, Q \subseteq C$. Then

$$D\left(\pi\left(\tilde{R}_P\right), \pi\left(\tilde{R}_Q\right)\right) = \frac{1}{n^2} \sum_{i=1}^{n} \left(\left|[x_i]_{\tilde{P}} \cup [x_i]_{\tilde{Q}}\right| - \left|[x_i]_{\tilde{P}} \cap [x_i]_{\tilde{Q}}\right|\right) \tag{4}$$

is a distance between fuzzy partitions $\pi\left(\tilde{R}_P\right)$ and $\pi\left(\tilde{R}_Q\right)$.

Proof. It is clear that $D\left(\pi\left(\tilde{R}_P\right), \pi\left(\tilde{R}_Q\right)\right) \geq 0$ and $D\left(\pi\left(\tilde{R}_P\right), \pi\left(\tilde{R}_Q\right)\right) = D\left(\pi\left(\tilde{R}_Q\right), \pi\left(\tilde{R}_P\right)\right)$. Next, we have to prove triangle inequality $D\left(\pi\left(\tilde{R}_P\right), \pi\left(\tilde{R}_Q\right)\right) + D\left(\pi\left(\tilde{R}_P\right), \pi\left(\tilde{R}_S\right)\right) \geq D\left(\pi\left(\tilde{R}_Q\right), \pi\left(\tilde{R}_S\right)\right)$ for any $\pi\left(\tilde{R}_P\right), \pi\left(\tilde{R}_Q\right), \pi\left(\tilde{R}_S\right)$. According to Proposition 1, for any $x_i \in U$ we have: $D\left([x_i]_{\tilde{P}}, [x_i]_{\tilde{Q}}\right) + D\left([x_i]_{\tilde{P}}, [x_i]_{\tilde{S}}\right) \geq D\left([x_i]_{\tilde{Q}}, [x_i]_{\tilde{S}}\right)$. Then

$$D\left(\pi\left(\tilde{R}_P\right), \pi\left(\tilde{R}_Q\right)\right) + D\left(\pi\left(\tilde{R}_P\right), \pi\left(\tilde{R}_S\right)\right)$$

$$= \frac{1}{n^2} \sum_{i=1}^{n} \left(\left|[x_i]_{\tilde{P}} \cup [x_i]_{\tilde{Q}}\right| - \left|[x_i]_{\tilde{P}} \cap [x_i]_{\tilde{Q}}\right|\right)$$

$$+ \frac{1}{n^2} \sum_{i=1}^{n} \left(\left|[x_i]_{\tilde{P}} \cup [x_i]_{\tilde{S}}\right| - \left|[x_i]_{\tilde{P}} \cap [x_i]_{\tilde{S}}\right|\right)$$

$$= \frac{1}{n^2} \sum_{i=1}^{n} d\left([x_i]_{\tilde{P}}, [x_i]_{\tilde{Q}}\right) + \frac{1}{n^2} \sum_{i=1}^{n} d\left([x_i]_{\tilde{P}}, [x_i]_{\tilde{S}}\right) \geq \frac{1}{n^2} \sum_{i=1}^{n} d\left([x_i]_{\tilde{Q}}, [x_i]_{\tilde{S}}\right)$$

$$= D\left(\pi\left(\tilde{R}_Q\right), \pi\left(\tilde{R}_S\right)\right)$$

It's easy to see that $D\left(\pi\left(\tilde{R}_P\right), \pi\left(\tilde{R}_Q\right)\right)$ achieves the minimum value of 0 if and only if $\pi\left(\tilde{R}_P\right) = \pi\left(\tilde{R}_Q\right)$ and $D\left(\pi\left(\tilde{R}_P\right), \pi\left(\tilde{R}_Q\right)\right)$ achieves the maximum value of 1 if and only if $\pi\left(\tilde{R}_P\right) = \pi\left(\tilde{\omega}\right)$ and $\pi\left(\tilde{R}_Q\right) = \pi\left(\tilde{\delta}\right)$ (or $\pi\left(\tilde{R}_P\right) = \pi\left(\tilde{\delta}\right)$ and $\pi\left(\tilde{R}_Q\right) = \pi\left(\tilde{\omega}\right)$, so $0 \leq D\left(\pi\left(\tilde{R}_P\right), \pi\left(\tilde{R}_Q\right)\right) \leq 1$.

Proposition 3. Given $DS = (U, C \cup D)$ where $U = \{x_1, x_2, ..., x_n\}$ and $\tilde{R}$ is a fuzzy equivalence relation. Then fuzzy partition distance between C and $C \cup D$ is defined as

$$D\left(\pi\left(\tilde{R}_C\right), \pi\left(\tilde{R}_{C \cup D}\right)\right) = \frac{1}{n^2} \sum_{i=1}^{n} \left(\left|[x_i]_{\tilde{C}}\right| - \left|[x_i]_{\tilde{C}} \cap [x_i]_{\tilde{D}}\right|\right) \tag{5}$$

Proof. According to Proposition 2 we have $D\left(\pi\left(\tilde{R}_C\right), \pi\left(\tilde{R}_{C \cup D}\right)\right) = \frac{1}{n^2} \sum_{i=1}^{n}$

$\left(\left|[x_i]_{\tilde{C}} \cup [x_i]_{\widetilde{C \cup D}}\right| - \left|[x_i]_{\tilde{C}} \cap [x_i]_{\widetilde{C \cup D}}\right|\right)$

$= \frac{1}{n^2} \sum_{i=1}^{n} \left(\left|[x_i]_{\tilde{C}} \cup \left([x_i]_{\tilde{C}} \cap [x_i]_{\tilde{D}}\right)\right| - \left|[x_i]_{\tilde{C}} \cap [x_i]_{\tilde{D}}\right|\right)$

$= \frac{1}{n^2} \sum_{i=1}^{n} \left(\left|[x_i]_{\tilde{C}}\right| - \left|[x_i]_{\tilde{C}} \cap [x_i]_{\tilde{D}}\right|\right)$

We have $0 \leq D\left(\pi\left(\tilde{R}_C\right), \pi\left(\tilde{R}_{C \cup D}\right)\right) \leq 1 - \frac{1}{n}$. $D\left(\pi\left(\tilde{R}_C\right), \pi\left(\tilde{R}_{C \cup D}\right)\right) = 0$ when $\pi\left(\tilde{R}_C\right) \preceq \pi\left(\tilde{D}\right)$ and $D\left(\pi\left(\tilde{R}_C\right), \pi\left(\tilde{R}_{C \cup D}\right)\right) = 1 - \frac{1}{n}$ when $\pi\left(\tilde{R}_C\right) = \pi\left(\tilde{\delta}\right)$ and $[x_i]_D = \{x_i\}$ where $1 \leq i \leq n$.

Proposition 4. Given $DS = (U, C \cup D)$ where $U = \{x_1, x_2, ..., x_n\}$, $B \subseteq C$ and $\tilde{R}$ is a fuzzy equivalence relation. Then

$$D\left(\pi\left(\tilde{R}_B\right), \pi\left(\tilde{R}_{B \cup D}\right)\right) \geq D\left(\pi\left(\tilde{R}_C\right), \pi\left(\tilde{R}_{C \cup D}\right)\right)$$

Proof: From $B \subseteq C$, according to [15] we have $\pi\left(\tilde{R}_C\right) \preceq \pi\left(\tilde{R}_B\right)$, that is $[x_i]_{\tilde{C}} \subseteq [x_i]_{\tilde{B}}$ where $1 \leq i \leq n$, so $\left|[x_i]_{\tilde{C}}\right| \leq \left|[x_i]_{\tilde{B}}\right|$ where $1 \leq i \leq n$. Consider the object $x_i \in U$ we have:

$$\left|[x_i]_{\tilde{C}}\right| - \left|[x_i]_{\tilde{C}} \cap [x_i]_{\tilde{D}}\right| = \sum_{j=1}^{n} \mu_{[x_i]_{\tilde{C}}}(x_j) - \sum_{j=1}^{n} \min\left\{\mu_{[x_i]_{\tilde{C}}}(x_j), \mu_{[x_i]_{\tilde{D}}}(x_j)\right\}$$

$$\left|[x_i]_{\tilde{B}}\right| - \left|[x_i]_{\tilde{B}} \cap [x_i]_{\tilde{D}}\right| = \sum_{j=1}^{n} \mu_{[x_i]_{\tilde{B}}}(x_j) - \sum_{j=1}^{n} \min\left\{\mu_{[x_i]_{\tilde{B}}}(x_j), \mu_{[x_i]_{\tilde{D}}}(x_j)\right\}$$

(1) For $x_j \in [x_i]_D$ we have $\mu_{[x_i]_{\tilde{D}}}(x_j) = 1$, so $\left|[x_i]_{\tilde{C}}\right| - \left|[x_i]_{\tilde{C}} \cap [x_i]_{\tilde{D}}\right| = 0 = \left|[x_i]_{\tilde{B}}\right| - \left|[x_i]_{\tilde{B}} \cap [x_i]_{\tilde{D}}\right|$

(2) For $x_j \notin [x_i]_D$ we have $\mu_{[x_i]_{\widetilde{D}}}(x_j) = 0$, so $\left| [x_i]_{\widetilde{C}} \right| - \left| [x_i]_{\widetilde{C}} \cap [x_i]_{\widetilde{D}} \right| = \left| [x_i]_{\widetilde{C}} \right| \leq \left| [x_i]_{\widetilde{B}} \right| = \left| [x_i]_{\widetilde{B}} \right| - \left| [x_i]_{\widetilde{B}} \cap [x_i]_{\widetilde{D}} \right|$.

From (1), (2) we have $\left| [x_i]_{\widetilde{B}} \right| - \left| [x_i]_{\widetilde{B}} \cap [x_i]_{\widetilde{D}} \right| \geq \left| [x_i]_{\widetilde{C}} \right| - \left| [x_i]_{\widetilde{C}} \cap [x_i]_{\widetilde{D}} \right|$

$\Leftrightarrow \frac{1}{n^2} \sum_{i=1}^{n} \left(\left| [x_i]_{\widetilde{B}} \right| - \left| [x_i]_{\widetilde{B}} \cap [x_i]_{\widetilde{D}} \right| \right) \geq \frac{1}{n^2} \sum_{i=1}^{n} \left(\left| [x_i]_{\widetilde{C}} \right| - \left| [x_i]_{\widetilde{C}} \cap [x_i]_{\widetilde{D}} \right| \right)$

$\Leftrightarrow D\left(\pi\left(\widetilde{R}_B \right), \pi\left(\widetilde{R}_{B \cup D} \right) \right) \geq D\left(\pi\left(\widetilde{R}_C \right), \pi\left(\widetilde{R}_{C \cup D} \right) \right)$.

The equality $D\left(\pi\left(\widetilde{R}_B \right), \pi\left(\widetilde{R}_{B \cup D} \right) \right) = D\left(\pi\left(\widetilde{R}_C \right), \pi\left(\widetilde{R}_{C \cup D} \right) \right)$ if and only if $\left| [x_i]_{\widetilde{B}} \right| = \left| [x_i]_{\widetilde{C}} \right|$ for any $x_i \in U$.

Zhang et al. [23] indicated that fuzzy conditional entropy does not satisfy monotonicity with the cardinality of conditional attribute set in inconsistent fuzzy decision tables. Thus, fuzzy entropy based attribute reduction methods in [6,12,13,23] are limited by the use of fuzzy conditional entropy to evaluate the criterion for selecting attributes. Proposition 4 shows that the fuzzy partitioning distance satisfies the monotonicity with the cardinality of the conditional attribute set, that is, the smaller the cardinality of condition attribute set, the greater the fuzzy partition distance. Thus, the fuzzy partitioning distance can be used as the criterion for selecting attributes in a heuristic algorithm, as shown in the following section.

4   Fuzzy Partition Distance Based Attribute Reduction in Decision Tables

First of all, we present the traditional method of finding reduct using proposed fuzzy partition distance according to the filter approach. The proposed method consists of the following steps: defining a reduct, defining the importance of the attribute, and constructing a heuristic algorithm to find a reduct.

Definition 1. Given a decision table $DS = (U, C \cup D)$ where $B \subseteq C$, $\widetilde{R}_B, \widetilde{R}_C$ is two fuzzy equivalence relations on B, C. If

(1) $D\left(\pi\left(\widetilde{R}_B \right), \pi\left(\widetilde{R}_{B \cup D} \right) \right) = D\left(\pi\left(\widetilde{R}_C \right), \pi\left(\widetilde{R}_{C \cup D} \right) \right)$

(2) $\forall b \in B$, $D\left(\pi\left(\widetilde{R}_{B-\{b\}} \right), \pi\left(\widetilde{R}_{\{B-\{b\}\} \cup D} \right) \right) \neq D\left(\pi\left(\widetilde{R}_C \right), \pi\left(\widetilde{R}_{C \cup D} \right) \right)$

then B is a reduct of C based on fuzzy partition distance.

Definition 2. Given a decision table $DS = (U, C \cup D)$ where $B \subset C$ and $b \in C - B$. The attribute significance of b with respect to B is defined as

$$SIG_B(b) = D\left(\pi\left(\widetilde{R}_B \right), \pi\left(\widetilde{R}_{B \cup D} \right) \right) - D\left(\pi\left(\widetilde{R}_{B \cup \{b\}} \right), \pi\left(\widetilde{R}_{B \cup \{b\} \cup D} \right) \right)$$

By Proposition 4 we have $SIG_B(b) \geq 0$. $SIG_B(b)$ characterizes the classification quality of the attribute b with respect to D and it used as the attribute selection criteria for the following heuristic algorithm.

Algorithm. F_FPDAR (Filter - Fuzzy Partition Distance based Attribute Reduction)

Input: A decision table $DS = (U, C \cup D)$, a fuzzy equivalence relation $\widetilde{R}$
Output: Reduct B of C

1. $B \leftarrow \emptyset$; $D\left(\pi\left(\widetilde{R}_B\right), \pi\left(\widetilde{R}_{B\cup D}\right)\right) = 1$;

2. Compute fuzzy partition distance $D\left(\pi\left(\widetilde{R}_C\right), \pi\left(\widetilde{R}_{C\cup D}\right)\right)$;

3. While $D\left(\pi\left(\widetilde{R}_B\right), \pi\left(\widetilde{R}_{B\cup D}\right)\right) \neq D\left(\pi\left(\widetilde{R}_C\right), \pi\left(\widetilde{R}_{C\cup D}\right)\right)$ do

4. Begin

5. $Foreach\ a \in C - B$ compute
$$SIG_B(a) = D\left(\pi\left(\widetilde{R}_B\right), \pi\left(\widetilde{R}_{B\cup D}\right)\right) - D\left(\pi\left(\widetilde{R}_{B\cup\{a\}}\right), \pi\left(\widetilde{R}_{B\cup\{a\}\cup D}\right)\right)$$

6. $Select\ a_m \in C - B$ satisfying $SIG_B(a_m) = \underset{a \in C-B}{Max}\{SIG_B(a)\}$;

7. $B = B \cup \{a_m\}$;

8. End;

Return B;

Assume that $D = \{d\}$ and $|C|$, $|U|$ are the cardinality of C, D respectively. The time complexity of computing fuzzy partition distance in command line 2 is $O\left(|C| * |U|^2\right)$. The time complexity of While loop from command line 3 to 8 is $O\left(|C|^2 * |U|^2\right)$. Therefore, the time complexity of algorithm F_FPDAR is $O\left(|C|^2 * |U|^2\right)$.

Let us consider the decision table $DS = (U, C \cup D)$ where $C = \{a_1, a_2, ..., a_m\}$. Let $\omega = D\left(\pi\left(\widetilde{R}_C\right), \pi\left(\widetilde{R}_{C\cup D}\right)\right)$, according to algorithm F_FPDAR, asumme that the attributes $a_{i_1}, a_{i_2}, ...$ are added to the empty set by the maximum value of the attribute significance until there exists $t \in \{1, 2, ...m\}$ satisfying $D\left(\pi\left(\widetilde{R}_{\{a_{i_1}, a_{i_2}, ..., a_{i_t}\}}\right), \pi\left(\widetilde{R}_{\{a_{i_1}, a_{i_2}, ..., a_{i_t}\}\cup D}\right)\right) = \omega$. When the algorithm terminates, we obtain the reduct $B = \{a_{i_1}, a_{i_2}, ..., a_{i_t}\}$. The classification accuracy of attribute sets does not compute in the process of finding reduct. Therefore, F_FPDAR is a filter algorithm.

On the other hand, according to Proposition 4 we have

$$D\left(\pi\left(\widetilde{R}_{\{a_{i_1}\}}\right), \pi\left(\widetilde{R}_{\{a_{i_1}\}\cup D}\right)\right) \geq D\left(\pi\left(\widetilde{R}_{\{a_{i_1}, a_{i_2}\}}\right), \pi\left(\widetilde{R}_{\{a_{i_1}, a_{i_2}\}\cup D}\right)\right) \geq$$
$$... \geq D\left(\pi\left(\widetilde{R}_{\{a_{i_1}, ..., a_{i_t}\}}\right), \pi\left(\widetilde{R}_{\{a_{i_1}, ..., a_{i_t}\}\cup D}\right)\right) = \varepsilon.$$ For given threshold $\varepsilon > \omega$, let $B_k = \{a_{i_1}, ..., a_{i_k}\}$ satisfying $D\left(\pi\left(\widetilde{R}_{B_k}\right), \pi\left(\widetilde{R}_{B_k\cup D}\right)\right) \geq \varepsilon$ and

$$D\left(\pi\left(\widetilde{R}_{B_k\cup\{a_{i_{k+1}}\}}\right), \pi\left(\widetilde{R}_{B_k\cup\{a_{i_{k+1}}\}\cup D}\right)\right) < \varepsilon.$$ Then, B_k is called a ε-approximate reduct.

For the purpose of finding approximate reduct with the best classification accuracy, we proposed a hybrid filter-wrapper approach, in which the filter phase searches for approximation reduct, the wrapper phase searches for the approximate reduct with the best classification accuracy. However, the execution time of the filter-wrapper algorithm will be larger compared with filter algorithm.

Our filter-wrapper algorithm finds approximate reduct using fuzzy partitioning distance as follows:

Algorithm. FW_FPDAR (Filter-Wrapper Fuzzy Partition Distance based Attribute Reduction).

Input: A decision table $DS = (U, C \cup D)$, a fuzzy equivalence relation $\widetilde{R}$

Output: The best reduct B_{best}

1. $B \leftarrow \emptyset$; $T \leftarrow \emptyset$; $D\left(\pi\left(\widetilde{R}_B\right), \pi\left(\widetilde{R}_{B \cup D}\right)\right) = 1$;

// Filter phase: find approximate reducts as candidates for the best reduct

2. Compute fuzzy partition distance $D\left(\pi\left(\widetilde{R}_C\right), \pi\left(\widetilde{R}_{C \cup D}\right)\right)$;

3. While $D\left(\pi\left(\widetilde{R}_B\right), \pi\left(\widetilde{R}_{B \cup D}\right)\right) \neq D\left(\pi\left(\widetilde{R}_C\right), \pi\left(\widetilde{R}_{C \cup D}\right)\right)$ do

4. Begin

5. $Foreach\ a \in C - B$

   $SIG_B(a) = D\left(\pi\left(\widetilde{R}_B\right), \pi\left(\widetilde{R}_{B \cup D}\right)\right) - D\left(\pi\left(\widetilde{R}_{B \cup \{a\}}\right), \pi\left(\widetilde{R}_{B \cup \{a\} \cup D}\right)\right)$

6. $Select\ a_m \in C - B$ satisfying $SIG_B(a_m) = \underset{a \in C - B}{Max} \{SIG_B(a)\}$;

7. $B = B \cup \{a_m\}$;

8. $T = T \cup \{B\}$;

9. End;

// Wrapper phase: find the reduct with the best classification acurracy

10. Let $t = |T|$ //t is the number of elemenst of T, T contains the selected attribute strings, that is $T = \{\{a_{i_1}\}, \{a_{i_1}, a_{i_2}\}, ..., \{a_{i_1}, a_{i_2}..., a_{i_t}\}\}$;

11. Let $T_1 = \{a_{i_1}\}, T_2 = \{a_{i_1}, a_{i_2}\}, ..., T_t = \{a_{i_1}, a_{i_2}, ..., a_{i_t}\}$

12. For $j = 1$ to t

13. Begin

14. Compute the classification accuracy of T_j by a classifier and use the 10-fold cross validation;

15. End

16. $B_{best} = T_{jo}$ where T_{jo} has the best classification accuracy. Return B_{best}

The time complexity of filter phase is $O\left(|C|^2 * |U|^2\right)$. The time complexity of wrapper phase depends on the time complexity of the classifier. Assume that the time complexity of the classifier is $O(T)$, then the time complexity of wrapper phase is $O(|C| * T)$. Therefore, the time complexity of algorithm FW_FPDAR is $O\left(|C|^2 * |U|^2\right) + O(|C| * T)$.

5   Experiments

The objective of our experiment is to compare the proposed algorithm FW_FPDAR with algorithm FEBAR [23] and F_DBAR [1]. The proposed filter-wrapper algorithm FW_FPDAR finds the best approximate reduct based on fuzzy partition distance, while filter-wrapper algorithm FEBAR [23] finds the best approximate reduct based on λ-fuzzy entropy and filter algorithm F_DBAR

[1] finds a reduct based on fuzzy Jaccard distance. The comparison is based on two criteria: classification accuracy and execution time.

The experiments was performed on 8 datasets from Machine Learning Repository (UCI) [26] (see Table 1). On each dataset, for each real value attribute, we normalized data domain in [0, 1] by using the following formula

$$a_0(x_i) = \frac{a(x_i) - \min(a)}{\max(a) - \min(a)}$$

where max(a), min(a) is the maximal and the minimal value on the value domain of attribute a respectively. We use the following fuzzy equivalence relation on the attribute a

$$\widetilde{R}_a(x_i, x_j) = 1 - |a(x_i) - a(x_j)| \text{ where } x_i, x_j \in U$$

For $a \in C$ has nominal or binary value, we use the following equivalence relation where $x_i, x_j \in U$

$$R_a = \begin{cases} 1, & a(x_i) = a(x_j) \\ 0, & otherwise \end{cases}$$

We use the equivalence relation $R_{\{d\}}$ on the decision attribute d. Partition $U/R_{\{d\}} = \left\{ [x]_{\{d\}} \mid x \in U \right\}$ where $[x]_{\{d\}} = \left\{ y \in U \mid R_{\{d\}}(x, y) = 1 \right\}$ is an equivalence class. Then, the equivalence class $[x]_d$ can be seen as a fuzzy equivalence class, denoted as $[x]_{\widetilde{d}}$, where membership fuction $\mu_{[x]_{\widetilde{d}}}(y) = 1$ if $y \in [x]_d$ and $\mu_{[x]_{\widetilde{d}}}(y) = 0$ if $y \notin [x]_d$.

For the filter-wrapper algorithm FW_FPDAR and FEBAR [23], we use the CART classifier to compute classification accuracy in the wrapper phase. For the filter algorithm F_DBAR [1], we also use the CART classifier to evaluate the classification accuracy after finding the reduct. We used the 10-fold cross validation method, which means that the original dataset was divided into 10 equal parts, randomly one part as a test data set, and the remainder as the training data set. The process is repeated 10 times. Classification accuracy is expressed by $v \pm \sigma$, in which v is the mean and σ is the standardized error. The implementation tool is Matlab. The experimental environment is a PC with Intel (R) Core (TM) i7-3770CPU @ 3.40 GHz configuration, running Windows 7, 32 bit.

The classification accuracy of three algorithms are described in Table 2. In which, $|C|$ is the cardinality of attributes of the original dataset, $|B|$ is the cardinality of attributes of obtained reduct. The results in Table 2 show that, the cardinality of reduct of proposed filter-wrapper FW_FPDAR is much smaller compared with filter fuzzy Jaccard distance based algorithm F_DBAR [1], especially for Horse, Heart, Credit, German data sets. Meanwhile, the accuracy of FW_FPDAR and F_DBAR is approximately equal. Therefore, the execution time and the generalization of classification rules of FW_FPDAR are much higher than F_DBAR. For filter-wrapper algorithm FEBAR based on λ-fuzzy entropy [23], the cardinality of reduct of FW_FPDAR is approximately equal to FEBAR

Table 1. Data sets for experiment

No	Data sets	Description	Table size ($\#obj \times \#attr$)	Type of attributes Nominal	Real-valued	Number of classes
1	Lympho	Lymphography	148 × 18	18	0	2
2	Wine	Wine	178 × 13	0	13	3
3	Libra	Libras movement	360 × 90	0	90	15
4	WDBC	Wisconsin diagnostic	569 × 30	0	30	2
5	Horse	Horse colic	368 × 22	15	7	2
6	Heart	Statlog (heart)	270 × 13	7	6	2
7	Credit	Credit approval	690 × 15	9	6	2
8	German	German credit data	1000 × 20	13	7	2

Table 2. The classification acurracy of algorithms

No	Data sets	Original data set $|C|$	Accuracy	FW_FPDAR $|B|$	Accuracy	FEBAR $|B|$	Accuracy	F_DBAR $|B|$	Accuracy
1	Lympho	18	0.776 ± 0.008	4	0.768 ± 0.085	4	0.768 ± 0.085	6	0.788 ± 0.062
2	Wine	13	0.910 ± 0.066	5	0.893 ± 0.072	5	0.893 ± 0.072	7	0.908 ± 0.058
3	Libra	90	0.566 ± 0.137	7	0.658 ± 0.077	8	0.605 ± 0.103	26	0.556 ± 0.205
4	WDBC	30	0.924 ± 0.037	4	0.968 ± 0.058	3	0.952 ± 0.027	6	0.925 ± 0.644
5	Horse	22	0.829 ± 0.085	5	0.806 ± 0.052	4	0.788 ± 0.066	12	0.836 ± 0.058
6	Heart	13	0.744 ± 0.072	3	0.803 ± 0.074	3	0.803 ± 0.074	12	0.752 ± 0.055
7	Credit	15	0.826 ± 0.052	3	0.865 ± 0.028	2	0.846 ± 0.048	14	0.820 ± 0.078
8	German	20	0.692 ± 0.030	6	0.716 ± 0.029	5	0.702 ± 0.043	11	0.725 ± 0.024

Table 3. The execution time of algorithms (s)

No	Data sets	FW_FPDAR Filer phase	Wrapper phase	Total	FEBAR Filer phase	Wrapper phase	Total	F_DBAR
1	Lympho	0.32	0.50	0.82	0.38	0.52	0.90	0.34
2	Wine	0.46	1.21	1.67	0.51	1.18	1.69	0.48
3	Libra	46.28	86.18	132.46	55.12	88.26	143.38	48.48
4	WDBC	20.15	8.74	28.89	26.38	8.22	34.60	22.32
5	Horse	4.85	2.68	7.53	5.26	2.65	7.91	4.98
6	Heart	1.22	1.52	2.74	1.45	1.78	3.23	1.26
7	Credit	16.58	3.42	20.00	19.26	3.98	23.24	18.02
8	German	52.48	8.64	61.12	71.22	8.28	79.50	54.65

(more on some datasets). The classification accuracy of FW_FPDAR is approximately or slightly higher than FEBAR on some datasets. However, the execution time of FW_FPDAR is smaller than FEBAR. The reason is that FEBAR must calculate the value of λ based on the fuzzy positive region and calculate fuzzy entropies contained logarithm formulas. This is shown in Table 3.

The results of the comparison of the execution time in Table 3 show that FW_FPDAR has a significantly less execution time than FEBAR [23], mainly filter phase. However, filter-wrapper algorithms FW_FPDAR and FEBAR have a greater execution time than the filter algorithm F_DBAR [1] because they have to implement the classifier to compute the classification accuracy of approximate reduct in wrapper phase.

6   Conclusions

The objective of the attribute reduction is to find the smallest subset of attributes to improve the efficiency of classification models. On the obtained reduct, the generalizability of classification rules is higher. The classification accuracy of fuzzy rough set based attribute reduction algorithms is higher that of traditional rough set based attribute reduction algorithms since they execute directly on original decision tables without preprocessing data. However, most of them are filter algorithms which finds a reduct preserving the given measure, does not compute the classification accuracy on candidates reduct in the process of finding reduct. So, the obtained reduct is not optimal for the cardinality of attributes and classification accuracy. In this paper, we proposed the filter-wrapper algorithm FW_FPDAR to find the best approximate reduct using a fuzzy partition distance. Experimental results on some data sets show that the proposed algorithm is more efficient than filter algorithms on classification accuracy and the cardinality of obtained reduct, typically the filter algorithm F_DBAR [1]. Moreover, the execution time of proposed algorithm is less than that of the filter-wrapper algorithm FEBAR [23] using λ-fuzzy entropy. The next research direction is to propose some incremental algorithms to find reduct in dynamic decision tables.

References

1. Cao, C.N., Vu, D.T., Nguyen, L.G., Tan, H.: Fuzzy distance based attribute reduction in decision tables. J. Inf. Commun. Technol. Res. Dev. Inf. Commun. Technol. Vietnam, 2, 16 (36), 104–111 (2016)
2. Dübois, D., Prade, H.: Rough fuzzy sets and fuzzy rough sets. Int. J. Gen. Syst. **17**, 191–209 (1990)
3. Chen, D.G., Hu, Q.H., Yang, Y.P.: Parameterized attribute reduction with Gaussian kernel based fuzzy rough sets. Inf. Sci. **181**(23), 5169–5179 (2011)
4. Chen, D.G., Zhang, L., Zhao, S.Y., Hu, Q.H., Zhu, P.F.: A novel algorithm for finding reducts with fuzzy rough sets. IEEE Trans. Fuzzy Syst. **20**(2), 385–389 (2012)
5. Tsang, E.C.C., Chen, D.G., Yeung, D.S., Wang, X.Z., Lee, J.W.T.: Attributes reduction using fuzzy rough sets. IEEE Trans. Fuzzy Syst. **16**, 1130–1141 (2008)
6. Dai, J., Xu, Q.: Attribute selection based on information gain ratio in fuzzy rough set theory with application to tumor classification. Appl. Soft Comput. **13**, 211–221 (2013)
7. Liang, J.Y., Li, R., Qian, Y.H.: Distance: a more comprehensible perspective for measures in rough set theory. Knowl. Based Syst. **27**, 126–136 (2012)

8. Nguyen L.G.: Metric based attribute reduction in decision tables. In: Federated Conference on Computer Science and Information System (FEDCSIS), pp. 311–316. IEEE, Wroclaw (2010)

9. Nguyen, L.G., Nguyen, H.S.: Metric based attribute reduction in incomplete decision tables. In: Ciucci, D., Inuiguchi, M., Yao, Y., Ślęzak, D., Wang, G. (eds.) RSFDGrC 2013. LNCS (LNAI), vol. 8170, pp. 99–110. Springer, Heidelberg (2013). https://doi.org/10.1007/978-3-642-41218-9_11

10. Nguyen, L.G., Cao, C.N., Nguyen, Q.H., Nguyen, T.L.H., Nguyen, N.C., Tran, A.T.: About a fuzzy distance and application in attribute reduction in decision tables. In: Proceedings of 20th National Conference on Information Technology and Telecommunication, pp. 404–409. Quy Nhon, Vietnam (2017)

11. Nguyen, V.T., Nguyen, L.G., Nguyen, N.S.: Fuzzy rough set based attribute reduction in numeric domain decision tables. J. Inf. Commun. Technol. Res. Dev. Inf. Commun. Technol. Vietnam, V-2, 16 (36), 40–49 (2016)

12. Hu, Q.H., Yu, D.R., Xie, Z.X., Liu, J.F.: Fuzzy probabilistic approximation spaces and their information measures. IEEE Trans. Fuzzy Syst. **14**(2), 191–201 (2006)

13. Hu, Q., Yu, D.R., Xie, Z.X.: Information-preserving hybrid data reduction based on fuzzy-rough techniques. Pattern Recognit. Lett. **27**(5), 414–423 (2006)

14. Hu, Q., Xie, Z.X., Yu, D.R.: Hybrid attribute reduction based on a novel fuzzy-rough model and information granulation. Pattern Recogn. **40**, 3509–3521 (2007)

15. Qian, Y.H., Liang, J.Y., Wu, W.Z., Dang, C.Y.: Information granularity in fuzzy binary GrC model. IEEE Trans. Fuzzy Syst. **19**(2), 253–264 (2011)

16. Hu, Q.H., Xie, Z.X., Yu, D.R.: Comments on fuzzy probabilistic approximations spaces and their information measures. IEEE Trans. Fuzzy Syst. **16**, 549–551 (2008)

17. Jensen, R., Shen, Q.: Semantics-preserving dimensionality reduction: rough and fuzzy-rough-based approaches. IEEE Trans. Knowl. Data Eng. **16**(12), 1457–1471 (2004)

18. Jensen, R., Shen, Q.: Fuzzy-rough attribute reduction with application to web categorization. Fuzzy Sets Syst. **141**, 469–485 (2004)

19. Jensen, R., Shen, Q.: Fuzzy-rough sets assisted attribute reduction. IEEE Trans. Fuzzy Syst. **15**(1), 73–89 (2007)

20. Jensen, R., Shen, Q.: New approaches to fuzzy-rough feature selection. IEEE Trans. Fuzzy Syst. **17**(4), 824–838 (2009)

21. Bhatt, R.B., Gopal, M.: On fuzzy-rough sets approach to feature selection. Pattern Recognit. Lett. **26**, 965–975 (2005)

22. Vu, V.D., Vu, D.T., Ngo, Q.T., Nguyen, L.G.: Partition distance based attribute reduction in incomplete decision tables. J. Inf. Commun. Technol. Res. Dev. Inf. Commun. Technol. Vietnam V-2, 14(34), pp. 23–32 (2015)

23. Zhang, X., Mei, C., Chen, D.G., Li, J.: Feature selection in mixed data: a method using a novel fuzzy rough set-based information entropy. Pattern Recogn. **56**, 1–15 (2016)

24. Qian, Y.H., Wang, Q., Cheng, H.H., Liang, J.Y., Dang, C.Y.: Fuzzy-rough feature selection accelerator. Fuzzy Sets Syst. **258**, 61–78 (2015)

25. Pawlak, Z.: Rough Sets: Theoretical Aspects of Reasoning about Data. Kluwer Academic Publisher, London (1991)

26. The UCI machine learning repository, October 2017. http://archive.ics.uci.edu/ml/datasets.html., https://sourceforge.net/projects/weka/

Dynamic and Discernibility Characteristics of Different Attribute Reduction Criteria

Dominik Ślęzak[1(✉)] and Soma Dutta[2]

[1] Institute of Informatics, University of Warsaw, Warsaw, Poland
slezak@mimuw.edu.pl
[2] School of Information Engineering, Vistula University, Warsaw, Poland
somadutta9@gmail.com

Abstract. We investigate different notions of decision superreducts, their interrelations, their way of dealing with inconsistent data and their so-called discernibility characteristics. We refer to superreducts understood as attribute subsets that are aimed at maintaining – when compared to original sets of attributes – unchanged rough set approximations of decision classes, positive regions and generalized decision values. We also include into our studies superreducts that maintain the same data-driven conditional probability distributions (known as rough membership functions), as well as those which let discern all pairs of objects belonging to different decision classes that are also distinguishable using all available attributes. We compare strengths of the corresponding attribute reduction criteria when applied to the whole data sets, as well as families of their subsets (which is an idea inspired by so-called dynamic reducts). We attempt to put together mostly known mathematical results concerning the considered criteria and prove several new facts to make overall picture more complete. We also discuss about importance of developing attribute reduction criteria for inconsistent data sets from the perspectives of machine learning and knowledge discovery.

Keywords: Rough sets · Inconsistent decision tables
Dynamic decision reducts
Discernibility characteristics of decision reducts

1 Introduction

The theory of rough sets is a tool to formalize vague, imprecise concepts [1]. Its approach towards approximating a concept can be easily explained through tabular data sets, usually referred as information systems. Any information system (U, A) describes a set of objects (U) with respect to a set of attributes (A). Based on (U, A) two objects become indistinguishable, with respect to a set of attributes $B \subseteq A$, if they have the same descriptions or values for each attribute of B. Thus, U is partitioned into disjoint classes. Respective equivalence relation

© Springer Nature Switzerland AG 2018
H. S. Nguyen et al. (Eds.): IJCRS 2018, LNAI 11103, pp. 628–643, 2018.
https://doi.org/10.1007/978-3-319-99368-3_49

is known as *indiscernibility relation* generated by B, denoted as $Ind(B)$. Any subset $X \subseteq U$ is now approximated by two sets. One is the union of all classes that are completely included in X (surely belonging to X). The other is the union of all classes that have some overlap with X (possibly belonging to X). This pair of sets gives an approximation of X with respect to $Ind(B)$.

Being such a simple concept, rough set approximations have a great impact in decision making with incomplete information and knowledge representation. However, representing knowledge of a set of objects or a concept greatly depends on a chosen set of attributes. Often, in practice we need to deal with a big set of data described by a large set of attributes. Based on this large set of attributes, a set of objects falling into a decision class, i.e., with a specific decision value, is approximated by its sure and possible cases. As dealing with a large set of attributes is not convenient from different practical perspectives, finding a suitably smaller $B \subseteq A$ generating (almost) *the same description of decision classes* as the whole set A, has a lot of significance in research. In rough set literature, such a set of attributes that can suitably replace the original set of attributes *without losing significant information* is known as a decision superreduct. If there is no proper subset of such a set satisfying a property of *not losing significant information*, then it is called a decision reduct [2].

There can be different ways of understanding phrases *the same description of decision classes* and *without losing significant information*. Based on different interpretations of these phrases different definitions of decision superreducts are available in the literature. In [3][1], we can find a comparative study among different ways of attribute reduction following classical rough set methods, including rough set approximations, rough membership functions and generalized decision functions. In [5], one can find more detailed information about models and mechanisms based on generalized decision functions, with their comparison to classical notions known from the theory of relational databases. In [6], there is more background on superreducts based on rough membership functions, including the first steps toward specifying *approximate attribute reduction* criteria allowing for paying no attention to a need of distinguishing between *almost the same* probability distributions. Let us refer, e.g., to [7] for further examples how to handle rough memberships – or, in other words, conditional probabilities of decision classes derived from the data – in attribute reduction processes.

One may say that all variations of attribute reduction criteria mentioned above (as well as others discussed in this paper) are meaningful only when dealing with *inconsistent decision tables* $(U, A \cup \{d\})$, where two elements indiscernible by A can be distinguished by decision attribute d. Indeed, for *consistent decision tables* all considered decision superreduct formulations are equivalent to each other.

---

[1] The paper [3] is a highly valuable source of information about different ways of specifying data reduction criteria. However, we cannot refer to this in context of "knowledge reduction", as by finding superreducts and reducts we extend – rather than reduce – our knowledge about analyzed data sets. This is analogous to the tasks of reducing complexity – or in other words, searching for simpler solutions – in other fields of data exploration and modeling [4].

Thus, one may say that this kind of study is of no practical use as data tables considered for learning purposes are usually consistent. However, quite often, we need to operate with inconsistent data, especially for the tasks that are actually not related to prediction or classification of specific decision values [4,6]. Moreover, inconsistencies may need to be handled even for consistent data sets, e.g., in the area of feature selection, where people often construct inconsistent subtables composed of relatively small subsets of attributes and then, they need some criteria to reduce them by removing redundant attributes [8,9]. Finally, a goal may not be to learn a single model providing *the same description of decision classes* as the whole set of attributes; but rather to learn a collection of weaker models, each of which able to manage its inconsistencies, that communicate with each other in order to reach to a joint knowledge about approximated concepts [5,10].

The paper is organized as follows. In Sect. 2, we recall different variants of decision superreducts in inconsistent decision tables. In Sect. 3, interrelations among these variants are established. In Sect. 4, we introduce new characteristics – inspired by dynamic decision reducts [11] – of a classical attribute reduction criterion according to which all pairs of objects belonging to different decision classes need to remain distinguished, if only it is so with respect to the whole set of attributes A. In Sect. 5, for each considered variant of decision superreduct a way of constructing respective consistent decision table is presented and a one-to-one connection between a specific definition of decision superreduct and its respective way of translating inconsistent decision table to a consistent one is shown. We also discuss various aspects of practical meaning of such translations, e.g., from the perspective of adaptation of popular attribute reduction algorithms [12]. In Sect. 6, we conclude the paper.

For conceptual and mathematical perspectives, the following paper's fragments are worth special attention. First, in Theorem 1, equivalence of variants D2-D5 seems to be common knowledge but there was no single publication that would gather all these criteria together. Moreover, the fragment D7⇒D6 was partially formulated in [6] but it has never been shown explicitly. Further, Theorem 2 provides brand new characteristics, although to some extent it refers to [6] from the perspective of criterion D6. Finally, Theorem 3 gathers more or less known facts except its fragment devoted to criterion D7 that has never been stated before. All other results are already proved in earlier papers, although we recall (and sometimes re-polish) their proofs for completeness.

2   Rough Set Criteria for Attribute Reduction in Decision Tables

As already outlined, different notions of decision superreducts, for different purposes, are available in the literature. Let us first present a preliminary background for the existing definitions and then explore interrelations among them. Some of those interrelations are already well-known for researchers specialized in the rough set theory [3] while others are delivered as new observations. Actually, paper [3] is a good starting point for analyzing mathematical properties of decision superreducts recalled and introduced herein. Readers are also referred to [2,5,6] for more detailed information about discernibility criteria, generalized decision functions and rough membership functions, respectively. Certainly, we realize that a list of cited publications should be far longer. We attempt to refer to just a few of them given space constraints.

Let $(U, A \cup \{d\})$ be a decision table where U is a set of objects, A is a set of conditional attributes and d is a decision attribute. For each $a \in A \cup \{d\}$ there is a set of values V_a such that $a : U \mapsto V_a$, where elements of V_d will be called decision values (determining decision classes). For any $B \subseteq A$, we write $B(u)$ to denote a vector of values that u receives under each attribute of B. We now present some basic prerequisites to characterize different formulations of decision superreducts as a next step.

Definition 1. *Given $(U, A \cup \{d\})$, $B \subseteq A$ and $X \subseteq U$, lower and upper approximations of X induced by B, denoted as $\underline{X}_B$ and $\overline{X}_B$, are defined as sets $\cup\{[u]_B : [u]_B \subseteq X\}$ and $\cup\{[u]_B : [u]_B \cap X \neq \emptyset\}$, respectively.*

Definition 2. *Given $(U, A \cup \{d\})$ and $B \subseteq A$, positive region induced by B is defined as $POS(B) = \{u \in U : \forall_{u' \in [u]_B} d(u) = d(u')\}$ or, equivalently, a set-theoretic sum of lower approximations of decision classes $X_i = \{u \in U : d(u) = v_i\}$, $i = 1, \ldots, |V_d|$.*

Definition 3. *Given $(U, A \cup \{d\})$ and $B \subseteq A$, generalized decision function induced by B is defined as $\partial_B(u) = \{d(u') : u' \in [u]_B\}$ for each $u \in U$.*

Definition 4. *Given $(U, A \cup \{d\})$ and $B \subseteq A$, rough membership function induced by B is defined as $\mu_B^i(u) = \frac{|\{u' \in [u]_B : d(u') = v_i\}|}{|[u]_B|}$ for each $u \in U$ and $i = 1, \ldots, |V_d|$.*

Definition 5. *Given $(U, A \cup \{d\})$, a subset $B \subseteq A$ is said to be a decision superreduct, if and only if $Ind(B) \subseteq Ind(\{d\})$. Additionally, if there is no proper subset of B that satisfies analogous inclusion, then B is called a decision reduct.*

The last out of the above definitions specifies the background for rough set approaches to data exploration. There are various useful representations (including Boolean representations) of problems of searching for decision reducts in large decision tables [2]. Many heuristic algorithms are also designed to cope with the corresponding computational problems with reasonable time complexity [12]. However, in many practical situations, such reducts do not exist. Below we present seven modified variants of the notion of decision superreduct that can replace condition $Ind(B) \subseteq Ind(\{d\})$.

Definition 6. *For decision table $(U, A \cup \{d\})$ and $B \subseteq A$, consider the following criteria:*

Variant D1. *B should generate the same positive region as A: $POS(B) = POS(A)$, i.e., the same lower approximations of decision classes $X_1, \ldots, X_{|V_d|}$ as A.*

Variant D2. B should generate the same upper approximation of each decision class as A: $\cup\{[u]_B : [u]_B \cap X_i \neq \emptyset\} = \cup\{[u]_A : [u]_A \cap X_i \neq \emptyset\}$ *for each* $i = 1, \ldots, |V_d|$.

Variant D3. B should generate the same lower and upper approximations of each decision class as A: $\cup\{[u]_B : [u]_B \subseteq X_i\} = \cup\{[u]_A : [u]_A \subseteq X_i\}$ *and* $\cup\{[u]_B : [u]_B \cap X_i \neq \emptyset\} = \cup\{[u]_A : [u]_A \cap X_i \neq \emptyset\}$.

Variant D4. B should generate the same lower and upper approximations of each set-theoretic sum of decision classes as A: $\cup\{[u]_B : [u]_B \cap Y \neq \emptyset\} = \cup\{[u]_A : [u]_A \cap Y \neq \emptyset\}$ *as well as* $\cup\{[u]_B : [u]_B \subseteq Y\} = \cup\{[u]_A : [u]_A \subseteq Y\}$ *where* $Y = \cup_{i=j_1}^{j_n} X_i$ *for* $j_1, \ldots, j_n \in \{1, \ldots, |V_d|\}$.

Variant D5. B should generate the same values of generalized decision function as A: *for every* $u \in U$, *there is* $\partial_B(u) = \partial_A(u)$, *i.e.,* $\{d(u') : u' \in [u]_B\} = \{d(u') : u' \in [u]_A\}$.

Variant D6. B should generate the same values of rough membership function as A: *for every* $u \in U$, *there is* $\vec{\mu}_B(u) = \vec{\mu}_A(u)$ *where each i-th component of vector* $\vec{\mu}_B(u)$ *is given by rough membership* $\mu_B^i(u)$.

Variant D7. B should discern the same pairs of objects with different decision values as A: *for every* $u, u' \in U$, *there is* $d(u) \neq d(u') \wedge A(u) \neq A(u') \Rightarrow B(u) \neq B(u')$.

When $(U, A \cup \{d\})$ is consistent, which means that $Ind(A) \subseteq Ind(\{d\})$ (or, in other words, every pair of objects belonging to different decision classes is discerned with respect to A), then all variants D1–D7 are equivalent to classical formulation given in Definition 5. However, in the next sections we show that for inconsistent decision tables these variants may differ from each other, in quite surprising ways.

Table 1. Example of decision table $(U, A \cup \{d\})$, where $U = \{o_1, \ldots, o_9\}$ and $A = \{a_1, a_2, a_3\}$.

	a_1	a_2	a_3	d
o_1	Average	Close	Moderate	High
o_2	Average	Close	Moderate	High
o_3	Average	Close	Moderate	High
o_4	More than average	Far	High	Moderate
o_5	More than average	Far	High	Low
o_6	More than average	Far	Low	Low
o_7	Average	Close	Moderate	High
o_8	More than average	Far	Low	Low
o_9	More than average	Far	Low	High

3   Interrelations Between Different Decision Superreduct Variants

Below we gather together interrelations that are mostly well-known in the rough set community. Majority of them were reported in [3], although some of properties were published earlier, e.g. in [6], while others were not formulated up to now. Perhaps the most interesting aspects of all these interrelations are the following:

- There is a substantial difference between the requirement of holding lower approximations versus the requirement of holding both lower and upper approximations of decision classes during attribute reduction. This fact is often forgotten as this difference does not exist for decision tables with only two decision classes. It arises only when there are three or more decision classes.
- Requirement D7, which is widely applied in the (rough set [9], but not only rough set [8]) literature, is actually more restrictive than others, including even D6 (which can be compared to the notion of Markov blanket [4]). This may seem to be quite counterintuitive as D6 operates with probability estimates that are usually unexpected to hold fully precisely while removing conditional attributes, while D7 – which means that all pairs of objects with different decision values that can be discerned by full set of attributes need to remain discerned also by the considered subset – has been always perceived as quite a natural criterion [2].

Theorem 1. *1. Criteria D2, D3, D4 and D5 are equivalent to each other.*
2. D1 is implied by D2-D5, but not the converse.[2]
3. D6 implies D5, but not the converse.
4. D7 implies D6, but not the converse.

Proof. 1. Below we omit cases D3$\Rightarrow$D2, D4$\Rightarrow$D2 and D4$\Rightarrow$D3, as they are obvious.

D2$\Rightarrow$D3. We assume D2. Let $u' \in \cup\{[u]_B : [u]_B \subseteq X_i\}$. Then $u' \in [u]_B$ for some $[u]_B \subseteq X_i$. As $B \subseteq A$, we have $[u]_A \subseteq [u]_B$. Hence, there is $[u]_A \subseteq X_i$. If $u' \in [u]_A$ we are done. If not, then let us consider $[u']_A$. We know $[u']_A \subseteq [u']_B = [u]_B$ as $u' \in [u]_B$. So, as $[u]_B \subseteq X_i$ we have a class $[u']_A \subseteq X_i$. So, $u' \in \cup\{[u]_A : [u]_A \subseteq X_i\}$. Hence, $\cup\{[u]_B : [u]_B \subseteq X_i\} \subseteq \cup\{[u]_A : [u]_A \subseteq X_i\}$.
Conversely, let $u' \in \cup\{[u]_A : [u]_A \subseteq X_i\}$. So, $u' \in [u]_A$ for some $[u]_A \subseteq X_i$. So, $[u]_A \cap X_j = \emptyset$ for any $j \neq i$. Now, as $[u]_A \subseteq [u]_B$, $[u]_B \cap X_i \neq \emptyset$. We claim that $[u]_B \subseteq X_i$. If not, then $[u]_B \cap X_j \neq \emptyset$ for some $j \neq i$. So, there is some $u'' \in [u]_B$ such that $u'' \in X_j$. Hence, $[u'']_A \cap X_j \neq \emptyset$. As $u'' \in [u]_B$, $u \in [u'']_B$. However, $u \notin [u'']_A$ because if $u \in [u'']_A$, then $[u'']_A \cap X_j = \emptyset$. So, $u \in \cup\{[u]_B : [u]_B \cap X_j \neq \emptyset\}$, but as $u \notin [u'']_A$ such that $[u'']_A \cap X_j \neq \emptyset$, $u \notin \cup\{[u]_A : [u]_A \cap X_j \neq \emptyset\}$. This contradicts D2. So,

---

[2] As already mentioned, D1 is equivalent to D2–D5 for decision tables with two decision classes. However, in this paper we consider the case of their arbitrary amount.

$[u]_B \subseteq X_i$. Thus, $u' \in [u]_A \subseteq X_i$ implies $u' \in [u]_B$ and $[u]_B \subseteq X_i$. Thus, we have other side, i.e., $\cup\{[u]_A : [u]_A \subseteq X_i\} \subseteq \cup\{[u]_B : [u]_B \subseteq X_i\}$.

D2⇒D4. Let us assume D2 and consider an $Y = X_i \cup X_j$ for $i \neq j$.

$$\begin{aligned}
\cup\{[u]_B : [u]_B \cap Y \neq \emptyset\} &= \cup\{[u]_B : [u]_B \cap (X_i \cup X_j) \neq \emptyset\} \\
&= \cup\{[u]_B : ([u]_B \cap X_i) \cup ([u]_B \cap X_j) \neq \emptyset\} \\
&= \cup\{[u]_B : [u]_B \cap X_i \neq \emptyset\} \cup \cup\{[u]_B : [u]_B \cap X_j \neq \emptyset\} \\
&= \cup\{[u]_A : [u]_A \cap X_i \neq \emptyset\} \cup \cup\{[u]_A : [u]_A \cap X_j \neq \emptyset\} \\
&= \cup\{[u]_A : [u]_A \cap Y \neq \emptyset\}
\end{aligned}$$

Now, let $u' \in \cup\{[u]_B : [u]_B \subseteq X_i \cup X_j\}$. Thus, $u' \in [u]_B$ for some $[u]_B$ such that either $[u]_B \subseteq X_i$ or $[u]_B \subseteq X_j$ or $[u]_B \cap X_i \neq \emptyset$, $[u]_B \cap X_j \neq \emptyset$ and $[u]_B \subseteq X_i \cup X_j$. For the first two cases, as D2 and D3 are equivalent, we have $u' \in \cup\{[u]_A : [u]_A \subseteq X_i\}$ and $u' \in \cup\{[u]_A : [u]_A \subseteq X_j\}$ respectively. For the third case we have that u' belongs to $\cup\{[u]_B : [u]_B \cap X_i \neq \emptyset\} \cap \cup\{[u]_B : [u]_B \cap X_j \neq \emptyset\}$, which is equal to $\cup\{[u]_A : [u]_A \cap X_i \neq \emptyset\} \cap \cup\{[u]_A : [u]_A \cap X_j \neq \emptyset\}$. Thus, $u' \in [u]_A$ where $[u]_A \cap X_i \neq \emptyset$, $[u]_A \cap X_j \neq \emptyset$ and $[u]_A \subseteq X_i \cup X_j$. So, by combining all above cases we have $u' \in \cup\{[u]_A : [u]_A \subseteq X_i \cup X_j\}$. Hence, we proved that $\cup\{[u]_B : [u]_B \subseteq X_i \cup X_j\} \subseteq \cup\{[u]_A : [u]_A \subseteq X_i \cup X_j\}$.

Now, let $u' \in \cup\{[u]_A : [u]_A \subseteq X_i \cup X_j\}$. Thus, $u' \in [u]_A$ for some $[u]_A$ such that either $[u]_A \subseteq X_i$, or $[u]_A \subseteq X_j$ or $[u]_A \cap X_i \neq \emptyset$, $[u]_A \cap X_j \neq \emptyset$ and $[u]_A \subseteq X_i \cup X_j$. For the first two cases, as before, we can show $u' \in \cup\{[u]_B : [u]_B \subseteq X_i\}$ and $u' \in \cup\{[u]_B : [u]_B \subseteq X_j\}$ respectively. For the third case, we want to prove that $[u]_B \subseteq X_i \cup X_j$. If not, then there is some $u'' \in [u]_B$ such that $u'' \in X_k$ where $k \neq i, j$. So, $[u'']_A \cap X_k \neq \emptyset$. Now as $u'' \in [u]_B$, $u \in [u'']_B$. However, $u \notin [u'']_A$ as $[u]_A \neq [u'']_A$ and $[u]_A \cap X_k = \emptyset$. So, this violates D2 as $u \in \cup\{[u'']_B : [u'']_B \cap X_k \neq \emptyset\}$ but $u \notin \cup\{[u'']_A : [u'']_A \cap X_k \neq \emptyset\}$. Hence $[u]_B \subseteq X_i \cup X_j$ and $u' \in \cup\{[u]_B : [u]_B \subseteq X_i \cup X_j\}$. Thus, it is proved that $\cup\{[u]_A : [u]_A \subseteq X_i \cup X_j\} = \cup\{[u]_B : [u]_B \subseteq X_i \cup X_j\}$. The case of set theoretic sum of finitely many decision classes can be shown similarly.

D2⇒D5. Let $\cup\{[u]_B : [u]_B \cap X_i \neq \emptyset\} = \cup\{[u]_A : [u]_A \cap X_i \neq \emptyset\}$ and let $v_i \in \partial_B(u)$. Thus, $[u]_B \cap X_i \neq \emptyset$. So, as $[u]_A \subseteq [u]_B \subseteq \cup\{[u]_B : [u]_B \cap X_i \neq \emptyset\} = \cup\{[u]_A : [u]_A \cap X_i \neq \emptyset\}$, we can conclude that $[u]_A \cap X_i \neq \emptyset$. Thus, $v_i \in \partial_A(u)$.

Conversely, let $v_i \in \partial_A(u)$. Thus, $[u]_A \cap X_i \neq \emptyset$ and – as $[u]_A \subseteq [u]_B$ – we have $[u]_B \cap X_i \neq \emptyset$. So, $v_i \in \partial_B(u)$. Hence, D5 is proved.

D5⇒D2. Let $\partial_B(u) = \partial_A(u)$ for all $u \in U$. First, we know that $\cup\{[u]_A : [u]_A \cap X_i \neq \emptyset\} \subseteq \cup\{[u]_B : [u]_B \cap X_i \neq \emptyset\}$ is immediate. To prove the other direction, let us assume $u' \in \cup\{[u]_B : [u]_B \cap X_i \neq \emptyset\}$. So, for some $u'' \in [u]_B$, we have $u'' \in X_i$, i.e., $d(u'') = v_i$. So, $v_i \in \partial_B(u) = \partial_A(u)$. Hence, $[u]_A \cap X_i \neq \phi$. Now if $u' \in [u]_A$ we have $u' \in \cup\{[u]_A : [u]_A \cap X_i \neq \emptyset\}$. If $u' \notin [u]_A$, we consider $[u']_A$. Since $[u']_A \subseteq [u']_B = [u]_B$, $[u']_B \cap X_i \neq \emptyset$. Hence, there is $v_i \in \partial_B(u') = \partial_A(u')$. So, we have $[u']_A \cap X_i \neq \emptyset$ and $u' \in \cup\{[u]_A : [u]_A \cap X_i \neq \emptyset\}$. Finally, we have $\cup\{[u]_B : [u]_B \cap X_i \neq \emptyset\} \subseteq \cup\{[u]_A : [u]_A \cap X_i \neq \emptyset\}$, so D2 is proved.

2. **D3⇒D1** is obvious. The converse is as follows:

   Example for D1 does not imply D2. Let $U = \{o_1, o_2, o_3, o_4, o_5, o_6,$ $o_7, o_8, o_9\}$ be a set of hotels and a set of attributes is given by $A = \{a_1, a_2, a_3\}$ where a_1 reflects amenities, a_2 – distance from public transport and a_3 – the price. Let d reflect hotel demand and $B = \{a_1, a_2\}$. Now, let us refer to Table 1. Partitions with respect to A and B are as follows: $U/A = \{\{o_1, o_2, o_3, o_7\}, \{o_4, o_5\}, \{o_6, o_8, o_9\}\}$, $U/B = \{\{o_1, o_2, o_3, o_7\}, \{o_4, o_5, o_6, o_8, o_9\}\}$. Let X_1, X_2, X_3 be decision classes corresponding to values high, low, moderate for decision attribute d, respectively, and d be given as $d(o_1) = d(o_2) = d(o_3) = d(o_7) = d(o_9) =$ high, $d(o_5) = d(o_6) = d(o_8) =$ low and $d(o_4) =$ moderate. So, $POS(A) = \{o_1, o_2, o_3, o_7\} = POS(B)$. Thus, D1 holds. However, $\cup\{[u]_A : [u]_A \cap X_3 \neq \emptyset\} = \{o_4, o_5\}$ and $\cup\{[u]_B : [u]_B \cap X_3 \neq \emptyset\} = \{o_4, o_5, o_6, o_8, o_9\}$. So, D2 does not hold.

3. **D6⇒D5:** Let $\mu_B(u) = \mu_A(u)$ for each $u \in U$. Thus, for each $v_j \in V_d$, $\mu_B^j(u) = \mu_A^j(u)$ for each $u \in U$. Now as $[u]_A \subseteq [u]_B$, $\partial_A(u) \subseteq \partial_B(u)$. So, we need to prove that $\partial_B(u) \subseteq \partial_A(u)$. Let $v_i \in \partial_B(u)$. Thus, for some $u' \in [u]_B$, there is $d(u') = v_i$. Hence $\mu_B^i(u) \neq 0$. Now as $\mu_B^i(u) = \mu_A^i(u)$, $v_i \in \partial_A(u)$. Thus, we obtain D5.

   Example for D5 does not imply D6. Let us consider a slightly changed Table 1. Let partitions with respect to A and B be the same as before, i.e., $U/A = \{\{o_1, o_2, o_3, o_7\}, \{o_4, o_5\}, \{o_6, o_8, o_9\}\}$, $U/B = \{\{o_1, o_2, o_3, o_7\}, \{o_4, o_5, o_6, o_8, o_9\}\}$. Let decision attribute be such that $d(o_i)$ is the same as in Table 1 for $i = 1, \ldots, 8$, but $d(o_9) =$ moderate. So, for $i = 1, 2, 3, 7$, $\partial_{d|A}(o_i) = \partial_{d|B}(o_i) = \{\text{high}\}$ and for $i = 4, 5, 6, 8, 9$, $\partial_{d|A}(o_i) = \{\text{low, moderate}\} = \partial_{d|B}(o_i)$. On the other hand, $\mu_{d|A}(o_4) = \langle \frac{1}{2}, \frac{1}{2}, 0 \rangle$ but $\mu_{d|B}(o_4) = \langle \frac{3}{5}, \frac{2}{5}, 0 \rangle$. So, D5 holds but D6 does not.

4. **D7⇒D6:** Following D7, if $u' \in [u]_B$, then $u' \in [u]_A$ or $d(u) = d(u')$. Now, $[u]_B$ can consist of elements from a single decision class or more than one decision classes. If $[u]_B$ is contained in a single decision class X_i, then so is $[u]_A$. Moreover, $\mu_B^i(u) = \mu_A^i(u) = 1$ and $\mu_B^j(u) = \mu_A^j(u) = 0$ for $j \neq i$. If $[u]_B$ contains elements of different decision classes, then $[u]_B = [u]_A$ and hence they have the same rough membership function with respect to B and A.

   Example for D6 does not imply D7. Let us consider $U = \{o_1, o_2, o_3, o_4\}$, where partitions with respect to A and $B \subseteq A$ are as follows: $U/A = \{\{o_1, o_2\}, \{o_3, o_4\}\}$ and $U/B = \{o_1, o_2, o_3, o_4\}$. Let values corresponding to decision attribute d be $d(o_1) = v_1$, $d(o_2) = v_2$, $d(o_3) = v_1$ and $d(o_4) = v_2$. So, for each $i = 1, 2, 3, 4$, $\mu_A(o_i) = \langle \frac{1}{2}, \frac{1}{2} \rangle$ and $\mu_B(o_i) = \langle \frac{2}{4}, \frac{2}{4} \rangle = \langle \frac{1}{2}, \frac{1}{2} \rangle$. So, D6 is satisfied. However, D7 does not hold as for $o_1, o_4 \in U$, though $[o_1]_A \neq [o_4]_A$ and $d(o_1) \neq d(o_4)$, $[o_1]_B = [o_4]_B$. □

4   Characterizations in Terms of Dynamic Superreducts

As we already know, criterion D7 is the strongest – most restrictive, allowing potentially to remove the smallest amount of attributes – out of all seven variants discussed in Sect. 3. Interestingly, the theorem below shows that D7 holds, if and only if any of those criteria hold for all subtables of a given decision table, created by taking arbitrary subsets of its universe. Thus, strict implications expressed by Theorem 1 become equivalences when required for all subuniverses. Moreover, in our opinion, these equivalences stand for an additional illustration that D7 is significantly stronger than all other considered criteria.

This kind of derivation was first formulated for D6 in [6]. More precisely, the following was shown:

– Criterion D6 holds for each decision subtable $(U', A \cup \{d\})$ obtained by taking any subset of objects U' from U, if and only if dependency $B \Rightarrow A \vee d$ holds in $(U, A \cup \{d\})$, i.e., for each $u \in U$ we have $[u]_B \subseteq [u]_A$ (which basically means equality $[u]_B = [u]_A$) or $[u]_B$ is fully contained in one of decision classes.

Although therein it was not noticed that $B \Rightarrow A \vee d$ is an equivalent formulation of D7, we can still think about paper [6] as the first step toward Theorem 2 below. Before proceeding further, let us just comment on the fact of using term *dynamic reducts* in the title of that paper. Actually, dynamic reducts were introduced in [11] as a tool for conducting more robust attribute reduction by comparing decision reducts obtained from a sample of different randomly selected subsets of the universe of objects. On the other hand, in [6], we followed this idea from a more theoretical than empirical perspective – by taking into account all subsets $U' \subseteq U$ instead of random samples.

Definition 7. *For each variant Di introduced in Definition 6, $i = 1, \ldots, 7$, we define:*

Variant D*i*'. For the considered decision table $(U, A \cup \{d\})$, $B \subseteq A$ should satisfy variant Di over all subtables $(U', A \cup \{d\})$, for all non-empty subsets $U' \subseteq U$.

Theorem 2. *For each $i = 1, \ldots, 7$, the following holds:*

– *Given $(U, A \cup \{d\})$, a subset $B \subseteq A$ satisfies D7, if and only if it satisfies Di'.*

Proof. First, note that D7 implies D7'. Second, by Theorem 1, we know that D7 implies D6, D6 implies D2-D5 and D2-D5 implies D1. Surely, it holds also at the level of subsets $U' \subseteq U$. Therefore, the only thing to show is that D1' implies D7.

Let D1 hold for any $U' \subseteq U$. We want to prove that $\forall_{u,u'}(d(u) \neq d(u') \wedge A(u) \neq A(u') \Rightarrow B(u) \neq B(u'))$. Consider $U' = \{u, u'\}$ such that $[u]_B = [u']_B$. We have $d(u) = d(u')$ or $d(u) \neq d(u')$. Let us denote by $POS'(A)$ and $POS'(B)$ positive regions induced by A and B in $(U', A \cup \{d\})$. If $d(u) \neq d(u')$, then $u, u' \notin POS'(B)$. Hence, $u, u' \notin POS'(A)$. Now, as $[u]_A \subseteq [u]_B$, either $[u]_A = [u]_B$ or $[u]_A \subsetneq [u]_B$. If $[u]_A = [u]_B$, then it is immediate that $[u]_A = [u']_A$.

If $[u]_A \subsetneqq [u]_B$, then $[u]_A = \{u\}$ and $[u']_A = \{u'\}$. So, being single element decision classes, $u, u' \in POS'(A)$. This contradicts that $u, u' \notin POS'(A)$. So, $[u]_A = [u']_A$. Hence, D7 is proved as for $B(u) = B(u')$, we have $d(u) = d(u')$ or $A(u) = A(u')$. $\qquad\qquad\qquad\qquad\qquad\qquad\qquad\qquad\qquad\qquad\qquad\qquad\qquad\quad\square$

5   Discernibility Characterizations of Decision Superreducts

Up to now, we examined interrelations among different criteria for decision superreducts in inconsistent decision tables. In this section, for each considered variant D1-D7, we present a translation of decision attribute d, for a given table $(U, A \cup \{d\})$, into a new decision attribute $d_i^{\#}$, $i = 1, \ldots, 7$, in such a way that:

- $d_i^{\#}$ agrees with d for all objects u for which equivalence class of $[u]_A$ is contained in a single decision class.
- Di holds for a given $B \subseteq A$ in $(U, A \cup \{d\})$, if and only if there is inclusion $Ind(B) \subseteq Ind(\{d_i^{\#}\})$ in consistent decision table $(U, A \cup \{d_i^{\#}\})$.

This kind of replacement of decision attribute, which actually makes a resulting decision table consistent as values of $d_i^{\#}$ are assigned to indiscernibility classes $[u]_A$, is a well-known mechanism in rough set literature [2,6]. From this perspective, such discernibility representations gathered in the theorem below are mostly common knowledge, although the fragment related to D7 is a brand new result.

Definition 8. *For decision table $(U, A \cup \{d\})$ and $B \subseteq A$, consider the following criteria:*

Variant D1$^{\#}$. *There is $Ind(B) \subseteq Ind(\{d_1^{\#}\})$, where $d_1^{\#}$ is defined as*

$$d_1^{\#}(u) = \begin{cases} d(u) & \text{if } u \in POS(A) \\ \# & \text{otherwise} \end{cases}$$

where $\# \notin V_d$ is a new value assigned to objects suffering from inconsistencies.
Variant D5$^{\#}$. *There is $Ind(B) \subseteq Ind(\{d_5^{\#}\})$, where $d_5^{\#}$ is defined as*

$$d_5^{\#}(u) = \partial_A(u)$$

Variant D6$^{\#}$. *There is $Ind(B) \subseteq Ind(\{d_6^{\#}\})$, where $d_6^{\#}$ is defined as*

$$d_6^{\#}(u) = \overrightarrow{\mu}_A(u)$$

Variant D7$^{\#}$. *There is $Ind(B) \subseteq Ind(\{d_7^{\#}\})$, where $d_7^{\#}$ is defined as*

$$d_7^{\#}(u) = \begin{cases} d(u) & \text{if } u \in POS(A) \\ \#_{m(u)} & \text{otherwise} \end{cases}$$

where $\#_{m(u)} \notin V_d$ are new decision values indexed by ordinal numbers of the corresponding indiscernibility classes $[u]_A$, such that $\#_{m(u)} \neq \#_{m(u')}$ if $[u]_A \neq [u']_A$.

Let us note that domains of decision values are modified in each of the above cases. For D1$^\#$ and D7$^\#$, V_d is changed to subsets (as some of original values and/or new values may not occur) of $V_d \cup \{\#\}$ and $V_d \cup \{\#_{m(u)} : u \in U\}$, respectively. In case of D5$^\#$ and D6$^\#$, we begin to operate with domains embedded into a family of non-empty subsets of V_d and a simplex of probability distributions over V_d, whereby original decision values are interpreted as singletons and zero-one distributions, respectively.

Table 2. New decisions constructed for Table 1. Column m denotes ordinal numbers of indiscernibility classes induced by the whole set of conditional attributes that objects belong to.

	m	d	$d_1^\#$	$d_5^\#$	$d_6^\#$	$d_7^\#$
o_1	1	High	High	{High}	$\langle 1,0,0 \rangle$	High
o_2	1	High	High	{High}	$\langle 1,0,0 \rangle$	High
o_3	1	High	High	{High}	$\langle 1,0,0 \rangle$	High
o_4	2	Moderate	#	{Medium, Low}	$\langle 0, \frac{1}{2}, \frac{1}{2} \rangle$	#$_2$
o_5	2	Low	#	{Medium, Low}	$\langle 0, \frac{1}{2}, \frac{1}{2} \rangle$	#$_2$
o_6	3	Low	#	{High, Low}	$\langle \frac{1}{3}, 0, \frac{2}{3} \rangle$	#$_3$
o_7	1	High	High	{High}	$\langle 1,0,0 \rangle$	High
o_8	3	Low	#	{High, Low}	$\langle \frac{1}{3}, 0, \frac{2}{3} \rangle$	#$_3$
o_9	3	High	#	{high, low}	$\langle \frac{1}{3}, 0, \frac{2}{3} \rangle$	#$_3$

Theorem 3. *For each $i = 1, 5, 6, 7$, the following holds:*

- *Given $(U, A \cup \{d\})$, a subset $B \subseteq A$ satisfies Di, if and only if it satisfies Di$^\#$.*

Proof. **D1$\Leftrightarrow$D1$^\#$.** Let $POS(B) = POS(A)$. We want to prove that if $u' \in [u]_B$, then $d_1^\#(u) = d_1^\#(u')$. Now for $u' \in [u]_B$ either $u' \in POS(B)$ or $u' \notin POS(B)$. If $u' \in POS(B)$, then $u' \in POS(A)$. Thus, $[u]_A$ is contained in a single decision class and hence $d_1^\#(u) = d_1^\#(u')$. If $u' \notin POS(B)$, $u' \notin POS(A)$. Then $[u]_A$ is not contained in any single decision class. So, $d_1^\#(u) = d_1^\# (u') = \#$.

Conversely, let $Ind(B) \subseteq Ind(\{d_1^\#\})$. As $B \subseteq A$, we know $[u]_A \subseteq [u]_B$. So, $POS(B) \subseteq POS(A)$. Now let $u \in POS(A)$. So, for all $u' \in [u]_A$, $d(u) = d(u')$ and hence $d_1^\#(u) = d_1^\#(u')$. We want to prove that $u \in POS(B)$. If not, then for some $u'' \in [u]_B$, there is $d(u) \neq d(u'')$. So, $u'' \notin [u]_A$. Hence $d_1^\#(u) \neq d_1^\#(u'')$. Thus, $B(u) \neq B(u'')$ as $Ind(B) \subseteq Ind(\{d_1^\#\})$. This contradicts with assumption that $u'' \in [u]_B$.

D5⇔D5$^{\#}$. Let us assume that D2 holds. We want to prove that if $u' \in [u]_B$, then $d_5^{\#}(u) = d_5^{\#}(u')$. Let $u' \in [u]_B$. Now we know that either $[u]_B \cap X_i \neq \emptyset$ or $[u]_B \cap X_i = \emptyset$. If $[u]_B \cap X_i \neq \emptyset$, then $u' \in \cup\{[u]_B : [u]_B \cap X_i \neq \emptyset\} = \cup\{[u]_A : [u]_A \cap X_i \neq \emptyset\}$. Now, as $u' \in [u]_B$, for all X_i's, $[u]_B \cap X_i \neq \emptyset$, if and only if $[u']_B \cap X_i \neq \emptyset$. Thus, for any X_i, $u \in \cup\{[u]_A : [u]_A \cap X_i \neq \emptyset\}$, if and only if $u' \in \cup\{[u]_A : [u]_A \cap X_i \neq \emptyset\}$. Hence $\partial_A(u) = \partial_A(u')$. Thus, $d_5^{\#}(u) = d_5^{\#}(u')$. Thus we have $Ind(B) \subseteq Ind(\{d_5^{\#}\})$.

Conversely, let $Ind(B) \subseteq Ind(\{d_5^{\#}\})$. Thus, for any $u' \in [u]_B$, $\partial_A(u) = \partial_A(u')$. Let $v_i \in \partial_A(u) = \partial_A(u')$. Then $[u]_A \cap X_i \neq \emptyset$ and $[u']_A \cap X_i \neq \emptyset$. As $[u]_A \subseteq [u]_B$, we know $\cup\{[u]_A : [u]_A \cap X_i \neq \emptyset\} \subseteq \cup\{[u]_B : [u]_B \cap X_i \neq \emptyset\}$. So, let us consider $u' \in \cup\{[u]_B : [u]_B \cap X_i \neq \emptyset\}$. Now, as for any $u' \in [u]_B$ such that $[u]_B \cap X_i \neq \emptyset$, $\partial_A(u) = \partial_A(u')$, $[u']_A \cap X_i \neq \emptyset$. So, $u' \in \cup\{[u]_A : [u]_A \cap X_i \neq \emptyset\}$. So, $\cup\{[u]_B : [u]_B \cap X_i \neq \emptyset\} \subseteq \cup\{[u]_A : [u]_A \cap X_i \neq \emptyset\}$. Thus, $\cup\{[u]_B : [u]_B \cap X_i \neq \emptyset\} = \cup\{[u]_A : [u]_A \cap X_i \neq \emptyset\}$.

D6⇔D6$^{\#}$. Let $\overrightarrow{\mu}_A(u) = \overrightarrow{\mu}_B(u)$ and $u' \in [u]_B$. We want to show $d_6^{\#}(u) = d_6^{\#}(u')$. As $u' \in [u]_B$, for any decision class X_i, we have $\mu_B^i(u) = \frac{|\{u'' \in [u]_B : d(u'') = v_i\}|}{|[u]_B|} = \mu_B^i(u')$. So, $\overrightarrow{\mu}_B(u) = \overrightarrow{\mu}_B(u')$. This is same as $\overrightarrow{\mu}_A(u) = \overrightarrow{\mu}_A(u')$ and hence $d_6^{\#}(u) = d_6^{\#}(u')$.

Conversely, let $Ind(B) \subseteq Ind(\{d_6^{\#}\})$. Thus, for any $u' \in [u]_B$, $d_6^{\#}(u) = d_6^{\#}(u')$, or in other words, $\overrightarrow{\mu}_A(u) = \overrightarrow{\mu}_A(u')$. We want to prove that $\overrightarrow{\mu}_A(u) = \overrightarrow{\mu}_B(u)$. We know $[u]_A \subseteq [u]_B$. Now, if $[u]_A = [u]_B$, then the proof is immediate. Let $[u]_A \subsetneq [u]_B$. Now, for any decision class X_i, if $[u]_B \cap X_i \neq \emptyset$, then $[u]_A \cap X_i \neq \emptyset$ as well, as for an $u \in [u]_A \subseteq [u]_B$ and for an $u' \in [u]_B \setminus [u]_A$, $\frac{|\{u'' \in [u]_A : d(u'') = v_i\}|}{|[u]_A|} = \frac{|\{u'' \in [u']_A : d(u'') = v_i\}|}{|[u']_A|}$. Thus, for any $u' \in [u]_B \setminus [u]_A$, $[u']_A \cap X_i \neq \emptyset$ also. So, there can be a number of disjoint equivalence classes $[u]_A, [u_1]_A, \ldots, [u_n]_A$ included in $[u]_B$, such that each has elements from decision class X_i and union of these classes covers $[u]_B$. Now, for simplicity let us assume that there are only two elements $u, u' \in [u]_B$ such that $[u]_A \cap [u']_A = \emptyset$ and $[u]_A \cup [u']_A = [u]_B$. Therefore, $\frac{|\{u'' \in [u]_A : d(u'') = v_i\}|}{|[u]_A|} = \frac{|\{u'' \in [u']_A : d(u'') = v_i\}|}{|[u']_A|}$ and we have the following relation: $\mu_B^i(u) = \frac{|\{u'' \in [u]_B : d(u'') = v_i\}|}{|[u]_B|} = \frac{|\{u'' \in [u]_A : d(u'') = v_i\}| + |\{u'' \in [u']_A : d(u'') = v_i\}|}{|[u]_A| + |[u']_A|} = \frac{|\{u'' \in [u]_A : d(u'') = v_i\}|}{|[u]_A|} = \mu_A^i(u)$. Hence, $\overrightarrow{\mu}_B(u) = \overrightarrow{\mu}_A(u)$.

D7⇔D7$^{\#}$. Assume that D7 is true. Consider $u' \in [u]_B$. Now $[u]_B$ can contain elements from a single decision class or more than one decision class. If $[u]_B$ contains element from a single decision class, then so do $[u]_A$ and $[u']_A$. Thus, $d(u) = d(u')$ and $u, u' \in POS(A)$. So, $d_7^{\#}(u) = d(u) = d(u') = d_7^{\#}(u')$. If $[u]_B$ contains elements from more than one decision class, then by D7, we have $[u]_A = [u]_B$. Thus, as $u' \in [u]_B$, there is $[u]_A = [u']_A$. So, $d_7^{\#}(u) = d_7^{\#}(u') = \#_{m(u)}$. Hence, $Ind(B) \subseteq Ind(\{d_7^{\#}\})$.

Conversely, let for any $u' \in [u]_B$ be $d_7^{\#}(u) = d_7^{\#}(u')$. Then, either $[u]_A$ and $[u']_A$ are both contained in the same decision class or $[u]_A = [u']_A$. This proves D7.   □

In Definition 8, we did not formulate the cases of D2–D4 as their are equivalent to D5, therefore, they are addressed by D5$^{\#}$. It is also worth pointing out that Theorem 3 provides an additional insight into interrelationships regarding strengths of criteria D1–D7. By conducting a simple comparative analysis of a nature of newly constructed decision values $d_1^{\#}$, $d_5^{\#}$, $d_6^{\#}$ and $d_7^{\#}$, one can design alternative proofs of components of Theorem 1, with some of "but not converse" examples displayed in Table 2.

Moreover, Theorem 3 is highly important from algorithmic perspective. This is because it allows us for adapting a variety of efficient methods for searching for most useful decision reducts developed originally for consistent decision tables [2] to deal with all considered variants of inconsistencies. This ability – and in particular ability of encoding specifications of different decision superreduct criteria in terms of so-called discernibility characteristics – is essential to take advantage of various techniques of accelerating computations, available in different frameworks [7].

Let us also recall that many rough-set-based attribute reduction methods implicitly assume that decision superreduct criteria are in some sense monotonic with respect to set-theoretic inclusion, i.e., if $B \subseteq A$ is a decision superreduct, then its superset C $(B \subseteq C \subseteq A)$ is a decision superreduct too [12]. Theorem 3 delivers this kind of property for all variants of attribute reduction considered in this paper. If we recall the aforementioned analogy between decision superreducts and Markov blankets, known from probability-based graphical models [4], then this kind of monotonicity could be interpreted as so-called *weak union* of conditional independence statements. Indeed, the facts that some subsets $B \subseteq A$ are decision superreducts for particular variants of handling inconsistencies in decision tables may be rephrased as *saturated conditional independences* [13]. We already know that properties of some of decision superreduct formulations are actually quite similar to those of classical probabilistic independence statements [5]. Nevertheless, this knowledge is still incomplete. In future, we intend to investigate further interrelations between the property of weak union and the property of discernibility characteristics for different models of superreducts.

In the remainder of this section, we focus on better understanding how new decision attributes $d_i^{\#}$ encode inconsistencies inherited from original decision tables. Indeed, deriving from each given (potentially inconsistent) $(U, A \cup \{d\})$ a consistent $(U, A \cup \{d_i^{\#}\})$ seems to be something more than just a technical trick. This is transformation of criteria formulated by means of preserving particular kinds of decision representations into criteria aimed at comparing representations of indiscernibility classes that can be merged during attribute reduction. Below we provide some comments, case by case:

– Following D1, a decision superreduct B is such that $POS(A) = POS(B)$. Herein, different forms of inconsistency are simply ignored by putting a dummy decision value $\#$ for all elements, indiscernible by $Ind(A)$ but having different decision values. The only aspect that matters is to distinguish between inconsistent and consistent cases, so positive regions do not decrease while reducing attributes.

- In case of D2 to D4, B has to generate the same lower and/or upper approximations of decision classes or their set theoretic sums as in case of A. As for D5, the set of all decision values of an equivalence class with respect to $Ind(A)$ should be the same as with respect to $Ind(B)$. In all these variants, inconsistency occurring within a given $[u]_A$ is handled by assigning a set of all possible decision values to each element of that class. Thus, if for some $u \in U$, there is $u \notin POS(A)$, then each element of $[u]_A$ would have new decision value $\partial_A(u)$. Thus, in contrast to D1, herein values corresponding to inconsistent elements are not completely erased from the new decision table. Those cases are rather grouped together with regard to how diverse sets of decision values particular indiscernibility classes can assume.
- As for D6, decision superreduct B should generate the same rough membership function as A. Thus, if objects u and u' are indistinguishable with respect to $Ind(A)$ but have different decision values, then they are unified with probability distribution. In a consistent decision table, no information regarding different decision values of a particular equivalence class is lost. They are now encoded with their probabilities. Therefore, we have analogous groupings of different indiscernibility classes as in case of D5, although now encoded information is richer.
- In case of D7, B should preserve discernibility of elements belonging to different decision classes, whenever it is possible with a usage of all attributes in A. Herein, each of equivalence classes $[u]_A$ for which inconsistency in decision arises, is assigned to a unique decision $\#_{m(u)}$. Thus, in contrast to D1, for each element of an inconsistent equivalence class $[u]_A$, a dummy value, indexed by ordinal number of that particular equivalence class, is assigned. Therefore, in the new consistent decision table, though a dummy value is assigned to each inconsistent element, the value still reflects an origin of a given element. Actually, one can try to compare the meanings of dummy decision values $\#$ and $\#_{m(u)}$ with analogous differences in handling unknown values of conditional attributes in incomplete information systems [14]. Therein, two undetermined values could be – among other strategies – regarded as potentially the same (which is an analogy to $\#$) or potentially different (which is an analogy to $\#_{m(u)}$). Like in our case, such two approaches to interpreting undetermined values can lead toward totally different results.

6   Conclusion

The study of decision reducts gives a practical way of abstracting significant knowledge and reason about the data, ignoring redundant attributes. In this paper, we attempted to gather together the most popular formulations of rough-set-based attribute reduction criteria, summarizing their interdependencies, outlining their alternative representations and providing some missing mathematical results that were not proved or sufficiently exposed up to now. From this perspective, one may pay a special attention on Theorem 2, as well as the part of Theorem 3 that corresponds to criterion D7. However, we also believe that an overall picture delivered by Theorems 1–3 can be helpful.

Redundancy of attributes can have different senses as it is clear from different considered ways of obtaining decision reducts. From case by case analysis, it can be noticed that by following criterion D1 (or equivalently D1$^\#$) a way of obtaining a decision superreduct B is the simplest one, as it completely ignores a nature of inconsistent cases. For variants D2–D5, the whole set of decision values that an equivalence class can take is assigned to every element of that class, but how frequently those values are taken by its different elements is ignored. As pointed out in [5], this strategy of dealing with data-driven information is analogous to a way of formulating multi-valued dependencies in relational databases. In case of D6/D6$^\#$, information about probabilities with which different decision values are taken by elements of particular equivalence classes is nicely encoded in respective consistent decision table $(U, A \cup \{d_6^\#\})$. Thus, from a point of view of probabilities, no information from original decision table is lost, though from computational perspective this approach might be more complex than others. For D7 (or equivalently D7'/D7$^\#$), respective consistent decision table seems to be more informative than in case of D1 as a dummy value assigned to each element of a given inconsistent equivalence class also carries an index specifying ordinal number of that class. These differences, somewhat similar to different ways of working with unknown values as summarized in [14], can have a huge impact on results of attribute reduction. One might even claim that D7 represents the most rigoristic criterion out of possible formulations (not only those stated as D1–D6) of decision superreducts.

Certainly, being the most rigoristic does not need to mean being inappropriate. As an example, let us discuss a situation of decision superreduct B, selected by following D6, that does not reflect a finer distinction when criterion D7 is applied. Consider U that is partitioned into three equivalence classes $[u_1]_A$, $[u_2]_A$, $[u_3]_A$ under $Ind(A)$. Imagine that there are only two decision values – v_1 and v_2 – and their probability distributions are given by $\vec{\mu}(u_1) = \langle \frac{1}{3}, \frac{2}{3} \rangle$, $\vec{\mu}(u_2) = \langle \frac{1}{3}, \frac{2}{3} \rangle$ and $\vec{\mu}(u_3) = \langle \frac{1}{2}, \frac{1}{2} \rangle$. Now, if we take one element u_1' from $[u_1]_A$ and one element u_2' from $[u_2]_A$ such that $d(u_1') = v_1$ and $d(u_2') = v_2$, then – when following D7 – u_1' and u_2' must be distinguished. However, if superreduct B is to be obtained using D6, then – as both u_1' and u_2' have the same vectors of rough membership degrees – these two elements do not need to kept in separate indiscernibility classes while removing attributes. This example shows that a deeper analysis of advantages and disadvantages of using different notions of decision superreduct is required, as it is not so obvious in what situations one can truly agree to think about such cases as potentially indistinguishable. Thus, one of our further research directions is to explore a meta-theoretic investigation regarding which kind of attribute reduction criterion would be suitable for which practical context.

References

1. Pawlak, Z.: A treatise on rough sets. In: Peters, J.F., Skowron, A. (eds.) Transactions on Rough Sets IV. LNCS, vol. 3700, pp. 1–17. Springer, Heidelberg (2005). https://doi.org/10.1007/11574798_1
2. Nguyen, H.S.: Approximate Boolean reasoning: foundations and applications in data mining. In: Peters, J.F., Skowron, A. (eds.) Transactions on Rough Sets V. LNCS, vol. 4100, pp. 334–506. Springer, Heidelberg (2006). https://doi.org/10.1007/11847465_16
3. Kryszkiewicz, M.: Comparative study of alternative types of knowledge reduction in inconsistent systems. Int. J. Intell. Syst. 16(1), 105–120 (2001)
4. Betliński, P., Ślęzak, D.: The problem of finding the sparsest Bayesian network for an input data set is NP-hard. In: Chen, L., Felfernig, A., Liu, J., Raś, Z.W. (eds.) ISMIS 2012. LNCS (LNAI), vol. 7661, pp. 21–30. Springer, Heidelberg (2012). https://doi.org/10.1007/978-3-642-34624-8_3
5. Ślęzak, D.: On generalized decision functions: reducts, networks and ensembles. In: Yao, Y., Hu, Q., Yu, H., Grzymala-Busse, J.W. (eds.) RSFDGrC 2015. LNCS (LNAI), vol. 9437, pp. 13–23. Springer, Cham (2015). https://doi.org/10.1007/978-3-319-25783-9_2
6. Ślęzak, D.: Searching for dynamic reducts in inconsistent decision tables. In: Proceedings of IPMU 1998 Part I I, 1362–1369 (1998)
7. Widz, S.: Introducing NRough framework. In: Polkowski, L., et al. (eds.) IJCRS 2017. LNCS (LNAI), vol. 10313, pp. 669–689. Springer, Cham (2017). https://doi.org/10.1007/978-3-319-60837-2_53
8. Dash, M., Liu, H.: Consistency-based search in feature selection. Artif. Intell. 151(1–2), 155–176 (2003)
9. Janusz, A., Ślęzak, D.: Computation of approximate reducts with dynamically adjusted approximation threshold. In: Esposito, F., Pivert, O., Hacid, M.-S., Raś, Z.W., Ferilli, S. (eds.) ISMIS 2015. LNCS (LNAI), vol. 9384, pp. 19–28. Springer, Cham (2015). https://doi.org/10.1007/978-3-319-25252-0_3
10. Chakraborty, M.K., Banerjee, M.: Rough dialogue and implication lattices. Fundamenta Informaticae 75(1–4), 123–139 (2007)
11. Bazan, J.G., Skowron, A., Synak, P.: Dynamic reducts as a tool for extracting laws from decisions tables. In: Raś, Z.W., Zemankova, M. (eds.) ISMIS 1994. LNCS, vol. 869, pp. 346–355. Springer, Heidelberg (1994). https://doi.org/10.1007/3-540-58495-1_35
12. Yao, Y., Zhao, Y., Wang, J.: On reduct construction algorithms. In: Gavrilova, M.L., Tan, C.J.K., Wang, Y., Yao, Y., Wang, G. (eds.) Transactions on Computational Science II. LNCS, vol. 5150, pp. 100–117. Springer, Heidelberg (2008). https://doi.org/10.1007/978-3-540-87563-5_6
13. Gyssens, M., Niepert, M., Van Gucht, D.: On the completeness of the semigraphoid axioms for deriving arbitrary from saturated conditional independence statements. Inf. Process. Lett. 114(11), 628–633 (2014)
14. Clark, P.G., Gao, C., Grzymala-Busse, J.W.: Rule set complexity for incomplete data sets with many attribute-concept values and "do not care" conditions. In: Flores, V. (ed.) IJCRS 2016. LNCS (LNAI), vol. 9920, pp. 65–74. Springer, Cham (2016). https://doi.org/10.1007/978-3-319-47160-0_6

A New Trace Clustering Algorithm Based on Context in Process Mining

Hong-Nhung Bui[1,2(✉)], Tri-Thanh Nguyen[1], Thi-Cham Nguyen[1,3],
and Quang-Thuy Ha[1]

[1] Vietnam National University, Hanoi (VNU), VNU-University of Engineering
and Technology (UET), 144, Xuan Thuy, Cau Giay, Hanoi, Vietnam
nhungbth@hvnh.edu.vn, {ntthanh, thuyhq}@vnu.edu.vn,
nthicham@hpmu.edu.vn
[2] Banking Academy of Vietnam, 12, Chua Boc, Dong Da, Hanoi, Vietnam
[3] Hai Phong University of Medicine and Pharmacy, 72A, Nguyen Binh Khiem,
Ngo Quyen, Haiphong, Vietnam

Abstract. In process mining, trace clustering is an important technique that attracts the attention of researchers to solve the large and complex volume of event logs. Traditional trace clustering often uses available data mining algorithms which do not exploit the characteristic of processes. In this study, we propose a new trace clustering algorithm, especially for the process mining, based on the using trace context. The proposed clustering algorithm can automatic detects the number of clusters, and it does not need a convergence iteration like traditional ones like K-means. The algorithm takes two loops over the input to generate the clusters, thus the complexity is greatly reduced. Experimental results show that our method also has good results when compared to traditional methods.

Keywords: Event log · Process mining · Trace context · Clustering algorithm

1 Introduction

Most today's modern information systems have collection of data that describes all the events of the user occur during the execution of the software system so-called event logs. Event logs play an important role in modern software systems, they record information about the system in real-time including a set of events that contain several information, e.g., *case id*, *event id*, *timestamp*, *activity*, etc., Table 1 introduces some examples about an event log. The events in the same *case* are ordered by *timestamp* and have the same "*case id*". These are valuable data for managers to analyze and evaluate the company's business processes.

Process mining includes three tasks process discovery, conformance checking and enhancement is the field that allows the use of the event log data for analysis and improvement of the processes.

© Springer Nature Switzerland AG 2018
H. S. Nguyen et al. (Eds.): IJCRS 2018, LNAI 11103, pp. 644–657, 2018.
https://doi.org/10.1007/978-3-319-99368-3_50

The target of process discovery is to generate a process model that captures all of the behaviors found in the event log [23]. The generated model can be used to analyze what is actually applied in daily activities of the company. It can be used to verify whether the formal process is strictly followed, or to enhance the formal process.

The event log quality is an important factor in process model generation. If the event log is homogeneous and small enough, the process model is easy to analyze as one example in Fig. 1a. However, real-life event logs are extremely huge with diverse characteristics, thus, the discovered process model may be diffuse and very hard to understand as an example in Fig. 1b. To overcome this problem, clustering a complex event log into sub-logs/clusters is one of the most widely used solution. The generated model from an event sub-log will have much lower complexity [5, 7, 9–11, 15–18, 21].

Table 1. A fragment of the event log [23]

Case id	Event id	Properties				
		Timestamp	Activity	Resource	Cost	...
1	4423	30-12-2010:11.02	Register request	Pete	50	
	4424	31-12-2010:10.06	Examine thoroughly	Sue	400	
	4425	06-01-2011:15.12	Check ticket	Mike	100	
	4426	07-01-2011:11.18	Decide	Sara	200	
	4427	07-01-2011:14.24	Reject request	Pete	200	
2	4483	30-12-2010:11.32	Register request	Mike	50	
	4485	30-12-2010:12.12	Check ticket	Mike	100	
	4487	30-12-2010:14.16	Examine casually	Pete	400	
	4488	06-01-2011:11.22	Decide	Sara	200	
	4489	08-01-2011:12.06	Pay compensation	Ellen	200	
3	4521	30-12-2010:14.32	Register request	Pete	50	
	4522	30-12-2010:15.06	Examine casually	Mike	400	
	4524	30-12-2010:16.34	Check ticket	Ellen	100	
	4525	06-01-2011:09.18	Decide	Sara	200	
	4526	08-01-2011:12.18	Reinitie request	Sara	200	
	...					

Traditional approaches use the data mining clustering algorithms such as Agglomerative Hierarchical Clustering, K-Means, K-Modes, etc., to cluster event logs. These algorithms are designed and used in the field of data mining, they do not exploit the specific characteristics of business processes.

In this paper, we propose a new trace clustering algorithm based on a specific characteristic of process, i.e., the context of traces in a process. The contribution of the paper includes: (1) defining a new trace context; (2) introducing a context tree; (3) giving a new event log clustering algorithm. The proposed algorithm can automatically detect the suitable number of clusters, and it does not need a convergence iteration which takes lot of time. The experimental results show that our method has significant contributions to improving the efficiency and the performance time of the process discovery task.

The rest of this article is organized as follows: First, we give an overview of the process discovery. Section 3 introduces the trace context in process mining and the new trace clustering. The experimental evaluation is described in Sect. 4. Section 5 introduces the related work. Conclusions and future work are shown in the last section.

2 The Brief Summary of Process Discovery Task in Process Mining

Event Logs

An event log is the starting point of process mining. Table 1 shows a fragment of the event log related to the handling of compensation requests of an airline. There are three cases corresponds to three compensation requests. The case 1 has five events with *id* from 4423 to 4427 that are ordered by execution time, i.e., property timestamp. For example, event 4423 executes activity "register request" at "30-12-2010:11.02" occurs before event 4424 which executes activity "examine thoroughly" at "31-12-2010:10.06". Each event in event log also is described by *resources* property, i.e., the persons executing the activities or the cost of the activity.

In process mining, the "*case id*" and "*activity*" are minimum properties that can be used to represent a case. For example, *case* 1 is represented by a sequence of five activities Register request, Examine thoroughly, Check ticket, Decide, Reject request. Such a sequence of activities is called a *trace*. For the sake of simplicity for computation, each activity name is assigned by a distinct letter label, e.g., *a* denotes activity register request. Hence, the event log in Table 1 has a more compact representation shown in Table 2, e.g., *case* 1 is represented by a trace $\langle a, b, d, e, h \rangle$. This representation is used for computation, such as clustering. For example, in K-means a trace is converted into a vector as the input to the algorithm.

Table 2. The trace in an event log (where a = "register request", b = "examine thoroughly", c = "examine casually", d = "check ticket", e = "decide", f = "reinitiate request", g = "pay compensation", h = "reject request")

Case id	Trace
1	$\langle a, b, d, e, h \rangle$
2	$\langle a, d, c, e, g \rangle$
3	$\langle a, c, d, e, f, b, d, e, g \rangle$
4	$\langle a, d, b, e, h \rangle$
5	$\langle a, c, d, e, f, d, c, e, h \rangle$
6	$\langle a, c, d, e, g \rangle$
...	...

Process Discovery Task

Process discovery is the first task of process mining. It takes an event log as an input data and produces a model represented in a process modeling language, e.g., Petri net (Fig. 1), which describes the behaviors recorded in the event log by applying a process discovery algorithm, e.g., α-algorithm [23].

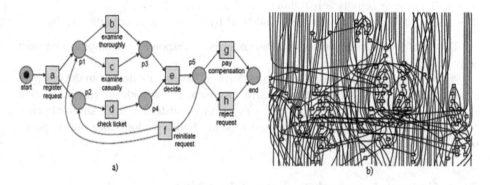

Fig. 1. The process model discovered from the event log by the α-algorithm

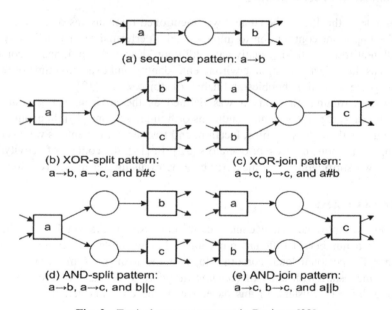

(a) sequence pattern: a→b

(b) XOR-split pattern:
a→b, a→c, and b#c

(c) XOR-join pattern:
a→c, b→c, and a#b

(d) AND-split pattern:
a→b, a→c, and b‖c

(e) AND-join pattern:
a→c, b→c, and a‖b

Fig. 2. Typical process patterns in Petri net [23]

α-Algorithm

The α-algorithm was one of the first process discovery algorithms. It generates the process model by reconstructing causality from a set of sequences of events in the event logs.

Given an event log of a business process L, α-algorithm scans L to find the relationships between activities based on the execution order. There are four ordering relations, e.g., *direct succession, causality, parallel, choice*. Let a, b are two activities in L.

1. Direct succession $a > b$: if some case a is followed by b.
2. Causality $a \rightarrow b$: if activity a is followed by b but b is never followed by a.
3. Parallel $a\|b$: if activity a is followed by b and b is followed by a.
4. Choice $a\#b$: if activity a is never followed by b and b is never followed by a.

To reflect those dependencies, the Petri net has corresponding notations to connect activities as illustrated in Fig. 2.

As mentioned above, to mine easy-to-understand process models from the complex event log, the trace clustering is the effective approach. The key idea of trace clustering algorithms is to create clusters that the traces within a cluster are more similar to each other than the traces in the different clusters. Next section we introduce our proposed trace clustering algorithm.

3   A Context Approach to Trace Clustering

3.1   Context in Process Mining

In the middle of the 1990s, the context was mentioned by many researchers [2, 3, 14]. It had the important contribution to improving the performance of practical systems. Each different research fields usually have different ideas and definitions of context. It is defined as the object's location, environment, identity and execution time or object's emotional state as well as hobbies and habits of objects, etc. [12].

In process mining, the context was defined as the environment surrounding a business process, e.g., the weather conditions or holiday seasons [13]. In another study, the context was defined as the time, location, and frequency of events as well as related communication, tools, devices, or operators [22]. In [19], the context of activity a was the set of two surrounding activities xy, i.e., xay, by using 3-g in an event log.

3.2   Trace Context

In this paper we introduce a new context definition based on the fact that each business process has a number of different procedures. For example, the credit process has procedures for personal loan, corporate loan, home loan, consumer loan, etc. Each procedure may start with a set of *common activities* which are the clue to separate traces into different clusters. In this paper we define common activities as the trace contexts.

Definition 1. Let $L = \{t_0, t_1, \ldots\}$ be an event log, where t_i is a trace. Let p be the longest common prefix p of a trace subset, i.e., $SP = \{t \in L | t = p|d\}$, such that $|SP| > 1$, where d is a sequence of activities, notation '|' in $p|d$ denotes sequence concatenation operation, then p is called as a *trace context*.

3.3 Context Tree

Since the common prefix of traces can be represented by a prefix tree, to efficiently identify the context, we introduce a *Context-tree* based on the idea of *frequent pattern tree* (FP-tree) [8].

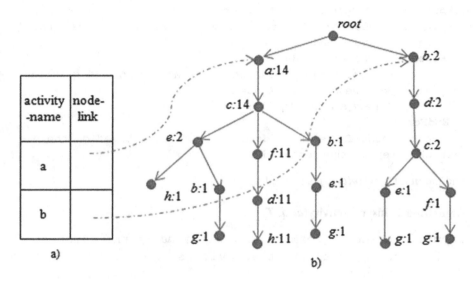

Fig. 3. (a) Header table; (b) The context-tree

Definition 2. A context tree is a tree that has:

1. One root labeled as "*root*" to form a complete tree.
2. A header table helps to access the tree faster during tree construction and traversal. Each entry in the Context-tree header table consists of two fields, (1) *activity-name*, and (2) head of *node-link* which points to the first node below the root carrying this activity.
3. Each node in the context tree consists of three fields except for the root node:

   activity-name: registers which activity is represented by the node;
   count: the number of traces that travel to this node;
   node-link: the pointers to its children, or null if there is none.

4. A trace in the event log is placed on a certain branch of the tree with the top- down fashion. Traces with the same prefix share a chunk of branch from the root node.

The idea is to map traces with the same prefix into the same chunk of tree branch as depicted in Fig. 3. The context tree construction procedure is described as follows:

Algorithm 1. ContextTreeConstruction.

```
Input:   An event log L

Output:  A corresponding context tree T
1. Create a node of a Context-tree T and label it as "root",
      i.e., the root node and T = root.
2. Foreach trace t in L do
        Let t=ac|q, where ac is the first activity, and q is the
        rest of the activity sequence
        call insert_activity(ac|q, T);
   EndFor
3. Create HeaderTable and update the node-link based on the di-
rect children of the root node.
```

And the *insert_activity(.)* is defined as:

Algorithm 2. insert_activity(ac|q, T)

```
Input: A trace in term of ac|q where ac is the first ac-
tivity, and q is the rest activities
     T is a tree node
Output: T is updated with new activities
1.  If T has a child N such that N.activity-name=ac then
        Increase N's count by 1
    Else
        Create a new node N, with its count = 1,
        Create a new node-link linked from T to N.
    EndIf
2.  If q is nonempty then
        call insert_activity(q, N) recursively.
    EndIf
```

Let L = [<*aceh*>, <*acfdh*>[10], <*acebg*>, <*acbeg*>, <*bdceg*>, <*bdcfg*>] be an event log, which includes 15 traces, the trace $\langle acfdh \rangle$ appears 10 times. The corresponding context-tree is illustrated in Fig. 3.

Mapping the context tree with the Definition 1 it is clear that, for each trace on the tree, the longest common prefix is the sequence of activities that have *count* > 1. From the context-tree in Fig. 3, the set of trace contexts of L is $\{ace, acfdh, ac, bdc\}$.

If a trace is distinct from the others, then it has no context. The following procedure is responsible for identifying the context of a given trace.

Algorithm 3. ContextDetection(*ac|q, T, context*)

Input: A trace in term of *ac|q* where *ac* is the first ac-
tivity, and *q* is the rest activities
        T is a context tree node
Output: The *context* of the trace
1. **If** *T* is root **then**
        context={};
        Get the node *N* pointed by *node-link* from the
        HeaderTable of *T* in the entry corresponding to *ac*;
    Else
        Find the child node *N* of *T* that has the label *ac*;
    EndIf
2. **If** the node *N* has *count* > 1 **then**
        Context = context|*ac*; //Concatenate a sequence
        If *q* is nonempty **then**
            call *ContextDetection(q, N, context)*;
        EndIf
    EndIf

3.4   Context Trace Clustering Algorithm

A new trace clustering algorithm called ContextTracClus which aims at creating
clusters of traces based on contexts is proposed. The algorithm consists of two distinct
phases: (1) Determining trace contexts and Building clusters; (2) Adjusting clusters.

The first phase, *Determining trace contexts and Building clusters*, includes two
steps.

Step 1 builds a compact data structure called the Context-tree that stores quantitative
information about activities of each trace in a event log. *Step 2* traverses the Context-tree
for each trace to find its trace context, and assigns the trace to the cluster corresponding
to this context. Based on the Context-tree construction process, for any trace *t* in event
log, there exists a path *p* in the Context-tree starting from the root. The trace context of
this trace is the sequence of nodes of *p* that have *count* $\geq$ 2. In case a trace has no
context, a new cluster is created for storing this trace for later adjustment in Phase 2.

The second phase, *Adjusting clusters*, handles the case where small clusters are
generated. If a cluster size, i.e., the number of traces in the cluster, is smaller than a
given minimum cluster size threshold *mcs* (e.g., each cluster size should be at least 10%
of the number of traces in the event log), this cluster will be added to its closest cluster.
The distance between to clusters is defined as the distance between two corresponding
trace contexts. In the case that a trace has no context, it will be added to the cluster
whose trace context includes the maximum number of duplicate activities with this
trace. The pseudo-code of the proposed algorithm, denoted ContextTracClus, is shown
in Algorithm 4.

Algorithm 4. ContextTracClus.

Input: An event log L
          A minimum cluster size threshold mcs
Output: The complete set of clusters C
Phase 1: Determining trace contexts and Building clusters
1. $C = \{\}$;
2. $T =$ ContextTreeConstruction(L); //T is the context tree
3. **Foreach** trace t in event log L **do**
    ContextDetection$(t, T, context)$;
    If context is empty **then**
        Create a new cluster c; //c has no label
        Add t to c; //This cluster has only one trace
        $C = C \cup c$;
    Else
      If C has no cluster labeled $context$ **then**
        Create a new cluster c labeled $context$;
        Add t to c;
        $C = C \cup c$;
      Else
        Add t to the cluster labeled $context$;
      EndIf
    EndIf
  EndFor
Phase 2: Adjusting clusters
4. **Foreach** cluster c in C **do**
    If size$(c) < mcs$ **then**
        Merge c to its closest cluster in C;
    EndIf
  EndFor

Our algorithm can automatically detect a suitable number of clusters. Unlike traditional clustering algorithms which need convergence loops, our algorithm takes only one loop to identify the clusters, and one loop to merge small clusters.

In K-means algorithm, it randomly selects some data points as the initial center of clusters, and the quality of clustering greatly depends on this selection, especially on event log, where a same trace can occur several times as depicted in Fig. 3, where the trace *acfdh* repeats 10 times. The repeated traces with a big number of times should be a cluster candidate. One more advantage of the algorithm is the ability to put repeated traces into a cluster candidate and removes the uncertainty of random.

The proposed algorithm needs one loop for context tree construction, one loop for clustering. Thus, its complexity is much less than that of traditional clustering algorithms such as K-means, K-modes. Furthermore, the proposed algorithm does not need to transform trace in an intermediate representation (e.g., binary, k-gram, maximal pair, maximal repeat, super maximal repeat and near super maximal repeat, etc.), convert this representation into vector, since it works directly with the traces, then the pre-processing time is greatly reduced.

3.5   An Application Framework for ContextTracClus Algorithm

In process discovery application, we propose a framework as described in Fig. 4, which consists of 5 steps.

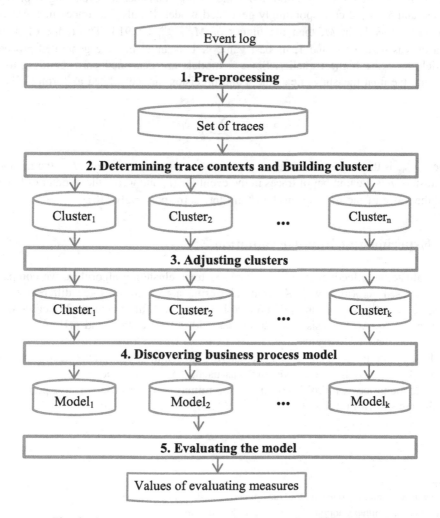

Fig. 4. An application framework of the ContextTracClus algorithm

The Pre-processing step transforms the input event log into a list of traces, i.e., merger all the events with the same *caseid* in the event log into a sequence of activities (sorted by recorded time) to form a trace [20, 23].

Step 2 and 3 use ContextTracClus algorithm to determine the contexts that appear in the event log, and generate n clusters. After adjustment, the number of clusters is k, where $k \leq n$. Each cluster is used to create a sub-log for process discovery.

In step 4, the α-algorithm is used to generate the sub-process models corresponding to each event sub-log.

The Evaluating model step evaluates the quality of each generated process models by two *Fitness* and *Precision*. The fitness measure determines whether all traces in the log can be replayed by the model from beginning to end. The precision measure determines whether the model has behavior very different from the behavior seen in the event log. Additional explanation about the fitness: consider an event log L of 600 traces, and M is the correspondingly generated model. If only 548 traces in L can be replayed correctly in M, then the fitness of M is $\frac{548}{600} = 0.913$. The range of those measures is between 0 and 1, its best value is 1 meaning that the generated process models have the highest quality. Since k models are generated corresponding to k clusters, the final measures, i.e., fitness and precision, are calculated as formula (1).

$$w_{avg} = \sum_{1}^{k} \frac{n_i}{n} w_i \qquad (1)$$

where w_{avg} is the aggregated value of the fitness or precision measure, k is the number of clusters, n is the number of traces in the event log, n_i and w_i are the number of traces and the value of the measure in the ith cluster, correspondingly [18].

4   Experimental Result Evaluation

To evaluate the effectiveness of the proposed trace clustering algorithm, we compare our proposed algorithm with K-means clustering algorithm, on three different event logs, i.e., Lfull[1], prAm6[2] and prHm6 (see Footnote 2). Lfull includes 1391 cases with 7539 events; prAm6 consists of 1200 cases with 49792 events; and prHm6 contains 1155 cases with 1720 events.

In the experiment with K-means clustering algorithm, the k-grams trace representation $(k = 1, 2, 3)$ for binary vectors was used. To generate the process model and evaluate the processes, ProM 6.6[3], a process mining tool, was used. The experimental results are shown in Table 3.

---

[1] www.processmining.org/event_logs_and_models_used_in_book/Chapter7.zip

[2] http://data.3tu.nl/repository/uuid:44c32783-15d0-4dbd-af8a-78b97be3de49

[3] http://www.processmining.org/prom/start

Table 3. Results of K-means and ContextTracClus trace clustering algorithm

Algorithm	Event log					
	Lfull		prAm6		prHm6	
	Fitness	Precision	Fitness	Precision	Fitness	Precision
Scenario 1: Using K-means algorithm						
1-g	**0.991**	0.754	0.968	0.809	0.902	0.66
2-g	0.951	0.958	0.968	0.809	0.902	0.66
3-g	0.955	0.962	0.968	0.809	0.902	0.66
Scenario 2: Using ContextTracClus algorithm						
	0.982	**1**	**0.975**	**0.904**	**0.922**	**0.673**

The experimental results show that ContextTracClus always has a higher precision, i.e., it ensures that the generated process model has the least behaviors not seen in the event log. This is because the traces in a cluster have the same context, i.e., they have the same set of actions so the generated model will have at least superfluous behaviors.

In the scenario 1, we found out the most suitable number of clusters for the data set is 3 after trying with different numbers of clusters, such as 2, 3, 4, 5. The scenario 2 automatically detected the number of clusters based on the input size threshold.

5   Related Work

Greco et al. [4] proposed a clustering solution on traces in event log using bag-of-activities trace representation for K-means algorithm.

Song et al. [11] presented a trace clustering approach based on log profiles which captured the information typically available in event logs e.g., activity profile, originator profile. In their approach, the K-means, Quality Threshold, Agglomerative Hierarchical Clustering, and SelfOrganizing Maps clustering algorithms were used.

Jagadeesh Chandra Bose et al. [20] proposed a trace representation method based on using some control-flow context information e.g., Maximal Pair, Maximal Repeat, Super Maximal Repeat and Near Super Maximal Repeat. They used some of the clustering algorithms such as Agglomerative Hierarchical Clustering, K-means.

Weerdt et al. [6] proposed the ActiTraC algorithm, a three-phase algorithm for clustering an event log into a collection of sub-logs to increase the quality of the process discovery task. The ActiTraC algorithm includes three phases: Selection, Look ahead, and Residual trace resolution. The important idea of this algorithm is the sampling strategy, i.e., a trace is added to the current cluster if and only if it does not decrease the process model accuracy too much.

Ha et al. [18] provided a trace representation solution based on the distance graph model for K-Modes, K-means clustering algorithms. In this representation, it can describe the ordering and the relationship between the activities in a trace. Distance graphs order k of a trace describe the activity pairs which has distance at most k activities in the trace.

Baldauf et al. [12] presented a survey on an architecture of context-aware systems, which includes the design principles, the common context models. They introduced the existent context-aware systems and discussed their advantages and disadvantages. Their paper mentioned a number of different definitions of "context" such as location, identities of nearby people, objects and changes to those objects (Schilit and Theimer 1994); The user's location, environment, identity and time (Ryan et al. 1997); The user's emotional state, focus of attention, location and orientation, date and time, as well as objects and people in the user's environment (Dey 1998); The aspects of the current situation (Hull et al. 1997). The elements of the user's environment which the computer knows about (Brown 1996).

Becker et al. [22] introduced the support of context information in analyzing and improving processes in logistics. They defined the context as time, location, and frequency of events, tools, devices, or operators. In the experiments, they used the frequency of a process and its overall cycle time as the context data. In addition, they used K-Medoids clustering algorithm for the identification of process groups and for the evaluation of context information.

Bolt et al. [1] presented an unsupervised technique to detect relevant process variants in event logs by applying existing data mining techniques. This technique splits a set of instances based on dependent and independent attributes.

Leyer [13] presented a new approach to identify the effect of context factors on business process performance in the aspect of processing time. They proposed a two-stage approach to identify the relevant data and to determine the context impact by applying the statistical methods.

6 Conclusions and Future Work

This paper proposed a definition of context in business process and a new trace clustering algorithm base on contexts. A context tree was introduced to make the complexity of the algorithm is reduced with two loops over the input for finding clusters, and one small loop over the clusters for adjustment. The ability to work directly with the traces without transforming to an immediate representation is an additional advantage of the algorithm. Another ability to automatically detect the optimal number of clusters makes algorithm to remove the disadvantage of traditional clustering algorithms and produce determined results. As future work, we plan to study the impact of the context in other tasks of the process mining.

References

1. Bolt, A., van der Aalst, W.M.P., de Leoni, M.: Finding process variants in event logs. In: Panetto, H., et al. (ed.) On the Move to Meaningful Internet Systems. OTM 2017 Conferences. OTM 2017. LNCS, vol. 10573, pp. 45–52. Springer, Cham (2017). https://doi.org/10.1007/978-3-319-69462-7_4
2. Dey, A.K.: Context-aware computing: the CyberDeskProject. In: Proceedings of the AAAI, Spring Symposium on Intelligent Environments, pp. 51–54 (1998)

3. Schilit, B.N., Adams, N., Want, R.: Context-aware computing applications. In: WMCSA, pp. 85–90 (1994)
4. Greco, G., Guzzo, A., Pontieri, L., Saccà, D.: Discovering expressive process models by clustering log traces. IEEE Trans. Knowl. Data Eng. **18**, 1010–1027 (2006)
5. Fischer, I., Poland, J.: New methods for spectral clustering. In: Proceedings of ISDIA (2004)
6. Weerdt, J.D., vanden Broucke, S.K.L.M., Vanthienen, J., Baesens, B.: Active trace clustering for improved process discovery. IEEE Trans. Knowl. Data Eng. **25**(12), 2708–2720 (2013)
7. Poland, J., Zeugmann, T.: Clustering the Google distance with eigenvectors and semidefinite programming. Knowl. Media Technol. **21**, 61–69 (2006)
8. Han, J., Pei, J., Yin, Y.: Mining frequent patterns without candidate generation. In: SIGMOD Conference, pp. 1–12 (2000)
9. Weerdt, J.D.: Business process discovery_new techniques and applications. Runner up Ph.D. thesis (2014)
10. Evermann, J., Thaler, T., Fettke, P.: Clustering traces using sequence alignment. In: Reichert, M., Reijers, Hajo A. (eds.) BPM 2015. LNBIP, vol. 256, pp. 179–190. Springer, Cham (2016). https://doi.org/10.1007/978-3-319-42887-1_15
11. Song, M., Günther, Christian W., van der Aalst, Wil M.P.: Trace clustering in process mining. In: Ardagna, D., Mecella, M., Yang, J. (eds.) BPM 2008. LNBIP, vol. 17, pp. 109–120. Springer, Heidelberg (2009). https://doi.org/10.1007/978-3-642-00328-8_11
12. Baldauf, M., Dustdar, S., Rosenberg, F.: A survey on context aware systems. IJAHUC **2**(4), 263–277 (2007)
13. Leyer, M.: Towards a context-aware analysis of business process performance. In: PACIS, vol. 108 (2011)
14. Ryan, N., Pascoe, J., Morse, D.: Enhanced reality fieldwork: the context-aware archaeological assistant. In: Proceeding of the 25th Anniversary Computer Applications in Archaeology (1997)
15. Vitányi, P.M.B.: Information distance: new developments. CoRR abs_1201.1221 (2012)
16. De Koninck, P., De Weerdt, J., vanden Broucke, S.K.L.M.: Explaining clusterings of process instances. Data Min. Knowl. Discov. **31**(3), 774–808 (2017)
17. Koninck, P.D., Weerdt, J.D.: Determining the number of trace clusters_a stability-based approach. In: ATAED@Petri Nets_ACSD, pp. 1–15 (2016)
18. Ha, Q.-T., Bui, H.-N., Nguyen, T.-T.: A trace clustering solution based on using the distance graph model. In: Nguyen, N.-T., Manolopoulos, Y., Iliadis, L., Trawiński, B. (eds.) ICCCI 2016. LNCS (LNAI), vol. 9875, pp. 313–322. Springer, Cham (2016). https://doi.org/10.1007/978-3-319-45243-2_29
19. Jagadeesh Chandra Bose, R.P., van der Aalst, W.M.P.: Context aware trace clustering: towards improving process mining results. In: SDM 2009, pp. 401–412 (2009)
20. Jagadeesh Chandra Bose, R.P.: Process mining in the large preprocessing, discovery, and diagnostics. Ph.D. thesis, Eindhoven University of Technology (2012)
21. Thaler, T., Ternis, S.F., Fettke, P., Loos, P.: A comparative analysis of process instance cluster techniques. Wirtschaftsinformatik **2015**, 423–437 (2015)
22. Becker, T., Intoyoad, W.: Context aware process mining in logistics. Procedia CIRP **63**, 557–562 (2017)
23. Van der Aalst, W.M.P.: Process Mining: Discovery, Conformance and Enhancement of Business Processes. Springer, Heidelberg (2011). https://doi.org/10.1007/978-3-642-19345-3

Author Index

Printed in the United States
By Bookmasters